EMIS Datareviews Series No. 25

Series Advisor: Professor B. L. Weiss

PHYSICAL PROPERTIES OF
Liquid Crystals: Nematics

ELECTRONIC MATERIALS INFORMATION SERVICE

Other books in the EMIS Datareviews Series from INSPEC:

Details of these and forthcoming books in the EMIS Datareviews Series are available at the following address on the World Wide Web:

http: //www.iee.org.uk/publish/books/emis.html

PHYSICAL PROPERTIES OF
Liquid Crystals: Nematics

Edited by

DAVID DUNMUR
University of Southampton, UK

ATSUO FUKUDA
Shinshu University, Japan

and

GEOFFREY LUCKHURST
University of Southampton, UK

Published by: INSPEC, The Institution of Electrical Engineers, London, United Kingdom

British Library Cataloguing in Publication Data

Physical properties of liquid crystals: nematics.–(EMIS Datareviews series; no.25)
1. Liquid crystals
I. Dunmar, D.A. II. Fukuda, A. III. Luckhurst, G. R.
530.4'29

ISBN 0 85296 784 5

Printed in England by Short Run Press Ltd., Exeter

Contents

DEDICATION

This book is dedicated to the late Frank M Leslie FRS FRSE

in recognition of his major contributions to the field of liquid crystals

Foreword

I am delighted to have been asked to write the foreword to the Datareview volume on the physical properties of nematic liquid crystals. The public perception of liquid crystals has changed dramatically over the last thirty years. Initially they were relatively unknown and regarded as strange materials of interest to only a few academic researchers. Now display devices based on nematic liquid crystals form a huge international industry dominating the flat panel display market and rivalling the traditional cathode ray tube market. Liquid crystal displays are now part of the every day life of most people and the acronym LCD is universally recognised.

Early LCDs were simple devices with a low information content based on the twisted nematic (TN) mode still found in products such as pocket calculators and digital watches. As the technology evolved displays were expected to show increasingly large amounts of information. The early TN mode evolved into the supertwisted nematic (STN) and thin film transistor (TFT) addressed TN modes found in today's lap top and desktop computers, and TV screens. Very recently we have seen the emergence of a number of newer modes based on alternative device configurations of nematic liquid crystals designed to improve various aspects of display performance. Alternative flat panel display technologies based on different materials keep emerging, but it is clear that none will rival the supremacy of LCDs in the short term. The inherent versatility and variety of liquid crystals phases and device configurations suggest that LCDs will remain a key flat panel display technology for many years.

The operating principles behind most LCDs are deceptively simple and can be explained to a non-technical audience. However this is a veneer hiding the tremendous depth and quality of the underlying science and technology. The physical properties of nematic liquid crystals are a key part of this underlying knowledge and determine the detailed operation of the various modes. A knowledge of the physical properties of nematic liquid crystals, and understanding of the way in which they are related to the chemical composition and how they influence the device performance has been, and will remain, absolutely crucial to the success of LCD technology.

Physical Properties of Liquid Crystals: Nematics provides an invaluable data source and insight into the physical properties of nematic liquid crystals governing the performance of liquid crystal display devices. I applaud the editors for collecting together contributions from a truly distinguished group of experts and the authors themselves for writing a set of definitive contributions. It is however sad to note the death of two distinguished contributors before publication and I think it entirely appropriate that the book has been dedicated to the late friend and colleague of so many of us in the liquid crystal field, Professor Frank Leslie, FRS. I strongly commend this volume to you and I look forward to it playing a key part in helping keep LCDs as the pre-eminent flat panel display technology well into the future.

Peter Raynes, FRS
Oxford
September 2000

Introduction

Liquid crystals have now joined a wide range of materials, loosely described as electronic materials, which are at the core of the electronics industry. Over the past twenty-five years there has been a rapid growth of liquid crystal technology in the displays sector of the industry. However, liquid crystals have applications in other areas, and there continues to be considerable research and development directed to non-display applications. This Datareview volume on the physical properties of liquid crystals will be of benefit to scientists and engineers involved with applications as well as those concerned with fundamental research.

It is about one hundred years since liquid crystals were identified as a state of matter having a structure intermediate between that of the crystal and liquid. They are now recognised as anisotropic fluids in which varying degrees of orientational order of the constituent molecules impart macroscopic anisotropy to the physical properties of the phases. Liquid crystals, like solid crystals, exhibit polymorphism, and a number of phase types have been identified with varying degrees of structural organisation between the extremes of complete order, as in a perfect crystal, and zero order in an isotropic liquid. These phases can be categorised as nematic, with liquid-like translational order, smectic having one degree of translational order which results in the formation of a wealth of layered structures, and columnar in which there are two degrees of translational order, with molecules stacked liquid-like in columns, and the columns having a solid-like organisation. Many possible applications of both nematic and smectic liquid crystals have been demonstrated, including fast switching displays based on ferroelectric smectic phases, but this book is concerned only with nematic liquid crystals. A feature is the inclusion of reviews on the theoretical aspects of the physical properties as well as their measurement. In a number of areas, there is a dearth of reliable data to test the theories and this impedes further development of applications and understanding. It is hoped that these reviews will stimulate further research on the properties, so that the full range of physical phenomena identified in liquid crystals may be exploited.

The Datareviews encompass the broad classes of nematic liquid crystals and give data on the properties of the most common thermotropic nematics. The contributions cover structure/property relations, thermodynamic, electric, magnetic, optical, elastic and viscous properties, theories and computer modelling. The latter has played an important role in developing an understanding of the properties of liquid crystals, and now computer modelling is extensively used to improve the operational characteristics of displays. The authors are all acknowledged experts in the topics covered, and have been selected from both industry and academic institutions around the world. Tragically two of our distinguished contributors, Professor Pier Luigi Nordio and Professor Frank Leslie, died before the publication of this Datareview. Both have made seminal contributions to our understanding of liquid crystals.

The area of liquid crystals involves complex chemistry and physics, demanding theories and a wide range of difficult experimental techniques, and the preparation of this volume of the EMIS Datareview series has presented new challenges to the publishing team led by John Sears, managing editor of the EMIS series. The editors are most grateful to him, Pamela Sears and Karen Arthur for their help, expertise and patience, the outcome of which is hopefully a volume of lasting benefit to liquid crystal science and technology.

David Dunmur
University of Southampton, UK.
Atsuo Fukuda
Shinshu University, Japan.
Geoffrey Luckhurst
University of Southampton, UK.

September 2000

Contributing Authors

P.J. Alonso
6.4

Universidad de Zaragoza, Department of Physics,
Facultad de Ciencias, Zaragoza 50009, Spain

R.J. Atkin
5.1, 8.5

University of Sheffield, Department of Applied Mathematics,
Sheffield, S3 7RH, UK

M.A. Bates
12.1

University of Southampton, Department of Chemistry and
Southampton Liquid Crystal Institute, Southampton,
SO17 1BJ, UK

V.V. Belyaev
8.4

Central R&D Institute Kometa,
Velozavodskaya Street, Moscow, 109280, Russia

G.-H. Chen
10.1

Technical University of Berlin, Institute for Technical Chemistry,
Fachgebeit Makromolecular Chemie, Strasse des 17 Juni 135,
D-10623 Berlin, Germany

S.J. Clark
2.4

University of Durham, Department of Physics,
Science Laboratories, South Road, Durham, DU1 3LE, UK

D.J. Cleaver
12.4

Sheffield Hallam University, Division of Applied Physics,
Pond Street, Sheffield, S1 1WB, UK

G.P. Crawford
5.3

Brown University, Division of Engineering, 182 Hope Street,
Box D, Providence, RI 02912, USA

S. Day
12.5

University College London, Department of Electronic and
Electrical Engineering, Torrington Place, London,
WC1E 7JE, UK

E. Dorr
7.2

Universitat Kaiserslautern, Fachbereich Chemie,
Erwin-Schrödinger Strasse, D-67663 Kaiserslautern, Germany

D.A. Dunmur
5.2, 7.1

University of Southampton, Department of Chemistry and
Southampton Liquid Crystal Institute, Southampton,
SO17 1BJ, UK

F.A. Fernandez
12.5

University College London, Department of Electronic and
Electrical Engineering, Torrington Place, London,
WC1E 7JE, UK

A. Ferrarini
2.3

Universita Degli Studi di Padova, Dipartimento di Chimica
Fisica, via Loredan, 2-35131 Padova, Italy

A.M. Figueiredo Neto
6.5

University of Sao Paulo, Institute of Physics,
Caixa Postal 66318, 05315-970 Sao Paulo, Brazil

W. Haase
6.3

Technical University of Darmstadt, Institut fur Physikalische
Chemie, Petersenstrasse 20, 64287 Darmstadt, Germany

M. Hara
10.3

RIKEN, The Institute of Physical & Chemical Research,
2-1 Hirosawa, Wako-shi, Saitama, 351-0198 Japan

M. Heckmeier
11.4

Merck KGaA, Postfach 64271, Darmstadt Frankfurter
Strasse 250, 64293 Darmstadt, Germany

M. Hird
1.1

University of Hull, School of Chemistry,
Hull, HU6 7RX, UK

K. Hori
4.3

Ochanomizu University, Department of Chemistry,
Otsuka, Bunkyo-ku, Tokyo 112, Japan

C.T. Imrie
1.3

University of Aberdeen, Department of Chemistry,
Meston Walk, Aberdeen, AB9 24E, Scotland, UK

T. Ishikawa
5.4

Kent State University, Liquid Crystal Institute,
Kent, OH 44242, USA

Y. Kawamura
3.4

RIKEN, The Institute of Physical & Chemical Research,
2-1 Hirosawa, Wako-shi, Saitama, 351-0198 Japan

I.C. Khoo
7.4

Penn State University, Department of Electrical Engineering,
University Park, PA 16802, USA

L. Kramer
8.6

University of Bayreuth, Theoretical Physics II,
Postfach 101251, D-95440 Bayreuth, Germany

H. Kresse
6.2

Martin Luther University, Institute for Physical Chemistry,
Muhlpforte 1, Halle Saale D-06108, Germany

H.-G. Kuball
7.2

Universitat Kaiserslautern, Fachbereich Chemie,
Erwin-Schrödinger Strasse, D-67663 Kaiserslautern, Germany

O.D. Lavrentovich
5.4

Kent State University, Liquid Crystal Institute,
Kent, OH 44242, USA

F.M. Leslie*
8.1, 8.5

University of Strathclyde, Department of Mathematics,
Glasgow, G1 1XH, Scotland, UK

G.R. Luckhurst
2.1

University of Southampton, Department of Chemistry and
Southampton Liquid Crystal Institute, Southampton,
SO17 1BJ, UK

J.I. Martinez
6.4

Universidad de Zaragoza, Department of Physics,
Facultad de Ciencias, Zaragoza 50009, Spain

A.F. Martins
8.3

Universidade Nova de Lisboa, FCT-Dept. Ciência dos Materiais,
2825-114 Monte da Caparica, Portugal

S. Miyajima
9.1

Okazaki National Research Institute, Department of. Molecular Assemblies, Institute for Molecular Science, Myodaiji, Okazaki 444-8585, Japan

H. Molsen
11.5

Sharp Laboratories of Europe Ltd., Oxford Science Park, Edmund Halley Road, Oxford, OX4 4GA, UK

G.J. Moro
2.3

Universita Degli Studi di Padova, Dipartimento di Chimica Fisica, via Loredan, 2-35131 Padova, Italy

K.J. Moscicki
8.2

Jagellonian University, Institute of Physics, Reymonta 4, 30-059 Krakow, Poland

S. Naemura
11.2

Merck Japan Limited, Atsugi Technical Center, 4084 Nakatsu, Aikawa-machi, Kanagawa-ken 243-0303, Japan

P.L. Nordio*
2.3

Universitata Degli Studi di Padova, Dipartimento di Chimica Fisica, via Loredan, 2-35131 Padova, Italy

H. Ohnishi
11.1

Dainippon Ink & Chemicals Inc., 4472-1 Ooazakomuro, Inamachi, Kitaadachi-gun, Saitama, 362-8577 Japan

T. Ohtsuka
11.1

Dainippon Ink & Chemicals Inc., 4472-1 Ooazakomuro, Inamachi, Kitaadachi-gun, Saitama, 362-8577 Japan

K. Okano
3.4

University of Tokyo
contact address: 3-19-4-209 Kamishakujii, Nerima-ku, Tokyo 177-0044, Japan

W. Pesch
8.6

University of Bayreuth, Theoretical Physics II, Postfach 101251, D-95440 Bayreuth, Germany

A.G. Petrov
5.5

Bulgarian Academy of Sciences, Institute of Solid State Physics (Biomolecular Layers Department), 2 Tzarigradsko chaussee, BG-1784 Sofia, Bulgaria

S.J. Picken
2.2

Delft University of Technology, Department of Materials Science and Technology, Julianalaan 136, 2628 BL Delft, The Netherlands

K. Praefcke
1.2

Technical University of Berlin, J4, Strasse des 17 Juni 124, D-10623 Berlin, Germany

R.M. Richardson
4.2, 9.3

University of Bristol, Department of Physics, Tyndall Avenue, Bristol, BS8 1TL, UK

R. Righini
9.2

University of Florence, Department of Chemistry, LENS, Via G. Capponi, 9-50121 Florence, Italy

H. Saito
11.3

Chisso Petrochemical Corporation, Speciality Chemicals Research Center, 5-1 Goikaigan Ichihara-shi, Chiba-ken 290, Japan

J.R. Sambles
7.3

University of Exeter, Department of Physics,
Stocker Road, Exeter, Devon, EX4 4QL, UK

M. Sandmann
3.3

Ruhr University, Physical Chemistry II, Universitatsstrasse 150,
D-44780 Bochum, Germany

Y.R. Shen
7.5

University of California, Berkeley, Department of Physics,
Berkeley, CA 94720-7300, USA

M. Sorai
3.2

Osaka University, Research Centre for Molecular
Thermodynamics, Graduate School of Science, Toyonaka,
Osaka 560, Japan

J. Springer
10.1

Technical University of Berlin, Institute for Technical Chemistry,
Fachgebeit Makromolecular Chemie, Strasse des 17 Juni 135,
D-10623 Berlin, Germany

I.W. Stewart
5.1

University of Strathclyde, Department of Mathematics,
Glasgow, G1 1XH, Scotland, UK

A. Sugimura
10.2

Osaka Sangyo University, Department of Information
Systems Engineering, 3-1-1 Nakagaito, Osaka, 574, Japan

H. Takatsu
11.1

Dainippon Ink & Chemicals Inc., 4472-1 Ooazakomuro,
Inamachi, Kitaadachi-gun, Saitama, 362-8577 Japan

K. Tarumi
11.4

Merck KGaA, Postfach 64271, Darmstadt Frankfurter
Strasse 250, 64293 Darmstadt, Germany

G. Ungar
4.1

University of Sheffield, Department of Engineering Materials,
Robert Hadfield Building, Mappin Street, Sheffield, S1 3JD, UK

S. Urban
6.1

Jagellonian University, Institute of Physics,
Reymonta 4, 30-059 Krakow, Poland

G.R. Van Hecke
3.1

Harvey Mudd College, Department of Chemistry,
Claremont, CA 91711, USA

M.R. Wilson
12.3

University of Durham, Department of Chemistry,
South Road, Durham, DH1 3LE, UK

A. Würflinger
3.3

Ruhr University, Physical Chemistry II, Universitatstrasse 150,
D-44780 Bochum, Germany

F. Yang
7.3

University of Exeter, Department of Physics,
Stocker Road, Exeter, Devon, EX4 4QL, UK

C. Zannoni
12.2

Universita di Bologna, Dipartimento di Chimica Fisica
ed Inorganica, viale Risorgimento 4, Bologna, 40136, Italy

*deceased

Abbreviations

AA	all-atom force field
AC	alternating current
AMBER	force field
AR	abnormal roll
BS	Schiff's base
BTN	bistable twisted nematic
CD	circular dichroism
CDF	cylindrical distribution function
CMOS	complementary metal oxide semiconductor
CW	continuous wave
DC	direct current
DNMR	deuterium nuclear magnetic resonance (see also NMR)
DSC	differential scanning calorimetry
DSM	dynamic scattering mode
DTA	differential thermal analysis
EC	electroconvection
ECB	electrically controlled birefringence
EISF	elastic incoherent structure factor
EMS	extended Maier-Saupe
EPR	electron paramagnetic resonance
ER	escaped-radial
ESR	electron spin resonance
ET	escaped-twisted
FG	field gradient
GB	Gay-Berne
HAN	hybrid-aligned nematic
HB	Hopf bifurcation
HFF	force field
HLGW	half-leaky guided wave
HOPG	highly oriented pyrolytic graphite
HTP	helical twisting power
IPS	in-plane switching
IQENS	incoherent quasi-elastic neutron scattering
IR	infrared
LC	liquid crystal
LCD	liquid crystal display
LJ	Lennard-Jones
LRO	long range order
LS	light scattering spectrum

MC	Monte Carlo
MD	molecular dynamics
MM2	force field
MM3	force field
MM4	force field
MO	molecular orbital
MPP	Martin-Parodi-Pershan
N	nematic
N*	chiral nematic
N_b	biaxial nematic
N_D	discotic nematic
N_{col}	columnar nematic
N^*_D	chiral discotic nematic
N^*_{col}	chiral columnar nematic
N_u	uniaxial nematic
NEMD	non-equilibrium molecular dynamics
NI	nematic-isotropic
NMR	nuclear magnetic resonance (see also DNMR)
NR	normal roll
OCB	optically compensated birefringence
OPLS	force field
ORD	optical rotatory dispersion
PC	polycarbonate
PD	pendant drop method
PDLC	polymer dispersed liquid crystal
PFG	pulsed field gradient
PGSE	pulsed field gradient spin echo
PUT	positionally unbiased thermostat
pVT	pressure-volume-temperature
RC	radii of curvature
RF	radio frequency
R-LCD	reflective liquid crystal display
SCN	surface controlled nematic
SD	sessile drop method
SM	standard model
SmA	smectic A
SQUID	supraconducting quantum interference device
SRA	straight ray approximation
STM	scanning tunnelling microscopy
STN	supertwisted nematic
TBM	thermobarometric method
TE	transverse electric
TFT	thin film transistor
TM	transverse magnetic
TN	twisted nematic

UA	united atom force field
UV	ultraviolet
VA	vertically aligned
WEM	weak electrolyte model
ZBD	zenithal bistable device
ZZ	zig-zag

Acronyms of chemical substances
(Alternative acronyms are sometimes used, and the
more common acronym is indicated in brackets)

nAB	4,4′-dialkylazoxybenzene
5AB	4,4′-dipentylazoxybenzene
APAPA	4-methoxybenzylidene-4′-acetoxybenzene *or* anisilidene-4-amino phenylacetate
AZA9	poly(4,4′-dioxy-2,2′-dimethoxyazoxybenzene-nonanediyl) *(see also DDA9)*
Azpac	Pd-complex of 4,4′-dihexyloxyazoxybenzene
BBBA	4-butyloxybenzilidene-4′-butylaniline
BMAB	4-methoxyphenylazoxy-4′-butylbenzene
BMAOB	4-butyl-4′-methoxyazoxybenzene
6BAP3	1-(4-propylphenyl)-2-(4-hexyl[2,2,2-bicyclooctyl])ethane
6BAP(F)	1-(4-propoxy-3-fluorophenyl)-2-(4-hexyl[2,2,2-bicyclooctyl])ethane
55-BBCO	4,4′-dipentylbibicyclo[2,2,2]octane
CBAn	α,ω-bis(4-cyanobiphenyl-4′-yloxy)alkanes
CBn	*see nCB*
nCB	4-alkyl-4′-cyanobiphenyl *the less common IUPAC systematic name is* 4-(4-alkylphenyl)benzenecarbonitrile
5CB	4-pentyl-4′-cyanobiphenyl
5CB-βd$_2$	4-pentyl[β,β′-^{2}H]-4′-cyanobiphenyl
6CB	4-hexyl-4′-cyanobiphenyl
7CB	4-heptyl-4′-cyanobiphenyl
8CB	4-octyl-4′-cyanobiphenyl
8CB-βd$_2$	4-octyl[β,β′-^{2}H]-4′-cyanobiphenyl
9CB	4-nonyl-4′-cyanobiphenyl
11CB	4-undecyl-4′-cyanobiphenyl
CB15	4-[(2-methyl)butyl]-4′-cyanobiphenyl
3CBCN	4-(trans-4-propylcyclohexyl)benzonitrile (PCH3)
CBOBP	4-cyanobenzoyloxy-4′-octylbenzoyloxy-p-phenylene
CBOOA	4-cyanobenzylidene-4′-octyloxyaniline *the less common IUPAC systematic name is* 1-[(E)-2-aza-2-(4-octyloxyphenyl)vinyl]benzenecarbonitrile
7CCF	4-fluorophenyl-trans-4′-heptylcyclohexanecarboxylate
nCCH	*see CCHn*
CCHn	trans,trans-4′-alkylbicyclohexyl-4-carbonitrile *the less common IUPAC systematic name is* 4-(4-alkylcyclohexyl)cyclohexanecarbonitrile
CCH5	trans,trans-4′-pentylbicyclohexyl-4-carbonitrile
CCH7	trans,trans-4′-heptylbicyclohexyl-4-carbonitrile
6CHBT	4-(trans-4′-hexylcyclohexyl)isothiocyanatobenzene
5CT	4-pentyl-4″-cyano-p-terphenyl
DABT	aramid polymer
DaCl	decylammonium chloride
DDA9	poly(4,4′-dioxy-2,2′-dimethoxyazoxybenzene-dodecanediyl) *(see also AZA9)*
D-EABAC	deuteriated 4-(4′-acetoxybenzylidene)aminocinnamate
D-IBPBAC	deuteriated 4-isobutyl(4′-phenylbenzylidene)aminocinnamate
DOBAMBC	4-decyloxybenzylidene-4′-amino 2-methyl butyl cinnamate
D-5CB	deuteriated 4-pentyl-4′-cyanobiphenyl

EBBA	4-ethoxybenzylidene-4′-butylaniline *the less common IUPAC systematic name is* 4-[(1E)-2-aza(4-butylphenyl)vinyl]-1-ethoxybenzene
EPPC	4-ethoxyphenyl trans-4-propylcyclohexylcarboxylate
E7	commercial mixture (Merck UK, formerly BDH Chemicals)
F-5CB	monofluorinated 4-pentyl-4′-cyanobiphenyl
HBAB	4-hexyloxybenzylidene-4′-aminobenzonitrile
HBA	4-hydroxybenzoic acid
HCPB	4-heptyl-4′-cyanophenylbenzoate
HOAB	4,4′-dihexyloxyazoxybenzene
HOACS	4-hexyloxy-4′-amyl-α-cyanostilbene
HOBA	4-heptyloxybenzoic acid
HOT	4,4′-dihexyloxytolane
5HexPhCN	1-pentyl-4(4′-cyanophenyl)cyclohexane (PCH5)
7HexPhCN	1-heptyl-4(4′-cyanophenyl)cyclohexane (PCH7)
10HexPhCN	1-decyl-4(4′-cyanophenyl)cyclohexane (PCH10)
5HexHexCN	4-pentyl-4′-cyanobicyclohexane (CCH5)
ITO	indium-tin-oxide
I52	4-pentyl(cyclohexyl)-4-ethyl-2-fluorobiphenylethane
KL	potassium laurate
MBBA	4-methoxybenzylidene-4′-butylaniline *the less common IUPAC systematic name is* 1-[(E)-2-aza(4-butylphenyl)vinyl]-1-methoxybenzene
MDCB	metadichlorobenzene
Mixture A	(4-butyl-4′-methoxyazoxybenzene + 4-butyl-4′-heptanoyloxy-azoxybenzene, 2:1)
nOAB	4,4′-dialkoxyazoxybenzene *the less common IUPAC systematic name is* bis(4-alkoxyphenyl)diazeneoxide
nOCB	4-alkoxy-4′-cyanobiphenyl *the less common IUPAC systematic name is* 4-(4-alkoxyphenyl)benzenecarbonitrile
ODCB	orthodichlorobenzene
nOmPCH	4-alkoxy-1-(4′-alkylcyclohexyl)benzene
OHMBBA	2-hydroxy-4-methoxybenzylidene-4′-butylaniline
OOBA	4-octyloxybenzoic acid
5OCB	4-pentyloxy-4′-cyanobiphenyl
6OCB	4-hexyloxy-4′-cyanobiphenyl
7OCB	4-heptyloxy-4′-cyanobiphenyl
8OPCBOB	4-octyloxyphenyl-4-(4-cyanobenzyloxy)benzoate
4O.8	4-butoxybenzylidene-4′-octylaniline
5O.7	4-pentyloxybenzylidene-4′-heptylaniline
PAA	4,4′-dimethoxyazoxybenzene *(see nOAB)*
PAP	4,4′-diethoxyazoxybenzene *(see nOAB)*
PB	quaternary mixture of phenylbenzoates
nPBCm	4-alkylphenyl-(4′-alkyl)bicyclohexane
PBLG	poly(γ-benzyl-L-glutamate)
PC	polycarbonate
PCPB	4-pentylphenyl-2-chloro-4-(4-pentylbenzoyloxy)benzoate
nPCH	*see PCHn*

PCHn	trans-4-alkyl(4'-cyanophenyl)cyclohexane *the less common IUPAC systematic name is 4-(4-alkylphenyl)benzenecarbonitrile*
PCH5	trans-4-pentyl(4'-cyanophenyl)cyclohexane
PCH7	trans-4-heptyl(4'-cyanophenyl)cyclohexane
PEBAB	4-ethoxybenzylidene-4'-benzonitrile
pHB	phenyl-4-(4-benzoyloxy)benzoyloxybenzoate
PMMA	poly(methylmethacrylate)
PPTA	poly(p-phenylene-terephthalamide)
PS	polystyrene
Psi4	poly[(2,3,5,6-tetradeuterio-4-methoxyphenyl-4'-butanoxybenzoate)-methylsiloxane]
PVCi	polyvinylcinnamate
5PhHexCN	1-cyano-4(4'-pentylphenyl)cyclohexane (PCH5)
6PCHNCS	4-(4-hexylcyclohexyl)benzeneisothiocyanate *(see also 6CHBT)*
ROCP-7037	5-heptyl-2-(4-cyanophenyl)-pyrimidine (Roche Company)
R6G	rhodamine dye
SiO	silicon oxide
TAPA	2-((2,4,5,7)-tetranitro-9-fluorenylideneaminooxypropionic acid
TBBA	terephthalylidene-bis(4-butylaniline)
THE5	hexakis(pentyloxy)triphenylene
TMS	tetramethylsilane
TPB10	poly[1,10-decylene-1-(4-hydroxy-4'-biphenylyl-2-(4-hydroxyphenyl)butane]
nTOm	4-alkyl-4'-alkoxytolane
5TO1	4-pentyl-4'-methoxytolane

CHAPTER 1

MOLECULAR STRUCTURE AND ORGANISATION

1.1   Relationship between molecular structure and transition
       temperatures for calamitic structures
1.2   Relationship between molecular structure and transition
       temperatures for organic materials of a disc-like
       molecular shape in nematics
1.3   Relationship between molecular structure and transition
       temperatures for unconventional molecular structures

1.1 Relationship between molecular structure and transition temperatures for calamitic structures

M. Hird

September 1998

A INTRODUCTION

There can surely be no better way of beginning a review of the physical properties of nematic liquid crystals than by discussing the effects of changing molecular structure on the melting point, transition temperatures and mesophase morphology. Such parameters are the most fundamental physical properties of liquid crystalline materials, and yet they are frequently undervalued or simply taken for granted.

The nematic mesophase has a structure similar to that of an isotropic liquid (no long range positional and orientational ordering), except for a statistically parallel arrangement of the constituent molecules which confers long range orientational order (see FIGURE 1). With this one degree of order, the nematic phase is distinct from the more ordered smectic phases, which additionally have one-dimensional translational ordering of the molecules to give a lamellar structure. There are even more ordered crystalline mesophases which have positional ordering of the molecules in three dimensions [1].

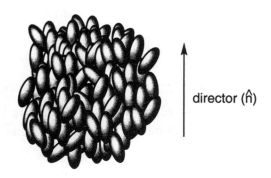

director ($\hat{n}$)

FIGURE 1 The structure of the nematic phase.

FIGURE 2 shows a general template of a rod-like molecule that can be used to represent typical calamitic systems that exhibit nematic and smectic phases. A and B represent core units (see FIGURE 3(a)) consisting of linearly linked aromatic or alicyclic rings that sometimes contain heteroatoms. X, Y and Z are linking groups (see FIGURE 3(b)) which are often absent (direct links), but when present consist of moieties which maintain the long lath-like structure and enhance the longitudinal polarisability. R and R′ are terminal moieties (see FIGURE 3(c)) which usually consist of moderately long alkyl or alkyloxy chains (sometimes branched), or often the cyano group. In FIGURE 3(d) M and N are the lateral substituents which are not present for all systems.

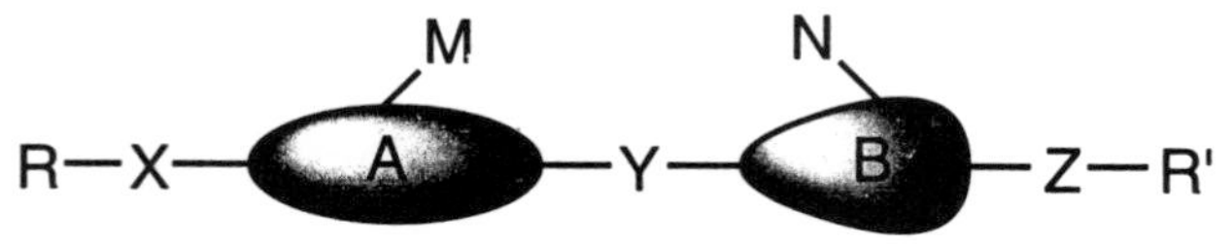

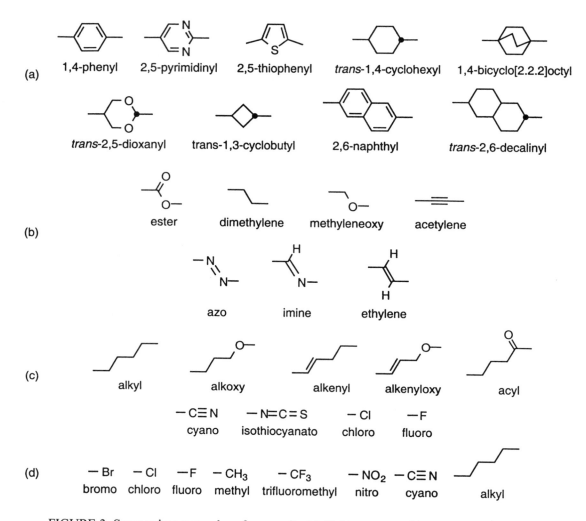

FIGURE 2 A general structural template for calamitic liquid crystals.

(a) 1,4-phenyl   2,5-pyrimidinyl   2,5-thiophenyl   *trans*-1,4-cyclohexyl   1,4-bicyclo[2.2.2]octyl

trans-2,5-dioxanyl   trans-1,3-cyclobutyl   2,6-naphthyl   *trans*-2,6-decalinyl

(b) ester   dimethylene   methyleneoxy   acetylene

azo   imine   ethylene

(c) alkyl   alkoxy   alkenyl   alkenyloxy   acyl

−C≡N   −N=C=S   −Cl   −F
cyano   isothiocyanato   chloro   fluoro

(d)   −Br   −Cl   −F   −CH₃   −CF₃   −NO₂   −C≡N
bromo   chloro   fluoro   methyl   trifluoromethyl   nitro   cyano   alkyl

FIGURE 3 Some prime examples of core units (a), linking groups (b), terminal moieties (c)
and lateral substituents (d) employed in mesogenic molecules.

Many books and review articles have been written on liquid crystalline structures and their mesomorphic behaviour [1-7].

B   CORE UNITS

There are a great many combinations of core units that have been used in the generation of liquid crystals. However, not all will solely provide the nematic phase; some will give only smectic phases and others will allow for both phase types to be exhibited. *[Note on diagram captions in this Datareview: all phase transition temperatures are in ℃.]*

1
Cr 60.0 E 72.5 B 73.5 [N 50] I

2
Cr 37.0 N 52.0 I

3
Cr 70.0 SmA 90.5 N 140.5 I

4
Cr 39.9 N 59.7 I

The smectic phases are lamellar in structure and more ordered than the nematic phase; accordingly the molecular structure required to generate a smectic phase must allow for lateral intermolecular attractions. In terms of core units, the smectic phases are favoured by materials composed of a wholly aromatic core (e.g. compound **1**) [8] or a wholly alicyclic core (e.g. compound **3**) [9]: see also other compounds in later discussions. The incompatibility of mixed cores, e.g. compounds **2** [10] and **4** [9], tends to disrupt lamellar packing and favour the generation of the nematic phase. If the core is broad then a layered structure is also less favoured, a feature illustrated by the phase morphology of compound **3** which exhibits a smectic A phase and a high nematic phase stability despite a wholly alicyclic core. Increasing the length of the core with a unit similar to those present tends to increase the transition temperatures significantly and to increase the melting point to a lesser extent; hence a wider mesophase range results, but generally the mesophase morphology is not altered: compare compounds **5** [11] and **6** [12].

5
Cr –0.8 (Sm –8.0 N –5.0) I

6
Cr 13.0 SmA 164.0 N 166.0 I

Very few compounds with compatible core units without linking groups or lateral substituents exist that solely exhibit the nematic phase. Compound **7e** [13-15] 4-pentyl-4′-cyanobiphenyl (5CB or K15) is perhaps the most famous nematogen because it is widely used in mixtures for displays, but the reason for the generation of the nematic phase is more a consequence of the short, polar terminal cyano group than the choice of core unit. The phenylcyclohexane (PCH) materials such as compound **8** [16] were not expected to be liquid crystalline because of a reduced anisotropy in the polarisability, but compound **8** is a very useful low birefringence, low viscosity nematogen for use in mixtures for displays. However, where the cyclohexane ring separates the phenyl ring and the cyano group (compound **9**) [17] the nematic phase stability is very low. In fact, compound **10** [18] (CCH) has an even higher nematic phase stability than compound **8**. Hence longitudinal polarisability is not the only factor which determines transition temperatures; shape, polarity and flexibility are also important.

Structures of compounds 7e, 8, 9, 10, 11, 12, 13, 14, 15:

8
Cr 31.0 N 55.0 I

7e
Cr 24.0 N 35.0 I

9
Cr < 20 (N –25.0) I

10
Cr 62.0 N 85.0 I

11
Cr 130.0 N 239.0 I

12
Cr 68.0 N 130.0 I

13
Cr 71.0 (N 52.0) I

14
Cr 56.0 (N 52.0) I

15
Cr 36.0 N 61.0 I

Increasing the size of the core with another phenyl ring provides a nematogen (compound **11**, 4-pentyl-4″-cyano-p-terphenyl (5CT)) [14] with an increased melting point and a much higher nematic phase stability which is used in nematic mixtures for displays in order to enhance the nematic-isotropic transition temperature. Extension of the core with a naphthalene ring (compound **12**) [19] also broadens the core and so the transition temperatures are intermediate between those for compounds **7e** and **11**. Core units that contain heteroatoms are quite common in the structures of liquid crystals and some quite useful modifications to physical properties can be generated by them. The heteroatoms in compound **13** [20] reduce the interannular twisting of the biphenyl unit and the nematic-isotropic transition temperature is higher than for compound **7e**. However, the enhanced polarity of the nitrogens produces a rather high melting point, which is analogous to the effect on melting point of the heterocyclic oxygens in compound **14** when compared with compound **8**. The nematic-isotropic transition temperature of compound **14** [21] is little different from the parent system (**8**) because polarisability is not an issue with the saturated ring system. Phenylpyrimidines with terminal alkyloxy and alkyl chains are more common as smectic C ferroelectric host materials; however, the short chain homologues (e.g. compound **15**) [22] are nematogens of low melting point. Interestingly, the biphenyl analogues (e.g. compound **16**) [11] are strong smectogens; seemingly the heterocyclic nitrogens must disrupt the lamellar packing.

Importantly, this liquid crystalline behaviour for the various core units can change dramatically through the influence of other structural variations, such as linking groups, terminal moieties and lateral substituents. The possibility of such wide variations in structure leads to some interesting, and often confusing, trends in melting points, transition temperatures and mesophase morphology.

C   LINKING GROUPS

16
Cr 26.0 Sm 47.6 S 52.2 I

17
Cr 34.8 (N 26.0) I

18
Cr 62.0 [N −24] I

19
Cr 30.0 N 51.0 I

In general, linking groups tend to reduce the lateral attractions and hence disrupt lamellar packing and so favour the nematic phase. For example, compound **16** [11] is strongly smectogenic, but when the core is separated by an ester linkage lateral attractions are disrupted and a nematogen (compound **17**) [23] results. However, smectic phases are supported in many materials with an ester linkage due to the polarity providing some tendency for lateral attractions.

20
Cr 64.5 (N 55.5) I

21
Cr 79.5 (N 70.5) I

22
Cr 40.0 SmB 110.4 I

23
Cr 52.0 (N 50.0) I

24
Cr 22.0 N 47.0 I

25
Cr 22.0 N 65.0 I

The effect of linking groups on mesophase morphology and transition temperatures is very dependent upon the core units that are linked. For example, where a non-conjugative linking group (e.g. dimethylene) separates two aromatic core units the liquid crystal phase stability is drastically reduced; compare compounds **7e** and **18** [24]. However, if the same linking group does not separate two compatible cores then liquid crystalline phases are well supported; compare compounds **8** and **19** [25]. An ester linking group between two aromatic core units increases the molecular length and enhances the longitudinal polarisability and planarity and so increases the nematic phase stability despite the generation of broader molecules due to the zig-zag structure; compare compounds **7e** and **20** [26]. The acetylenic (tolane) linking group between two aromatic core units maintains linearity, extends the molecular length and enhances longitudinal polarisability, and hence the nematic phase stability is significantly enhanced; compare compounds **7e** and **21** [27]. Cyclohexane rings in most systems are very prone to generate smectic phases due to excellent lamellar packing (compound **22**) [28], but the use of an acetylene linking group (compound **23**) [29] eliminates the smectic phase to leave a monotropic nematic and because the linking group is incompatible with the two cyclohexane rings the nematic-isotropic transition temperature is drastically reduced. Imine (**24**) [30] and azoxy

(25) linking groups enhance molecular length and longitudinal polarisability to give nematogens of reasonably high nematic-isotropic transition temperature; however, they proved unsuitable for display applications because the imine group is very susceptible to moisture and the azoxy unit suffers from severe photochemical instability.

D   TERMINAL MOIETIES

Of course, all of the generalisations so far vary depending on the exact structure, i.e. the effect of linking group changes depends on the core units, the effect of changing the core units depends on terminal chains, and so on. Speaking of terminal chains, it is probable that the terminal chains have the greatest influence on the relative smectic and nematic character, but yet again other factors are important. Short terminal chains tend to disfavour lamellar packing, but allow the orientational order necessary for the generation of the nematic phase. As the chain length increases then entanglement occurs and the molecules aggregate in layers and hence smectic phases become prevalent. As the terminal chain length increases the nematic tendency often eventually decreases because the flexible chains generate many conformers and the orientational order is lost without lamellar stabilisation. So typically short chain homologues will exhibit the nematic phase with no smectic tendency (e.g. compound 26), but if the core system is strongly supportive of smectic character then even very short chain homologues will be smectogenic. As the chains increase in length the nematic phase stability may initially increase and the smectic tendency increases (e.g. compound 27); eventually all nematic character will be lost and smectogenic materials will be produced (e.g. compound 28) [31].

R^1—⬡—C(=O)—O—⬡—R^2

26; $R^1 = C_3H_7$, $R^2 = C_3H_7$;   Cr 22.8 N 36.6 I
27; $R^1 = C_5H_{11}$, $R^2 = C_3H_7$;   Cr 24.0 SmB 37.5 N 52.5 I
28; $R^1 = C_5H_{11}$, $R^2 = C_5H_{11}$;   Cr 52.0 SmB 72.0 I

Odd-even effects in the nematic phase stability occur for many systems. Assuming an all-trans conformation of the terminal chains, the addition of a CH_2 to provide an odd chain enhances the molecular length without increasing the breadth whereas the additional CH_2 unit to give an even chain would increase the molecular breadth; hence the nematic phase stability for odd chain homologues is slightly higher than comparable even chain homologues based on the expected trend (see TABLE 1 for examples). Eventually, as the length of the terminal chain becomes very long the odd-even effect tends to diminish.

Many other structural possibilities for the terminal chains can have a great influence on melting point, transition temperatures and mesophase morphology. For example, branched chains are common, usually for the purpose of generating a chiral centre for chiral nematic or chiral smectic C liquid crystals. The effect of the branch is to broaden the molecules and hence the transition temperatures are usually depressed significantly, often with the largest reduction in the smectic phase stability, but melting points are often not so much affected (compare compounds 29 and 30) [32].

TABLE 1 Transition temperatures for alkylcyanobiphenyl homologues [13,14,27].

R—⬡—⬡—CN

Compound		Transition temperatures (°C)			
No	R	Cr	SmA	N	I
7a	CH_3	• 109.0	-	[• 45.0]	•
7b	C_2H_5	• 75.0	-	[• 22.0]	•
7c	C_3H_7	• 66.0	-	(• 25.5)	•
7d	C_4H_9	• 48.0	-	(• 16.5)	•
7e	C_5H_{11}	• 24.0	-	• 35.0	•
7f	C_6H_{13}	• 14.5	-	• 29.0	•
7g	C_7H_{15}	• 30.0	-	• 43.0	•
7h	C_8H_{17}	• 21.5	• 33.5	• 40.5	•
7i	C_9H_{19}	• 42.5	• 48.0	• 49.5	•
7j	$C_{10}H_{21}$	• 44.0	• 50.5	-	•
7k	$C_{11}H_{23}$	• 53.0	• 57.5	-	•
7l	$C_{12}H_{25}$	• 48.0	• 58.5	-	•

However, if the branch unit is polar (e.g. fluoro), then smectic phases are often supported to the exclusion of the nematic phase due to enhanced lateral attractions (e.g. compound **31** [33] compared with compound **15**). In fact, superfluoro and perfluoro chains are very stiff which facilitates excellent lamellar packing and so exceptionally high smectic phase stability is generated (e.g. compound **32**) [34].

29
Cr 65.5 SmI 74.5 SmC 118.5 SmA 135.0 N 137.0 I

30
Cr 64.0 SmC 86.0 N 100.5 I

31
Cr 53.0 SmC 67.0 SmA 82.0 I

32
Cr 65.0 (SmC 64.0) SmA 138.0 I

Thus such polar units in the terminal chains are best avoided in the generation of nematic materials, unless other structural features are included to disrupt the lamellar attractions (e.g. lateral substituents). The use of an alkyloxy terminal chain with liquid crystals (compounds **33**) based on aromatic core units tends to generate higher phase stability than compounds with analogous alkyl chains (compounds **9**).

Typical differences of transition temperatures between alkyl and alkyloxy terminal chains can be seen by comparing the compounds in TABLES 1 and 2; together with short chains the alkyloxy unit enhances phase stability by about 60°C, but for longer chains the increase is only about 30°C. In addition to a large increase in the nematic phase stability, the polar nature of the ether oxygen can also cause, for example, the introduction of a tilted smectic C phase (compound **35**) [35] not present in the analogous alkyl homologues (compound **34**) [35].

TABLE 2 Transition temperatures for alkyloxycyanobiphenyl homologues [13,14,27].

$$R \text{—} \langle \rangle \text{—} \langle \rangle \text{— CN}$$

Compound		Transition temperatures (°C)			
No	R	Cr	SmA	N	I
33a	CH_3O	• 105.0	-	(• 85.5)	•
33b	C_2H_5O	• 102.0	-	(• 90.5)	•
33c	C_3H_7O	• 74.5	-	(• 64.0)	•
33d	C_4H_9O	• 78.0	-	(• 75.5)	•
33e	$C_5H_{11}O$	• 48.0	-	• 68.0	•
33f	$C_6H_{13}O$	• 57.0	-	• 75.5	•
33g	$C_7H_{15}O$	• 54.9	-	• 74.0	•
33h	$C_8H_{17}O$	• 54.5	• 67.0	• 80.0	•
33i	$C_9H_{19}O$	• 64.0	• 77.5	• 80.0	•
33j	$C_{10}H_{21}O$	• 59.5	• 84.0	-	•
33k	$C_{11}H_{23}O$	• 71.5	• 87.5	-	•
33l	$C_{12}H_{25}O$	• 70.0	• 90.0	-	•

34
Cr 51.5 SmB 62.0 SmA 109.5 N 136.5 I

35
Cr 65.5 SmC 96.5 N 172.5 I

36
Cr –5.0 SmB 67.0 I

37
Cr 10.0 N 17.0 I

38
Cr 38.0 (SmC 35.2) N 106.6 I

The use of an alkyloxy chain in liquid crystals does not always enhance phase stability. When an alkyloxy chain is used in a cyclohexane ring, the incompatibility of the oxygen causes a significant reduction in the nematic-isotropic transition temperature, and because of the severe disruption in lamellar packing the smectic B phase is replaced by the less ordered nematic phase (compare compounds **36** and **37**) [36]. Additionally, an oxygen atom away from the core within an alkyl terminal chain can disrupt the molecular packing through structural incompatibility to give nematic liquid crystals with lower melting points and much reduced smectic phase stability (e.g. compound **38**).

39
Cr 16.0 N 58.5 I

40
Cr 60.0 N 73.7 I

41
Cr 30.0 (N 10.2) I

Alkenic units within the terminal chains have been employed to generate nematic liquid crystals [37], and the transition temperatures depend critically on the position of the alkenic unit. For example, when compared to the saturated parent system (compound **8**), compound **39** has a much lower melting point with a slightly higher nematic-isotropic transition temperature, compound **40** has a much higher nematic-isotropic transition temperature, but the melting point is also much higher, and compound **41** has a similar melting point but a very much reduced nematic-isotropic transition temperature.

C_5H_{11} —◯—◯—◯— X
42, X = CF$_3$ Cr 123.0 N 124.2 I
43, X = OCF$_3$ Cr 43.0 Sm 128.0 N 147.0 I

C_3H_7 —◯—◯—◯— X
44, X = F   Cr 90.0 N 158.0 I
45, X = Cl   Cr 70.0 Sm 79.0 N 193.0 I

C_5H_{11} —◯—◯—◯—
46
Cr 45.2 N 125.0 I

C_3H_7 —◯—◯—◯—
47
Cr 36.1 N 105.2 I

The terminal cyano group provides good nematogens for simple displays, but for active matrix displays nematic liquid crystals of high resistivity are required. This requirement has been met through compounds that incorporate fluoro and/or chloro substituents as part of the terminal unit (e.g. **42-47**) [38-40]. The use of short terminal chains, linking groups and incompatible core units ensures low melting points and a lack of smectic phases for many compounds. Appropriate compounds are blended into mixtures with optimised physical properties for displays. Such materials for active matrix displays have been reviewed in detail by Petrov [41].

C_3H_7 —◯—◯—
48
Cr 54.0 SmB 91.0 N 103.8 I

C_5H_{11} —◯—◯—
49
Cr 9.2 N 60.2 I

C_5H_{11} —◯—◯—≡— Cl
50
Cr 66.0 N 70.0 I

C_5H_{11} —◯—◯—≡— CN
51
Cr 50.0 N 129.0 I

Some rather unusual fluoro-substituted terminal units have been used to provide nematic liquid crystals. Compound **48** [42] has a wholly alicyclic structure and hence gives a high smectic B phase stability in addition to a short nematic temperature range. Compound **49** [43] has a low melting point and a room temperature nematic phase; however, it is likely to be very unstable. Likewise, acetylenic compounds **50** [44] and **51** [45] are interesting nematogens, but may have stability problems.

E   LATERAL SUBSTITUENTS

The use of lateral substituents in liquid crystals has proved to be very important, initially in nematic material and later in smectic C materials. Clearly, anything that sticks off the side of a rod-like molecule will tend to reduce the liquid crystal phase stability, and generally the larger the lateral substituent the greater the reduction in liquid crystal phase stability. Usually, the smectic phase stability is much more affected than that of the nematic phase, especially by larger substituents because of the obvious reduction in lateral attractions, but increased lateral attractions associated with polar substituents cause a smaller reduction in smectic phase stability (see compounds **52-57**) [46].

52, X = H   Cr 50.0 Sm 196.0 I
53, X = F   Cr 61.0 Sm 79.2 N 142.8 I
54, X = Cl   Cr 46.1 N 96.1 I
55, X = CH$_3$ Cr 55.5 N 86.5 I
56, X = Br   Cr 40.5 N 80.8 I
57, X = CN   Cr 62.8 (Sm 43.1) N 79.5 I

Generally, fluorine is the most used lateral substituent because it is the next smallest to hydrogen (1.47 Å versus 1.20 Å), and the combination of small size and high electronegativity can provide nematic materials with low melting points, reasonably high nematic-isotropic transition temperatures and some beneficial physical properties.

58
Cr 180 E 200 SmB 214 SmA 218 I

59
Cr 50.0 N 140.6 I

60
Cr 61.0 SmA 99.5 N 141.5 I

The precise influence of lateral substituents on melting points, transition temperatures and mesophase morphology is more complex and depends strongly on other structural moieties, and more subtly on their location. Terphenyls (e.g. compound **58**) have very high melting points and show a strong smectic tendency; however fluoro substitution in a lateral position has a dramatic effect on mesomorphic behaviour, with compound **59** having a low melting point, a wide nematic range and no smectic phases. The subtlety of structural changes on mesomorphic behaviour is illustrated by compound **60** which is isomeric to compound **59**, but it exhibits a higher melting point and a smectic A phase to high temperature, yet the nematic phase stabilities are very similar [35]. Many other terphenyls with lateral fluoro substituents have been prepared and a wide variety of mesomorphism can be generated depending on the position and number of fluoro substituents and the nature and length of the terminal chains [32,47-49]. The central location of two fluoro substituents, in combination with reasonably short terminal chains (compound **61**), ensures a lack of smectic phases. However, the isomeric terphenyl with the two fluoro substituents in an outer ring (compound **62**) generates strong smectic C and smectic A tendency in addition to a much higher nematic phase stability than that of compound **61**. The longitudinal polarisability of compound **61** is reduced by the two inter-annular twists caused by the two fluoro substituents, and the overall liquid crystal phase stability is lower than for compound **62** which only has one such twisting. The high smectic tendency of compound **62** is due to the outer fluoro substituent which serves to fill vacant space and enhance the lateral attractions [32].

61
Cr 60.0 N 120.0 I

62
Cr 81.0 SmC 115.5 SmA 131.5 N 142.0 I

A range of fluoro-substituted biphenyls with a dimethylene-linked cyclohexane unit have provided nematogens (I compounds, e.g. **65**) with very low melting points that have been used in nematic mixtures for displays. Once again the fluoro substituent is far more destructive of smectic phases than of the nematic phase, but the location of the fluoro substituent is important (see compounds **63-65**) [50].

63, X = H, Y = H   Cr 67.0 SmA 119.0 N 144.0 I
64, X = H, Y = F   Cr 59.0 (SmA 34.0) N 108.0 I
65, X = F, Y = H   Cr 40.0 N 108.0 I

Esters are very common liquid crystal compounds, and lateral fluoro substitution has provided some interesting modifications to melting points, transition temperatures, mesophase morphology and physical properties. Compounds such as **66** [26] allow for the generation of high positive dielectric anisotropy, but the fluoro substituent has caused a large reduction in the nematic phase stability when compared with compound **20**, due to the reduction in antiparallel correlations. However, the fluoro substituent in compound **67** [26] is somewhat shielded by the zig-zag nature of the ester linkage; hence the nematic-isotropic transition temperature is identical to that of compound **20**. The lateral fluoro substituent in compound **68** [51] is not as shielded by the zig-zag structure and the nematic-isotropic transition temperature is much reduced, but not to the same extent as for compound **66**.

66
Cr 30.5 (N 24.5) I

67
Cr 53.5 N 55.0 I

68
Cr 65.5 (N 32.0) I

Lateral fluoro substituents are also valuable in providing nematic materials of negative dielectric anisotropy (see compound **61**), which is required for the electrically controlled birefringence (ECB) displays. The two-ring esters (**69-71**) have low clearing points, but lateral fluoro substitution reduces the melting point by more than the clearing point giving a wider nematic range than for the parent system [52].

13

69
Cr 37.0 (SmA 29.0) N 47.0 I

70
Cr 17.0 N 37.0 I

71
Cr 13.0 N 29.0 I

72
Cr 110.0 N 253.0 I

73
Cr 84.0 N 229.0 I

A very wide nematic range is found in the lateral fluoro-substituted tolane (**73**) which has a high birefringence and high negative dielectric anisotropy. Reductions in melting point and nematic-isotropic transition temperature of compound **73** when compared with the parent system (**72**) are both modest at about 25°C [52].

74
Cr 25.0 SmB 30.0 N 66.0 I

75
Cr 76.0 N 126.0 I

More unusual forms of lateral substitution are seen in compounds **74** and **75**. Compound **74** uses an axial cyano group to act as a lateral substituent which tends to reduce the strong smectic tendency of the parent system (**22**) and to introduce a nematic phase with a remarkably low melting point for such a polar compound [36]. Many compounds with a lateral carbonyl group tend to generate smectic phases because of the strong lateral polarity; however, use of short terminal chains and a dimethylene linking group ensures that compound **75** is a nematogen [53].

76
Cr 37.5 N 114.5 I

77
Cr 68.5 B 70.5 N 122.0 I

Lateral fluoro substitution in an aromatic environment tends to cause large reductions in melting points and/or smectic phase stability to create compounds with wide nematic ranges (compare compound **76** with compound **6**) [54]. Much less reduction of smectic character is caused by the corresponding lateral fluoro substitution in an alicyclic environment (compound **77**), but the nematic phase is generated [12].

F   CONCLUSION

Despite the essential restriction to a rod-like architecture, the structural variations that generate calamitic nematic liquid crystals are enormous. The precise structural fingerprint of each compound determines the various physical properties, and this Datareview has shown the significant changes in melting point, transition temperatures and mesophase morphology that can be achieved with seemingly very minor alterations to the molecular structure.

Of course, the vast majority of nematic liquid crystals do not have the required properties to be useful in applications, but many have contributed towards the ultimate goal of commercially-successful materials. In future the knowledge gained from structure-property relationships will be invaluable in the design of novel materials to satisfy new, advanced applications.

REFERENCES

[1]   P.J. Collings, M. Hird [*Introduction to Liquid Crystals Chemistry and Physics* (Taylor & Francis, London, 1997)]

[2]   G.W. Gray, P.A. Winsor [*Liquid Crystals and Plastic Crystals, vol.1&2* (Ellis Horwood, Chichester, 1974)]

[3]   D. Demus, H. Demus, H. Zaschke [*Flussige Kristalle in Tabellen* (Deutscher Verlag für Grundstoffindustrie, Leipzig, 1974)]

[4]   D. Demus, H. Zaschke [*Flussige Kristalle in Tabellen vol.2* (Deutscher Verlag für Grundstoffindustrie, Leipzig, 1984)]

[5]   P.J. Collings, J.S. Patel [*Handbook of Liquid Crystal Research* (Oxford University Press, Oxford, 1997)]

[6]   D. Demus, J. Goodby, G.W. Gray, H.-W. Spiess, V. Vill [*Handbook of Liquid Crystals* (Wiley-VCH, Weinheim, 1998)]

[7]   K.J. Toyne [in *Thermotropic Liquid Crystals* Ed. G.W. Gray (Wiley, Chichester, 1987)]

[8]   R. Dabrowski, E. Zytynski [*Mol. Cryst. Liq. Cryst. (UK)* vol.87 (1982) p.109]

[9]   M. Petrzilka, K. Schleich [*Helv. Chim. Acta (Switzerland)* vol.65 (1982) p.1242]

[10]   R. Dabrowski, J. Dziaduszek, T. Szczucinski [*Mol. Cryst. Liq. Cryst. (UK)* vol.102 (1984) p.155]

[11]   M.A. Osman [*Z. Nat.forsch. A (Germany)* vol.38 (1983) p.693]

[12]   M. Hird, K.J. Toyne, A.J. Slaney, J.W. Goodby, G.W. Gray [*J. Chem. Soc., Perkin Trans. 2 (UK)* (1993) p.2337]

[13]   G.W. Gray, K.J. Harrison, J.A. Nash [*Electron. Lett. (UK)* vol.9 (1973) p.130]

[14]   G.W. Gray, K.J. Harrison, J.A. Nash [*J. Chem. Soc., Chem. Commun. (UK)* (1974) p.431]

[15]   G.W. Gray, A. Mosley [*J. Chem. Soc., Perkin Trans. 2 (UK)* (1976) p.97]

[16]   R. Eidenschink, D. Erdmann, J. Krause, L. Pohl [*Agnew. Chem., Int. Ed. Engl. (Germany)* vol.16 (1977) p.100]

[17]   A. Villiger, A. Boller, M. Schadt [*Z. Nat.forsch. B (Germany)* vol.34 (1979) p.1535]

[18]   R. Eidenschink, D. Erdmann, J. Krause, L. Pohl [*Angew. Chem., Int. Ed. Engl. (Germany)* vol.17 (1978) p.133]

[19]   M. Hird, K.J. Toyne, G.W. Gray, S.E. Day, D.G. McDonnell [*Liq. Cryst. (UK)* vol.15 (1993) p.123]

[20]   A. Boller, M. Cereghetti, M. Schadt, H. Scherrer [*Mol. Cryst. Liq. Cryst. (UK)* vol.42 (1977) p.215]

[21]   D. Demus, H. Zaschke [*Mol. Cryst. Liq. Cryst. (UK)* vol.63 (1981) p.129]

[22]   H. Zaschke [*J. Prakt. Chem. (Germany)* vol.317 (1975) p.617]

[23]   M.E. Neubert, L.T. Carlino, D.L. Fishel, R.M. D'Sidock [*Mol. Cryst. Liq. Cryst. (UK)* vol.59 (1980) p.253]

[24]   N. Carr, G.W. Gray [*Mol. Cryst. Liq. Cryst. (UK)* vol.124 (1985) p.27]

[25]   N. Carr, G.W. Gray, D.G. McDonnell [*Mol. Cryst. Liq. Cryst. (UK)* vol.97 (1983) p.13]

[26]   S.M. Kelly [*Helv. Chim. Acta (Switzerland)* vol.67 (1984) p.1572]

[27]   G.W. Gray, A. Mosely [*Mol. Cryst. Liq. Cryst. (UK)* vol.37 (1976) p.213]

[28]   M.A. Osman, T. Huynh-Ba [*Mol. Cryst. Liq. Cryst. (UK)* vol.116 (1984) p.141]

[29]   K. Praefcke, D. Schmidt, G. Heppke [*Chem.-Ztg. (Germany)* vol.104 (1980) p.269]

[30]   H. Kelker, B. Scheurle [*Angew. Chem., Int. Ed. Engl. (Germany)* vol.8 (1969) p.884]

[31]   M.A. Osman [*Mol. Cryst. Liq. Cryst. (UK)* vol.72 (1981) p.291]

[32]   G.W. Gray, M. Hird, D. Lacey, K.J. Toyne [*J. Chem. Soc., Perkin Trans. 2 (UK)* (1989) p.2041]

[33]   S. Saito et al [*Ferroelectrics (UK)* vol.147 (1983) p.367]

[34]   T. Geelhaar, H. Lannert, B. Littwitz, A. Pausch, V. Reiffenrath, A.E.F. Wachtler [presented at the *13th Int. Liquid Crystal Conf.* Vancouver, Canada, 1990, FER-13-O-Fri]

[35]   L.K.M. Chan, G.W. Gray, D. Lacey [*Mol. Cryst. Liq. Cryst. (UK)* vol.123 (1985) p.185]

[36]   R. Eidenschink [*Mol. Cryst. Liq. Cryst. (UK)* vol.123 (1985) p.57]

[37]   M. Schadt, M. Petrzilka, P.R. Gerber, A. Villiger [*Mol. Cryst. Liq. Cryst. (UK)* vol.122 (1985) p.241]

[38]   E. Bartmann, D. Dorsch, U. Finkenzeller, H.A. Kurmeier, E. Poetsch [presented at the *19th Freiburger Arbeitstagung Flussigkristalle* Freiburg, Germany, 1990, talk 8]

[39]   H.J. Plach, E. Bartmann, E. Poetsch, S. Naemura, B. Rieger [*SID Digest (USA)* (1992) p.13]

[40]   Y. Goto, T. Ogawa, S. Sawada, S. Sugimori [*Mol. Cryst. Liq. Cryst. (UK)* vol.209 (1991) p.1]

[41]   V.F. Petrov [*Liq. Cryst. (UK)* vol.19 (1995) p.729]

[42]   U. Finkenzeller, E. Poetsch, K. Tarumi [presented at the *14th Int. Liquid Crystal Conf.* Pisa, Italy, 1992, A-P94]

[43]   M. Ushioda, M. Ushida, T. Suzuki [*Eur. Pat. Appl.* (1988) EP 0325796]

[44]   S.T. Wu, D. Coates, E. Bartmann [*Liq. Cryst. (UK)* vol.10 (1991) p.635]

[45]   M. Petrzilka [*Mol. Cryst. Liq. Cryst. (UK)* vol.111 (1984) p.329]

[46]   M.A. Osman [*Mol. Cryst. Liq. Cryst. (UK)* vol.128 (1985) p.45]

[47]   G.W. Gray, M. Hird, K.J. Toyne [*Mol. Cryst. Liq. Cryst. (UK)* vol.195 (1991) p.221]

[48]   G.W. Gray, M. Hird, K.J. Toyne [*Mol. Cryst. Liq. Cryst. (UK)* vol.204 (1991) p.43]

[49]   M. Hird, K.J. Toyne, G.W. Gray, D.G. McDonnell, I.C. Sage [*Liq. Cryst. (UK)* vol.18 (1995) p.1]

[50]   P. Balkwill, D. Bishop, A. Pearson, I. Sage [*Mol. Cryst. Liq. Cryst. (UK)* vol.123 (1985) p.1]

[51]   G.W. Gray, M. Hird, D. Lacey, K.J. Toyne [*Mol. Cryst. Liq. Cryst. (UK)* vol.172 (1989) p.165]

[52]   V. Reiffenrath, J. Krause, H.J. Plach, G. Weber [*Liq. Cryst. (UK)* vol.5 (1989) p.159]

[53]   V. Bezborodov, R. Dabrowski, J. Dziaduszek, K. Czuprynski, Z. Raszewski [*Liq. Cryst. (UK)* vol.20 (1996) p.1]

[54]   M. Hird, K.J. Toyne, A.J. Slaney, J.W. Goodby [*J. Mater. Chem. (UK)* vol.5 (1995) p.423]

1.2 Relationship between molecular structure and transition temperatures for organic materials of a disc-like molecular shape in nematics

K. Praefcke

October 1999

A INTRODUCTION

The possible existence of discotic liquid crystals and their columnar stacking was discussed by a German research group [1] nearly eighty years ago; however, their discovery only occurred half a century later in India [2]. Shortly thereafter, the first nematogen of a discotic molecular shape was described by French researchers [3]. Up to now, the number of thermotropic nematic mesogens of more or less disc-like molecular architectures has grown to two hundred and thirty two [4]. They represent a very small but nonetheless important subgroup among the thirty-three thousand [4] pure organic compounds showing a nematic phase on change of temperature.

This particular type of mesophase can also occur on heating mixtures of certain organic solids of which at least one has a disc-like molecular structure, as we shall see. Furthermore, it should be mentioned here that suitable discotic compounds dissolved in various solvents, thus constituting binary lyotropic systems, may also give rise to a nematic phase.

Comparable to the nematic (N) phase of rod-like compounds the least ordered (usually highest temperature) mesophase exhibited by disc-like molecules is also the nematic (discotic, N_D) phase; the index D simply refers to their molecular shape. Both nematic phases are of the same symmetry and identical types of defects are seen in both cases [5]; they exhibit similar fluid Schlieren textures [3,6]. However, the nematic phases of these two low-molar-mass liquid crystals are not miscible [3,6] and phase separation occurs due to fundamental differences in their molecular structures.

The disc-like molecules of an N_D phase have also only one degree of order, the orientational order, however, with the director (**n**) perpendicular to their molecular plane, i.e. preferring the molecular short axis or the disc's normal (FIGURE 1 [7], centre). Unlike the usual nematic phase of calamitic molecules (FIGURE 1 [7], left), the N_D phase is optically and diamagnetically negative and of much higher viscosity.

B SPECIAL REMARKS

Prior to discussing the structural features of discotic mesogens in Section C, some other distinguishing features of these fascinating, discotic - for many still exotic - materials will be discussed here: induction of nematic phases and the possibility of thermotropic phase biaxiality envisaged for them. Along with thermotropic N_D phases of pure compounds, further fundamental modes of formation of nematic phases of disc-like molecules have become known.

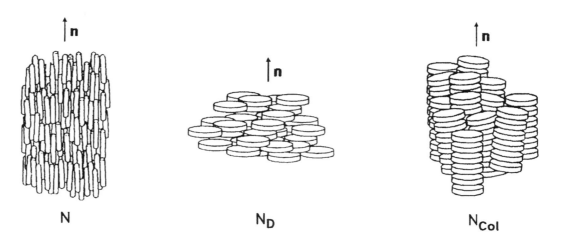

FIGURE 1   Schematic representations [7] of the nematic phase (N, left) of rod-like molecules compared with the discotic nematic (N_D, centre) and the non-tilted columnar nematic (N_{Col}, right) phases of disc-like molecules.   D = discotic, Col = columnar, **n** = director.

B1   Chirality, Thermo- and Lyotropic Mixtures

Regarding the introduction of chirality into disc-like nematogens, two different paths toward this important goal have been successfully pursued [8] following the discovery [2] of discotic liquid crystals.   These early examples of realising thermotropic chiral nematic phases ($N*_D$) were, first [8], the synthesis of pure chiral triphenylene hexaesters carrying asymmetric centres in their covalently bound ester chains, and, secondly [8], the occurrence of such an $N*_D$ phase in mixtures consisting of analogous, but achiral hexaesters doped with a small amount of a chiral one [8].   The latter situation constitutes the first example of the induction of the chiral nematic discotic phase [8,9].   However, in the few cases reported then [8], the $N*_D$ phase appears at rather high temperatures.   Only the extension of this activity into the well-known family of disc-like radial multiynes [10-15] yielded progress by obtaining enantiotropic chiral nematic discotic phases at moderate temperatures ([14-17], and see some examples in TABLE 3) and, furthermore, the phenomenon of selective reflection as well as a temperature induced helix inversion [15].   FIGURE 2 [15] illustrates the arrangements of rod- and disc-like molecules in their chiral nematic phases N* or $N*_D$, respectively.

A third and very easy approach to the induction of thermotropic nematic phases of disc-like compounds, in particular, exists by mixing suitable substances (electron donor materials, mesomorphic or not!) with one or two strong (non-discotic, non-mesomorphic!) electron acceptor compounds [9,14,17,18], for instance, with an achiral or chiral multi(nitro)fluorenone derivative as shown in TABLES 4 and 6.   Such binary or even ternary electron-donor-acceptor mixtures can yield nematic columnar (N_{Col}, FIGURE 1 [7], right) or chiral nematic columnar ($N*_{Col}$) phases, respectively, in which the components stack alternately on top of each other upon heating due to π- and lone-pair electron interactions between the sufficiently large, functionalised aromatic regions of the molecules (see Section C3 and FIGURE 3 in comparison with FIGURE 1).

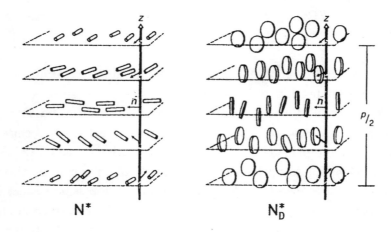

FIGURE 2 Schematic representation for the organisation of (classical) rod- and disc-like molecules in their chiral nematic phases with the mean orientations of their symmetry axes where the director field forms a helical array of molecules characterised by the magnitude of the pitch, p, and the screw sense. In both sketches (from [15], by permission of the publishers), the director, **n**, turns in the same way, i.e. the circles in the right half of this figure are nearly in edge-on orientation. N* = chiral nematic phase, D = discotic, **n** = director, p = pitch, z = z-axis.

FIGURE 3 Schematic representation of a thermotropic columnar nematic (N_{Col}) phase induced by electronic interactions between donor and acceptor molecules or parts of molecules (see TABLES 4 to 6) stacked coin-like alternately on top of each other displaying, for instance, the donor molecules in black and others shaded (see FIGURE 1).

In this new type of nematic phase reported about ten years ago [18,19], its columnar superstructure acts as calamitic molecules do in their nematic phase and also displays a Schlieren texture, which is responsible for its name. X-ray diffraction studies [18,19] have proved the nematic array of columns. This method of mesophase induction constitutes an attractive and novel principle [9,14,17,18] for the design of tailor-made thermotropic nematogens from discotic starting materials by spontaneous supramolecular self-organisation.

Lyotropic nematic phases (see Section A) can also be produced by preparing, for instance, binary or ternary mixtures of organic disc-like compounds in suitable solvents such as hydrocarbons [20]. In linear saturated [20,21] or, as found recently [21], even better in cyclic saturated hydrocarbons, preferably cyclohexane [21], alone or in such a solvent plus an achiral or a chiral electron acceptor compound, induction of lyotropic N_{Col} or N^*_{Col} phases, respectively, can occur. Sometimes, an N_{Col} phase can be formed in addition to a columnar phase [21]. Furthermore, it has also been observed that even two different N_{Col} phases can be induced in that way in the same system [22,23] showing a nematic-nematic phase transition [22-24] due to a difference in the construction of their columns. In one of these N_{Col} phases the constituent discs of the columns spontaneously formed are tilted with respect to the column axis, but in the second, parallel N_{Col} phase they are untilted [22,23]. However, reliable data about the length of the columns in N_{Col} phases do not yet seem to exist.

An alternative way [25] to obtain a chiral nematic discotic phase is simply the use of a chiral organic solvent, e.g. the hydrocarbon R-(+)-limonene, in which, for example, an achiral, slightly folded disc-like di-palladium organyl is dissolved.

For further information and suggestions about lyotropic systems of some more disc-like compounds generated in aqueous or various organic solvents, especially about those exhibiting nematic behaviour, recent review papers and monographs on organic lyotropics quoted here [26] are useful. Also a detailed review [27] on inorganic lyotropic liquid crystals with a very extensive section on nematics (water or water-organic dispersions of certain non-organic substances) is recommended. This represents a new and unexplored area, in spite of the fact that this chemical type of lyomesogen has also been known about for a hundred years.

B2   Biaxial Nematics

Another hot topic which has entwined itself around nematic discotic materials for about fifteen years is the growing interest in thermotropic biaxial nematic phases of low molecular mass compounds, especially since the lyotropic analogue was well established by Saupe et al [28] some years earlier. It has been presumed that the chance of obtaining the thermotropic nematic biaxial phase (N_b) should be good with molecules which combine the structural features of a rod and a disc (the latter at least in part) [29-31]. However, although the synthetic efforts in several research groups have been and are still high [30,32], the few promising cases worked out so far [32,33a] remain controversial.

A new and effective method for studying biaxial nematics with X-rays has been described recently [33b] with the classic inorganic (aqueous) lyotropic liquid crystal, vanadium pentoxide [33c], exhibiting now a biaxial nematic gel phase. Undoubtedly, biaxial nematogens would indeed be of great importance for the system of phase types, the theory of liquid crystals, and the understanding of the molecular dynamics of mesogens, and for the development or discovery of new possibilities of applications for such novel pure materials.

C   STRUCTURAL FEATURES

In contrast to the vast number of calamitic liquid crystals having a great variation in their chemical constitutions (Datareview 1.1 of this volume), the relatively few discotic molecular structures including those associated with nematic phenomena consist, in almost all cases, either of a quite flat, highly unsaturated rigid core (or in part), for instance present in the hexaester/-ether series **1** and **2** or **4**, respectively, in the cyclic type of multiyne **3g**, and in the metal complex **8**, or of a flexible, so-called [34] super-disc-like core type as indicated by a grey background for the hexaester series **1** and the radial multiyne families **3a-3f**, and **6**, as well as for the dimetallomesogens of type **7**. The structural formulas **1** to **8** are each depicted at the bottom of the following six tables. Based on Dreiding models, the diameters of these 'super-discs' are about 1.5 nm for **7** but in most cases between ~1.95 nm as for the radial hexayne **3a** and ~2.4 nm for the largest one (**3f**) of this hexayne type. At their peripheries, each of them carries mostly six (up to twelve in certain families of metallomesogens [22,23,35]) radially located covalently bound alkyl, chalcogenoalkyl or -alkanoyl chains. Their inner parts are either disc-/star- (e.g. see **2** and **3a**) or hoop-like (see **3g**) in shape; they are sketched and further described in FIGURE 4.

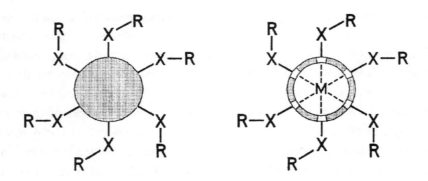

FIGURE 4 Sketches of the two general molecular structures consisting of a disc-/star-like (left) or hoop-like
core with or without a metal centre (right) and mostly six peripheral chains, found in liquid crystals,
exhibiting the discotic nematic or, as a result of spontaneous stacking of such individual molecules,
the columnar nematic phase (see FIGURE 1). R = alkyl (in left cases only), substituted phenyl or
aroyl groups, X = methylene or oxygen, □: ring position of a defined segment of carbons for a
metal-free, hollow core or a heteroelement (e.g. nitrogen) in metal-centred molecules (e.g. **8**).

The formation of the N_D phase directly from the crystalline state of discotic materials on heating or above a columnar type of phase (in general, the latter by far more typical for them) seems to be controlled to a greater extent by the length of the outer moieties [6].

Likewise, as for instance analogously substituted members of the radial multiyne families **3a** and **3f**, compiled in TABLE 3, demonstrate, it seems that the wider their molecular core, the more stable and broader the N_D phase formed on heating [12]. Re-entrant occurrences of an N_D phase are also known, i.e. the nematic phase appears at temperatures both above and below the columnar phase. Phase sequences also occur in which the high temperature nematic phase is absent: these are referred to as inverse phases. Examples of these are hexaesters of the truxene types of multicycles **2a-2c** [36-41] (see TABLE 2) or with two 4-alkoxybenzoyloxytruxene hexaesters [42] of type **2a** (see the footnote to TABLE 2), respectively. The correct chemical names of these ring systems **2a-2c** according to the CA System of the Ring Index are also given in the footnote. Some radial multiynyl hydrocarbons of type **3e** [11] derived from naphthalene (see TABLE 3) also exhibit the N_D phase thermotropically in an inverse manner.

The presentation of the phase transition temperatures jointly with the appropriate basic molecular formula(s) of the discotic material(s) in question, known so far, is given in six tables. Each of the pure discotic compounds (see TABLES 1 to 3, and, in part, TABLE 6) and their donor-acceptor complexes (see TABLES 4, 5, and, in part, 6) listed here shows a very small enthalpy value for the nematic $\rightarrow$ isotropic transition, i.e. (in most cases) considerably below but seldom above 1 kJ/mol. This behaviour is characteristic for this transition on heating [10,11,17,18,43-45] and so in order to save space, these standard values of discotic nematics have been omitted from the tables discussed in the following four sections.

C1   Carbo- and Heterocycles

Probably, the best known carbocyclic core system in the field of discotic liquid crystals is triphenylene. Multisubstituted derivatives (esters, also ethers) of, for instance, 2,3,6,7,10,11-hexa(hydroxy)triphenylene are among the first thermomesogens of disc-like molecular shape synthesised about two decades ago [3,6] but still of some interest. Other chalcogeno-(thio- or seleno-)bridged analogues followed later [46]. Particular attention was given to certain hexabenzoates (**1**) of these triphenylene families [3,6] because their N_D phases were among the first to be observed. Together with their structural formula, the series **1a-1j** of monosubstituted hexabenzoates, selected here [6], is presented in TABLE 1. Unfortunately the temperature ranges of their enantiotropic N_D phases are very high, above columnar phases, evidently, independent of the para-alkyl or -alkoxy substitution. However, increasing the chain length of R tends to reduce the transition temperatures slightly, but more pronounced effects [47,48] are observed when additional ortho- or meta-methyl groups are introduced into the six benzoyl functions. These latter triphenylene derivatives also exhibit the N_D phase [47,48]: their transition temperatures are a little lower than those of the compounds in TABLE 1 and the phase sequence shows changes, obviously, on account of steric influences resulting from the higher degree of substitution of the benzoyl groups compared to those of series **1** in TABLE 1.

Trans-cyclohexane carboxylates [49] (in which the benzoyl groups of **1** are hydrogenated) and also hexakisalkanoates [6,50,51], analogous to the structure **1**, exhibit liquid crystalline phases but not nematics in their pure state.

2,3,7,8,12,13-hexa(alkanoyloxy)derivatives (esters) of another, second carbocycle, truxene (**2a**), and its two hetero analogues **2b** and **2c** are also well known [36-42] discotic liquid crystals each with very interesting polymorphism including the N_D inverse phase, mostly below columnar phases: see TABLE 2 and compare four other cases of the radial multiyne series **3e** in TABLE 3. The inverse nematic

TABLE 1 Transition temperatures (°C, on heating) of ten 4-substituted
2,3,6,7,10,11-hexa(benzoyloxy)triphenylene derivatives with the
super-disc core (diameter ≈2.05 nm) shown in formula **1**
each exhibiting an N_D phase.

Triphenylene hexaester structure	R	Phase transition temperatures							Ref
		Cr		M		N_D		I	
1a	C_8H_{17}	•	208	-		•	210	•	[6]
1b	C_9H_{19}	•	175	•	183	•	192	•	[6]
1c	OC_4H_9	•	257	-		•	>300	•	[6]
1d	OC_5H_{11}	•	224	-		•	298	•	[6]
1e	OC_6H_{13}	•	186	•	193	•	274	•	[6]
1f	OC_7H_{15}	•	168	-		•	253	•	[6]
1g	OC_8H_{17}	•	152	•	163	•	244	•	[6]
1h	OC_9H_{19}	•	154	•	183	•	227	•	[6]
1i	$OC_{10}H_{21}$	•	142	•	191	•	212	•	[6]
1j	$OC_{11}H_{23}$	•	145	•	179	•	185	•	[6]

M = different types of columnar phases.

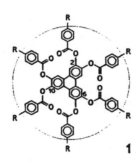

1

phase of members of series **2a** and **2c** (upper half) develops enantiotropically on average above 64°C (series **2a**) or about 10°C higher for the other; their nematic ranges are relatively small and fluctuate between 16 and 27 or 6 and 38°C, respectively. Surprisingly, all of the hexaesters of series **2b** exhibit a monotropic N_D phase. These and other differences in the thermomorphic behaviour of the alkanoate derivatives of **2a**, **2b**, and **2c** clearly stem from their structural differences caused by X in positions 5, 10 and 15 of formula **2**.

The four analogous 4-substituted hexa(benzoyloxy)derivatives in the lower part of series **2c** are monomesomorphic showing very broad nematic N_D ranges (between 145 and 217°C), thus offering great advantages over the N_D behaviour of similar hexaesters (**1c-1j**) of the triphenylene series (see TABLE 1).

TABLE 2 Transition temperatures (°C, on heating) of twenty-three hexaesters in four homologous groups of 2,3,7,8,12,13-hexa-(hydroxy)-truxene[a] (**2a**), -trioxatruxene[b] (**2b**), and -trithiatruxene[c] (**2c**) exhibiting mostly an inverse N_D phase.

Truxene type structure	X	R (ester groups)	Cr		N_D		M				I	Ref
2a	CH_2	$O\text{-}CO\text{-}C_9H_{19}$[d]	•	68	•	85	•		138/280		•	[36,37]
	CH_2	$O\text{-}CO\text{-}C_{10}H_{21}$	•	62	•	89	•		118/250		•	[36,37]
	CH_2	$O\text{-}CO\text{-}C_{11}H_{23}$[e]	•	64	•	84	•		130/≈250		•	[36,37]
	CH_2	$O\text{-}CO\text{-}C_{12}H_{25}$[e]	•	57	•	84	•		≈107/249		•	[36,37]
	CH_2	$O\text{-}CO\text{-}C_{13}H_{27}$	•	61	•	84	•		112/235		•	[36,37]
	CH_2	$O\text{-}CO\text{-}C_{14}H_{29}$	•	64	•	82	•		95/241		•	[37,38]
	CH_2	$O\text{-}CO\text{-}C_{15}H_{31}$	•	69	•	84	•		95/210		•	[37,38]
2b	O	$O\text{-}CO\text{-}C_7H_{15}$	•	96	{•	89}	-		194		•	[39]
	O	$O\text{-}CO\text{-}C_8H_{17}$	•	90	{•	75}	-		197		•	[39]
	O	$O\text{-}CO\text{-}C_9H_{19}$	•	82	{•	67.5}	-		192		•	[39]
	O	$O\text{-}CO\text{-}C_{10}H_{21}$	•	76	{•	61.5}	-		194		•	[39]
	O	$O\text{-}CO\text{-}C_{11}H_{23}$	•	78	{•	58}	{•	64}	/-/184		•	[39]
	O	$O\text{-}CO\text{-}C_{12}H_{25}$	•	78	{•	59}	{•	64}	/172/177		•	[39]
	O	$O\text{-}CO\text{-}C_{13}H_{27}$	•	74	{•	58}	{•	71}	/158/166		•	[39]
	O	$O\text{-}CO\text{-}C_{15}H_{31}$	•	80	{•	60}	•	84	/138/152		•	[39]
2c	S	$O\text{-}CO\text{-}C_9H_{19}$	•	87	•	93	•		185/210		•	[40,41]
	S	$O\text{-}CO\text{-}C_{10}H_{21}$	•	62	•	98	•		151/193		•	[40]
	S	$O\text{-}CO\text{-}C_{11}H_{23}$	•	64	•	93	•		146/149/179		•	[40]
	S	$O\text{-}CO\text{-}C_{12}H_{25}$	•	79	•	87	•		136/191		•	[41]
	S	$O\text{-}CO\text{-}C_6H_4OC_{10}H_{21}$	•	83	•	>300	-				•	[41]
	S	$O\text{-}CO\text{-}C_6H_4OC_{12}H_{25}$	•	81	•	295	-				•	[41]
	S	$O\text{-}CO\text{-}C_6H_4OC_{13}H_{27}$	•	86	•	280	-				•	[41]
	S	$O\text{-}CO\text{-}C_6H_4OC_{16}H_{33}$	•	96	•	241	-				•	[41]

N_D = inverse nematic discotic phase, M = two, in four cases three, types of columnar phases, { } = monotropic phase transition. Nomenclature: [a]truxene ≡ 10,15-dihydro-10,15-dihydro-5H-diindenol;[1,2-a:1′,2′-c]fluorene, [b]the so-called trioxatruxene ≡ benzo[1,2-b:3,4-b′:5,6-b″]trisbenzofuran, and [c]the so-called trithiotruxene ≡ benzo[1,2-b:3,4-b′:5,6-b″]-tris[1]benzothiophene. [d]The three shorter homologues to this one exhibit the N_D phase monotropically between 85 and 96°C [36,37]. [e]An analogous 4-alkoxybenzoyloxytruxene derivative with this alkyl chain exhibits a sequence of thermotropic mesophases which are the first examples of a reentrant N_D occurrence among disc-like compounds [42].

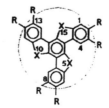

2a: X = CH_2
b: X = O
c: X = S

TABLE 3 Transition temperatures (°C, on heating) of forty-six derivatives of six different radial multiynes (type 3a to 3f) and of the cyclic multiyne type 3g with super-disc cores of different diameters each exhibiting mostly an N_D phase which in four cases of group 3e are inverse. Five of the radial multiynes of the types 3b, 3c, and 3e exhibit the chiral nematic discotic (N_D^*) phase.

3a 3b 3c

Multiyne structure	R	R'	Cr	M	N_D/N_D^*	I	Ref
3a	C_5H_{11}	-	• 170	-	• 186[a]	•	[12]
	C_6H_{13}	-	• 124	-	• 142	•	[12]
	C_7H_{15}	-	• 99	-	• 132	•	[12]
	OC_7H_{15}	-	• 109	-	• 193	•	[53]
	C_8H_{17}	-	• 80	-	• 96	•	[10]
	C_9H_{19}	-	• 71	-	• 88	•	[54]
3b	CH_3	CH_3	• 261	-	• 270	•	[14]
	C_7H_{15}	CH_3	• 49/63[c]	-	{• 2.5}	•	[14]
	CH_3	OC_5H_{11}	• 263	-	• ≈283	•	[14,55]
	CH_3	OC_9H_{19}	• 200	-	• ≈259	•	[14,55]
	CH_3	$OC_{11}H_{22}$-OH	• ≈172	-	• ≈227	•	[44]
	CH_3	$OC_{13}H_{27}$	• 168	-	• 222	•	[14,55]
	CH_3	$OC_{16}H_{33}$	• 162	-	• 193	•	[17]
	C_5H_{11}[b]	OC_9H_{19}	• 86	-	• 110	•	[14,30]
	C_5H_{11}	$OC_{10}H_{21}$[#]	• 78	-	#• 98	•	[15]
	C_5H_{11}	OC_9H_{18}-CH=CH$_2$	• 77	-	• 101	•	[30]
	C_5H_{11}[b]	$OC_{10}H_{20}$-COOH	• ≈83	-	• 91	•	[56]
	C_5H_{11}[b]	$OC_{10}H_{20}$-COOEt[d]	• ≈57	-	• 69	•	[56]
	C_5H_{11}	$OC_{11}H_{23}$	• 75	-	• 101	•	[57]
	C_5H_{11}[b]	$OC_{11}H_{22}$-OH	• 67	-	• ≈94	•	[56]
	C_5H_{11}[b]	$OC_{16}H_{33}$	• 88	-	• 100	•	[15]
	OC_5H_{11}[#]	$OC_{16}H_{33}$	• T_g = –36	-	#• ≈23	•	[15]
	$OC_{10}H_{21}$[#]	$OC_{16}H_{33}$	• 171	-	#• ≈250[a]	•	[15]
	C_6H_5[b]	$OC_{16}H_{33}$	-	• ≈230	• 250[a]	•	[13]
	4-$H_{17}C_8C_6H_4C\equiv C$	$OC_{16}H_{33}$	• 58	• 251[a]	• 265[a]	•	[58]

25

TABLE 3 continued.

Multiyne structure	R	R'	Cr	M	N_D/N*_D	I	Ref
3c	H	$OC_{16}H_{33}$	• 173	-	• 177	•	[58]
	$OC_{10}H_{21}$ #	$OC_{16}H_{33}$	• 55	-	## • 156	•	[58]
3d	C_5H_{11}	$O(CH_2)_8O$	• 127	-	• 129	•	[30]
	C_5H_{11}	$O(CH_2)_9O$	• 131	-	• 113	•	[30]
	C_5H_{11}	$O(CH_2)_{10}O$	• 129	-	• 154	•	[30]
	C_5H_{11}	$O(CH_2)_{11}O$	• 118	-	• 141	•	[30]
	C_5H_{11}	$O(CH_2)_{12}O$	• 121	-	• 155	•	[30]
	C_6H_{13}	$O(CH_2)_{12}O$	• 93	-	• 107	•	[15]
	C_7H_{15}	$O(CH_2)_{12}O$	• 76	-	• 85	•[a]	[54]
3e	OC_5H_{11}	-	• 144	-	• ≈250	•[a]	[11]
	C_5H_{11}	-	• ≈122	• ≈157/≈260	• ≈90	•	[11]
	C_6H_{13}	-	• ≈69	• ≈134/≈245[a]	• ≈98	•	[11]
	C_7H_{15}	-	• ≈49	• 159/≈230[a]	• ≈95	•	[11]
	C_8H_{17}	-	• ≈60	• ≈137/≈200	• ≈113	•	[11]
	C_9H_{19}	-	• ≈68	-	• ≈110	•	[11]
	$OC_{10}H_{21}$ #	-	(oil) • 39	-	## • 108	•	[58]
	$4\text{-}C_9H_{19}C_6H_4$	-	• 58	• ≈180	• ≈222	•[a]	[13]
3f	C_5H_{11}	-	• 156	-	• 237[a]	•	[12]
	C_7H_{15}	-	• 121	-	• 175	•	[12]
3g	OC_6H_{13}	-	• 216[e]	-	• 233[e]	•	[59]
	OC_7H_{15}	-	• 168[e]	-	• 192[e]	•	[59]
	$O\text{-}CO\text{-}C_7H_{15}$	-	• 121[e]	-	• 241[e]	•	[59]

Tg = glass transition temperature, M = different columnar phases, in part their structures are as yet unknown, { } = monotropic phase transition; [a] = decomposition, [b] the parent radial pentayne (R = H) with this particular R' is non-thermomesomorphic, see its melting point in [17,44,56,60], [c] two crystalline modifications, [d] Et = ethyl group, [e] this transition temperature is reversible and was obtained on cooling the isotropic melt, [#] chiral molecule carrying five (= R) (S)-2-methylbutyloxy, respectively one (= R') or five (= R) (S)-3,7-dimethyloctyloxy substituents, [##] chiral nematic discotic (N*_D) phase, [□]: data given in opposite order assigning inverse phase sequences.

The broadest and most stable N_D phase is formed by the sulphur containing heterocycle with six ester functions in the positions shown in formula **2**; the only complication with them lies in their relatively long and very tedious chemical synthesis which is given in [41]. Benzotrisfuran is a third disc-/star-like heterocycle of which one, the 2,3,5,6,8,9-hexa(octyloxybenzoyloxyphenyl)-substituted, derivative (a hexaester) has been prepared and found to give a nematic discotic phase above its melting point [52]; however, it decomposes rather quickly at higher temperatures which excludes further detailed investigation.

C2   Radial and Cyclic Multiynes

Forty-three derivatives of six radial multiyne families (**3a-3f**) and three of the one cyclic multiyne family **3g** are shown in TABLE 3 together with the formulas of their super-disc core structures of which that of **3g** is hollow and hoop-like (see FIGURE 3). With only a few exceptions the multiyne derivatives are a collection of chemically as well as thermally stable and mostly coloured monomesomorphic, nematic discotic liquid crystals. Their N_D phases have different ranges and occur at different temperatures; half the compounds enter the N_D phase on heating below 100°C, even (a chiral) one with the super-disc core **3b** below room temperature being among them. As expected, homologues of the families **3a**, **3d**, **3e** (in part), and **3g** prove that the extension of the chains R reduces their transition temperatures. Five radial multiynes with the super-disc cores **3b**, **3c**, and **3e** are chiral and exhibit a chiral nematic discotic phase starting (in part) far below 90°C.

Obviously, the structurally biaxial members of the naphthalene family **3e** do not behave in line with the other, benzene or triphenylene centred, multiynes and in four cases there is an inversion of the phase sequence (see TABLE 3 and compare the examples in TABLE 2).

The syntheses of the multiynes are straightforward and are described in the references quoted in TABLE 3.

C3   Inter- and Intramolecular Electron Donor-Acceptor Complexes

The synthesis of new single discoid molecules with special liquid crystalline properties is very costly and time-consuming, requiring tedious purification procedures. Another, completely different and in principle very easy method [9] for preparing thermomesomorphic materials has been applied and found to be fruitful. This is a technique through which mesophase stabilisations/manipulations or even inductions occur simply by mixing electron donor molecules disc-like in shape with electron acceptor compounds of which one component (**5a**, **5b** or **5c**) or even both (in seven mixtures of TABLE 4 also the respective donor compound) are not liquid crystalline. The organisation of the donor and acceptor molecules in such binary mixtures is illustrated in FIGURE 3.

Mostly, the induction of phases on heating means an increase of the order of such molecules by spontaneous complexation, i.e. formation of dimers at the beginning, thereafter of stacks, longer single columns which can give rise either to a clearly more viscous columnar nematic phase or for sufficiently strong intercolumnar interactions to still higher ordered phases, for instance to the very common hexagonal columnar (Col_h) phase.

The eleven examples of TABLE 4 based on electron-donor compounds of family **3b** show that induction processes starting from either non-mesomorphic or nematic discotic precursors can indeed give the ('intermediate' kind of) stable columnar nematic phase including the three known cases of chiral nematic phases. Unfortunately, for application purposes the temperature ranges of the nematic phases of the intermolecular electron donor-acceptor complexes (TABLE 4) may be much too high!

The two compositions of the hexaether **4** which exhibits a columnar phase with the non-mesomorphic fluorenone derivative **5c** are the known examples [18] for reduced order phase inductions on heating of low molar mass mesogens resulting not in the complete destruction of the previous highly ordered mesophase, but the formation of the columnar nematic phase with its reduced degree of order. Apparently, this process arises from a combination of steric and electron donor-acceptor interactions within the complexes.

A further development of such complexes is the synthesis of the electron donor-acceptor twins **6** of which seven representatives with three different kinds of bridging groups XY (one is chiral, see **6c**, **6d** and **6e**, TABLE 5) have not only been prepared but show desired nematic properties. Their columnar nematic phases have a much higher viscosity at low temperatures, in some cases considerably below room temperature. Three of these interesting twin compounds (**6c**, **6d** and **6e**) exhibit chiral nematic phases.

C4   Metal Complexes

The research into metal-containing liquid crystals although still quite young has already brought to light an enormous number of fascinating metallomesogens [61] among which are also the very first examples (see TABLE 6) of disc-like molecular shape exhibiting the discotic/columnar types of nematic phase.

This collection contains one platinum and four palladium organyls (**7a** and **7b-7e**, respectively) as well as an azacyclic cobalt(II) complex (**8**). Whereas the five pure metal organyls **7a-7e** melt between 73 and 96°C and show only a monotropic N_D phase, the cobalt(II) compound **8** does exhibit the N_{Col} phase enantiotropically at lower temperatures, between 30 and 60°C. The metal-free precursor ligand of **8** is not mesomorphic [63]!

Interestingly, equimolar binary mixtures of the two palladomesogens **7d** and **7e** with the non-mesomorphic acceptor compound **5a**, referred to in TABLE 4, lead not only to a decrease of their melting points by somewhat more than 10°C but simultaneously to a significant stabilisation and widening of their nematic range, which under these conditions turns into the more viscous N_{Col} phase and becomes even enantiotropic for the complex consisting of **5a** and **7e**.

TABLE 4 Transition temperatures (°C, DSC data) of intermolecular donor-acceptor complexes formed spontaneously when heating equimolar mixtures of electron donor molecules (e.g. radial pentaynes of type **3b**, R = H, not liquid crystalline, or the 2,3,6,7,10,11-triphenylene hexaether **4**) and the non-liquid crystalline multi(nitro)fluorenone derivatives **5a**, **5b** or **5c** as electron acceptors yielding preferably by induction the nematic or chiral nematic columnar (N_{Col} or N^*_{Col}, respectively) phases.

Basic donor structure	R	R' (ether groups)	Mesophase of donor compound	Acceptor: multinitrofluorenone[a-c]	Cr	Col_{ho}	N_{Col}	I	Ref
3b	H	OC_9H_{19}	-	**5a** (50 mol %)	• 124	• 176	-	• 180	[19]
	H	$OC_{10}H_{21}$-COOH	-	**5a** (50 mol %)	• 120	-	•	• 145	[56]
	H	$OC_{10}H_{20}$-COOEt	-	**5a** (50 mol %)	• 104	-	•	• 124	[56]
	H	$OC_{11}H_{22}$-OH	-	**5a** (50 mol %)	• 131	-	•	• 158	[44]
	CH_3	$OC_{11}H_{22}$-OH	$N_D^{ø}$	**5a** (50 mol %)	• ≈120	-	•	• ≈219	[44]
	CH_3	$OC_{11}H_{22}$-OH	N_D	**5b** (15 mol %)	• ≈170	-	## •	• ≈214	[44]
	H	$OC_{13}H_{27}$	-	**5a** (50 mol %)	• 119	-	•	• 153	[19]
	H	$OC_{16}H_{33}$	-	**5a** (50 mol %)	• 118	-	•	• 140	[17]
	H	$OC_{16}H_{33}$	-	**5a/5b** (38/12 mol %)	• 109	-	## •	• 118	[17]
	CH_3	$OC_{16}H_{33}$	$N_D^{ø}$	**5a** (50 mol %)	• 129	-	•	• 184	[17]
	CH_3	$OC_{16}H_{33}$	N_D	**5b** (20 mol %)	• 156	-	## •	• 170	[17]
4 [6b]	C_5H_{11}	-	Col_{ho}	**5c** (40 mol %)	• 323	-	•	• 324	[18]
	C_5H_{11}	-	Col_{ho}	**5c** (50 mol %)	• 332	-	{ •	.330}	[18]

Col_{ho} = columnar hexagonal ordered phase, { } = monotropic transition, observed by polarising microscopy on cooling; [a]2,4,7-trinitrofluorenone (TNF, **5a**), [b]the chiral (-)-2-(2,4,5,7-tetranitro-9-fluorenylideneaminooxy)propionic acid ((-)-TAPA, **5b**, commercially available) or [c](2,4,7-trinitro-9-fluorenylidene)malonic bishexadecylester (**5c**) [18]. [ø]The transition temperatures of this donor are given in TABLE 3. [##]Here, due to the chiral TAPA an induced chiral nematic phase.

N_D phase

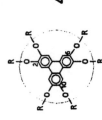

5a (Cr 176 °C Iso)   **5b** (Cr 196-9 °C Iso)   **5c** (Cr 109 °C Iso)

4 (R = pentyl, Cr 69 Col_{ho} 122°C Iso)

3b

TABLE 5 Transition temperatures (°C) of four intramolecular donor-acceptor complexes of structure **6** with bridging groups XY, different in composition and length, between the radial multiyne donor section (see TABLE 3) and the 2,4,7-trinitrofluorenone oxime acceptor part exhibiting by internal induction the columnar nematic phase.

Intramolecular donor-acceptor complex	R	XY (bridging groups)	Phase transition temperatures					Ref
			Cr	M		N_{Col}	I	
6a	H	$O(CH_2)_{11}O\text{-}CO\text{-}(CH_2)_2\text{-}O$	-	T_g ≈60 •	≈125 •	134 •	•	[44]
6b	C_5H_{11}	$O(CH_2)_{11}O\text{-}CO\text{-}(CH_2)_2\text{-}O$	-	T_g ≈-15 •	50[a,b] •	59[a] •	•	[44]
6c	C_5H_{11}	$O(CH_2)_{11}O\text{-}CO\text{-}CH(CH_3)\text{-}O^\#$	-	T_g ≈59 -		## ≈95 •	•	[44]
6d	CH_3	$O(CH_2)_{11}O\text{-}CO\text{-}CH(CH_3)\text{-}O^\#$	-	T_g ≈74 -		## 195 •	•	[44]
6e	C_5H_{11}	$O(CH_2)_{11}O\text{-}CO\text{-}CH(CH_3)\text{-}O^\#$	-	T_g ≈11 -		## ≈77 •	•	[44]
6f	H	$O(CH_2)_{11}O\text{-}CO\text{-}⬡\text{-}N{=}N\text{-}⬡\text{-}O\text{-}CO\text{-}(CH_2)_2\text{-}O$	-	T_g ≈69 -		≈105 •	•	[45]
6g	CH_3	$O(CH_2)_{11}O\text{-}CO\text{-}⬡\text{-}N{=}N\text{-}⬡\text{-}O\text{-}CO\text{-}(CH_2)_2\text{-}O$	-	T_g ≈80 -		≈147 •	•	[45]

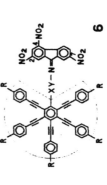

6

aObserved on cooling by polarising microscopy; bthis value is ≈67°C on heating, the sample does not show then an N_{Col} phase. #Chiral bridging group synthetically introduced by applying (-)-TAPA, see **5b** in the footnotes of TABLE 4. ##Here, due to the chiral TAPA as part of the starting material **6c**, **6d** or **6e**, respectively, an induced chiral columnar nematic phase.

TABLE 6 Transition temperatures (°C) of eight disc-shaped metallorganic compounds of the molecular structures **7** or **8** of which **7a-7e** contain a super-disc core with the greatest diameter of ≈1.5 nm, two platinum or two palladium atoms, and different bridging units. Included are the binary mixtures of **7d** and **7e** with 2,4,7-trinitrofluorenone (**5a**) each exhibiting by induction the columnar nematic phase.

Basic donor structure	Metal M	R	Bridging element/group	Acceptor: trinitrofluorenone[a]	Phase transition temperatures			Ref
					Cr	N_D	I	
7a	Pt	C_6H_{13}	Cl	-	• 92	{• 54}	•	[62]
7b	Pd	C_6H_{13}	Cl	-	• 79	{• 44}	•	[43]
7c	Pd	C_6H_{13}	Br	-	• 73	{• 28}	•	[43]
7d	Pd	C_6H_{13}	I	-	• 97	{• 27}	•	[43]
	Pd	C_6H_{13}	I	**5a** (50 mol %)	• 84	{N_{Col} 73}	•	[43]
7e	Pd	C_6H_{13}	SCN	-	• 96	{• ≈45}	•	[43]
	Pd	C_6H_{13}	SCN	**5a** (50 mol %)	• 85	N_{Col} 112	•	[43]
8	Co(NO$_3$)$_2$	$OC_{14}H_{29}$	-		• 30	N_{Col} 60	•	[63]

{ } = monotropic transition, observed by polarising microscopy on cooling. [a]See its name in the footnote to TABLE 4.

7

8

D   CONCLUDING REMARKS

The development of organic nematic liquid crystals of other than only calamitic molecular structures has brought forward valuable materials of several disc-/star- or hoop-like shapes. The more than one hundred examples discussed here and compiled in six tables provide evidence for the successful research activities in this branch of scientific interest during the past twenty-five years.

In contrast to the single kind of fluid nematic phase known to be exhibited by calamitic mesogens, the disc-like mesogens described in this Datareview can form two nematic phases, each showing a Schlieren texture: a discotic nematic or a columnar nematic type of phase; both differ in their supra-molecular array of the disc-like molecules and viscosity.

Details of their molecular architectures and mesomorphic properties have been discussed, and also, the exotic nature of these materials has prompted special comments about the following topics: (i) ways to introduce chirality, (ii) the question of biaxial nematics, (iii) thermo- and lyotropic mesomorphism based on mixtures of meso- or non-mesomorphic organic compounds, and (iv) mesophase manipulations and inductions through doping with suitable chemicals.

However, in contrast to their fascinating flat molecular structures very little attention has yet been paid to possible applications. Probably, more discotic materials with still lower phase transition temperatures must be designed for such purposes. An interesting optically controlled electro-optic effect in discotic nematic radial multiynes has been observed very recently [57], and hopefully, this encouraging experience will induce more and systematic work on these materials.

ACKNOWLEDGEMENT

The author thanks Mr. A. Eckert for his kind help during the preparation of this manuscript.

REFERENCES

[1]   D. Vorländer [*Z. Phys. Chem. (Germany)* vol.105 (1923) p.211; *Chem. Abstr. (USA)* vol.18 (1924) p.1072]. See also P.G. de Gennes [*The Physics of Liquid Crystals* (Oxford University Press, Oxford, 1974)]

[2]   S. Chandrasekhar, B.K. Sadashiva, K.A. Suresh [*Pramana (India)* vol.7 (1977) p.471; *Chem. Abstr. (USA)* vol.88 (1977) p.30566y]. The first columnar (hexagonal) phase is formed by three disc-/star-shaped chemically defined, pure hexaalkanoates of hexahydroxybenzene. NB: Similarly, the layered structure of graphite causes the formation of the so-called carbonaceous phase on heating; see J.D. Brooks, G.H. Taylor [*Carbon (UK)* vol.3 (1965) p.185 and reference [3]].

[3]   H.T. Nguyen, C. Destrade, H. Gasparoux [*Phys. Lett. A (Netherlands)* vol.72 (1979) p.251]

[4]   Source: version LiqCryst 3.2a (status of September 1999) of *Liq. Cryst 3.2 - Database of Liquid Crystalline Compounds for Personal Computers* V. Vill, Fujitsu Kyushu System (FQS) Ltd, Fukuoka, Japan, 1998; LCI Publisher, Hamburg, Germany, 1999; http://liqcryst.chemie.uni-hamburg.de.

[5]   S. Chandrasekhar, G.S. Ranganath [*Rep. Prog. Phys. (UK)* vol.53 (1990) p.57]

[6]   (a) H.T. Nguyen, H. Gasparoux, C. Destrade [*Mol. Cryst. Liq. Cryst. (UK)* vol.68 (1981) p.101]; (b) C. Destrade, H.T. Nguygen, H. Gasparoux, J. Malthête, A.M. Levelut [*Mol. Cryst. Liq. Cryst. (UK)* vol.71 (1981) p.111]

[7]   K. Praefcke, D. Blunk [unpublished results from D. Blunk, PhD Thesis, Technische Universität Berlin, Germany, 1999, ISBN 3-934445-00-4]

[8]   C. Destrade, H.T. Nguyen, J. Malthête, J. Jacques [*Phys. Lett. A (Netherlands)* vol.79 (1980) p.189]

[9]   K. Praefcke, D. Singer [in *Handbook of Liquid Crystals* Eds. D. Demus, J. Goodby, G.W. Gray, H.-J. Spiess, V. Vill (Wiley-VCH, Weinheim, 1998) vol.2B ch.XVI p.945 and references therein]

[10]   B. Kohne, K. Praefcke [*Chimia (Switzerland)* vol.41 (1987) p.196]

[11]   K. Praefcke, B. Kohne, K. Gutbier, N. Johnen, D. Singer [*Liq. Cryst. (UK)* vol.5 (1989) p.233]

[12]   K. Praefcke, B. Kohne, D. Singer [*Angew. Chem., Int. Ed. Engl. (Germany)* vol.29 (1990) p.177]

[13]   K. Prafecke, D. Singer, B. Gündogan, K. Gutbier, M. Langner [*Ber. Bunsenges. Phys. Chem. (Germany)* vol.98 (1994) p.118]

[14]   K. Praefcke, D. Singer [unpublished results from D. Singer, PhD Thesis, Technische Universität Berlin, Germany, 1994, ISBN 3-929937-79-4]

[15]   M. Langner, K. Praefcke, D. Krüerke, G. Heppke [*J. Mater. Chem. (UK)* vol.5 (1995) p.693]

[16]   D. Krüerke, H.-S. Kitzerow, G. Heppke [*Ber. Bunsenges. Phys. Chem. (Germany)* vol.97 (1993) p.1371]

[17]   K. Praefcke, D. Singer, A. Eckert [*Liq. Cryst. (UK)* vol.16 (1994) p.53]. The chiral dopant applied here is (-)-2-(2,4,5,7-tetranitro-9-fluorenylidenaminooxy)propionic acid, (-)-TAPA.

[18]   H. Bengs, O. Karthaus, H. Ringsdorf, C. Baehr, M. Ebert, J.H. Wendorff [*Liq. Cryst. (UK)* vol.10 (1991) p.161]

[19]   K. Praefcke, D. Singer, B. Kohne, M. Ebert, A. Liebmann, J.H. Wendorff [*Liq. Cryst. (UK)* vol.10 (1991) p.147]

[20]   N. Usol'tseva, K. Praefcke, D. Singer, B. Gündogan [*Liq. Cryst. (UK)* vol.16 (1994) p.617]

[21]   N. Usol'tseva, K. Praefcke, A. Smirnova, D. Blunk [*Liq. Cryst. (UK)* vol.26 (1999) p.1723]

[22]   N. Usol'tseva, G. Hauck, H.D. Koswig, K. Praefcke, B. Heinrich [*Liq. Cryst. (UK)* vol.20 (1996) p.731 and earlier papers cited therein]

[23]   K. Praefcke, B. Bilgin, N. Usol'tseva, B. Heinrich, D. Guillon [*J. Mater. Chem. (UK)* vol.5 (1995) p.2257]

[24]   A.V. Kasnatscheev, K. Praefcke, A.S. Sonin, N.V. Usol'tseva [*Kristallografiya (Russia)* vol.42 (1997) p.744 or *Crystallogr. Rep. (Russia)* vol.42 (1997) p.683]

[25]   N. Usol'tseva, P. Espinet, J. Buey, K. Praefcke, D. Blunk [*Mol. Cryst. Liq. Cryst. (Switzerland)* vol.299 (1997) p.457]

[26]   N.V. Usol'tseva [*Lyotropic Liquid Crystals: Chemical and Supramolecular Structure* (IvGU, Ivanovo, Russia, 1994, ISBN 5-230-02212-4)]; T. Sierra [in *Metallomesogens - Synthesis, Properties and Applications* Ed. J.L. Serrano (VCH, Weinheim, 1996, ISBN 3-527-29296-9) ch.2 p.29]; M.R. Kuzma, A. Saupe [in *Handbook of Liquid Crystal Research* Eds. P.J. Collings, J.S. Patel (Oxford University Press, 1997, ISBN 0-19-508442-X) ch.7 p.237]; J. Lydon [in *Handbook of Liquid Crystals* Eds. D. Demus, J. Goodby, G.W. Gray, H.-J. Spiess, V. Vill (Wiley-VCH, Weinheim, 1998, ISBN 3-527-29491-0) vol.2B ch.XVIII p.981]; D.

Blunk, K. Praefcke, V. Vill [in *Handbook of Liquid Crystals* Eds. D. Demus, J. Goodby, G.W. Gray, H.-J. Spiess, V. Vill (Wiley-VCH, Weinheim, 1998, ISBN 3-527-29272-1) vol.3 ch.VI p.305]; A. Petrov [*The Lyotropic State of Matter - Molecular Physics and Living Matter Physics* (Gordon & Breach Science Publishers, 1999, ISBN 90-5699-638-X)]

[27]   A.S. Sonin [*J. Mater. Chem. (UK)* vol.8 (1998) p.2557]

[28]   L.Y. Yu, A. Saupe [*Phys. Rev. Lett. (USA)* vol.45 (1980) p.1000]

[29]   S. Chandrasekhar [*Mol. Cryst. Liq. Cryst. (UK)* vol.124 (1985) p.1]

[30]   K. Praefcke et al [*Mol. Cryst. Liq. Cryst. (UK)* vol.198 (1991) p.393]

[31]   B.K. Sadashiva [in *Handbook of Liquid Crystals* Eds. D. Demus, J. Goodby, G.W. Gray, H.-J. Spiess, V. Vill (Wiley-VCH, Weinheim, 1998) vol.2B ch.XV p.933 and references cited therein]

[32]   K. Praefcke et al [*Mol. Cryst. Liq. Cryst. (Switzerland)* vol.323 (1998) p.231 and references cited therein]

[33]   (a) See a report by 23 participants for the 'Oxford Workshop on Biaxial Nematics', Oxford, UK, 20-22 December 1996 [*Liq. Cryst. Today (UK)* vol.7 (1997) p.13]; D.W. Bruce, G.R. Luckhurst, D.J. Photinos (Eds.) [*Mol. Cryst. Liq. Cryst. (Switzerland)* vol.323 (1998) p.154-65 and p.261-93]; S.M. Fan et al [*Chem. Phys. Lett. (Netherlands)* vol.204 (1993) p.517]; J.R. Hughes et al [*J. Chem. Phys. (USA)* vol.107 (1997) p.9252] (b) O. Pelletier, C. Bourgaux, O. Diat, P. Davidson, J. Livage [*Eur. Phys. J. B (France)* vol.12 (1999) p.541] (c) For a recent review on 'Inorganic Liquid Crystals' see A.S. Sonin [*J. Mater. Chem. (UK)* vol.8 (1998) p.2557]

[34]   B. Kohne, K. Praefcke [*Chem.-Ztg. (Germany)* vol.109 (1985) p.121]

[35]   B. Heinrich, K. Praefcke, D. Guillon [*J. Mater. Chem. (UK)* vol.7 (1997) p.1363 and earlier papers cited therein]

[36]   C. Destrade, H. Gasparoux, A. Babeau, H.T. Nguyen [*Mol. Cryst. Liq. Cryst. (UK)* vol.67 (1981) p.37 and an earlier paper cited therein]

[37]   H.T. Nguyen, P. Foucher, C. Destrade, A.M. Levelut, J. Malthête [*Mol. Cryst. Liq. Cryst. (UK)* vol.111 (1984) p.277]

[38]   H.T. Nguyen, J. Malthête, C. Destrade [*Mol. Cryst. Liq. Cryst. (UK)* vol.64 (1981) p.291]

[39]   L. Mamlok, J. Malthête, H.T. Nguyen, C. Destrade, A.M. Levelut [*J. Phys. Lett. (France)* vol.43 (1982) p.L641]; C. Destrade, H.T. Nguyen, L. Mamlok, J. Malthête [*Mol. Cryst. Liq. Cryst. (UK)* vol.114 (1984) p.139]

[40]   H.T. Nguyen, R. Cayuela, C. Destrade [*Mol. Cryst. Liq. Cryst. (UK)* vol.122 (1984) p.141]

[41]   R. Cayuela, H.T. Nguyen, C. Destrade, A.M. Levelut [*Mol. Cryst. Liq. Cryst. (UK)* vol.177 (1989) p.81]

[42]   H.T. Nguyen, J. Malthête, C. Destrade [*J. Phys. Lett. (France)* vol.42 (1981) p.L417]; C. Destrade, P. Foucher, J. Malthête, H.T. Nguyen [*Phys. Lett. A (Netherlands)* vol.88 (1982) p.187]; W.K. Lee et al [*Liq. Cryst. (UK)* vol.4 (1989) p.87]

[43]   K. Praefcke, D. Singer, B. Gündogan [*Mol. Cryst. Liq. Cryst. (UK)* vol.223 (1992) p.181]; D. Singer, A. Liebmann, K. Praefcke, J.H. Wendorff [*Liq. Cryst. (UK)* vol.14 (1993) p.785]

[44]   D. Goldmann et al [*Liq. Cryst. (UK)* vol.24 (1998) p.881]

[45]   S. Mahlstedt, D. Janietz, C. Schmidt, A. Stracke, J.H. Wendorff [*Liq. Cryst. (UK)* vol.26 (1999) p.1359]

[46]   B. Kohne, W. Poules, K. Praefcke [*Chem.-Ztg. (Germany)* vol.108 (1984) p.113]; B. Kohne, K. Praefcke, T. Derz, W. Frischmut, C. Gansau [*Chem.-Ztg. (Germany)* vol.108 (1984) p.408]

[47]   T.J. Phillips, J.C. Jones, D.G. McDonnell [*Liq. Cryst. (UK)* vol.15 (1993) p.203]

[48]   P. Hindmarsh, M. Hird, P. Styring, J.W. Goodby [*J. Mater. Chem. (UK)* vol.3 (1993) p.1117]

[49]   D.R. Beattie, P. Hindmarsh, J.W. Goodby, S.D. Haslam, R.M. Richardson [*J. Mater. Chem. (UK)* vol.2 (1992) p.1261]

[50]   D.M. Collard, C.P. Lillya [*J. Org. Chem. (USA)* vol.56 (1991) p.6064]

[51]   C. Destrade, M.C. Mondon, J. Malthête [*J. Phys. Colloq. (France)* vol.C3 (1979) p.17]

[52]   C. Destrade, H.T. Nguyen, H. Gasparoux, L. Mamlok [*Liq. Cryst. (UK)* vol.2 (1987) p.229]
The systematic name (CA system) of the heterocyclic type of the discotic liquid crystal core studied here is benzo(1,2-b;3,4-b';5,6-b'')trisfuran; the three oxygen positions of this heterocycle are 1, 4 and 7.

[53]   M. Ebert, D.A. Jungbauer, R. Kleppinger, J.H. Wendorff, B. Kohne, K. Praefcke [*Liq. Cryst. (UK)* vol.4 (1989) p.53]

[54]   J.S. Patel, K. Praefcke, D. Singer, M. Langner [*Appl. Phys. B (Germany)* vol.60 (1995) p.469]

[55]   K. Praefcke, J.D. Holbrey [*J. Inclusion Phen., Molec. Recogn. Chem. (Netherlands)* vol.24 (1996) p.19]

[56]   D. Janietz, K. Praefcke, D. Singer [*Liq. Cryst. (UK)* vol.13 (1993) p.247]

[57]   D. Sikharulidze, G. Chilaya, K. Praefcke, D. Blunk [*Liq. Cryst. (UK)* vol.23 (1997) p.439]

[58]   K. Praefcke, M. Langner [unpublished results from M. Langner, PhD Thesis, Technische Universität Berlin, Germany, 1996]

[59]   J. Zhang, J.S. Moore [*J. Am. Chem. Soc. (USA)* vol.116 (1994) p.2655]; O.Y. Mindyuk, M.R. Stetzer, P.A. Heiney, J.C. Nelson, J.S. Moore [*Adv. Mater. (Germany)* vol.10 (1998) p.1363]

[60]   K. Praefcke et al [*Mol. Cryst. Liq. Cryst. (UK)* vol.215 (1992) p.121]

[61]   The reader is referred to the book by J.L. Serrano (Ed.), cited here in [26], and to M. Giroud-Godquin [in *Handbook of Liquid Crystals* Eds. D. Demus, J. Goodby, G.W. Gray, H.-J. Spiess, V. Vill (Wiley-VCH, Weinheim, 1998) vol.2B ch.XIV p.901]

[62]   K. Praefcke, B. Bilgin, J. Pickardt, M. Borowski [*Chem. Ber. (Germany)* vol.127 (1994) p.1543]

[63]   A. Liebmann, C. Mertesdorf, T. Plesnivy, H. Ringsdorf, J.H. Wendorff [*Angew. Chem., Int. Ed. Engl. (Germany)* vol.30 (1991) p.1375]

1.3 Relationship between molecular structure and transition temperatures for unconventional molecular structures

C.T. Imrie

September 1999

A INTRODUCTION

The overwhelming majority of low molar mass liquid crystals may be described as conventional calamitic liquid crystals and consist of molecules composed of a semi-rigid rod-like core, normally consisting of phenyl rings linked through short unsaturated linkages, attached to which are one or two alkyl chains [1]. By comparison, a conventional discotic liquid crystal consists of molecules which have essentially planar cores, for example triphenylene, attached to which are either six or eight alkyl chains [2]. This Datareview focuses on the relationship between molecular structure and the nematic-isotropic transition temperature, T_{NI}, for mesogens which do not fall into these classifications and thus may be termed unconventional structures.

B DIMERS AND OLIGOMERS

Liquid crystal dimers consist of molecules comprising two mesogenic units linked via a flexible spacer, normally an alkyl chain, containing between 3 and 12 methylene units [3,4]. In the majority of dimers reported in the literature the two mesogenic units are identical and rod-like; these are termed symmetric calamitic dimers. The characteristic behaviour of such compounds can be illustrated using the α,ω-bis(4′-cyanobiphenyl-4-yloxy) alkanes, and TABLE 1 lists their nematic-isotropic transition

temperatures [5]. A very large odd-even effect is apparent in T_{NI} on varying the length and parity of the spacer in which the even members of the series exhibit the higher values, although this alternation is attenuated on increasing the spacer length. This odd-even effect is very much more pronounced than that exhibited by T_{NI} on varying the terminal chain length of conventional calamitics. This behaviour has been interpreted most often in terms of how the molecular shape is governed by the parity of the spacer in the all-trans conformation. Specifically in an even member the mesogenic groups are antiparallel whereas in an odd membered dimer they are inclined. More realistic interpretations allow for the flexible nature of the spacer [3].

In general, the effects on T_{NI} for a liquid crystal dimer on varying the chemical structure of the mesogenic units [6,7] or on increasing the length of a terminal alkyl chain [8] tend to mirror those seen for conventional calamitics.

36

TABLE 1 Nematic-isotropic transition temperatures for the α,ω-bis (4'-cyanobiphenyl-4-yloxy) alkanes [5]. n refers to the number of methylene units in the alkyl spacer and ( ) indicates a monotropic transition.

n	1	2	3	4	5	6	7	8	9	10	11	12
$T_{NI}/°C$	(124)	265	(170)	250	186	221	181	201	172	184	164	169

Liquid crystal dimers in which two differing rod-like mesogenic units are linked via a spacer are termed non-symmetric calamitic dimers. The effect on T_{NI} of varying the length of the spacer for a series of non-symmetric dimers is analogous to that already seen for symmetric dimers. TABLE 2 lists the nematic-isotropic transition temperatures for the α-(4-cyanobiphenyl-4'-yloxy)-ω-(4-ethylanilinebenzylidene-4'-oxy) alkanes [9], T_{NI} exhibits a pronounced odd-even effect as the length and parity of the spacer is varied.

TABLE 2 Nematic-isotropic transition temperatures for the α-(4-cyanobiphenyl-4'-yloxy)-ω-(4-ethylanilinebenzylidene-4'-oxy) alkanes [9]. n denotes the number of methylene units in the spacer. ( ) indicates a monotropic transition.

n	3	4	5	6	7	8	9	10	11	12
$T_{NI}/°C$	(130)	236	165	206	165	177	162	177	158	168

The mesogenic units in a liquid crystal dimer may also be linked in lateral positions. TABLE 3 lists the T_{NI}s of the α,ω-bis-[2,5-bis(4-octyloxybenzoyloxy)-benzamido] alkanes [10]. Again T_{NI} exhibits a pronounced alternation as the length and parity of the spacer is varied but this attenuates quicker than for terminally linked liquid crystal dimers. Similar behaviour has been reported for other laterally linked dimers [11] although there is a strong tendency for this class of materials to exhibit exclusively smectic behaviour [12,13].

TABLE 3 Nematic-isotropic transition temperatures for the α,ω-bis-[2,5-bis(4-octyloxybenzoyloxy)-benzamido] alkanes [10]. n represents the number of methylene units in the flexible spacer. ( ) indicates a monotropic transition.

n	3	4	5	6	7	8	9	10	12
$T_{NI}/°C$	(141)	172.5	154	160	144.5	142	132.5	(127)	(123)

Liquid crystal dimers have been reported in which one mesogen is attached via a lateral position while the other is connected via a terminal site. TABLE 4 reveals how T_{NI} increases as the bulky lateral group is systematically changed into a mesogenic structure [14]. Again there is a strong tendency for these lateral-terminal or T-shaped dimers to exhibit exclusively smectic behaviour although nematogenic examples have been reported [10,15]; see TABLE 5.

TABLE 4 Nematic-isotropic transition temperatures
for compounds containing lateral bulky groups [14].

R	T_{NI}/°C
H	98
OC_8H_{17}	104
[phenyl]—OC_8H_{17}	151.5
—O.OC—[phenyl]—O.OC—[phenyl]—OC_8H_{17}	197

TABLE 5 Nematic-isotropic transition temperatures for T-shaped liquid crystal dimers [15].
n represents the number of methylene units in the flexible alkyl spacer.

n	5	6	7	8	9	10	12
T_{NI}/°C	190	193	178	182	164	168	147

Dimers consisting of two discotic units linked via a flexible spacer have a strong tendency not to exhibit nematic behaviour [16]. An example of a nematogenic discotic dimer has been reported by Praefcke et al which was based on two discotic multialkylnyl units [17],

Cr 121.4°C N 156.2°C I

Conoscopic studies of this nematic phase have suggested that it is in fact biaxial although this assignment has still to be verified using other techniques. The search for the thermotropic biaxial nematic phase prompted the synthesis of dimers consisting of one calamitic and one discotic mesogenic unit [18]. A single compound exhibited a strongly monotropic nematic phase, and the

Cr 125°C N (79°C) I

authors attributed the general absence of liquid crystalline behaviour in this class of mesogens to the extreme difficulty in packing the disc-like and rod-like units simultaneously. A related class of dimers consists of molecules containing a pentayne electron rich moiety attached to an electron deficient accepting group: see TABLE 6 [19]. It is widely believed that in these and similar compounds, a charge-transfer interaction between the mesogenic discotic unit and the rod-like, but not mesogenic, 2,4,7-trinitrofluoren-9-one group is the driving force responsible for the observation of liquid crystalline behaviour. It has been shown by computer simulation studies, however, that the driving force for phase formation involves an electrostatic quadrupolar interaction between the unlike components whose quadrupole moments differ in sign [20].

TABLE 6 Nematic-isotropic transition temperatures
for pentayne-based donor-acceptor dimers [19].

R	T_{NI}/°C
H	134
C_5H_{11}	59

U-shaped dimers have been reported in which a central unit introduces a kink into the molecular structure. TABLE 7 lists the nematic-isotropic transition temperatures of the benzene-1,2-di-(4-carboxyalkoxybenzylidene-4'-alkylaniline)s [21],

On increasing the terminal chain length for an odd-membered spacer, T_{NI} alternates in a fashion similar to that observed for conventional calamitics. Compounds containing even-membered spacers are exclusively smectogenic.

TABLE 7 Nematic-isotropic transition temperatures of the benzene-1,2-di-(4-carboxyalkoxybenzylidene-4'-alkylaniline)s [21]. n refers to the number of methylene units in the alkyl spacers and m to the number of carbon atoms in the terminal alkyl chains. ( ) denotes a monotropic transition.

m	n = 3 T_{NI} / °C	n = 5 T_{NI} / °C
1	(47)	(54)
2	-	(49)
3	(66)	(68)
4	(67)	(62)
5	(83)	(75)

Liquid crystal trimers contain molecules comprising three mesogenic groups and two flexible spacers. In a linear calamitic trimer these groups are attached at terminal sites. TABLE 8 lists the nematic-isotropic transition temperatures of the 4,4'-bis[ω-(4-cyanobiphenyl-4'-yloxy)alkoxy] biphenyls [22],

T_{NI} exhibits a dramatic odd-even effect as the length and parity of the spacer is varied, in which the even members exhibit the higher values. This behaviour is interpreted in terms of how the spacer controls the average molecular shape. The magnitude of the alternation in T_{NI} for the trimers is only marginally larger than that found for the corresponding dimers, suggesting that the mesogenic units are correlated to the same extent in the dimers and trimers.

TABLE 8 Nematic-isotropic transition temperatures of the 4,4'-bis[ω-(4-cyanobiphenyl-4'-yloxy)alkoxy] biphenyls [22a]. n denotes the number of methylene units in the flexible alkyl spacers.

n	3	4	5	6	7	8	9	10	11	12
T_{NI}/°C	202	297	215	262	206	231	196	213	184	194

The addition of a fourth mesogenic unit via a third alkyl spacer yields liquid crystal tetramers. TABLE 9 lists the nematic-isotropic transition temperatures of a homologous series of tetramers in which the length and parity of the outer flexible spacers is varied [23]. A dramatic odd-even effect in T_{NI} is apparent on varying the outer spacer length and again this is attributed to how the spacers control the average molecular shape. Non-linear tetramers have also been described in the literature; for example, molecules have been reported with tetrahedral symmetry although these have all been exclusively smectic in behaviour [24,25]. These structures may be thought of as zeroth order dendrimers. A 1,3,5,7-tetrakiscubane-based nematogen has been reported [26] but in general substituted caged structures have a strong tendency to exhibit smectic behaviour [25].

TABLE 9 Nematic-isotropic transition temperatures for the liquid crystal tetramers [23]. n denotes the number of methylene units in the outer flexible spacers.

n	3	4	5	6	7	8	9	11
T_{NI}/°C	246	307	250	279	241	258	235	210

A laterally linked trimer, 2',5'-bis[3-(4,4''-diethoxy-p-terphenyl-2'-yl)-2-oxaprop-1-yl]-4,4''-didecyloxy-p-terphenyl, has been shown to exhibit nematic behaviour but extending the terminal chains promotes smectic behaviour [27].

Cr 162°C N (155°C) I

Three mesogenic units have been attached to a central 1,3,5-triazine unit yielding the nematogenic tris[4-(4-alkyloxyphenylazo)phenylamino]-1,3,5-triazines [28] (see TABLE 10) although the removal of one mesogenic unit and replacement by a methoxy group produces exclusively smectic or columnar materials.

TABLE 10 Nematic-isotropic transition temperatures for the tris[4-(4-alkyloxyphenylazo)phenylamino]-1,3,5-triazines [28]. n is the number of carbon atoms in the terminal alkyloxy chains.

n	10	12	16
T_{NI}/°C	240.7	239.2	203.4

The introduction of flexible spacers between the central 1,3,5-substituted phenyl ring and the mesogenic groups yields nematogenic materials for short spacer lengths (see TABLE 11) but exclusively smectic materials on increasing the spacer length [29]. Structurally analogous materials containing disc-like mesogenic units do not exhibit nematic behaviour [30]. The corresponding 1,3,5-cyclohexanetricarboxylic acid derivatives containing three calamitic mesogenic groups also exhibit nematic behaviour for short spacer lengths [26].

As many as six mesogenic groups have been attached to a central ring. Thus, the phosphazene-based cyclic trimer [31] exhibits a nematic phase.

TABLE 11 Nematic-isotropic transition temperatures for trimers derived from benzene-1,3,5-tricarboxylic acid [29]. m denotes the number of carbon atoms in the terminal alkyl chains.

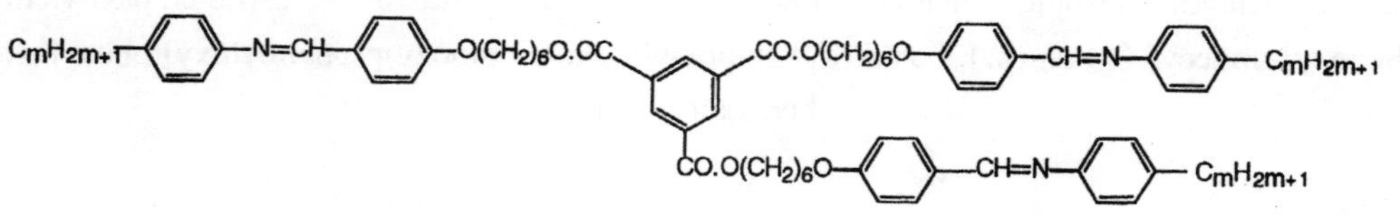

m	2	3	4	5	6
T_{NI}/°C	102	101	103	108	99

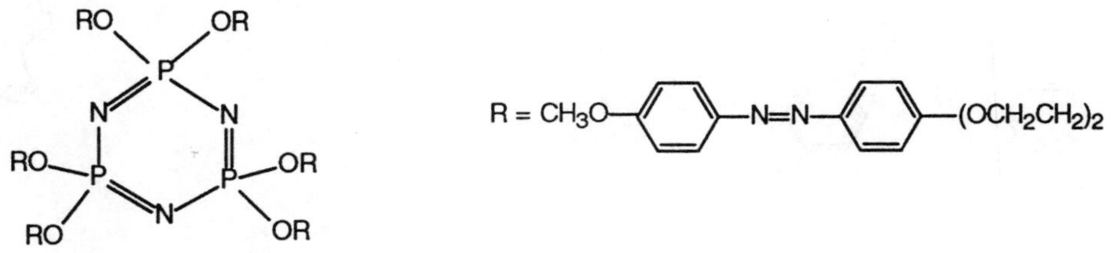

Cr 197°C N (192°C) I

Eight cyanobiphenyl groups have been attached to a cyclotetrabenzylene central core yielding a nematic material [32],

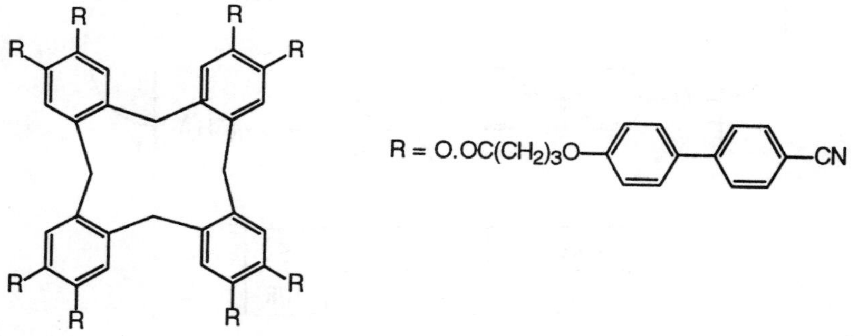

Cr 222°C N (220°C) I

Linear trimeric structures have been reported consisting of a discotic electron rich mesogen, a rod-like mesogen and an electron deficient acceptor unit each separated by flexible alkyl spacers: see TABLE 12 [33]. Preliminary conoscopic investigations suggested that the nematic phase was not biaxial.

Cyclic liquid crystal oligomers based on 1-(4-hydroxy-4′-biphenyl)-2-(4-hydroxyphenyl) butane with 1,7-dibromoheptane have been reported: see TABLE 13 [34]. The cyclic trimer, tetramer and pentamer exhibit nematic behaviour and their clearing temperatures are higher than that of the

43

analogous linear polymer. This suggests that the cyclic oligomers adopt elongated conformations in the nematic phase and for a cyclic trimer a biaxial nematic phase has been proposed [35].

TABLE 12 Nematic-isotropic transition temperatures of the 11-[pentakis(4-substituted phenylethynyl) phenoxy]-undecyl 4{4-[3-(2,4,7-trinitro-9-fluorenylidene-aminooxy)propanoyloxy] phenylazo} benzoates [33].

R	H	CH$_3$
T$_{NI}$/°C	105.4	146.6

TABLE 13 Nematic-isotropic transition temperatures for the cyclic oligomers where x + y = ring size [34].

Ring size	3	4	5
T$_{NI}$/°C	81	115	108

A nematogenic cyclic dimer has also been reported [36],

Cr 172°C (N 142°C) I

Cyclic oligomers based on triethylene glycol [37] and on a siloxane core [38] also exhibit nematic behaviour; see TABLES 14 and 15, respectively.

TABLE 14 Nematic-isotropic transition temperatures of the cyclic oligomers where z is the number of repeat units [37].

z	4	5	6	7	8	9	10
T_{NI}/°C	112	57	92	68	82	66	74

TABLE 15 Nematic-isotropic transition temperatures for the cyclic oligomers where m and n are shown on the structure [38].

m, n	0,0	1,1	3,1
T_{NI}/°C	135	175	175

C   LIQUID CRYSTAL DENDRIMERS

Liquid crystal dendrimers can be divided into two classes, one in which mesogenic units are found in each generation of the dendritic molecule, or alternatively, those in which mesogenic units are found only on the surface layer [39]. For the latter class of dendrimers only smectic behaviour has been observed [25]. Nematic behaviour has been observed, however, for dendrimers containing mesogens in each generation as well as for the monodendrons used in their synthesis; see TABLE 16 [40]. The observation of liquid crystalline behaviour for dendritic molecules strongly suggests that they must adopt rod-like shapes in the liquid crystal environment.

TABLE 16 Nematic-isotropic transition temperatures for a monodendron
and the corresponding dendrimer [40].

Monodendron: N 110°C I

TABLE 16 continued

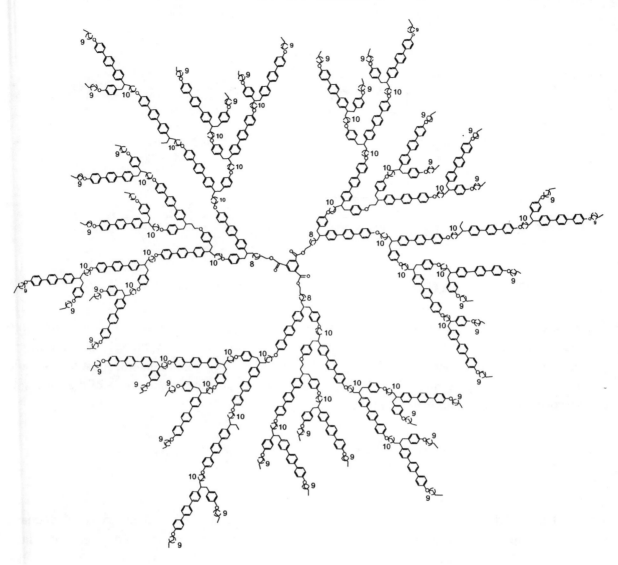

Dendrimer: N 108°C I

D   BANANA-SHAPED COMPOUNDS

Banana-shaped liquid crystals consist of molecules containing a bent semi-rigid core attached to which are one or two terminal alkyl chains. This class of mesogen is primarily of interest because they can exhibit novel smectic phases [41]. Banana-shaped compounds tend not to exhibit nematic behaviour but for series which do, T_{NI} tends to decrease with little or no alternation: see TABLE 17 [42]. This behaviour is analogous to that observed for conventional calamitic nematogens.

TABLE 17 Nematic-isotropic transition temperatures of the banana-shaped compounds [42].
n is the number of carbon atoms in the terminal alkyl chains.

n	8	9	10	11	12	13
T_{NI}/°C	179	174	173	168	167	163

E   LONG CHAIN LATERALLY SUBSTITUTED COMPOUNDS

It was widely believed for many years that the incorporation of large lateral substituents into mesogenic structures would destroy liquid crystalline behaviour. This view changed dramatically in the early 1980s with the discovery that compounds containing large flexible lateral substituents could indeed exhibit liquid crystallinity [43]. The nematic-isotropic transition temperatures of the 1,4-bis(4-octyloxybenzoyloxy)-2-alkyl benzenes [44]

are listed in TABLE 18. The initial increase in the length of the alkyl chain causes a dramatic reduction in the nematic-isotropic transition temperature. Further increases, however, have little effect on T_{NI}. This is thought to imply that the lateral chain adopts conformations in which it lies coparallel to the molecular long axis [45].

TABLE 18 The nematic-isotropic transition temperatures of the 1,4-bis(4-octyloxybenzoyloxy)-2-alkyl benzenes [44]. n denotes the number of carbon atoms in the lateral alkyl chain.

n	0	1	2	3	4	5	6	7	8	9	10	11	12	16
T_{NI}/°C	195	156	119	98	91	84.5	81.5	79	77.5	77	74	72	69.5	69.5

Two lateral alkyl chains can be attached to the same core. For example, TABLE 19 lists the nematic-isotropic transition temperatures of the di-alkylketoximino 2,5-bis[4-octyloxy-benzoyloxy] benzoates [46],

The increase in T_{NI} for long chain lengths supports the view that the lateral chains tend to lie coparallel to the molecular long axis.

TABLE 19 The nematic-isotropic transition temperatures for the di-alkylketoximino 2,5-bis[4-octyloxybenzoyloxy] benzoates [46]. n denotes the number of carbon atoms in the lateral alkyl chains. ( ) indicates a monotropic transition.

n	1	2	3	4	5	6	7	8	9	10	11
T_{NI}/°C	(69.5)	(56.5)	(46.5)	(44.5)	(42)	(40)	(39)	(41)	(42.2)	(43)	(43)

The two lateral alkyl chains can also be attached at different sites on the mesogenic core; for example, the four ring oximester has a T_{NI} of 148°C [47].

F   SWALLOW-TAILED LIQUID CRYSTALS

Swallow-tailed liquid crystals are composed of molecules containing a branched terminal alkyl chain in which the branch itself is a long alkyl chain [43]. TABLE 20 lists the nematic-isotropic transition temperatures of the di-alkyl 4-(4'-(4''-octyloxybenzoyloxy) benzoyloxy) benzylidene malonates [48],

As with lateral long alkyl chains, T_{NI} tends to approach a plateau value on increasing the chain length. Unlike mesogens containing lateral alkyl chains, however, swallow-tailed compounds often exhibit smectic behaviour.

TABLE 20 The nematic-isotropic transition temperatures of the di-alkyl 4-(4'-(4''-n-octyloxybenzoyloxy) benzoyloxy) benzylidene malonates [48]. n denotes the number of carbon atoms in the branched alkyl chains.

n	1	2	3	4	5	6	7	8	9	10	11	12	16
T_{NI}/°C	162	144	130	115	107	101	97	94	94	91	90	89	86

Liquid crystals consisting of molecules containing branched alkyl chains in both terminal positions are referred to as double-swallow-tailed mesogens, see TABLE 21. Double-swallow-tailed mesogens tend to exhibit complex behaviour typically involving smectic C, cubic phases and nematic phases for shorter chains and columnar phases on increasing the chain length.

TABLE 21 The nematic-isotropic transition temperatures for the double-swallow-tailed compound [49]. n denotes the number of carbon atoms in the alkyl chains.

n	5	6	7	8
T_{NI}/°C	378	355	334	312

G   POLYCATENAR LIQUID CRYSTALS

Polycatenar liquid crystals are composed of molecules containing a conventional rod-like core attached to which are two to six flexible chains around the terminal rings [50]. Bicatenar mesogens include not only conventional calamitic liquid crystals but also molecules in which both terminal chains are on the same ring, occupying the para and meta positions, see TABLE 22. This class of mesogens are also referred to as forked mesogens and can be considered analogous to swallow-tailed mesogens.

TABLE 22 The nematic-isotropic transition temperatures of the bicatenar or forked mesogens, the 4-cyano-4'-biphenylyl 3″, 4″-di-alkyloxybenzoates [45]. n denotes the number of carbon atoms in the alkyloxy chains. ( ) indicates a monotropic transition.

n	1	2	3	4
T_{NI}/°C	191	(167)	(142)	(143)

Tricatenar compounds containing two chains in para positions and one in a meta position do exhibit nematic behaviour (see TABLE 23) and moving the meta-chain into an ortho position promotes nematic tendencies [51]. Nematic phases have not been observed for tricatenar compounds with two chains in the meta-positions of one ring and the third in the para position of the other terminal ring.

TABLE 23 The nematic-isotropic transition temperatures for the tricatenar liquid crystals [51]. n denotes the number of carbon atoms in the alkyloxy chains. ( ) indicates a monotropic transition.

n	4	8	12
T_{NI}/°C	93	(88)	(86)

Four types of tetracatenar liquid crystal have been studied [50]. Molecules containing chains in the para and meta positions of both terminal rings exhibit nematic behaviour, see TABLE 24. These compounds are also referred to as biforked mesogens and are structurally analogous to the double-swallow-tailed mesogens.

TABLE 24 The nematic-isotropic transition temperatures for the tetracatenar liquid crystals [52]. n denotes the number of carbon atoms in the alkyloxy chains.

n	6	7	8	9
$T_{NI}/°C$	217	205.5	196.5	183.5

Special classes of tetracatenar liquid crystals exhibit nematic behaviour; for example,

exhibits a T_{NI} of 89°C [53] and this was assigned as a biaxial nematic phase. Subsequent studies, however, revealed that this was not the case [54-56]. The 1,2,4,5-tetra(4'-alkoxybenzoyloxy) benzenes are also nematogenic [57,58] (see TABLE 25) and the nematic phase is uniaxial. These materials are referred to as cross-shaped liquid crystals and the analogous materials containing three substituents, the so-called λ-mesogens, are also nematogens, see TABLE 26 [59].

TABLE 25 The nematic-isotropic transition temperatures of the 1,2,4,5-tetra(4'-alkoxybenzoyloxy) benzenes [57,58]. n denotes the number of carbon atoms in the alkyloxy chains and ( ) indicates a monotropic transition.

n	3	4	5	6	7	8
$T_{NI}/°C$	(115.0)	125.0	(103.0)	(107.8)	(104.5)	(105.4)

n	9	10	11	12	13	14
$T_{NI}/°C$	(100.9)	102.0	(99.6)	(100.2)	100.0	100.0

TABLE 26 The nematic-isotropic transition temperatures of the λ-shaped 1,2,4-benzenetricarboxylates [59]. n denotes the number of carbon atoms in the alkyloxy chains.

n	1	3	4	6	8	12
T_{NI}/°C	258	245	189	166	172	190

Only a small number of pentacatenars have been reported and these do not exhibit nematic behaviour [50]. Similarly hexacatenars containing alkyl chains in the para- and meta-positions do not exhibit nematic behaviour [50]. Both classes of compounds tend to exhibit columnar phases. Hexacatenars containing alkyl chains in the ortho-, meta- and para-positions of both terminal rings do, however, exhibit nematic behaviour; see TABLE 27 [60].

TABLE 27 The nematic-isotropic transition temperatures of the hexacatenars [60]. n is the number of carbon atoms in the alkyloxy chains and ( ) indicates a monotropic transition.

n	1	4	6
T_{NI}/°C	268.1	(144.2)	(116.3)

H   CONCLUSION

This review has served to highlight the surprising richness of structures now known to exhibit nematic behaviour. We are still at an early stage in the development of empirical structure-property relationships for nematogens having unconventional molecular structures and much work now needs to be done. In order to understand these empirical relationships and hence be able to design novel materials in a rational manner, we must develop, in parallel, molecular theories for nematic liquid crystals and the materials described here provide an important and demanding test bed for such theories. This synergistic approach between experiment and theory will result in the discovery of new classes of nematic liquid crystals and may also lead to the design of materials that exhibit the much sought after but elusive thermotropic biaxial nematic phase.

REFERENCES

[1]   M. Hird [Datareview in this book: *1.1 Relationship between molecular structure and transition temperatures for calamitic structures in nematics*]

[2]   K. Praefcke [Datareview in this book: *1.2 Relationship between molecular structure and transition temperatures for organic materials of a disc-like molecular shape in nematics*]

[3]   C.T. Imrie, G.R. Luckhurst [in *Handbook of Liquid Crystals vol.2B* Eds. D. Demus, J. Goodby, G.W. Gray, H.-W. Spiess, V. Vill (Wiley-VCH, Weinheim, 1998) p.801-33]

[4]   C.T. Imrie [*Struct. Bond. (Germany)* vol.95 (1999) p.149]

[5]   J.W. Emsley, G.R. Luckhurst, G.N. Shilstone, I. Sage [*Mol. Cryst. Liq. Cryst. Lett. (UK)* vol.102 (1984) p.223]

[6]   J.-I. Jin [*Mol. Cryst. Liq. Cryst. (Switzerland)* vol.267 (1995) p.249]

[7]   C.T. Imrie [*Liq. Cryst. (UK)* vol.6 (1989) p.391]

[8]   R.W. Date, C.T. Imrie, G.R. Luckhurst, J.M. Seddon [*Liq. Cryst. (UK)* vol.12 (1992) p.203]

[9]   G.S. Attard et al [*Liq. Cryst. (UK)* vol.16 (1994) p.529]

[10]   W. Weissflog, D. Demus, S. Diele, P. Nitschke, W. Wedler [*Liq. Cryst. (UK)* vol.5 (1989) p.111]

[11]   S.M. Huh, J.-I. Jin, M.F. Achard, F. Hardouin [*Liq. Cryst. (UK)* vol.25 (1998) p.285 and vol.26 (1998) p.919]

[12]   J. Andersch, C. Tschierske [*Liq. Cryst. (UK)* vol.21 (1996) p.51]

[13]   J. Andersch, C. Tschierske, S. Diele, D. Lose [*J. Mater. Chem. (UK)* vol.6 (1996) p.1297]

[14]   W. Weissflog, D. Demus, S. Diele [*Mol. Cryst. Liq. Cryst. (UK)* vol.191 (1990) p.9]

[15]   J.W. Lee, X.L. Piao, Y.K. Yun, J.-I. Jin, Y.S. Kang, W.C. Zin [*Liq. Cryst. (UK)* vol.26 (1999) p.1671]

[16]   S. Zamir et al [*Liq. Cryst. (UK)* vol.23 (1997) p.689]

[17]   K. Praefcke, B. Kohne, D. Singer, D. Demus, G. Pelzl, S. Diele [*Liq. Cryst. (UK)* vol.7 (1990) p.589]

[18]   I.D. Fletcher, G.R. Luckhurst [*Liq. Cryst. (UK)* vol.18 (1995) p.175]

[19]   D. Goldmann et al [*Liq. Cryst. (UK)* vol.24 (1998) p.881]

[20]   M.A. Bates, G.R. Luckhurst [*Liq. Cryst. (UK)* vol.24 (1998) p.229]

[21]   G.S. Attard, A.G. Douglass [*Liq. Cryst. (UK)* vol.22 (1997) p.349]

[22]   (a) C.T. Imrie, G.R. Luckhurst [*J. Mater. Chem. (UK)* vol.8 (1998) p.1339]; (b) H. Furuya, K. Asahi, A. Abe [*Polym. J. (Singapore)* vol.18 (1986) p.779]; (c) N.V. Tsvetkov, V.V. Zuev, V.N. Tsvetkov [*Liq. Cryst. (UK)* vol.22 (1997) p.245]

[23]   C.T. Imrie, D. Stewart, C. Rémy, D.W. Christie, I.W. Hamley, R. Harding [*J. Mater. Chem. (UK)* vol.9 (1999) p.2321]

[24]   R. Eidenschink, F.H. Kreuzer, W.H. de Jeu [*Liq. Cryst. (UK)* vol.8 (1990) p.879]

[25]   J.W. Goodby et al [*Chem. Commun. (UK)* (1998) p.2057]

[26]   S.H. Chen, J.C. Mastrangelo, T.N. Blanton, A. Bashir-Hashemi, K.L. Marshall [*Liq. Cryst. (UK)* vol.21 (1996) p.683]

[27]   J. Andersch, S. Diele, D. Lose, C. Tschierske [*Liq. Cryst. (UK)* vol.21 (1996) p.103]

[28]   D. Goldmann, D. Janietz, C. Schmidt, J.H. Wendorff [*Liq. Cryst. (UK)* vol.25 (1998) p.711]

[29]   G.S. Attard, A.G. Douglass, C.T. Imrie, L. Taylor [*Liq. Cryst. (UK)* vol.11 (1992) p.779]

[30]   S. Kumar, M. Manickam [*Liq. Cryst. (UK)* vol.26 (1999) p.939]

[31]   H.R. Allcock, C. Kim [*Macromolecules (USA)* vol.22 (1989) p.2596]

[32]   R. Lunkwitz, C. Tschierske, S. Diele [*J. Mater. Chem. (UK)* vol.7 (1997) p.2001]

[33]   S. Mahlstedt, D. Janietz, C. Schmidt, A. Stracke, J.H. Wendorff [*Liq. Cryst. (UK)* vol.26 (1999) p.1359]

[34]   V. Percec, M. Kawasumi [*Liq. Cryst. (UK)* vol.13 (1993) p.83]

[35]   J.F. Li, V. Percec, C. Rosenblatt, O.D. Lavrentovich [*Europhys. Lett. (Switzerland)* vol.25 (1994) p.199]

[36]   B. Neumann, T. Hegmann, R. Wolf, C. Tschierske [*Chem. Commun. (UK)* (1998) p.105]

[37]   V. Percec, P.J. Turkaly, A.D. Asandei [*Macromolecules (USA)* vol.30 (1997) p.943]

[38]   K.D. Gresham, C.M. McHugh, T.J. Bunning, R.L. Crane, H.E. Klei, E.T. Samulski [*J. Polym. Sci. A, Polym. Chem. (USA)* vol.32 (1994) p.2039]

[39]   S.A. Ponomarenko, E.A. Rebrov, A.Yu. Bobrovsky, N.I. Boiko, A.M. Muzafarov, V.P. Shibaev [*Liq. Cryst. (UK)* vol.21 (1996) p.1]

[40]   V. Percec, P. Chu, G. Ungar, J. Zhou [*J. Am. Chem. Soc. (USA)* vol.117 (1995) p.11441]

[41]   G. Pelzl, S. Diele, W. Weissflog [*Adv. Mater. (Germany)* vol.11 (1999) p.707]

[42]   H.T. Nguyen, J.C. Rouillon, J.P. Marcerou, J.P. Bedel, P. Barois, S. Sarmento [*Mol. Cryst. Liq. Cryst. (Switzerland)* vol.328 (1999) p.177]

[43]   W. Weissflog [in *Handbook of Liquid Crystals vol.2B* Eds. D. Demus, J. Goodby, G.W. Gray, H.-W. Spiess, V. Vill (Wiley-VCH, Weinheim, 1998) p.835]

[44]   W. Weissflog, D. Demus [*Cryst. Res. Technol. (Germany)* vol.18 (1983) p.K21 and vol.19 (1984) p.55]

[45]   C.T. Imrie, L. Taylor [*Liq. Cryst. (UK)* vol.6 (1989) p.1]

[46]   W. Weissflog, D. Demus [*Mol. Cryst. Liq. Cryst. (UK)* vol.129 (1985) p.235]

[47]   W. Weissflog, A. Wiegeleben, D. Demus [*Mater. Chem. Phys. (Switzerland)* vol.12 (1985) p.461]

[48]   W. Weissflog, A. Wiegelben, S. Diele, D. Demus [*Cryst. Res. Technol. (Germany)* vol.19 (1984) p.583]

[49]   W. Weissflog, G. Pelzl, I. Letko, S. Diele [*Mol. Cryst. Liq. Cryst. (Switzerland)* vol.260 (1995) p.157]

[50]   H.T. Nguyen, C. Destrade, J. Malthête [in *Handbook of Liquid Crystals vol.2B* Eds. D. Demus, J. Goodby, G.W. Gray, H.-W. Spiess, V. Vill (Wiley-VCH, Weinheim, 1998) p.865]

[51]   J. Malthête, H.T. Nguyen, C. Destrade [*Liq. Cryst. (UK)* vol.13 (1993) p.171]

[52]   H.T. Nguyen, C. Destrade, J. Malthête [*Liq. Cryst. (UK)* vol.8 (1990) p.797]

[53]   J. Malthête, L. Liébert, A.-M. Levelut, Y. Galerne [*C.R. Acad. Sci. (France)* vol.303 (1986) p.1073]

[54]   I.G. Shenouda, Y. Shi, M.E. Neubert [*Mol. Cryst. Liq. Cryst. (Switzerland)* vol.257 (1994) p.209]

[55]   J. Malthête, P. Davidson [*Bull. Soc. Chim. Fr. (France)* vol.131 (1994) p.812]

[56]   J.R. Hughes et al [*J. Chem. Phys. (USA)* vol.107 (1997) p.9252]

[57]   S. Berg, V. Krone, H. Ringsdorf, U. Quotschalla, H. Paulus [*Liq. Cryst. (UK)* vol.9 (1991) p.151]

[58]   W.D.J.A. Norbert, J.W. Goodby, M. Hird, K.J. Toyne, J.C. Jones, J.S. Patel [*Mol. Cryst. Liq. Cryst. (Switzerland)* vol.260 (1995) p.339]

[59]   D. Braun, M. Reubold, L. Schneider, M. Wegmann, J.H. Wendorff [*Liq. Cryst. (UK)* vol.16 (1994) p.429]

[60]   K. Praefcke et al [*Mol. Cryst. Liq. Cryst. (UK)* vol.198 (1991) p.393]

CHAPTER 2

ORDER PARAMETERS AND MOLECULAR PROPERTIES

2.1 Orientational order: distribution functions and order parameters

G.R. Luckhurst

July 2000

A   INTRODUCTION

The defining characteristic of a liquid crystal is its long range orientational order. That is, particular axes set in the molecules tend to be correlated over large distances and are on average parallel to the director. It is the long range orientational order which is responsible for the anisotropy in the properties of a nematic phase, both static and dynamic. A knowledge and an understanding of the orientational order of a nematic is, therefore, of prime importance. Here we introduce the formal definitions of the singlet orientational distribution function and order parameters starting with the simplest molecular systems and building up to the complexity of real mesogenic molecules. In addition we shall also describe certain molecular field theories of nematics and their predictions of the distribution function and order parameters.

B   UNIAXIAL MOLECULES IN UNIAXIAL PHASES

Nematogenic molecules are usually depicted as being rigid with $D_{\infty h}$ symmetry even though real molecules never attain this ideal. We begin by assuming such a symmetry and then gradually increase the molecular complexity to that of real nematogens. Within a uniaxial nematic phase the relevant orientational coordinate is the single angle, β, between the director and the molecular symmetry axis. The most complete description of the orientational order is provided by the singlet orientational distribution function, $f(\beta)$, which gives the probability density of finding a molecule at a given orientation to the director.

By definition the distribution function is normalised, that is,

$$\int_{o}^{\pi} f(\beta) \sin \beta d\beta = 1 \tag{1}$$

In general, provided the distribution function is well-behaved it can be expanded in a complete basis set of functions spanning the relevant orientational space. For uniaxial molecules in a uniaxial nematic phase the appropriate functions are the Legendre polynomials, $P_L(\cos\beta)$ [1], where L is a positive integer. The first few Legendre polynomials, of particular relevance in liquid crystal science, are given on the following page.

$$P_0(\cos\beta) = 1$$

$$P_1(\cos\beta) = \cos\beta$$

$$P_2(\cos\beta) = (3\cos^2\beta - 1)/2$$

$$P_3(\cos\beta) = (5\cos^3\beta - 3\cos\beta)/2$$

$$P_4(\cos\beta) = (35\cos^4\beta - 30\cos^2\beta + 3)/8$$

$$P_5(\cos\beta) = (63\cos^5\beta - 70\cos^3\beta + 15\cos\beta)/8$$

$$P_6(\cos\beta) = (693\cos^6\beta - 945\cos^4\beta + 315\cos^2\beta - 15)/48 \qquad (2)$$

The polynomials are orthogonal but not normalised [2], that is

$$\int_o^\pi P_L(\cos\beta)P_{L'}(\cos\beta)\sin\beta d\beta = 2\delta_{LL'}/(2L+1) \qquad (3)$$

where $\delta_{LL'}$ is the Kronecker delta function. The singlet orientational distribution function is then written as [3]

$$f(\beta) = \sum_L f_L P_L(\cos\beta) \qquad (4)$$

However, this can be simplified by noting that for nematics which are non-polar the probability densities $f(\beta)$ and $f(\pi-\beta)$ are identical whereas the Legendre polymials change sign if L is odd but not if L is even. Accordingly the summation in EQN (4) must be restricted to even terms.

To proceed further we need to identify the expansion coefficients, f_L, which can be done by using the orthogonality of the Legendre polynomials. This gives, for L equal to L',

$$f_{L'} = <P_{L'}>(2L' + 1)/2 \qquad (5)$$

where the angular brackets denote an ensemble average over the orientations of the molecules. This average is related to the singlet orientational distribution function by

$$<P_{L'}> = \int_o^\pi P_{L'}(\cos\beta)f(\beta)\sin\beta d\beta \qquad (6)$$

and defines a set of orientational order parameters which we shall consider shortly. The first few terms in the distribution function are then

$$f(\beta) = \frac{1}{2} + \left(\frac{5}{2}\right)<P_2>P_2(\cos\beta) + \left(\frac{9}{2}\right)<P_4>P_4(\cos\beta) + \left(\frac{13}{2}\right)<P_6>P_6(\cos\beta)... \qquad (7)$$

Although the order parameters usually satisfy the inequalities $<P_2> > <P_4> > <P_6>$ the series for $f(\beta)$ does not converge rapidly partly because of the growth in the numerical factors of $^5/_2$, $^9/_2$, $^{13}/_2$ ... [3]. An alternative representation of the singlet orientational distribution which generally has a better convergence is obtained by writing $f(\beta)$ as

$$f(\beta) = Q^{-1} \exp\{A(\beta)\} \tag{8}$$

where Q is given by

$$Q = \int_o^\pi \exp\{A(\beta)\} \sin\beta d\beta \tag{9}$$

to ensure that the distribution function is normalised. Now it is the function $A(\beta)$ which is expanded in a basis of Legendre polynomials so that

$$f(\beta) = Q^{-1} \exp\left\{\sum_L a_L P_L (\cos\beta)\right\} \tag{10}$$

where the summation is again restricted to even values of L. However, unlike the expansion coefficients in EQN (4) the a_L are not directly related to the order parameters $<P_L>$. They are, of course, connected indirectly through the defining EQN (6).

The $<P_L>$ form an infinite set of order parameters which are necessary, in principle, to provide a complete description of the orientational order of a nematic. The limiting values of all of these are particularly convenient; thus for perfect order when the symmetry axes of the molecules are parallel to the director they all take the value of unity. At the other extreme when the system is disordered, that is in the isotropic phase, the order parameters all vanish. Another limiting case occurs when the molecular symmetry axes are perpendicular to the director so that the nematic phase remains uniaxial. Now the limiting values of the order parameters depend on L and for the first three $<P_2> = -^1/_2$, $<P_4> = {}^3/_8$ and $<P_6> = -5/16$. Under the same limiting conditions the singlet orientational distribution functions are given by

$$f(\beta) = \delta(\beta - 0) \tag{11}$$

for complete order, where $\delta(\beta - 0)$ is the Dirac delta function

$$f(\beta) = \frac{1}{2} \tag{12}$$

in the isotropic phase where this value follows from the normalisation constraint, and

$$f(\beta) = \delta\left(\beta - \frac{\pi}{2}\right) \tag{13}$$

when the molecules are orthogonal to the director.

Other notation has been used for certain of these orientational order parameters. Thus Tsvetkov [4] first introduced the second rank orientational order parameter for nematics although he denoted this with the letter S. This notation can be extended to include the higher rank orientational order parameters by using S_2 for $<P_2>$, S_4 for $<P_4>$ and in general S_L for $<P_L>$.

In concluding this section we explore the relationship between the anisotropic properties of a nematic and the orientational order parameters. It is convenient to do this using irreducible spherical tensors since they have particularly straightforward transformation properties under rotation, as we shall discover. We begin with the tensor components set in the molecular frame which are necessarily independent of the molecular orientation in the laboratory frame. For a tensor of rank L there are $(2L + 1)$ components which are distinguished by the label m, taking values L, L - 1, ... -L + 1, -L; these are denoted by T_{mol}^{Lm} [2]. They are simply linear combinations of the components of the symmetric Cartesian tensor of the same rank. The components of the tensor in the laboratory frame are T_{lab}^{Ln} and are related to T_{mol}^{Lm} by

$$T_{lab}^{Ln} = \sum_m D_{nm}^{L*}(\Omega) T_{mol}^{Lm} \tag{14}$$

where $D_{nm}^{L*}(\Omega)$ is a Wigner rotation matrix and Ω donates the Euler angles, $\alpha\beta\gamma$, connecting the two frames [2]. The observed quantities are normally the components in the laboratory frame averaged over the orientations adopted by the molecules. These averages are denoted by a tilde and so

$$\widetilde{T}_{lab}^{Ln} = \sum_m \left\langle D_{nm}^{L*} \right\rangle T_{mol}^{Lm} \tag{15}$$

In the special case which we are considering of uniaxial molecules in a uniaxial phase the averaging of the Wigner rotation matrix gives

$$\left\langle D_{nm}^{L*} \right\rangle = \left\langle P_L \right\rangle \delta_{0n} \delta_{0m} \tag{16}$$

and so

$$\widetilde{T}_{lab}^{L0} = \left\langle P_L \right\rangle T_{mol}^{L0} \tag{17}$$

provided the director is taken to define the Z axis. We see, therefore, that measurement of the partially averaged tensorial property of rank L in a uniaxial liquid crystal phase leads to the orientational order parameter of the same rank.

In principle this result should provide a route to a complete set of order parameters, $<P_L>$. In practice this is not the case because the majority of properties transform under rotation like tensors of rank two. Consequently only $<P_2>$ is readily available experimentally and since this is usually the largest of the order parameters it is normally used alone to characterise the long range orientational order in a liquid crystal. To illustrate this we consider the relationship between the anisotropic magnetic susceptibility [5] of a nematic and the second rank orientational order parameter. The molecular property of relevance is the magnetic polarisability κ and its non-zero partially averaged component in the laboratory frame is obtained from EQN (17) as

$$\widetilde{\kappa}_{lab}^{20} = <P_2> \kappa_{mol}^{20} \tag{18}$$

The magnetic susceptibility is obtained from this simply by multiplying by N_A, the Avogadro constant, which gives

$$\widetilde{\chi}_{lab}^{20} = N_A <P_2> \kappa_{mol}^{20}$$

$$= <P_2> \chi_{mol}^{20} \tag{19}$$

The irreducible component $\widetilde{\chi}_{lab}^{20}$ is related to the Cartesian components which are usually the quantities measured in an experiment. Since the director is taken to define the Z axis then

$$\widetilde{\chi}^{20} = \sqrt{\frac{2}{3}}(\widetilde{\chi}_{\parallel} - \widetilde{\chi}_{\perp}) \tag{20}$$

where the subscript $\parallel$ indicates the component parallel to the director and $\perp$ denotes the perpendicular component. The factor of $\sqrt{2/3}$ results from this transformation of the magnetic susceptibility tensor from the irreducible spherical to the Cartesian form. The subscript lab has been removed to simplify the notation. This should not lead to any confusion since the tilde already indicates the value for the liquid crystal phase. Thus the anisotropic magnetic susceptibility of a uniaxial nematic composed of uniaxial molecules is given by

$$\widetilde{\chi}_{\parallel} - \widetilde{\chi}_{\perp} = <P_2>(\chi_l - \chi_t) \tag{21}$$

where χ_l is the magnetic susceptibility parallel to the molecular symmetry axis and χ_t that perpendicular to it.

C   BIAXIAL MOLECULES IN UNIAXIAL PHASES

For biaxial molecules, that is those deviating from uniaxial or cylindrical symmetry, two angles are needed to define their orientation with respect to the director. These are taken to be the spherical polar angles, $\beta\gamma$, made by the director in a frame set in the molecule. Again the most complete description of the long range orientational order at the single molecule level is provided by the singlet orientational distribution function, $f(\beta\gamma)$, which is necessarily a function of the two angles. It is normalised, that is,

$$\int_0^{2\pi} \int_0^{\pi} f(\beta\gamma) \sin\beta \, d\beta \, d\gamma = 1 \tag{22}$$

Provided, as is usually the case, the distribution function is well-behaved it can be expanded in an appropriate set of functions. Here the modified spherical harmonics, $C_{Lm}(\beta\gamma)$ [1,3], are especially convenient. The subscript L is an integer denoting the rank of the spherical harmonic and m takes values L, L - 1, ... -L + 1, -L giving a total of 2L + 1 components. The component with m equal to zero is independent of the azimuthal angle γ and, in fact, is equal to the Legendre polynomial of the same rank, that is

$$P_L(\cos\beta) = C_{Lm}(\beta\gamma)\delta_{0m} \tag{23}$$

The other components have a particularly simple dependence on γ as may be seen from the first few modified spherical harmonics

$$C_{00}(\beta\gamma) = 1$$

$$C_{10}(\beta\gamma) = \cos\beta$$

$$C_{1\pm1}(\beta\gamma) = \sqrt{\frac{1}{2}}\sin\beta\exp(\mp i\gamma)$$

$$C_{20}(\beta\gamma) = (3\cos^2\beta - 1)/2$$

$$C_{2\pm1}(\beta\gamma) = \pm\sqrt{\frac{3}{2}}\sin\beta\cos\beta\exp(\mp i\gamma)$$

$$C_{2\pm2}(\beta\gamma) = \sqrt{\frac{3}{8}}\sin^2\beta\exp(\mp i2\gamma)$$

$$C_{40}(\beta\gamma) = (35\cos^4\beta - 30\cos^2\beta + 3)/8$$

$$C_{4\pm1}(\beta\gamma) = \pm\sqrt{\frac{5}{16}}\sin\beta\cos\beta(7\cos^2\beta - 3)\exp(\mp i\gamma)$$

$$C_{4\pm2}(\beta\gamma) = \sqrt{\frac{5}{32}}\sin^2\beta(7\cos^2\beta - 1)\exp(\mp i2\gamma)$$

$$C_{4\pm3}(\beta\gamma) = \pm\sqrt{\frac{7}{8}}\sin^3\beta\cos^2\beta\exp(\mp i3\gamma)$$

$$C_{4\pm4}(\beta\gamma) = \sqrt{\frac{35}{128}}\sin^4\beta\exp(\mp i4\gamma) \tag{24}$$

The orientational distribution function is then

$$f(\beta\gamma) = \sum_{Lm} f_{Lm}C_{Lm}(\beta\gamma) \tag{25}$$

where the summation is again restricted to even values of L because of the symmetry of the phase. The expansion coefficients can be found by exploiting the orthogonality of the modified spherical harmonics [2], namely,

$$\int_o^{2\pi}\int_o^{\pi}C_{Lm}(\beta\gamma)C^*_{L'm'}(\beta\gamma)\sin\beta d\beta d\gamma = \{4\pi/(2L'+1)\}\delta_{LL'}\delta_{mm'} \tag{26}$$

This gives the expansion coefficients as

$$f_{Lm} = \{(2L+1)/4\pi\}\left\langle C^*_{Lm}\right\rangle \tag{27}$$

where the $<C^*_{Lm}>$ are components of an Lth rank orientational ordering tensor defined as the average

$$\left\langle C^*_{Lm}\right\rangle = \int_o^{2\pi}\int_o^{\pi}C^*_{Lm}(\beta\gamma)f(\beta\gamma)\sin\beta d\beta d\gamma \tag{28}$$

The formal expansion of the singlet orientational distribution function is then

$$f(\beta\gamma) = \frac{1}{4\pi} + \left(\frac{5}{4\pi}\right)\sum_m\left\langle C^*_{2m}\right\rangle C_{2m}(\beta\gamma) + \left(\frac{9}{4\pi}\right)\sum_m\left\langle C^*_{4m}\right\rangle C_{4m}(\beta\gamma) + ... \tag{29}$$

The combination of the modified spherical harmonic with the average of its complex conjugate ensures that the distribution function is real, as it must be. The series is normally slow to converge even though the ordering tensors decrease with increasing rank, L. As for uniaxial molecules it is usually possible to obtain a more rapidly converging expansion for the distribution function by using the exponential form

$$f(\beta\gamma) = Q^{-1}\exp\left\{\sum_{Lm}a_{Lm}C_{Lm}(\beta\gamma)\right\} \tag{30}$$

Here Q is given by

$$Q = \int_o^{2\pi}\int_o^{\pi}\exp\left\{\sum_{Lm}a_{Lm}C_{Lm}(\beta\gamma)\right\}\sin\beta d\beta d\gamma \tag{31}$$

which guarantees that $f(\beta\gamma)$ is normalised. Again the expansion coefficients a_{Lm} are related to the orientational order parameters but not in such a direct manner as the f_{Lm} (see EQN (28)).

The use of the formal expansion for $f(\beta\gamma)$ shows that, in principle, an infinite set of order parameters, $<C^*_{Lm}>$, albeit with L even are needed to provide a complete description of the orientational order. These order parameters take relatively simple limiting forms. Thus in the isotropic phase they vanish while in a completely orientationally ordered liquid crystal phase

$$\left\langle C_{Lm}^{*} \right\rangle = \delta_{0m} \qquad (32)$$

This result follows from the definition of the modified spherical harmonics given in EQN (24). For perfect order the molecular z axis is aligned parallel to the director so that $< C_{L0}^{*} >$ is just unity. The other components $< C_{Lm}^{*} >$ with m not equal to zero vanish because they are all proportional to $\sin\beta$. The components of the ordering tensors, $< C_{Lm}^{*} >$, also depend on the choice of the molecular frame and its relation to the molecular symmetry [6]. For most mesogenic molecules there is no symmetry and so it is not possible to simplify the ordering tensors in any formal sense. This is not the case for solutes dissolved in liquid crystals which may often be of high symmetry [7]. Thus for a molecule with D_{2h} symmetry, such as anthracene, there are just two independent components of the second rank tensor, namely $<C_{20}>$ and $<C_{22}>$ $(\equiv < C_{22}^{*} >)$, while for the fourth rank ordering tensor there are three independent components $<C_{40}>$, $<C_{42}>$ $(\equiv < C_{42}^{*} >)$ and $<C_{44}>$ $(\equiv < C_{44}^{*} >)$.

Although the irreducible spherical tensor approach is valuable for the definition of the ordering tensors and for their manipulation under rotation it does not always provide a ready understanding of the physical significance of the various components. This is sometimes available from a Cartesian representation of the ordering tensor. The most familiar example is the Saupe ordering matrix which represents the orientational ordering at the second rank level [8]. It is defined by

$$S_{\alpha\beta} = \left\langle \left(3\ell_{\alpha}\ell_{\beta} - \delta_{\alpha\beta} \right)/2 \right\rangle \qquad (33)$$

where $\alpha\beta$ denote the molecular axes and ℓ_{α} is the direction cosine between the α axis and the director. Since there are three axes the matrix contains nine elements but these are not independent. The matrix is symmetric which follows from its definition and, in addition, the properties of the direction cosines mean that it is traceless,

$$\sum_{\alpha} S_{\alpha\alpha} = 0 \qquad (34)$$

There are then five independent components of **S** in keeping with the five elements of $< C_{2m}^{*} >$. The relationship between these two descriptions is as follows:

$$S_{zz} = <C_{20}>$$

$$S_{xx} - S_{yy} = \sqrt{\frac{3}{2}} \left(\left\langle C_{22} \right\rangle + \left\langle C_{2-2} \right\rangle \right)$$

$$S_{xy} = i\sqrt{\frac{3}{8}} \left(\left\langle C_{2-2} \right\rangle - \left\langle C_{22} \right\rangle \right)$$

$$S_{xz} = \sqrt{\frac{3}{8}} \left(\left\langle C_{2-1} \right\rangle - \left\langle C_{21} \right\rangle \right)$$

$$S_{yz} = i\sqrt{\frac{3}{8}}\left(\langle C_{21}\rangle + \langle C_{2\text{-}1}\rangle\right) \tag{35}$$

Since the ordering matrix is real and symmetric it is always possible to cast it in a diagonal form irrespective of the molecular symmetry. The three principal elements of **S** then correspond to the second rank order parameters for the associated principal axes. By convention these are labelled so that z is the axis about which the ordering matrix approximates to cylindrical symmetry while x and y are selected to make the difference $(S_{xx} - S_{yy})$ positive. In other words $S_{zz} > (S_{xx} - S_{yy}) > 0$. Since the ordering matrix is traceless there are only two independent principal components and these are usually selected to be S_{zz} the major order parameter and $(S_{xx} - S_{yy})$ which is the biaxiality in **S**. We can now see that it is not necessary for the molecule to have a particular symmetry to be able to reduce the elements of the second rank ordering tensor to just $\langle C_{20}\rangle$ and $\langle C_{22}\rangle$ $(\equiv <C_{2\text{-}2}>)$. It is important to recognise that $(S_{xx} - S_{yy})$ measures the biaxiality in the ordering of a molecule in a uniaxial phase and does not reflect the phase biaxiality.

Although the Cartesian analogue of the second rank ordering tensor is relatively straightforward to manipulate the same is not true for the fourth rank ordering matrix [9] or rather supermatrix. This is defined by

$$S_{\alpha\beta\gamma\delta} = <\{35\ell_\alpha\ell_\beta\ell_\gamma\ell_\delta - 5(\ell_\alpha\ell_\beta\delta_{\gamma\delta} + \ell_\alpha\ell_\gamma\delta_{\beta\delta} + \ell_\alpha\ell_\delta\delta_{\beta\delta} + \ell_\beta\ell_\gamma\delta_{\alpha\delta}$$

$$+ \ell_\beta\ell_\delta\delta_{\alpha\gamma} + \ell_\gamma\ell_\delta\delta_{\alpha\beta}) + (\delta_{\alpha\beta}\delta_{\gamma\delta} + \delta_{\alpha\gamma}\delta_{\beta\delta} + \delta_{\alpha\delta}\delta_{\beta\gamma})\}/8 > \tag{36}$$

Unlike its second rank counterpart there is no transformation which will cast $S_{\alpha\beta\gamma\delta}$ into a diagonal form unless the molecule has the appropriate symmetry. This can be seen most readily in terms of the ordering tensor, $\langle C_{4m}\rangle$. For this to be diagonal it is required that

$$<C_{42}> = <C_{4\text{-}2}>$$

$$<C_{44}> = <C_{4\text{-}4}>$$

$$<C_{41}> = <C_{4\text{-}1}> = 0$$

and

$$<C_{43}> = <C_{4\text{-}3}> = 0 \tag{37}$$

There are then six constraints which the rotation must satisfy but as there are only three Euler angles in the transformation this is not possible. To satisfy these conditions the molecule must have at the minimum C_{2v}, D_2 or D_{2h} symmetry [6].

In completing this section we return to the anisotropy in the magnetic susceptibility and its relation to the second rank ordering tensor. From EQN (15) the zeroth component of $\tilde{\chi}^{2n}$ in the uniaxial liquid crystal phase is

$$\widetilde{\chi}^{20} = \sum_{m} \left\langle C_{2m}^{*} \right\rangle \chi^{2m} \tag{38}$$

where χ^{2m} denotes the susceptibility tensor in the molecular frame. Here both tensors are expressed in the same axis system and if this is chosen to be the principal axis system for the ordering tensor then

$$\widetilde{\chi}^{20} = \left\langle C_{20} \right\rangle \chi^{20} + 2 \left\langle C_{22} \right\rangle \left(\chi^{22} + \chi^{2\text{-}2} \right) \tag{39}$$

The anisotropy in the magnetic susceptibility is then

$$\Delta\widetilde{\chi} = \left(\frac{3}{2} \right)^{\frac{1}{2}} \left\{ \left\langle C_{20} \right\rangle \chi^{20} + 2 \left\langle C_{22} \right\rangle \left(\chi^{22} + \chi^{2\text{-}2} \right) \right\} \tag{40}$$

which in terms of Cartesian tensors becomes

$$\Delta\widetilde{\chi} = S_{zz} \chi_{zz}' + \left(\frac{1}{2} \right) \left(S_{xx} - S_{yy} \right) \left(\chi_{xx}' - \chi_{yy}' \right) \tag{41}$$

Here the prime denotes the anisotropic part of the tensor, that is

$$\chi_{\alpha\alpha}' = \chi_{\alpha\alpha} - \left(\frac{1}{3} \right) \sum_{\alpha} \chi_{\alpha\alpha} \tag{42}$$

The anisotropy in the magnetic susceptibility like that in other properties depends not only on the major order parameter but also on the biaxiality parameter $(S_{xx} - S_{yy})$. We shall consider the likely magnitude of this new contribution in the final section. The converse of the result in EQN (41) is that measurement of the anisotropy in a single property is clearly insufficient to determine the two order parameters even if the principal axes are known.

D   NON-RIGID MOLECULES IN UNIAXIAL PHASES

So far the mesogenic molecules have been taken to be rigid and in this section this final constraint is removed. From a formal point of view this is especially important because the vast majority of mesogenic molecules are non-rigid primarily because they possess one or more flexible alkyl chains but also because of the rotation of aromatic rings in the mesogenic groups. The description of the molecular conformation can be described in terms of a variety of internal coordinates depending on the nature of the conformational change. In order to simplify the analysis the mesogenic molecule is taken to be composed of a set of rigid fragments linked by bonds about which rotation can occur, as in an alkyl chain. The conformation of the molecule can then be described in terms of a set of torsional angles which we denote by $\{\varphi_i\}$ where the subscript i denotes the bond about which the torsional angle is defined. If this set $\{\varphi_i\}$ together with the orientation of just one rigid fragment with respect to the director are known then the orientation of all fragments can be obtained [10].

The orientational and conformational state of a single molecule is then described by $f(\beta\gamma\{\varphi_i\})$ which is the probability density of finding the molecule in a given conformation and orientation with respect to the director. As we have seen in the previous sections this distribution function can be expanded in an appropriate set of basis functions. For the orientational coordinates we use the modified spherical harmonics, as for biaxial molecules, while for each of the torsional angles we use the Fourier function, $\exp(-iq\varphi)$, where q is an integer taking values, $0, \pm1, \pm2 \ldots$ [10]. The distribution is then written as the sum of products of the basis functions

$$f(\beta\gamma\{\varphi_i\}) = \sum_{Lm\{q_i\}} f_{Lm\{q_i\}} C_{Lm}(\beta\gamma)\prod_i \exp(-iq_i\varphi_i) \tag{43}$$

which is normalised

$$\int_0^{2\pi} \int_0^{2\pi} \int_0^{2\pi} f(\beta\gamma\{\varphi_i\}) \sin\beta d\beta d\gamma d\{\varphi_i\} = 1 \tag{44}$$

It is possible to derive two sub-distribution functions by projecting out the unwanted coordinates from the complete distribution. Thus the pure torsional distribution, $f(\{\varphi_i\})$, which gives the probability of the molecule having a conformation irrespective of its orientation, is obtained from the complete distribution by integrating over β and γ. Thus

$$f_{conf}(\{\varphi_i\}) = \int_0^{2\pi} \int_0^{\pi} f(\beta\gamma\{\varphi_i\}) \sin\beta d\beta d\gamma$$

$$= \sum_{\{q_i\}} f_{00\{q_i\}} \prod_i \exp(-iq_i\varphi_i) \tag{45}$$

The orientational distribution for the selected rigid fragment independent of the molecular conformation, $f_{orien}(\beta\gamma)$, is obtained from the complete distribution by integration over all of the torsional angles

$$f_{orien}(\beta\gamma) = \int_0^{2\pi} f(\beta\gamma\{\varphi_i\}) d\{\varphi_i\}$$

$$= \sum_{Lm} f_{Lm\{0\}} C_{Lm}(\beta\gamma) \tag{46}$$

As we have come to expect, the expansion coefficients in the distribution are proportional to the order parameters for the system. These order parameters are obtained by exploiting the orthogonality of the modified spherical harmonics (see EQN (26)) and the Fourier functions

$$\int_0^{2\pi} \exp(-iq\varphi)\exp(iq'\varphi) d\varphi = 2\pi\delta_{qq'} \tag{47}$$

This gives the expansion coefficients as

$$f_{Lm\{q_i\}} = \{(2\pi)^{n+1}/(2L+1)\}\left\langle C^*_{Lm}(\beta\gamma)\prod_i \exp(iq_i\varphi_i)\right\rangle \qquad (48)$$

where n is the number of bonds linking the rigid sub-units in the molecule. The set of order parameters can be sub-divided in the following way. If all of the q_i are zero then

$$f_{Lm\{0\}} = \{2\pi/(2L+1)\}\left\langle C^*_{Lm}(\beta\gamma)\right\rangle \qquad (49)$$

is a purely orientational order parameter, which is to be expected because only these coefficients occur in the orientational sub-distribution function. The orientational ordering tensor, $<C^*_{Lm}(\beta\gamma)>$, provides a measure of the orientational order of a given rigid sub-unit in the molecule independently of the molecular conformation. At the other extreme when L is zero so that m also vanishes then

$$f_{00\{q_i\}} = (2\pi)^n\left\langle \prod_i \exp(iq_i\varphi_i)\right\rangle \qquad (50)$$

These are the coefficients occurring in the sub-distribution function for the molecular conformation, irrespective of the orientation of the rigid fragment with respect to the director. The

$$\left\langle \prod_i \exp(iq_i\varphi_i)\right\rangle$$

are purely conformational or internal order parameters which, unlike the $<C^*_{Lm}(\beta\gamma)>$, do not vanish in the isotropic phase. The difference between the values in the isotropic and nematic phases provides a measure of the influence of the long range orientational order on the conformational distribution. There are various pure conformational order parameters depending on n and the number of q_i which are non-zero. For example, if all of the q_i vanish except q_1 then $<\exp(iq_1\varphi)>$ provides a measure of the conformational order about a single bond. If two of the q_i are non-zero, say q_1 and q_3, then $<\exp(iq_1\varphi_1)\exp(iq_3\varphi_3)>$ reflects both the conformational order about these two bonds and the correlation between these conformations. Finally, when L is non-zero and at least one of the q_i does not vanish we have the mixed orientational-conformational order parameters

$$\left\langle C^*_{Lm}(\beta\gamma)\prod_i \exp(iq_i\varphi_i)\right\rangle$$

These provide a measure of the correlation between the orientational and conformational degrees of freedom.

To appreciate the significance of these numerous order parameters it is helpful to see how they are related to the anisotropic properties of a nematic. As an example we consider the anisotropy in the magnetic susceptibility, $\Delta\tilde{\chi}$, as we have for previous systems. The starting point is the magnetic polarisability expressed in a molecular frame set in a rigid fragment of the molecule. This tensor $\kappa^{2m}_{mol}(\{\varphi_i\})$ now depends, in general, on the conformation of the molecule, a fact which we have made explicit. A functional dependence of the tensor on the torsional angles $\{\varphi_i\}$ defining the conformation can be obtained via the Fourier expansion

$$\kappa_{mol}^{2m}(\{\varphi_i\}) = \sum_{\{q_i\}} \kappa_{mol}^{2m}(\{q_i\}) \prod_i \exp(-iq_i\varphi_i) \tag{51}$$

where the expansion coefficients, $\kappa_{mol}^{2m}(\{q_i\})$, are second rank tensor components depending on the set of integers $\{q_i\}$. Transforming to the laboratory frame now gives

$$\kappa_{lab}^{2n} = \sum_{m\{q_i\}} \kappa_{mol}^{2m}(\{q_i\}) D_{nm}^{2*}(\Omega) \prod_i \exp(-iq_i\varphi_i) \tag{52}$$

and to obtain the anisotropic magnetic susceptibility this expression must be averaged over both orientational and conformational coordinates. This gives

$$\Delta\widetilde{\chi} = N_A \sqrt{\frac{3}{2}} \sum_{m\{q_i\}} \kappa_{mol}^{2m}(\{q_i\}) \left\langle C_{Lm}^*(\beta\gamma) \prod_i \exp(iq_i\varphi_i) \right\rangle \tag{53}$$

so that the anisotropy in $\widetilde{\chi}$ depends on the mixed second rank orientational-conformational order parameters and the pure second rank orientational ordering tensor; it is independent of the pure conformational order parameters. The variation of the magnetic anisotropy with the mixed order parameters appears somewhat involved. The extent of this complexity is determined by the number of terms needed in the Fourier expansion of the tensor $\kappa_{mol}^{2m}(\{\varphi_i\})$. However, there is an especially simple limiting case when $\kappa_{mol}^{2m}(\{\varphi_i\})$ is independent of the molecular conformation, that is when the only contribution to $\kappa_{mol}^{2m}(\{\varphi_i\})$ originates from the rigid sub-unit used to define the molecular frame. Then

$$\Delta\widetilde{\chi} = N_A \sqrt{\frac{3}{2}} \sum_m \kappa_{mol}^{2m}(\{0\}) \left\langle C_{2m}^*(\beta\gamma) \right\rangle \tag{54}$$

where $<C_{2m}^*(\beta\gamma)>$ is a pure second rank orientational order parameter for the rigid fragment averaged over the various conformations. Although such an approximation is unlikely to be good for the magnetic susceptibility it will be valid for interactions in NMR spectroscopy, such as the dipolar interaction between nuclei in the same rigid fragment [11] but not for those in different fragments [12].

There is an alternative set of order parameters which can be used to describe the long range orientational order and conformational order of a nematic composed of non-rigid molecules. To understand these it is again useful to consider an expansion of the singlet orientational and conformational distribution function, $f(\beta\gamma\{\varphi_i\})$. This is now written as the product of the probability, $f_{conf}(\{\varphi_i\})$, that the molecule has a given conformation and the conditional probability, $f(\{\varphi_i\}|\beta\gamma)$, that in this conformation the director has an orientation, $\beta\gamma$, in a frame set in a rigid fragment of the molecule. The conditional probability can be expanded in a complete basis set of modified spherical harmonics

$$f(\{\varphi_i\}|\beta\gamma) = \sum_{Lm} f_{Lm}(\{\varphi_i\})C_{Lm}(\beta\gamma) \tag{55}$$

where now the expansion coefficients are dependent on the molecular conformation, $\{\varphi_i\}$. As for rigid molecules the summation is restricted to even values of L and the expansion coefficients can be obtained from the orthogonality of the $C_{Lm}(\beta\gamma)$. This gives

$$f_{Lm}(\{\varphi_i\}) = \{(2L+1)/4\pi\}\langle C_{Lm}^* \rangle_{\{\varphi_i\}} \tag{56}$$

where the subscript $\{\varphi_i\}$ indicates that the ordering tensor is that for the molecule in the conformation defined by the set of torsional angles $\{\varphi_i\}$. This gives the composite orientational-conformational distribution function as

$$f(\beta\gamma\{\varphi_i\}) = f_{conf}(\{\varphi_i\})\sum_{Lm}\{(2L+1)/4\pi\}\langle C_{Lm}^* \rangle_{\{\varphi_i\}} C_{Lm}(\beta\gamma) \tag{57}$$

The orientational order for the nematic is now defined in terms of the orientational ordering tensors for each conformational state. The conformational order is defined by the singlet distribution function, $f_{conf}(\{\varphi_i\})$, which as we have seen can be expanded in a basis of Fourier functions, but we shall not pursue this aspect here.

The expansion of the conditional probability, $f(\{\varphi_i\}|\beta\gamma)$, is not expected to be rapidly convergent by analogy with the behaviour of rigid biaxial molecules. However, as we have seen for this system the exponential form

$$f(\{\varphi_i\}|\beta\gamma) = Q(\{\varphi_i\})^{-1}\exp\left\{\sum_{Lm}a_{Lm}(\{\varphi_i\})C_{Lm}(\beta)\right\} \tag{58}$$

should converge more rapidly (cf. EQN (30)). In the summation the expansion coefficients are those for a particular conformation, $\{\varphi_i\}$. The total single molecule orientational-conformational distribution function is then given by

$$f(\beta\gamma\{\varphi_i\}) = f_{conf}(\{\varphi_i\})Q(\{\varphi_i\})^{-1}\exp\left\{\sum_{Lm}a_{Lm}(\{\varphi_i\})C_{Lm}(\beta\gamma)\right\} \tag{59}$$

A similar result has been obtained previously but by using less general arguments [13].

In this approach based on characterising the orientational order in terms of a set of ordering tensors for each conformational state the anisotropy in properties such as the magnetic susceptibility is obtained in the following manner. The starting point is the expression for the zeroth component of the magnetic polarisability in the laboratory frame

$$\kappa_{lab}^{20} = \sum_m \kappa_{mol}^{2m}(\{\varphi_i\})C_{2m}(\beta\gamma) \tag{60}$$

where the director defines the Z axis. This component of the tensor must now be averaged over all orientations and conformations via the product distribution function

$$\widetilde{\kappa}_{lab}^{20} = \sum_m \int_o^{2\pi} \int_o^{2\pi} \int_o^{\pi} \kappa_{mol}^{2m}(\{\varphi_i\}) C_{2m}(\beta\gamma) f_{conf}(\{\varphi_i\}) f(\{\varphi_i\} | \beta\gamma) \sin\beta d\beta d\gamma d\{\varphi_i\}$$

$$= \int \sum_m \kappa_{mol}^{2m}(\{\varphi_i\}) \langle C_{2m} \rangle_{\{\varphi_i\}} d\{\varphi_i\}$$

$$= \left\langle \sum_m \kappa_{mol}^{2m}(\{\varphi_i\}) \langle C_{2m} \rangle_{\{\varphi_i\}} \right\rangle_{conf}$$

$$= \left\langle \kappa_{lab}^{20}(\{\varphi_i\}) \right\rangle_{conf} \tag{61}$$

This gives the anisotropy in the magnetic susceptibility as

$$\Delta\widetilde{\chi} = N_A \sqrt{\frac{3}{2}} \left\langle \kappa_{lab}^{20}(\{\varphi_i\}) \right\rangle_{conf} \tag{62}$$

that is, each conformational state makes a contribution to $\Delta\widetilde{\chi}$ which depends on the orientational ordering tensor for that state. The contribution is then weighted with the conformational probability and the total is obtained by integrating over all conformations [13].

E   BIAXIAL MOLECULES IN BIAXIAL PHASES

As Freiser [14] has demonstrated one consequence of the deviation of nematogenic molecules from uniaxial symmetry is the possibility that in addition to a uniaxial nematic phase a biaxial nematic with D_{2h} symmetry should be formed. In the biaxial phase second rank tensorial properties such as the magnetic susceptibility, $\widetilde{\chi}$, should have three independent principal components, $\widetilde{\chi}_{XX} \neq \widetilde{\chi}_{YY} \neq \widetilde{\chi}_{ZZ}$. The principal axes give the orientations of the three orthogonal directors characterising the phase, **l**, **m** and **n**. Unlike the situation for the uniaxial nematic where the director is uniquely associated with the symmetry axis of any second rank property the labelling of the axes in a biaxial nematic is a matter of convention. For example, the Z axis could be identified as that about which $\widetilde{\chi}$ approximates to cylindrical symmetry; that is $|\widetilde{\chi}_{ZZ}| > \widetilde{\chi}_{XX} - \widetilde{\chi}_{YY}$ where X and Y are chosen so that the biaxiality in $\widetilde{\chi}$ is positive. With this choice **n** would be identified with Z, X with **m** and Y with **l**. However, it is important to recognise that this labelling of the three directors will be dependent on the tensorial property selected. At a molecular level the directors would be associated with the preferred orientations of three orthogonal axes set in the molecule.

To see the nature of the orientational order parameters needed to characterise the biaxial nematic we start with the singlet orientational distribution function. Since the molecules are taken to be rigid the distribution is a function of the three Euler angles, $\alpha\beta\gamma$, which we denote by Ω connecting the director frame with an axis system set in the molecule. Again this is expanded in an appropriate set of basis functions which are taken to be the Wigner rotation matrices [3]

$$f(\Omega) = \sum_{Lmn} f_{Lnm} D_{nm}^L(\Omega) \tag{63}$$

For a non-polar biaxial nematic the summation will be restricted to even values of L. The $D_{nm}^L(\Omega)$ are functions of the three Euler angles and are related to the modified spherical harmonics. Thus the subset $D_{0m}^L(\Omega)$ is independent of α and equal to $C_{Lm}(\beta\gamma)$. The expansion coefficients which will form the order parameters are obtained via the orthogonality of the Wigner rotation matrices

$$\int D_{nm}^L(\Omega) D_{n'm'}^{L'*}(\Omega) d\Omega = \{8\pi^2/(2L'+1)\}\delta_{LL'}\delta_{mm'}\delta_{nn'} \tag{64}$$

This gives an expansion coefficient as

$$f_{Lnm} = \{(2L+1)/8\pi^2\}\left\langle D_{nm}^{L*}\right\rangle \tag{65}$$

and it is the averages, $<D_{nm}^{L*}>$, which constitute the set of orientational order parameters needed to characterise the biaxial nematic completely. For a given rank L only certain components m and n are non-zero in the biaxial nematic. Thus $<D_{00}^L>$ is non-zero in both the uniaxial and biaxial nematic phases, measuring as it does the orientational order of the molecular z axis with respect to the **n** director. Similarly the order parameters $<D_{0m}^L>$ are non-zero in both uniaxial and biaxial nematic phases because $<D_{0m}^L>$ is independent of the Euler angle, α, which is analogous to the azimuthal angle for the molecular z axis in the director frame. The order parameters $<D_{0m}^L>$ reflect, therefore, the biaxiality in the ordering of the molecule with respect to the director, **n**. In contrast the order parameters $<D_{n0}^L>$ with n not zero vanish in the uniaxial nematic phase and are non-zero in the biaxial nematic. They provide a measure of the biaxial ordering of the molecular z axis in the director frame. Finally the order parameters $<D_{nm}^L>$ where both n and m are non-zero vanish in the uniaxial nematic but not in the biaxial nematic.

To illustrate further the role of the orientational order parameters in a biaxial nematic we consider the case of the second rank ordering tensor, $<D_{nm}^2>$, both because it is available from experiment and because it provides the simplest description of the ordering. In principle $<D_{nm}^2>$ has twenty-five independent components because both n and m take values from 2 to -2 in steps of unity. However, it is possible to reduce this number to just four by assuming D_{2h} symmetry for both the molecule and the nematic phase. The four principal second rank order parameters are $<D_{00}^2>$, $<D_{02}^2>$, $<D_{20}^2>$ and $<D_{22}^2>$. The first two of these are non-zero in both uniaxial and biaxial nematic phases; they are equivalent to those which we encountered in Section C for a biaxial molecule in a uniaxial phase. As we have seen, in the limit of perfect order when the molecular z axis is parallel to the director **n** the major order parameter $<D_{00}^2>$ is unity while the molecular biaxial order parameter, $<D_{02}^2>$, vanishes. Of the other two order parameters, which are zero in the uniaxial phase but not in the biaxial nematic one, $<D_{20}^2>$, vanishes in the limit of complete order (that is $<D_{00}^2>=1$) whereas the other order parameter $<D_{22}^2>$ takes the value of $\sqrt{3/8}$. Since the major order parameter is expected to be high in a biaxial nematic phase it would seem that a property with a strong dependence on $<D_{22}^2>$ is needed in order to observe the phase biaxiality.

To explore this we shall now see how the components of a tensorial property are related to the orientational order parameters. Thus for the magnetic susceptibility $\widetilde{\underline{\chi}}$ the irreducible spherical components measured in the laboratory frame are given by

$$\widetilde{\chi}^{2n} = N_A \sum_m \left\langle D_{nm}^{2*} \right\rangle \kappa_{mol}^{2m} \tag{66}$$

(see EQN (15)). The major component of $\underline{\chi}$ in the principal axis system for both the phase and the molecule is given by

$$\widetilde{\chi}^{20} = N_A \left\{ \left\langle D_{00}^2 \right\rangle \kappa_{mol}^{20} + 2Re\left(\left\langle D_{02}^2 \right\rangle \kappa_{mol}^{2-2} \right) \right\} \tag{67}$$

which is independent of the biaxial ordering of the phase but does depend on the biaxial ordering of the molecules. This is, of course, not the case for the components $\widetilde{\chi}^{2\pm2}$ which reflect the phase biaxiality. In the principal axis system for the biaxial nematic both of these components are equal and use of this equality gives

$$\widetilde{\chi}^{22} = N_A \left[Re\left\langle D_{20}^2 \right\rangle \kappa_{mol}^{20} + Re\left\{ \left(\left\langle D_{22}^2 \right\rangle + \left\langle D_{-22}^2 \right\rangle \right) \kappa_{mol}^{2-2} \right\} \right] \tag{68}$$

As we have seen the dominant order parameter is expected to be $Re(<D_{22}^2> + <D_{-22}^2>)$ and so for the magnetic susceptibility to have a large biaxiality it is necessary for the biaxial components of magnetic polarisability in the molecular frame, $\kappa_{mol}^{2\pm2}$, also to be large. This is an important consideration in magnetic resonance studies of the phase symmetry where the magnetic interactions approximate to cylindrical symmetry. Then if the symmetry axis for the interaction is parallel to the molecular z axis the biaxial component is zero and the observed biaxiality in the partially averaged interaction is determined by $Re<D_{20}^2>$ which is likely to be small. Ideally the symmetry axis of the magnetic interaction should be orthogonal to the molecular z axis thus maximising its biaxiality in the molecular frame and so enhancing the biaxiality in the partially averaged magnetic interaction.

The orientational ordering in a biaxial nematic phase can also be described using a Cartesian representation of the order parameters. At the second rank level this description is provided by the supermatrix

$$S_{\alpha\beta}^{AB} = \left\langle \left(3\ell_{A\alpha} \ell_{B\beta} - \delta_{AB}\delta_{\alpha\beta} \right)/2 \right\rangle \tag{69}$$

where the capital letters denote laboratory axes and the lower case letters are for axes set in the molecule [15]. In general the supermatrix contains eighty-one elements which contrasts with the twenty-five independent components of the ordering tensor $<D_{nm}^2>$. This emphasises the compactness of the notation based on Wigner rotation matrices to describe the orientational order in liquid crystal phases. In the principal axes of both the phase and the molecule only the diagonal elements, $S_{\alpha\alpha}^{AA}$, are non-zero. There are nine diagonal elements but these are not independent because as we have seen there are just four independent components of $<D_{nm}^2>$ in the principal axis systems. In fact the five relationships linking the $S_{\alpha\alpha}^{AA}$ result from the properties of the direction cosines which give

$$\sum_\alpha S_{\alpha\alpha}^{XX} = 0$$

$$\sum_\alpha S_{\alpha\alpha}^{YY} = 0$$

$$\sum_\alpha S_{\alpha\alpha}^{ZZ} = 0$$

$$\sum_A S_{xx}^{AA} = 0$$

and

$$\sum_A S_{yy}^{AA} = 0 \tag{70}$$

The final vanishing trace,

$$\sum_A S_{zz}^{AA}$$

is already contained in the five other traces. Physically the $S_{\alpha\alpha}^{AA}$ are the ordering matrices for the principal molecular axes with respect to the three principal axes **l**, **m** and **n** of the biaxial nematic phase. The four independent components of $S_{\alpha\alpha}^{AA}$ [16] used to describe the second rank orientational ordering within a biaxial nematic phase are:

$$S_{zz}^{ZZ} = \left\langle \left(3\ell_{Zz}^2 - 1\right)/2 \right\rangle$$

$$S_{xx}^{ZZ} - S_{yy}^{ZZ} = \left\langle 3\left(\ell_{Zx}^2 - \ell_{Zy}^2\right)/2 \right\rangle$$

$$S_{zz}^{XX} - S_{zz}^{YY} = \left\langle 3\left(\ell_{Xz}^2 - \ell_{Yz}^2\right)/2 \right\rangle$$

and

$$\left(S_{xx}^{XX} - S_{xx}^{YY}\right) - \left(S_{yy}^{XX} - S_{yy}^{YY}\right) = \left\langle 3\left[\left(\ell_{Xx}^2 - \ell_{Yx}^2\right) - \left(\ell_{Xy}^2 - \ell_{Yy}^2\right)\right]/2 \right\rangle \tag{71}$$

The first two of these order parameters are non-zero in both the uniaxial and the biaxial nematic phases. They are the major and biaxial order parameters for the molecule measured with respect to the director **n** (see Section C). The remaining two order parameters vanish in the uniaxial nematic phase but are non-zero in the biaxial nematic. One of these, $S_{zz}^{XX} - S_{zz}^{YY}$, measures the biaxiality in the ordering of the molecular z axis with respect to the laboratory X and Y axes. It is, therefore, the phase analogue of the molecular biaxial order parameter, $S_{xx}^{ZZ} - S_{yy}^{ZZ}$. The final quantity is the difference in the phase biaxial order parameter for the molecular x and y axes. The dependence of these four order parameters on the direction cosines, $\ell_{A\alpha}$, allows the limiting forms to be easily identified. Thus for a completely ordered system with the molecular axes x, y and z parallel to the laboratory axes X, Y and Z, respectively,

$$S_{zz}^{ZZ} = 1$$

$$S_{xx}^{ZZ} - S_{yy}^{ZZ} = 0$$

$$S_{zz}^{XX} - S_{zz}^{YY} = 0$$

$$\left(S_{xx}^{XX} - S_{xx}^{YY}\right) - \left(S_{yy}^{XX} - S_{yy}^{YY}\right) = 3 \tag{72}$$

Finally, we note the relationship between the principal components of the ordering matrix $S_{\alpha\alpha}^{AA}$ and the ordering tensor $< D_{nm}^{2} >$,

$$S_{zz}^{ZZ} = \left\langle D_{00}^{2} \right\rangle$$

$$S_{xx}^{ZZ} - S_{yy}^{ZZ} = \sqrt{6} \, \mathrm{Re} \left\langle D_{02}^{2} \right\rangle$$

$$S_{zz}^{XX} - S_{zz}^{YY} = \sqrt{6} \, \mathrm{Re} \left\langle D_{20}^{2} \right\rangle$$

$$\left(S_{xx}^{XX} - S_{xx}^{YY}\right) - \left(S_{yy}^{XX} - S_{yy}^{YY}\right) = 3\mathrm{Re}\left(\left\langle D_{22}^{2} \right\rangle + \left\langle D_{-22}^{2} \right\rangle\right) \tag{73}$$

The combinations of order parameters based on the Wigner rotation matrices are proportional to the four order parameters introduced by Straley [17].

F   **MOLECULAR FIELD THEORIES OF NEMATICS**

Various analytic theories have been proposed for nematics but some of the most successful have been based on the molecular field approximation. In this the many body problem which bedevils theories of soft condensed matter is solved in a fairly dramatic manner. For the case of nematics the direct orientational correlations are ignored and the probability of finding two molecules with particular orientations is replaced by the product of finding each molecule with a particular orientation to the director [18]. The effect of this approximation is to replace the many interactions between one molecule and its neighbours by an effective or molecular field. There are several ways in which to derive the form of the molecular field or potential of mean torque, each with their strengths and weaknesses. One method starts with the pair potential and then averages over the coordinates of one of the particles [18]. The strength of this method is that the parameters occurring in the potential of mean torque can be related to specific molecular interactions. The weakness is that the pair potential cannot always be written in a convergent form or one in which the molecular orientational coordinates are separable. The variational procedure described by de Gennes [19] avoids this problem by starting with the dominant order parameters for the phases of interest. This means that the formulation of the pair potential is not necessary; however the weakness of the method is that the relationship between the parameters in the potential of mean torque and the molecular interactions is unknown. Nonetheless both methods normally yield the same form for the potential of mean torque and it is this with which we shall be primarily concerned. We shall see that once this potential is known it provides the singlet orientational distribution function, the orientational order parameters and the Helmholtz free energy from which the transition temperatures can be determined. In the following sub-sections we shall describe the molecular field theories for the systems which were encountered in the previous sections and will give some of the theoretical predictions for these.

F1   Uniaxial Molecules in a Uniaxial Nematic

The seminal molecular field theory of nematics was developed by Maier and Saupe [20] who showed that the potential of mean torque for a uniaxial molecule in a nematic is given by

$$U(\beta) = -\varepsilon <P_2> P_2(\cos\beta) \tag{74}$$

where ε is a parameter which in their theory is related to the anisotropy in the molecular polarisability. This potential has been re-derived in a more general way and it is seen that ε should be treated more as an adjustable parameter [18]. In fact evaluation of the molar Helmholtz free energy

$$A = N_A \varepsilon <P_2>^2/2 - N_A k_B T \ln Q \tag{75}$$

where Q is the orientational partition function

$$Q = \int_0^\pi \exp\{(\varepsilon/k_B T) <P_2> P_2(\cos\beta)\} \sin\beta d\beta \tag{76}$$

shows that ε is directly proportional to the nematic-isotropic transition temperature

$$\varepsilon = 0.2203 k_B T_{NI} \tag{77}$$

This proportionality has important consequences for certain predictions of the Maier-Saupe theory, as we shall see.

The singlet orientational distribution function is obtained directly from the potential of mean torque as

$$f(\beta) = Q^{-1}\exp\{(\varepsilon/k_B T)<P_2>P_2(\cos\beta)\} \tag{78}$$

This has the same form as that given in EQN (10) with the series truncated at the first non-trivial term, namely $L = 2$. However, one important difference is that the Maier-Saupe theory gives an explicit dependence of the expansion coefficient a_2 on the orientational order parameters. We should also note that the form of the distribution function is equivalent to that obtained from an information theory approach when the dominant or only order parameter known is $<P_2>$ [21]. There are relatively few experimental techniques which are able to determine the singlet orientational distribution function and so it is difficult to test this prediction of the theory. However, computer simulation studies of model systems, which conform to the molecular symmetry assumed in the theory, are able to provide the singlet orientational distribution function. For a wide range of these, including the Gay-Berne mesogen [22], the distribution function is found to be in surprisingly good agreement with that predicted by the Maier-Saupe theory. Essentially perfect agreement can be obtained by extending the Maier-Saupe theory to include fourth rank terms in the potential of mean torque, that is

$$U(\beta) = -\varepsilon_2 <P_2> P_2(\cos\beta) - \varepsilon_4 <P_4> P_4(\cos\beta) \tag{79}$$

as in the Humphries-James-Luckhurst theory [18].

The major information available experimentally concerning the orientational order of nematics is the second rank order parameter which is given by the Maier-Saupe theory as

$$<P_2> = Q^{-1} \int_0^\pi P_2(\cos\beta) \exp\{(\varepsilon/k_B T) <P_2> P_2(\cos\beta)\} \sin\beta \, d\beta \tag{80}$$

Solution of this consistency equation gives $<P_2>$ as a function of the scaled temperature, $k_B T/\varepsilon$, which can be converted into a reduced temperature, T/T_{NI}, via EQN (77). Since this scaling removes the only unknown, ε, the theory predicts that the order parameter $<P_2>$ should be a universal function of T/T_{NI}, in good agreement with experiment [23]. The predicted dependence is shown in FIGURE 1. Although the potential of mean torque does not depend on the fourth rank order parameter it can, of course, be used to calculate this. The results of such calculations are shown in FIGURE 1 and as expected $<P_4>$ is significantly smaller than $<P_2>$ at the same reduced temperature in the vicinity of the phase transition. By the same token the predicted values of $<P_6>$ are still smaller (see FIGURE 1) which makes their determination experimentally especially challenging. It is also possible to present the results for the order parameters in a way which does not depend on the specific dependence of the coefficient multiplying $P_2(\cos\beta)$ in the distribution function on temperature or order, unlike the results in FIGURE 1. This is achieved by plotting one order parameter, say $<P_4>$, against another, say $<P_2>$. The results of such a presentation are shown in FIGURE 2 where the line ends at the phase transition. This representation is especially valuable for comparing experiment with theory [23].

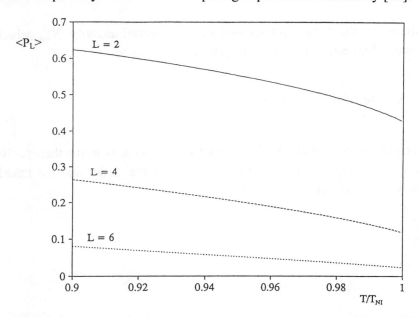

FIGURE 1 The dependence of the orientational order parameters $<P_2>$, $<P_4>$ and $<P_6>$ on the reduced temperature, T/T_{NI}, predicted by the Maier-Saupe theory.

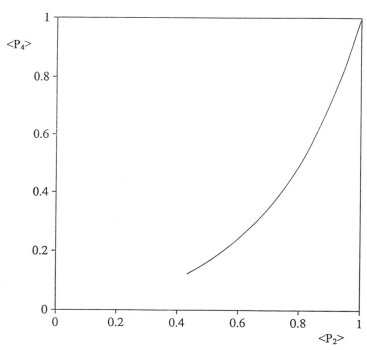

FIGURE 2 The variation of the fourth rank orientational order
parameter, $<P_4>$, with the second rank $<P_2>$.

F2   Biaxial Molecules in Uniaxial Nematics

The Maier-Saupe theory has been extended to include the influence of molecular biaxiality in the
potential of mean torque which now takes the form [18]

$$U(\beta) = -\sum_m (-)^m X_{2m} C_{2-m}(\beta\gamma) \tag{81}$$

The strength of the molecular field is represented by the tensorial quantity, X_{2m}, which being second
rank can be cast into a diagonal form so that the potential reduces to

$$U(\beta\gamma) = -X_{20}(3\cos^2\beta - 1)/2 - \sqrt{\frac{3}{2}} X_{22} \sin^2\beta \cos 2\gamma \tag{82}$$

It is clearly the second term which allows for the molecular biaxiality in the theory. As in the Maier-
Saupe theory the tensor components depend on the orientational order and the Luckhurst-Zannoni-
Nordio-Segre theory [24] gives this dependence as

$$X_{2m} = \sum_n \varepsilon_{2nm} \langle C_{2n} \rangle \tag{83}$$

Thus

$$X_{20} = \varepsilon_{200}<C_{20}> + 2\varepsilon_{220}Re<C_{22}> \tag{84}$$

and

$$X_{22} = \varepsilon_{220}\langle C_{20}\rangle + 2\varepsilon_{222}Re\langle C_{22}\rangle \qquad (85)$$

since for single component systems ε_{2nm} is equal to ε_{2mn}. The parameters ε_{2nm} depend on the molecular structure although in a way which is not given by the variational derivation of the molecular field theory. In the principal axis system for the molecule there are still three parameters in the theory but this number can be reduced to just two by assuming the geometric mean approximation [24]

$$\varepsilon_{220} = (\varepsilon_{200}\varepsilon_{222})^{\frac{1}{2}} \qquad (86)$$

With this approximation the ratio of X_{22}/X_{20} is independent of the orientational order and given by

$$X_{22}/X_{20} = \varepsilon_{220}/\varepsilon_{200}; \qquad (87)$$

it is denoted by λ and provides a measure of the molecular biaxiality. The range of this parameter, λ, is from zero for uniaxial particles to $1/\sqrt{6}$ which corresponds to the maximum biaxiality, for prolate molecules.

The theory gives the singlet orientational distribution function as

$$f(\beta\gamma) = Q^{-1}\exp\left\{\sum_m (-)^m (X_{2m}/k_BT)C_{2-m}(\beta\gamma)\right\} \qquad (88)$$

which has the same form as that encountered previously (see EQN (30)) but with the expansion truncated at the second rank term. Again this form of the distribution function is equivalent to that obtained from information theory when the dominant order parameters are the $\langle C_{2m}\rangle$. Use of the distribution function then gives the coupled consistency equations for the major and biaxial order parameters

$$\langle C_{20}\rangle = Q^{-1}\int_0^{2\pi}\int_0^{\pi} C_{20}(\beta\gamma)\exp\left[(X_{20}/k_BT)\{C_{20}(\beta\gamma) + 2\lambda ReC_{22}(\beta\gamma)\}\right]\sin\beta d\beta d\gamma \qquad (89)$$

and

$$\langle C_{22}\rangle = Q^{-1}\int_0^{2\pi}\int_0^{\pi} C_{22}(\beta\gamma)\exp\left[(X_{20}/k_BT)\{C_{20}(\beta\gamma) + 2\lambda ReC_{22}(\beta\gamma)\}\right]\sin\beta d\beta d\gamma \qquad (90)$$

where the angles defining the director orientation have been written in the principal axis system for X_{2m}. Having evaluated the second rank order parameters it is then possible to determine their fourth rank equivalents. The theory contains two parameters, ε_{200}, equivalent to the Maier-Saupe ε, and λ, the molecular biaxiality parameter; consequently the order parameters are not expected to exhibit a universal behaviour when plotted as a function of the reduced temperature. This is manifestly

79

apparent from the values of the order parameters evaluated at the nematic-isotropic transition. This temperature is determined from the molar Helmholtz free energy

$$A = N_A \sum_m (-)^m X_{2m} \langle C_{2-m} \rangle / 2 - N_A k_B T \ln Q \tag{91}$$

The values of the scaled transition temperature, $k_B T_{NI}/\varepsilon_{200}$, together with the transitional values of the order parameters $\langle C_{20} \rangle_{NI}$, $\langle C_{22} \rangle_{NI}$ and $\langle C_{40} \rangle_{NI}$ are listed in TABLE 1 for several values of the biaxiality parameter, λ. As the molecular biaxiality increases so the major order parameter at the transition decreases; indeed when λ is 0.3 $\langle C_{20} \rangle_{NI}$ has decreased to about half that for a uniaxial molecule. In contrast although as expected the biaxial order parameter $\langle C_{22} \rangle_{NI}$ increases with λ the value is extremely small. It would seem, therefore, that the molecular biaxiality has the greatest effect on $\langle C_{20} \rangle_{NI}$ and that $\langle C_{22} \rangle_{NI}$ is essentially negligible. The influence of λ on the second rank order parameter is mimicked by its fourth rank counterpart $\langle C_{40} \rangle_{NI}$; indeed when λ is 0.3 this order parameter is four times smaller than that for uniaxial molecules. For λ of 0.4 all of the transitional order parameters are extremely small, hinting at the approach of a second order transition. This occurs when λ is $1/\sqrt{6}$ but now the transition from the isotropic phase is directly to a biaxial and not a uniaxial nematic phase, as we shall see.

TABLE 1 The scaled nematic-isotropic transition
temperature and transitional order parameters.

λ	$k_B T_{NI}/\varepsilon_{200}$	$\langle C_{20} \rangle_{NI}$	$\langle C_{22} \rangle_{NI}$	$\langle C_{40} \rangle_{NI}$
0	0.2203	0.429	0	0.120
0.1	0.2220	0.408	0.017	0.109
0.2	0.2280	0.341	0.035	0.075
0.3	0.2404	0.207	0.041	0.029
0.4	0.2540	0.016	0.006	0.001

The definitive order parameter for a biaxial molecule in a uniaxial nematic is $\langle C_{22} \rangle$ which demonstrates unambiguously the role of the molecular biaxiality. However, this order parameter depends not only on the molecular biaxiality but also on the major order parameter. In order to allow for this it is helpful to examine the dependence of $\langle C_{22} \rangle$ on $\langle C_{20} \rangle$ for given values of λ. Molecular field theory predictions for this dependence are shown in FIGURE 3 for λ equal to 0.1, 0.2, 0.3 and 0.4. These results show the limiting values discussed in Section C, that is the biaxial order parameter vanishes when the major order parameter is both zero and unity. In between these limiting values $\langle C_{22} \rangle$ increases, passes through a maximum and then decreases again. It is clear that any comparison of biaxial order parameters must be made at the same value of the major order parameter. The termination of the curves in FIGURE 3 occurs at the nematic-isotropic transition and as we have seen $\langle C_{20} \rangle_{NI}$ decreases significantly with increasing λ. The molecular field theory predicts that $\langle C_{20} \rangle$ increases monotonically with decreasing temperature so that these plots reveal an intriguing temperature dependence for $\langle C_{22} \rangle$. Thus for λ of 0.1 the biaxial order parameter is predicted to decrease with decreasing temperature. In contrast for the larger values of λ, 0.2, 0.3 and 0.4, the biaxial order parameter first increases with decreasing temperature, passes through a maximum and then decreases.

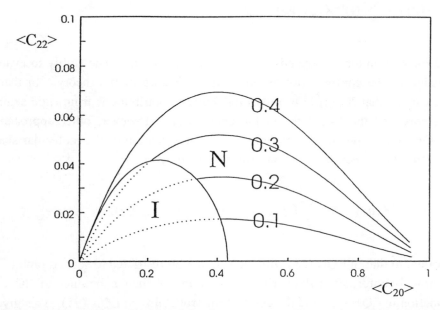

FIGURE 3 The dependence of the biaxial order parameter $<C_{22}>$ on the major order parameter $<C_{20}>$ predicted for a selection of values of the molecular biaxiality parameter, λ. The solid lines end at the nematic-isotropic transition.

F3 Flexible Molecules in Uniaxial Nematics

The Maier-Saupe molecular field theory was extended to allow for the flexibility of the constituent molecules by Marcelja [25]. This theory was refined by Counsell et al [26] and a variational derivation of their theory has been given by Luckhurst [27]. The starting point of the theory is that the nematic can be treated as a multicomponent mixture with each component being a conformer. The number of components can be minimised by using Flory's rotameric state model [28] to describe the conformations of the flexible alkyl chains of the nematogenic molecules; it is possible to relax this simplification but at some computational expense [29]. In the Flory model the conformation about a given bond in the chain is either trans, gauche$^+$ or gauche$^-$, and so the number of conformations is simply 3^{N-2} for a chain with N groups. In the isotropic phase the probability of a particular conformation, denoted by the set of integers {n}, is just

$$f_{conf}(\{n\}) = Z^{-1}\exp\{-U_{int}(\{n\})/k_BT\} \tag{92}$$

where the normalisation factor is given by

$$Z = \sum_{\{n\}}\exp\{-U_{int}(\{n\})/k_BT\} \tag{93}$$

Here U_{int} is the conformational energy in the Flory model and Z is given by a sum because there is a discrete number of states. The conformational distribution changes in the nematic phase because it favours the elongated conformers, as we shall see.

Each conformer is usually biaxial and so the potential of mean torque for them is analogous to that in EQN (81)

$$U_{ext}(\beta\gamma\{n\}) = -\sum_m (-)^m X_{2m}(\{n\})C_{2-m}(\beta\gamma) \qquad (94)$$

where now the interaction tensor depends on the conformation $\{n\}$. The ability to evaluate $X_{2m}(\{n\})$ from a knowledge of the conformation is an essential feature of the theory. In that proposed by Marcelja it is assumed that $X_{2m}(\{n\})$ is a tensorial sum of contributions from rigid segments, such as the mesogenic core and the C-C links in the chain [25]. However, other approaches have been proposed which provide a potentially more realistic representation of the molecular shape [30]. The orientational distribution function for a conformer is given by

$$f(\{n\}|\beta\gamma) = Q(\{n\})^{-1}\exp\left[\sum_m (-)^m (X_{2m}(\{n\})/k_BT)C_{2-m}(\beta\gamma)\right] \qquad (95)$$

which is equivalent to that in EQN (58) with the summation restricted to second rank terms. The total singlet orientational-conformational distribution function is then a product of the conformational distribution function in EQN (92) and the conditional probability in EQN (95). This gives

$$f(\beta\gamma\{n\}) = Z^{-1}\exp\left[-U_{int}(\{n\})/k_BT\right]\exp\left[\sum_m (-)^m (X_{2m}(\{n\})/k_BT)C_{2-m}(\beta\gamma)\right] \qquad (96)$$

where the normalisation factor is now

$$Z = \sum_{\{n\}}\exp\left[-U_{int}(\{n\})/k_BT\right]Q(\{n\}) \qquad (97)$$

Integration of the expression for the combined distribution function over all orientations gives the probability of finding a molecule in a particular conformation irrespective of its orientation (see EQN (45)) as

$$f(\{n\}) = Z^{-1}\exp[-U_{int}(\{n\})/k_BT]Q(\{n\}) \qquad (98)$$

We see that the anisotropic environment of the nematic is predicted to modify the conformational probability according to the orientational partition function of the conformer.

The orientational order parameters for each conformer are obtained by solving the coupled equations

$$\langle C_{2m}(\{n\})\rangle = Q(\{n\})^{-1}\int_o^{2\pi}\int_o^\pi C_{2m}(\beta\gamma)\exp\left[\sum_m \{(-)^m X_{2m}(\{n\})/k_BT\}C_{2-m}(\beta\gamma)\right]si\beta d\beta d\gamma \quad (99)$$

These equations are coupled because as we have seen for rigid molecules the interaction tensor depends on the orientational order. This is determined by the order of each conformer averaged over all conformers. Thus for a segment of type i making a contribution of X_i to the total interaction tensor

$$X_i = \sum_j \varepsilon_{ij}\varphi_j v_j^{-1}\langle\langle P_2^j\rangle\rangle \qquad (100)$$

This expression is analogous to that for a multicomponent mixture [31] and in it ε_{ij} is a second rank interaction parameter between segments of type i and type j, φ_j is the volume fraction of segment j, v_j is its volume and $\ll P_2^j \gg$ is the order parameter for segment j, $< P_2^j >$ averaged over all conformations. These order parameters $\ll P_2^j \gg$ are important not only because of their occurrence in the theory but also because they are available from NMR experiments [11]. In particular deuterium NMR spectroscopy is able to yield the order parameters for the C-D bonds along a suitably deuterated alkyl chain in a nematogenic molecule.

The predicted dependence of the C-D bond order parameter on its position in the hexyloxy chain of 4-hexyloxy-4′-cyanobiphenyl (6OCB) is shown in FIGURE 4. The order parameters $\ll P_2^{CD} \gg$ are negative because the C-D bonds tend to be orthogonal to the director when the long molecular axis is parallel to **n**. We see that as the position moves away from the mesogenic core so the order parameter falls as a result of enhanced conformational averaging. In addition the order parameter exhibits a characteristic variation with a small drop followed by a large one. This predicted variation is found to be in good agreement with experiment [32]. The theory also predicts how the conformational probability changes on passing from the isotropic to the nematic phase. An example of this behaviour is shown in TABLE 2 for 6OCB which lists the probabilities for some of the most common conformers in the two phases. The most dramatic change occurs for the all-trans conformation whose

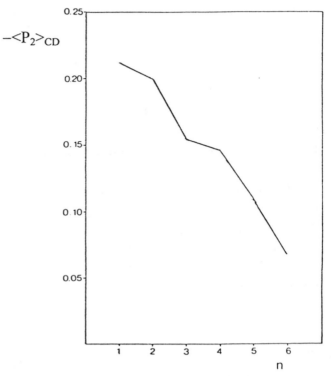

FIGURE 4 The variation of the order parameter for the C-D bonds along the chain in 4-hexyloxy-4′-cyanobiphenyl predicted by the Marcelja-like molecular field theory.

probability essentially doubles in the nematic phase. This clearly occurs because of the elongated form of this conformer. Within the version of the Flory model employed in the calculations the conformational energy for a single gauche link is independent of its position in the hexyloxy chain. This is apparent from the results for the isotropic phase. However, this degeneracy is removed in the nematic phase which favours the more elongated conformers. These occur when the gauche link is at an even position in the chain. For the gauche link at an odd position the molecular anisotropy is reduced, as is the orientational partition function and hence the probability of that conformer. This does not occur, however, when the bond is at the end of the chain for which there is a small increase in the probability.

TABLE 2 The predicted probabilities of the eleven most common conformers for
4-hexyloxy-4'-cyanobiphenyl in the isotropic and nematic phases.

Conformer*	% Probability f({n})	
	Isotropic	Nematic
ttttt	12.72	20.02
g^+tttt	3.25	2.10
tg^+ttt	3.25	4.17
ttg^+tt	3.25	2.68
$tttg^+$t	3.25	4.51
$ttttg^\pm$	3.25	3.63

*The first letter denotes the conformation of the
chain segment nearest to the cyanobiphenyl group.

F4   Biaxial Molecules in Biaxial Nematics

The first molecular field theory of biaxial nematics was presented by Freiser [14]; indeed it was his prediction which stimulated the hunt for thermotropic biaxial nematics. An alternative version of the theory was then given by Straley [17] and, although not primarily concerned with liquid crystals, Boccara et al [33] have presented a theory applicable to uniaxial and biaxial nematic phases. The key feature of these theories is the potential of mean torque which is written using a second rank interaction as

$$U(\Omega) = -\sum_{pn} X_{2pn} D^2_{-np}(\Omega) \tag{101}$$

where X_{2pn} is a supertensor determining the strength of the molecular field. In this expression the laboratory and molecular axis systems are taken to be the principal axes of the phase and of the molecule. Accordingly the summation is taken over p and n equal to 0 and ± 2. When n is confined to zero the system reverts to that of a uniaxial nematic composed of biaxial molecules and EQN (101) for the potential of mean torque necessarily reduces to EQN (81). The supertensor, X_{2pn}, is determined by the orientational order and in the theory this is confined to the second rank order parameters $<D^2_{nm}>$. Their relationship to the strength supertensor is, according to the theory,

$$X_{2pn} = \sum_m \varepsilon_{2pm} \langle D^2_{nm} \rangle \tag{102}$$

where the expansion coefficients are those which we encountered for biaxial molecules in a uniaxial nematic (see EQN (83)). Again if n is set equal to zero we recover EQN (83) for the dependence of the strength tensor on the second rank ordering tensor, $<C_{2m}>$, for a nematic phase composed of biaxial molecules.

The principal components of the ordering tensor $<D^2_{00}>$, $<D^2_{02}>$, $<D^2_{20}>$ and $<D^2_{22}>$ are obtained by solving the coupled consistency equations

$$\left\langle D_{nm}^2 \right\rangle = Q^{-1} \int D_{nm}^2(\Omega) \exp\left\{ \sum_{mnp} (\epsilon_{2pm} / k_B T) \left\langle D_{nm}^2 \right\rangle D_{-np}^2(\Omega) \right\} d\Omega \tag{103}$$

where the orientational partition function is

$$Q = \int \exp\left\{ \sum_{mnp} (\epsilon_{2pm} / k_B T) \left\langle D_{nm}^2 \right\rangle D_{-np}^2(\Omega) \right\} d\Omega \tag{104}$$

The second rank supertensor ϵ_{2pm} is related to the molecular anisotropy and, as we have seen, in the principal axis system of the molecule there are just three independent components ϵ_{200}, ϵ_{220} and ϵ_{222}. These can be related via the geometric mean approximation (see EQN (86)) so that a single molecular parameter $\lambda = \epsilon_{220}/\epsilon_{200}$ enters the theory. The temperature dependence of the four principal order parameters predicted for λ of 0.38 is shown in FIGURE 5; the scaled temperature, T^*, used in these plots is $k_B T/\epsilon_{200}$. We begin with the major order parameter $< D_{00}^2 >$; this jumps at T^* of about 0.258 to a value of approximately 0.05 indicating a weak transition to what proves to be a uniaxial nematic. The weakness of the transition is to be expected, as we saw in Section F2, from the large molecular biaxiality. This order parameter then increases with decreasing temperature until at T^* of about 0.178 there is a discontinuity in the gradient at the transition from the uniaxial to the biaxial nematic phase. In contrast to $< D_{00}^2 >$ the order parameter $< D_{02}^2 >$ reflecting the biaxiality in the molecular order is small even though λ is close to its maximal value of $1/\sqrt{6}$. Following the initial jump at the uniaxial nematic-isotropic transition the order parameter increases, passes through a maximum and then decreases, as discussed in Section F2. After the transition to the biaxial nematic $< D_{02}^2 >$ continues to decrease but relatively slowly. The other two order parameters, $< D_{20}^2 >$ and $< D_{22}^2 >$, are definitive of the biaxial nematic phase and zero in the uniaxial nematic. At the transition to the biaxial nematic both change continuously from zero in accord with the second order character predicted for the biaxial nematic-uniaxial nematic phase transition. The order parameter $< D_{20}^2 >$ is found to be small as we had anticipated in Section E. In addition it has a relatively weak temperature dependence. In marked contrast the order parameter $< D_{22}^2 >$ increases rapidly with decreasing temperature in keeping with expectation (see Section E). The large value predicted for $< D_{22}^2 >$ suggests that the biaxiality in the properties of a biaxial nematic can reach large values in keeping with NMR studies of a lyotropic biaxial nematic [34].

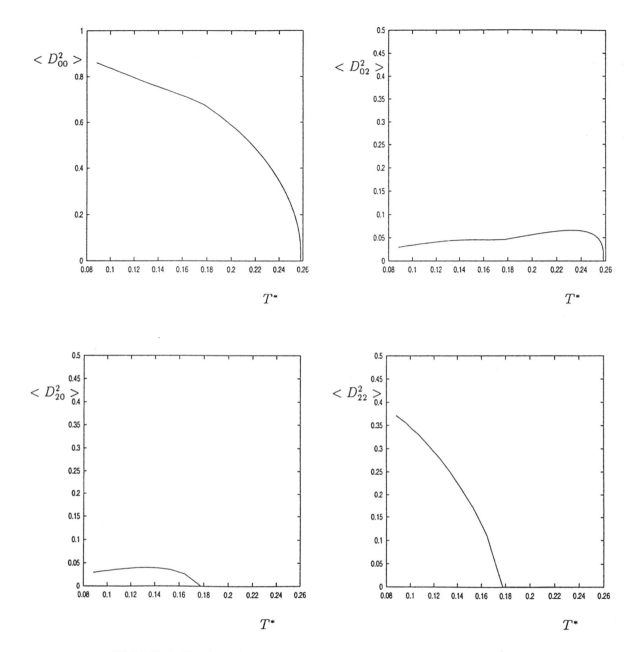

FIGURE 5 The dependence of the four principal order parameters $< D^2_{00} >$, $< D^2_{02} >$, $< D^2_{20} >$ and $< D^2_{22} >$ on the scaled temperature, $k_B T/\varepsilon_{200}$, predicted for a molecular biaxiality, λ, of 0.38.

REFERENCES

[1]   M. Abramowitz, I.A. Stegun (Eds.) [*Handbook of Mathematical Functions* (Dover, 1964)]

[2]   M.E. Rose [*Elementary Theory of Angular Momentum* (Wiley, New York, 1957)]

[3]   C. Zannoni [in *The Molecular Dynamics of Liquid Crystals* Eds. G.R. Luckhurst, C.A. Veracini (Kluwer Academic Publishers, Dordrecht, 1994) ch.2]

[4]   V. Tsvetkov [*Acta Physioch. URSS (USSR)* vol.10 (1939) p.557]

[5]   W.H. de Jeu [*Physical Properties of Liquid Crystalline Materials* (Gordon and Breach Science Publishers, New York, 1980)]

[6]   C. Zannoni [in *The Molecular Physics of Liquid Crystals* Eds. G.R. Luckhurst, G.W. Gray (Academic Press, London, 1979) ch.3]

[7]   J.W. Emsley, R. Hashim, G.R. Luckhurst, G.N. Shilstone [*Liq. Cryst. (UK)* vol.1 (1986) p.437]

[8]   A. Saupe [*Angew. Chem., Int. Ed. Engl. (Germany)* vol.7 (1968) p.97]

[9]   A.D. Buckingham [*Discuss. Faraday Soc. (UK)* vol.43 (1967) p.205]

[10]   C. Zannoni [in *Nuclear Magnetic Resonance of Liquid Crystals* Ed. J.W. Emsley (Reidel, 1985) ch.1]

[11]   See, for example, J.W. Emsley (Ed.) [*Nuclear Magnetic Resonance of Liquid Crystals* (Reidel, 1985)]

[12]   L. Di Bari, C. Forte, C.A. Veracini, C. Zannoni [*Chem. Phys. Lett. (Netherlands)* vol.143 (1987) p.263]

[13]   J.W. Emsley, G.R. Luckhurst [*Mol. Phys. (UK)* vol.41 (1980) p.19]

[14]   M.J. Freiser [*Phys. Rev. Lett. (USA)* vol.24 (1970) p.1041]

[15]   P.G. de Gennes [*The Physics of Liquid Crystals* (Clarendon Press, Oxford, 1974) p.27]; D.W. Allender, M.A. Lee [*Mol. Cryst. Liq. Cryst. (UK)* vol.110 (1984) p.331]

[16]   D.A. Dunmur, K. Toriyama [in *Physical Properties of Liquid Crystals* Eds. D. Demus, J. Goodby, G.W. Gray, H.-W. Spiess, V. Vill (Wiley-VCH, Weinheim, 1999) ch.IV.1]

[17]   J.P. Straley [*Phys. Rev. A (USA)* vol.10 (1974) p.1881]

[18]   G.R. Luckhurst [in *The Molecular Physics of Liquid Crystals* Eds. G.R. Luckhurst, G.W. Gray (Academic Press, 1979) ch.4]

[19]   P.G. de Gennes [*The Physics of Liquid Crystals* (Clarendon Press, Oxford, 1974) p.42]

[20]   W. Maier, A. Saupe [*Z. Nat.forsch. A (Germany)* vol.13 (1958) p.564, vol.14 (1959) p.882 and vol.15 (1960) p.287]

[21]   D.I. Bower [*J. Polym. Sci. (USA)* vol.19 (1981) p.93]

[22]   A.P.J. Emerson, R. Hashim, G.R. Luckhurst, S. Romano [*Mol. Phys. (UK)* vol.76 (1992) p.241]

[23]   G.R. Luckhurst [in *Dynamics and Defects in Liquid Crystals: A Festschrift in Honour of Alfred Saupe* Eds. P.E. Cladis, P. Palffy-Muhoray (Gordon and Breach Science Publishers, Amsterdam, 1998)]

[24]   G.R. Luckhurst, C. Zannoni, P.L. Nordio, U. Segre [*Mol. Phys. (UK)* vol.30 (1975) p.1345]

[25]   S. Marcelja [*J. Chem. Phys. (USA)* vol.60 (1974) p.3599]

[26]   C.J.R. Counsell, J.W. Emsley, N.J. Heaton, G.R. Luckhurst [*Mol. Phys. (UK)* vol.54 (1985) p.847]

[27]   G.R. Luckhurst [*Mol. Phys. (UK)* vol.82 (1994) p.1063]

[28]   P.J. Flory [*Statistical Mechanics of Chain Molecules* (Interscience, New York, 1969)]

[29]   A. Ferrarini, G.R. Luckhurst, P.L. Nordio [*Mol. Phys. (UK)* vol.85 (1995) p.131]

[30]   D.J. Photinos, E.T. Samulski, H. Toriumi [*J. Phys. Chem. (USA)* vol.94 (1990) p.4688 and p.4694]; A. Ferrarini, G.J. Moro, P.L. Nordio, G.R. Luckhurst [*Mol. Phys. (UK)* vol.77 (1992) p.1]

[31]   D.E. Martire [in *The Molecular Physics of Liquid Crystals* Eds. G.R. Luckhurst, G.W. Gray (Academic Press, London, 1979) ch.11]

[32]   C.J.R. Counsell, J.W. Emsley, G.R. Luckhurst, H.S. Sachdev [*Mol. Phys. (UK)* vol.63 (1988) p.33]

[33]   N. Boccara, R. Mejdani, L. De Seze [*J. Phys. (France)* vol.38 (1977) p.149]

[34]   F.P. Nicoletta, G. Chidichimo, A. Golemme, N. Picci [*Liq. Cryst. (UK)* vol.10 (1991) p.665]

2.2   Measurements and values for selected order parameters

S.J. Picken

January 1999

A   SOME REMARKS ON THE MEASUREMENT AND CALCULATION OF ORDER PARAMETERS

It is not particularly easy to give a comprehensive overview of the available experimental methods that can be used to determine the order parameters of nematics. The reason for this is that nearly all physical properties of nematics are in some way influenced by the orientational order. Therefore, we will restrict ourselves to the methods used to determine $<P_2>$ that are relatively simple and are readily accessible to most laboratories. In addition, we will discuss some more specialised techniques that allow the determination of $<P_4>$ and the biaxial order parameter D.

It should be stated from the outset that there really are no experimental methods that provide unambiguous absolute values for the order parameters. This is related to the fact that the theory that relates the measured property to the molecular orientational order is not exact and/or that there are unknown parameters in the theory. The values for these parameters can often only be obtained by fitting the data to a model for the order parameter or by comparing data from various experimental techniques. Finally, there is a limit to what can be done since each method strictly provides data on a different order parameter which is due to the fact that different time and/or length scales are probed. In addition each method will tend to highlight the orientational order of different parts of the molecule.

Despite these problems the various experimental methods normally do give numbers for the order parameter that are rather similar in value. This means that if the details are disregarded it is easy to obtain quite a good estimate for the order parameter. However, if we wish to be exact then extreme care has to be taken to eliminate the various experimental problems, and when comparing the results to theory it is necessary to examine in detail whether the theory is appropriate.

If the reader finds this state of affairs somewhat disappointing it should be realised that contrary to popular belief the nematic phase is still not understood in detail. This is due to its amorphous structure which causes a variety of experimental and theoretical problems, and makes exact results somewhat difficult to obtain.

B   BIREFRINGENCE AND DICHROISM

The main characteristic of the nematic phase is the birefringence, as is apparent from the use of polarising microscopy to distinguish crystal and liquid crystal phases from the isotropic phase. Therefore, it will not come as a surprise that birefringence is one of the most convenient methods to obtain the value of $<P_2>$. The birefringence can be measured directly from uniform planar samples. Boundary conditions for uniform planar alignment can be conveniently obtained by using glass slides coated with a rubbed poly-imide film (e.g. Merck Liquicoat-PI); alternatively oblique sputtered SiO_x and various other coating techniques may be used. Assuming that the director is aligned at angle 0° and the polariser and analyser are at +45° and -45° (see FIGURE 1) then the transmitted intensity is given by

$$I = \frac{I_0}{4} \sin^2(2\alpha) \sin^2(\phi(T)) = I_0 \sin^2\left(\frac{\pi d \Delta \tilde{n}(T)}{\lambda}\right) \tag{1}$$

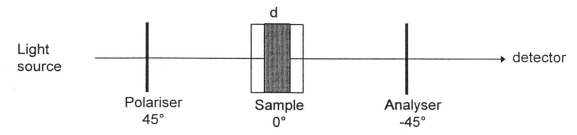

FIGURE 1 Experimental geometry for birefringence measurements.

Here d is the layer thickness, λ is the wavelength, α is the angle of 45° between the analyser and the director, and

$$\Delta \tilde{n}(T) = \tilde{n}_\parallel(T) - \tilde{n}_\perp(T)$$

is the temperature dependent birefringence. I_0 is the intensity of the incident light corrected for the transmission and extinction characteristics of the polarising elements and the cell containing the sample.

By examining the transmitted intensity versus sample temperature (preferably also for different wavelengths) it is possible to invert EQN (1) so as to obtain the phase angle $\phi(T)$, which is strictly proportional to the birefringence $\Delta \tilde{n}(T)$, if the layer thickness is assumed to be constant. There is a certain ambiguity in obtaining the absolute values of the phase angle due to the periodic nature of EQN (1). This can be eliminated by comparing results on samples of various thickness or from using various wavelengths.

An alternative but less accurate method is to determine the depolarisation colour associated with EQN (1) using a polarising microscope equipped with a hot-stage. The depolarisation colour can be transformed into an optical path difference $\Gamma = d\Delta\tilde{n}$ by visual comparison using a colour chart (or by using a compensator) and if the layer thickness is known the $\Delta\tilde{n}$ value is easily obtained. Note that this works best for thin samples with optical path differences around 1 - 2 λ and so requires d values in the range of 3 - 10 μm, assuming a typical $\Delta\tilde{n}$ value of 0.1 - 0.2. Making such samples certainly is feasible but may be found to be inconvenient. Here again, it is not necessary to have a precise value for the layer thickness as the optical path difference itself can be used as a figure to obtain the $<P_2>$ values (assuming d is constant). An interesting variation on this type of method is to make a wedge shaped sample where the layer thickness varies in a controlled fashion and the absolute value of $\Delta\tilde{n}$ is obtained from the distance between the coloured fringes (again via EQN (1)).

Finally, the temperature dependence of the birefringence can be determined using an Abbé refractometer. With some practice it is possible to measure $\tilde{n}_\perp(T)$ in the nematic phase and $n_{iso}(T)$ in the isotropic phase. Observing the angle of total reflection associated with $\tilde{n}_\perp(T)$ can be aided by placing a polariser in the eyepiece of the refractometer and rotating it until the contribution from the extraordinary wave is eliminated. Also, care has to be taken with the temperature calibration as there may be a considerable difference between the actual sample temperature and the temperature of the thermostated oil-bath used to heat the prisms of the refractometer. The temperature calibration can be performed by determining the clearing temperature of various standard samples using the refractometer. Then a figure showing the experimentally determined values of $\tilde{n}_\perp^2$ and $n_{iso}^2(T)$ may be used to estimate the value of $\tilde{n}_\parallel^2(T)$. This is done by linear extrapolation of $n_{iso}^2(T)$ into the nematic region ($n_{iso}^2(T)^*$) and using the relation:

$$n_{iso}^2(T)^* = \frac{2\tilde{n}_\perp^2(T) + \tilde{n}_\parallel^2(T)}{3} \tag{2}$$

In performing the extrapolation we assume that the change in density at the nematic-isotropic transition is small and that the thermal expansion of the nematic and the isotropic phase is the same. Typical values for the density change at the transition are about 0.3% and the thermal expansion typically is about 10^{-4} K^{-1} or 1% per 100°C. Of course, this procedure for compensating for the effect of thermal expansion can be improved by using measured density values, but this is not easily accomplished as a function of temperature. Only a few density measurements have been reported in the literature, e.g. see Dunmur [1]. It should be noted that the use of n^2 for the extrapolation is preferred, as it is the polarisability $\alpha = n^2$ that is roughly proportional to the density. In practice we may find that straightforward extrapolation of $\tilde{n}(T)$ gives similar results and the complication of taking the squared values may be avoided. Indeed, the birefringence measuring set-up shown in FIGURE 1 or the method using the polarisation colour chart only provides $\Delta\tilde{n}(T)$ or $\Gamma(T)$ and it is not possible to obtain $\Delta\tilde{\alpha}(T)$ by this method. Observe that the expression

$$\Delta\tilde{\alpha}(T) = \tilde{n}_\parallel^2(T) - \tilde{n}_\perp^2(T) = \left(2\bar{n} + \frac{1}{3}\Delta\tilde{n}\right)\Delta\tilde{n} \approx 2\bar{n}(T)\Delta\tilde{n}(T)$$

shows that $\Delta\tilde{n}$ and $\Delta\tilde{\alpha}$ should be roughly proportional. This is demonstrated in FIGURE 2 where results for 4-pentyl-4'-cyanobiphenyl (5CB) obtained using an Abbé refractometer and the phase angle method from FIGURE 1 are compared. FIGURES 3 and 4 show some birefringence results for 4,4'-dimethoxyazoxybenzene (PAA) (Chatelain [2]) and other 4,4'-dialkoxyazoxybenzenes (Hanson and Shen [3]).

To obtain values for the $<P_2>$ order parameter from such measurements a model is required. The curve drawn in FIGURE 2 is from extended Maier-Saupe theory, which can conveniently be calculated with an accuracy of about 1% using [4,5]:

$$<P_2> = 0.1 + 0.9\left[1 - 0.99\left(\frac{T}{T_{NI}}\right)^\varepsilon\right]^{\frac{1}{4}} \text{ for } T < T_{NI} \text{ and } <P_2> = 0 \text{ for } T > T_{NI} \tag{3}$$

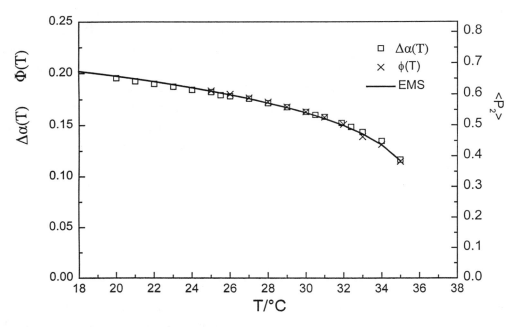

FIGURE 2 Measured anisotropy of the polarisability $\Delta\tilde{\alpha}$ values for 5CB using an Abbé refractometer
(closed squares) and $\phi(T)$ using the set-up from FIGURE 1 (x). The $\phi(T)$ data are scaled
to the $\Delta\tilde{\alpha}$ values. The solid curve is from the extended Maier-Saupe theory
using EQN (3) with $\varepsilon = 3$ and $T_{NI} = 35°C$.

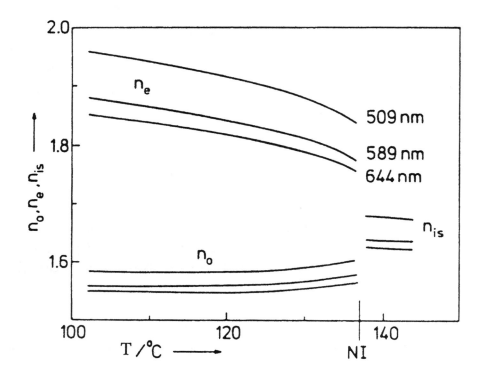

FIGURE 3 The refractive indices of PAA at various wavelengths [2].

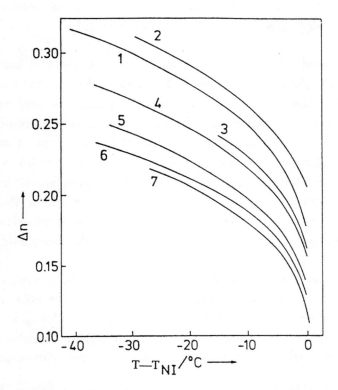

FIGURE 4 The birefringence of 4,4′-dialkoxyazoxybenzenes [3] with the number
indicating the length of the alkoxy chains.

where the exponent $\varepsilon = 1$ for standard Maier-Saupe theory and higher values for ε of 2 - 3 seem to be more appropriate for polymeric liquid crystals. EQN (3) can be interpreted as the Maier-Saupe equivalent of a stretched exponential, where the temperature dependence contains an additional exponent that allows improved fitting to the data. FIGURE 5 shows some birefringence data for a

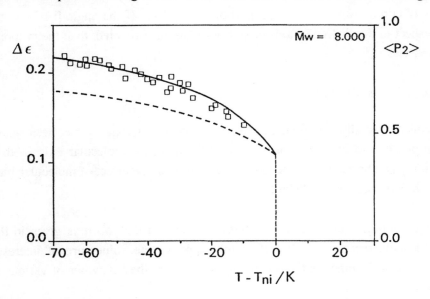

FIGURE 5 The birefringence of DABT aramid polymer in nematic solution, from Abbé refractometer measurements [5]. The dashed curve is from the standard Maier-Saupe theory ($\varepsilon = 1$ in EQN (3)) and the solid curve is obtained using $\varepsilon = 3$.

nematic aramid polymer (DABT) in sulphuric acid solution [5] obtained using an Abbé refractometer. The curves show the extended model with ε of 1 and 3 (see EQN (3)).

Using the Maier-Saupe model for $\langle P_2 \rangle$, or any other appropriate model, and using the expression $\Delta \tilde{n} = \Delta n_0 \langle P_2 \rangle$ or more preferably $\Delta \tilde{\alpha} = \Delta \alpha_0 \langle P_2 \rangle$, we can quickly obtain a reasonable estimate for $\langle P_2 \rangle$. Again, the density change via thermal expansion is not taken into account which leads to errors of the order of several percent. Despite the approximations in this procedure, substantial effort is required to improve on the $\langle P_2 \rangle$ results obtained from this simple type of measurement. Another method with which to extract the order parameter is by using a Haller plot [6], where it is assumed that if the birefringence data are plotted as $\log(\Delta \tilde{n}(T))$ versus T a straight line results and the birefringence Δn_0 at $T = 0$ is obtained by extrapolation. The order parameter can then be obtained directly from the experiments using $\Delta \tilde{n} = \Delta n_0 \langle P_2 \rangle$. Of course, this method can be used to evaluate the order parameter from other anisotropic properties, e.g. polarisability and diamagnetic susceptibility.

In addition to birefringence, in some cases it is possible to obtain the dichroic ratio by measuring the extinction coefficient parallel and perpendicular to the director. For coloured mesogens, such as nitrostilbene or azoxy compounds, the absorption bands are in the visible region of the spectrum, but in most cases UV or IR [7] spectroscopy is required as many nematic samples are transparent in the visible region which allows their use in electro-optic displays. Whatever the method, by calculating the dichroic ratio R from:

$$R = \tilde{\alpha}_{\parallel} / \tilde{\alpha}_{\perp} = \log(T_{\parallel})/\log(T_{\perp})$$

and using

$$\langle P_2 \rangle_{dichr.} = (R - 1)/(R + 2)$$

we immediately obtain a number that is proportional to $\langle P_2 \rangle$. If the angle β of the transition dipole moment with respect to the molecular axis is known or can be estimated, then the proportionality factor can be eliminated via:

$$\langle P_2 \rangle_{dichr.} = P_2(\cos \beta) \langle P_2 \rangle \tag{4}$$

In this expression the usually rather small effect of molecular biaxiality has been ignored altogether. Also it is often possible to identify transitions that lie along the molecular axis so that the effect of molecular biaxiality can be reduced. For more information on the effect of molecular biaxiality and the biaxial order parameter see the article by Korte [8].

Summarising, the simplified expressions for birefringence and dichroism as given in this section will provide quite adequate results for $\langle P_2 \rangle$ if mainly approximate values are of interest. Also, these methods can be used with some confidence simply to compare the behaviour of various samples, e.g. a homologous series.

C ANISOTROPY OF THE DIAMAGNETIC SUSCEPTIBILITY

Better values for the $<P_2>$ order parameter can be obtained from diamagnetic susceptibility measurements. Here, as for the Abbé refractometer method, we obtain values of χ in the isotropic and in the nematic phase. When using a Faraday balance the magnetic alignment of the director will usually be along the field leading to measurement of $\widetilde{\chi}_{\parallel}$ in the nematic phase, and $\Delta\widetilde{\chi}(T)$ is again obtained by extrapolation of the isotropic results into the nematic region. In a Faraday balance the sample is suspended in an inhomogeneous magnetic field and the diamagnetic susceptibility will give rise to an additional force acting on the sample that can be measured using a microbalance (see FIGURE 6). The relation between the measured force F_Z and the diamagnetic susceptibility is given by:

$$F_Z = \mu_0^{-1}\widetilde{\chi}_{\parallel}^{\,m}\,m\left(\mathbf{B}\frac{\partial\mathbf{B}}{\partial Z}\right) \tag{5}$$

where $\widetilde{\chi}_{\parallel}^{\,m}$ is the specific susceptibility and m is the sample mass.

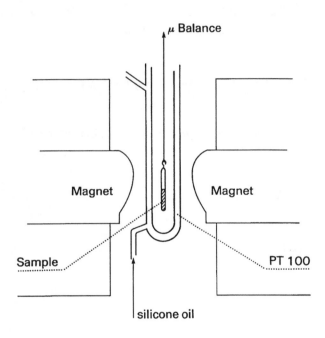

FIGURE 6 Schematic set-up of a Faraday balance.

Here, as for the birefringence measurements using an Abbé refractometer, the anisotropy can be obtained by extrapolation of the isotropic results into the nematic phase. The macroscopic anisotropy of the diamagnetic susceptibility can be extracted from the diamagnetic susceptibility tensor:

$$\widetilde{\chi}_{ij} = \begin{pmatrix} \widetilde{\chi}_{\perp} & 0 & 0 \\ 0 & \widetilde{\chi}_{\perp} & 0 \\ 0 & 0 & \widetilde{\chi}_{\parallel} \end{pmatrix} \tag{6a}$$

and

$$\widetilde{\chi}_{ij} - \overline{\chi}\delta_{ij} = \Delta\widetilde{\chi} \begin{pmatrix} -\frac{1}{3} & 0 & 0 \\ 0 & -\frac{1}{3} & 0 \\ 0 & 0 & \frac{2}{3} \end{pmatrix} \qquad\qquad (6b)$$

with

$$\Delta\widetilde{\chi} = \widetilde{\chi}_{\parallel} - \widetilde{\chi}_{\perp}$$

and

$$\overline{\chi} = \frac{1}{3}\widetilde{\chi}_{\parallel} + \frac{2}{3}\widetilde{\chi}_{\perp}$$

Therefore, there is only one anisotropy parameter on the macroscopic level because of the uniaxial nature of the nematic phase. Then, relating this to the molecular susceptibility components κ_{ij} (assumed to be diagonal on the η,ξ,ζ molecular coordinates) we find [9]:

$$\Delta\widetilde{\chi} = N_A\left(\Delta\kappa S + \delta\kappa D\right) = N_A\left[\left(\kappa_{zz} - \frac{1}{2}(\kappa_{xx} + \kappa_{yy})\right)S + \frac{1}{2}(\kappa_{xx} - \kappa_{yy})D\right] \qquad (7)$$

This expression shows that the macroscopic anisotropy contains information on both the major order parameter

$$<P_2> = \frac{1}{2}\left\langle 3\cos^2\theta - 1\right\rangle$$

and the biaxial order parameter

$$D = \frac{3}{2}\left\langle\sin^2\theta\cos 2\varphi\right\rangle$$

Here D provides information on whether the molecules deviate away from the director with a preferred alignment around the molecular long axis z. Usually it is assumed that both D and $\delta\kappa$ are small and so the second term is often ignored ($|D|$ is typically 0.1 or less: see Section D on NMR spectroscopy).

The amount of effort required to set up a Faraday balance that is accurate and sufficiently stable to perform diamagnetic susceptibility measurement should by no means be underestimated, e.g. see de Jeu and Claassen [10] for some details. Even with a field of 2 T the rather small forces require microgram resolution and an excellent mechanical stability of the equipment. In addition, the sample temperature has to be controlled and the change in buoyancy due to expansion of the surrounding gas can easily lead to large experimental errors.

Some results for the diamagnetic susceptibility of 4-heptyl-4′-cyanobiphenyl (7CB) are shown in FIGURE 7 (Leenhouts et al [11]), and similar results have also been reported by Sherrell and Crellin [12]. Somewhat inaccurate measurements on main chain aramid polymers in nematic solution have been given by Picken [5]. FIGURE 8 shows data for <P₂> determined from diamagnetic susceptibility, birefringence and NMR spectroscopy, which is the topic of the next section.

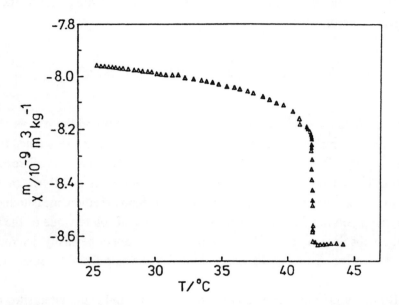

FIGURE 7 The diamagnetic susceptibility of 7CB [11] in the nematic phase and the isotropic phase (above 42°C) giving $\widetilde{\chi}_{\parallel}$ and χ_{iso} respectively.

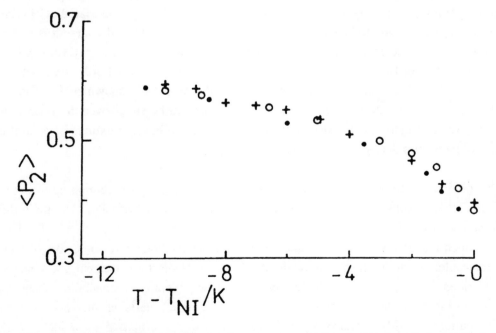

FIGURE 8 <P₂> values of 5CB from NMR spectroscopy (closed circles), $\Delta\widetilde{\chi}$ (open circles) and Raman depolarisation [21] (+) measurements.

D   NMR SPECTROSCOPY

A useful method with which to determine order parameters is NMR spectroscopy. One of the more frequently used techniques is deuterium NMR where the mesogenic compounds have to be specifically deuteriated to reduce measuring time. Then assuming that the director aligns along the magnetic field the anisotropy of the nematic phase shows up as a quadrupolar splitting $\Delta\tilde{\nu}$ of the NMR signal for each set of equivalent deuterons [9]:

$$\Delta\tilde{\nu} = \frac{e^2qQ}{h}\left\{\left[\mu_z^2 - \frac{1}{2}\left(\mu_x^2 + \mu_y^2\right)\right]S + \frac{1}{2}\left(\mu_x^2 - \mu_y^2\right)D\right\}. \qquad (8)$$

Here the prefactor e^2qQ/h is the static quadrupole coupling constant and the μ_i's are the components of the electric field gradient along the principal axes of the ordering matrix. It is noted that the splitting is proportional to S or $<P_2>$ if the effect of the biaxial order parameter is disregarded and so the quadrupolar splitting immediately gives a result that is proportional to $<P_2>$. In FIGURE 8 some results for $<P_2>$ are shown from NMR, $\Delta\tilde{\chi}$ and Raman depolarisation measurements on 5CB. It should be realised that again there is some uncertainty as to the absolute value of the various constants in EQN (8) so that an absolute value for the order parameters cannot be easily determined using NMR: in FIGURE 8 the $<P_2>$ values from $\Delta\tilde{\chi}$ have been scaled to the data from the other methods.

Using proton NMR and by judicious choice of the hydrogenous and deuteriated sites in the mesogen, it is possible to practically eliminate the first term in the proton NMR equivalent of EQN (8) (this requires the vector between the proton pairs to be close to the 'magic angle' with the molecular major axis, i.e. when P_2 (cos θ) = 0 at 54.3°) and then the proton NMR signal will highlight the contribution of the D order parameter. This type of study has been done on partially deuteriated 5CB (Emsley et al [13]) and the D value is found to be small with S = 0.62 and D = -0.02. Of course it is to be expected that the relative importance of the biaxial order parameter and indeed its sign may vary from sample to sample, but the results of Emsley et al indicate that disregarding the biaxial order parameter altogether is not unreasonable. The temperature dependence of $<P_2>$ and D is shown in FIGURE 9, in ^{15}N enriched 4,4′-diheptylazobenzene, where the D value is found to be small over the entire temperature range and only varies slightly [14]. Also, work has been done on binary mixtures of biaxial molecules [15] and on solute probes dissolved in a nematic solvent [16].

Despite the obvious success of the NMR technique some words of caution are required when comparing the NMR results to order parameters obtained from other methods. The main difficulty is that NMR necessarily only gives information on the orientational order of a specific rigid part of the molecule in which the observed nuclei reside, while the other methods give some sort of average over the entire molecule. In addition the timescale at which the NMR measurement takes place is in the microsecond range and therefore the NMR results will not contain a contribution from the slower collective modes (some of these can be observed visually using a polarising microscope in the form of director fluctuations). NMR spectroscopy therefore provides information on a local scale and the major order parameter values tend to be somewhat higher than those found for instance from birefringence or diamagnetic susceptibility measurements, unless as is done in FIGURES 8 and 9 the various $<P_2>$ values are scaled. The use of stimulated-echo three pulse sequences has been reported to allow investigation of order-fluctuations on longer timescales [17].

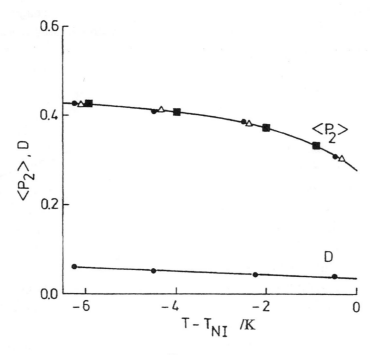

FIGURE 9 <P$_2$> and D from NMR studies on ^{15}N enriched 4,4'-diheptylazobenzene as a function of temperature (closed circles). Additional <P$_2$> data are shown from birefringence (closed squares) and diamagnetic susceptibility (open triangles) measurements.

E   RAMAN DEPOLARISATION MEASUREMENTS

The techniques described so far only provide information on the second rank order parameter <P$_2$>. However, there are a few methods that allow the determination of higher moments of the singlet orientational distribution function. In this section polarised Raman spectroscopy [18] that allows determination of the fourth rank order parameter <P$_4$> is discussed. The reason why <P$_4$> can be determined is that the response is proportional to the probability to excite a Raman state, which depends on the cosine of the angle of polarisation with respect to the mesogen (or to be more precise to a certain Raman active molecular vibration), times the decomposition of the decay along the axis of the analyser which also contains geometric factors such as cos θ. Therefore, the polarised Raman intensities will contain terms up to order cos^4 θ and the Raman depolarisation can be expressed in terms of both <P$_2$> and <P$_4$>. The measurement involves determining the depolarisation ratios (e.g. of a cyano-stretch vibration at 2227 cm^{-1} in cyanobiphenyls) for a variety of sample geometries, with the director parallel and perpendicular to the incident polarisation (giving ratios $R_1 = I_{XY}/I_{XX}$ and $R_2 = I_{YY}/I_{YX}$) and along the light path for a homeotropically oriented sample (giving $R_3 = I_{zy}/I_{zx}$). Additionally, the depolarisation ratio in the isotropic phase is required giving R_{iso}. In each case the depolarisation ratio is the quotient of the measured Raman intensities with the analyser position respectively parallel and perpendicular to the incident light. In this manner four depolarisation ratios (R_1, R_2, R_3, R_{iso}) are obtained and these are used to solve the expressions relating the Raman intensities to <P$_2$> and <P$_4$> and the two unknown components of the molecular Raman tensor. The details of these equations can be found in [18] and [19]. In the case that the Raman active molecular vibration is not along the molecular long axis the effect of the biaxial order parameter has to be considered as well and for this reason it is desirable to choose a Raman band that lies along the major axis of the mesogen. For example, in the case of cyanobiphenyls these are the cyano-stretch vibration at 2227 cm^{-1} and the symmetric phenyl stretch vibration at 1597 cm^{-1}.

Some results are shown in FIGURES 10 and 11 [20,21]. One of the most interesting features is the observation that for some compounds $<P_4>$ may be negative (see FIGURE 11 for 5CB and 6CB). This is not allowed by Maier-Saupe theory and various explanations for this behaviour have been proposed. One idea is that in strongly polar compounds such as the cyanobiphenyls there is a substantial degree of antiparallel association of the molecules causing the long axis of a 'dimer' to be tilted with respect to the measured CN stretch vibration. This idea was also studied by comparing molecular dynamics simulations on 5CB, with and without a dipole moment, and the inclusion of molecular charges indeed lowers the $<P_4>$ value [22]. These simulations were performed on relatively small systems with only 64 molecules, while including the full molecular structure with all

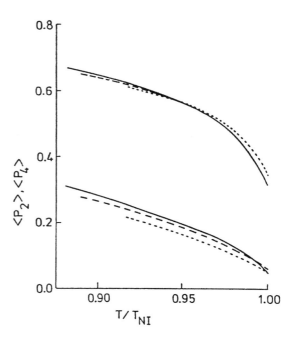

FIGURE 10 $<P_2>$ and $<P_4>$ versus temperature from polarised Raman scattering of 7PCH (solid line), 5PCH (dashed line) and 3PCH (dotted line).

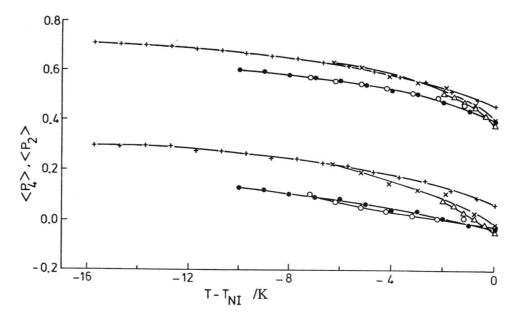

FIGURE 11 $<P_2>$ and $<P_4>$ versus temperature from polarised Raman scattering on 5CB (closed circles), 6CB (open circles), 7CB (+), 8CB (x), and 9CB (open triangles).

internal degrees of freedom, and therefore the noise on the MD simulation data is rather large. Alternatively, there are claims that the negative $<P_4>$ phenomenon is due to the anisotropic local electric fields influencing the Raman depolarisation ratios [20]. This problem is not unique to Raman scattering and also occurs in the interpretation of birefringence measurements and other optical techniques such as polarised fluorescence spectroscopy, which is one of the other optical methods used to measure $<P_4>$ albeit for probe molecules. With polarised fluorescence again use is made of an excitation and decay process to obtain the higher moments of the orientational distribution. As an example see [23] for measurements on stilbene probe molecules dissolved in 5CB, 7CB and PCH7.

It should also be noted that neutron scattering experiments also provide a route, in principle, to order parameters of all ranks although in practice only $<P_2>$ and $<P_4>$ have been determined. This technique is discussed in more detail in Datareview 4.2.

F   OTHER EXPERIMENTAL TECHNIQUES

In addition to the methods described here there are various other experimental techniques to determine the order parameters. These will be described very briefly and some more recent references are given.

When the nematic solvent does not show a clear dichroism it may be useful to investigate the dichroism of solute dye molecules. In these guest-host systems the dichroic dye may be considered to be a probe that allows measurement of the order parameters of the solute dye molecules in the nematic environment. Although this method provides exact information on the degree of alignment of the dye molecules it only provides indirect results on the order parameter of the nematic solvent. Some references are Herba et al [24] and Bauman [25]. Analogously, NMR and ESR spectroscopy can also be performed on probe molecules dissolved in a nematic solvent, and again the same interpretation problems occur.

Dielectric spectroscopy has also been reported as a method to determine $<P_2>$, which in addition provides information on the molecular dynamics in the nematic phase. The idea behind the method is similar to using birefringence, which may be regarded as performing dielectric spectroscopy at (high) optical frequencies. However, for dielectric studies the problems of interpretation are compounded by having to consider the various timescales of the molecular motions and the effect of the permanent dipole moments. Some recent articles on this very interesting method are Würflinger [26] and Urban et al [27]. Vertogen and de Jeu [9] also give detailed information on this technique, and Haws et al [28] have written an overview of this method as applied to liquid crystal polymers.

Finally, it should be noted that in addition to the even rank order parameters $<P_2>$ and $<P_4>$ occurring in the bulk there is a substantial literature on the odd rank order parameters that provide information on the structure of the nematic phase at interfaces, where the up-down symmetry is broken, and on odd rank order parameters from the bulk of DC poled nematic polymer samples. The information that is obtained on these odd rank order parameters is via $<\cos^3(\theta)>$ from second harmonic generation or Pockels effect measurements: see for example [30] and [31] for some experimental results and [32] for the Maier-Saupe theory which includes the effect of external DC fields.

REFERENCES

[1]   D.A. Dunmur, W.H. Miller [*J. Phys. Colloq. (France)* vol.40 (1979) p.141]

[2]   P. Chatelain, M. Germain [*C.R. Hebd. Seances Advan. Sci. (France)* vol.259 (1964) p.127]

[3]   E.G. Hanson, Y.R. Shen [*Mol. Cryst. Liq. Cryst. (UK)* vol.36 (1976) p.193]

[4]   S.J. Picken, L. Noirez, G.R. Luckhurst [*J. Chem. Phys. (USA)* vol.109 (1998) p.7612]

[5]   S.J. Picken [*Macromolecules (USA)* vol.23 (1990) p.467-70]

[6]   I. Haller, H.A. Huggins, H.R. Lilienthal, T.R. McGuire [*J. Phys. Chem. (USA)* vol.77 (1973) p.950]

[7]   R. Kiefer, G. Baur [*Mol. Cryst. Liq. Cryst. (UK)* vol.174 (1989) p.101]

[8]   E.H. Korte [*Mol. Cryst. Liq. Cryst. (UK)* vol.100 (1983) p.127]

[9]   G. Vertogen, W.H. de Jeu [*Liquid Crystals, Fundamentals* (Springer, Berlin, 1988)]

[10]   W.H. de Jeu, W.A.P. Claassen [*J. Chem. Phys. (USA)* vol.68 (1978) p.102]

[11]   F. Leenhouts, W.H. de Jeu, A.J. Dekker [*J. Phys. (France)* vol.40 (1979) p.989]

[12]   P.L. Sherrell, D.A. Crellin [*J. Phys. Colloq. (France)* vol.3 (1979) p.211]

[13]   J. Emsley, G.R. Luckhurst, G.W. Gray, A. Mosley [*Mol. Phys. (UK)* vol.35 (1978) p.1499]

[14]   L.G.P. Dalmolen [PhD Thesis, Groningen, 1984]

[15]   G.L. Hoatson, J.M. Goetz, P. Palffy-Muhoray, G.P. Crawford, J.W. Doane [*Mol. Cryst. Liq. Cryst. (UK)* vol.203 (1991) p.45]

[16]   M.Y. Kok, A.J. Van der Est, E.E. Burnell [*Liq. Cryst. (UK)* vol.3 (1988) p.485]

[17]   F. Grinberg, R. Kimmich [*J. Chem. Phys. (USA)* vol.103 (1995) p.365]

[18]   S. Jen, N.A. Clark, P.S. Pershan, E.B. Priestley [*J. Chem. Phys. (USA)* vol.66 (1977) p.4635]

[19]   L.G.P. Dalmolen, W.H. de Jeu [*J. Chem. Phys. (USA)* vol.78 (1983) p.7353]

[20]   R. Seeliger, H. Haspeklo, F. Noack [*Mol. Phys. (UK)* vol.49 (1983) p.1039]

[21]   L.G.P. Dalmolen, S.J. Picken, A.F. de Jong, W.H. de Jeu [*J. Phys. (France)* vol.46 (1985) p.1443]

[22]   S.J. Picken, W.F. van Gunsteren, P.Th. van Duynen, W.H. de Jeu [*Liq. Cryst. (UK)* vol.6 (1989) p.357]

[23]   E. Wolarz, D. Bauman [*Mol. Cryst. Liq. Cryst. (UK)* vol.197 (1991) p.1]

[24]   H. Herba, B. Zywucki, G. Czechowski, D. Bauman, A. Wasik, J. Jadzyn [*Acta Phys. Pol. A (Poland)* vol.87 (1995) p.985]

[25]   D. Bauman [*Proc. SPIE - Int. Soc. Opt. Eng. (USA)* vol.1845 (1993) p.354]

[26]   A. Würflinger [*Z. Naturforsch. A, Phys. Sci. (Germany)* vol.53 (1998) p.141-4]

[27]   S. Urban, D. Busing, A. Würflinger, B. Gestblom [*Liq. Cryst. (UK)* vol.25 (1998) p.253]

[28]   C.M. Haws, M.G. Clark, G.S. Attard [*Side Chain Liquid Crystal Polymers* Ed. C.B. McArdle (Blackie, Glasgow, 1989) ch.7]

[29]   S. Sen, K. Kali, S.K. Roy, S.B. Roy [*Mol. Cryst. Liq. Cryst. (UK)* vol.126 (1985) p.269]

[30]   B. Jérôme [*Mol. Cryst. Liq. Cryst. (UK)* vol.212 (1992) p.21]

[31]   P. le Barny et al [*Proc. Soc. Photo-Opt. Instrum. Eng. (USA)* vol.682 (1986) p.56]

[32]   C.P.J.M. van der Vorst, S.J. Picken [*J. Opt. Soc. Am. B (USA)* vol.7 (1990) p.320]

2.3 Quantitative representation of molecular shape and liquid crystal properties

A. Ferrarini, G.J. Moro and P.L. Nordio[*]

May 1999

A INTRODUCTION

It is well-known that the organisation of liquid crystal phases, as well as their existence, is strongly dependent on the shape of the constituent molecules and this has stimulated the search for empirical correlations between molecular shape and mesomorphic properties. Typical mesogenic molecules are made of a rigid moiety, in most cases an aromatic or alicyclic core, and one or more flexible chains. For molecules belonging to a homologous series with chains of differing lengths, strong odd-even effects are observed in the liquid crystal properties. In addition, the extension of flexible and rigid parts affects the relative stability of nematic and smectic phases. For chiral mesogenic molecules, twisted nematic and smectic phases are found.

Such a variety of behaviour can be traced back to the intermolecular short-range interactions, which are modulated by the molecular shape. This has been confirmed by theoretical investigations and simulations performed on model systems of simple shape, interacting with simplified anisotropic potentials. A complete theory for the correlation between molecular and phase properties, which would be useful also for application purposes, would require a detailed description of the attractive and repulsive interactions for realistic molecular structures. Starting from the recognition of the relevance of the molecular shape, as a first approximation, this correlation has been investigated using phenomenological approaches. Thus, ignoring the details of the intermolecular potentials, the anisotropy of the interactions in liquid crystal phases has been modelled by molecular field potentials accounting for the relevant features of the molecular shape. The first attempts in this direction were the segmental models proposed for flexible molecules, in which the single particle pseudopotential or potential of mean torque is written as a sum of contributions deriving from the rigid moieties constituting the molecule [1-3]. Subsequently other models, based on a more accurate description of the molecular shape, have been formulated (for a recent review see [4]). These phenomenological approaches start from different assumptions; thus, if the role of the repulsive interactions is emphasised, the single particle pseudopotential has an excluded-volume form [5], while a dependence on the molecular surface appears when the analogy with the attractive interactions as the origin of anchoring effects [6] or when the elastic response to deformations in liquid crystals [5] is exploited. Despite the differences, all of these models have been successfully used to predict orientational order parameters of solutes of different shape dissolved in nematic phases. The reason for the success can be attributed to the introduction of a specific shape dependence into the anisotropic pseudopotential in all of these models. In the following we shall present a summary of the results obtained with the surface tensor model, which has been used for various classes of molecules to predict, in addition to orientational order parameters, transition properties of nematics and the helical pitch of chiral nematics.

---

[*] Deceased, October 20th 1998

B   MODEL AND COMPUTATIONAL PROCEDURE

A summary of the relevant expressions of the model and its implementation, necessary to understand the reported results, will be presented here. A complete description and discussion of the theory can be found in [6-9].

The surface tensor model is based on the assumption that each surface element ds of a molecule tends to align parallel to the mesophase director **n**. In analogy with the expression for the anchoring free energy of nematics, the potential acting on a surface element is written as

$$dU = k_B T \varepsilon P_2 (\mathbf{n} \cdot \mathbf{s}) ds \tag{1}$$

where ε is a parameter controlling the strength of the orienting interactions, **s** is a unit vector perpendicular to the surface element and $P_2 (\cos\beta)$ is the second Legendre polynomial. After integration over the whole molecular surface, by using the addition theorem for spherical harmonics, the potential experienced by a molecule with orientation Ω can be written as

$$U(\Omega) = - k_B T \varepsilon \sum_m T^{(2,m)*} D_{0m}^2 (\Omega) \tag{2}$$

where $D_{0m}^2 (\Omega)$ are Wigner matrix components and $T^{(2,m)}$ are irreducible spherical components of a second rank tensor, called the surface tensor, defined as

$$\mathbf{T} = - 6^{-1/2} \int \rho_S(\omega_S)(3\mathbf{s} \otimes \mathbf{s} - 1) d\omega_S \tag{3}$$

In this expression $\rho_S(\omega_S)$ is the density of surface elements with orientation ω_S. It follows from the definition that the surface tensor is traceless and its elements measure the anisotropy of the surface elements' distribution. Thus, all elements of **T** vanish for a spherical surface; in the case of axially symmetric particles **T** is diagonal, with $T_{zz} = -2 T_{xx} = -2 T_{yy}$ in a frame with the z axis parallel to the C_∞ axis, and T_{zz} positive or negative for prolate (rod-like) and oblate (disc-like) shapes, respectively.

In chiral nematics there is a position dependent director field $\mathbf{n}(\mathbf{R})$, corresponding to a helical structure characterised by its handedness and pitch p, or by the wave vector **q** of magnitude $q = 2\pi/p$. In the limit $q \to 0$, the orientational potential experienced by a molecule in twisted nematics can be approximated as [7,8]

$$U(\Omega) = - k_B T \varepsilon \sum_m \left[T^{(2,m)*} - q Q^{(2,m)*} \right] D_{0m}^2 (\Omega) \tag{4}$$

where Ω is the molecular orientation in a laboratory frame with its Z axis along the local director. EQN (4) differs from EQN (2) because of the presence of $Q^{(2,m)*}$, which are irreducible spherical components of the second rank pseudo tensor **Q** called the helicity tensor, defined as

$$\mathbf{Q} = (3/8)^{1/2} \int_S \rho_S(\omega_S) \left[\mathbf{s} \otimes (\mathbf{s} \times \mathbf{r}) + (\mathbf{s} \times \mathbf{r}) \otimes \mathbf{s} \right] d\omega_S \tag{5}$$

where **r** is the vector position of the surface element ds in the molecular frame. Again **Q** is a traceless tensor, whose diagonal element Q_{ii} can be interpreted as the helicity of the molecular surface along the i axis.

A measurable property directly determined from the surface tensor **T** is the Saupe ordering matrix **S** [10]

$$\mathbf{S} = \frac{\frac{1}{2}\int(3\mathbf{n}\otimes\mathbf{n}-\mathbf{1})\exp\left[-U(\Omega)/k_BT\right]d\Omega}{\int\exp\left[-U(\Omega)/k_BT\right]d\Omega} \tag{6}$$

with the components of the unit vector **n** expressed in the molecular frame. Thus, by using the orientational potential in EQN (2) it is possible to predict the orientational order of different solutes in the same nematic solvent at a given temperature, that is at a given value of the parameter ε. This parameter in turn can be related to the orientational order parameters, and then to the shape of the solvent molecules, through a molecular field interpretation of the orientational potential in EQN (2).

The macroscopic pitch in twisted nematic phases is determined by the competition between bulk elastic forces and a chiral distortion contribution, associated with the term proportional to q in EQN (4). An expression for the helical twisting power β of chiral dopants in nematic solvents, defined as the inverse pitch for unit concentration of enantiomeric pure dopant [11], is obtained by minimising the q-dependent free energy which includes elastic and chiral contributions [7,8]

$$\beta = \frac{RT\varepsilon Q}{2\pi K_2 V_m} \tag{7}$$

where K_2 and V_m are the twist elastic constant and molar volume of the solvent, and Q, the chirality order parameter, is the average value of the helicity tensor

$$Q = -\sqrt{2/3}\,\mathbf{Q}\cdot\mathbf{S} \tag{8}$$

An expression analogous to EQN (8) can be used for the inverse pitch of pure phases, in which case K_2 and V_m are the twist elastic constant and molar volume of the chiral nematics.

EQNS (6) and (8) for orientational and chirality order parameters are appropriate for rigid molecules described by a single structure. But this is not the case for mesogenic species, which usually contain flexible moieties and can adopt various conformations. EQNS (6) and (8) can be generalised for such molecules, provided that order parameters are defined as averages over all possible conformers, with weights corresponding to the conformer populations.

Calculation of the surface and helicity tensors, and the consequent prediction of liquid crystal properties, requires the definition of the molecular surface. First, the molecular structure is needed: this can be derived from crystal structure data or calculated using molecular mechanical and quantum mechanical methods. For a given structure, the simplest choice of surface is obtained by considering the outer part of an array of van der Waals spheres centred at the nuclear position [8,9]. A more realistic description of the surface experienced by the surrounding molecules is provided by the rolling sphere algorithm through identification of the contours drawn by a sphere rolling on the van der Waals envelope [12,13].

C   RESULTS

In the following, molecular surface properties, as described by the **T** and **Q** tensors, and the corresponding liquid crystal properties, such as orientational and chirality order parameters, will be reported for some typical molecules.   Both mesogenic and non-mesogenic molecules will be considered: in the latter case order parameters should be viewed as appropriate for the molecules dissolved in ordinary calamitic phases.

For systems such as benzene and biphenyl, a simple molecular structure with regular hexagons and ordinary bond lengths is assumed; for more complex systems the structures are obtained by geometry optimisation with ab initio or semi-empirical methods, as reported in the quoted references.   In all cases the molecular surface was defined using standard van der Waals radii for atoms [14] and the rolling sphere algorithm with a sphere radius of 3 Å.   The surface obtained in this way for anthracene and two conformers of 4-pentyl-4'-cyanobiphenyl (5CB) is shown in FIGURE 1.

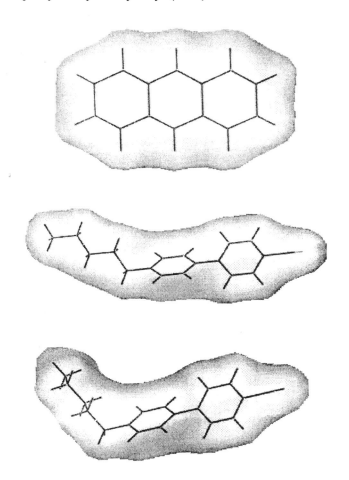

FIGURE 1 The surface obtained with the rolling sphere algorithm (with a sphere radius of 3 Å) for anthracene and for two isomers of 5CB with the alkyl chain in the ttt and g$^+$tt conformation.

Unless explicitly stated, the molecular axes are labelled in such a way that z and y correspond to the largest and the smallest diagonal elements of the **T** and **S** tensors, respectively. According to the definition of **T**, this means that y and z are the directions of highest and lowest surface density (recalling that the direction of a surface element is identified by its normal); from the point of view of **S** these are the directions of greatest and least tendency to alignment parallel to the director.

First of all, three simple rigid aromatic molecules will be considered: benzene, naphthalene and anthracene, whose principal **T** and **S** components are reported in TABLE 1. The order parameters of these molecules, as well as those of all other molecules listed in TABLE 1, were calculated by taking $\varepsilon = 0.04$ Å^{-2}, a value reasonable for typical nematic liquid crystals. As expected from the molecular shape, a continuous increase in the magnitude of the **T** tensor components and in the difference between the components perpendicular to the molecular plane is observed along the series. Correspondingly, a change in the ordering behaviour is predicted. Benzene can be approximated as a disk; it tends to align with the molecular plane parallel to the director, with no difference between the x and z axes. There is a significant difference between these two directions for the more elongated homologues, which show a net increase in order of the molecular long axis. The table also shows the **T** and **S** components for meta- and ortho-dichlorobenzene (MDCB and ODCB, respectively). The replacement of two hydrogens with bulkier chlorine atoms has the effect of increasing the magnitude of the tensor components and breaking the quasi-axial symmetry of benzene, especially in the case of the meta derivative. It can be seen that for this molecule the principal elements of **T**, if scaled with a factor of about 1.2, are close to those of naphthalene, with the main alignment axis lying in the molecular plane, perpendicular to the C_2 symmetry axis. In contrast, for ODCB, the axis of preferential alignment is parallel to the C_2 symmetry axis. Finally, the table reports the **T** and **S** tensors for the two 5CB conformers shown in FIGURE 1. Larger components with nearly axial symmetry are found in both cases. On the other hand, significant differences between the two conformers are observed, as could be expected from their molecular shapes.

TABLE 1 Principal components of the **T** tensor
and of the ordering matrix **S** for selected systems.

	$T_{xx}/\text{Å}^2$	$T_{yy}/\text{Å}^2$	$T_{zz}/\text{Å}^2$	S_{xx}	S_{yy}	S_{zz}
Benzene	9.7	-19.4	9.7	0.081	-0.162	0.081
Naphthalene	10.9	-36.3	25.7	0.030	-0.259	0.229
Anthracene	12.8	-54.8	42.0	-0.009	-0.281	0.290
MDCB	8.7	-31.5	22.7	0.026	-0.235	0.209
ODCB	12.3	-32.2	19.8	0.069	-0.237	0.168
5CB (ttt)	-18.9	-57.4	76.3	-0.308	-0.370	0.678
5CB (g^+tt)	-21.7	-29.8	57.5	-0.254	-0.279	0.533

If experimental order parameters are available, a comparison with the theoretical predictions can be made [6,12,15]. FIGURE 2 displays the difference $(S_{xx} - S_{yy})$ versus S_{zz} for MDCB and ODCB, as obtained from calculations [12] and from NMR measurements [16]. Order parameters for the C-D bonds of 5CB, derived from NMR experiments on deuterated derivatives [17], are compared with theoretical predictions in TABLE 2. In this case, the order parameters were obtained by averaging over 27 conformers, with structure and weights obtained from quantum mechanical calculations; ε was used as a fitting parameter [18].

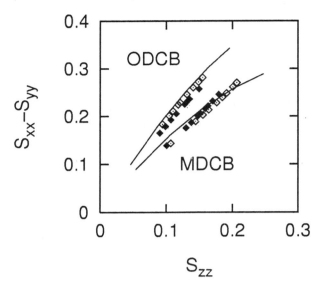

FIGURE 2 Calculated (continuous line) and experimental (squares) orientational order parameters for ortho-dichlorobenzene and meta-dichlorobenzene. Open and filled squares denote values measured in two different nematic solvents [16].

TABLE 2 S_{CD} order parameters for deuterons in the aromatic rings (R) and the alkyl chain (1-5 positions) of 5CB.

	R	1	2	3	4	5
Expt. [17]	-0.044	-0.202	-0.137	-0.147	-0.099	-0.072
Calc. [18]	-0.041	-0.200	-0.132	-0.144	-0.105	-0.078

The surface model can also be used, in the context of a molecular field approach, to predict properties at the nematic-isotropic (NI) transition [19]. The Helmholtz free energy is written as a sum of an energy contribution, according to the orientational potential in EQN (2), and an entropy term determined from the orientational distribution function. The transition is identified by equal values of the nematic and isotropic free energies.

FIGURE 3 shows orientational order parameters, entropy changes and transition temperatures for the nematic-isotropic transition in two series, α,ω-bis(4'-cyanobiphenyl-4-yloxy)alkanes and α,ω-bis(4'-cyanobiphenyl-4-yl)alkanes [19]. These are examples of the so-called liquid crystal dimers, or twin systems, which are characterised by a common structure with two mesogenic units joined by a flexible spacer. These compounds show unusual transitional properties and strong odd-even effects. Calculations were performed by averaging over the rotational isomeric state conformers, which totalled 35,000 for the ether dimer with ten groups in the spacer (N = 10). In order to appreciate the results, it should be remembered that the difference between ether and methylene dimers lies in the geometry of the connection between flexible spacers and aromatic rings: the C_{phenyl}-O-CH_2 bonds lie in the plane of the ring with a bond angle close to 125°, while the C_{phenyl}-CH_2-CH_2 segment is perpendicular to the ring with a bond angle of about 112°. It should also be noted that both biaxiality and flexibility effects are taken into account by the model, but the packing entropy contribution is neglected. The results can be compared with the Maier-Saupe model, which predicts $S_{zz}^{NI} = 0.429$ and $\Delta S_{NI}/R = 0.417$ [10].

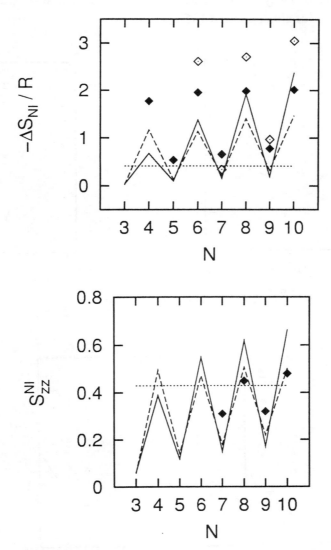

FIGURE 3 Entropy change and order parameter S_{zz} at the nematic-isotropic transition for α,ω-bis(4'-cyanobiphenyl-4-yloxy)alkanes (dashed line: calculated; filled diamonds: experimental) and α,ω-bis(4'-cyanobiphenyl-4-yl)alkanes (solid line: calculated; open diamonds: experimental) as functions of the number of units in the spacer [19]. Predictions of the Maier-Saupe theory are represented by the horizontal line. The z axis is taken along the CN bond.

In order to discuss chiral properties and their relation to the molecular shape, it is worth considering in some detail the biphenyl molecule, which is chiral when the twist angle between aromatic rings is different from 0° and 90°. FIGURE 4 shows the components of the Saupe matrix **S**, calculated with $\varepsilon = 0.04$ Å^{-2}, of the chirality tensor **Q**, and of the chirality order parameter Q, as functions of the twist angle ϑ. The angle is defined in such a way that positive values correspond to P (right-handed) configurations for $\vartheta < 90°$ (M, left-handed, for $\vartheta > 90°$). It can be seen from the figure that the molecule tends to align with the para axis parallel to the director; the two perpendicular axes have a different tendency to align, unless the twist angle is equal to 90°. A strong angular dependence of the helicity tensor components is observed. The x and y components of the **Q** tensor are opposite in sign and very close in magnitude. The component Q_{zz}, which represents the helicity along the para axis, is always small and changes sign at 90°, in keeping with the change of configuration from P to M. The chirality order parameter is antisymmetric with respect to reflection at 90°, and reaches its greatest magnitude for twist angles of about 45° and 135°.

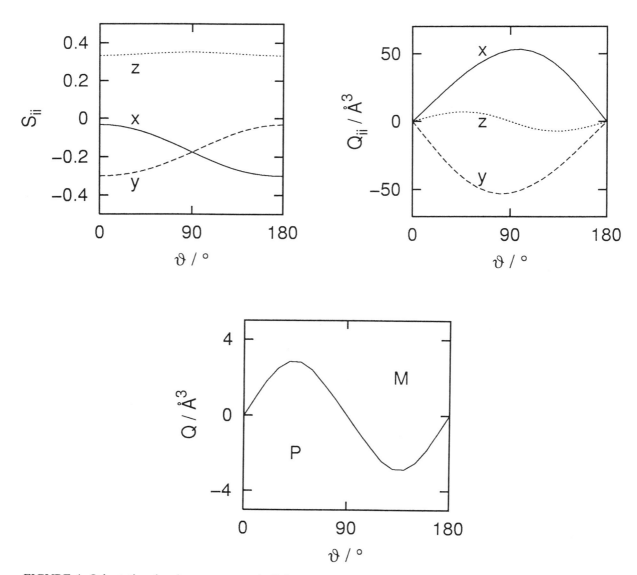

FIGURE 4 Orientational order parameters, helicity tensor components and chirality order parameter calculated for biphenyl as functions of the twist angle between the aromatic rings. The z and y axes are taken along the para direction and parallel to the line bisecting the twist angle, respectively.

Biphenyl is a simple example which can be used for illustrative purposes; however, it does not behave as a chiral molecule, because enantiomeric forms are present with equal weight, unless the twist angle is constrained by introducing bulky substituents or by connecting the rings with a bridge. The helical pitch of various twisted nematic phases has been successfully predicted by the surface model [9,20-22]. A comparison between theoretical results and experimental data for some selected chiral dopants is presented in TABLE 3. Because of their simple rigid structure and high dissymmetry, these are suitable examples with which to illustrate the correlation between molecular shape and liquid crystal chirality. The correlation would certainly not appear so clearly from the analysis of chiral mesogens, which are mixtures of conformers often not strongly dissymmetric. The data reported in the table are measured helical twisting powers and calculated chirality order parameters. According to EQN (7), the prediction of the twisting power requires knowledge of solvent properties which is often not available, especially for commercial mixtures. The proportionality factor between β and Q can be reasonably estimated to be of the order of unity for common nematics under ordinary conditions, and in particular it should not change significantly for similar solvents to those used for the reported measurements. Therefore, the trend of Q values is expected to reflect that of the corresponding twisting powers. All of the reported data refer to P enantiomers (the twist angle of biphenyl and binaphthol derivatives is about 50° - 60°). Although no general relation exists between absolute configuration of a dopant and handedness of the induced chiral nematic phase, molecules with very similar structure and the same configuration are expected to produce chiral nematic phases with the same sense of twist. As appears from the table, this occurs for the binaphthol derivatives **IV** and **V**. In contrast, in spite of their close similarity, the biphenyl derivatives **I**, **II** and **III** have a completely different behaviour. It can be seen that the surface tensor model, taking into account the complex interplay of ordering and chiral effects, is able to explain the experimental findings and to provide good estimates for the helical twisting power.

TABLE 3 Calculated chirality order parameters and experimental
twisting powers for chiral dopants (P configuration).

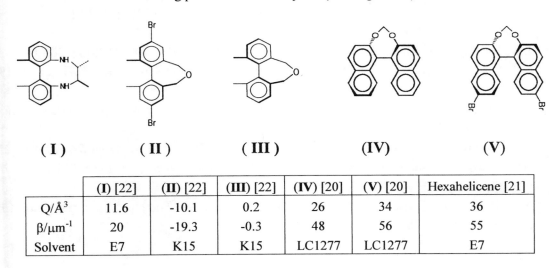

| (I) | (II) | (III) | (IV) | (V) |

	(I) [22]	**(II)** [22]	**(III)** [22]	**(IV)** [20]	**(V)** [20]	Hexahelicene [21]
Q/Å³	11.6	-10.1	0.2	26	34	36
$\beta/\mu m^{-1}$	20	-19.3	-0.3	48	56	55
Solvent	E7	K15	K15	LC1277	LC1277	E7

D   CONCLUSION

The surface model describes the anisotropy of the interactions of a molecule with a liquid crystal environment in terms of the second rank surface and helicity tensors, which account for the anisometry and the helicity of the molecular surface, respectively. In this way it is possible to relate the molecular ordering and transitional properties of nematics, as well as the helical pitch of pure and induced chiral nematics, to the structural features of the molecule.

REFERENCES

[1]   S. Marcelja [*Nature (UK)* vol.241 (1973) p.451]

[2]   J.W. Emsley, G.R. Luckhurst, C.P. Stockley [*Proc. R. Soc. Lond. A (UK)* vol.381 (1982) p.117]

[3]   D.J. Photinos, E.T. Samulski, H. Toriumi [*J. Phys. Chem. (USA)* vol.94 (1990) p.4688]

[4]   E.E. Burnell, C.A. de Lange [*Chem. Rev. (USA)* vol.98 (1998) p.2359 and references therein]

[5]   A.F. Terzis et al [*J. Am. Chem. Soc. (USA)* vol.118 (1996) p.2226]

[6]   A. Ferrarini, G.J. Moro, P.L. Nordio, G.R. Luckhurst [*Mol. Phys. (UK)* vol.77 (1992) p.1]

[7]   A. Ferrarini, G.J. Moro, P.L. Nordio [*Phys. Rev. E (USA)* vol.53 (1996) p.681]

[8]   A. Ferrarini, G.J. Moro, P.L. Nordio [*Mol. Phys. (UK)* vol.87 (1996) p.485]

[9]   L. Feltre, A. Ferrarini, F. Pacchiele, P.L. Nordio [*Mol. Cryst. Liq. Cryst. (Switzerland)* vol.290 (1996) p.109]

[10]   C. Zannoni [in *The Molecular Physics of Liquid Crystals* Eds. G.R. Luckhurst, G.W. Gray (Academic Press, London, 1979) p.51]

[11]   G. Solladié, R. Zimmermann [*Angew. Chem., Int. Ed. Engl. (Germany)* vol.23 (1984) p.348]

[12]   A. Ferrarini, F. Janssen, G.J. Moro, P.L. Nordio [*Liq. Cryst. (UK)* vol.26 (1999) p.201]

[13]   M.L. Connolly [*Science (USA)* vol.221 (1983) p.709]

[14]   A. Bondi [*J. Phys. Chem. (USA)* vol.68 (1964) p.441]

[15]   G. Celebre, G. De Luca, A. Ferrarini [*Mol. Phys. (UK)* vol.92 (1997) p.1039]

[16]   T. Chandrakumar, E.E. Burnell [*Mol. Phys. (UK)* vol.90 (1997) p.303]

[17]   J.W. Emsley, G.R. Luckhurst, C.P. Stockley [*Mol. Phys. (UK)* vol.44 (1981) p.565]

[18]   C.J. Adam, A. Ferrarini, M.R. Wilson, G.J. Ackland, J. Crain [*Mol. Phys. (UK)* vol.97 (1999) p.541]

[19]   A. Ferrarini, G.R. Luckhurst, P.L. Nordio, S.J. Roskilly [*J. Chem. Phys. (USA)* vol.100 (1994) p.1460]

[20]   A. Ferrarini, P.L. Nordio, P.V. Shibaev, V.P. Shibaev [*Liq. Cryst. (UK)* vol.24 (1998) p.219]

[21]   A. Ferrarini, G. Gottarelli, P.L. Nordio, G.P. Spada [*J. Chem. Soc. Perkin Trans. II (UK)* (1999) p.411]

[22]   N. Dal Mas, A. Ferrarini, P.L. Nordio, P. Styring, S.M. Todd [*Mol. Cryst. Liq. Cryst. (Switzerland)* vol.328 (1999) p.391]

2.4 Measurements and calculation of dipole moments, quadrupole moments and polarisabilities of mesogenic molecules

S. Clark

February 1999

A INTRODUCTION

Measurements and calculations concerning the dipole and quadrupole moments of a selection of molecules that form nematic liquid crystals are presented here. We report results on the more common nematics, namely, cyanobiphenyls, phenylcyclohexanes, bicyclohexanes, terphenyls and phenylbenzoates. Their polarisabilities and anisotropic polarisabilities are also given. In addition we give a brief overview of the theoretical and experimental methods used to obtain these quantities.

FIGURE 1 illustrates the structure of the molecules that are reviewed here and introduces the nomenclature that will be used. In each molecule considered, it will have an alkyl tail C_nH_{2n+1} that contains n carbon atoms. In the majority of cases the molecule will contain a cyano group which is the main source of dipole moment in mesogenic molecules, although halide substituted hydrogens also contribute to the overall electric multipole moments of the system.

B MOLECULAR DIPOLE MOMENTS

The dipole moment of a molecule is a measure of the difference in the relative position of nuclear and electronic charge. If the centre of positive and negative charge (an analogous definition to that of centre of mass) is at r_+ and r_- respectively, then the molecular dipole is defined as

$$\mu = q(r_+ - r_-)$$

However, molecular dipole measurements of anything but the smallest gas phase molecules are extremely difficult. In fact, the molecular dipole of a molecule in a bulk system such as a nematic liquid crystal cannot be measured directly, but instead it has to be inferred from various optical and electrical measurements of the bulk system. Several assumptions have to be made in order to do this because the electrostatic interactions are large in nematic systems. They cause considerable dipole-dipole interactions, which are a major contribution to the internal field of the bulk materials, making it extremely difficult to obtain the dipole moment of a single molecule.

Knowledge of the dipole moment of mesogenic molecules is, however, extremely important since it is a major contributing factor to the stability of the liquid crystal phase. This is demonstrated in many computer simulations [1-8] of model liquid crystal materials based on ellipsoids and spherocylinders. These use parametrised potentials (probably the most common being the Gay-Berne potential) that include dipole-dipole interactions.

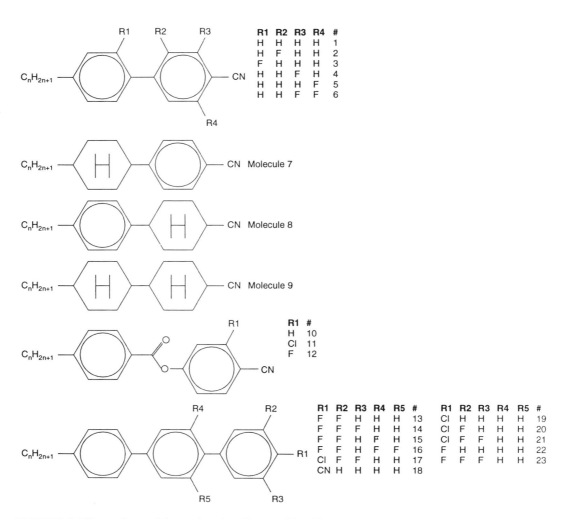

FIGURE 1 Illustrations of the molecules discussed in this Datareview. From top to bottom they are cyanobiphenyls, phenylcyclohexanes, bicyclohexanes, phenylbenzoates and terphenyls. For brevity, they will be referenced by the number shown and by the length of the alkyl chain n.

There are several methods which relate the molecular dipole moment to dielectric permittivities and refractive indices of the bulk material; these include the Guggenheim-Smith [9] method where the system under consideration is dissolved in a solvent at concentration (weight fraction) W. The dipole is obtained from

$$\mu = \frac{3k_B T}{5\pi N_A} \frac{M}{\rho(\varepsilon+2)^2} \left(\frac{\varepsilon_1 - \varepsilon}{W} - \frac{n_1^2 - n^2}{W} \right)$$

where ρ is the sample density, ε and ε_1 are the dielectric permittivities, T is the absolute temperature, M is the sample molecular weight, and n and n_1 are refractive indices of the solution and solvent, respectively. Similar measurements [10] can also be used to give an estimate of the molecular dipole using the Kirkwood correlation factor, g, which relates the molecular dipole to the effective dipole,

$$\mu_{eff}^2 = \mu^2 g$$

114

where

$$\mu_{\text{eff}}^2 = \frac{\left(\varepsilon - n^2\right)\left(2\varepsilon + n^2\right)}{\varepsilon\left(n^2 + 2\right)^2} \frac{9\varepsilon_0 k_B T}{(\rho/M)N_A}$$

Again, dipoles are determined from dilute solutions of the liquid crystalline material. A fuller description of the theory can be found in [11]. In several of the articles discussed here, the dipole moments were determined by extrapolating refractive index and permittivity measurements of dilute solutions of the molecules under consideration using the method of Raynes [12].

A fairly accurate measurement of the dipole can also be obtained using the Maier-Meier formulae [13] where the mean effective values of the dipole moment in the parallel and perpendicular directions to the director can be obtained from the dielectric functions, density and refractive indices. This has been done for a range of nematogens in [14], and is reported in TABLES 2, 3 and 4.

In each of these experimental methods to determine the molecular dipole, the errors involved in measuring the required permittivities and refractive indices are relatively small compared to the approximations that are required to relate the measurements to the dipole. Each method described is an approximate model of the dipole interactions that are involved within the actual material; this is the main source of error in determining the molecular dipole moment.

An alternative method of determining the molecular dipole moment of mesogenic materials is to use various computational techniques to obtain the molecular charge distribution where the dipole can then be calculated directly. From the atomic positions within a molecule, the electronic distribution must be determined. Since the molecules that form liquid crystals tend to have a highly complex electronic structure, a reasonably sophisticated method must be used - these range from semi-empirical calculations [15,16] through to the highly computationally intensive first principle ab initio methods [17]. In these methods, an attempt to solve the many-body Schrödinger equation for the molecule is made. An introduction to these methods can be found in [18] and [19]. These theoretical methods have shown varying degrees of success depending on the approximations of the model used. The most popular methods of calculating molecular dipoles are at the semi-empirical level using various commercial packages such as MOPAC. These give a reasonable measure of the dipole moment [20] for only a modest use of computer time. However, the electronic structure of molecules which form liquid crystals is rather complicated and therefore more accurate ab initio methods are often employed [18].

In TABLE 1 we report the dipole moments of several cyanobiphenyl molecules that have been obtained using these methods. Possibly the most commonly studied nematogen is 4-pentyl-4'-cyanobiphenyl (5CB). The results illustrate the problems of obtaining the dipole for such a large molecule in a complex fluid. It can be seen that the reported dipole moments vary over a considerable range.

TABLE 1 The molecular dipole moments of selected cyanobiphenyl molecules obtained by experimental and theoretical techniques. 1 Debye (D) = 3.33564 × 10^{-30} C m.

Molecule	Alkyl tail (n)	μ/D	Ref
1 (5CB)	5	4.8	[10]
1 (5CB)	5	5.2	[22]
1 (5CB)	5	4.7	[2]
1 (5CB)	5	3.9	[16]
1 (5CB)	5	3.21	[14]
1 (6CB)	6	3.24	[14]
1 (7CB)	7	3.15	[14]
1 (8CB)	8	3.09	[14]
2 (F-5CB)	5	4.2	[10]
3 (F-5CB)	5	5.2	[10]
4 (F-5CB)	5	5.6	[10]
5 (F-5CB)	5	4.8	[16]
6 (F-5CB)	5	5.4	[16]

The dielectric properties of 5CB have also been studied as a function of pressure [21]. Good results are obtained by assuming that the molecular dipole remains unchanged as the pressure varies.

In TABLE 1, the dipoles for the nCB homologous series are reported for n = 5, ..., 8. The cyano group is the main contributor to the molecular dipole with the electronic charge displaced towards the N atom. However, alkyl chains can usually be thought of as weak electron donors when bonded to phenyl groups and therefore it may be expected that the dipole moment should increase slowly with increasing chain length. However, that is not found to be the case [14] where no particular trend is shown. It is likely that the expected increase could be smaller than the accuracy of the measurements.

McDonnel et al [10] report the dipole moments of several fluorinated alkyl-cyanobiphenyls which are also given in TABLE 1. Fluorine is highly electronegative which creates a localised dipole along the C-F bond. The effect on the overall dipole moment of the molecule is to either increase or decrease it as expected, depending on the relative orientations of the C-F and C≡N bonds. The molecular dipoles of many other fluorinated liquid crystal forming heterocyclic and benzoic esters are also reported in [10].

In TABLES 2, 3 and 4 we show similar quantities for the cyclohexane substituted molecules, the phenylbenzoates and the terphenyls, respectively. Again the halide substitutions contribute to the overall molecular dipole moment in an obvious sense, but there is no specific trend through a particular homologous series.

TABLE 2 The dipole moments for a selection of nematic
liquid crystals containing cyclohexane cores.

Molecule	Alkyl tail (n)	μ/D	Ref
7 (5HexPhCN)	5	4.14	[2]
7 (5HexPhCN)	5	2.76	[14]
7 (7HexPhCN)	7	2.61	[14]
7 (10HexPhCN)	10	2.70	[14]
8 (5PhHexCN)	5	3.63	[2]
9 (5HexHexCN)	5	3.60	[2]

TABLE 3 The dipole moments for a selection of phenylbenzoate
molecules that form nematic liquid crystals.

Molecule	Alkyl tail (n)	μ/D	Ref
10	6	4.47	[14]
10	7	4.44	[14]
10	8	4.44	[14]
10	9	4.32	[14]
10	10	3.96	[14]
10	3	5.8	[10]
11	3	6.4	[10]
12	3	6.3	[10]
12	3	7.2	[16]
12	5	7.1	[16]
12	2	6.6	[10]
12	3	6.6	[10]
12	5	6.6	[10]
12	7	6.6	[10]

TABLE 4 The dipole moments for a selection of terphenyl molecules.

Molecule	Alkyl tail (n)	μ/D	Ref
13	3	3.4	[16]
14	3	4.2	[16]
15	3	3.9	[16]
16	3	4.7	[16]
17	5	3.6	[16]
18	3	3.9	[16]
19	3	1.7	[16]
20	3	2.8	[16]
21	3	3.2	[16]
22	3	2.0	[16]
23	3	3.8	[16]

C   QUADRUPOLES

Recently it has become clear that higher order multipolar interactions, such as the molecular quadrupole moments, in liquid crystals play an important role in determining the phase behaviour of the materials.   The quadrupole moment of a molecule essentially measures the deviation from spherical symmetry of the charge distribution.   There is no unique definition of quadrupole moment (for dipolar molecules) since it is origin dependent.   However, here we will use the most commonly used definition

$$\mathbf{Q} = \frac{1}{2}\begin{pmatrix} \sum_i q_i\left(3x_i^2 - r_i^2\right) & 3\sum_i q_i x_i y_i & 3\sum_i q_i x_i z_i \\ 3\sum_i q_i x_i y_i & \sum_i q_i\left(3y_i^2 - r_i^2\right) & 3\sum_i q_i y_i z_i \\ 3\sum_i q_i x_i z_i & 3\sum_i q_i y_i z_i & \sum_i q_i\left(3z_i^2 - r_i^2\right) \end{pmatrix}$$

where the index i runs over all nuclear and electronic charges, q_i, located at coordinates x_i, y_i, z_i, a distance r_i from a chosen origin.   It is usual to diagonalise the quadrupole tensor and report the eigenvalues Q_{xx}, Q_{yy} and Q_{zz}, which is then the quadrupole moment with respect to the molecular axes given by the eigenvectors.

It is not yet possible, experimentally, to determine the higher order multipoles of molecules that form mesophases due to their large size and complexity.   However, there have been a large number of computer simulations of such phases including quadrupolar interactions.   These usually take the form of an extension to the standard Gay-Berne model by including quadrupolar interactions on spherocylinders or soft spheres usually considered in such models.   It has been shown [23], for example, that the quadrupolar interaction in nematics shifts the isotropic to ferroelectric phase and ferroelectric to solid phase transitions to higher densities compared to results which only include dipole interactions.   There is now growing interest in constructing Gay-Berne models beyond the simple dipole-dipole approximation to include higher order multipolar terms [4,24-26].   However, since there is very little experimental data on molecular quadrupoles of mesogenic molecules many of the values used have been crudely estimated by adding quadrupoles of smaller molecules.   The importance of quadrupolar interactions in nematics has been studied in detail by Burnell and de Lange [37] primarily through measurements of solute orientational order.

Recently however, the quadrupole of 5CB [18] has been reported, where it has been calculated from the electronic structure obtained from first principles calculations, as described previously.   It was found that the quadrupole of 5CB, with the origin taken at the centre of mass, in units of 10^{-40} C m^2, is

$$Q_{5CB} = \begin{pmatrix} -159.0 & -124 & 0.2 \\ -12.4 & 106.8 & -5.8 \\ 0.2 & -5.8 & 52.3 \end{pmatrix}$$

Bates and Luckhurst [26] have successfully performed a molecular dynamics simulation using quadrupolar interactions of discotic systems where the (uniaxial) quadrupole was estimated from benzene and naphthalene quadrupoles measured at -30×10^{-40} C m^2 and -45×10^{-40} C m^2 respectively.

D   POLARISABILITIES

A static electric field, **E**, when applied across a system changes its electronic structure inducing a dipole $\mu_1 - \mu_0$, where μ_1 and μ_0 are the molecular dipoles with and without the field, respectively. This induced dipole can be expressed as a power series

$$\mu_1 - \mu_0 = \alpha_{ij} E_j + \frac{1}{2} \gamma_{ijk} E_j E_k + \dots$$

where α_{ij} is known as the polarisability tensor and the indices i, j and k run over the Cartesian x, y and z axes. The higher order terms, γ, are known as hyperpolarisabilities, which we will not discuss here. In a similar manner to the quadrupole moment, the polarisability is usually expressed in the molecular frame of reference, which is obtained from the eigenvectors of the polarisability tensor with eigenvalues (principal polarisabilities) α_{xx}, α_{yy} and α_{zz}.

Polarisability measurements are more straightforward than that of the molecular dipole. However, the full polarisability tensor cannot be measured. The trace

$$\alpha_T = \frac{1}{3}\left(\alpha_{xx} + \alpha_{yy} + \alpha_{zz}\right)$$

is readily obtained from refractive index measurements on isotropic fluids.

We can assume that the molecule under consideration has a long molecular axis in the direction of α_{xx} and two shorter molecular axes in the orthogonal directions (or one short and two long axes for discotic materials). In nematic materials, it is often assumed that the polarisabilities along two directions are equal and so the polarisability tensor is

$$\alpha = \begin{pmatrix} \alpha_{xx} & 0 & 0 \\ 0 & \alpha_{yy} & 0 \\ 0 & 0 & \alpha_{yy} \end{pmatrix}$$

with respect to the principal molecular axes. This is usually an adequate approximation since in nematic liquid crystals there is one preferred direction along which the molecule aligns, leaving it free to rotate about that axis. For this reason it is not usually possible to experimentally distinguish between the α_{yy} and α_{zz} directions but only obtain an average.

In recent years it has been common practice to obtain the polarisabilities of mesogenic molecules by assuming that they are equal to the addition of bond or fragment polarisabilities [27]. In this, it is assumed that each bond contributes a small amount to the overall molecular polarisability but is not affected by its surrounding environment. For example, in TABLE 5 we give a selection of polarisabilities that can be used in this way to calculate the molecular polarisability.

TABLE 5 Bond or fragment polarisabilities from [27] which can be used to calculate
the overall molecular polarisability of many mesogenic molecules.

Bond or fragment	$\alpha_{xx}/\text{Å}^3$	$\alpha_{yy}/\text{Å}^3$	$\alpha_{zz}/\text{Å}^3$
C-H	0.65	0.65	0.65
C-C	0.97	0.26	0.26
C-C≡N	40.3	1.54	1.54
C_6H_6	11.15	11.15	7.44
$C_6H_5C≡N$	16.27	11.34	8.30

A computational determination of the molecular polarisability can be used to find the total tensor. There are many methods available ranging from simple empirical charge displacements, summing the polarisabilities of individual bonds [27], up to the more reliable but computationally intensive ab initio methods [17].

Experimentally, it is possible to extract the anisotropy defined as

$$\Delta\alpha = (\alpha_{xx} - \alpha_{yy})$$

using methods such as Rayleigh scattering [28], where local fluctuations in the anisotropic part of the molecular polarisability tensor give rise to a depolarised component of the scattered light.

Molecular polarisabilities cannot be measured directly, but can only be calculated using theoretical techniques that model the internal field of the material. For this reason there is a wide variety of different values leading to a rather confusing situation. In the results given here we present experimental values obtained with a range of models.

There are many alternative methods available where the molecular polarisabilities can be extracted from various experimental measurements of optical properties. An extensive review of optical properties can be found in [29]. Similar to the case of dipole measurement, the main sources of error involved are the approximations that are required to construct a model which relates the molecular polarisabilities to measured quantities such as the refractive index or dielectric functions [30-32]. According to the Saupe-Maier theory, the polarisability of the bulk liquid crystal depends on the individual components α_{xx} and α_{yy} of the molecular polarisability and the orientational order of the molecules, described by the order parameter, S, through the relations [31]

$$\tilde{\alpha}_\| = \overline{\alpha} + \frac{2}{3}(\alpha_{xx} - \alpha_{yy})S$$

$$\tilde{\alpha}_\perp = \overline{\alpha} - \frac{1}{3}(\alpha_{xx} - \alpha_{yy})S$$

where $\tilde{\alpha}_\|$ and $\tilde{\alpha}_\perp$ are the bulk polarisability components parallel and perpendicular to the director, respectively, and $\overline{\alpha}$ is the mean molecular polarisability. The average polarisability can then be expressed by

2.4 Dipole moments, quadrupole moments and polarisabilities of mesogenic molecules

$$\overline{\alpha} = \frac{\left(\widetilde{\alpha}_\parallel + 2\widetilde{\alpha}_\perp\right)}{3} = \frac{\left(\alpha_{xx} + 2\alpha_{yy}\right)}{3}$$

such that

$$S = \frac{\left(\widetilde{\alpha}_\parallel - \widetilde{\alpha}_\perp\right)}{\left(\alpha_{xx} - \alpha_{yy}\right)}$$

However, one of the more popular methods of determining the molecular polarisabilities is the Vuks [33] model. This relates the main refractive indices $\tilde{n}_e$ and $\tilde{n}_o$ to $\widetilde{\alpha}_\parallel$ and $\widetilde{\alpha}_\perp$ by

$$\frac{\tilde{n}_e^2 - 1}{n^2 + 2} = \frac{4}{3}\pi N\widetilde{\alpha}_\parallel$$

$$\frac{\tilde{n}_o^2 - 1}{n^2 + 2} = \frac{4}{3}\pi N\widetilde{\alpha}_\perp$$

where N is the number of molecules per unit volume and

$$n^2 = \frac{\tilde{n}_e^2 + 2\tilde{n}_o^2}{3}$$

A full description of the theoretical background to these methods can be found in [31].

In TABLE 6 we present a selection of molecular polarisabilities for nematogens. It can be seen that there is still a large discrepancy in the values reported using various methods, but certain trends can be recognised. As the phenyl core groups are replaced by cyclohexane we see that the polarisability only decreases slightly. Also, as the alkyl tails increase in length, the polarisabilities rise by only a modest amount.

There have been very few calculations to obtain the full polarisability tensor of mesogenic molecules; however, recently results have been obtained from ab initio calculations [17]. The polarisability for 5CB, in $Å^3$, is

$$\alpha_{5CB} = \begin{pmatrix} 66.15 & -2.65 & -0.90 \\ -2.65 & 26.67 & -2.07 \\ -0.90 & -2.07 & 27.64 \end{pmatrix}$$

where the long axis is in the x direction. It also shows that the approximation of making the y and z polarisabilities equal is reasonable in this case. These values agree well with experimental values in TABLE 5 with $\mathrm{Tr}(\alpha)/3 = 40.1 \ Å^3$.

TABLE 6 Molecular polarisabilities. In some references the polarisability is presented over a range of temperatures. In those cases, the average value is given. It is found that the polarisabilities vary by less than 1% over the nematic range.

Molecule	Alkyl tail (n)	$\alpha/\text{Å}^3$	$\Delta\tilde{\alpha}/\text{Å}^3$	Ref
1	7	109.3	-	[34]
8	7	108.0	-	[34]
9	7	107.9	-	[34]
1	5	33.8	17.5	[2]
7	5	32.6	14.4	[2]
8	5	32.6	11.0	[2]
9	5	32.0	11.3	[2]
10	3	32.4	17.6	[35]
10	4	34.3	18.0	[35]
10	5	36.1	20.4	[35]
10	6	37.9	21.2	[35]
10	7	39.8	21.4	[35]
10	8	41.6	20.3	[35]
1	5	33.7	17.6	[36]
1	7	38.1	15.8	[36]

E   CONCLUSION

Experimentally, the dipole moments and molecular polarisabilities are not measured directly, but instead a model is constructed that relates dipoles and polarisabilities to optical properties such as permittivities and refractive indices of the bulk liquid crystal. These models are the main source of disagreement between various experimental values.

It should also be noted that in making a comparison between experiment and theory we have to be careful in comparing like with like. For example, experimentally in the bulk material the molecules are at finite temperature and are able to explore a large part of their conformational phase space. In the theoretical calculations, when constructing the electronic charge density a single molecular conformation is usually taken (and in all cases considered here, the all-trans conformation is the one that is reported). However, with the onset of large-scale, accurate ab initio calculations, experiment and theory are reporting similar results.

REFERENCES

[1]   S.C. McGrother, A. Gil-Villegas, G. Jackson [*J. Phys., Cond. Mat. (UK)* vol.8 (1996) p.9649]

[2]   D.A. Dunmur, A.E. Tomes [*Mol. Cryst. Liq. Cryst. (UK)* vol.97 (1983) p.241]

[3]   K. Toriyama, D.A. Dunmur, S.E. Hunt [*Liq. Cryst. (UK)* vol.5 (1989) p.1001]

[4]   C. Vega, S. Lago [*J. Chem. Phys. (USA)* vol.100 (1994) p.6727]

[5] J.J. Weis, D. Levesque, G.J. Zarragiocoechea [*Mol. Phys. (UK)* vol.80 (1993) p.1077]

[6] P.G. Kusalik [*Mol. Phys. (UK)* vol.81 (1994) p.199]

[7] S. Hauptmann, T. Mosell, S. Reiling, J. Brickmann [*Chem. Phys. (Netherlands)* vol.208 (1996) p.57]

[8] S.C. McGrother, A. Gil-Villegas, G. Jackson [*Mol. Phys. (UK)* vol.95 (1998) p.657]

[9] E.A. Guggenheim, J.W. Smith [*Trans. Faraday Soc. (UK)* vol.45 (1949) p.349]

[10] D.G. McDonnel, E.P. Raynes, R.A. Smith [*Liq. Cryst. (UK)* vol.6 (1989) p.515]

[11] D.A. Dunmur, P. Palffy-Muhoray [*Mol. Phys. (UK)* vol.76 (1992) p.1015]

[12] P.E. Raynes [*Mol. Cryst. Liq. Cryst. Lett. (UK)* vol.3 (1985) p.69]

[13] W. Maier, G. Meier [*Z. Nat.forsch A (Germany)* vol.16 (1961) p.262]

[14] Z. Rasewski et al [*Mol. Cryst. Liq. Cryst. (UK)* vol.251 (1994) p.357]

[15] M. Kodaka, S.N. Shah, T. Tomohiro, N.K. Chudgar [*J. Phys. Chem. (USA)* vol.102 (1998) p.1219]

[16] M. Klasen, M. Bremer, A. Gotz, A. Manabe, S. Maemura [*Jpn. J. Appl. Phys. (Japan)* vol.37 (1998) p.L945]

[17] S.J. Clark, G.J. Ackland, J. Crain [*Europhys. Lett. (France)* vol.44 (1998) p.578]

[18] S.J. Clark, C.J. Adam, G.J. Ackland, J. White, J. Crain [*Liq. Cryst. (UK)* vol.22 (1997) p.469]

[19] S.J. Clark, C.J. Adam, D.J. Cleaver, J. Crain [*Liq. Cryst. (UK)* vol.22 (1997) p.477]

[20] D.A. Dunmur, M. Grayson, B.T. Pickup, M.R. Wilson [*Mol. Phys. (UK)* vol.90 (1997) p.179]

[21] H.G. Kreul, S. Urban, A. Würflinger [*Phys. Rev. A (USA)* vol.45 (1992) p.8624]

[22] H. Furuya, S. Okamoto, A. Abe, G. Petekidis, G. Fytas [*J. Phys. Chem. (USA)* vol.99 (1995) p.6483]

[23] D. Wei [*Mol. Cryst. Liq. Cryst. (Switzerland)* vol.269 (1995) p.89]

[24] V.M. Kaganer, M.A. Osipov [*J. Chem. Phys. (USA)* vol.109 (1998) p.2600]

[25] M.P. Neal, A.J. Parker, C.M. Care [*Mol. Phys. (UK)* vol.91 (1997) p.603]

[26] M.A. Bates, G.R. Luckhurst [*Liq. Cryst. (UK)* vol.24 (1998) p.229]

[27] S.Y. Yakovenko, A.A. Muravski, F. Eikelschulte, A. Geiger [*J. Chem. Phys. (USA)* vol.105 (1996) p.10766]

[28] G. Floudas, A. Patowski, G. Fytas, M. Ballauff [*J. Chem. Phys. (USA)* vol.94 (1990) p.3215]

[29] S. Elston, R. Sambles (Eds.) [*The Optics of Thermotropic Liquid Crystals* (Taylor & Francis Ltd, 1998)]

[30] M. Mitra, S.S. Roy, T.P. Majumder, S.K. Roy [*Mol. Cryst. Liq. Cryst. (Switzerland)* vol.301 (1997) p.363]

[31] A. Hauser, G. Pelzl, C. Selbmann, D. Demus [*Mol. Cryst. Liq. Cryst. (UK)* vol.91 (1983) p.97]

[32] D.A. Dunmur [in *The Optics of Thermotropic Liquid Crystals* Eds. S. Elston, R. Sambles (Taylor & Francis Ltd, 1998) ch.2]

[33] M.F. Vuks [*Opt. Spektrosk. (USSR)* vol.20 (1966) p.644]

[34] P. Adamski [*Mol. Cryst. Liq. Cryst. (UK)* vol.201 (1991) p.87]

[35] A. Hauser, D. Demus [*Wiss. Z. Univ. (Germany)* vol.37 (1988) p.137]

[36] S.R. Sharma [*Mol. Phys. (UK)* vol.78 (1993) p.733]

[37] E.E. Burnell, C.A. de Lange [*Chem. Rev. (USA)* vol.98 (1998) p.2359]

CHAPTER 3

THERMODYNAMICS

3.1 Phase transitions and the effects of pressure

G.R. Van Hecke

December 1998

A INTRODUCTION

The effects of pressure on the nematic-isotropic transition temperature as measured by a variety of experimental techniques are here reviewed and summarised. A wide variety of liquid crystal compounds are covered.

B EXPERIMENTAL METHODS

The experimental methods of measuring transition temperatures, whether for crystal to mesophase, mesophase to mesophase, or mesophase to isotropic liquid, fall into four categories. A high pressure cell with optical detection is a common method [1,2]. High pressure differential thermal analysis (DTA) systems have been built [3]. In some cases DTA has been combined with optical cells for visual as well as thermal detection [4]. A variation of thermal detection of phase transitions is the high pressure differential scanning calorimetry system [5]. The technique that provides pVT data is based on measuring volume or density as a function of pressure for various isotherms [6-10]. A unique method is based on the thermobarometer developed by Busine [11].

C SUMMARY OF AVAILABLE DATA

This Datareview has been organised in the following manner. TABLE 1 presents an overview of the data listed according to the type of measurement used: optical pressure temperature; high pressure DTA or DSC; density or volume versus pressure at constant temperature (a method which gives not only p-T phase transition data but also the volume change at the phase transition); the thermobarometric method. The entries in TABLE 1 give references to the data available obtained by the specific type of measurement for the compounds covered here. The compounds are grouped by generic classes and specific compounds are designated by the commonly used abbreviations. Each line in TABLE 1 represents a class of compound for which further data are provided in the Datareview. The additional data are of two kinds. If the original literature provided tabulated pressure and temperature those data are presented here. The parameters for a fit of temperature as a second order polynomial in the pressure are provided for every compound discussed in this review; the fitting equation is

$$T/K = a + b \, (p/bar) + c \, (p/bar)^2$$

Bar is the pressure unit used since one bar is very close to one atmosphere and has a simple relationship to a Pascal, the SI unit of pressure: 1 bar = 10^5 Pa. For those cases where tabulated

TABLE 1 Sources of pressure - temperature data by compound class and type of measurement.

Substance by class		p-T[a] optical	p-T[b] DTA	p-T[c] pVT	p-T[d] TBM	ΔV at N-I
Cyanobiphenyl	nCB n = 5 - 8	[12]	[13-17]	[18-23]		[21-23]
	nOCB n = 5, 7, 8	[24]		[10,20]		
Cyclohexylphenyl	nPCH n = 3, 5, 8, 12	[25]	[4,25]	[21,26-29]		
	3OnPCH n = 2, 4		[25]			
	nPCH-NCS n = 6		[14]			
Cyclohexyl cyclohexane	nCCH n = 3, 5, 7		[25,30]			
Schiff bases	MBBA homologues	[1,2,31]	[3]	[7,8]		
	R = alkyl, R' = alkyloxy	[1]				
	R = alkyloxy, R' = alkyl	[1]				
	R = alkyloxy, R' = CN	[2]				
Azoxybenzenes	nOAB n = 1 - 9	[1,2,31,32]	[3]	[6,9,33-35]		[9]
Azobenzenes	R = alkyl	[1,6]				[1]
	R = alkyloxy	[1]				
Hydroquinone	NB		[36]			
Cholesteryl	ChOC R = oleyl carbonate	[1,2]	[36]			
	ChN R = geranil carbonate	[1,2]	[36]			
Truxenes	nHATX n = 10 - 12				[11]	[11]

[a]High pressure optical observations
[b]High pressure differential thermal analysis measurements
[c]Measurements based on high pressure - volume isotherms
[d]Thermobarometric apparatus for measuring pressure - volume isotherms

TABLE 2 Thermodynamic properties for the nematic-isotropic transition in homologous 4-alkyl-4'-cyanobiphenyls and 4-octyloxy-4'-cyanobiphenyl. Abbreviations and references are given in TABLE 3.

5CB			6CB			7CB			8CB			8OCB		
T_{NI}/K	p/bar	ΔV/cm^3 mol^{-1}	T_{NI}/K	p/bar	ΔV/cm^3 mol^{-1}	T_{NI}/K	p/bar	ΔV/cm^3 mol^{-1}	T_{NI}/K	p/bar	ΔV/cm^3 mol^{-1}	T_{NI}/K	p/bar	ΔV/cm^3 mol^{-1}
308.2	1	0.45	302.5	1	0.34	315.7	1	0.62	313.2	1	0.57	363.1	303	0.055
320.2	284	0.54	305.2	120	0.66	320.2	183	0.75	315.2	35	0.93	373.2	606	0.52
325.2	442	0.75	310.2	226	0.54	325.2	281	0.72	320.2	182	0.93	383.2	924	0.40
330.2	559	0.69	315.2	375	0.59	330.2	433	0.93	325.2	337	0.87	393.2	1255	0.25
335.2	709	0.79	320.2	475	0.46	335.2	583	0.93	330.2	494	0.82			
340.2	811	0.64	325.2	628	0.59	340.2	685	0.85	335.2	680	0.44			
345.2	961	0.70	330.2	753	0.28	345.2	834	0.89	340.2	824	0.85			
350.2	1115	0.79	335.2	902	0.29	350.2	985	0.87	345.2	926	0.85			
355.2	1241	0.89	340.2	1002	0.46	355.2	1132	0.69	350.2	1103	0.90			
360.2	1370	0.85	345.2	1106	0.23	360.2	1259	0.49	355.2	1258	0.82			
365.2	1479	1.03	350.2	1206	0.19	365.2	1382	0.51	360.2	1458	0.82			
370.2	1677	0.85	355.2	1402	0.38	370.2	1584	0.67	365.2	1613	0.82			
			360.2	1506	0.46				370.2	1762	0.73			
			365.2	1609	0.43									
			370.2	1804	0.35									

original data were available, the tabulated data were fitted. For those cases where the literature only presented figures showing the dependence of p on T or T on p, data were read off the figure and fitted to the second order polynomial in pressure. TABLES 2, 4, 7, 8, 9, 11, 13 and 16 present tabular data while TABLES 3, 5, 6, 10, 12, 14, 15, 17 and 18 give the fitting parameters together with the IUPAC compound names.

D PRESSURE - TEMPERATURE DATA FOR SELECTED COMPOUNDS

D1 Cyanobiphenyls

Cyanobiphenyl compounds are well-known for their widespread practical utility in displays. The data in TABLE 2 are based on pVT measurements which provide estimates of the volume discontinuity at the nematic-isotropic transition as well as the pressure and temperature points of phase equilibrium. The general range of ΔV is from 0.2 to 1.0 cm^3/mol. The fitting constants presented in TABLE 3 were evaluated by a least squares fit to the second order polynomial defined above. The constant term is approximately the room pressure transition temperature. The data fit the polynomial well showing a mild curvature which increases at higher pressures. No obvious trends are apparent from the magnitudes and signs of the parameters.

TABLE 3 Fitting equation parameters for nematic-isotropic pressure-temperature transition lines for various homologous cyanobiphenyl compounds. Fitting equation is: $T/K = a + b(p/bar) + c(p/bar)^2$.

Compound	a	b/10^{-2}	c/10^{-6}	Ref
4-(4-Alkylphenyl)benzenecarbonitriles				
Alkyl = pentyl, hexyl, heptyl, octyl				
5CB	308.28	4.24	-3.20	[21]
6CB	301.23	3.50	-1.89	[22]
7CB	314.61	3.70	-84	[22]
8CB	313.81	3.55	-2.19	[23]
4-(4-Octyloxyphenyl)benzenecarbonitrile				
8OCB	352.6	3.538	-2.42	[10]

D2 Cyclohexylphenyls

Here pVT data were also available to present in TABLE 4 transition volume changes together with the basic pressure temperature data. Worth noting is the larger size of the ΔV values. These values are, however, overestimates for although the volume of the isotropic phase was well-defined at the nematic-isotropic transition, the volume of the nematic phase was not. The values presented in TABLE 4 used the last value measured for the nematic phase which is likely to be too small making ΔV an overestimate. The fitting parameters are given in TABLE 5. Since only figures presenting p-T data were available for the alkyloxy-propylcyclohexylbenzenes and the isothiocyanate compounds,

TABLE 4 Thermodynamic properties for the nematic-isotropic transition for homologous 4-(4-alkylcyclohexyl)benzenecarbonitriles. Abbreviations and references are given in TABLE 5.

3PCH			5PCH			7PCH			8PCH		
T_{NI}/K	p/bar	ΔV/cm^3 mol^{-1}	T_{NI}/K	p/bar	ΔV/cm^3 mol^{-1}	T_{NI}/K	p/bar	ΔV/cm^3 mol^{-1}	T_{NI}/K	p/bar	ΔV/cm^3 mol^{-1}
323.2	80	1.87	333.2	110	2.24	333.2	40	1.63	328.5	1	0.82
333.2	290	1.51	343.2	335	2.63	343.2	275	1.98	330.2	43	1.08
343.2	505	2.73	353.2	560	2.01	353.2	520	2.73	335.2	167	1.07
353.2	726	2.47	363.2	800	2.69	363.2	777	1.65	340.2	307	1.15
363.2	955	2.04			(a)			(a)	345.2	415	1.18
		(a)							350.2	571	1.26
									355.2	696	1.23
									360.2	866	1.05
									365.2	992	0.88
									370.2	1166	1.37
									375.2	1290	1.31
									380.2	1465	1.14

(a) Estimated from original data using the last tabulated value for nematic density.
The values are overestimates.

TABLE 5 Fitting equation parameters for nematic-isotropic pressure-temperature phase transition lines for homologous cyclohexylphenyl derivatives. Fitting equation is: T/K = a + b(p/bar) + c(p/bar)^2.

Compound	a	b/10^{-2}	c/10^{-6}	Ref
4-(4-n-Alkylcyclohexyl)benzenecarbonitrile				
Alkyl = propyl, pentyloxy, heptyloxy, octyloxy, dodecyloxy				
3PCH	319.3	4.885	-3.01	[26]
5PCH	328.1	4.639	-3.10	[26]
7PCH	331.5	4.374	-3.71	[26]
8PCH	328.28	4.12	-3.613	[21]
12PCH	335.1	3.53	-1.838	[25]
4-Alkyloxy-1-(4-propylcyclohexyl)benzene				
Alkyloxy = ethoxy, butyloxy				
3O2PCH	312.4	3.518	-1.345	[25]
3O4PCH	307.4	3.204	-1.198	[25]
4-(4-Hexylcyclohexyl)benzeneisothiocyanate				
6PCH-NCS	317.0	4.266	-4.515	[14]

the fits are somewhat dependent on the ability to read data from the published figures. In all cases smooth quadratic lines fit the data very well and the parameters fall in the same range of values as those for the cyanobiphenyls. Even the isothiocyanate compound follows the same trends.

D3   Cyclohexylcyclohexanes

Only two examples of cyclohexylcyclohexanes were found and the data were presented as pressure-temperature figures. The data presented in TABLE 6 are then just the fitting parameters obtained from reading the figures. Again the magnitudes of the parameters and trends of the fitting lines follow those of previous compounds.

TABLE 6 Fitting equation parameters for nematic-isotropic pressure-temperature phase transition lines for homologous cyclohexylcyclohexanes. Fitting equation is: $T/K = a + b(p/bar) + c(p/bar)^2$.

Compound	a	$b/10^{-2}$	$c/10^{-6}$	Ref
4-(4-Alkylcyclohexyl)cyclohexanecarbonitrile Alkyl = propyl, heptyl				
3CCH	354.5	5.141	-3.124	[25]
7CCH	357.1	4.72	-4.1	[30]

D4   Schiff's Bases

TABLE 7 Thermodynamic properties for the nematic-isotropic transition in homologous Schiff's bases: 4-[(1E)-2-aza-(4-butylphenyl)vinyl]-1-alkyloxybenzenes. Abbreviations and references are given in TABLE 10.

	-methoxy	-ethoxy	-propyloxy	-butyloxy	-pentyloxy	-nonyloxy
p/bar	T_{NI}/K	T_{NI}/K	T_{NI}/K	T_{NI}/K	T_{NI}/K	T_{NI}/K
1	317.9	351.9	331.4	348.3	342.5	355.2
100	321.4	355.5	334.8	352.1	345.9	357.9
300	328.6	362.9	341.7	359.0	352.6	363.4
500	335.5	370.0	348.4	365.4	359.0	368.8
700	342.3	376.9	354.8	371.8	365.2	374.0
900	349.1	383.4	361.1	377.9	371.2	379.1
1100	355.7	389.8	367.4	383.9	377.3	384.2
1300	362.3	396.0	373.3	389.7	382.9	389.2
1500	368.6	401.9	379.1	395.2	388.3	393.9
1700	374.9	407.6	384.7	400.4	393.4	398.3
1900	380.9	412.8	390.3	405.5	398.6	402.5
2100	386.7	418.1	395.6	410.5	403.5	406.5
2300	392.5		400.8	415.5	407.9	410.1
2500	397.9		405.8		412.2	414.0

132

TABLE 8 Thermodynamic properties for the nematic-isotropic transition in homologous Schiff's bases: 1-{(1E)-2-Aza-2-[4-(2-methylpropyl)phenyl]vinyl}-4-alkyloxybenzenes. Abbreviations and references are given in TABLE 10.

p/bar	-methoxy T_{NI}/K	-ethoxy T_{NI}/K	-propyloxy T_{NI}/K	-butyloxy T_{NI}/K	-pentyloxy T_{NI}/K
1	324.2	353.5	331.4	344.7	337.7
100	327.7	357.1	334.8	348.2	340.6
300	334.8	364.0	341.2	354.7	346.8
500	341.7	370.8	347.8	360.9	353.0
700	348.3	377.4	354.2	366.9	358.9
900	354.8	383.9	360.4	373.0	364.5
1100	361.2	390.3	366.3	378.8	370.3
1300	367.7	396.3	372.1	384.5	375.9
1500	373.9	402.2	377.9	389.9	380.3
1700	380.0	407.9	383.5	394.9	386.4
1900	386.2	413.6	389.0	399.9	391.5
2100	392.1		394.1	404.8	396.3
2300	397.9		399.3	409.6	400.9
2500	403.4		404.1	414.5	405.4

TABLE 9 Thermodynamic properties for the nematic-isotropic transition in homologous Schiff's bases: 1-[(1E)-2-Aza-2-(4-alkylphenyl)vinyl]-4-methoxybenzenes. Abbreviations and references are given in TABLE 10.

p/bar	-ethyl T_{NI}/K	-propyl T_{NI}/K	-butyl T_{NI}/K	-pentyl T_{NI}/K	-hexyl T_{NI}/K
1	306.2	335.7	317.9	336.2	326.2
100	309.9	339.5	321.4	339.9	329.7
300	317.4	347.0	328.5	347.3	337.0
500	324.9	354.5	335.5	354.4	344.2
700	332.4	362.0	342.4	361.7	351.1
900	339.6	369.0	349.1	368.5	357.6
1100	346.7	376.4	355.7	375.2	364.1
1300	353.6	383.5	362.3	381.7	370.6
1500	360.6	390.2	368.6	388.4	376.9
1700	367.5	396.8	374.9	394.8	382.9
1900	374.3	403.0	380.9	401.0	388.9
2100	381.0	409.1	386.7	407.0	394.8
2300	387.8	415.0	392.5	412.4	400.6
2500	393.8	420.9	397.9	417.3	

The tabular data available for Schiff's bases were derived from pressure-temperature measurements. Homologues of 4-methoxybenzylidine-4'-butylaniline (MBBA) are presented in TABLE 7 where the R' of the figure in TABLE 10 is butyl and R is an alkoxy chain of varying length. TABLE 8 presents data for the homologues of R' being 2-methylpropyl while R varied again as alkoxy chains of varying length. The third group of homologues presented in TABLE 9 is for the compounds with R' being methoxy while R is an alkyl chain from ethyl to hexyl. A fourth homologous group is presented in TABLE 10 which contains the fitting parameters. The fourth group has a terminal cyano group for R'

TABLE 10 Fitting equation parameters for nematic-isotropic pressure-temperature phase transition lines for homologous Schiff's bases. Fitting equation is: $T/K = a + b(p/bar) + c(p/bar)^2$.

Compound	a	$b/10^{-2}$	$c/10^{-6}$	Ref
4-[(1E)-2-Aza-(4-butylphenyl)vinyl]-1-alkyloxybenzene Alkyloxy = methoxy, ethoxy, butyloxy, pentyloxy, nonyloxy				
-methoxy	317.8	3.650	-1.749	[1]
	319.3	3.791	-1.595	[3]
-ethoxy	351.8	3.790	-3.019	[1]
	353.2	3.856	-2.076	[3]
-propyloxy	331.4	3.494	-2.071	[1]
-butyloxy	348.5	3.500	-2.597	[1]
-pentyloxy	342.4	3.447	-2.611	[1]
	343.0	3.409	-2.019	[3]
-nonyloxy	354.9	2.906	-2.164	[1]
1-{(1E)-2-Aza-2-[4-(2-methylpropyl)phenyl]vinyl}-4-alkyloxybenzene Alkoxy = methoxy, ethoxy, propyloxy, butyloxy, pentyloxy				
-methoxy	324.2	3.530	-1.440	[1]
-ethoxy	353.5	3.577	-2.186	[1]
-propyloxy	331.3	3.392	-1.914	[1]
-butyloxy	344.8	3.324	-2.190	[1]
-pentyloxy	337.6	3.165	-1.783	[1]
1-[(1E)-2-Aza-2-(4-alkylphenyl)vinyl]-4-methoxybenzene Alkyl = ethyl, propyl, butyl, pentyl, hexyl				
-ethyl	306.1	3.822	-1.227	[1]
-propyl	317.7	3.653	-1.763	[1]
-butyl	335.4	3.963	-2.166	[1]
-pentyl	336.0	3.819	-2.189	[1]
-hexyl	326.2	3.661	-1.876	[1]
1-[(E)-2-Aza-2-(4-alkyloxyphenyl)vinyl]benzenecarbonitrile Alkyloxy = ethoxy, octyloxy				
PEBAB, ethoxy	401.7	5.487	-4.202	[3]
CBOOA, octyloxy	377.1	3.866	-2.405	[2]

134

while R is an alkoxy group. For this latter group no tabular data were available which meant that the fitting was done to data read from literature figures. Here also the constants are of the same magnitude and the fitted lines all show gentle quadratic behaviour.

D5   Azoxybenzenes

Azoxybenzenes are among the oldest known nematic liquid crystals. Derivatives have mostly been symmetrically disubstituted with either alkyl or alkyloxy groups. TABLE 11 presents the available tabular data while in TABLE 12 the fitting parameters are given. Only the nonyl azoxybenzene was fitted to data read from a literature figure. The data available for the propyloxy and butyloxy derivatives were over such a short pressure range that a quadratic was not judged useful and instead a simple linear fit was used. Even for the linear fit the constant term and the second coefficient b are the same orders of magnitude as the corresponding parameters for the quadratic fits. What this implies is that the quadratic dependence on pressure of the transition temperature is weak for all of these compounds.

TABLE 11 Thermodynamic properties for the nematic-isotropic
transition for homologous 4-4'-dialkyloxyazoxybenzenes.
Abbreviations and references are given in TABLE 12.

	PAA	PAP		propyloxy		butyloxy		pentyl	hexyl	heptyl
p/bar	T_{NI}/K	T_{NI}/K	p/bar	T_{NI}/K	p/bar	T_{NI}/K	p/bar	T_{NI}/K	T_{NI}/K	T_{NI}/K
1	408.9	439.0	117.20	402.20	144.80	414.40	1	341.4	327.7	344.6
500	433.3	462.1	189.00	405.20	355.90	422.30	100	345.4	331.1	348.2
1000	453.8	481.8	258.60	408.10	583.40	430.20	300	352.9	337.9	355.1
1500	471.7	499.2					500	360.2	344.4	361.7
2000	488.2	514.8					700	367.2	350.9	368.2
							900	373.8	357.3	374.5
							1100	380.1	363.6	380.7
							1300	386.4	370.0	386.8
							1500	392.3	376.0	392.5
							1700	397.9	381.9	398.1
							1900	403.5	387.6	403.4
							2100	408.9	393.3	408.8
							2300	414.3	398.6	413.9

TABLE 12 Fitting equation parameters for nematic-isotropic pressure-temperature phase transition lines for homologous azoxybenzene derivatives. Fitting equation is: $T/K = a + b(p/bar) + c(p/bar)^2$.

Compound	a	$b/10^{-2}$	$c/10^{-6}$	Ref
bis(4-Alkyloxyphenyl)diazene oxides				
Alkyloxy = methoxy, ethoxy, propyloxy, butyloxy				
PAA methoxy	439.1	4.770	-4.970	[3]
PAP ethoxy	409.1	4.999	-5.285	[3]
propyloxy	397.3	4.173	-	[35]
butyloxy	409.3	3.601	-	[35]
bis(4-Alkylphenyl)diazene oxides				
Alkyl = pentyl, hexyl, heptyl, nonyl				
pentyl	341.6	3.832	-2.977	[1]
hexyl	327.6	3.453	-1.563	[1]
heptyl	344.6	3.527	-2.249	[1]
nonyl	355.2	2.761	-1.298	[36]

D6   Azobenzenes

While azobenzenes also have a long history of numerous studies and synthetic variations, relatively few nematic-isotropic pressure-temperature measurements exist in the literature. Data for three disubstituted alkyl derivatives are tabulated in TABLE 13 while TABLE 14 presents the fitting

TABLE 13 Thermodynamic properties for the nematic-isotropic transition for homologous azobenzenes: references are given in TABLE 14.

	pentyl	hexyl	heptyl
p/bar	T_{NI}/K	T_{NI}/K	T_{NI}/K
1	316.0	300.0	321.4
100	319.5	303.0	324.8
300	326.5	309.6	331.6
500	333.5	316.0	338.3
700	340.3	322.5	344.5
900	346.8	328.7	350.5
1100	353.1	335.0	356.4
1300	359.3	341.2	362.1
1500	365.1	347.0	367.5
1700	370.7	352.5	373.0
1900	376.0	357.9	378.4
2100	381.2	363.3	383.5
2300	386.2	368.6	388.6
2500	391.0	373.8	393.6

parameters for the 4,4′-dialkylazobenzenes. The values and trends in the fitting lines are in line with other nematic isotropic data.

TABLE 14 Fitting equation parameters for nematic-isotropic pressure-temperature phase transition lines for homologous azobenzene derivatives. Fitting equation is: $T/K = a + b(p/bar) + c(p/bar)^2$.

Compound	a	$b/10^{-2}$	$c/10^{-6}$	Ref
4,4′-dialkylazobenzenes				
Alkyl = pentyl, hexyl, heptyl				
pentyl	315.8	3.691	-2.738	[1]
hexyl	299.7	3.402	-1.747	[1]
heptyl	321.6	3.387	-2.060	[1]

D7   Hydroquinones

TABLE 15 presents the parameters used to fit data drawn from a figure in the work of Rein and Demus [36]. The goal of these authors was to test the influence of large lateral substituents bound to the central core of the mesogenic molecule. The example presented in TABLE 15 is one of several, all of whose pressure-temperature data and trends are extremely similar. The quadratic fit to the data was excellent offering an alternative treatment to the fitting based on the Simon-Glatzel equation used by the authors. The second order polynomial fit was used here to allow better comparison with the fitting parameters of the other substances reviewed.

TABLE 15 Fitting equation parameters for nematic-isotropic pressure-temperature phase transition lines for a hydroquinone derivative. Fitting equation is: $T/K = a + b(p/bar) + c(p/bar)^2$.

Compound	a	$b/10^{-2}$	$c/10^{-6}$	Ref
6-(4-hexylphenylcarbonyloxy)tricyclo[6.2.0<2,7>]undeca-2(7),3,5,9-tetraen-3-yl 4-hexyl				
R = $C_6H_{13}C_6H_4COO-$	382.7	3.322	-1.889	[37]

D8   Chiral Nematics

Tabular pressure-temperature data are available for two derivatives of cholesterol carbonate (TABLE 16). The maximum pressure used to study these derivatives was several hundred bar more than for any other compounds reported in this Datareview yet the trends in parameters are still similar. What is different is that at high pressures the p-T curves almost coalesce implying that at high enough pressure the structure of the phase itself is most important and the nature of the molecule involved is unimportant as long as it exhibits a chiral nematic phase.

TABLE 16 Thermodynamic properties for the chiral nematic-isotropic transition for cholesterol derivatives. Abbreviations and references are given in TABLE 17.

	Oleyl carbonate	Geranil carbonate
p/bar	T_{N*I}/K	T_{N*I}/K
1	311.9	325.0
100	315.1	329.1
200	318.4	333.2
400	324.8	341.1
600	331.4	349.0
800	337.6	356.8
1000	343.4	364.4
1200	349.3	371.4
1400	354.9	378.2
1600	360.5	384.9
1800	366.0	391.4
2000	371.5	397.7
2200	377.0	403.6
2400	382.0	409.3
2600	387.0	414.8
2800	391.5	420.0
3000	396.0	425.0

TABLE 17 Fitting equation parameters for nematic-isotropic pressure-temperature phase transition lines for homologous cholesterol derivatives. Fitting equation is: $T/K = a + b(p/bar) + c(p/bar)^2$.

Compound	a	$b/10^{-2}$	$c/10^{-6}$	Ref
14-(1,5-dimethylhexyl)-2,15-dimethyltetraacyclo [8.7.0.0<2,7>.0<11,15>] heptadec-7-en-5-yl ((9E)octadec-9-enyloxy)formate. R1 = ((9E)octadec-9-enyloxy)				
R1 (oleyl)	311.8	3.330	-1.727	[1]
14-(1,5-dimethylhexyl)-2,15-dimethyltetraacyclo [8.7.0.0<2,7>.0<11,15>] heptadec-7-en-5-yl ((2E,6E)-3,7-dimethylocta-2,6-dienyloxy)formate. R2 = ((2E,6E)-3,7-dimethylocta-2,6-dienyloxy)				
R2 (geranil)	324.8	4.235	-2.982	[1]

D9   Truxenes

Even though no tabular data are available for truxene derivatives, inclusion in this Datareview is important because these materials exhibit discotic nematic phases. The truxene phase data presented by Busine and co-workers [11] are for the transition of the discotic nematic phase to the disordered rectangular columnar phase (TABLE 18). Over the phase range studied which was up to 900 bar, the relationship between T and p is most conveniently represented as linear. Once again the magnitudes of the fitting parameters are consistent with those for the other compounds reviewed here. These materials also exhibit columnar-columnar transitions that are not mentioned here. The method of study is unique in the pressure-temperature literature because it involves the thermobarometric technique pioneered by Busine and co-workers. This technique also provides estimates for the density, isothermal compressibility and thermal expansivity coefficient using quite small sample sizes, the order of microlitres.

TABLE 18 Fitting equation parameters for discotic nematic-disordered rectangular columnar pressure-temperature phase transition lines for homologous truxenes. Fitting equation is: $T/K = a + b(p/bar)$.

Compound	a	$b/10^{-2}$	Ref
2,3,7,8,12,13-Hexaalkyloxyindeno[3,2-a]indeno[3,2-c]fluorene Alkyloxy = decyloxy, undecyloxy, dodecyloxy, tridecyloxy			
decyloxy	361	3.7	[11]
undecyloxy	356	5.5	[11]
dodecyloxy	358	4.1	[11]
tridecyloxy	357	4.3	[11]

E   REMARKS

This Datareview has concentrated on the nematic-isotropic transition except for the case of the truxenes where the discotic nematic to a columnar phase was considered. No smectic or crystal phase data have been considered here even though the literature reviewed often contained pressure-temperature data involving these phases. Data for the crystal-nematic transition are almost always included in the studies, but only sometimes even when the compound exhibits a smectic phase are such data included. No surprises exist in the body of pressure-temperature data for the nematic-isotropic transition. This transition for any nematogen could be described by the polynomial fitting equation with surprisingly small uncertainties in the average fitting parameters: for example

$$T_{NI}(p)/K = (T_{NI})_0 + (3.8 \pm 0.6)\ 10^{-2}\ (p/bar) - (2.5 \pm 1.0)\ 10^{-6}\ (p/bar)^2$$

where $(T_{NI})_0$ is the nematic-isotropic transition temperature at normal atmospheric pressure.

REFERENCES

[1]   M. Feyz, E. Kuss [*Ber. Bunsenges. Phys. Chem. (Germany)* vol.78 (1974) p.834]

[2]   P.H. Keyes, H.T. Weston, W. Lin, W.B. Daniels [*J. Chem. Phys. (USA)* vol.63 (1975) p.5006]

[3]   W. Spratte, G.M. Schnieder [*Ber. Bunsenges. Phys. Chem. (Germany)* vol.80 (1976) p.886]

[4]   G.M. Schnieder, A. Bartelt, J. Friedrich, H. Reisig, A. Rothert [*Physica B & C (Netherlands)* vol.139-140 (1986) p.616]

[5]   R. Sandrock, M. Kamphausen, G.M. Schnieder [*Mol. Cryst. Liq. Cryst. (UK)* vol.45 (1978) p.257]

[6]   V. Baskakov, V.K. Semenchenko, N.A. Nedostup [*Sov. Phys. Crystallogr. (USA)* vol.19 (1974) p.112]

[7]   A.C. Zawisza, M. Stecki [*Solid State Commun. (USA)* vol.19 (1976) p.1173]

[8]   E. Kuss [*Mol. Cryst. Liq. Cryst. (UK)* vol.47 (1978) p.71]

[9]   P.H. Keyes, W.B. Daniels [*J. Phys. (France)* vol.44 (1979) p.51]

[10]   C.S. Johnson, P.J. Collings [*J. Chem. Phys. (USA)* vol.79 (1983) p.4056]

[11]   J.M. Busine, R. Cayuela, C. Destrade, N.H. Tinh [*Mol. Cryst. Liq. Cryst. (UK)* vol.144 (1987) p.137]

[12]   R. Shashidar, G. Venkatesh [*J. Phys. Colloq. (France)* vol.40 (1979) p.396]

[13]   H.G. Kruel, S. Urban, A. Wurflinger [*Phys. Rev. A, Gen. Phys. (USA)* vol.45 (1992) p.8624]

[14]   S. Urban, A. Wurflinger [*Liq. Cryst. (UK)* vol.12 (1992) p.931]

[15]   T. Bruckert, A. Wurflinger, S. Urban [*Ber. Bunsenges. Phys. Chem. (Germany)* vol.97 (1993) p.1209]

[16]   S. Urban, T. Bruckert, A. Wurflinger [*Liq. Cryst. (UK)* vol.15 (1993) p.919]

[17]   S. Urban, T. Bruckert, A. Wurflinger [*Z. Nat.forsch. A (Germany)* vol.49 (1994) p.552]

[18]   T. Shirakawa, T. Inoue, T. Tokuda [*J. Phys. Chem. (USA)* vol.86 (1982) p.1700]

[19]   T. Shirakawa, T. Hayakawa, T. Tokuda [*J. Phys. Chem. (USA)* vol.87 (1983) p.1406]

[20]   R.A. Orwoll, V.J. Sullivan, G.C. Campbell [*Mol. Cryst. Liq. Cryst. (UK)* vol.149 (1987) p.121]

[21]   M. Sandmann, F. Hamann, A. Wurflinger [*Z. Nat.forsch. A (Germany)* vol.52 (1997) p.739]

[22]   M. Sandmann, A. Wurflinger [*Z. Nat.forsch. A (Germany)* vol.53 (1998) p.233]

[23]   M. Sandmann, A. Wurflinger [*Z. Nat.forsch. A (Germany)* vol.53 (1998) p.787]

[24]   P.E. Cladis, D. Guillon, F.R. Bouchet, P.L. Finn [*Phys. Rev. A, Gen. Phys. (USA)* vol.23 (1981) p.2594]

[25]   A. Bartelt, G.M. Schnieder [*Mol. Cryst. Liq. Cryst. (UK)* vol.173 (1989) p.75]

[26]   E. Kuss [*Mol. Cryst. Liq. Cryst. (UK)* vol.76 (1981) p.199]

[27]   T. Shirakawa, M. Arai, T. Tokuda [*Mol. Cryst. Liq. Cryst. (UK)* vol.104 (1984) p.131]

[28]   T. Shirakawa, H. Ichimura, I. Ikemoto [*Mol. Cryst. Liq. Cryst. (UK)* vol.142 (1987) p.101]

[29]   T. Bruckert, D. Busing, A. Wurflinger, S. Urban [*Mol. Cryst. Liq. Cryst. Sci. Technol., Sec. A (Germany)* vol.262 (1995) p.1497]

[30]   D. Busing, T. Bruckert, A. Wurflinger, B. Gestblom [*Proc. SPIE (USA)* vol.3318 (1998) p.213]

[31]   N.A. Tikhomirova, L.K. Vistin', V.N. Nosov [*Sov. Phys.-Crystallogr. (USA)* vol.17 (1973) p.878]

[32]   B. Deloche, B. Cabane, D. Jerome [*Mol. Cryst. Liq. Cryst. (UK)* vol.15 (1971) p.197]

[33]   W. Klement, L.H. Cohen [*Mol. Cryst. Liq. Cryst. (UK)* vol.27 (1974) p.359]

[34]   R.V. Tranfield, P.J. Collings [*Phys. Rev. A, Gen. Phys. (USA)* vol.25 (1982) p.2744]

[35]   M.W. Lampe, P.J. Collings [*Phys. Rev. A, Gen. Phys. (USA)* vol.34 (1986) p.524]

[36]   C. Rein, D. Demus [*Liq. Cryst. (UK)* vol.16 (1994) p.323]

3.2   Calorimetric measurements in nematics

M. Sorai

April 1999

A   INTRODUCTION

We review here calorimetric studies on nematics performed mainly by the adiabatic method, focusing on phase transitions and the glassy state.

B   HEAT CAPACITY CALORIMETRY

Among various thermodynamic measurements heat capacity calorimetry is an extremely useful tool with which to investigate thermal properties of liquid crystals [1,2]. The heat capacity is usually measured under constant pressure and designated as C_p. It is defined as the enthalpy, H, required to raise the temperature of one mole of a given substance by 1 K. From this definition, $C_p = (\partial H/\partial T)_p$, the enthalpy increment is determined by integration of C_p with respect to temperature, that is

$$ \tag{1} $$

Since C_p is alternatively defined as $C_p = T (\partial S/\partial T)_p$, the entropy S can also be obtained by integration of C_p with respect to lnT:

$$ \tag{2} $$

The availability of both enthalpy and entropy from C_p measurements means that we can estimate the Gibbs energy G from

$$ G = H - TS \tag{3} $$

The heat capacity is a very valuable and efficient physical quantity in that three fundamental thermodynamic quantities (H, S and G) can be determined simultaneously solely from heat capacity measurements.

The heat capacity is sensitive to a change in the degree of both the long range and short range order. Thus, heat capacity calorimetry is the most reliable experimental tool to detect the existence of a phase transition originating in the onset of a long range ordering. For a discussion of the nature of the phase transition, the thermal contribution gained by the relevant degrees of freedom should be accurately separated from the observed heat capacity as the excess heat capacity ΔC_p beyond a normal (or lattice) heat capacity. The excess enthalpy ΔH and entropy ΔS arising from the phase transition are determined by integration of ΔC_p with respect to T and lnT, respectively. Based on these

141

quantities, we can gain an insight into the mechanism of the phase transition mainly from the viewpoint of energy and entropy.

Phase transitions occurring between liquid crystalline states are often of a continuous type and so they are regarded as critical phenomena. For these phase transitions, the excess heat capacities ΔC_p in the vicinity of the critical temperature T_C are given by the power law:

$$\Delta C_p \propto \varepsilon^{-\alpha} \ (\varepsilon > 0) \quad \text{or} \quad \Delta C_p \propto (-\varepsilon)^{-\alpha'} \ (\varepsilon < 0) \tag{4}$$

where $\varepsilon \equiv (T - T_C)/T_C$, and α and α' are critical exponents.

C   EXPERIMENTAL METHODS

Various experimental methods have so far been employed to investigate the thermal properties of liquid crystals [3]. The most commonly used thermal technique for studying liquid crystals is differential scanning calorimetry (DSC). The phase sequence of a given substance is recorded in a short time by this method. The transition temperatures and qualitative enthalpy changes associated with the phase transitions are easily determined with commercially available instruments. The amount of material necessary for DSC is usually 10 mg. As an extremely high-resolution experimental technique, there is alternating current calorimetry (AC calorimetry) [4-8]. This is a powerful method for the study of phase transitions, in particular for the determination of critical exponents and comparison with theoretical models. By making measurements with different oscillating frequencies, we can use an AC calorimeter as a spectrometer and gain insight into time-dependent kinetics at the phase transition. However, the shortcoming of this calorimetry is its failure to detect a latent heat inherent in a first order phase transition. The absolute value of the heat capacity cannot be obtained by this method. Relaxation calorimetry [8,9] complements AC calorimetry in that it can sense the latent heat of a first order phase transition.

The most reliable method for the determination of the enthalpy and hence the heat capacity as a function of temperature is traditional adiabatic calorimetry [10-13], in which precise temperature measurement is carried out by a stepwise addition of an accurately measured small increment of electrically supplied heat. When the thermal properties of liquid crystals are discussed in terms of entropy, this traditional calorimetry plays a crucial role. However, this method needs a large sample (1 to 10 g) and is very time-consuming. The reason for the large sample size is based on the fact that the calorimeter cell for the material cannot be reduced owing to the large size of the working thermometer, while the reason for the long times is that adiabatic calorimetry is carried out under the condition of thermal equilibrium, the time necessary for which cannot be significantly shortened artificially.

D   PHASE TRANSITIONS RELATING TO THE NEMATIC STATE

Although 4-methoxybenzilidene-4′-butylaniline (MBBA) is the epoch-making pure nematogen and its heat capacity has been measured by adiabatic calorimetry [14,15], this compound is not so stable chemically. So as the representative heat capacity of a nematogenic compound, FIGURE 1 demonstrates the molar heat capacity of 2-hydroxy-4-methoxybenzilidene-4'-butylaniline (OHMBBA) [16]. The crystal of OHMBBA melts at T_{CrN} = 314.30 K and the nematic to isotropic liquid transition occurs at T_{NI} = 333.65 K. In order to determine the excess heat capacities ΔC_p associated with the phase transitions, normal heat capacity curves, that is the base lines, should be estimated. In case of melting, the base lines are determined by simple extrapolation of the heat capacity of the crystal and that of the nematic state. On the other hand, for the nematic-isotropic transition, the normal heat capacity is usually approximated by a curve smoothly connecting the heat capacities in the isotropic liquid and those in the low temperature part of the nematic state. The enthalpy and entropy gains at the melting, obtained by integration of ΔC_p, are ΔH_{CrN} = 22.41 kJ mol^{-1} and ΔS_{CrN} = 71.43 J K^{-1} mol^{-1}, while those for the nematic-isotropic transition are ΔH_{NI} = 0.887 kJ mol^{-1} and ΔS_{NI} = 2.69 J K^{-1} mol^{-1}. Another example of the nematic-isotropic transition is given in FIGURE 2 for 4-(trans-4-propylcyclohexyl)benzonitrile (3CBCN) [17]. This compound melts into the nematic phase at 316.33 K. Since the nematic phase persists only over a short range and changes into the isotropic liquid phase at 319.09 K, it is difficult to determine precisely the enthalpy and entropy change at the nematic-isotropic transition. However, as shown in FIGURE 2 by solid triangles, the nematic phase is easily supercooled to 25 K below the melting point when the cooling rate is rather high, whereas further cooling brings about a monotropic transformation to the solid phase. This supercooling makes it possible to determine the enthalpy and entropy of the nematic-isotropic transition. The enthalpy and entropy gained at fusion are 20.4 kJ mol^{-1} and 64.4 J K^{-1} mol^{-1}, while those for the nematic-isotropic transition are 1.5 kJ mol^{-1} and 4.7 J K^{-1} mol^{-1}, respectively. Many liquid crystal compounds have their enthalpy changes at the nematic-isotropic transition in the range (0.08 → 0.962) kJ mol^{-1} [18]. They can be thought of as the energies necessary for long range orientational disordering of the molecules.

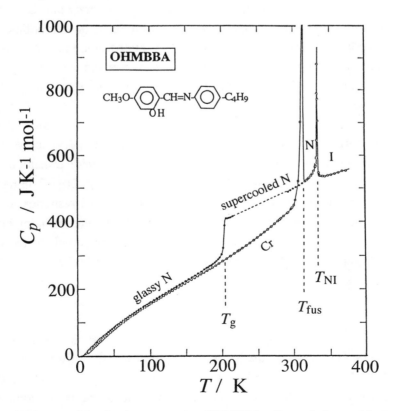

FIGURE 1 Molar heat capacities for the nematogen OHMBBA. Open circles: stable crystal, nematic and isotropic liquid phases. Solid circles: the glassy nematic and the supercooled nematic states. From [16].

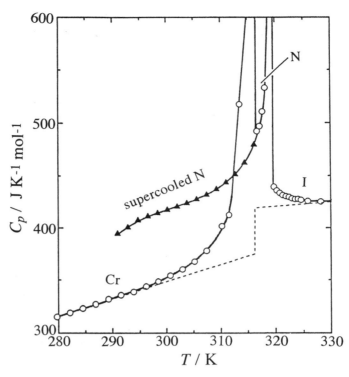

FIGURE 2 Molar heat capacities of 3CBCN in the melting and mesomorphic phase transition regions.
Open circles: heat capacities in the stable states. Broken lines: the normal heat capacity.
Solid triangles: heat capacities of the supercooled nematic state. From [17].

The nematic-isotropic transition is weakly first order due to a cubic invariant in the free energy [19]. Although the existence of latent heat and/or thermal hysteresis (supercooling or superheating phenomenon) is the definitive evidence for a first order phase transition, the usual isotropic to nematic transition does not show any obvious supercooling effect even when the cooling rate is extremely high. In contrast to this, since the latent heat at the nematic-isotropic transition is very small, pseudo critical behaviour is expected, in which various physical properties are described by power laws with appropriate critical exponents. As seen in FIGURES 1 and 2, pre- and post-transitional heat capacities arising from these fluctuation effects are dominant. FIGURE 3(a) shows a similar example of the temperature dependence of C_p near the nematic-isotropic transition for 4-hexyl-4'-cyanobiphenyl (6CB) determined by adiabatic scanning calorimetry [3,20]. The narrow box around the arrow marking the nematic-isotropic transition temperature T_{NI} corresponds to the nematic and isotropic coexistence region characteristic of a first order transition. The enthalpy change near T_{NI} is shown in FIGURE 3(b) for 6CB [3,20]; the nearly vertical part corresponds to the latent heat of the nematic-isotropic transition. Although the latent heat seems large in FIGURE 3(b), it is two orders of magnitude smaller than the latent heat at the melting transition.

The SmA-N transition is often encountered in liquid crystals and has been extensively studied both theoretically and experimentally. Based on the analogy with the superconductor to normal metal transition, de Gennes [19,21] classified the SmA-N transition in the isotropic three-dimensional XY universality class. Actually, however, deviation from the isotropic 3D-XY behaviour occurs to give the cross-over from second order to first order behaviour via a tricritical point due to coupling between smectic and nematic order parameters. This type of deviation has been predicted for the

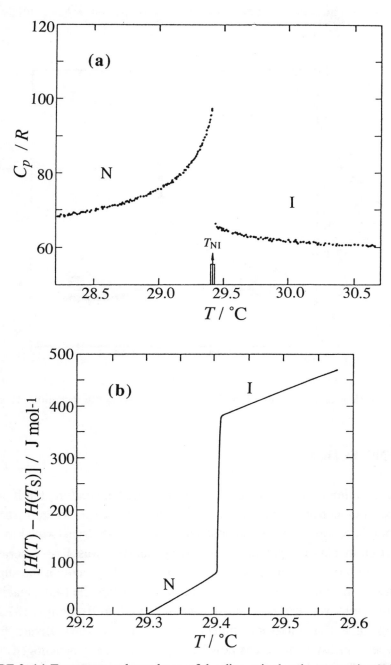

FIGURE 3 (a) Temperature dependence of the dimensionless heat capacity C_p/R near the nematic-isotropic transition for 6CB. The width of the arrow on the temperature axis shows the two-phase region. (b) The enthalpy curve of 6CB near the nematic-isotropic transition with respect to the value at 29.3°C. From [3,20].

SmA_m-N [19], SmA_d-N [19] and SmA_2-N [22] transitions, where SmA_m denotes the monolayer structure for non-polar compounds, while SmA_d and SmA_2 imply the interdigitated bilayer and the bilayer smectic structures respectively for polar compounds. For the SmA_1-N transition, where SmA_1 denotes the monolayer smectic structure for polar liquid crystals, 3D-XY critical behaviour has been expected theoretically [23,24]. FIGURE 4 [25] shows a typical example of 3D-XY heat capacity behaviour determined by AC calorimetry for the compound 4-octyloxyphenyl-4-(4-cyanobenzyloxy)-

benzoate (8OPCBOB). The smooth curves represent a fit to the 3D-XY model and show good agreement.

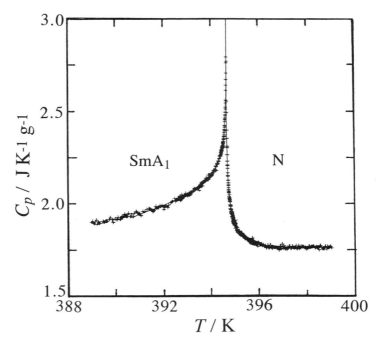

FIGURE 4   Heat capacity of 8OPCBOB near the SmA_1-N phase transition.   From [25].

E   GLASSY NEMATICS

The nematic phase can often be supercooled below the melting temperature, but in many cases further cooling brings about a monotropic transition to the crystalline phase. When the cooling rate is high enough, say 10 K min^{-1}, the supercooled nematic state is eventually transformed to a non-equilibrium thermodynamic frozen-in glassy nematic. This glassy nematic was first reported for OHMBBA [16,26]. As shown in FIGURE 1, the glassy nematic is realised by cooling the supercooled nematic phase below the glass transition temperature T_{NgN} at 204 K, at which the heat capacity jumps abruptly. The magnitude of the jump ΔC_p relating to the quenched degrees of freedom is 107 J K^{-1} mol^{-1}. In the glass transition region, a characteristic temperature drift is observed arising from the enthalpy relaxation. This type of glassy nematic has now been found for many nematogens [27,28]. It should be remarked here that although glassy smectics are also possible, their thermal behaviour around the glass transition region is quite different from that of the glassy nematic. For glassy SmG, double glass transitions have been found [29-31].

FIGURE 5(a) shows the enthalpy diagram for the various phases of OHMBBA. Since the transition enthalpy at the nematic-isotropic transition ΔH_{NI} is extremely small in comparison to the latent heat at melting ΔH_{CrN}, the former cannot clearly be discriminated on this scale of diagram. The enthalpy difference at 0 K and thus the energy difference between the vitreous state (the glassy nematic) and the crystal is 11.38 kJ mol^{-1}. FIGURE 5(b) shows the entropy diagram. One of the most important thermodynamic quantities of a glass is the residual entropy at 0 K, which approximates the configurational entropy frozen-in at the glass transition. The residual entropy of the glassy nematic of OHMBBA is $S_0 = (12.7 \pm 0.16)$ J K^{-1} mol^{-1}. This value is smaller than the residual entropies for ordinary glassy liquids but larger than those for glassy crystals. This reflects the fact that the degree of orientational order in liquid crystals is higher than that in ordinary liquids but lower than that in orientationally disordered crystals.

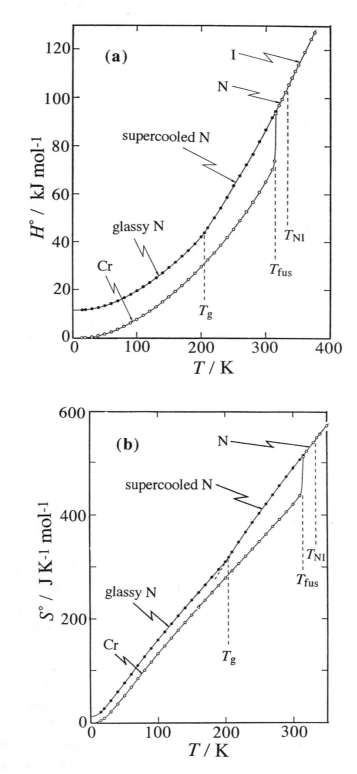

FIGURE 5 (a) Enthalpy and (b) entropy diagrams for the various phases of OHMBBA. From [16].

As for amorphous solids, the heat capacity of a glassy nematic is remarkably large at low temperatures in comparison to that of the crystal. FIGURE 6 shows the excess heat capacities ΔC_p [= C_p(nematic glass) - C_p(crystal)] of OHMBBA below T_{NgN} [16]. From 80 K upwards the heat capacity of the glassy nematic is about 2.5 - 3% higher than that of the crystal. When the temperature is lowered below 80 K, this gradually increases to 20% at 30 K and finally to 50% at 13 K, the lowest temperature studied. The ΔC_p peak centred around 25 K may be attributed to the low-lying excitation due to localised vibrations, which has been interpreted in terms of the two-level or tunnelling model [32] or the soft potential model [33,34].

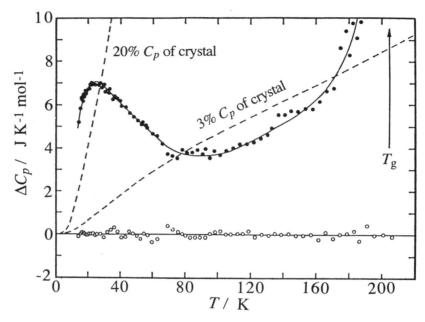

FIGURE 6 Excess heat capacities ΔC_p [= C_p(nematic glass) - C_p(crystal)] of OHMBBA below T_{NgN}. Base line: smoothed value of crystal. Open circles: crystal. Solid circles: glassy liquid crystal. From [16].

F   RE-ENTRANT NEMATIC

The orientational order and viscosity of the nematic phase is lower than that of the smectic A phase and so the nematic phase has long been thought to appear on the higher temperature side of the smectic. However, Cladis [35] reported in 1975 the re-entrant behaviour of the nematic phase in a binary mixture of liquid crystals. Since then, extensive studies have been performed and this phenomenon has been found to occur widely not only in mixtures, but also in single component systems. Moreover, multiple re-entrancy has been also reported; the enantiotropic doubly re-entrant compound first reported is 4-cyanobenzoyloxy-4′-octylbenzoyloxy-p-phenylene (CBOBP) [36]. Its thermotropic phase sequence is Cr-SmA$_1$-N$_{re}$-SmA$_d$-N-I. FIGURE 7 shows the experimental heat capacities determined by adiabatic calorimetry [37]. Literature values for the phase transitions are marked approximately by the arrows for comparison. These heat capacity results show the presence of the SmA$_1$-N$_{re}$ transition at 413 K, the N$_{re}$-SmA$_d$ at 425 K, and the SmA$_d$-N at 463 K. Since the enthalpy and entropy changes at the latter two phase transitions are extremely small, it may be concluded that although the molecular aggregation is quite different in the nematic and smectic states, the degrees of molecular disorder in those two phases reflected by the entropy resemble each other.

148

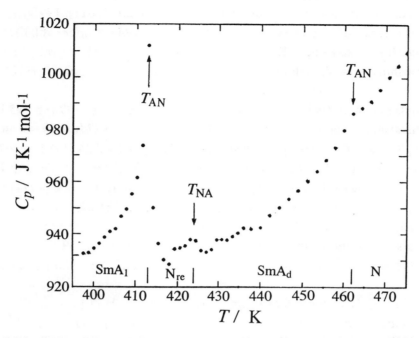

FIGURE 7 Molar heat capacity versus temperature for the doubly re-entrant CBOBP. The literature transition temperatures are marked approximately by the arrows. From [37].

G   CONCLUSION

Heat capacities of the nematic-isotropic transition are primarily characterised by dominant pre- and post-transitional effects arising from the fluctuation of molecular alignment, and are described by power laws with appropriate critical exponents.   However, in accordance with the theoretical prediction that the nematic-isotropic transition is weakly first order consistent with a cubic invariant in the free energy [19], a very small latent heat has been observed in the vicinity of the phase transition. For the smectic A-nematic transition, the heat capacities essentially follow the 3D-XY behaviour. Actually, however, deviation from the 3D-XY behaviour occurs to give the cross-over from second order to first order behaviour via a tricritical point due to coupling between smectic and nematic order parameters.   When the cooling rate is high enough, the nematic phase is often supercooled below the melting temperature and eventually transformed to a non-equilibrium thermodynamic frozen-in glassy nematic.

REFERENCES

[1]   M. Sorai [*Thermochim. Acta. (Netherlands)* vol.88 (1985) p.1]

[2]   P. Barois [in *Handbook of Liquid Crystals* Eds. D. Demus, J. Goodby, G.W. Gray, H.-W. Spiess, V. Vill (Wiley-VCH, Weinheim, 1998) vol.1 p.281]

[3]   J. Thoen [in *Handbook of Liquid Crystals* Eds. D. Demus, J. Goodby, G.W. Gray, H.-W. Spiess, V. Vill (Wiley-VCH, Weinheim, 1998) vol.1 p.310]

[4]   C.A. Schantz, D.L. Johnson [*Phys. Rev. A (USA)* vol.17 (1978) p.1504]

[5]   C.C. Huang, J.M. Viner, J.C. Novack [*Rev. Sci. Instrum. (USA)* vol.56 (1985) p.1390]

[6]   G.B. Kasting, K.J. Lushington, C.W. Garland [*Phys. Rev. B (USA)* vol.22 (1980) p.321]

[7]   I. Hatta, A.J. Ikushima [*Jpn. J. Appl. Phys. (Japan)* vol.20 (1981) p.1995]

[8]   K. Ema, T. Uematsu, A. Sugata, H. Yao [*Jpn. J. Appl. Phys. (Japan)* vol.32 (1993) p.1846]

[9]   D. Djurek, J. Baturic-Rubcic, K. Franulovic [*Phys. Rev. Lett. (USA)* vol.33 (1974) p.1126]

[10]   M.A. Anishimov, A.V. Voronel, T.M. Ovodova [*Sov. Phys.-JETP (USA)* vol.35 (1972) p.536]

[11]   J.T.S. Andrews, W.E. Bacon [*J. Chem. Thermodyn. (UK)* vol.6 (1974) p.515]

[12]   M. Yoshikawa, M. Sorai, H. Suga, S. Seki [*J. Phys. Chem. Solids (UK)* vol.44 (1983) p.311]

[13]   M. Sorai, K. Kaji, Y. Kaneko [*J. Chem. Thermodyn. (UK)* vol.24 (1992) p.167]

[14]   J. Mayer, T. Waluga, J.A. Janik [*Phys. Lett. A (Netherlands)* vol.41 (1972) p.102]

[15]   T. Shinoda, Y. Maeda, H. Enokido [*J. Chem. Thermodyn. (UK)* vol.6 (1974) p.921]

[16]   M. Sorai, S. Seki [*Mol. Cryst. Liq. Cryst. (UK)* vol.23 (1973) p.299]

[17]   S. Asahina, M. Sorai, R. Eidenschink [*Liq. Cryst. (UK)* vol.24 (1998) p.201]

[18]   D. Marzotko, D. Demus [*Pramana (India)* suppl. no.1 (1975) p.189]

[19]   P.G. de Gennes, J. Prost [*The Physics of Liquid Crystals 2nd Edition* (Clarendon, Oxford, 1993)]

[20]   J. Thoen [*Int. J. Mod. Phys. B (Singapore)* vol.9 (1995) p.2157]

[21]   P.G. de Gennes [*Solid State Commun. (USA)* vol.10 (1972) p.753]

[22]   J. Prost, P. Barois [*J. Chim. Phys. (France)* vol.80 (1983) p.65]

[23]   T.C. Lubensky [*J. Chim. Phys. (France)* vol.80 (1983) p.31]

[24]   J. Prost [*Adv. Phys. (UK)* vol.33 (1984) p.1]

[25]   C.W. Garland, G. Nounesis, K.J. Stine, G. Heppke [*J. Phys. (France)* vol.50 (1989) p.2291]

[26]   M. Sorai, S. Seki [*Bull. Chem. Soc. Jpn. (Japan)* vol.44 (1971) p.2887]

[27]   W. Wedler, D. Demus, H. Zaschke, K. Mohr, W. Schäfer, W. Weissflog [*J. Mater. Chem. (UK)* vol.1 (1991) p.347]

[28]   W. Wedler, P. Hartmann, U. Bakowsky, S. Diele, D. Demus [*J. Mater. Chem. (UK)* vol.2 (1992) p.1195]

[29]   H. Yoshioka, M. Sorai, H. Suga [*Mol. Cryst. Liq. Cryst. (UK)* vol.95 (1983) p.11]

[30]   M. Sorai, K. Tani, H. Suga [*Mol. Cryst. Liq. Cryst. (UK)* vol.97 (1983) p.365]

[31]   M. Sorai, H. Yoshioka, H. Suga [in *Liquid Crystals and Ordered Fluids* Eds. A.C. Griffin, J.F. Johnson (Plenum, New York, 1984) vol.4 p.233]

[32]   P.W. Anderson, B.I. Halperin, C.M. Varma [*Philos. Mag. (UK)* vol.25 (1972) p.1]

[33]   U. Buchenau, Yu.M. Galperin, V.L. Gurevich, H.R. Schober [*Phys. Rev. B (USA)* vol.43 (1991) p.5039]

[34]   L. Gil, M.A. Ramos, A. Bringer, U. Buchenau [*Phys. Rev. Lett. (USA)* vol.70 (1993) p.182]

[35]   P.E. Cladis [*Phys. Rev. Lett. (USA)* vol.35 (1975) p.48]

[36]   N.H. Tinh, C. Destrade [*Nouv. J. Chim. (France)* vol.5 (1981) p.337]

[37]   G.R. Van Hecke, M. Sorai [*Liq. Cryst. (UK)* vol.12 (1992) p.503]

3.3 Equations of state for nematics

A. Würflinger and M. Sandmann

January 1999

A INTRODUCTION

Pressure - volume - temperature (pVT) data are indispensable for any discussion of the inter-molecular potential in condensed phases and provide an understanding of the roles of attractive and repulsive molecular interactions. This situation obtains because the behaviour of physical properties along isochores, isotherms and isobars can be distinguished. Such an analysis has been applied to the order parameter [1-3] and nematic potential in nematic phases [4-6]. Similarly it has been found in the framework of dielectric relaxation studies of nematic phases that the activation energy (at constant volume) is only half of the activation enthalpy (at constant pressure) [6-9]. Very often dilatometry is combined with X-ray studies for a better understanding of the molecular organisation in smectic layers or columnar phases [10-12]. Few accounts of high-pressure studies on liquid crystals mention the equation of state or pVT behaviour [13-16]. Nonetheless the equation of state of nematics has attracted considerable attention by theoreticians, because it is the least ordered liquid crystalline phase [17-20]. A wealth of atmospheric pressure volume data has been reported, from which we mention only some reviews [21-26], in which a detailed correlation of various physical properties is presented (packing fraction, molecular length to breadth ratios, intermolecular potential parameters).

B EXPERIMENTAL METHODS

Pollmann [14], Wedler [14], Guillon et al [27] and Grasso et al [28] have given short descriptions of experimental methods. For studies at normal pressure pycnometers [29,30] and vibrating tube density meters [24,31] are widely used. The former method is, however, questionable, because it is difficult to avoid wetting the glass tube wall and to define the meniscus. The latter requires a careful calibration with substances of known density. The capillary method can be improved by separating the sample from the surroundings with mercury [21,32]. It was applied by Orwoll et al to slightly elevated pressures to allow the determination of thermal pressure coefficients [25,33]. Jadzyn et al employed the hydrostatic displacement method based on Archimedes principle [34]. For studies at high pressures the volume change is essentially measured by three methods: (a) piezometer with mercury separation [35,36], (b) a bellows cell [37-39] or (c) the displacement of a piston [40,41]. Ter Minassian has developed a piezothermal method for the measurements of high-pressure expansivity [42]. Horn et al have derived pVT data of 4-pentyl-4′-cyanobiphenyl (5CB) and 4-methoxybenzylidene-4′-butylaniline (MBBA) from high-pressure refractive index measurements [43].

C RESULTS

In TABLE 1 we list some liquid crystals, for which volume data at higher pressures have been reported, including atmospheric pressure work for comparison. TABLE 1 is constrained to typical homologous series of nematics; more compounds have been reviewed by Pollmann and Wedler [14]. Only a few authors present their volume data in the form of comprehensive tables or equations of states, which is denoted in TABLE 1 by 'T' or 'E', respectively. As an example we show in FIGURE 1 the specific volume as a function of temperature at 1 atm for various 4-alkyl-4'-cyanobiphenyls (nCB) [44-47], trans-4-alkyl-(4'-cyanophenyl)-cyclohexanes (nPCH) [35], and trans,trans-4'-alkylbicyclohexyl-4-carbonitriles (nCCH) [31,48]. The plots clearly reveal the volume step corresponding to the nematic-isotropic transition. Comparing the specific volumes of 7CB, 7PCH and 7CCH at atmospheric pressure, we note an increase in this order. In the same order we also find an increase in the activation volumes for the low-frequency dielectric relaxation in the nematic phase [49]. Apparently these common trends are connected with the larger volume and flexibility of the cyclohexyl ring, which has replaced the phenyl ring. At higher pressures the same volume behaviour is observed for the nCBs and nPCHs [41]; no high-pressure densities have been reported for nCCHs.

TABLE 1 High-pressure volumetric studies on selected liquid crystals (F,T,E: data are presented in Figures or Tables or Equations, respectively; $\Delta S_{NI,V}$: calculation of constant-volume entropy).

Substance	Studies at elevated pressures			Studies at 1 atm
	Pressure	Remarks	Author(s)	
nCB: 4-alkyl-4'-cyanobiphenyls				
5CB	300 MPa	F, T, $\Delta S_{NI,V}$, $\gamma = 5.3$	Sandmann et al [44,45]	[24,34,69]
	30 MPa	F, $\gamma = 7.6$	Shirakawa et al [52]	
	5.8 MPa	F, T, $\Delta S_{NI,V}$	Orwoll et al [25]	
	500 MPa	E, $\Delta S_{NI,V}$, $\gamma = 6.0$	Horn et al [43]	
6CB	300 MPa	F, T, $\Delta S_{NI,V}$, $\gamma = 6.3$	Sandmann et al [46,41]	[24,34]
	30 MPa	F, $\gamma = 6.1$	Shirakawa et al [52]	
7CB	300 MPa	F, T, $\Delta S_{NI,V}$, $\gamma = 4.7$	Sandmann et al [46,41]	[24,31,34,69]
	30 MPa	F, $\gamma = 5.15$	Shirakawa et al [52]	
	5.8 MPa	F, T, $\Delta S_{NI,V}$	Orwoll et al [25]	
8CB	300 MPa	F, T, $\Delta S_{NI,V}$, $\gamma = 4.0$	Sandmann et al [47]	[24,34,70,71]
	30 MPa	F, $\gamma = 4.3$	Shirakawa et al [53]	
9CB	90 MPa	F, $\gamma = 4.26$	Shirakawa et al [54]	[24]
nPCH: trans-4-alkyl-(4'-cyanophenyl)-cyclohexanes				
3PCH	200 MPa	T, E	Kuss [35]	
	60 MPa	F, $\gamma = 8.3$	Shirakawa et al [55]	
4PCH	40 MPa	F	Shirakawa et al [56]	
5PCH	200 MPa	T, E	Kuss [35]	
	20 MPa	F, $\gamma = 5.24$	Ichimura et al [57]	
7PCH	200 MPa	F, T, E	Kuss [35]	[31]
		$\gamma = 3.32$	Table 1 in [58]	
8PCH	300 MPa	F, T, $\Delta S_{NI,V}$, $\gamma = 3.4$	Sandmann et al [44]	
nCCH: trans,trans-4'-alkyl-bicyclohexyl-4-carbonitriles				
n = 2, 3, 4, 5, 7				[48]
7CCH				[31]

TABLE 1 continued

Substance	Studies at elevated pressures			Studies at 1 atm
	Pressure	Remarks	Author(s)	
nOCB: 4-alkyloxy-4'-cyanobiphenyls				
5OCB	5.8 MPa	F, T, $\Delta S_{NI,V}$	Orwoll et al [25]	[69]
7OCB				[69]
8OCB	50 MPa	F	Shashidhar et al [59]	[69,72]
	160 MPa	F, $\gamma = 3.2$	Johnson et al [60]	
	5.8 MPa	F, T, $\Delta S_{NI,V}$	Orwoll et al [25]	
nAB: 4,4'-dialkylazoxybenzenes				
n = 7 - 9	250 MPa	F, $\gamma = 3; 2.7; 2.8$	Lampe et al [64]	
5AB				[75]
nOAB: 4,4'-dialkyloxyazoxybenzenes				
n = 1 - 6	123 MPa	F, T, $\gamma = \sim 4 ... 2$	Tranfield et al [37]	
PAA (n = 1)	64 MPa	F, $\gamma = 4$	McColl et al [1]	[21,73,74]
	40 MPa	F	Baskakov et al [61]	
	110 MPa	F, T	Kachinskii et al [62]	
5OAB				[73,74]
7OAB	260 MPa	F, $\gamma = 1.64$	Keyes et al [39]	[73]
	160 MPa	F, T, $\gamma = 2.07$	Johnson et al [60]	
8OAB	90 MPa	F	Baskakov et al [63]	
4-methoxyphenylazoxy-4'-butylbenzene				
BMAB	223 MPa	F, T	Kachinskii et al [62]	
4-alkyloxybenzylidene-4'-butylanilines				
MBBA	200 MPa	T, E	Kuss [36]	[76]
	260 MPa	F, $\gamma = 4.76$	Keyes et al [39]	
	200 MPa	F, E, $\gamma = 2.6$	Horn et al [43]	
	700 MPa	F	Lewis et al [65]	
	14 MPa	F	Kim et al [66]	
	7 MPa	F	Zawisza et al [67]	
EBBA	200 MPa	T, E	Kuss [36]	[77,78]
	315 MPa	T	Kachinskii et al [62]	
4-butyloxybenzylidene-4'-butylaniline				[79]
p-pentyloxybenzylidene-p-butylaniline				
	100 MPa	F, $\gamma = 3.6$	Hanawa et al [40]	[80]
N-(4-butyloxybenzylidene)-4-octylaniline				
	600 MPa	F, T	Nefedov et al [68]	
4-cyanobenzylidene-4'-octyloxyaniline				
CBOOA	260 MPa	F, $\gamma = 3.23$	Keyes et al [39]	[81,82]
α,ω-bis [4-cyanobiphenyl-4'-yloxy]alkanes				
CBA -9, -10	100 MPa	F, $\Delta S_{NI,V}$	Abe et al [38]	

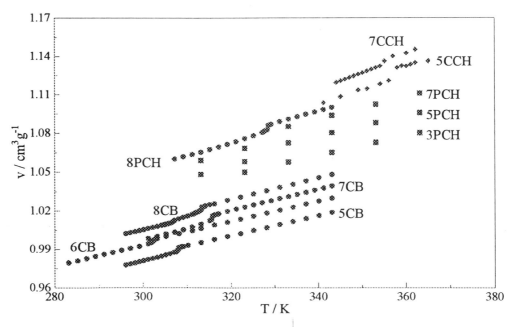

FIGURE 1 Specific volumes of nCBs [44-47], nPCHs [35] and nCCHs [31,48] at 1 atm.

Kiefer et al [26] report that terminally cyano-substituted liquid crystals show smaller molar volumes than nematogens which possess a non-polar terminal group. They argue that terminally polar molecules form dimers and therefore achieve a higher packing density. Belyaev et al [23] calculated the molecular packing coefficient

$$K_\rho = N_A V_{van} \rho/M \tag{1}$$

where V_{van} is the intrinsic van der Waals volume of the molecule and M is the molar mass. The authors find that increasing the length of the alkyl substituent or replacing the phenyl by a cyclohexyl ring lowers K_ρ. Demus et al [22] investigated thermotropic liquid crystals with lateral substituents and found that the branched molecules are less densely packed than the basic molecule. The critical packing fraction which is necessary for the occurrence of the nematic state must be higher, the lower the molecular length-to-breadth ratio. Dunmur et al [24] used densities for nCBs to employ a van der Waals theory of liquid crystals resulting in an unrealistically small ratio of the anisotropic to isotropic potential parameters.

Kuss expressed the pVT data of MBBA, 4-ethoxybenzylidene-4'-butylaniline (EBBA) [36] and the nPCHs (n = 3, 5, 7) [35] with the semi-empirical Tait equation in the form

$$(V_o - V)/V_o = C \log[(B + p)/(B + p_o)] \tag{2}$$

where V_o is the volume at atmospheric pressure, p_o. The constant C is independent of temperature whereas B depends on temperature and molecular structure. Tsykalo et al [50] fitted Kuss's data for EBBA to the polynomial

$$p(T) = A(T)\rho + B(T)\rho^3 + C(T)\rho^5 \tag{3}$$

154

where ρ is the density.

Many dilatometric studies focus on the volumetric behaviour along the phase boundaries of the nematic phase. For example, the volume data can be used to separate the entropy change at the nematic - isotropic transition into a constant-volume (or configurational) part, and a dilation part

$$\Delta S_{NI} = \Delta S_{NI,V} + \Delta S_{dil} = \Delta S_{NI,V} + (\partial p/\partial T)_V \, \Delta V_{NI} \tag{4}$$

or

$$\Delta S_{NI,V} \equiv \Delta S_{conf} = [(\partial p/\partial T)_{NI} - (\partial p/\partial T)_V] \, \Delta V_{NI} \tag{5}$$

This procedure is reviewed in detail by Orwoll et al [25] and Sandmann et al [44]. For the calculation it is necessary to determine the slopes $(\partial p/\partial T)_V$ along the phase boundary.

In FIGURE 2 we present the corresponding plots for some nCBs showing that (a) $(\partial p/\partial T)_V$ is different from the slope $(\partial p/\partial T)_{NI}$ for the nematic-isotropic transition, that is the density is not constant along the phase boundary, and (b) the slopes $(\partial p/\partial T)_V$ are not too different in the nematic and isotropic states. It was applied by many workers to various liquid crystals (see TABLE 1) with the result that the constant-volume entropy (or entropy of disordering [25]) is found to be about half of the total entropy change. For 5CB and 8PCH Sandmann et al found a small increase of the ratio $\Delta S_{conf}/\Delta S_{NI}$ with increasing pressure [44], whereas Horn et al report a small decrease [43]. Abe et al determined constant-volume entropies for the liquid crystal dimers α,ω-(4-cyanobiphenyl-4′-yloxy)alkanes (CBA-9 and CBA-10) and compared them with conformation entropies derived from deuterium quadrupolar splitting data [38]. Although the agreement is fairly good the authors point out that this might be fortuitous; the conformational entropy should decrease during the compression required to achieve the constant-volume disorder.

Another interesting quantity is the thermodynamic potential parameter

$$\gamma = -(\partial \ln T_{NI}/\partial \ln V_{NI}) \tag{6}$$

which can be derived from the slope of a log-log plot of the nematic-isotropic transition temperature against the molar volume at the transition. Values of γ for several liquid crystals are gathered in TABLE 1. In every case γ is significantly larger than 2, showing that the volume dependence of the interaction coefficient

$$\varepsilon = \varepsilon_o \, V^{-\gamma} \tag{7}$$

cannot be described solely by attractive forces after Maier-Saupe [51]. Inspecting the different homologous series (nCB, nPCH, nOAB) shows that γ decreases with increasing alkyl chain length. Shirakawa et al [58] compared 7CB, 7PCH, and 4-fluoro-phenyl-trans-4′-heptylcyclohexanecarboxylate (7CCF with γ of 3) and attributed the differences in γ to the flexibility or softness of the molecular cores. For further discussion of γ and the related quantity

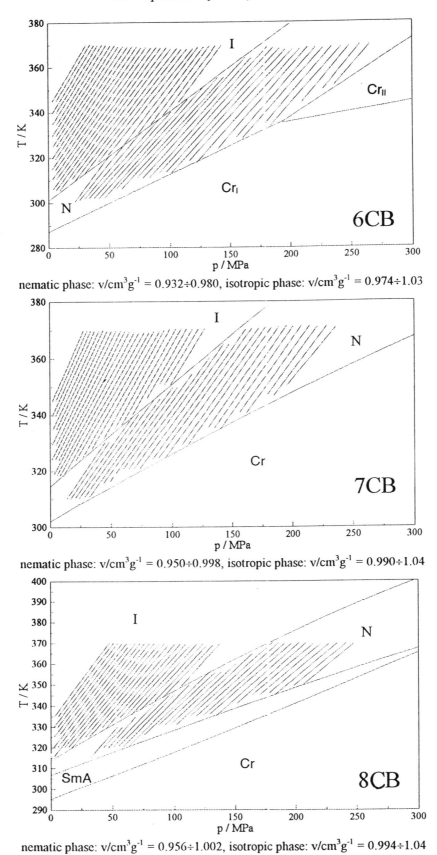

nematic phase: $v/cm^3g^{-1} = 0.932 \div 0.980$, isotropic phase: $v/cm^3g^{-1} = 0.974 \div 1.03$

nematic phase: $v/cm^3g^{-1} = 0.950 \div 0.998$, isotropic phase: $v/cm^3g^{-1} = 0.990 \div 1.04$

nematic phase: $v/cm^3g^{-1} = 0.956 \div 1.002$, isotropic phase: $v/cm^3g^{-1} = 0.994 \div 1.04$

FIGURE 2 T-p diagrams for 6CB, 7CB and 8CB with isochores in steps of 0.002 cm^3 g^{-1}.

$$\Gamma = - (\partial \ln T / \partial \ln V) \overline{P}_2 \qquad \qquad (8)$$

see [1,7,83].

The low-temperature phase boundary of the nematic phase can be either a crystal or a smectic phase. In the case of the smectic A - nematic transition considerable effort has been devoted to the analysis of its order, either first or second [84,85]. In previous high-pressure measurements on 4-octyloxy-4'-cyanobiphenyl (8OCB) there was no indication of a volume change at the SmA-N transition [59]. Zywocinski et al conclude from careful volume measurements that even at ambient pressure this transition is always second order [71]. They changed the temperature in steps of 5 mK and found an S-shaped dependence of the molar volume in the vicinity of the SmA-N transition. We have recently investigated 8CB both with a high-pressure dilatometer and with a vibrating density meter [41,47]. At atmospheric pressure we find a discontinuity in the molar volume in agreement with other authors [24,70]. However, it is difficult to avoid an inclusion of pretransitional volume changes within the limits of the experimental temperature control. At moderate pressures a discontinuity can be detected with the vibrating density meter, which decreases with increasing pressure. A similar behaviour has been observed in heat capacity measurements at high pressure [86].

Dilatometric studies are clearly useful to establish phase diagrams and in this respect they are complementary to high-pressure DTA or DSC investigations [15,44,87]. Thus a new solid phase has been discovered recently in 4-hexyl-4'-cyanobiphenyl (6CB) [46]. Shirakawa et al investigated trans-4-alkyl-(4'-cyanophenyl)-cyclohexane (4PCH), which exhibits a monotropic nematic-isotropic transition that becomes enantiotropic with increasing pressure [56]. Stabilisation of nematic phases with increasing pressure has also been observed for laterally aryl branched liquid crystals [88]. Their mesophase structures are apparently connected with latent holes between the wedge-shaped molecules [89], which should be sensitive to density changes. Therefore, we expect that high-pressure dilatometry gives more insight into packing conditions and phase behaviour.

ACKNOWLEDGEMENT

W. Weissflog and A. Würflinger thank the Deutsche Forschungsgemeinschaft for financial support.

REFERENCES

[1]   J.R. McColl, C.S. Shih [*Phys. Rev. Lett. (USA)* vol.29 (1972) p.85]; J.R. McColl [*Phys. Lett. A (Netherlands)* vol.38 (1972) p.55]

[2]   J.W. Emsley, G.R. Luckhurst, S.W. Smith [*Mol. Phys. (UK)* vol.70 (1980) p.967]; J.W. Emsley, G.R. Luckhurst, B.A. Timini [*J. Phys. (France)* vol.48 (1987) p.473]

[3]   S. Chandrasekhar, N.V. Madhusudana [*Mol. Cryst. Liq. Cryst. (UK)* vol.24 (1973) p.179]

[4]   A. Würflinger [*Z. Nat.forsch. A (Germany)* vol.53 (1998) p.141]; A. Würflinger, S. Urban [*Z. Nat.forsch. A (Germany)* vol.53 (1998) p.883]

[5]   S. Urban, D. Büsing, A. Würflinger, B. Gestblom [*Liq. Cryst. (UK)* vol.25 (1998) p.253]

[6]   S. Urban, A. Würflinger, D. Büsing, T. Brückert, M. Sandmann, B. Gestblom [*Pol. J. Chem. (Poland)* vol.72 (1998) p.241]

[7]   S. Urban, A. Würflinger [*Adv. Chem. Phys. (USA)* vol.98 (1997) p.143]

[8]   A. Würflinger [*Int. Rev. Phys. Chem. (UK)* vol.12 (1993) p.89]

[9]   T. Brückert, D. Büsing, S. Urban, A. Würflinger [*Proc. SPIE - Int. Soc. Opt. Eng. (USA)* vol.3318 (1998) p.198]

[10]   D. Guillon, A. Skoulios, J.J. Benattar [*J. Phys. (France)* vol.47 (1986) p.133]; D. Guillon, A. Skoulios [*J. Phys. Colloq. (France)* vol.3 (1976) p.83]

[11]   H. Allouchi [Thesis d'ordre 1125, University of Bordeaux, France, 1994]

[12]   B.R. Ratna, C. Nagabhushana, V.N. Raja, R. Shashidhar, S. Chandrasekhar, G. Heppke [*Mol. Cryst. Liq. Cryst. (UK)* vol.138 (1986) p.245]

[13]   W.B. Daniels, P.E. Cladis, P. Keyes [in *High Pressure Sci. Technol., Proc. 7th Int. AIRAPT Conf.* 1979, Eds. B. Vodar, P. Marteau (Pergamon, Oxford, UK, 1980) vol.2 p.655]

[14]   P. Pollmann [in *Handbook of Liquid Crystals, Volume 1: Fundamentals* Eds. D. Demus, J. Goodby, G.W. Gray, H.W. Spiess, V. Vill (Wiley-VCH, Weinheim, Germany, 1998) p.355]; W. Wedler [in *Handbook of Liquid Crystals, Volume 1: Fundamentals* Eds. D. Demus, J. Goodby, G.W. Gray, H.W. Spiess, V. Vill (Wiley-VCH, Weinheim, Germany, 1998) p.334]

[15]   S. Chandrasekhar, R. Shashidhar [*Adv. Liq. Cryst. (USA)* vol.4 (1979) p.83]

[16]   D. Demus, A. Hauser [in *Selected Topics in Liquid Crystal Research* Ed. H.D. Koswig (Akademie-Verlag Berlin, 1990) p.39]

[17]   G. Vertogen, W.H. de Jeu [*Springer Ser. Chem. Phys. (Germany)* vol.45 (1988)]

[18]   S.D.P. Flapper, G. Vertogen [*Phys. Lett. A (Netherlands)* vol.79 (1980) p.87; *J. Chem. Phys. (USA)* vol.75 (1981) p.3599; *Phys. Rev. A (USA)* vol.24 (1981) p.2089]

[19]   J.G.J. Ypma, G. Vertogen [*Phys. Rev. A (USA)* vol.17 (1978) p.1490]

[20]   J. Liu, L. Lin [*Mol. Cryst. Liq. Cryst. (UK)* vol.89 (1982) p.275 and p.259]

[21]   B. Bahadur [*J. Chim. Phys. Phys.-Chim. Biol. (France)* vol.73 (1976) p.255; *Mol. Cryst. Liq. Cryst. (UK)* vol.35 (1976) p.83]

[22]   D. Demus, S. Diele, A. Hauser, I. Latif, C. Selbmann, W. Weissflog [*Cryst. Res. Technol. (Germany)* vol.20 (1985) p.1547]; D. Demus, A. Hauser, C. Selbmann, W. Weissflog [*Cryst. Res. Technol. (Germany)* vol.19 (1984) p.271]

[23]   V. Belyaev, M.F. Grebenkin, V.F. Petrov [*Zh. Fiz. Khim. (Russia)* vol.64 (1990) p.958 and p.963]; V. Belyaev [*Izv. Akad. Nauk SSSR Ser. Fiz. (Russia)* vol.60 (1996) p.12]

[24]   D.A. Dunmur, W.H. Miller [*J. Phys. Colloq. (France)* vol.C3 no.40 (1979) p.141]

[25]   R.A. Orwoll, V.J. Sullivan, G.C. Campbell [*Mol. Cryst. Liq. Cryst. (UK)* vol.149 (1987) p.121]

[26]   R. Kiefer, G. Baur [*Mol. Cryst. Liq. Cryst. (UK)* vol.188 (1990) p.13]

[27]   D. Guillon, A. Skoulios [*Mol. Cryst. Liq. Cryst. (UK)* vol.39 (1977) p.139]

[28]   D. Grasso, S. Fasone, G. Di Pasquale, F. Castelli [*Thermochim. Acta (Netherlands)* vol.140 (1989) p.31]

[29]   A. Hauser, R. Rettig, C. Selbmann, W. Weissflog, J. Wulf, D. Demus [*Cryst. Res. Technol. (Germany)* vol.19 (1984) p.261]; D. Demus, R. Rurainski [*Z. Phys. Chem. (Germany)* vol.253 (1973) p.53]; H. Sackmann, F. Sauerwald [*Z. Phys. Chem. (Germany)* vol.195 (1950) p.295]

[30]   N.S.V. Rao, V.G.K.M. Pisipati, J.S.R. Murthy, P. Bhaskar Rao, P.R. Alapati [*Liq. Cryst. (UK)* vol.5 (1989) p.539]

[31]   I.H. Ibrahim, W. Haase [*Mol. Cryst. Liq. Cryst. (UK)* vol.66 (1981) p.189]

[32]   S. Torza, P.E. Cladis [*Phys. Rev. Lett. (USA)* vol.32 (1974) p.1406]

[33]   R.A. Orwoll, P.J. Flory [*J. Am. Chem. Soc. (USA)* vol.89 (1967) p.6814]

[34]   J. Jadzyn, W. Labno [*Chem. Phys. Lett. (Netherlands)* vol.73 (1980) p.307]; J. Jadzyn, J. Malecki [*Roczn. Chem. (Poland)* vol.48 (1974) p.531]; W. Labno, J. Jadzyn [*Pr. Kom. Mat.-Przyr., Poznan, Tow. Przyj. Nauk, Fiz. Dielektr. Radiospektrosk. (Poland)* vol.12 (1981) p.75]

[35]   E. Kuss [*Mol. Cryst. Liq. Cryst. (UK)* vol.76 (1981) p.199]

[36]   E. Kuss [*Mol. Cryst. Liq. Cryst. (UK)* vol.47 (1978) p.71]

[37]   R.V. Tranfield, P.J. Collings [*Phys. Rev. A (USA)* vol.25 (1982) p.2744]

[38]   A. Abe, Su.Y. Nam [*Macromolecules (USA)* vol.28 (1995) p.90; *Macromolecules (USA)* vol.29 (1996) p.3337]; A. Abe, H. Furuya, R.N. Shimizu, Su.Y. Nam [*Macromolecules (USA)* vol.28 (1995) p.96]

[39]   P.H. Keyes, W.B. Daniels [*J. Phys. Colloq. (France)* vol.C3 no.40 (1979) p.380]

[40]   C. Hanawa, T. Shirakawa, T. Tokuda [*Chem. Lett. (Japan)* vol.10 (1977) p.1223]

[41]   M. Sandmann [Doctoral thesis, University of Bochum, Germany, 1998]

[42]   L. Ter Minassian, P. Pruzan [*J. Chem. Thermodyn. (UK)* vol.9 (1977) p.375]; R. Shashidhar, L. Ter Minassian, B.R. Ratna, A.N. Kalkura [*J. Phys. Lett. (France)* vol.43 (1982) p.239]

[43]   R.G. Horn [*J. Phys. (France)* vol.39 (1978) p.167]; R.G. Horn, T.E. Faber [*Proc. R. Soc. Lond. A (UK)* vol.368 (1979) p.199]

[44]   M. Sandmann, F. Hamann, A. Würflinger [*Z. Nat.forsch. A (Germany)* vol.52 (1997) p.739]

[45]   M. Sandmann, D. Büsing, T. Brückert, A. Würflinger, S. Urban, B. Gestblom [*Proc. SPIE - Int. Soc. Opt. Eng. (USA)* vol.3318 (1998) p.223]

[46]   M. Sandmann, A. Würflinger [*Z. Nat.forsch. A (Germany)* vol.53 (1998) p.233]

[47]   M. Sandmann, A. Würflinger [*Z. Nat.forsch. A (Germany)* vol.53 (1998) p.787]

[48]   S. Sen, K. Kali, S.K. Roy [*Bull. Chem. Soc. Jpn. (Japan)* vol.61 (1988) p.3681]

[49]   D. Büsing, T. Brückert, A. Würflinger, B. Gestblom [*Proc. SPIE - Int. Soc. Opt. Eng. (USA)* vol.3318 (1998) p.213]

[50]   A.L. Tsykalo, V.A. Vasserman [*Zh. Fiz. Khim. (Russia)* vol.64 (1990) p.1118]

[51]   W. Maier, A. Saupe [*Z. Nat.forsch. A (Germany)* vol.14 (1959) p.982 and vol.15 (1960) p.287]

[52]   T. Shirakawa, T. Hayakawa, T. Tokuda [*J. Phys. Chem. (USA)* vol.87 (1983) p.1406]

[53]   T. Shirakawa, T. Inoue, T. Tokuda [*J. Phys. Chem. (USA)* vol.86 (1982) p.1700]

[54]   T. Shirakawa, Y. Kikuchi, T. Seimiya [*Thermochim. Acta (Netherlands)* vol.197 (1992) p.399]

[55]   T. Shirakawa, M. Arai, T. Tokuda [*Mol. Cryst. Liq. Cryst. (UK)* vol.104 (1984) p.131]

[56]   T. Shirakawa, H. Ichimura, I. Ikemoto [*Mol. Cryst. Liq. Cryst. (UK)* vol.142 (1987) p.101]

[57]   H. Ichimura, T. Shirakawa, T. Tokuda, T. Seimiya [*Bull. Chem. Soc. Jpn. (Japan)* vol.56 (1983) p.2238]

[58]   T. Shirakawa, H. Eura, H. Ichimura, T. Ito, K. Toi, T. Seimiya [*Thermochim. Acta (Netherlands)* vol.105 (1986) p.251]

[59]   R. Shashidhar, P.H. Keyes, W.B. Daniels [*Mol. Cryst. Liq. Cryst. Lett. (UK)* vol.3 (1986) p.169]

[60]   C.S. Johnson, P.J. Collings [*J. Chem. Phys. (USA)* vol.79 (1983) p.4056]

[61]   V.Ya. Baskakov, V.K. Semenchenko, N.A. Nedostup [*Kristallografiya (USSR)* vol.19 (1974) p.185]

[62] V.N. Kachinskii, V.A. Ivanov, A.N. Zisman, S.M. Stishov [*Zh. Eksp. Teor. Fiz. (USSR)* vol.75 (1978) p.545]; S.M. Stishov, V.A. Ivanov, V.N. Kachinskii [*Pis'ma Zh. Eksp. Teor. Fiz. (USSR)* vol.24 (1976) p.329]

[63] V.Y. Baskakov, V.K. Semenchenko, V.M. Byankin [*Zh. Fiz. Khim. (USSR)* vol.50 (1976) p.200]

[64] W.M. Lampe, P.J. Collings [*Phys. Rev. A (USA)* vol.34 (1986) p.524]

[65] E.A.S. Lewis, H.M. Strong, G.H. Brown [*Mol. Cryst. Liq. Cryst. (UK)* vol.53 (1979) p.89]

[66] Y.B. Kim, R. Gakujutsu [*Chosen Shogakkai (Japan)* vol.9 (1979) p.114]; Y.B. Kim, K. Ogino [*Mol. Cryst. Liq. Cryst. (UK)* vol.53 (1979) p.122]

[67] A.C. Zawisza, J. Stecki [*Solid State Commun. (USA)* vol.19 (1976) p.1173]

[68] S.N. Nefedov, A.N. Zisman, S.M. Stishov [*Zh. Eksp. Teor. Fiz. (USSR)* vol.86 (1984) p.125]

[69] S. Sen, P. Brahma, S.K. Roy, D.K. Mukherjee, S.B. Roy [*Mol. Cryst. Liq. Cryst. (UK)* vol.100 (1983) p.327]

[70] A.J. Leadbetter, J.L.A. Durrant, M. Rugman [*Mol. Cryst. Liq. Cryst. Lett. (UK)* vol.34 (1977) p.231]

[71] A. Zywocinski, S.A. Wieczorek [*J. Phys. Chem. B (USA)* vol.101 (1997) p.6970; *Phys. Rev. A (USA)* vol.31 (1985) p.479; *Phys. Rev. A (USA)* vol.36 (1987) p.1901]

[72] K. Czuprynski, R. Dabrowski, J. Baran, A. Zywocinski, J. Przedmojski [*J. Phys. (France)* vol.47 (1986) p.1577]

[73] R.M. Stimpfle, R.A. Orwoll, M.E. Schott [*J. Phys. Chem. (USA)* vol.83 (1979) p.613]

[74] O. Phaovibul, K. Pongthana-Ananta, I.M. Tang [*Mol. Cryst. Liq. Cryst. (UK)* vol.62 (1980) p.25]

[75] K.W. Sadowska, A. Zywocinski, J. Stecki, R. Dabrowski [*J. Phys. (France)* vol.43 (1982) p.1673]

[76] E. Gulari, B. Chu [*J. Chem. Phys. (USA)* vol.62 (1975) p.795]

[77] B. Bahadur, S. Chandra [*J. Phys. C (UK)* vol.9 (1976) p.5]

[78] H.S. Subramhanyam, J.S. Prasad [*Mol. Cryst. Liq. Cryst. (UK)* vol.37 (1976) p.23]

[79] J.V. Rao, N.V.S. Rao, V.G.K.M. Pisipati, C.R.K. Murty [*Ber. Bunsenges. Phys. Chem. (Germany)* vol.84 (1980) p.1157]

[80] J.V. Rao, L.V. Choudary, K.R.K. Rao, P. Venkatacharyulu [*Acta Phys. Pol. A (Poland)* vol.72 (1987) p.517]

[81] A.K. Jaiswal, G.L. Patel [*Acta Phys. Pol. A (Poland)* vol.69 (1986) p.723]

[82] D. Guillon, P. Seurin, A. Skoulios [*Mol. Cryst. Liq. Cryst. (UK)* vol.51 (1979) p.149]; D. Guillon, P.E. Cladis, D. Aadsen, W.B. Daniels [*Phys. Rev. A (USA)* vol.21 (1980) p.658]

[83] G.R. Luckhurst [in *The Molecular Physics of Liquid Crystals* Eds. G.R. Luckhurst, G.W. Gray (Academic Press, 1979) ch.4]

[84] J. Thoen [*Int. J. Mod. Phys. B (Singapore)* vol.9 (1995) p.2157]; H. Marynissen, J. Thoen, W. van Dael [*Mol. Cryst. Liq. Cryst. (UK)* vol.97 (1983) p.149]

[85] P.R. Alapati, D. Saran, S.V. Raman [*Mol. Cryst. Liq. Cryst. (Switzerland)* vol.287 (1996) p.239]

[86] G.B. Kasting, C.W. Garland, K.J. Lushington [*J. Phys. (France)* vol.41 (1980) p.879]; G.B. Kasting, K.J. Lushington, C.W. Garland [*Phys. Rev. B (USA)* vol.22 (1980) p.321]

[87] C. Schmidt, M. Rittmeier-Kettner, H. Becker, J. Ellert, R. Krombach, G.M. Schneider [*Thermochim. Acta (Netherlands)* vol.238 (1994) p.321]

[88]   C.R. Ernst, G.M. Schneider, A. Würflinger, W. Weissflog [*Ber. Bunsenges. Phys. Chem. (Germany)* vol.102 (1998) p.1870]; W. Weissflog, D. Demus [*Liq. Cryst. (UK)* vol.3 (1988) p.275]

[89]   S. Haddawi, S. Diele, H. Kresse, G. Pelzl, W. Weissflog [*Cryst. Res. Technol. (Germany)* vol.29 (1994) p.745]

3.4 Measurements and interpretation of acoustic properties in nematics

K. Okano and Y. Kawamura

February 1999

A HYDRODYNAMIC BEHAVIOUR OF NEMATICS

We consider the propagation of small oscillations in an otherwise uniform (equilibrium) nematic with temperature, T, and pressure, p. We assume that the director in equilibrium **n** is directed along the z-axis. A complete set of hydrodynamic equations of motion of a nematic comprises the equation of the director as well as the conservation laws of mass, energy and momentum, supplemented by the relevant constitutive equations. (For the details of the following discussion refer to [1-5].)

As for the constitutive relation that expresses the dissipative (viscous) stress tensor σ_{ik}' in terms of the velocity gradient tensor

$$v_{ik} = (1/2)(\nabla_i v_k + \nabla_k v_i)$$

we employ the following form given by Martin et al [1] (i.e. MPP notation of viscosity coefficients), which is particularly suited for the analysis of wave propagation:

$$
\begin{pmatrix}
\sigma_{xx}' \\
\sigma_{yy}' \\
\sigma_{zz}' \\
\sigma_{yz}' \\
\sigma_{zx}' \\
\sigma_{xy}'
\end{pmatrix}
=
\begin{pmatrix}
\eta_2 + \eta_4 & \eta_4 - \eta_2 & \eta_5 & & & \\
\eta_4 - \eta_2 & \eta_2 + \eta_4 & \eta_5 & & & \\
\eta_5 & \eta_5 & \eta_1 & & & \\
& & & 2\eta_3 & & \\
& & & & 2\eta_3 & \\
& & & & & 2\eta_2
\end{pmatrix}
\cdot
\begin{pmatrix}
v_{xx} \\
v_{yy} \\
v_{zz} \\
v_{yz} \\
v_{zx} \\
v_{xy}
\end{pmatrix}
\tag{1}
$$

In considering the propagation of small oscillations we linearise these equations with respect to the oscillating parts of the variables: the velocity **v** and the deviation of the director $\delta\mathbf{n}$ from its equilibrium value. We assume that the oscillating variables have the form of a plane wave:

$$
\begin{pmatrix}
\mathbf{v} \\
\delta\mathbf{n} \\
\cdot \\
\cdot \\
\cdot
\end{pmatrix}
\sim \exp(i\omega t - i\mathbf{q} \cdot \mathbf{r})
\tag{2}
$$

Then the linearised equations of motion yield the sets of dispersion relations described in A1 and A2.

A1   Longitudinal Mode in which v is Parallel to q

This mode corresponds to the ordinary sound in a nematic and the dispersion relation is

$$\omega = cq + \Delta\omega \tag{3}$$

where

$$c = (\partial p/\partial \rho)_s^{1/2} \tag{4}$$

and

$$\Delta\omega = (iq^2/2\rho) \, [(\eta_2 + \eta_4)\sin^2\theta$$

$$+ (2\eta_5 - \eta_4 + 4\eta_3 - \eta_2 - \eta_1)\sin^2\theta\cos^2\theta + \eta_1\cos^2\theta \tag{5}$$

$$+ (\{1/c_v\} - \{1/c_p\})(\tilde{\kappa}_\parallel\cos^2\theta + \tilde{\kappa}_\perp\sin^2\theta)]$$

In these equations, ρ is the equilibrium density, and the differentiation is taken at constant entropy (i.e. under adiabatic conditions), θ being the angle between the equilibrium director (z-axis) and the direction of wave propagation $\mathbf{q}$. c_v and c_p are respectively the constant volume and the constant pressure heat capacity per unit volume of nematic. $\tilde{\kappa}_\parallel$ and $\tilde{\kappa}_\perp$ are the thermal conductivities in the direction of the unperturbed director and perpendicular to the director, respectively.

For an experiment in which speed and attenuation of the sound wave are measured at a constant ω, we can rewrite the factor $\exp(i\omega t - i\mathbf{q} \cdot \mathbf{r})$ as

$$\exp(i\omega t - i\mathbf{q} \cdot \mathbf{r}) = \exp[i\omega(t - \mathbf{e} \cdot \mathbf{r}/c) - \alpha\mathbf{e} \cdot \mathbf{r}], \ (\mathbf{e} = \mathbf{q}/q) \tag{6}$$

Thus c is the speed of sound and α is the attenuation coefficient of the sound wave:

$$\alpha\rho c^3/\omega^2 = -i(\Delta\omega/c)(\rho c^3/\omega^2)$$

$$= (1/2) \, [\eta_1 - (\eta_1 - \eta_2 - \eta_4)\sin^2\theta - (1/4) \, (n_1 + \eta_2 - 4\eta_3 + \eta_4 - 2\eta_5) \times \sin^2 2\theta \tag{7}$$

$$+ (\{1/c_v\} - \{1/c_p\}) \, (\tilde{\kappa}_\parallel\cos^2\theta + \tilde{\kappa}_\perp\sin^2\theta)]$$

Note that the speed of the ordinary sound wave in nematics is isotropic, whereas its attenuation coefficient is anisotropic. In organic liquids including nematics the last term in EQN (7) due to the conduction of heat is generally negligibly small compared to the viscosity terms.

A2   Transverse Mode in which v is Perpendicular to q

In such oscillations the fluid can be regarded as incompressible: $\nabla \cdot \mathbf{v} = 0$. Because $\mathbf{q} \cdot \mathbf{v} = 0$, the vector $\mathbf{v}$ and $\delta\mathbf{n}$ have two independent components each. Thus the equations of motion for $\mathbf{v}$ and $\delta\mathbf{n}$

constitute a set of four linear equations which define four oscillation modes in each of which the velocity and the director undergo coupled oscillation. Usually, however, the situation is considerably simplified by the fact that the dimensionless ratio

$$\mu = (K\rho/\eta^2) \tag{8}$$

is very small (in a typical case, $\mu \sim 10^{-4}$); here K and η denote the order of magnitude of the Frank constants and the viscosity coefficients. Because $\mu \ll 1$, the so-called fast mode with dispersion relations of the type

$$\omega_f \sim i(\eta/\rho)q^2 \tag{9}$$

associated with the shear velocity oscillations, is decoupled from the slow modes with dispersion relations of the type

$$\omega_s \sim i(K/\rho)q^2 \tag{10}$$

associated with the oscillation of the director. Note that

$$\omega_s/\omega_f \sim \mu \ll 1 \tag{11}$$

and hence the adjectives fast and slow.

We now discuss the general feature of an oscillation having the dispersion relation of the form

$$\omega = iAq^2, \text{ A: real positive} \tag{12}$$

In an acoustic experiment in which the measurement is performed at a constant frequency ω, q becomes complex:

$$q = (\omega/2A)^{1/2} (1 - i) \tag{13}$$

and

$$\exp(i\omega t - i\mathbf{q} \cdot \mathbf{r}) = \exp[i\omega\{t - \mathbf{e} \cdot \mathbf{r}/(2A\omega)^{1/2}\} - (\omega/2A)^{1/2} \mathbf{e} \cdot \mathbf{r}], \ (\mathbf{e} = \mathbf{q}/q) \tag{14}$$

The wavelength λ and the attenuation coefficient α are given by

$$\lambda = (2A/\omega)^{1/2} \tag{15}$$

and

$$\alpha = (\omega/2A)^{1/2} \tag{16}$$

Defining the penetration depth δ by

$$\alpha\delta = 1$$

we have

$$\delta = (2A/\omega)^{1/2} \tag{17}$$

and so

$$\delta = \lambda \tag{18}$$

Thus an eigenmode having a dispersion relation of the form given in EQN (12) represents a strongly damped oscillation as a diffusive wave.

We now turn to the modes of shear oscillations or the fast modes in nematics. We have two dispersion relations for the fast shear modes. For oscillations **v** perpendicular to both **q** and **n**, we have

$$\omega = (iq^2/\rho)(\eta_2 \sin^2\theta + \eta_3 \cos^2\theta) \tag{19}$$

and for **v** in the plane through **q** and **n**

$$\omega = (iq^2/\rho)[(\eta_1 + \eta_2 + \eta_4 - 2\eta_5)\sin^2\theta \cos^2\theta + \eta_3 \cos^2 2\theta] \tag{20}$$

As for the director oscillation modes (the slow modes), since they are not genuine acoustic waves, we shall not enter into any detail here.

In the experimental study of the strongly damped shear oscillations having dispersion relations of the form of EQNS (19) and (20), the shear mechanical impedance, Z, defined as the negative ratio of the shear stress to the velocity, is conveniently measured instead of the speed and attenuation of the wave. Detailed expressions are given later.

For the shear oscillation of **v** orthogonal to the plane through **q** and **n**, whose dispersion relation is given by EQN (19), we have

$$Z_\perp = (i\omega\rho\eta_\perp)^{1/2} \tag{21}$$

where

$$\eta_\perp = \eta_2 \sin^2\theta + \eta_3 \cos^2\theta \tag{22}$$

The real and imaginary parts of $Z_\perp$ are, respectively

$$R_\perp = \mathrm{Re}Z_\perp = \left(\frac{1}{2}\omega\rho\eta_\perp\right)^{1/2} \tag{21a}$$

and

$$X_\perp = \mathrm{Im} Z_\perp = \left(\frac{1}{2}\omega\rho\eta_\perp\right)^{1/2} \tag{21b}$$

For the oscillation of **v** polarised in the plane through **q** and **n**, whose dispersion relation is given by EQN (20), we have

$$Z_\parallel = (i\omega\rho\eta_\parallel)^{1/2} \tag{23}$$

where

$$\eta_\parallel = (\eta_1 + \eta_2 + \eta_4 - 2\eta_5)\sin^2\theta\cos^2\theta + \eta_3\cos^2 2\theta \tag{24}$$

The real and imaginary parts of $Z_\parallel$ are, respectively

$$R_\parallel = \mathrm{Re} Z_\parallel = \left(\frac{1}{2}\omega\rho\eta_\parallel\right)^{1/2} \tag{23a}$$

and

$$X_\parallel = \mathrm{Im} Z_\parallel = \left(\frac{1}{2}\omega\rho\eta_\parallel\right)^{1/2} \tag{23b}$$

Finally we summarise briefly the relations between the viscosity coefficients appearing in other formulations of nematodynamics.

In the ultrasonic literature the notation by Forster et al [6] is frequently used. The relation to that of Martin et al [1] is

$$\eta_1 - \eta_2 + \eta_4 - 2\eta_5 = 2v_1$$
$$\eta_2 = v_2, \; \eta_3 = v_3, \; \eta_4 = v_4, \; \eta_5 = v_5 \tag{25}$$

Note that the factor $\eta_1 + \eta_2 - 4\eta_3 + \eta_4 - 2\eta_5$ in the third term of EQN (7) is equal to $2v_1 + 2v_2 - 4v_3$, which is a shear viscosity coefficient.

In the case of incompressible nematics the Leslie viscosity coefficients $\alpha_1 \sim \alpha_6$ have traditionally been used. Due to the Parodi relation

$$\alpha_6 - \alpha_5 = \alpha_3 + \alpha_2 \tag{26}$$

which is a consequence of the Onsager reciprocal relation of irreversible thermodynamics, there are five independent coefficients among the six Leslie coefficients. The Leslie coefficients are related to those of Martin et al [1] by

$$\eta_2 = \frac{1}{2}\alpha_4$$

$$\eta_1 - \eta_2 + \eta_4 - 2\eta_5 = \alpha_1 + \alpha_4 + \alpha_5 + \alpha_6 = 2\nu_1 \tag{27}$$

$$\eta_3 = \frac{1}{2}(\alpha_4 + \alpha_5 + \lambda\alpha_2), \ \lambda = -(\alpha_3 + \alpha_2)/(\alpha_3 - \alpha_2)$$

B   MEASUREMENT TECHNIQUES

In measurements of the acoustic properties of nematics, a variety of ultrasonic measurement techniques, developed for application to conventional liquids, have been used. For the details of the methods refer to the references cited together with [2] and [7].

For longitudinal waves, which can propagate through nematic liquids, a pulse propagation method is applicable. In this method the propagation time and the attenuation in amplitude of pulsed ultrasonic waves after travelling through a known path length are measured, from which we can obtain the speed, c, and the attenuation, α, at the measuring frequency, $f(= \omega/2\pi)$, typically in the MHz region and above.

In the lower frequency region, e.g. below about 5 MHz, the resonance cavity method is used since the lower the frequency the longer the wavelength approaches to the thickness of the acoustic path of the sample. In this method using a resonance cavity with known path length, l, the resonance frequencies (i.e. the frequencies at which standing waves are excited along the path) are measured by changing the frequency continuously together with the width at half maximum of the resonance curve at nth resonance (where n is an integer), Δf. c and α are then given by

$$c = 2 \ l \ | \ f_n - f_{n-1} \ |, \ \ \alpha = \pi\Delta f/\lambda \ f_n \tag{28}$$

where f_n is the nth resonance frequency and λ is the wavelength.

For transverse (shear) waves, which are strongly damped diffusive waves as described in the preceding section, a propagation method for acoustic measurements is not feasible due to the high attenuation of shear waves in nematics. Under such a condition the shear mechanical impedance is measured to obtain the acoustic properties of nematics.

At higher frequencies (typically f > 1 MHz), the shear wave reflectance method is used, in which the shear mechanical impedance is determined by measuring the reflection coefficient, r, and the phase retardation of the reflected wave caused by the liquid, Θ, for shear waves reflected at a solid-nematic liquid interface.   The reflection coefficient and the phase retardation measured by the shear reflectance method are related to the shear impedance by

$$Z/\cos \phi = Z_0[(1 - r^2 + 2ir\sin \Theta)/(1 + r^2 + 2r\cos \Theta)] \tag{29}$$

167

where ϕ is the incident angle of sound waves to the reflection surface and Z_0 is the shear mechanical impedance of the buffer solid, on the surface of which the oriented nematic sample is placed (see FIGURE 6).

C   ACOUSTIC PROPERTIES MEASURED BY LONGITUDINAL WAVES

C1   Effect of Relaxation Processes

Sound propagation in nematics is usually accompanied by non-hydrodynamic behaviour (relaxation processes) and the purely hydrodynamic behaviour mentioned in Section A is rarely observable. Two distinct relaxation processes are known to occur: one is the intramolecular relaxation process associated with the internal rotation (trans-gauche isomerisation) of the alkyl chains attached to the aromatic core, and the other is due to the critical fluctuations of the nematic ordering. The former is predominant at temperatures well below the nematic-isotropic transition temperature, T_{NI}, whereas the latter becomes pronounced in the vicinity of T_{NI}. The effect of these relaxation processes can be formally expressed by replacing the bulk components of the MPP viscosity coefficients by complex viscosities,

$$\eta_k \rightarrow \eta_k(\omega) = \eta'_k(\omega) - i\eta''_k(\omega), \; k = 1, 4, 5 \tag{30}$$

The replacement of the viscosity coefficients in the dispersion relation EQN (5) means that a frequency dependent anisotropic elasticity is induced by the relaxation processes. As a result the speed of sound becomes slightly anisotropic, which, in fact, has been observed.

C2   Hydrodynamic Behaviour

4,4'-dimethoxyazoxybenzene (PAA) is a unique compound in which no rotational isomerism can occur in the end groups. Thus at temperatures sufficiently below T_{NI}, this compound is expected to exhibit dispersion-less hydrodynamic behaviour. In fact, Kemp and Letcher [8] have shown that the speed of sound is isotropic and the attenuation constant varies as ω^2 in the frequency range of 3 to 18 MHz. Also, as shown in FIGURE 1, the angular dependence of the attenuation constant α is well fitted by EQN (7) neglecting the

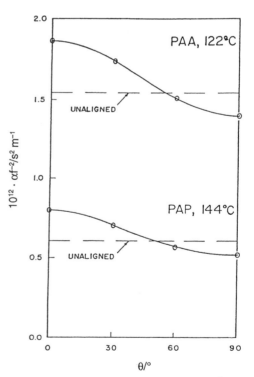

FIGURE 1 Angular dependence of α/f^2 for 4,4'-dimethoxyazoxybenzene (PAA) and 4,4'-diethoxyazoxybenzene (PAP) in the nematic phase. Solid curves are least squares fits of the data points (circles) to EQN (7). From Kemp and Letcher [8].

last term describing the sound attenuation by heat conduction, which is about four orders of magnitude smaller than the viscosity terms.

C3   Critical Behaviour

Since the nematic-isotropic phase transition is weakly first order, both sound speed and attenuation constant α at low frequencies (0.5 ~ 5 MHz) show critical anomalies in the vicinity of T_{NI}. Again PAA, which has no intramolecular relaxation, exhibits a pronounced critical effect as shown in FIGURES 2 and 3 (Thiriet and Martinoty [9]). For comparison, the temperature dependence of the attenuation of 4-pentyl-4'-cyanobiphenyl (5CB) at low (0.5 MHz) and high (115 MHz) frequencies is also shown (FIGURE 4) (Nagai et al [10]). In the low frequency range the angular dependence of the attenuation constant just below T_{NI} can be fitted to the form

$$\alpha(\theta) = \alpha(0)(1 - \delta \sin^2 \theta) \tag{31}$$

with $\delta \sim 0.1$. FIGURE 5 shows the angular dependence of the low frequency (1 MHz) attenuation constant of 5CB for various temperatures close to T_{NI} (Nagai et al [10]). A comparison between EQN (7) and EQN (31) indicates that at low frequencies the bulk viscosity $\eta'_1(\omega) - \eta'_4(\omega)$ dominates the attenuation near T_{NI}.

D   ACOUSTIC PROPERTIES MEASURED BY TRANSVERSE (SHEAR) WAVES

The shear mode in nematics is purely viscous and diffusive as described in Section A2. Experimentally there are three independent cases in the configuration geometry of **n**, **q** and **v** for shear wave propagation: i.e. (a) **n** $\perp$ **q** and **n** $\perp$ **v**, (b) **n** $\perp$ **q** and **n** $\parallel$ **v** and (c) **n** $\parallel$ **q** and **n** $\perp$ **v** (see FIGURE 6). However, from EQN (24) the effective viscosities $\eta_{\parallel}(\theta)$ of cases (b) and (c) are identical: i.e. $\eta_{\parallel}(\pi/2) = \eta_{\parallel}(0) = \eta_3$. In the Leslie formulation, this identity is guaranteed by the Parodi relation EQN (26).

By use of the shear wave reflectance method at 15 MHz, Kiry and Martinoty [11] have shown for all the three configurations described that the real and the imaginary parts of the shear mechanical impedance of 5CB in its nematic phase are, in fact, equal to each other as predicted by hydrodynamic theory (see EQNS (21a) and (21b), and (23a) and (23b)) (see FIGURE 7). They have also shown that the real parts of the shear mechanical impedance for configurations (b) and (c) are identical. This observation means that the Parodi relation has been proved experimentally since the effective viscosity η for each configuration is related to the shear mechanical impedance by

$$\eta = 2RX/\omega\rho \tag{32}$$

with $R = X$ for a purely viscous liquid. The effective viscosity thus obtained for configuration (b) or (c) compares well with the capillary viscosity implying that the capillary viscosity includes the effect of flow alignment and also no relaxation of the viscosity coefficient occurs, at least up to 15 MHz for 5CB. The relaxation processes observed by longitudinal waves are induced by the temperature variation associated with the volume change imposed by the waves. In the isotropic phase just above

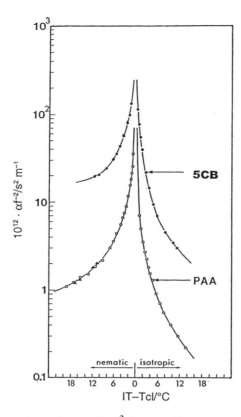

FIGURE 2 Temperature dependence of α/f^2 for 4,4'-dimethoxyazoxybenzene (PAA) and
4-pentyl-4'-cyanobiphenyl (5CB) in the nematic and isotropic phases. In the nematic
phase the data for $\theta = 90°$ are shown. For PAA α/f^2 is frequency independent in
the frequency range of 0.8 to 5 MHz investigated. The results for 5CB
at 0.5 MHz are also shown. From Thiriet and Martinoty [9].

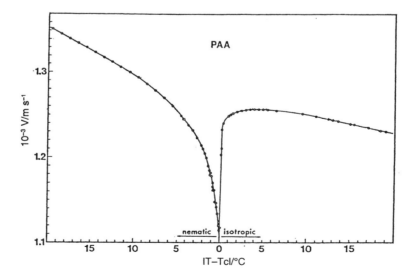

FIGURE 3 Temperature dependence of the speed of sound for 4,4'-dimethoxyazoxybenzene (PAA),
which showed no angular dependence and no frequency dispersion in the 1 to 5 MHz
range investigated. From Thiriet and Martinoty [9].

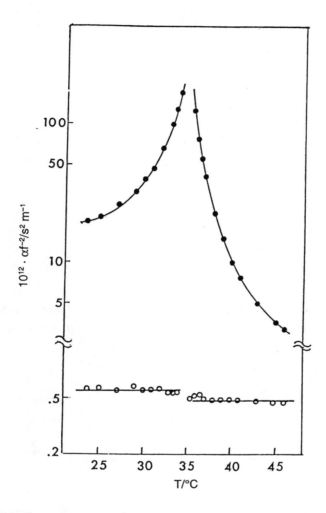

FIGURE 4 Temperature dependence of α/f^2 for 4-pentyl-4'-cyanobiphenyl (5CB): filled circles are for 0.5 MHz and open circles for 115 MHz. The nematic-isotropic phase transition temperature, T_{NI}, is 35°C. Note that for 5CB α/f^2 varies with the measuring frequency due to relaxation processes. Note also that α/f^2 at 115 MHz is almost temperature independent suggesting the relaxation frequency is sufficiently lower than the measuring frequency. (From Nagai et al [10]).

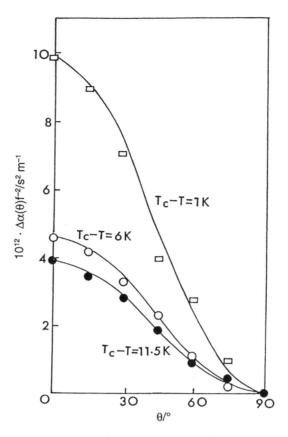

FIGURE 5 Angular dependence of $\Delta\alpha(\theta)/f^2 = [\alpha(\theta) - \alpha(\pi/2)]/f^2$ for 4-pentyl-4'-cyanobiphenyl (5CB) at 1 MHz for several selected temperatures near T_{NI}. Solid curves are fits to the data by EQN (31). From Nagai et al [10].

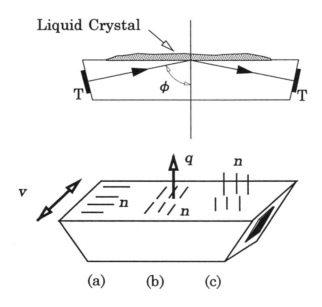

FIGURE 6 Schematic representation of the shear wave reflectance method and the geometrical configuration of **n**, **q** and **v**. T: shear transducer, ϕ: the incident angle.

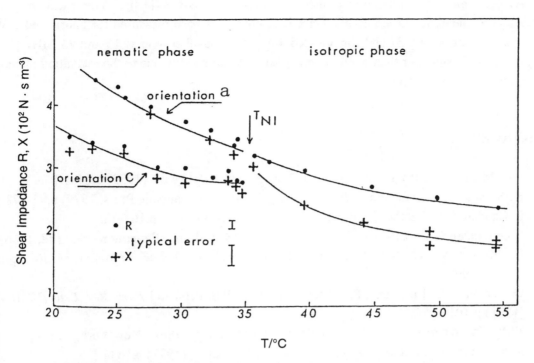

FIGURE 7 The real (R) and imaginary (X) parts of the shear mechanical impedance of 4-pentyl-4′-cyanobiphenyl (5CB) at 15 MHz in the nematic and isotropic phases. Note that in the nematic phase R ~ X but in the isotropic phase R > X. From Kiry and Martinoty [11].

T_{NI}, R is not equal to X as shown in FIGURE 7 suggesting the existence of a finite shear rigidity. Since the shear field can couple with the tensor order parameter as pointed out by de Gennes, relaxation phenomena associated with the phase transition (pretransitional effects) manifest themselves and non-hydrodynamic effects such as the occurrence of the imaginary part of a viscosity coefficient (i.e. the finite shear rigidity) appear [12].

E   CONCLUSION

Acoustic properties of nematics measured under well-defined orientational conditions were very useful probes to confirm the validity of linear hydrodynamic theories of nematics. Combined with other experiments such as conventional capillary viscometry, they have been used to determine the anisotropic viscosity coefficients of nematics.

Dynamical behaviour associated with the phase transition in the nematic phase near T_{NI} has been investigated extensively using both oriented and unoriented nematics by use of longitudinal waves. Although in the low frequency range below, say, about 100 MHz, the intramolecular relaxation processes manifest themselves in the nematic phase of most nematogens, still the remarkable anomaly in the speed and absorption has been useful to make clear the phase transitional behaviour associated with the order parameter fluctuation.

Near the smectic A phase of nematics a similar anomaly associated with the smectic A-nematic phase transition has been observed by longitudinal waves for 4-cyanobenzylidene-4′-octyloxyaniline

(CBOOA) [13] and terephthalylidene-bis(4-butylaniline) (TBBA) [14]. The shear mechanical impedance in the nematic phase near the smectic A phase has been measured for unoriented CBOOA between 50 and 150 kHz [15] and for oriented 4,4′-diheptylazobenzene at 15 and 25 MHz [16]. In this region again R is greater than X giving an apparent shear rigidity due to the relaxation phenomena associated with the phase transition.

REFERENCES

[1]   P.C. Martin, O. Parodi, P.S. Pershan [*Phys. Rev. A (USA)* vol.6 (1972) p.2401]

[2]   K. Miyano, J.B. Ketterson [*Physical Acoustics* vol.14 (Academic Press, 1979) p.93-178]

[3]   S. Candau, S.V. Letcher [*Adv. Liq. Cryst. (USA)* vol.3 (1978) p.167-225]

[4]   L.D. Landau, E.M. Lifshitz [*Theory of Elasticity* 3rd edition (Pergamon Press Ltd, 1986)]

[5]   E.I. Kats, V.V. Lebedev [*Fluctuational Effects in the Dynamics of Liquid Crystals* (Springer-Verlag, 1994)]

[6]   D. Forster, T.C. Lubensky, P.C. Martin, J. Swift, P.S. Pershan [*Phys. Rev. Lett. (USA)* vol.26 (1971) p.1016]

[7]   W.P. Mason (Ed.) [*Physical Acoustics* vol.1A (Academic Press, New York, 1964)]

[8]   K.A. Kemp, S.V. Letcher [*Phys. Rev. Lett. (USA)* vol.27 (1971) p.1634]

[9]   Y. Thiriet, P. Martinoty [*J. Phys. (France)* vol.40 (1979) p.789]

[10]   S. Nagai, P. Martinoty, S. Candau [*J. Phys. (France)* vol.37 (1976) p.769]

[11]   F. Kiry, P. Martinoty [*J. Phys. (France)* vol.38 (1977) p.153]

[12]   See, for example, P.G. de Gennes, J. Prost [*The Physics of Liquid Crystals* 2nd edition (Oxford University Press, New York, 1993)]

[13]   F. Kiry, P. Martinoty [*J. Phys. (France)* vol.39 (1978) p.1019]

[14]   S. Bhattacharya, B.K. Sarma, J.B. Ketterson [*Phys. Rev. B (USA)* vol.5 (1981) p.2397]

[15]   Y. Kawamura, K. Okano [*Jpn. J. Appl. Phys. (Japan)* vol.22 (1983) p.1749]

[16]   F. Kiry, P. Martinoty [*J. Phys. (France)* vol.38 (1977) p.L389]

CHAPTER 4

STRUCTURAL STUDIES

4.1 X-ray studies of nematic systems

G. Ungar

July 2000

A   INTRODUCTION

This Datareview describes the types of information on nematic liquid crystals obtainable from X-ray scattering and illustrates some of these on selected examples. The basic theory of amorphous scattering is introduced first for isotropic liquids, and subsequently for uniaxial amorphous systems such as nematics.

B   SCATTERING FROM AN ISOTROPIC LIQUID

FIGURE 1 shows an arbitrary object with the incident X-ray beam defined by vector $\mathbf{k_0}$ and a beam defined by $\mathbf{k}$, scattered through angle ϕ. The path difference between two waves scattered by electrons at arbitrary points O and P is $\mathbf{r \cdot k} - \mathbf{r \cdot k_0}$. The phase difference for radiation of wavelength λ is $(2\pi/\lambda) \, \mathbf{r \cdot (k - k_0)} = \mathbf{r \cdot Q}$, where $\mathbf{Q}$, the wavevector, is defined as $\mathbf{Q} = (2\pi/\lambda)(\mathbf{k - k_0})$.

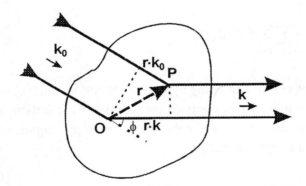

FIGURE 1 A scattering object.

On a relative scale, the complex number describing the amplitude and phase of the wave scattered on a small volume element at P with electron density ρ is $\rho\exp(i\mathbf{r \cdot Q})$. The resultant wave scattered by the whole object in the direction $\mathbf{k}$ is therefore given by

$$F(\mathbf{Q}) = \int\rho(\mathbf{r}) \exp(i\mathbf{r \cdot Q}) \, d\mathbf{r} \tag{1}$$

If integration is over all space then F is the Fourier transform of electron density, providing the link between the real ($\mathbf{r}$) space and the reciprocal ($\mathbf{Q}$) space. The scattered intensity at a particular point in $\mathbf{Q}$-space is

$$I(\mathbf{Q}) = F(\mathbf{Q}) \, F^*(\mathbf{Q}) \tag{2}$$

and is the square of the amplitude (modulus) of F(**Q**).

To obtain the scattered intensity from a molecule we have to add the contributions from each atom, taking account of phase differences. Thus

$$I(\mathbf{Q}) = \sum_j f_j(\mathbf{Q}) \exp(i\mathbf{r}_j\mathbf{Q}) \sum_k f_k(Q) \exp(-i\mathbf{r}_k \cdot \mathbf{Q})$$

$$= \sum_j \sum_k f_j(Q) f_k(Q) \exp[i(\mathbf{r}_j - \mathbf{r}_k) \cdot \mathbf{Q}] \tag{3}$$

where f_j and f_k are atomic scattering factors.

In studies of isotropic and anisotropic liquids, key to relating the diffraction pattern to the scatterers in real space are two properties of Fourier transforms: additivity and rotation [1]. Thus the transform of a sum is equal to the sum of the transforms; also rotation in real space causes an equal rotation in reciprocal space. The reciprocal space, i.e. F(**Q**), will therefore accurately reflect the averaging over the irradiated volume of the specimen. For a spherically averaged rigid molecule the scattered intensity is

$$I(Q) = \sum_j \sum_k f_j(Q) f_k(Q) \frac{\sin(Qr_{jk})}{Qr_{jk}}$$

$$= \sum_j f_j^2(Q) + \sum_j \sum_{k \neq j} f_j(Q) f_k(Q) \frac{\sin(Qr_{jk})}{Qr_{jk}} \tag{4}$$

where r_{jk} is the modulus of $\mathbf{r}_j - \mathbf{r}_k$. Scattering by a gas of N non-interacting molecules is simply NI(Q). However, in a liquid intermolecular interactions play a role. An equation analogous to EQN (4) can be written to describe scattering by molecules in an isotropic liquid, with F being the Fourier transform of the molecular electron density

$$I(Q) = \sum_j F_j^2(Q) + \sum_j \sum_{k \neq j} F_j(Q) F_k(Q) \frac{\sin(Qr_{jk})}{Qr_{jk}} \tag{5}$$

For a liquid of N identical molecules this can be written as

$$I(Q) = NF^2(Q)S(Q) \tag{6a}$$

$$S(Q) = 1 + \frac{1}{N} \sum_j \sum_{k \neq j} \frac{\sin(Qr_{jk})}{Qr_{jk}} \tag{6b}$$

where S(Q) is the interference function.

The function in real space which gives the probability of finding a molecule at a distance r, i.e. between r and r + δr, from a molecule at the origin is the radial distribution function g(r). The probability of finding a given molecule is $(1/V) 4\pi r^2 \delta r\, g(r)$. Hence the interference function can be written as

$$S(Q) = 1 + n_0 \int_0^\infty 4\pi r^2\, g(r) \frac{\sin(Qr)}{Qr}\, dr \qquad (7)$$

where n_0 is the number density. The oscillations in g(r) tend to die out and the function approaches a constant at large r, corresponding to the overall average electron density ρ_0. If the function is normalised so that the constant is 1 and 1 is then subtracted we obtain the molecular correlation function

$$h(r) = g(r)-1$$

Since h(r) and S(Q) (more precisely S(Q)-1) are related by Fourier transformation (cf. EQN (7)), S(Q) will reflect any periodic oscillation in h(r) and show it as a peak at the corresponding Q. The damped oscillations in probability of finding a molecule in the first, second and further shells around a molecule at the origin in a liquid will result in a broad peak in S(Q); hence the ubiquitous amorphous halo with a radius corresponding to 2π/Q between 0.4 and 0.5 nm in most liquids and glasses. When long-range order is present, such as in a crystal, sharp Bragg peaks occur in S(Q).

C   SCATTERING BY AN ANISOTROPIC LIQUID

For amorphous systems with cylindrical symmetry, such as aligned uniaxial nematic liquid crystals or oriented amorphous polymers, the equivalent of the radial distribution function is the cylindrical distribution function (CDF). This function can be written as [2]

$$C(x,z) = \rho_0^2 + V^{-1} W(x,z)$$

where

$$W(x,z) = \frac{2r}{\pi} \int_0^\infty \int_{-\infty}^\infty \chi I(\chi, \varsigma) J_0(2\pi\chi x) \cos(2\pi\varsigma z)\, d\varsigma\, d\chi \qquad (8)$$

Here x, z and χ, ς are cylindrical co-ordinates in real and reciprocal space, respectively (see FIGURE 2). ρ_0 is the average value (scaled here to the electron density) which C approaches at long distances, while J_0 is the zero order Bessel function.

It can be more useful to use spherical co-ordinates [3] (see FIGURE 2). Any cylindrically symmetrical function with a centre of inversion can be developed into a series of even-order Legendre polynomials P_{2n}, in a fashion similar to the way a periodic function is depicted by a Fourier series. The scattered intensity distribution over the reciprocal space can thus be represented as

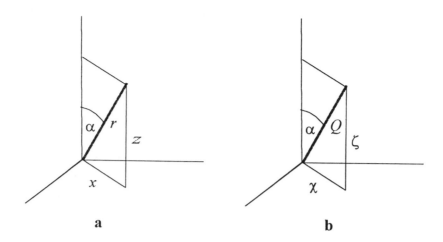

FIGURE 2 Cylindrical and spherical co-ordinates in (a) real and (b) reciprocal space.

$$I(Q,\alpha) = \sum_{n=0}^{\infty} I_{2n}(Q)P_{2n}(\cos\alpha)$$

where

$$I_{2n}(Q) = (4n+1)\int_{0}^{\pi/2} I(Q,\alpha)P_{2n}(\cos\alpha)\sin\alpha\, d\alpha \tag{9}$$

Similarly the cylindrical distribution function can be expanded in real space as [4]

$$W(r,\alpha) = \sum_{0}^{\infty} W_{2n}(r)P_{2n}(\cos\alpha) \tag{10a}$$

with the coefficients

$$W_{2n}(r) = (-1)^{n}\frac{2r}{\pi}\int_{0}^{\infty} Q^{2}I_{2n}(Q)j_{2n}(Qr)\,dQ \tag{10b}$$

Here the j_{2n} are spherical Bessel functions. The first three even-order Legendre polynomials in $\cos\alpha$ are

$$P_{0}(\cos\alpha)=1, \quad P_{2}(\cos\alpha)=\frac{1}{2}(3\cos^{2}\alpha-1), \quad P_{4}(\cos\alpha)=\frac{1}{8}(35\cos^{4}\alpha-30\cos^{2}\alpha+3) \tag{11}$$

Thus the first term in the series (EQN (10a)), i.e. for $n=0$, is independent of α and is in fact the radial distribution function; cf. EQN (7) and note that

$$j_{0}(Qr)=(Qr)^{-1}\sin(Qr)$$

The second and higher terms in EQN (10a) become the more prominent the higher the anisotropy in the phase.

The cylindrical distribution function is effectively the cylindrically symmetric Patterson or autocorrelation function and it contains all of the information experimentally available from X-ray scattering by a uniaxial nematic or oriented amorphous material. Complementary structural information could be obtained from a CDF derived, for example, from neutron scattering where the correlations are weighted differently, i.e. not by electron density but by the scattering lengths of the atomic nuclei. Alternatively, the lack of phase information may be overcome partly through isomorphous replacement techniques, often used in single crystal structure determination in conjunction with Patterson maps. Whereas heavy metal atoms are employed as labels by crystallographers, substitution of one or several selected hydrogen atoms by halogens may be applied in the study of thermotropic liquid crystals. A peak in CDF at **r** means that translation by **r** produces good overlap of electron densities. However, the cause of the overlap, i.e. the nature and shape of the repeating motif (a whole molecule, a part of it) may remain ambiguous. If the scattering pattern is dominated by a strongly scattering atom or group, the position of strong peaks in the autocorrelation function may be equated with translation vectors between such groups.

The CDF (cylindrical distribution function) can be used productively in studies of molecular conformation of oriented amorphous polymer or nematic liquid crystals, although the intermolecular correlations as well as the orientational distribution can also be obtained. The orientational distribution is usually determined in reciprocal space, i.e. directly from the scattering function, as will be discussed later. However, to aid in the interpretation of the CDF it is useful to reduce or eliminate the effect of the orientational distribution function by employing a sharpening procedure [4,5]. Neglecting intermolecular interference, the overall scattering function $I(Q,\alpha)$ can be represented as a convolution of scattering on a cylindrically averaged molecule $I^m(Q,\alpha)$ and of the orientational distribution $D(\alpha)$. Deas [4] has shown that the coefficients in the Legendre series expansion of the three functions are related through

$$I_{2n}(Q) = \frac{2\pi}{4n+1} D_{2n} I_{2n}^m(Q) \tag{12}$$

and that $I^m(Q,\alpha)$ can be written as

$$I^m(Q,\alpha) = \sum_{n=0}^{\infty} \langle P_{2n} \rangle_D^{-1} I_{2n}(Q) P_{2n}(\cos\alpha). \tag{13a}$$

Here

$$\langle P_{2n} \rangle_D = \frac{\displaystyle\int_0^{\pi/2} D(\alpha) P_{2n}(\cos\alpha)\sin\alpha\, d\alpha}{\displaystyle\int_0^{\pi/2} D(\alpha)\sin\alpha\, d\alpha} \tag{13b}$$

are the orientational order parameters.

Through EQN (13a) we can obtain the orientationally desmeared molecular scattering function. The experimental cylindrical distribution function can be sharpened to the same extent as the scattering function. This follows from the general equivalence of rotation in real and reciprocal spaces. It is also evident from the fact that each term $W_{2n}(r)$ in the expansion of the CDF (EQN (10b)) is independently related to the equivalent term $I_{2n}(Q)$.

A good example of the use of this distribution in determining molecular conformation and a step-by-step description of the procedure is given by Mitchell and Lovell [3] who determine the sharpened CDFs for two oriented amorphous polymers, atactic poly(methylmethacrylate) (PMMA) and atactic polystyrene (PS). CDFs for selected energetically viable conformations were also calculated. This was achieved by creating two-dimensional histograms of vectors connecting all pairs of atoms in a model, weighted by their number of electrons. Experimental CDF, derived from $I(Q)$, is shown in FIGURE 3, before (a) and after sharpening (b). FIGURE 4 shows the calculated cylindrically averaged single-chain CDFs for two model conformations of syndiotactic PS: $(tttt)_5$ (FIGURE 4(a)) and $(ttgg)_5$ (FIGURE 4(b)). There is reasonable similarity between FIGURE 3(b) and FIGURE 4(a), suggesting the presence of all-trans syndiotactic sequences in the atactic polymer. In the calculated single-chain CDFs there are no correlations with a normal-to-chain component larger than 10 Å. However, such correlations are observed in the experimental CDF and are thus clearly intermolecular. Note that a large part of the experimental $I(Q)$ data is recorded at Q-values beyond the range normally recorded in typical X-ray experiments on liquid crystals using flat film or an area detector. In order to study intramolecular correlations, i.e. molecular structure and conformation, it is necessary to explore a wide range of reciprocal space; this involves measurements up to high scattering angles and tilting the sample axis. Ideally a texture goniometer or a three- or four-circle diffractometer should be used.

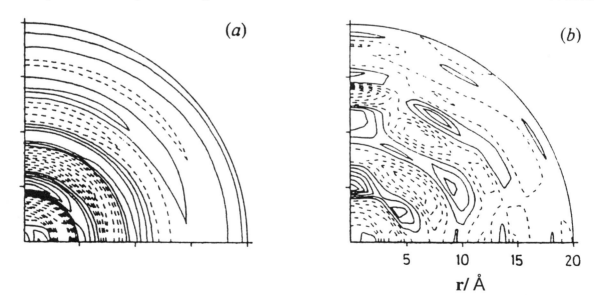

FIGURE 3 (a) Cylindrical distribution function for atactic PS obtained from $I(Q,\alpha)$ using EQN (10).
(b) Sharpened CDF. Orientational distribution $D(\alpha)$ was approximated by a function of the form
$\cos^3\alpha$ which closely describes the azimuthal distribution of the halo with the narrowest profile.
Dashed contours represent negative, i.e. lower than average, values. The alignment axis
is vertical. From [3] with permission.

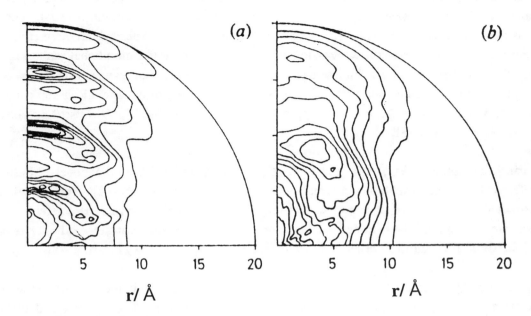

FIGURE 4 Calculated CDF for a model single chain of syndiotactic PS in $(tttt)_5$ conformation (a) and $(ttgg)_5$ conformation (b). The two-dimensional histograms described in the text were smoothed. From [3] with permission.

D   MEASUREMENT OF ORIENTATIONAL ORDER

Unlike other methods, X-ray scattering allows measurement of the whole orientational distribution $D(\alpha)$. As mentioned earlier, the reduced intensity distribution $I(Q,\alpha)$ is a convolution of the molecular scattering function and $D(\alpha)$. This results in EQN (12) for the Legendre coefficients

$$\langle P_{2n} \rangle = \langle P_{2n} \rangle_D \langle P_{2n} \rangle_m \tag{14}$$

for the amplitudes of the spherical harmonics. However, this simple situation is strictly valid only for a hypothetical nematic gas, whereas in real nematic liquids there is intermolecular scattering to consider. If a purely intramolecular scattering feature can be identified, then it can be used to extract $<P_{2n}>_D$ using EQN (14). This is usually possible in main-chain liquid crystal polymers, particularly those with rigid chains [6]. It is customary for low-molar mass liquid crystals or side-chain polymers to base $D(\alpha)$ on the measurement of azimuthal intensity profile of the strong diffuse equatorial arc mostly occurring around $Q \approx 1.2 - 1.5$ Å^{-1}. However, as this maximum arises primarily through intermolecular interactions, methods based on its measurement require further qualification.

The purely intramolecular scattering pattern can be calculated from the model molecular cylindrical distribution function such as those in FIGURE 4, or directly from the molecular model, containing N atoms, through [7]

$$NI_{2n}^m(Q) = (-1)^n (4n+1) \sum_{k=1}^{N} \sum_{l=1}^{N} f_k(Q) f_l(Q) j_{2n}(r_{kl}Q) P_{2n}(\cos \alpha_{kl}) \tag{15}$$

The selected scattering peak can be meridional or off-meridional, but it should be reasonably strong and have a small intrinsic azimuthal spread. The Q-value of the peak maximum is usually chosen for the measurement of $I(\alpha)$. The way $I(\alpha)$ is obtained will depend on experimental geometry. Tilting the sample axis, i.e. a symmetric transmission geometry, is essential if a meridional scattering peak is chosen. Only if Q is very small can one use a fixed sample with the director normal to the incident beam. The approximation is then that $I(\alpha) = I(\beta)$, where β is the azimuthal angle in the plane of the film or area detector. However, it is only appropriate to use peaks with low Q for stiff polymers with a large persistence length [8].

The intrinsic azimuthal spread, expressed as $<P_{2n}>_m$, can be calculated from the model. Alternatively, if a reference high order parameter sample can be obtained, where the value of $<P_{2n}>_D$ can be approximated as 1, then $<P_{2n}>_m$ can be determined experimentally. Order parameters $<P_{2n}>_D$ of different samples can then be obtained via EQN (14). Hereafter the subscript D will be removed from the symbol for the orientational order parameters.

At large enough Q all scattering is intramolecular and thus potentially suitable for measurement of the global orientational order of molecules. However, since such scattering is weak and difficult to separate from the background, it has rarely been used for the measurement of the order parameters. Some work has been done on nematics using high-Q neutron scattering and $<P_2>$ and $<P_4>$ were determined [9]. Neutron scattering does not suffer like X-ray scattering from the rather steep decrease in atomic scattering factor f with increasing Q.

Most measurements of the order parameters using X-rays are based on the azimuthal spread of the equatorial arc in the Q-region 1.2 - 1.5 Å^{-1} [10,11]: see FIGURE 5. The advantages are that this scattering peak is strong and that, as long as the scattering is mainly equatorial, the fixed-sample flat-film recording geometry is adequate. However, since this scattering is a feature of the intermolecular interference function it does not directly reflect the orientation of individual molecules but rather that of molecular clusters or correlated domains. From the width of the peak $\Delta Q \approx 0.5$ Å^{-1} the lateral extent of the (positionally) correlated domain can be estimated as about three molecular diameters, or

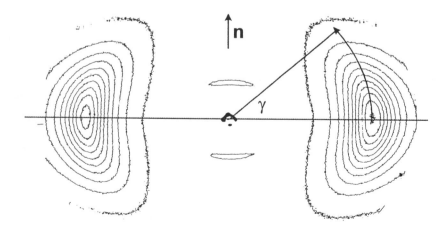

FIGURE 5 Low-to-medium-Q scattering pattern of a nematic side-chain polymer aligned in the magnetic field (field axis vertical). The pattern features a strong diffuse equatorial arc and a weak diffuse meridional streak, both attributed to intermolecular interference.

approximately ten molecules: cf. FIGURE 8. It should be recalled that the preceding discussion neglected the intermolecular interference function introduced in EQN (6). We could improve our simplified treatment and take account of the intermolecular interference function $S(Q, \alpha)$ by introducing a third factor into EQN (12). The global order parameter can then be written as

$$<P_{2n}> = <P_{2n}>' <P_{2n}>'' \qquad (16)$$

where $<P_{2n}>'$ and $<P_{2n}>''$ are, respectively, the local and the interdomain order parameter; the former is obtained by expansion of the interference function.

A problem with the method based on equatorial scattering is that $<P_{2n}>'$ is not known. However, the following two approximations can be made: (i) that the local orientational order is perfect, and (ii) that the scattering objects are infinitely long rods. In that case the interference function allows a finite intensity only for scattering which is confined strictly to the equatorial plane with respect to the local director. It is estimated [11,12] that for practical purposes condition (ii) is fulfilled if molecules are longer than 20 Å. An additional approximation is that molecules are cylindrically averaged individually, i.e. that there is no local biaxiality. Accepting these assumptions, we can proceed in different ways, as we shall see.

Mitchell and co-workers [6,13] suggested expanding the equatorial scattering in the 1.5 Å^{-1} region into spherical harmonics and dividing each $<P_{2n}>$ by the corresponding idealised $<P_{2n}>'$, according to EQN (16). The values of such $<P_{2n}>'$, describing the purely equatorial scattering on infinite parallel rods, are given by [13]

$$\left\langle P_{2n} \right\rangle' = \frac{(2n!)}{(-1)^n 2^{2n} (n!)^2} \qquad (17)$$

Thus, experimentally obtained $<P_2>$ should be divided by -1/2, and $<P_4>$ by 3/8.

Alternatively, Leadbetter and Norris [12] have derived a direct relationship between the orientational distribution function $D(\alpha)$ and the azimuthal distribution of the equatorial intensity $I(\gamma)$ (FIGURE 5):

$$I(\gamma) = \int\limits_{\gamma}^{\pi/2} \frac{D(\alpha)\sin\alpha}{\cos^2\gamma\sqrt{\tan^2\alpha - \tan^2\gamma}} d\alpha \qquad (18)$$

This expression takes account of the fact that clusters with a range of different polar angles α contribute to the intensity scattered at the same γ due to the different azimuthal orientation of their local directors around the global axis. In order to obtain $D(\alpha)$, EQN (18) can be inverted numerically [14]. If only the order parameter $<P_2>$ is required, the expression,

$$<P_2> = 1 - \frac{3}{2} \left(\int\limits_0^{\pi/2} I(\gamma)d\gamma \right)^{-1} \int\limits_0^{\pi/2} I(\gamma)[\sin^2\gamma + (\sin\gamma\cos\gamma)\ln\frac{1+\sin\gamma}{\cos\gamma}]d\gamma \qquad (19)$$

derived by Deutsch [15], can be used.

Davidson and co-workers [16] have shown that an inversion of EQN (18) can be achieved by expanding $D(\alpha)$ as a series of even-order cosines:

$$D(\alpha) = \sum_{n=0}^{\infty} D_{2n} \cos^{2n} \alpha \qquad (20)$$

EQN (18) then becomes

$$I(\gamma) = \sum_{n=0}^{\infty} \int_{\gamma}^{\pi/2} \frac{D_{2n} \cos^{2n} \alpha \sin \alpha}{\cos^2 \gamma \sqrt{\tan^2 \alpha - \tan^2 \gamma}} d\alpha \qquad (21)$$

which, after transformation, results in the series

$$I(\gamma) = D_0 + \frac{2}{3} D_2 \cos^2 \gamma + \frac{8}{15} D_4 \cos^4 \gamma + \frac{16}{35} D_6 \cos^6 \gamma + ... + \frac{2^n n!}{(2n+1)!!} D_{2n} \cos^{2n} \gamma + ... \qquad (22)$$

Thus, in order to obtain the complete orientational distribution function $D(\alpha)$, all that is needed is finding the coefficients D_{2n} by fitting the experimental intensity distribution via EQN (22) and reconstructing $D(\alpha)$ using EQN (20). Unfortunately, the $\cos^{2n}$ series is a slowly convergent one and for anything but very weakly ordered materials this method is impracticable.

An alternative to direct evaluation of $D(\alpha)$ is to assume that the orientational distribution follows the prediction of the Maier-Saupe theory [17]:

$$D(\alpha) = \frac{1}{Z} \exp(m \cos^2 \alpha) \qquad (23)$$

Here Z is a normalisation constant and m is the Maier-Saupe nematic interaction parameter. Reasonably good agreement between experimental and theoretical forms of $D(\alpha)$ has been demonstrated for a number of examples [12,18]. Substituting EQN (23) into EQN (18) we obtain, after transformation [19]:

$$I(\gamma) = \frac{\sqrt{\pi}}{2} \frac{\exp(m \cos^2 \gamma)}{Z \sqrt{m} \cos \gamma} \operatorname{erf}(\sqrt{m} \cos \gamma) \qquad (24a)$$

where erf is the error function

$$\operatorname{erf}(u) = \frac{2}{\sqrt{\pi}} \int_{0}^{u} \exp(-x^2) dx$$

In order to make EQN (24) useful in practice, the error function can be expanded, yielding [16]:

$$ZI(\gamma) = 1 + \frac{2m}{3}\cos^2\gamma + \frac{4m^2}{15}\cos^4\gamma + \frac{8m^3}{105}\cos^6\gamma + \frac{16m^4}{945}\cos^8\gamma +$$

$$\text{(24b)}$$

$$... + \frac{2^n m^n}{(2n+1)!!}\cos^{2n}\gamma + ...$$

where $(2n + 1)!! = 1 \times 3 \times 5 \times 7 \times ... \times (2n + 1)$. Although this is again a slowly convergent series, all of the coefficients are linked by m which therefore remains the single adjustable parameter. The goodness of the fit thus becomes the measure of compliance with the Maier-Saupe theory and of the assumptions in [12]. The second rank order parameter $<P_2>$ can be calculated from

$$<\cos^2\alpha> = \frac{e^m - s}{2ms} \quad \text{where} \quad s = \sum_{i=0}^{\infty}\frac{m^i}{i!(2i+1)} \tag{25}$$

Davidson and co-workers [16] have applied the method based on EQN (24) to a range of mesogens exhibiting nematic and smectic A phases, including side-chain and main-chain polymers with flexible spacers. The agreement between the experimental I(γ) and that obtained by fitting to EQN (24) is illustrated in FIGURE 6; on the whole the match was good, except for samples with $<P_2>$ of 0.7 - 0.8 or higher. One of the reasons for the discrepancy in the latter case is the fact that higher orientational order usually implies higher positional and conformational order. As a result other inter- and intramolecular scattering features become prominent, creating an uneven background.

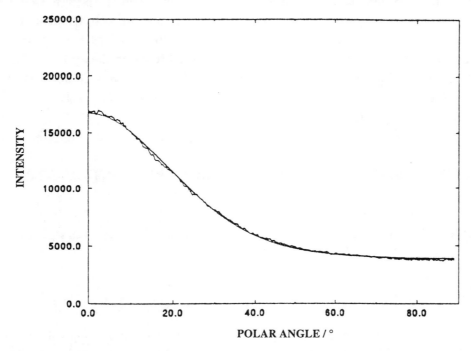

FIGURE 6 Scattered intensity I versus azimuthal angle γ for compound 4O.8 in the magnetically oriented nematic phase at 66°C. Experimental curve is shown together with that fitted to the Maier-Saupe distribution via EQN (24). $<P_2>$ = 0.65. From [16] with permission.

Results of different methods of determining $<P_2>$ for N-(4-butyloxybenzylidene)-4-octylaniline) (4O.8) are shown in TABLE 1 for a series of reduced temperatures T/T_{NI}.

TABLE 1 Results of different methods of determining <P$_2$> for 4O.8.

T/T$_{NI}$	<P$_2$> using EQN (24) (from [16])	<P$_2$> using EQN (18) (from [12])	<P$_2$> using EQN (19) (from [16])
0.997	0.40		0.67
0.991	0.50		0.64
0.989		0.60	
0.986	0.50		0.74
0.980		0.61	
0.974	0.60	0.63	0.79
0.963	0.65		0.82
0.960		0.67	
0.952	0.71		0.87
0.946		0.73	
0.935	0.78	0.75	0.91
0.920		0.76	

An extensive study on 4,4′-diheptyloxyazoxybenzene (HAB) has been carried out by Paul et al [20,21]. This compound shows nematic and smectic C phases. Pretransitional SmC-type fluctuations are observed in the nematic phase close to the transition temperature. This so-called skewed cybotactic nematic phase [22] is characterised by four diffuse spots at low angles. This may be compared with SmA-type fluctuations often observed as a meridional streak: see FIGURE 5. Both <P$_2$> and <P$_4$> have been measured for HAB and are shown as a function of temperature in FIGURE 7. For the nematic phase the agreement between X-ray and birefringence data for <P$_2$> is reasonably good, and they are broadly in line with the prediction of the McMillan theory [23]. The agreement however fails in the SmC phase. In fact we can notice a hint of a downward deviation in measured order parameters already in the nematic phase close to the transition; this is likely to be caused by the SmC pretransitional fluctuations.

E   OTHER INFORMATION FROM X-RAY SCATTERING ON NEMATICS

The transverse correlation length, ξ, has also been measured for HAB by fitting the equatorial intensity profile of the wide-angle intermolecular peak I(Q) to a Lorentzian lineshape [21]:

$$I(Q) = I_b(Q) + \frac{a}{b + (Q - Q_0)^2}$$

The correlation length is given by $\xi = 2\pi/b$ and is shown as a function of temperature in FIGURE 8. It is seen to diverge at the SmC-N transition, as expected for a second order phase transition.

Similarly the longitudinal correlation length in molecular position can be measured from the meridional spread of the small-angle scattering maximum arising from smectic-like fluctuations (FIGURE 5). The position Q_0 of such a peak has been used to estimate the apparent molecular length,

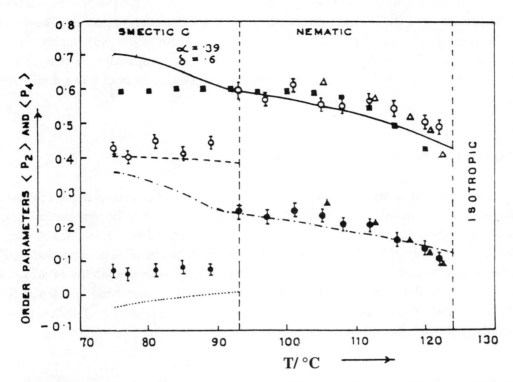

FIGURE 7 Order parameters <P$_2$> (open circles) and <P$_4$> (filled circles), obtained by X-ray via EQN (18), for HAB in the nematic and smectic C phases. Refractive index data for <P$_2$> are also shown (filled squares). Solid lines are predictions from McMillan's potential. X-ray data from Leadbetter et al [12] for <P$_2$> and <P$_4$>, respectively. From [20] with permission.

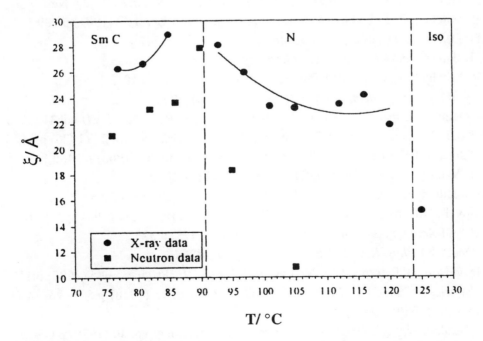

FIGURE 8 Transverse correlation length as a function of temperature for HAB obtained from the equatorial peak profile of X-ray (filled circles) and neutron (filled squares) scattering. From [21] with permission.

$\ell = 2\pi/Q_0$. Thus, for example, Sarkar et al [24] found $\ell = 24$ Å for 4-(trans-4'-hexylcyclohexyl) isothiocyanatobenzene (6CHBT) as against the molecular model length L of 22.2 Å. The discrepancy has been interpreted as indicating a degree of bimolecular association with highly overlapping molecules in a dimer. The ℓ/L ratio of 1.1 for 6CHBT should be compared to an equivalent ratio of 1.4 for the alkylcyanobiphenyls [25] (CB) where the two antiparallel molecules in a dimer overlap to a lesser degree.

F   CONCLUSION

The most common use of X-ray scattering in the field of nematics has been phase identification and measurement of orientational order parameters. Even when a birefringent optical texture is ambiguous (e.g. in some polymers), a nematic can be easily identified by the absence of sharp X-ray reflections. The capability of the X-ray technique to determine the complete orientational distribution function has been enhanced in recent years by the comparatively wide availability of area detectors. Much of the theoretical background work applicable to nematics has been developed for other oriented systems, notably oriented amorphous polymers. They have yet to be fully exploited in the field of liquid crystals, e.g. in determining the molecular conformation.

REFERENCES

[1]   D.W.L. Hukins [*X-Ray Diffraction by Disordered and Ordered Systems* (Pergamon Press, Oxford, 1981)]

[2]   L.E. Alexander [*X-ray Diffraction Methods in Polymer Science* (Wiley, New York, 1969)]

[3]   G.R. Mitchell, R. Lovell [*Acta Crystallogr. A (Denmark)* vol.37 (1981) p.189]

[4]   H.D. Deas [*Acta Crystallogr. (Denmark)* vol.5 (1952) p.542]

[5]   W. Ruland [*Colloid Polym. Sci. (Germany)* vol.255 (1977) p.833]

[6]   G.R. Mitchell, A.H. Windle [*Polymer (UK)* vol.24 (1983) p.1513]

[7]   R. Pynn [*J. Phys. Chem. Solids (UK)* vol.34 (1973) p.735]

[8]   J.A. Odell, G. Ungar, J.L. Feijoo [*J. Polym. Sci. Polym. Phys. Ed. (USA)* vol.31 (1993) p.141]

[9]   M. Kohli, K. Otnes, R. Pynn, T. Riste [*Z. Phys. B (Germany)* vol.24 (1976) p.147]

[10]   P. Delord, J. Falgueirettes [*C.R. Acad. Sci. (France)* vol.260 (1965) p.2468]

[11]   A. de Vries [*J. Chem. Phys. (USA)* vol.56 (1972) p.4489]

[12]   A.J. Leadbetter, E.K. Norris [*Mol. Phys. (UK)* vol.38 (1979) p.669]

[13]   R. Lovell, G.R. Mitchell [*Acta Crystallogr. A (Denmark)* vol.37 (1981) p.135]

[14]   R. Pynn [*Acta Crystallogr. A (Denmark)* vol.31 (1975) p.323]

[15]   M. Deutsch [*Phys. Rev. A (USA)* vol.44 (1991) p.8264]

[16]   P. Davidson, D. Petermann, A.M. Levelut [*J. Phys. II (France)* vol.5 (1995) p.113]

[17]   W. Maier, A. Saupe [*Z. Nat.forsch. A (Germany)* vol.13 (1958) p.564, vol.14 (1959) p.882 and vol.15 (1960) p.287]

[18]   A.J. Leadbetter, P.G. Wrighton [*J. Phys. Colloq. (France)* vol.40 (1979) p.C3-234]

[19]   A.S. Paranjpe, V.K. Kelkar [*Mol. Cryst. Liq. Cryst. (UK)* vol.102 (1984) p.289]; V.K. Kelkar, A.S. Paranjpe [*Mol. Cryst. Liq. Cryst. Lett. (UK)* vol.4 (1987) p.139]

[20]   B. Adhikari, R. Paul [*Mol. Cryst. Liq. Cryst. (Switzerland)* vol.261 (1995) p.241]

[21]   M.K. Das, B. Adhikari, R. Paul, S. Paul, K. Usha Deniz, S.K. Paranjpe [*Mol. Cryst. Liq. Cryst. (Switzerland)* vol.330 (1999) p.1]

[22]   A. de Vries [*Acta Crystallogr. A (Denmark)* vol.25 (1969) p.S135]

[23]   W.L. McMillan [*Phys. Rev. A (USA)* vol.6 (1972) p.936]

[24]   P. Sarkar et al [*Mol. Cryst. Liq. Cryst. (Switzerland)* vol.330 (1999) p.159]

[25]   A.J. Leadbetter, R.M. Richardson, C.N. Colling [*J. Phys. (France)* vol.Suppl.40 (1975) p.C1-37]

4.2   Neutron scattering studies of nematic mesophase structures

R.M. Richardson

November 1998

A   INTRODUCTION

Neutron scattering has been used to measure the orientational order parameters of nematic liquid crystals. In general, spectroscopic methods such as NMR can only measure the second rank order parameter $<P_2>$ although ESR and Raman scattering techniques also measure the fourth rank analogue $<P_4>$. It is the possibility of determining some of the higher rank order parameters (i.e. $<P_L>$ for $L > 2$) that makes neutron scattering unique.

B   MEASUREMENT TECHNIQUES

A conventional X-ray or neutron diffraction measurement on a single component liquid crystal contains information on both intermolecular and intramolecular correlations. The scattering maxima at $10 < Q/nm < 20$ (described in Datareview 4.1) result from intermolecular correlations. They appear as arcs in the diffraction pattern and this can be interpreted in terms of the orientational distribution of small groups of well ordered molecules [1]. Using neutron scattering from a mixture of different isotopically labelled molecules it is possible to identify a feature in the diffraction pattern that results from intramolecular correlations only. It is referred to as single molecule scattering. It contains information on the orientational distribution of the molecules and can be used to determine the order parameters $<P_2>$, $<P_4>$, $<P_6>$ etc.

The scattering vector, $\mathbf{Q}$, is the natural variable for describing the results from a scattering experiment. Its magnitude is given by

$$Q = \frac{4\pi\sin\theta}{\lambda} \tag{1}$$

and its direction bisects the incident and scattered beam. Single molecule scattering can be measured by utilising the very different scattering lengths of hydrogen and deuterium for neutrons. Hydrogen has an (unusual) negative value of -3.74×10^{-15} m while deuterium has a more typical value of 6.674×10^{-15} m. Since hydrogen and deuterium are chemically very similar, a mixture of hydrogenous and deuteriated versions of the same molecule will constitute a random mixture. The neutron scattering from such a mixture contains a new component which is single molecule scattering from an ensemble of well-separated so-called difference molecules [2]. A difference molecule may be regarded as one in which only the isotopically substituted sites contribute to the scattering. The order parameters may be determined from the scattering either by assuming the molecules to be cylindrical or (more rigorously) by using the molecular structure [3].

The measurements are made using neutron-scattering apparatus at facilities such as the Institut Laue Langevin, Grenoble, France, or Risø National Laboratory, Roskilde, Denmark. A monochromatic incident neutron beam is scattered by a magnetically aligned monodomain sample. An area detector registers the scattered neutrons so that the intensity is measured as a function of the magnitude of **Q** and its direction with respect to the nematic director. FIGURE 1 shows the scattering from a mixture of hydrogenous and deuterated 4,4′-dimethoxyazoxybenzene (PAA). Perfectly aligned long rods would give a horizontal streak of scattering but this is smeared into a more isotropic shape by the distribution of orientations in the nematic phase.

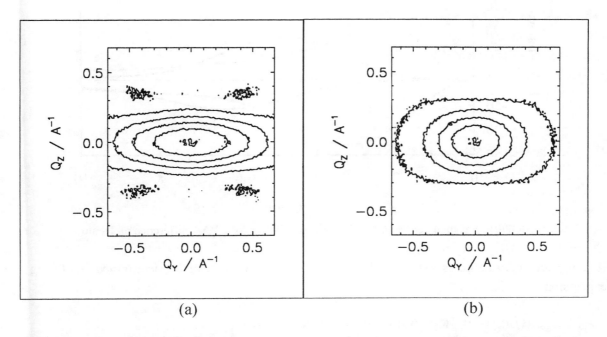

(a)　　　　　　　　　　　　　　　(b)

FIGURE 1 The single molecule scattering calculated for PAA difference molecules with a Maier-Saupe distribution of orientations (a) for $\overline{P}_2 = 1.0$ and (b) for $\overline{P}_2 = 0.5$. Q_Z is parallel to the director and Q_Y is perpendicular. Note $1 \, \text{Å}^{-1} \equiv 10 \, \text{nm}^{-1}$.

It has been shown [4] that the single molecule scattering depends on the order parameters, $<P_L>$, and a series of structure factors $S_L(Q)$, which act as coefficients for the terms in the series:

$$I_{SMS}(Q,\theta_Q) = \sum_{L \, even} S_L(Q) < P_L > P_L(\cos\theta_Q) \qquad (2)$$

where θ_Q is the angle between the scattering vector **Q** and the director. The $P_L(\cos\theta_Q)$ are the Legendre polynomial functions. The structure factors depend only on the structure of the difference molecule:

$$S_L(Q) = (2L+1)(-1)^{L/2} \sum_{i=1}^{N} \sum_{j=1}^{N} \Delta b_i \Delta b_j \, j_L(Qr_{ij}) \, P_L(\cos\beta_{ij}) \qquad (3)$$

where β_{ij} is the angle between the interatomic vector, r_{ij}, and the effective symmetry axis of the molecule. It is usual to choose this axis to be the principal axis of inertia of the normal molecule. The summations are over the N substituted H/D sites in the molecule.

FIGURE 2 shows the calculated values of the structure factors, $S_L(Q)$, for a PAA difference molecule.

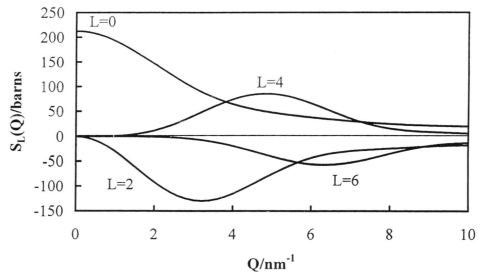

FIGURE 2 Calculated values of the structure factors for a PAA difference molecule.

If a molecular structure is assumed then the order parameters can be determined by fitting the expression

$$I_{SMS}(Q, \theta_Q) = B + K(S_0(Q) + <P_2>S_2(Q)P_2(\cos\theta_Q) +$$

$$<P_4>S_4(Q)P_4(\cos\theta_Q) + <P_6>S_6(Q)P_6(\cos\theta_Q) + ...)$$

(4)

to the experimental data. This requires a linear least squares fit to minimise the weighted deviation between experiment and theory with the flat background, B, the scaling factor, K, and the order parameters, $<P_L>$, as variable parameters.

C   ORIENTATIONAL ORDER PARAMETER RESULTS

An early application of neutron diffraction was to find an intramolecular interference peak at high Q where intermolecular effects have died away [5]. This is more practical with neutron diffraction than X-rays because the atomic form factor for X-rays tends to vanish at high Q. The angular distribution of such an intramolecular peak about the director can be interpreted in terms of the distribution of molecules about the director, $f(\beta)$. In fact there are difficulties in the scaling of a high Q peak which preclude the determination of individual order parameters but a value for the ratio of $<P_4>/<P_2>$ was measured for PAA. However, the value was rather larger than that obtained from later measurements using the single molecule scattering method and suggests that intermolecular effects were not completely negligible.

The single molecule scattering has its maximum at Q = 0 where $S_0(Q)$ is fixed by the number of substituted atoms in the molecule so the scaling parameter, K, is easily determined. This means that all of the order parameters may be determined individually. However, there are a number of experimental factors that can influence the accuracy of the order parameters determined by the single molecule scattering method.

(i)   The range of Q covered by the detector and the counting statistics. The effect of these experimental conditions has been assessed by applying the analysis method on simulated data. This has indicated the importance of measuring over the range of Q for which the structure factors are significant.

(ii)   The magnetic field used to create the monodomain should be sufficient to saturate the alignment of the director. Early experiments used a field of 0.07 T whereas 0.20 T was used more recently. It seems that imperfect alignment of the director with the lower field may have caused the order parameters to be low by a few percent.

(iii) A molecular structure has to be assumed in order to extract the order parameters from the scattering data. Assuming a cylindrical molecule of viable radius and height is clearly an approximation and it also suffers from the drawback of introducing more variable parameters (i.e. the radius and height) into the analysis. The later practice of assuming a realistic molecular structure in order to calculate the structure factors, $S_L(Q)$, avoids this. It will be most reliable for rigid molecules where it may safely be assumed that the molecular structure determined by modelling or X-ray crystallography is a reasonable representation of the molecular structure in the nematic phase.

For these reasons, the results presented here are for PAA, which is a fairly rigid molecule. The sample was aligned with a field of 0.20 T and the scattering was measured over a Q range of 0.8 to 6.0 nm^{-1}. FIGURE 3 shows the orientational order parameters that were determined [6] compared

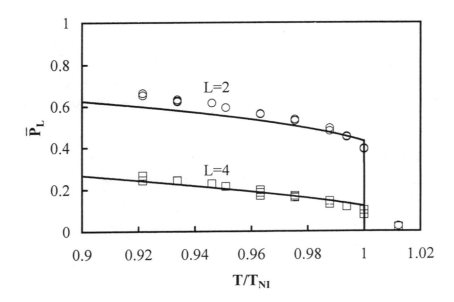

FIGURE 3 Orientational order parameters determined for PAA (points) compared
with the predictions of the Maier-Saupe theory (line).

with the predictions of the Maier-Saupe theory [7]. The accuracy of the order parameters here is about ±0.02. It was not possible to determine $<P_6>$ from this experiment because of insufficient data in the Q range where $S_6(Q)$ adopts significant values.

An experiment on the longer molecule, 4-pentyloxybenzylidine-4′-heptylaniline (5O.7), has given values for $<P_6>$ and $<P_8>$. This is probably because the regions of significance for $S_6(Q)$ and $S_8(Q)$ fell in the experimental range. However, this experiment [2] was made with the lower magnetic field and with a higher minimum Q so the values are not expected to be very accurate. In TABLE 1 we show the values for the order parameters for PAA measured with good and poor experimental conditions. The comparison shows the error that can be introduced and suggests that caution be exercised when using the results from 5O.7.

TABLE 1 Comparing order parameter results for PAA and 5O.7 interpolated to $T/T_{NI} = 0.966$.
Experimental uncertainty of the values is about ±0.02.

	Experimental conditions	$<P_2>$	$<P_4>$	$<P_6>$	$<P_8>$
PAA	Good	0.55	0.18	-	-
PAA	Poor	0.61	0.09	-	-
5O.7	Poor	0.44	0.17	0.10	0.00

D   CONCLUSION

Neutron scattering from mixtures of normal hydrogenous molecules with deuteriated versions of the same compound provides a route to the orientational order parameters of nematic phases. Reliable results exist for $<P_2>$ and $<P_4>$ in PAA. Higher order parameters have been determined in other materials but the experimental conditions used suggest that the values are not yet accurate.

REFERENCES

[1]   A.J. Leadbetter, E.K. Norris [*Mol. Phys. (UK)* vol.68 (1979) p.669]
[2]   R.M. Richardson, J.M. Allman, G.J. Mcintyre [*Liq. Cryst. (UK)* vol.7 (1990) p.701]
[3]   R.W. Date, I.W. Hamley, G.R. Luckhurst, J.M. Seddon, R.M. Richardson [*Mol. Phys. (UK)* vol.76 (1992) p.951]
[4]   P.G. de Gennes [*C.R. Acad. Sci. (France)* vol.B274 (1972) p.142]
[5]   M. Kohli, K. Otnes, R. Pynn, T. Riste [*Z. Phys. B (Germany)* vol.24 (1976) p.147]
[6]   I.W. Hamley et al [*J. Chem. Phys. (USA)* vol.104 (1996) p.10046]
[7]   W. Maier, W. Saupe [*Z. Nat.forsch. A (Germany)* vol.13 (1958) p.564]

4.3 Crystal structures of mesogens that form nematic mesophases

K. Hori

October 1998

A   INTRODUCTION

X-ray crystal structure analysis is the most powerful and widely used method for structure determination in chemistry. A brief survey of the principles of the method [1] and the characteristic features of typical examples of crystal structures of mesogens that form nematic phases are given here.

B   X-RAY CRYSTAL STRUCTURE ANALYSIS

X-rays are scattered by electrons. X-rays scattered from several points are summed up or cancelled by the superposition of waves. This phenomenon is called diffraction. The superposition of the scattered X-rays is represented by

$$E = \int \rho(\mathbf{r})\exp\{2\pi i(\nu t + \delta(\mathbf{r}))\}d\mathbf{r}$$

$$= \int \rho(\mathbf{r})\exp\{2\pi i\delta(\mathbf{r})\}d\mathbf{r}\,\exp(2\pi i\nu t) \tag{1}$$

where ν is the frequency of the X-ray and $\rho(\mathbf{r})$ and $\delta(\mathbf{r})$ are the electron density and the relative phase of the X-ray, respectively, at position $\mathbf{r}$. Since the intensity of the X-ray, I, is EE*, where E* is the complex conjugate of E,

$$I = \int \rho(\mathbf{r})\exp\{2\pi i\delta(\mathbf{r})\}d\mathbf{r}\,[\int \rho(\mathbf{r})\exp\{2\pi i\delta(\mathbf{r})\}d\mathbf{r}]^*$$

$$= FF^*, \tag{2}$$

where

$$F = \int \rho(\mathbf{r})\exp\{2\pi i\delta(\mathbf{r})\}d\mathbf{r}$$

is called the structure factor. The structure factor and the electron density distribution, i.e. a structure of matter, are related by a Fourier transform.

The X-ray diffraction pattern from the three-dimensional periodic array of a crystal has sharp spots with intensities that result from the structure of the unit cell, on the reciprocal lattice, which reflects the periodicity of the crystal. Here, the Fourier transform is represented as a sum instead of an integral,

$$F = \sum_j f_j T_j \exp\{2\pi i(hx_j + ky_j + lz_j)\} \tag{3}$$

where F is called the crystal structure factor with the scattering factor of the jth atom, f_j, T_j is a displacement parameter accounting for the thermal vibration of the atom, h, k, l are integers designating a reciprocal point, and atomic coordinates in the crystal are x_j, y_j, z_j.

Single crystal X-ray analysis consists of measuring diffraction intensities while rotating a crystal (to ensure the diffraction condition for as many reciprocal spots as possible), by using a computer-controlled automatic diffractometer and obtaining the electron density distribution $\rho(x, y, z)$ from the inverse relation of EQN (3),

$$\rho(x, y, z) = (1/V) \sum F(h\,k\,l) \exp\{-2\pi i(hx_j + ky_j + lz_j)\} \tag{4}$$

However, $\rho(x, y, z)$ cannot be derived automatically from the measured intensity data, because they lack information about the phases included in F(h k l). This is essential to crystal structure analysis and is called the phase problem. In order to use EQN (4), the phases in F(h k l) must be estimated. Recent developments of computers and programs have made this process much more successful than before. For a structure determined from EQN (4) based on the estimated phases, structure factors are calculated by using EQN (3). Next, the parameters of the estimated structure are adjusted little by little so that the discrepancies between observed and calculated structure factors, F_o and F_c, become smaller by least-squares fits. A measure of the reliability of the obtained structure is given by the R-factor defined as

$$R = \sum \left| |F_o| - |F_c| \right| / \sum |F_o|$$

The smaller the R-factor, the more reliable the obtained structure. For a molecule of moderate size with several tens of atoms other than hydrogen, several hundreds of parameters concerning the atomic coordinates (x, y, z, and six components of a displacement parameter tensor for each atom) are determined from several thousands of intensity data, giving precise and exact information on the structure. All of the published crystal structures of organic and organometallic compounds have been stored in the Cambridge Structural Database.

There are a few problems often encountered in crystal structure analysis of mesogens. First, it is sometimes very difficult to obtain single crystals suitable for crystal structure analysis, because the anisotropy of the molecules results in very thin plate- or needle-like crystals. Secondly, incompleteness of crystals such as disorder and large thermal motions of the molecules as well as structure modulation are occasionally found in crystals due to rather weak dispersive intermolecular forces. Some of the problems are solved by using strong X-ray sources such as synchrotron radiation and/or very sensitive area detectors at low temperatures [2]. More seriously, relationships between the crystal and mesophase structures are not straightforward due to the first order phase transition with its large enthalpy change. Nevertheless, crystal structure analysis gives precise and exact information about molecular conformations, and more importantly, it is almost the only way to show the mutual arrangements (packing modes) of molecules in bulk states.

C   CRYSTAL STRUCTURES OF NEMATOGENS

In a very early work in 1933, Bernal and Crowfoot estimated the crystal packings of several nematogens, such as 4,4′-dimethoxyazoxybenzene (PAA), and a smectogen, based on limited X-ray diffraction data in combination with the optical properties of single crystals. They showed that the nematogens crystallised in imbricated structures, where a terminal group of a molecule came nearly in contact with the central moiety of its neighbour, while the smectogen had a layer structure [3]. This was verified in 1970-71 for PAA [4-6], which has an imbricated structure, as shown in FIGURE 1.

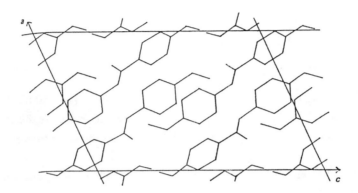

FIGURE 1 The crystal structure of PAA. Crystal data: P2$_1$/c, a = 1.09, b = 0.81, c = 1.56 nm, β = 114°, Z (the number of molecules in the unit cell) = 4. From [6].

Imbricated structures have been found for many typical nematogens, such as 2,2′-di-bromo-4,4′-bis-(4″-methoxybenzylideneamino)biphenyl [7], (4-ethoxybenzylidene)-4′-butylaniline (EBBA) [8], 4,4′-dihexyloxybenzalazine [9], 4-butylphenyl-4′-butylbenzoyl-oxybenzoate [10], propyloxyazoxybenzene [11], cholesteryl hexyloxybenzoate [12] and so on, and also for unconventional molecules with lateral chains on the core moieties [13].

Crystalline polymorphs (more than one crystal structure for a compound) are often obtained for mesogens, reflecting the subtle balance of intermolecular interactions. The single crystals obtained are not always thermodynamically the most stable phases. 4-methoxybenzylidene-4′-butylaniline (MBBA), the first room-temperature nematogen, has seven solid phases [14]. The structure of a phase, which was later assigned to be a metastable C4 phase [15], was determined for a single crystal obtained by a micro-zone-melting method at low temperature [16]. FIGURE 2 shows the crystal packing of MBBA. There are three crystallographically independent molecules, in which interplanar angles of the two phenyl rings are slightly different. Molecules are not arranged in an imbricated way, but they form a layer structure, where two molecules are parallel and the third is antiparallel to the former two.

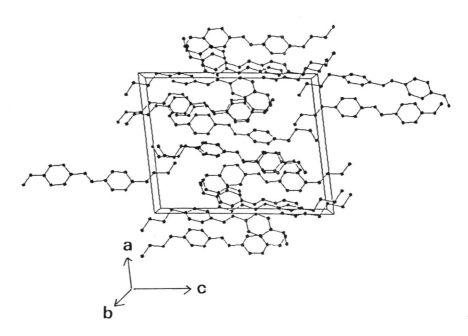

FIGURE 2 The crystal structure of MBBA. Crystal data: $P2_1$, a = 1.4908(4), b = 0.8391(3), c = 1.8411(4) nm, β = 96.40(2)°, Z = 6 at T = 110 K. From [16]. Figures in parentheses denote the estimated standard deviations in the smallest digit both here and for the other figures.

Mesophases of 4-alkyl-4′-cyanobiphenyl (nCB) and 4-alkyloxy-4′-cyanobiphenyl (nOCB) are supposed to be composed of highly overlapped molecules due to the strong dipole moment of the CN group [17]. Crystal structures of enantiotropically nematogenic 5CB [18], 6CB [19] and 7CB [19] are shown in FIGURES 3 - 5. 5CB has a unique molecular conformation with the alkyl chain almost perpendicular to the plane of the attached benzene ring. Intermolecular distances between CN groups and between a CN and a phenyl group are rather long with the least N ... C distance of 3.69 Å. For 6CB, N(CN) ... C(CN) distances are 3.56 and 3.67 Å and N(CN) ... C(phenyl) distances are 3.48 and 3.50 Å. The crystal of 7CB comprises highly overlapped molecules with CN-phenyl interaction, which is isomorphous with those of the smectogenic 9CB [20] and 11CB [21]. On the other hand, closely arranged CN-CN interactions are found in crystals of 5OCB, 6OCB and 7OCB. In 5OCB [22] (see FIGURE 6), one-dimensional chains of closely arranged antiparallel CN groups (C ... N, 3.30 and 3.34 Å) are formed along the b axis. In 6OCB [23] (see FIGURE 7), CN-CN interaction forms dimers (C ... N, 3.40 and 3.56 Å). Both crystals have imbricated structures of the associated units. 7OCB has four solid states [24]. The structures of square-plate [24] and needle crystals [23], which are metastable phases, were determined. Both crystals also show closely arranged CN-CN groups. In the needle crystal (see FIGURE 8), the CN-CN interaction forms tetramers (C ... N, 3.39 - 3.57 Å), while in the square-plate crystal (FIGURE 9), two-dimensional networks of CN-CN interaction (C ... N, 3.41 - 3.64 Å) are found. The needle crystal has a largely tilted layer structure, while the square-plate crystal has a less tilted layer structure.

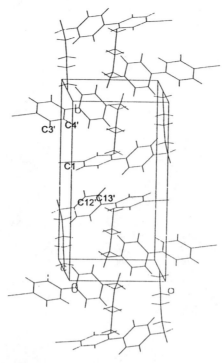

FIGURE 3 The crystal structure of 5CB. Crystal data: P2₁/a, a = 0.8249(5), b = 1.6022(4), c = 1.0935(3) nm, β = 95.09(3)°, Z = 4 at 253 K. From [18].

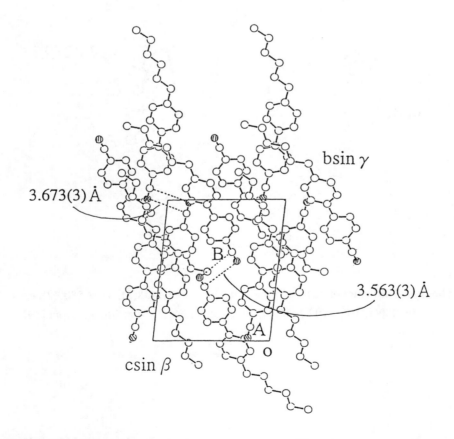

FIGURE 4 The crystal structure of 6CB. Crystal data: P$\overline{1}$, a = 1.2427(4), b = 1.2724(3), c = 1.0857(2) nm, α = 100.74(2), β = 112.54(2), γ = 75.89(2)°, Z = 4 at 200 K. From [19].

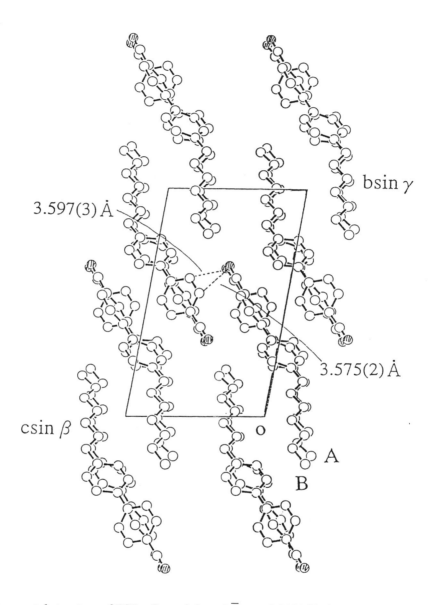

FIGURE 5 The crystal structure of 7CB. Crystal data: $P\bar{1}$, a = 1.1438(2), b = 1.5800(2), c = 0.9674(2) nm, α = 99.000(12), β = 107.164(11), γ = 91.062(11)°, Z = 4 at 240 K. From [19].

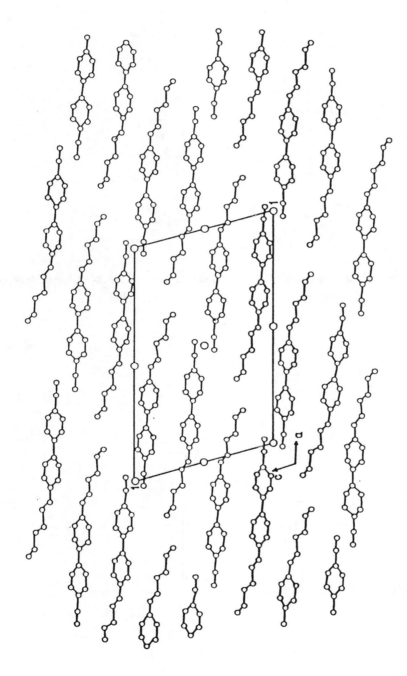

FIGURE 6 The crystal structure of 5OCB. Crystal data: P2₁/n, a = 2.1378(5), b = 0.5695(3), c = 1.2789(2) nm, β = 106.074(17)°, Z = 4. From [22].

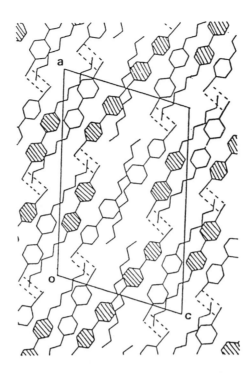

FIGURE 7 The crystal structure of 6OCB. Crystal data: P2$_1$/a, a = 2.676(1), b = 0.7609(3), c = 1.6636(7) nm, β = 105.52(3)°, Z = 8. From [23].

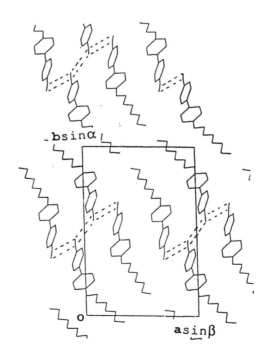

FIGURE 8 The crystal structure of the needle crystal of 7OCB. Crystal data: P$\bar{1}$, a = 1.26556(8), b = 1.9044(2), c = 0.73495(5) nm, α = 94.142(8), β = 100.108(5), γ = 91.036(7)°, Z = 4. From [23].

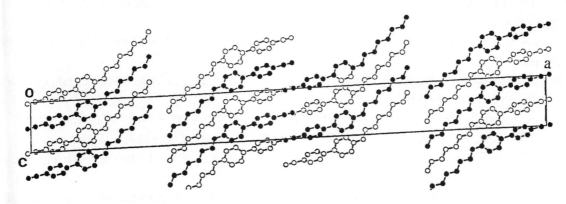

FIGURE 9 The crystal structure of the square-plate crystal of 7OCB. Crystal data: C2/c, a = 6.9458(11), b = 0.70442(8), c = 0.68686(5) nm, β = 92.359(9)°, Z = 8 at 243 K. From [24]. Black and white circles denote molecules at the front and back, respectively.

D   CONCLUSION

Good correlation has been generally found between crystal and nematic phase structures, i.e. crystals of nematogens have imbricated structures. However, there are exceptions which show smectic-like layer structures. In an extreme case, the opposite relation was found for two isomers, i.e. a largely tilted lamellar structure for a compound which has the smectic A phase and a distinct smectic-like layer structure for a nematogen [25]. Nevertheless, it is useful to know the crystal structures in order to understand the intermolecular interactions which control the mesophase structures. For example, a systematic comparison of the whole series of nCB and nOCB reveals that the former shows a remarkable odd-even effect, while the CN-CN interaction is more dominant in the latter [19]. It might be mentioned that results for the crystal structures of smectogens are more informative, first because smectic phases are nearer to crystalline phases and secondly there are many smectic phases where the intermolecular interactions play more important roles. Precise structural information at the atomic level would contribute to their understanding.

REFERENCES

[1]   See, for example, J.P. Glusker, M. Lewis, M. Rossi [*Crystal Structure Analysis for Chemists and Biologists* (VCH, New York, 1994)]

[2]   M. Helliwell, A. Deacon, K.J. Moon, A.K. Powell, M.J. Cook [*Acta Crystallogr. B (Denmark)* vol.53 (1997) p.231]. This paper describes the crystal structure of a discotic mesogen obtained with the most advanced techniques.

[3]   J.D. Bernal, D. Crowfoot [*Trans. Faraday Soc. (UK)* vol.29 (1933) p.1032]

[4]   W.R. Krigbaum, Y. Chatani, P.G. Barber [*Acta Crystallogr. B (Denmark)* vol.26 (1970) p.97]

[5]   A.L. Bednowitz [in *Crystallographic Computing* Eds. F.R. Ahmed, S.R. Hall, C.P. Huber (Munksgaard, Copenhagen, 1970) p.58]

[6]   C.H. Carlisle, C.H. Smith [*Acta Crystallogr. B (Denmark)* vol.27 (1971) p.1068]

[7] D.P. Lesser, A. de Vries, J.W. Reed, G.H. Brown [*Acta Crystallogr. B (Denmark)* vol.31 (1975) p.653]

[8] J.A.K. Howard, A.J. Leadbetter, M. Sherwood [*Mol. Cryst. Liq. Cryst. (UK)* vol.56 (1980) p.271]

[9] H. Astheimer, L. Walz, W. Haase, J. Loub, H.J. Müller, M. Gallardo [*Mol. Cryst. Liq. Cryst. (UK)* vol.131 (1985) p.343]

[10] W. Haase, H. Paulus, R. Pendzialek [*Mol. Cryst. Liq. Cryst. (UK)* vol.101 (1983) p.291]

[11] F. Romain, A. Gruger, J. Guilhem [*Mol. Cryst. Liq. Cryst. (UK)* vol.135 (1986) p.111]

[12] A.P. Polishuk, M.Yu. Antipin, T.V. Timofeeva, V.I. Kulishov, Yu.T. Strukov [*Kristallografiya (USSR)* vol.31 (1986) p.671]

[13] F. Perez, P. Judeinstein, J.P. Bayle, H. Allouch, M. Cotrait, E. Lafontaine [*Liq. Cryst. (UK)* vol.21 (1996) p.855]

[14] L. Rosta et al [*Mol. Cryst. Liq. Cryst. (UK)* vol.144 (1987) p.297]

[15] M. More, C. Gors, P. Derollez, J. Matavar [*Liq. Cryst. (UK)* vol.18 (1995) p.237]

[16] R. Boese, M.Yu. Antipin, M. Nussbaumer, D. Bläser [*Liq. Cryst. (UK)* vol.12 (1992) p.431]

[17] A.J. Leadbetter, R.M. Richardson, C.N. Colling [*J. Phys. (France)* vol.36 (1975) p.C1-37]

[18] T. Hanemann, W. Haase, I. Svoboda, H. Fuess [*Liq. Cryst. (UK)* vol.19 (1995) p.699]

[19] M. Kuribayashi, K. Hori [*Liq. Cryst. (UK)* vol.26 (1999) p.809]

[20] T. Manisekaran, R.K. Bamezai, N.K. Sharma, J. Shashidhara Prasad [*Liq. Cryst. (UK)* vol.23 (1997) p.597]

[21] T. Manisekaran, R.K. Bamezai, N.K. Sharma, J. Shashidhara Prasad [*Mol. Cryst. Liq. Cryst. (Switzerland)* vol.268 (1995) p.45]

[22] P. Mandal, S. Paul [*Mol. Cryst. Liq. Cryst. (UK)* vol.131 (1985) p.223]

[23] K. Hori, Y. Koma, A. Uchida, Y. Ohashi [*Mol. Cryst. Liq. Cryst. (Switzerland)* vol.225 (1995) p.15]

[24] K. Hori et al [*Bull. Chem. Soc. Jpn. (Japan)* vol.69 (1996) p.891]

[25] R. Centore, M.A. Ciajoro, A. Roviello, A. Sirigu, A. Tuzi [*Liq. Cryst. (UK)* vol.9 (1991) p.873]

CHAPTER 5

ELASTIC PROPERTIES

5.1   Torsional elasticity for mesophases

I.W. Stewart and R.J. Atkin

June 1999

A   INTRODUCTION

The free energy for nematic and chiral nematic liquid crystals based on torsional elasticity is presented in the most common mathematical forms.

·

B   NEMATICS

Nematic liquid crystals often consist of long molecules having an average alignment in space which can be described by the unit vector, $\mathbf{n}$, called the director. The bulk free energy per unit volume can be constructed in terms of the square of the spatial derivatives of the director. For non-chiral systems there are symmetry restrictions on the free energy which limit the possible terms. The terms must be invariant when $\mathbf{n}$ is replaced by $-\mathbf{n}$ and linear contributions in the derivatives cannot be allowed since they can change sign if the coordinate system is inverted (such terms, nevertheless, can occur in chiral nematic systems, considered later). The resulting free energy integrand has its origins in the works of Oseen [1,2], Zöcher [3] and Frank [4] and is usually called the Frank-Oseen free energy. It can be written in the usual vector notation as

$$F = \frac{1}{2}K_1(\nabla \cdot \mathbf{n})^2 + \frac{1}{2}K_2(\mathbf{n} \cdot \nabla \times \mathbf{n})^2 + \frac{1}{2}K_3(\mathbf{n} \times \nabla \times \mathbf{n})^2 +$$

$$\frac{1}{2}(K_2 + K_4)\nabla \cdot [(\mathbf{n} \cdot \nabla)\mathbf{n} - (\nabla \cdot \mathbf{n})\mathbf{n}]$$

(1)

where the K_i are the curvature elastic moduli, often referred to as the Frank elastic constants. The last term on the right-hand side of EQN (1), sometimes called the saddle-splay term, can be written in the equivalent form $\frac{1}{2}(K_2 + K_4)[\mathrm{tr}((\nabla \mathbf{n})^2) - (\nabla.\mathbf{n})^2]$ when the partial derivatives commute [6, p.96]. By considering states of minimum energy in which the director orientation is uniform and everywhere parallel, Ericksen [7] deduced that the elastic constants must satisfy

$$K_1 > 0, \quad K_2 > 0, \quad K_3 > 0$$

(2)

$$2K_1 > K_2 + K_4 > 0, \quad K_2 > |K_4|$$

(3)

and so whilst K_1, K_2, K_3 are all positive, K_4 can be either positive or negative.

The saddle-splay term is often omitted since it does not contribute to the bulk equilibrium equations or the boundary conditions for problems involving strong anchoring (see Section D). Being a divergence its contribution to

$$\int_V F\, dv$$

can be transformed into a surface integral over the boundary. Neglecting this surface term results in the most widely used form of the bulk energy density [5,6]

$$F_N = \frac{1}{2}K_1(\nabla \cdot \mathbf{n})^2 + \frac{1}{2}K_2(\mathbf{n} \cdot \nabla \times \mathbf{n})^2 + \frac{1}{2}K_3(\mathbf{n} \times \nabla \times \mathbf{n})^2 \qquad (4)$$

Each of the three terms on the right-hand side of EQN (4) represents a specific type of distortion of the director within a given sample of nematic. They are called, respectively, the splay, twist and bend terms, and FIGURE 1 schematically shows the possible orientation of $\mathbf{n}$, indicated by the lines, for each individual distortion: the director orientation is unchanged as the observer moves vertically through the page. The elastic constants in EQN (2) have been measured by many experimentalists and are approximately of the order of 10^{-6} dyn (10^{-11} N) [5, p.103-5] with K_3 often being two or three times larger than K_1 or K_2.

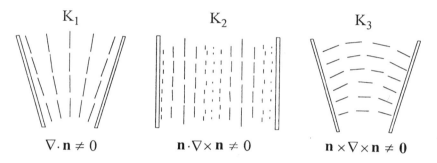

$$\nabla \cdot \mathbf{n} \neq 0 \qquad \mathbf{n} \cdot \nabla \times \mathbf{n} \neq 0 \qquad \mathbf{n} \times \nabla \times \mathbf{n} \neq 0$$

FIGURE 1 The three basic types of distortion in a nematic
liquid crystal: (1) splay, (2) twist, and (3) bend.

Sometimes, for example, when the relative values of the K_i are unknown or when the resulting equilibrium equations are rather complicated, the one constant approximation $K_1 = K_2 = K_3 = K$ is made. In this case the bulk energy F_N can be simplified further by using identities for unit vectors which results in the more amenable form [5, p.104]

$$F_N = \frac{1}{2}K\{(\nabla \cdot \mathbf{n})^2 + (\nabla \times \mathbf{n})^2\} \qquad (5)$$

In Cartesian component form this is equivalent to [8]

$$F_N = \frac{1}{2}K\sum_{i=1}^{3}\sum_{j=1}^{3}n_{i,j}n_{i,j} = \frac{1}{2}K\|\nabla\mathbf{n}\|^2 \qquad (6)$$

apart from a surface term.

Nehring and Saupe [9] proposed that the energy F can be extended to include linear terms involving the second derivatives of $\mathbf{n}$, arguing that such contributions ought to be of the same order as the quadratic terms in first derivatives. The resulting expression is

$$F_{NS} = \frac{1}{2} K_1^{'} (\nabla \cdot \mathbf{n})^2 + \frac{1}{2} K_2 (\mathbf{n} \cdot \nabla \times \mathbf{n})^2 + \frac{1}{2} K_3^{'} (\mathbf{n} \times \nabla \times \mathbf{n})^2$$

$$+ \frac{1}{2} (K_2 + K_4) \nabla \cdot [(\mathbf{n} \cdot \nabla)\mathbf{n} - (\nabla \cdot \mathbf{n})\mathbf{n}] + K_{13} \nabla \cdot ((\nabla \cdot \mathbf{n})\mathbf{n})$$

(7)

where $K_1^{'} = K_1 - 2K_{13}$ and $K_3^{'} = K_3 + 2K_{13}$, effectively rescaling K_1 and K_3. The K_{13} term in EQN (7) can clearly be converted to a surface integral, indicating, just as for the $K_2 + K_4$ term, that it does not contribute to the bulk equilibrium orientation of the director. However, in [9] it is reasoned on the basis of microscopic calculations that K_{13} is of a similar magnitude to the other Frank constants and should, therefore, be added to F_N. The interpretation of the K_{13} term is not fully resolved, but it has been suggested that it may be important in situations where samples are weakly anchored at the surface boundaries, or when investigating thin films: brief discussions of K_{13} can be found in [6, p.115] and [27, p.31].

C   CHIRAL NEMATICS

Chiral nematic liquid crystals possess a helical structure, the pitch P of which is defined to be the distance along the helix over which the director rotates through 2π radians. Since $\mathbf{n}$ and $-\mathbf{n}$ are indistinguishable the period of repetition L is half that of the pitch (see FIGURE 2). Typical values for L are in the range 300 nm [5, p.15]. For chiral nematics one additional term, linear in the derivatives, is added to EQN (1) leading to [6, p.97]

$$F_C = F + k_2 (\mathbf{n} \cdot \nabla \times \mathbf{n})$$

(8)

where F is as given in EQN (1) and k_2 is an elastic constant. The k_2 term changes sign when the axes are inverted, which is allowed in chiral phases. The pitch satisfies $P = 2L = 2\pi/|q_0|$ where q_0 corresponds to the wavevector. The sign of q_0 changes according to the type of helix present: q_0 is positive for a right-handed helix (for example, cholesterol chloride) and is negative for a left-handed helix (for example, many of the aliphatic esters of cholesterol) [5, p.264]. The expression for F_C readily yields some simple results for chiral nematics when a right-handed helical structure with $q_0 > 0$ is supposed. In the geometry of FIGURE 2 with the y-axis perpendicular into the page, the director rotates within the yz-plane as the observer travels along the x-axis in a right-handed helical structure when $\mathbf{n}$ takes the form $\mathbf{n} = (0, -\sin(2\pi x/P), \cos(2\pi x/P))$. The corresponding F_C evaluated over a unit volume yields

$$F_C = -k_2 2\frac{\pi}{P} + \frac{1}{2} K_2 \left(2\frac{\pi}{P} \right)^2$$

(9)

$$= -k_2 q_0 + \frac{1}{2} K_2 q_0^2$$

Setting the derivative of F_C with respect to q_0 equal to zero shows that F_C is minimised when

$$q_0 = \frac{k_2}{K_2} \qquad (10)$$

and so the wavevector and pitch of a chiral nematic are determined by the ratio of the two elastic constants k_2 and K_2. For this reason the bulk free energy density expression for F_C analogous to EQN (1) is often written in the convenient form

$$F_C = \frac{1}{2}K_1(\nabla \cdot \mathbf{n})^2 + \frac{1}{2}K_2(\mathbf{n} \cdot \nabla \times \mathbf{n} + q_0)^2 + \frac{1}{2}K_3(\mathbf{n} \times \nabla \times \mathbf{n})^2 +$$

$$\qquad (11)$$

$$\frac{1}{2}(K_2 + K_4)\nabla \cdot [(\mathbf{n} \cdot \nabla)\mathbf{n} - (\nabla \cdot \mathbf{n})\mathbf{n}]$$

where a constant contribution to the energy has been ignored. This form is generally correct provided $\nabla \mathbf{n}$ and q_0 are small on the molecular scale, which is the case in many practical simulations [5, p.287]. As for nematics, the $K_2 + K_4$ term is often omitted when considering the bulk alignment. The nematic energy F is recovered by setting $q_0 = 0$.

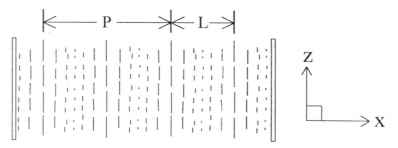

FIGURE 2 The pitch P over which the director rotates
through 2π radians in cholesterics; L = P/2.

D   SURFACE FREE ENERGY

Liquid crystals exhibit either strong or weak anchoring at boundaries or interfaces [5]. Strong anchoring indicates that the alignment is fixed and prescribed on the boundary; the most frequently occurring fixed directions for the director $\mathbf{n}$ are in the plane of the surface (planar alignment) or perpendicular to it (homeotropic alignment). Weak anchoring allows the alignment of $\mathbf{n}$ at the boundary or interface to vary subject to competing torques. Here, a surface energy per unit area is introduced and the form of this boundary condition, expressing a balance between the couple stress arising from the Frank-Oseen energy and the torque from the surface energy, has been discussed by Jenkins and Barratt [10]; for the explicit form of this condition see Datareview 8.1 by Leslie in this volume [EQN (52)]. In cylindrical samples the modulus K_4 can appear in the weak anchoring boundary condition: see Barratt and Duffy [11,12] who also give relevant references to the experimental determination of K_4 and suggest possible experiments when there is weak anchoring on at least part of the boundary. Barratt and Duffy [13] also discuss the possible determination of k_{13}.

The simplest surface energy is of the form first proposed by Rapini and Papoular ([14] and [23, p.289])

$$W_s = \tau_0 (1 + \omega (\mathbf{n} \cdot \mathbf{v})^2) \tag{12}$$

with $\tau_0 > 0$ and $\omega > -1$, where $\mathbf{v}$ is the outward unit normal to the boundary surface or interface. It has been discovered recently that the surface anchoring energy can be more complicated than that in EQN (12) since surface anchoring in nematics can be bistable [15]. Nobili and Durand [16] and Sergan and Durand [17] have attempted to model this bistability more accurately by introducing more complex forms for the surface energy.

E   OTHER FORMULATIONS

Some mention should be made of a more recent extension to the Frank-Oseen energy in EQNS (1) or (4) which has been proposed by Ericksen [18] in an attempt to improve the modelling of defects in nematic and chiral nematic liquid crystals. This is achieved by incorporating some variation in the degree of orientational order of the order parameter, leading to an energy of the form $F = F(S, \nabla S, \mathbf{n}, \nabla \mathbf{n})$ where S is a scalar function representing the extent of orientational order or alignment. For brevity we only state the energy for a commonly accepted approximation in the nematic case when it can be written as [19,20]

$$F = \kappa (k(\nabla S)^2 + S^2 \| \nabla \mathbf{n} \|^2) + \sigma_0 (S) \tag{13}$$

$$\sigma_0 (S) = \frac{1}{2} a S^2 - \frac{1}{3} b S^3 + \frac{1}{4} c S^4 \tag{14}$$

where κ and k are both positive constants while a, b, and c are non-negative constants that for thermotropic liquid crystals depend on the temperature: σ_0 is familiar from the work of de Gennes [21] and Doi [22]. When S takes the constant value $S_0 \neq 0$ then EQN (13) in the isothermal case reduces to

$$F = \kappa S_0^2 \| \nabla \mathbf{n} \|^2 \tag{15}$$

a constant multiple of the usual one constant approximation for nematics given by EQN (6), ignoring constant contributions to the energy. The construction of relevant energies for nematics and chiral nematics in this recently developed theory is discussed in detail by Ericksen and an account of some analyses and developments can be found in the book by Virga [23].

F   DISCUSSION

The general forms for the energies of nematics and chiral nematics have been introduced. The most employed expressions for nematics are given by EQNS (1), (4) and (5); for chiral nematics, EQN (11) is the most generally accepted form. The saddle-splay term is sometimes ignored although it is often considered to be important at surfaces and curved interfaces of liquid crystals, for example in lyotropic and membrane liquid crystals [24]; see also Datareview 5.3 by Crawford in this volume concerning the evaluation of the saddle-splay constant. Brief comments on surface energies have been made and the most widely accepted energy for weak anchoring of the director at a surface or interface is given by EQN (12). A simplified expression for the bulk energy in an extended theory developed by Ericksen has also been given in EQN (13) for nematics.

An accessible review of the Frank-Oseen energy for the bulk expressions is available in the book by Collings and Hird [25] while more detailed, yet concise, comments are available in the recent reviews of Leslie [26,27] and of Dunmur and Toriyama [28]. Fuller discussions and examples of applications are to be found in the reviews by Stephen and Straley [8] and Ericksen [29], and the books by de Gennes and Prost [5] and Chandrasekhar [6].

REFERENCES

[1]   C.W. Oseen [*Ark. Mat. Astron. Fys. (Sweden)* vol.19A (1925) p.1]

[2]   C.W. Oseen [*Trans. Faraday Soc. (UK)* vol.29 (1933) p.883]

[3]   H. Zöcher [*Trans. Faraday Soc. (UK)* vol.29 (1933) p.945]

[4]   F.C. Frank [*Discuss. Faraday Soc. (UK)* vol.25 (1958) p.19]

[5]   P.G. de Gennes, J. Prost [*The Physics of Liquid Crystals* (Clarendon, Oxford, 1993)]

[6]   S. Chandrasekhar [*Liquid Crystals* 2nd Ed. (CUP, Cambridge, 1992)]

[7]   J.L. Ericksen [*Phys. Fluids (USA)* vol.9 (1966) p.1205]

[8]   M.J. Stephen, J.P. Straley [*Rev. Mod. Phys. (USA)* vol.46 (1974) p.617]

[9]   J. Nehring, A. Saupe [*J. Chem. Phys. (USA)* vol.54 (1971) p.337]

[10]   J. Jenkins, P.J. Barratt [*Q. J. Mech. Appl. Math. (UK)* vol.27 (1974) p.111]

[11]   P.J. Barratt, B.R. Duffy [*Liq. Cryst. (UK)* vol.19 (1995) p.57]

[12]   P.J. Barratt, B.R. Duffy [*J. Phys. D, Appl. Phys. (UK)* vol.29 (1996) p.1551]

[13]   P.J. Barratt, B.R. Duffy [*Liq. Cryst. (UK)* vol.26 (1999) p.743]

[14]   A. Rapini, M. Papoular [*J. Phys. Colloq. (France)* vol.30 no.C4 (1969) p.54]

[15]   B. Jerome, P. Pieranski, M. Boix [*Europhys. Lett. (Switzerland)* vol.5 (1988) p.693]

[16]   M. Nobili, G. Durand [*Europhys. Lett. (Switzerland)* vol.25 (1994) p.527]

[17]   V. Sergan, G. Durand [*Liq. Cryst. (UK)* vol.18 (1995) p.171]

[18]   J.L. Ericksen [*Arch. Ration. Mech. Anal. (Germany)* vol.113 (1991) p.97]

[19]   D. Roccato, E.G. Virga [*Contin. Mech. Thermodyn. (Germany)* vol.4 (1992) p.121]

[20]   E.G. Virga [in *Nematics* Eds. J.-M. Coron, J.-M. Ghidaglia, F. Hélein (Kluwer, Dordrecht, 1991) p.371]

[21]   P.G. de Gennes [*Phys. Lett. A (Netherlands)* vol.30 (1969) p.454]

[22]   M. Doi [*J. Polym. Sci. (USA)* vol.19 (1981) p.229]

[23]   E.G. Virga [*Variational Theories for Liquid Crystals* (Chapman and Hall, London, 1994)]

[24]   J. Charvolin, J.F. Sadoc [*J. Phys. Chem. (USA)* vol.92 (1988) p.5787]

[25]   P.J. Collings, M. Hird [*Introduction to Liquid Crystals* (Taylor & Francis, London, 1997)]

[26]   F.M. Leslie [in *Theory and Applications of Liquid Crystals* IMA vol.5, Eds. J.L. Ericksen, D. Kinderlehrer (Springer, New York, 1987) p.211]

[27]   F.M. Leslie [in *Handbook of Liquid Crystals* vol.1, Eds. D. Demus, J. Goodby, G.W. Gray, H.-W. Spiess, V. Vill (Wiley-VCH, Weinheim, 1998) p.25]

[28]   D.A. Dunmur, K. Toriyama [in *Handbook of Liquid Crystals* vol.1, Eds. D. Demus, J. Goodby, G.W. Gray, H.-W. Spiess, V. Vill (Wiley-VCH, Weinheim, 1998) p.253]

[29]   J.L. Ericksen [in *Advances in Liquid Crystals* vol.2, Ed. G.H. Brown (Academic Press, New York, 1976) p.233]

5.2 Measurements of bulk elastic constants of nematics

D.A. Dunmur

July 2000

A   INTRODUCTION

One of the defining properties of liquid crystals is their ability to support torsional strain. The molecules in nematic liquid crystals have no positional organisation, and their centres of mass are randomly distributed as in isotropic liquids. However the orientational organisation of the molecules results in a macroscopic anisotropy for physical properties, and also gives rise to torsional or curvature elasticity. Deformation of the director from a uniform parallel alignment causes a torsional strain energy, which contributes to the free energy. The mathematical description of this has been given in the previous Datareview [1], which introduces the curvature elastic moduli for director deformations of splay (K_1), twist (K_2), bend (K_3) and saddle-splay ($K_2 + K_4$). Since the saddle-splay deformation does not contribute to the bulk equilibrium free energy, no further consideration will be given to it here. Elastic strain energy comes directly from the angle dependence of the intermolecular forces, and so the curvature elastic moduli are related to the angle dependent parameters of the intermolecular potential. Furthermore, any perturbation of the uniform alignment of the director of a liquid crystal can in principle be used to provide a measure of the elastic moduli. Perturbations include: the effect of external electric and magnetic fields; defects and surfaces, complicated by the need to include surface elastic terms; flow, complicated by the need to consider viscous forces; thermal fluctuations in the director orientation. Thus there are many possible techniques that can be used to measure elastic constants of liquid crystals, but they all depend on the detection and measurement of a secondary effect. This invariably means that the evaluation of the elastic constants from a physical measurement relies on the knowledge or additional measurement of another physical property of the liquid crystal. Clearly this introduces further uncertainties into the measurements, with the consequence that values reported for the elastic constants can vary quite widely from group to group depending not only on the techniques used but also on the values assumed for other properties required in the analysis of results. Methods have been devised to measure the elastic torque directly, but these do not escape other complications, because the strain is introduced by applying a magnetic field, and the detection of the torsion requires a surface interaction between the liquid crystal and the torsion balance.

B   METHODS

Any physical perturbation to the director configuration which produces a measurable response may be used to measure the elastic constants of nematic liquid crystals, and many techniques have been reported in the literature [2,3]. For simplicity of analysis it is preferable to use sample configurations, splay, twist or bend, that will give the principal elastic moduli, separately. In many situations changes in the director configuration involve contributions from combinations of splay, twist and bend deformations, and their separation to give individual elastic moduli can be delicate.

B1   Measurement of Elastic Constants using Electric or Magnetic Field-Induced Freedericksz Transitions

This has been the most commonly used technique, and it relies on measuring the response to external magnetic or electric fields of thin samples of surface-aligned nematics. The response of the director is usually probed by measurements of capacitance or birefringence changes, but any anisotropic physical property can be used. Depending on the surface alignment, parallel or perpendicular, and depending on the sign of the magnetic or electric susceptibility anisotropy and the direction of the applied field, the elastic constants for splay, twist and bend can be directly measured from the threshold fields. This technique using electric fields is described in Datareview 11.3 [4] in this volume. The accurate determination of the threshold field, or voltage in the case of electric Freedericksz transitions, is difficult, and so the response, capacitative or optical, as a function of applied voltage (electric) or field (magnetic) can be fitted to appropriate continuum equations [5].

The electric field-induced Freedericksz transition method has been used primarily to determine K_1 and K_3, although it is possible to obtain K_2 with reduced accuracy from the threshold voltage for twisted nematic samples. In order to obtain satisfactory values for K_2, it was necessary to use a magnetic field-induced Freedericksz transition, but recently [6] a new electric field method has been devised using an in-plane electrode configuration.

B2   Measurement of Elastic Constants using Light Scattering

Nematic liquid crystals in the bulk have a turbid appearance which is due to fluctuations in the refractive index over distances comparable with the wavelength of light. The variations in refractive indices are coupled to fluctuations of the director, and they persist even with uniformly aligned samples. In such samples, due to thermal excitation, the director orientation fluctuates about some average direction, and since deformations of the director are opposed by elastic forces, the amplitude of the fluctuations is a measure of the elastic moduli. Associated with the fluctuations of the director are corresponding fluctuations in the refractive indices, or more precisely the birefringence, and these cause incident light to be scattered and depolarised. Light scattering has long been used as a technique to study liquid crystals, starting with the studies of Chatelain [7], and many reviews have appeared [8-10]. The angle dependence of the intensity and depolarisation of scattered light can be analysed to give values for the elastic constants (quasi-static scattering). Additionally the spectrum of the scattered light is broadened by the dynamic fluctuations of the director (dynamic light scattering), and measurements of the line-width of scattered light can be used to obtain information on the viscoelastic properties of nematics (see Datareview 8.2 in this volume [11]).

B2.1   Quasi-static light scattering

In this technique, normally incident plane-polarised light is scattered by thin films (~20 μm thickness) of nematic aligned with the director either parallel or perpendicular to the incident light direction. The scattered light is detected as a function of angle, and for different polarisations of the incident (**i**) and scattered (**f**) beams. The differential scattering cross-section per unit volume is given by [12]:

$$\frac{d\sigma}{d\Omega} = \left(\frac{\pi \Delta \varepsilon_\lambda}{\lambda^2}\right)^2 k_B T \sum_{j=1,2} \frac{i_j f_z + i_z f_j}{K_3 q_\parallel^2 + K_j q_\perp^2 + \Delta \chi F^2} \tag{1}$$

where $\Delta \varepsilon_\lambda$ is the anisotropy in electric susceptibility for wavelength λ, and $\mathbf{q} = \mathbf{k}_i - \mathbf{k}_f$ is the wave-vector of the scattered light. The direction z is parallel to the optic axis of the sample, and components i_2 and i_1 are along directions perpendicular to z and $\mathbf{q}$, and z and i_2 respectively. The two contributions to the scattered light intensity arise from the normal-mode fluctuations of the director in $\mathbf{q}$-space, which correspond to splay-bend and twist-bend deformations. The inclusion of an external field term $\Delta \chi F^2$ indicates that application of a field parallel to the director (for materials of positive susceptibility anisotropy) will quench the fluctuations. This provides another variable, along with angle, which can be used to extract information on the elastic constants. Although many measurements have been made using this technique (see [10] for references), the values obtained for elastic constants are not always in accordance with results obtained on standard materials using other techniques, and a critical account of the technique has been published [13]. Thus the technique has not yet gained acceptance as a reliable method for obtaining precise values for elastic constants.

B2.2   Dynamic light scattering

Although apparently more complicated, the technique of dynamic light scattering has been developed to yield reliable values for the elastic constants. The technique also provides a useful method to obtain information on the viscoelastic properties, and this application is described in this volume by Moscicki [11]. Dissipative fluctuations of the director in nematics are paralleled by refractive index fluctuations which cause a broadening of the depolarised component of the Rayleigh scattered light. Photon correlation techniques are able to determine accurately the line-width of the scattered light, which in turn can be related to elastic and viscoelastic properties of the liquid crystal. A variety of different scattering geometries are possible, with the light propagation direction along the director or perpendicular to the director, giving different combinations of elastic and viscoelastic parameters in the expression for the line-width. As for the intensities, EQN (1) above, two fluctuation modes can be identified as splay-bend (1) and twist-bend (2), and the corresponding expressions for the line widths are [14]:

$$\Gamma_j = \frac{K_3 q_\parallel^2 + K_j q_\perp^2}{\eta_j(\mathbf{q})} \tag{2}$$

where η_j's are combinations of viscoelastic coefficients (see Datareview 8.2 in this book), which depend on the scattering geometry. Measurements of the line-widths as a function of scattering angle (varying q), and fitting the results to EQN (2), allows the determination of individual elastic constants [15]. Alternatively, it is possible to measure the change in line-width as a function of an applied electric or magnetic field. The corresponding line-widths become:

$$\Gamma_1 = \frac{K_3 q_\parallel^2 + K_1 q_\perp^2}{\eta_1(\mathbf{q})} + \frac{\varepsilon_o \varepsilon_{zz} \Delta\varepsilon E^2}{\eta_1(\mathbf{q})\varepsilon(\mathbf{q})}$$

(3)

$$\Gamma_2 = \frac{K_3 q_\parallel^2 + K_2 q_\perp^2}{\eta_2(\mathbf{q})} + \frac{\varepsilon_o \Delta\varepsilon E^2}{\eta_2(\mathbf{q})}$$

where ε_{zz} is the parallel component of the permittivity, $\Delta\varepsilon$ is the permittivity anisotropy, and $\varepsilon(\mathbf{q})$ is a function of the scattering geometry. By fitting the above equations to a quadratic function of the electric field, it is possible to obtain the elastic constants directly. The electric field dynamic light scattering method has been used to give accurate values for splay and twist elastic constants [9,16]. The accuracy of elastic and viscoelastic constants determined from electric field dynamic light scattering has been discussed in detail by Hasegawa et al [17], and they conclude that the accuracy of elastic constants measured using this technique is better than ±5%. This does of course rely on an accurate knowledge of the electric permittivity.

B3 Other Methods

B3.1 Measurement of elastic constants using the torsion pendulum

This provides a relatively direct method to determine the torsional elasticity by measuring the torque exerted on a plate immersed in a liquid crystal [18,19]. The torque is generated by an external magnetic field, and measured by observing the deflection of a torsion pendulum attached to the plate. By choosing different geometries it is possible to measure the three principal elastic constants independently. The response depends on the surface anchoring energy at the solid plate/liquid crystal interface, and on the magnetic susceptibility anisotropy of the liquid crystal. By making measurements as a function of magnetic field strength, it is possible to avoid any difficulties due to weak anchoring, and the anchoring energy can also be extracted from the measurements.

C RESULTS

Because of their importance in determining the behaviour of liquid crystals in devices, many measurements have been made of the elastic constants of a wide range of liquid crystalline compounds. In Datareview 11.3 in this volume [4] Saito lists measurements at a single temperature for many materials of importance for display mixtures. The results presented in this section are for liquid crystals widely used as standards for physical measurements, or as examples of the influence of molecular structure on elastic properties.

C1 Standard Materials

C1.1 Alkylcyanobiphenyls

These are probably the most widely studied of nematic liquid crystals, and results for the splay (K_1), twist (K_2) and bend (K_3) elastic constants for 4,4'pentylcyanobiphenyl (5CB) are given in FIGURE 1 and TABLE 1. The results for K_1 and K_3 have been taken from measurements of the Freedericksz transition [20] and have been averaged over two independent sets of measurements for both the electric and magnetic Freedericksz transitions. The results for K_2 have been taken from electric field dynamic light scattering [9]. As previously explained, the values listed depend on other measurements of electric and magnetic susceptibilities, but are expected to be accurate to better than 5%, and they are also in accord with other measurements in the literature [15].

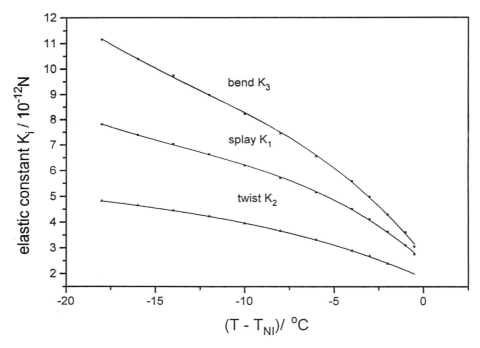

FIGURE 1 Splay, twist and bend elastic constants of
5CB as a function of shifted temperature.

TABLE 1 Splay, twist and bend elastic constants of
5CB as a function of shifted temperature.

$(T - T_{NI})/°C$	$K_1/10^{-12}$ N [20]	$K_2/10^{-12}$ N [9]	$K_3/10^{-12}$ N [20]
-0.5	2.8		3.1
-1.0	3.1		3.6
-2.0	3.6	2.4	4.3
-3.0	4.1	2.7	5.0
-4.0	4.5	2.9	5.6
-6.0	5.2	3.3	6.6
-8.0	5.7	3.7	7.5
-10.0	6.2	3.9	8.2
-12.0	6.6	4.2	9.0
-14.0	7.0	4.4	9.7
-16.0	7.4	4.6	10.4
-18.0	7.8	4.8	11.2

Results for other alkylcyanobiphenyls are gathered in TABLES 2 to 4.

TABLE 2 Splay, twist and bend elastic constants of
6CB as a function of shifted temperature.

$(T - T_{NI})/°C$	$K_1/10^{-12}$ N [20]	$K_2/10^{-12}$ N [9]	$K_3/10^{-12}$ N [20]
-2.0	2.9	2.0	3.2
-3.0	3.3	2.3	3.8
-4.0	3.8	2.4	4.3
-6.0	4.4	2.8	5.1
-8.0	5.0	3.1	5.8
-10.0	5.4	3.3	6.4
-12.0	5.8	3.5	7.0

TABLE 3 Splay, twist and bend elastic constants of 7CB as a function of shifted temperature (figures in brackets are from [20], but are from magnetic field Freedericksz transitions only, i.e. are not the average of measurements from the electric and magnetic Freedericksz transitions).

$(T - T_{NI})/°C$	$K_1/10^{-12}$ N [20]	$K_2/10^{-12}$ N [9]	$K_3/10^{-12}$ N [20]
-0.5	(2.9)		(3.1)
-1.0	(3.6)		(3.9)
-1.6*	3.3*	2.2*	4.1*
-2.0	(4.5)	2.7	(4.8)
-2.5*	3.9*	2.5*	4.6*
-3.0	4.8	3.0	5.3
-4.0	5.3	3.2	5.8
-4.7*	4.8*	3.0	5.7*
-6.0	6.1	3.7	6.8
-7.6*	5.9*	3.5*	7.0*
-8.0	6.8	4.1	7.7
-10.0	7.4	4.4	8.6
-12.0	8.0	4.7	9.3
-12.6	7.3*	4.2*	9.0*
-14.0	8.5	5.0	10.1
-16.0	9.0	5.2	10.8
-16.0*	8.6*	4.4*	10.2*
-18.0	9.5	5.4	11.5

*These values have been taken from [21], and were measured using the magnetic Freedericksz effect. They are lower than the other values quoted, which may reflect some uncertainty over the magnetic susceptibility anisotropy.

TABLE 4 Splay, twist and bend elastic constants of
8CB as a function of shifted temperature.

$(T - T_{NI})/°C$	$K_1/10^{-12}$ N [20]	$K_2/10^{-12}$ N [19]	$K_3/10^{-12}$ N [20]
-0.5	(2.9)	2.2	(2.8)
-1.0	3.2	2.3	3.1
-2.0	4.2	2.7	4.1
-3.0	4.9	3.0	4.8
-4.0	5.5	3.1	5.7
-5.0	6.1	3.3	6.8
-6.0	6.7	3.7	8.9
-6.2	6.8	4.3	9.6
-6.4	7.0	4.9	11.0

The measured elastic constants for 4,4′octylcyanobiphenyl (8CB) are also plotted as a function of temperature in FIGURE 2, where it can be seen that the elastic constants for twist and bend increase as the underlying smectic A phase is approached on cooling in the nematic phase. This is a general result, since the smectic A phase cannot support either a twist or bend torsional distortion.

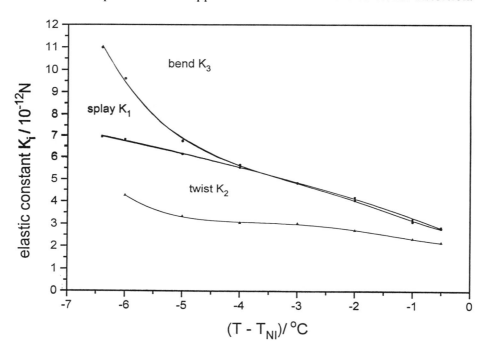

FIGURE 2 Splay, twist and bend elastic constants of
8CB as a function of shifted temperature.

Further results for alkoxycyanobiphenyls are given in [20].

C1.2   Phenylcyclohexanes, bicyclohexanes, Schiff's bases and azoxybenzenes

The alkylcyanophenylcyclohexanes (PCHn) and alkylcyanobicyclohexanes (CCHn) have a lower birefringence than the alkylcyanobiphenyls, and therefore do not scatter light so strongly. As a result they have not been used as standard materials in light scattering experiments. They have been, however, important materials for display applications, and their elastic properties have been measured by the magnetic Freedericksz effect [21,22]. Results for one example of the homologous series PCH7 and CCH7 are given in TABLES 5 and 6.

TABLE 5 Splay, twist and bend elastic constants of
PCH7 as a function of shifted temperature.

$(T - T_{NI})/°C$	$K_1/10^{-12}$ N [21]	$K_2/10^{-12}$ N [21]	$K_3/10^{-12}$ N [21]
-3.0	4.2	2.5	4.9
-5.0	5.0	3.0	7.3
-6.6	5.4	3.2	8.1
-7.9	5.8	3.4	8.6
-9.9	6.3	3.6	9.6
-12.9	6.9	4.0	10.9
-16.6	7.8	4.2	12.1
-22.9	8.6	4.7	14.3

TABLE 6 Splay, twist and bend elastic constants of
CCH7 as a function of shifted temperature.

$(T - T_{NI})/°C$	$K_1/10^{-12}$ N [21]	$K_2/10^{-12}$ N [21]	$K_3/10^{-12}$ N [21]
-1.4	2.1		3.3
-2.5	2.5	1.6	4.1
-4.3	3.2	1.9	5.3
-7.1	3.8	2.3	6.4
-10.4	4.5	2.5	7.6
-14.3	5.2	2.8	8.8
-17.9	5.9	3.1	9.85
-24.3	7.2	3.5	11.5

Many early studies of the physical properties of nematic liquid crystals were carried out on MBBA and PAA. Preferred values for these materials are listed in [23] and reproduced in TABLE 7. There do not appear to have been any more extensive and reliable measurements on these compounds than those reported in [23].

TABLE 7 Splay, twist and bend elastic constants of PAA and MBBA [23].

	$K_1/10^{-12}$ N	$K_2/10^{-12}$ N	$K_3/10^{-12}$ N
PAA for $(T - T_{NI}) = -13.5°C$	6.9	3.8	11.9
MBBA for $(T - T_{NI}) = -23°C$	7.1	4.0	9.2
MBBA for $(T - T_{NI}) = -18°C$	6.4	3.6	8.2

C2   Discotic Materials

There have been a few measurements reported of the elastic properties of disc-like mesogens, but no systematic studies have appeared. Therefore it is not yet possible to list reference data for the elastic constants of disc-shaped molecules which form nematic phases. Measurements have been reported [24] for the splay and bend elastic constants of hexakis(4-alkylphenylethynyl)benzene compounds. They were determined from capacitance measurements of the electric field-induced Freedericksz transition, which yielded K_3 from the threshold, and K_1 was obtained by fitting the field dependence of the capacitance above threshold. The values obtained were the same order of magnitude as those reported for rod-like mesogens. The viscoelastic properties of two hexa-substituted truxenes ((i) hexa-nonoyloxy- [25] and (ii) hexa-dodecoyloxytruxene [26]) have been measured using the technique of dynamic light scattering. Both materials exhibit low temperature nematic phases (i) 57°C - 84°C and (ii) 65°C - 84°C; however they are unusual in that the nematic phases are at lower temperatures than the more ordered columnar phases, which persist to over 200°C. Line-widths of the scattered light were measured as a function of angle (scattered wave-vector) for different geometries, and yielded values for the splay, twist and bend ratios K_i/η_i (see EQN (2)). The measured ratios are given in TABLE 8, where they are compared with specimen results for a nematic composed of rod-shaped molecules. The values for K_i/η_i for these nematics of disc-shaped molecules are nearly two orders of magnitude smaller than the values recorded for a nematic of rod-shaped molecules, but this is likely to be due to the much higher viscosities (η_i) for the low temperature nematic discotic phase.

TABLE 8 Viscoelastic properties of nematic phases of disc-shaped molecules.

	$K_1/\eta_{splay}/$ 10^{-12} m^2 s^{-1}	$K_2/\eta_{twist}/$ 10^{-12} m^2 s^{-1}	$K_3/\eta_{bend}/$ 10^{-12} m^2 s^{-1}
Truxene (i) 67°C [25]		0.4	0.6
Truxene (ii) 65°C [26]	0.6	0.4	
5CB 30°C [17]	87	55	54

As shown in Datareview 1.2 in this volume [27] it is possible to induce nematic phases in materials composed of disc-shaped molecules by addition of a charge-transfer dopant. Thus mixtures of tridecylpentakis(phenylethynyl)phenyl ether with the electron acceptor 2,4,7-trinitrofluorenone form nematic phases. The dielectric and elastic properties of the mixtures have been investigated [28], and the splay and bend elastic constants have been measured using the electric field Freedericksz transition. The values obtained for K_1 and K_3 were about ten times larger than is found for ordinary nematic discotic phases. This result has been attributed to the nature of the packing in the induced nematic phase, which is thought to consist of short columns of strongly correlated stacks of alternating molecules of the mixture components.

C3   Dimers and Trimers

If mesogenic molecular units are chemically linked through flexible alkyl chain spacers, then new liquid crystals are often formed. Depending on the number of mesogenic groups and the manner in which they are joined, the resulting molecules may be dimers, trimers, oligomers, polymers or dendrimers. In recent years there has been some progress in the characterisation of the physical properties of these materials, although there have not been any systematic studies. Connection of two mesogenic units with a flexible alkyl chain gives rise to dimers, and there is a pronounced odd-even effect in the physical properties of the dimers, depending on the odd or even number of methylene spacer units in the linking alkyl chain. Measurements on odd and even dimers of the monomer 4,4′-dipentyloxyphenylbenzoate have been reported [29,30] of the elastic properties as measured by the magnetic Freedericksz transition, and the viscoelastic properties as measured by dynamic light scattering [30,31].

The elastic constants of a structurally related trimer have also been measured [32] using the magnetic Freedericksz transition, and selected results are gathered in TABLE 9.

TABLE 9 Structures and elastic constants for dimers and trimers.

Monomer $T_{NI} = 81°C$	
Even dimer $T_{NI} = 149°C$	
Odd dimer $T_{NI} = 135°C$	
Trimer $T_{NI} = 155°C$	

	$K_1/10^{-12}$ N	$K_2/10^{-12}$ N	$K_3/10^{-12}$ N
Monomer [29] for			
$(T - T_{NI}) = -5°C$	5.7		
$(T - T_{NI}) = -10°C$	7.4		
Even dimer [30] for			
$(T - T_{NI}) = -5°C$	6.8	2.8	8.0
$(T - T_{NI}) = -10°C$	8.2	3.3	12.0
Odd dimer [30] for			
$(T - T_{NI}) = -5°C$	3.2	1.6	4.2
$(T - T_{NI}) = -10°C$	4.6	1.8	5.0
Trimer [32] for			
$(T - T_{NI}) = -5°C$	5.0	2.6	11.8
$(T - T_{NI}) = -10°C$	6.6	3.0	15.0

The results given for dimers refer to mesogenic units linked end to end; however it is possible to link mesogenic groups laterally. Some such materials have been investigated [33] from the point of view of elastic properties using 2,7-disubstituted fluorenes linked laterally through their 9-positions by a flexible alkyl chain.

D   DISCUSSION AND CONCLUSIONS

The elastic properties of liquid crystals are determined by intermolecular interactions, and so it is difficult to relate the elastic constants directly to molecular properties. The intermolecular potential plays a particularly important role in determining the torsional elastic properties of nematics, and molecular theories have been developed which relate the elastic constants to the pair distribution function [34]. A simple result of the molecular theories is that the elastic constants are predicted to lowest order to be proportional to the square of the order parameter $<P_2>$. Thus anything which reduces the orientational order, either through intermolecular interactions, or intramolecular flexibility, will reduce the elastic constants. The dependence of splay and twist elastic constants on $<P_2>^2$ is experimentally confirmed for many nematics [9,35], but often the bend elastic constant shows a more complex dependence on the order parameter. This is possibly because the bend elastic constant is especially sensitive to molecular shape and flexibility. Furthermore the development of pre-transitional smectic ordering in the nematic phase will strongly influence the twist and bend elastic constants. Hard particle theories predict a relatively simple dependence of elastic properties on molecular size, though in reality the shape and flexibility of molecular structures are as important as the size. It is fairly remarkable, given the wide range of possible molecular structures, that the elastic constants do not vary more extensively. Often the contributing factors of local order, shape and size act in a self-compensatory fashion.

The dependence of the elastic constants on the chemical constitution of the mesogens, i.e. the attached functional groups etc., has not been systematically investigated. There are a number of papers that discuss the relationship between the physical properties, including elastic constants, of nematics and the molecular structure of the mesogens [36,37]. It seems likely that the effect of changes of structure on the elastic constants is due to change of molecular shape, or through changes of intermolecular interactions. Thus it has been shown [17,35] that for a series of phenylbicyclohexane liquid crystals, the elastic constants for splay and twist show a dramatic change, when a longitudinal cyano-group is introduced in place of a fluoro-group. This behaviour is illustrated in FIGURE 3.

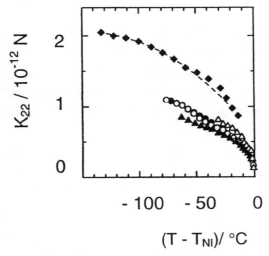

FIGURE 3 The twist elastic constants for PCBs, taken from [17] with permission.
The symbols are as given in TABLE 10.

The marked increase in splay and twist elastic constants due to the presence of a cyano-group is thought to be a consequence of the dramatic increase in molecular association that accompanies the increased longitudinal dipole moment. Thus the apparently strong influence of the dipole moment on the elastic constants is really a consequence of the enhanced intermolecular interactions [38].

TABLE 10 Molecular structures of phenylbicyclohexanes.

3PBC$_1$ (◆)		Cr - 54.7°C - N - 207.5°C - I
3PBC$_2$ (•)		Cr - 44.2°C - N - 118.0°C - I
3PBC$_3$ (○)		Cr - 47.0°C - N - 124.0°C - I
3PBC$_4$ (▲)		Cr - 88.6°C - N - 158.5°C - I
3PBC$_5$ (△)		Cr - 51.4°C - N - 91.0°C - I

For nematics formed from low molecular weight rod-shaped mesogens, the elastic constants are often found to be in the order $K_3 > K_1 > K_2$. This result can be rationalised by molecular theories, and is often found experimentally as the data presented in Section C1 illustrate. For disc-shaped molecules, the order of elastic constants is predicted [39] to be $K_2 > K_1 > K_3$, but as yet there are insufficient experimental data available to confirm this.

The data presented here have excluded nematic polymer liquid crystals. These present particular difficulties for the conventional methods of measurement, which require well-aligned samples. A magnetic resonance technique has been developed which is particularly suited to the measurement of the elastic and viscous properties of liquid crystalline polymers, and this technique and specimen results are described in Datareview 8.3 in this volume [40]. There have been some measurements on the precursors to polymers i.e. dimers and trimers (see Section C3), and these limited results do indicate the importance of molecular shape in determining the elastic properties. In particular, the extended even dimer has elastic constants almost twice those of the bent odd dimer. Thus in polymer liquid crystals, it is clear that the chain configuration will have a dramatic influence on the elastic properties.

REFERENCES

[1]   I.W. Stewart, R.J. Atkin [Datareview in this book: *5.1 Torsional elasticity for mesophases*]

[2]   W.H. de Jeu [*Physical Properties of Liquid Crystalline Materials* (Gordon and Breach, New York, 1980) p.76]

[3]   D.A. Dunmur, K. Toriyama [in *Handbook of Liquid Crystals* vol.1, Eds. D. Demus, J. Goodby, G.W. Gray, H.-W. Spiess, V. Vill (Wiley-VCH, Weinheim, 1998) p.266]

[4]   H. Saito [Datareview in this book: *11.3 Elastic properties of nematics for applications*]

[5]   H.J. Deuling [*Liquid Crystals* Sol. State Phys. Suppl. vol.14, Ed. L. Liebert (Academic Press, London, 1978) p.77]

[6]   K. Ikeda, H. Okada, H. Onnagawa, S. Sugimori [*J. Appl. Phys. (USA)* vol.86 (1999) p.5413]

[7]   P. Chatelain [*C.R. Acad. Sci. (France)* vol.218 (1944) p.652]

[8]   P.G. de Gennes, J. Prost [*The Physics of Liquid Crystals* (Oxford University Press, Oxford, 1993) p.139]

[9]   H.J. Coles [in *The Optics of Thermotropic Liquid Crystals* Eds. S. Elston, R. Sambles (Taylor and Francis, London, 1998) p.57]

[10]   H.F. Gleeson [in *Handbook of Liquid Crystals* vol.1, Eds. D. Demus, J. Goodby, G.W. Gray, H.-W. Spiess, V. Vill (Wiley-VCH, Weinheim, 1998) p.699]

[11]   J.K. Moscicki [Datareview in this book: *8.2 Measurements of viscosities in nematics*]

[12]   P.G. de Gennes, J. Prost [*The Physics of Liquid Crystals* (Oxford University Press, Oxford, 1993) p.146]

[13]   G.-P. Chen, H. Takezoe, A. Fukuda. [*Liq. Cryst. (UK)* vol.5 (1989) p.341]

[14]   P.G. de Gennes, J. Prost [*The Physics of Liquid Crystals* (Oxford University Press, Oxford, 1993) p.229]

[15]   D. Gu, A.M. Jamieson, C. Rosenblatt, D. Tomazos, M. Lee, V. Percec [*Macromolecules (USA)* vol.24 (1991) p.2385]

[16]   T. Toyooka, G.-P. Chen, H. Takezoe, A. Fukuda [*Jpn. J. Appl. Phys. (Japan)* vol.26 (1987) p.1959]

[17]   M. Hasegawa, K. Miyachi, A. Fukuda [*Jpn. J. Appl. Phys. (Japan)* vol.34 (1995) p.5694]

[18]   S. Faetti, M. Gatti, V. Palleschi [*Rev. Phys. Appl. (France)* vol.21 (1986) p.451]

[19]   S. Faetti, V. Palleschi [*Liq. Cryst. (UK)* vol.2 (1987) p.261]

[20]   M.J. Bradshaw, E.P. Raynes, J.D. Bunning, T.E. Faber [*J. Phys. (France)* vol.46 (1985) p.1513]

[21]   Hp. Schad, M.A. Osman [*J. Chem. Phys. (USA)* vol.75 (1981) p.880]

[22]   Hp. Schad, G. Baur, G. Meier [*J. Chem. Phys. (USA)* vol.70 (1979) p.2770]

[23]   W.H. de Jeu [*Physical Properties of Liquid Crystalline Materials* (Gordon and Breach, New York, 1980) p.88]

[24]   G. Heppke, A. Ranft, B. Sabaschus [*Mol. Cryst. Liq. Cryst. Lett. (UK)* vol.8 (1991) p.17]

[25]   N. Derbel, T. Othman, A. Gharbi [*Liq. Cryst. (UK)* vol.25 (1998) p.561]

[26]   T. Othman, M. Garbia, A. Gharbia, C. Destrade, G. Durand [*Liq. Cryst. (UK)* vol.18 (1995) p.839]

[27]   K Praefcke [Datareview in this book: *1.2 Relationship between molecular structure and transition temperatures for organic materials of a disc-like molecular shape in nematics*]

[28]   B. Sabaschus, D. Singer, G. Heppke, K. Praefcke [*Liq. Cryst. (UK)* vol.12 (1992) p.863]

[29]   G.A. DiLisi, C. Rosenblatt, A.C. Griffin, U. Hari [*Liq. Cryst. (UK)* vol.8 (1990) p.437]

[30]   G.A. DiLisi, E.M. Terentjev, A.C. Griffin, C. Rosenblatt [*J. Phys. II (France)* vol.2 (1992) p.1065]

[31]   G.A. DiLisi, E.M. Terentjev, A.C. Griffin, C. Rosenblatt [*J. Phys. II (France)* vol.3 (1993) p.597]

[32]   D. Kang et al [*Phys. Rev. E (USA)* vol.58 (1998) p.2041]

[33]   A.P. Filippov, V. Surendranath [*Liq. Cryst. (UK)* vol.26 (1999) p.817]

[34]   D.A. Dunmur, K. Toriyama [in *Handbook of Liquid Crystals* vol.1, Eds. D. Demus, J. Goodby, G.W. Gray, H.-W. Spiess, V. Vill (Wiley-VCH, Weinheim, 1998) p.274]

[35]   H. Ishikawa, A. Toda, H. Okada, H. Onnagawa, S. Sugimori [*Liq. Cryst. (UK)* vol.22 (1997) p.743]

[36]   M. Schadt, R. Buchecker, A. Villiger [*Liq. Cryst. (UK)* vol.7 (1990) p.519]

[37]   S.M. Kelly, M. Schadt, H. Seiberle [*Liq. Cryst. (UK)* vol.18 (1995) p.581]

[38]   K. Toriyama, S. Sugimori, K. Moriya, D.A. Dunmur, R. Hanson [*J. Phys. Chem. (USA)* vol.100 (1996) p.307]

[39]   K. Singh, N.S. Pandey [*Liq. Cryst. (UK)* vol.25 (1998) p.411]

[40]   A.F. Martins [Datareview in this book: *8.3 Measurement of viscoelastic coefficients for nematic mesophases using magnetic resonance*]

5.3   Measurement of surface elastic constants

G.P. Crawford

November 1998

A   INTRODUCTION

The effect of surface elastic constants on the nematic director configurations is of basic interest for the elastic theory of liquid crystals and plays a critical role in those device applications where the nematic is confined to a curved geometry. The saddle-splay surface elastic constant, K_{24}, and the splay-bend surface elastic constant, K_{13}, defied measurement for more than sixty years, since the pioneering work of Oseen, who made the first steps toward the elastic theory of liquid crystals.

B   ORIGIN OF SURFACE ELASTIC CONSTANTS

Measurements of the saddle-splay surface elastic constant, K_{24}, and the splay-bend surface elastic constant, K_{13}, were first introduced by Oseen [1] in 1933 from a phenomenological viewpoint, and later by Nehring and Saupe [2] from a molecular standpoint. These constants tend to be neglected in conventional elastic continuum treatments for fixed boundary conditions because they do not enter the Euler-Lagrange equation for bulk equilibrium. Experimental determination of the two surface elastic constants is undoubtedly a difficult task, since their effects are hard to discriminate from those of ordinary surface anchoring [3].

In order to appreciate and understand thoroughly the significance of the surface elastic constants, it is necessary to comprehend fully the phenomenology of bulk nematic ordering. The Landau-de Gennes approach is based on the tensor order parameter $\mathbf{Q}$ with components Q_{ij}. In the eigen frame defined by the nematic director, $\mathbf{n}$, the corresponding degrees of order are measured by eigenvalues S, -½(S + D), -½(S - D), where S is the nematic or major order parameter and D is the biaxial order parameter. Here, the free energy density, F_n, is expanded in terms of symmetry allowed powers of the order parameter and its derivatives. Keeping only invariant terms quadratic in the first derivatives and linear in the second derivatives [10,11] gives:

$$
\begin{aligned}
F_n = {} & F_o(T) + \frac{1}{2}a(T - T^*)Q_{ij}Q_{ji} - \frac{1}{2}bQ_{ij}Q_{jk}Q_{ki}/3 + \frac{1}{4}c_1(Q_{ij}Q_{ji})^2 + \\
& \frac{1}{4}c_2Q_{ij}Q_{jk}Q_{kl}Q_{li} + L_1^{(1)}Q_{ij,ij} + L_1^{(2)}Q_{jk,i}Q_{jk,i} + L_2^{(2)}Q_{ij,i}Q_{kj,k} + \\
& L_3^{(2)}Q_{jk,i}Q_{ik,j} + L_5^{(2)}Q_{ik,ij}Q_{jk} + L_6^{(2)}Q_{jk,ii}Q_{jk} + L_1^{(3)}Q_{ij}Q_{ij,k}Q_{kl,l} + \\
& L_2^{(3)}Q_{ij}Q_{ik,j}Q_{kl,l} + L_3^{(3)}Q_{ij}Q_{ik,k}Q_{jl,l} + L_4^{(3)}Q_{ij}Q_{ik,l}Q_{jk,l} + \\
& L_5^{(3)}Q_{ij}Q_{ik,l}Q_{jl,k} + L_6^{(3)}Q_{ij}Q_{ik,l}Q_{kl}
\end{aligned}
\tag{1}
$$

where $a(T - T^*)$, b, c_1 and c_2 are expansion coefficients. The supercooling limit temperature is T* and indices i, j, k and l represent any of our three coordinate axes. The expansion coefficients $L_j^{(i)}$ are temperature independent generalised elastic constants and the notation $Q_{ij,l}$ denotes the partial derivative with respect to the lth coordinate. The introduction of a homogeneous magnetic (or similarly electric) field **B** adds to the free energy density a contribution from the anisotropic susceptibility: $F_f = -\frac{1}{2}\mu_o^{-1}\Delta\chi_o\mathbf{BQB}$ where $\Delta\chi_o$ is the difference between the principal values of the susceptibility tensor in a completely ordered nematic phase.

For a uniaxial nematic phase, Q_{ij} is simply replaced by the scalar S, the nematic order parameter; although there are many orientational order parameters S is normally dominant. For a uniaxial nematic phase, the constant order parameter approximation leads to:

$$Q_{ij}(\mathbf{r}) = \frac{S}{2}\{3n_i(\mathbf{r})n_j(\mathbf{r}) - \delta_{ij}\} \tag{2}$$

where the nematic liquid crystal structure is described by $\mathbf{n}(\mathbf{r}) = (n_1(\mathbf{r}), n_2(\mathbf{r}), n_3(\mathbf{r}))$ with symmetry $\mathbf{n} = -\mathbf{n}$ [6]. The nematic free energy thus reduces to the Frank elastic free energy [2,5,7,8]:

$$F_d = \frac{1}{2}\int_V \{K_1(\nabla\cdot\mathbf{n})^2 + K_2(\mathbf{n}\cdot\nabla\times\mathbf{n})^2 + K_3(\mathbf{n}\times\nabla\times\mathbf{n})^2$$

$$- K_{24}\nabla\cdot[\mathbf{n}(\nabla\cdot\mathbf{n}) + \mathbf{n}\times\nabla\times\mathbf{n}] + K_{13}\nabla\cdot[\mathbf{n}(\nabla\cdot\mathbf{n})]\} \, dV \tag{3}$$

where the Frank elastic constants are related in the following way to the L_{ij} constants [9]:

$$K_1 = \frac{9S^2}{2}[2L_1^{(2)} + L_2^{(2)} + L_3^{(2)} - L_5^{(2)} - 2L_6^{(2)}] +$$

$$\frac{9S^3}{4}[-L_2^{(3)} + 2L_3^{(3)} + L_4^{(3)} + 2L_5^{(3)} - L_6^{(3)}]$$

$$K_2 = 9S^2[L_1^{(2)} - L_6^{(2)}] + \frac{9S^3}{4}L_4^{(3)}$$

$$K_3 = \frac{9S^2}{2}[2L_1^{(2)} + L_2^{(2)} + L_3^{(2)} - L_5^{(2)} - 2L_6^{(2)}] + \tag{4}$$

$$\frac{9S^3}{4}[2L_2^{(3)} - L_3^{(3)} + L_4^{(3)} - L_5^{(3)} + 2L_6^{(3)}]$$

$$K_{24} = 3SL_1^{(1)} + \frac{9S^2}{2}[2L_1^{(2)} + L_3^{(2)} - \frac{1}{3}L_5^{(2)} - 2L_6^{(2)}] +$$

$$\frac{9S^3}{2}[L_4^{(3)} + 2L_5^{(3)} - L_6^{(3)}]$$

$$K_{13} = 3SL_1^{(1)} + \frac{3S^2}{4}L_5^{(2)}$$

Here K_1, K_2 and K_3 correspond to the splay, twist, and bend bulk elastic constants, respectively. Further, the terms K_{24} and K_{13}, known as the saddle-splay surface elastic constant and mixed splay-bend surface elastic constant, respectively, are called surface elastic constants because they enter EQN (3) as divergences of a volume integral, which are converted to surface integrals via Green's theorem. It should be noted that the second derivative of the K_{24} term in EQN (3) is apparent; however, this is not true for the K_{13} term which is the only one with a second derivative of the director field. This so-called second-order elasticity causes problems when minimisation of the free energy is not performed carefully; there have been several suggestions about how this should be carried out [10-15].

In the single elastic constant approximation, where only the constant $L_1^{(2)}$ associated with pure quadratic terms in EQNS (4) is non-zero, we have $K_1 = K_2 = K_3 = K_{24} = 9S^2L_1^{(2)}$ and $K_{13} = 0$. Taking into account constants associated with other second-order terms in S, we find $K_1 = K_3 \neq K_2 \neq K_{24}$ and $K_{13} = 0$, where all non-zero constants are of the same order of magnitude. Including linear and quadratic terms with second derivatives, we find $K_{13} = 0$, and further including third-order terms yields $K_1 \neq K_3$.

In the chiral nematic case [17], the twist term has the form $\frac{1}{2}K_2(\mathbf{n} \cdot \nabla \times \mathbf{n} - \mathbf{q})^2$ with $2\pi/q$ as the pitch of the helical deformation. If q, K_{24} and K_{13} are all zero, all deformations enter quadratically into the free energy. Then a stable uniform solution requires that K_1, K_2 and K_3 are all positive. The non-zero values of q, K_{24} or K_{13} that are related to quadratic terms in the free energy can, in principle, stabilise a spontaneous deformation. The observations in bulk systems show that such deformations occur only in chiral systems with non-zero $\mathbf{q}$. Therefore, it can be shown that for non-chiral systems, $0 < K_{24} < 2 K_1$ or $2 K_2$, whichever is smaller [18].

C   MEASUREMENTS OF SURFACE ELASTIC CONSTANTS

The focus of this Datareview is on the measurement of surface elastic constants which we have discussed. Although the elastic theory of liquid crystals has been known for many years [19], the effects of the surface elastic constants, K_{24} and K_{13}, were usually neglected because their magnitude defied measurement. Such an approximation is reasonable in simple planar systems. Neglecting these terms in more complex systems (such as those arising from curved confining geometries), can result in large discrepancies between theory and experiment. The increasing interest in these curved systems in the 1990s has resulted in the first measurements of K_{24} and K_{13}. One of the primary driving forces behind the measurement of surface elastic constants is the polymer dispersed liquid crystal (PDLC) display. The value of the surface elastic constants was found to influence strongly the director distribution [5,9,20]. Although phenomenological modelling incorporated the K_{24} term (neglecting K_{13}), determining its value from measurement is very difficult.

The work on the surface elastic constants shifted to cylindrical geometries where its effect could be measured [3,9,21-24]. FIGURE 1 shows two configurations that are stable in cylindrical geometries: the escaped-radial (ER) [3,9,21,22] and the escaped-twisted (ET) [24]. These configurations are ideal candidates to measure the saddle-splay surface elastic constant because they are both sensitive to the surface parameter σ. For the escaped-radial structure, $\sigma = RW_\theta/K + K_{24}/K - 1$ ($K_1 = K_3 = K$), and for the escaped-twisted configuration, $\sigma = RW_\phi/K_2 + K_{24}/K_2 - 1$ when the effects of K_{13} are omitted. Note that σ also depends on the polar anchoring strength, W_θ, for the escaped-radial structure and on the azimuthal anchoring strength, W_ϕ, for the escaped-twisted structure. The free energy presented in EQN (3) is therefore supplemented by a surface anchoring term to arrive at these relations. The simplest expression for anchoring is the well-known Rapini-Papoular [25] form that is supplemented in the free energy density by $\frac{1}{2}W_\theta\sin^2\theta$ for homeotropic anchoring [25] and a similar form for

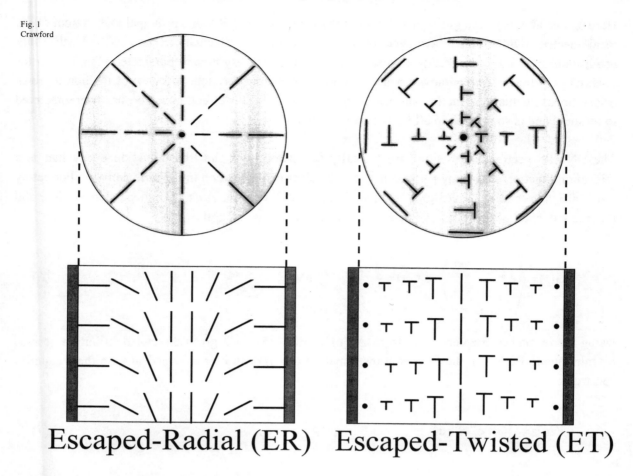

Fig. 1
Crawford

Escaped-Radial (ER) Escaped-Twisted (ET)

FIGURE 1 Schematic illustration of the escaped-radial (ER) and the escaped-twisted (ET) director distributions that are stable in cylindrical cavities. Both configurations sustain the K_{24} distortion.

monostable planar anchoring given by $\frac{1}{2}(W_\theta\cos^2\phi + W_\phi\sin^2\phi)\sin^2\theta$ where θ is the polar angle between the surface normal and the director at the surface and ϕ is the angle measuring the azimuthal direction on the surface [9]. It is obvious from the form of σ in both cases that for strong anchoring or very large radii, the effect of K_{24} is not measurable; in fact, the anchoring term will just wash out the effects of the saddle-splay elasticity.

There have been two techniques, nuclear magnetic resonance [3,9,21,22] and optical polarising microscopy [23], which have been successful in extracting the magnitude of K_{24} in the case when $K_{13} = 0$ is assumed in EQN (3).

Deuterium nuclear magnetic resonance (DNMR) is a very sensitive way to probe the structure if the mesogenic molecule has been selectively deuterated. DNMR has been extensively used to determine director distributions in small confining cavities [26-28] and director dynamics [29-31]. The observed quadrupole splitting for a selectively deuterated compound originating from molecules in a volume element at point $\mathbf{r}$ is given by:

$$\Delta\tilde{v} = \frac{1}{2}S(\mathbf{r})\Delta\tilde{v}_0\{(3\cos^2\theta_n - 1) + \tilde{\eta}(\mathbf{r})\sin^2\theta_n\cos[2\psi(\mathbf{r})]\} \tag{5}$$

Here $\theta_n(\mathbf{r})$ and $\psi_n(\mathbf{r})$ express the orientation of the magnetic field **B** in the principal axis system of the local electric field gradient averaged over fast orientational fluctuations and $\Delta\tilde{v}_0$ is the bulk quadrupole splitting of a perfectly aligned nematic phase. The asymmetry parameter $\tilde{\eta}(\mathbf{r})$ is zero for uniaxial nematics, but it can however have an effect in confined systems because of the non-uniform orientational fluctuations such as surface induced biaxiality. This effect, however, has been predicted to be small and is so ignored in all DNMR studies to date.

The DNMR spectral distributions are typically calculated from the free induction decay that is a relaxation function describing the attenuation of nuclear magnetisation in the time domain. For cavity sizes >0.1 mm, motional averaging is typically ignored; therefore, $\Delta\tilde{v}(\mathbf{r})$ is independent of time and the free induction decay, G(t), can be expressed as the volume integral over V:

$$G(t) = \left\langle e^{i\int_0^t \pi\Delta\tilde{v}[\mathbf{r}(t')]dt'} \right\rangle \Rightarrow G(t) = \frac{1}{V}\int_V e^{i\pi\Delta\tilde{v}(\mathbf{r})t}\, dV \tag{6}$$

Using the resonance frequency spectrum $\Delta\tilde{v}(\mathbf{r})$ in EQN (5) for a given director distribution derived by minimising EQN (3), the Fourier transformation of G(t) can be performed to give the frequency spectrum:

$$I(\Delta\tilde{v}) = \int_{-\infty}^{\infty} G(t)e^{-i\pi\Delta\tilde{v}t}\, dt \tag{7}$$

Line broadening is included by convoluting $I(\Delta\tilde{v})$ with a narrow Gaussian distribution function. If necessary, motional averaging can be included by considering instantaneous values of the quadrupole splitting frequency.

Both configurations presented in FIGURE 1 have been probed by DNMR. Results for the escaped-radial can be found in [21-23,28,32] and for the escaped-twisted in [24]. In order to perform a K_{24} experiment with DNMR, two fundamental parameters must be estimated to ensure that the experiment will yield results that are simple to analyse. The influence of the magnetic field on the director distribution can be estimated by calculating the magnetic coherence length, $\xi_m = (\mu_0 K / \Delta\tilde{\chi})^{1/2}/B$, which is approximately ~1.7 μm for the commonly used deuterated liquid crystal 4-pentyl-4′-cyanobiphenyl (5CB-βd$_2$) and a magnetic field strength of 4.7 T. The effect of motional averaging on the spectra during the time scale of the DNMR experiment is also a very important parameter. The distance a molecule diffuses on the DNMR time scale can be estimated from $d \sim (D/\Delta\tilde{v}_0)^{1/2}$ where D is the diffusion constant. For 5CB-βd$_2$, $D \sim 10^{-11}$ m^2 s^{-1} and $\Delta\tilde{v}_0 \sim 40$ kHz, and so the characteristic diffusion length is $d \sim 0.02$ μm. Therefore before doing a DNMR experiment, it is necessary to be confident that the size of the cylindrical containers is appreciably larger than the characteristic diffusion length and less than the magnetic coherence length, ξ_m. The inset in FIGURE 2 shows a scanning electron microscope photograph of a Nuclepore membrane, manufactured from thin 10 μm polycarbonate (PC) films, that has cylindrical pores etched throughout the thickness of the membrane. The pore diameters of Nuclepore membranes, 0.1 - 0.6 μm, fall well within this range found for d. To ensure that these simple estimates of d and ξ_m are a good rule-of-thumb, spectra were also simulated

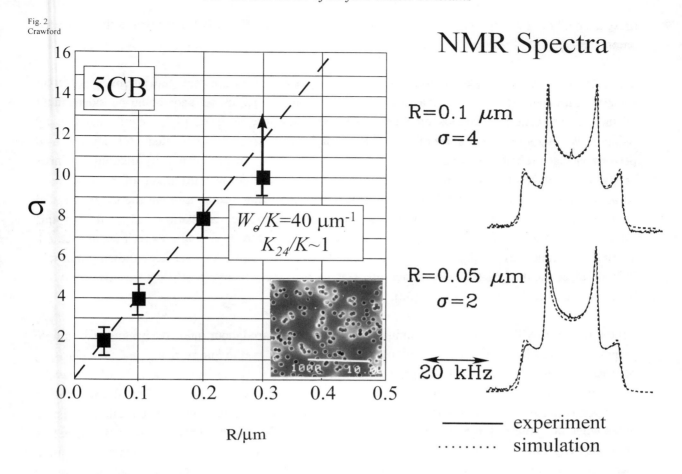

FIGURE 2 Measurement of K_{24} using deuterium nuclear magnetic resonance. Spectra are shown for the escaped-radial director distribution in the submicrometre pores of Nuclepore membranes where the pores are oriented perpendicular to the magnetic field and a simulation where σ is the fitting parameter (right). The measured values of σ from DNMR spectral fits are plotted as a function of the pore radius (left). The inset shows a scanning electron microscopy of a Nuclepore membrane.

to confirm that the magnetic field and diffusion had no appreciable effect on the experiment for the pore diameters in the range 0.1 - 0.6 μm [29]. The pores of the membranes are permeated with 5CB-βd₂, sliced into 5 mm × 20 mm strips, stacked on top of one another (~200) to ensure a sufficient signal-to-noise ratio for the DNMR experiment, and inserted into an NMR tube.

FIGURE 2 shows two experimental spectra (solid line) that were recorded and fitted to the predictions of the elastic theory (dashed line) derived from EQN (3). The sample was oriented such that the cylindrical axis of the pores was orthogonal to the magnetic field of the DNMR experiment. To remove any ambiguity in the fits, the situation where the cylindrical pores are parallel to the magnetic field was also probed to improve the quality of the fit and to confirm unambiguously the escaped-radial configuration [22,23,28,29]. The fitting parameter $\sigma = RW_\theta/K + K_{24}/K - 1$ (assuming $K_1 = K_3 = K$) is plotted in FIGURE 2 as a function of the pore radius. The slope of the line gives W_θ/K and the intercept yields K_{24}/K, so that information on the molecular anchoring strength and saddle-splay elasticity can be determined. From this particular experiment, it is found that $K_{24} = K$. This was the first measurement of the saddle-splay surface elastic constant which showed that its presence could not be ignored in EQN (3) for confined systems. TABLE 1 summarises the measured parameters.

These DNMR measurements led to subsequent studies on the same nematogen but using different surface treatments. Here it is emphasised that K_{24} is a material parameter and should be independent of surface interactions. Studies on lecithin treated Nuclepore surfaces [3,22] confirmed this by reducing W_θ by more than a factor of six, yet K_{24} remained unchanged, as expected. A study was also performed that probed the escaped-twisted configuration shown in FIGURE 1 by treating the inner cavity walls of Nuclepore membranes with polyimide, a well-known promoter of homogeneous parallel surface anchoring conditions. This situation is even more complex than the escaped-radial case because the azimuthal anchoring term also enters into the calculation [32]. It was this experiment that showed that W_θ, W_ϕ and K_{24} could be measured by a single experimental method. TABLE 1 summarises the data on the measurement of K_{24} for the various surfaces showing that it is indeed independent of the surface interaction. 4-octyl-4'-cyanobiphenyl (8CB-βd$_2$) was also shown to have a value of K_{24} of the order of K [32].

Although the DNMR is an elegant technique, the experiments are labour intensive because the sample preparation is tedious. It was for this reason that a similar methodology using optical polarising microscopy was developed [23]. The same averaging process as the order parameter S influences the refractive indices of nematic liquid crystals. Therefore, the optical axis parallel to the director corresponds to the extraordinary index of refraction, $\tilde{n}_e$, and the ordinary index of refraction, $\tilde{n}_o$, is the index when the polarised light is perpendicular to the optic axis. The birefringence $\Delta\tilde{n} = \tilde{n}_e - \tilde{n}_o \propto S$ is approximately 0.2. A liquid crystal, therefore, strongly influences the propagation of light. Depending on the scale of the homogeneous areas of the director distribution, L, as compared to the wavelength of light, λ, we can classify these situations into two distinct cases: $L > \lambda$ which covers most of optical polarising microscopy and allows for the use of geometrical optics, and $L \sim \lambda$ which is realised in light scattering experiments and requires wave optics. Here we are interested in $L > \lambda$.

Polarising microscopy has provided many of the clues to the structure of nematic droplets and nematic filled cylinders [33]. Here we describe how to use simulated optical polarising microscope textures of nematic filled capillary tubes to measure the value of the saddle-splay-surface elastic constant. A ray passing through a capillary tube filled with liquid crystal will experience relatively small variations in the refractive index; we can therefore neglect the diffraction of light. The direction of the ray passing through a capillary tube is assumed therefore to be undeviated. The intensity pattern at the location of interest is determined by the birefringence of the nematic which introduces a phase shift between the ordinary and extraordinary components of light. Within this straight ray approximation (SRA) the distribution of the transmitted light over a texture observed with a polarising microscope may be formally described by the Jones matrix. A matrix $\underline{T}(p)$ describes the evolution of the light polarisation vector as a ray k passing through the tube at a point p on the texture. Except for a very simple structure, $\underline{T}(p)$ cannot be calculated analytically; therefore the path of a ray through the tube is divided into small segments of length Δ' where the local field is assumed constant.

The continuous process of the polarisation changes is divided into the following steps: change of the polarisation vector caused by the rotation of the optic axis going from one interval to another and phase shifts within each interval. For N steps the relative intensity of the transmitted light can be expressed as:

TABLE 1 Summary of K_{24} measurements.

Liquid crystal[a]	Confinement {Surface} [Alignment]	W_θ K[-1]/μm[-1]	W_θ/10[-5] J m[-2b]	W_ϕ K[-1]/μm[-1]	W_ϕ/10[-5] J m[-2b]	K_{24}/K	Technique/Ref
5CB-βd₂	Nuclepore {Untreated PC} [Homeotropic]	40	20 using K = 5 × 10[-12] N			~1.0	DNMR [3;22]
5CB-βd₂	Nuclepore {Lecithin coat} [Homeotropic]	6	3 using K = 5 × 10[-12] N			~1.1	DNMR [3;22]
5CB-βd₂	Nuclepore {Untreated PC} [Homeotropic]	35	17.5 using K = 5 × 10[-12] N			~1.0	DNMR [28]
5CB-βd₂	Nuclepore {Polyimide coat} [Homogeneous]	13	6.5 using K = 5 × 10[-12] N	20	8 using K_2 = 4 × 10[-12] N	~1.2	DNMR [24]
8CB-βd₂	Nuclepore {Untreated PC} [Homeotropic]	100	70 using K_1 = 7 × 10[-12] N			~1.0	DNMR [32]
E7 Eutectic mixture	Glass capillary {Lecithin coat} [Homeotropic]	0.56	0.61 using K_1 = 1.1 × 10[-11] N			~2.6, 1.6 < K_{24}/K < 1.9[c]	Optical microscopy [23]
5CB	Glass capillary {Lecithin coat} [Homeotropic]	1.1	0.66 using K_1 = 6 × 10[-12] N			~3.1, 1.2 < K_{24}/K < 1.6[c]	Optical microscopy [23]

[a]The -βd₂ indicates that the materials are selectively deuterated in the second position on the hydrocarbon chain.

[b]The value of K denoted in the box was used to arrive at the value for W_θ or W_ϕ.

[c]Stability considerations to limit the possible values of K_{24} to a narrower interval using the Erickson inequality $0 < K_{24} < 2K_1$ (or K_2) whichever is smaller. See [18].

$$I(\mathbf{p}) = |\mathbf{e}_A T(\mathbf{p})\mathbf{e}_P|^2 = |\mathbf{e}_A \underline{R}_{N+1}\underline{P}_{N+1}\underline{P}_N\underline{R}_N \ldots \underline{P}_2\underline{R}_2\underline{P}_1\underline{R}_1\mathbf{e}_P|^2 \qquad (8)$$

Here $\mathbf{e}_P$ and $\mathbf{e}_A$ are the unit vectors defined by the polariser and analyser direction, $\underline{R}_i(\mathbf{p})$ are rotation matrices, and $\underline{P}_i(\mathbf{p})$ induces the appropriate phase shifts to ordinary and extraordinary components of light in the ith segment. More precisely, $\underline{R}_i(\mathbf{p})$ represents a simple rotation of the optical axis $\mathbf{n}_i$ around $\mathbf{k}$ going from the i-1 interval to the i interval. Introducing $\mathbf{n}_D$ as a direction of the symmetry axis of the structure we define the angle α_i between the reference $\mathbf{n}_D$-$\mathbf{k}$ plane and the $\mathbf{k}$-local optic axis ($\mathbf{n}_i$ or $\mathbf{e}_P$ or $\mathbf{e}_A$) plane and express the rotation matrix as:

$$\underline{R}_i = \begin{bmatrix} \cos(\alpha_i - \alpha_{i-1}) & -\sin(\alpha_i - \alpha_{i-1}) \\ \sin(\alpha_i - \alpha_{i-1}) & -\cos(\alpha_i - \alpha_{i-1}) \end{bmatrix} \qquad (9)$$

Values of the index i range from 1 to N and correspond to the segment in the capillary tube cross section. The end points i = 0 and i = N+1 represent the frames corresponding to the polariser and analyser, respectively. The $\underline{P}_i$ matrix introduces phase shifts in the i interval and is expressed as:

$$\underline{P}_i = \begin{bmatrix} e^{i n_o 2\pi\Delta'/\lambda} & 0 \\ 0 & e^{i n_e(\gamma_i')2\pi\Delta'/\lambda} \end{bmatrix} \qquad (10)$$

where λ is the wavelength of incoming light and $\tilde{n}_e(\gamma_i')$ is the effective extraordinary index in the i interval where $\gamma_i'(\mathbf{p})$ is the angle between the symmetry axis and the director $\mathbf{n}_i$ and the wavevector $\mathbf{k}$.

The extraordinary index is expressed as:

$$\tilde{n}_e(\gamma_e'(\mathbf{p})) = \frac{\tilde{n}_o\tilde{n}_e}{\sqrt{\tilde{n}_o^2 \sin^2 \gamma_i' + \tilde{n}_e^2 \cos^2 \gamma_i'}} \qquad (11)$$

To calculate the complete texture, the above procedure is repeated for several points $\mathbf{p}$ so that all details of the textures needed are represented. From EQN (8) the relative transmitted intensity at $\mathbf{p}$ can be calculated for values of $\tilde{n}_e$, $\tilde{n}_o$, λ, the tube radius R, and the angle α_o defined as the orientation of the polarisation vector $\mathbf{e}_P$ of incoming light and θ_o defined as the angle of the symmetry axis of the director distribution with respect to $\mathbf{k}$. For capillary tubes with a cylindrically symmetric director distribution, as is our case for the escaped-radial configuration, $\theta_o = 90°$ since the cylinder axis is restricted in the plane of the microscope stage, and so the symmetry axis is orthogonal to $\mathbf{k}$. Typically the microscope stage can be rotated through the angle α_o which is the angle that the symmetry axis of the escaped-radial configuration makes with the polarisation director.

FIGURE 3 shows the experimentally observed escaped-radial configuration in a capillary tube (R = 15 μm) oriented at $\alpha_o = 45°$ directly compared to the simulation. By matching the number of fringes and fringe position, the value of $\sigma = RW_\theta/K + K_{24}/K - 1$ (assuming $K_1 = K_3 = K$) can be determined [23] just as for the DNMR experiment. FIGURE 3 also shows $\sigma = RW_\theta/K + K_{24}/K - 1$ versus R, where the slope is W_θ/K and the intercept yields K_{24}/K. The final point on the graph in FIGURE 3 at R = 25 μm shows that we are approaching the strong anchoring regime and the texture

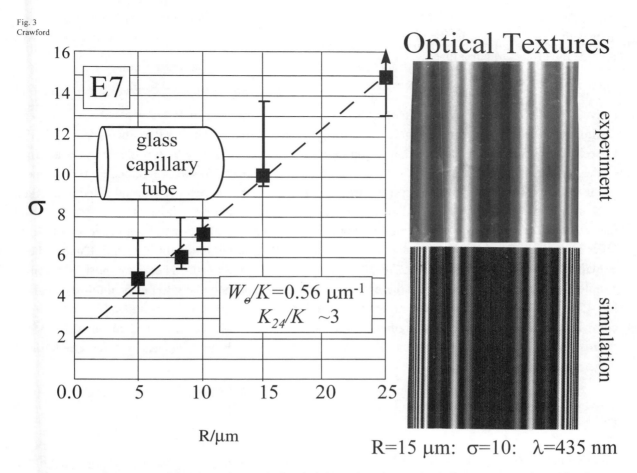

FIGURE 3 Measurement of K_{24} using optical polarising microscopy. Optical textures are shown for the escaped-radial director field in supramicrometre capillary tubes where the cylinder axis is oriented at 45° with respect to the polariser and analyser. The experimental observation is compared to simulation where σ is the fitting parameter (right). The measured values of σ from optical polarising microscopy fits are plotted as a function of the capillary radius (left).

is no longer sensitive to σ. TABLE 1 summarises the capillary tube experiments. Although the optical polarising microscope experiments are easier than DNMR experiments, the values determined for K_{24}/K are not as precise. This is because the capillary tubes are much larger than the pores of Nuclepore membranes. Therefore, the error in determining σ experimentally is larger. We have reverted to the theory of Erickson [18] that enables us to define a range of pore sizes for our optical experiments. As was the case with DNMR, the magnitude of K_{24} is significant.

These studies ignore the curvature dependence of the interfacial free energy; Yokoyama [34] has incorporated this effect. The curvature correction amounts to replacing the anchoring energy by the planar-equivalent effective anchoring energy:

$$W_{eff} = W_\theta \frac{d_e}{R \ln(1 + d_e / K)} \tag{12}$$

where $d_e = K/W_\theta$ is the extrapolation length. Yokayama argues that the values of K_{24} so obtained are subjected to a correction factor of -½. Clearly more precise determinations of K_{24} in the future will include the effects of curvature.

We have discussed two methods to determine the magnitude of K_{24}. There is another way in which to arrive at a reliable value of the surface elastic constants as pointed out by Pergamenshchik [11,35] and exploited by Sparavigna and co-workers [36-38], and by Lavrentovich and Pergamenshchik [17,39], to seek an orientational behaviour that is qualitatively different from anything possible in the absence of the surface-like constants. The most basic consequence of the existence of free energy terms, linear in the gradient, may be that the uniformly oriented state could be energetically destabilised, and taken over by a spontaneously modulated structure under an appropriate condition, as is known from the occurrence of a spontaneous twist in chiral nematics. Such an instability has been observed in a thin layer of nematic floating on a liquid substrate (known as hybrid alignment or stripes). This has been explained as the action of K_{24} [11,12,36-38] or K_{13} [16]. As mentioned earlier, care must be taken to handle the K_{13} term in EQN (3) and there have been several reports on how to perform this calculation [10-15]. The first measurements of both K_{24} and K_{13} came from these systems [16] where K_{24} compared well with earlier measurements and the value of K_{13}/K was reported to be -0.2. More details on the influence of K_{13} on the director distribution have been reviewed by Lavrentovich and Pergamenshchik [39].

D   CONCLUSION

The various ways to measure the saddle-splay surface elastic constant, K_{24}, have been presented. The Datareview finished with the effect of the mixed splay-bend term, K_{13}, on the stripe configuration. It should be emphasised that the determination of the surface-like elastic constants is not nearly as mature as the measurement of the bulk elastic constants K_1, K_2 and K_3. In fact, we can make the analogy that our current understanding of the surface-like elastic constants is similar to that of K_2 about thirty years ago when its value was unknown [18]. Our current understanding of the theory and supporting measurements to date, although the measurements are not as precise as those for the bulk elastic constants, does reveal that these surface-like elastic terms cannot be ignored in nematic systems (e.g. stripes, cylinders, droplets, hybrid aligned cells) that are known to sustain a surface-like distortion. By any objective opinion, this field is clearly still in its infancy and it will take researchers many more years to refine our current understanding of it.

REFERENCES

[1]   C.W. Oseen [*Trans. Faraday Soc. (UK)* vol.29 (1933) p.883]

[2]   J. Nehring, A. Saupe [*J. Chem. Phys. (USA)* vol.54 (1971) p.337]

[3]   D.W. Allender, G.P. Crawford, J.W. Doane [*Phys. Rev. Lett. (USA)* vol.67 (1991) p.1442]

[4]   G. Vertogen, W.H. de Jeu [*Thermotropic Liquid Crystals* (Springer, Berlin, 1988)]

[5]   S. Kralj, S. Zumer [*Phys. Rev. A (USA)* vol.45 (1992) p.2461]

[6]   P.G. de Gennes, J. Prost [*The Physics of Liquid Crystals* (Oxford University Press, Oxford, 1993)]

[7]   A. Saupe [*J. Chem. Phys. (USA)* vol.75 (1981) p.5118]

[8]   J. Nehring, A. Saupe [*J. Chem. Phys. (USA)* vol.56 (1972) p.5527]

[9]   S. Kralj, S. Zumer [*Phys. Rev. E (USA)* vol.51 (1995) p.366]

[10]   G. Barbero, A. Strigazzi [*Liq. Cryst. (UK)* vol.5 (1989) p.693]

[11]   V.M. Pergamenshchik [*Phys. Rev. E (USA)* vol.48 (1993) p.1254]

[12]   V.M. Pergamenshchik [*Phys. Rev. E (USA)* vol.49 (1994) p.934 (errata)]

[13]   H.P. Hinov [*Mol. Cryst. Liq. Cryst. (UK)* vol.148 (1987) p.197]

[14]   G. Barbero, G. Durand [*Phys. Rev. E (USA)* vol.48 (1993) p.1942]

[15]   S. Faetti [*Mol. Cryst. Liq. Cryst. (UK)* vol.241 (1994) p.131]

[16]   S. Zumer, S. Kralj, J. Bezic [*Mol. Cryst. Liq. Cryst. (UK)* vol.212 (1992) p.163]

[17]   O.D. Lavrentovich, V.M. Pergamenshchik [*Phys. Rev. Lett. (USA)* vol.73 (1994) p.979]

[18]   J.L. Erickson [*Phys. Fluids (USA)* vol.9 (1966) p.1205]

[19]   F.C. Frank [*Discuss. Faraday Soc. (UK)* vol.25 (1958) p.19]

[20]   S. Zumer, K. Kralj [*Liq. Cryst. (UK)* vol.12 (1995) p.613]

[21]   G.P. Crawford, D.W. Allender, J.W. Doane, M. Vilfan, I. Vilfan [*Phys. Rev. A (USA)* vol.44 (1991) p.2270]

[22]   G.P. Crawford, D.W. Allender, J.W. Doane [*Phys. Rev. A (USA)* vol.45 (1992) p.8693]

[23]   R.D. Polak, G.P. Crawford, B.C. Kostival, J.W. Doane, S. Zumer [*Phys. Rev. E (USA)* vol.49 (1994) p.R978]

[24]   R.J. Ondris-Crawford, G.P. Crawford, S. Zumer, J.W. Doane [*Phys. Rev. Lett. (USA)* vol.70 (1993) p.194]

[25]   A. Rapini, M. Papoular [*J. Phys. (France)* vol.30 (1959) p.C4]

[26]   A. Golemme, S. Zumer, J.W. Doane, M.E. Neubert [*Phys. Rev. A (USA)* vol.37 (1988) p.559]

[27]   A. Golemme, S. Zumer, D.W. Allender, J.W. Doane [*Phys. Rev. Lett. (USA)* vol.61 (1988) p.1937]

[28]   G.P. Crawford, M. Vilfan, J.W. Doane, I. Vilfan [*Phys. Rev. A (USA)* vol.43 no.2 (1991) p.835-42]

[29]   G.P. Crawford, D.K. Yang, S. Zumer, D. Finotello, J.W. Doane [*Phys. Rev. Lett. (USA)* vol.66 (1991) p.723]

[30]   N. Vrbancic, M. Vilfan, R. Blinc, J. Dolinsek, G.P. Crawford, J.W. Doane [*J. Chem. Phys. (USA)* vol.98 (1993) p.427]

[31]   M. Vilfan et al [*J. Chem. Phys. (USA)* vol.102 (1995) p.8726]

[32]   R.J. Ondris-Crawford, G.P. Crawford, S. Zumer, M. Vilfan, I. Vilfan, J.W. Doane [*Phys. Rev. E (USA)* vol.48 (1993) p.1998]

[33]   G.P. Crawford, J.W. Doane, S. Zumer [in *Handbook of Liquid Crystal Research* Eds. P.J. Collings, J.S. Patel (Oxford University Press, Oxford, 1997) ch.8]

[34]   H. Yokayama [in *Handbook of Liquid Crystal Research* Eds. P.J. Collings, J.S. Patel (Oxford University Press, Oxford, 1997) ch.6]

[35]   V.M. Pergamenshchik [*Phys. Rev. E (USA)* vol.47 (1993) p.1881]

[36]   A. Sparavigna, L. Komitov, B. Stebler, A. Strigazzi [*Mol. Cryst. Liq. Cryst. (UK)* vol.212 (1992) p.265]

[37]   A. Sparavigna, L. Komitov, O.D. Lavrentovich, A. Strigazzi [*J. Phys. II (France)* vol.2 (1992) p.1881]

[38]   A. Sparavigna, O.D. Lavrentovich, A. Strigazzi [*Phys. Rev. E (USA)* vol.49 (1994) p.1344]

[39]   O.D. Lavrentovich, V.M. Pergamenshchik [in *Liquid Crystals: In the Nineties and Beyond* Ed. S. Kumar (World Scientific, Singapore, 1995) ch.8]

5.4   Quantitative aspects of defects in nematics

T. Ishikawa and O.D. Lavrentovich

August 1998

A   INTRODUCTION

Defects in nematic liquid crystals are reviewed at three levels: (1) topological classification; (2) elastic features; (3) experimental observations.

B   TOPOLOGICAL CLASSIFICATION

Defects in uniaxial nematics are described in terms of the spatial distribution of the director $n(r)$. There are two types of functions $n(r)$: those containing singularities (at which n is not defined) and those without singularities. For three-dimensional nematics, the singular regions may be either zero-dimensional (points), one-dimensional (lines), or two-dimensional (walls). These are the defects. Whenever a non-homogeneous state cannot be eliminated by continuous variations of $n(r)$ (i.e. the homogeneous state $n(r)$ = const cannot be generated), it is called a topologically stable, or simply a topological defect. If the inhomogeneous state does not contain singularities, but nevertheless is not deformable continuously into a homogeneous state, the system is said to contain a topological configuration (or soliton).

Wall defects as singular defects are not topologically stable [1]. The energy per unit area of a singular wall $\sim U/a^2$ is defined by the energy U of molecular interactions and the molecular length scale a. If such a singularity is replaced by a smooth director reorientation over a macroscopic length $l >> a$, its energy reduces to $U/al \sim K/l$; here $K \sim U/a$ is an average value of the Frank elastic constants. Thus the singular walls are unstable and tend to smear out. The term 'wall' is often used to describe continuous reorientation of the director field by an angle π or 2π. When the macroscopic width l of the wall is fixed by some external factor, such as electromagnetic field or surface anchoring, the wall is a topological soliton.

Line defects can be topologically stable. Topological stability of defects is controlled by the order parameter space of the medium [2,3]. This space is the manifold of all possible values of the order parameter that do not alter the thermodynamical potentials of the system. In uniaxial nematics, the order parameter space is a sphere with pairs of diametrically opposite points being identical. Such a sphere is denoted as S^2/Z_2; every point of S^2/Z_2 represents a particular orientation of n. Any reorientation of the nematic as a whole leaves the thermodynamical potentials unchanged. In addition, since the nematics are non-polar, $n \equiv -n$, any two diametrically opposite points describe the same state.

Imagine now a singular line in a bulk nematic (FIGURE 1); the goal is to verify its topological stability. Let us surround the line by a loop γ; the only requirement is that γ does not approach the singular region too closely (the 'safe' distance is usually a few molecular lengths), so that the direction of n is well-defined at every point along γ. The function **n(r)** maps the points of a real space along γ into the order parameter space. When one goes around γ, **n(r)** draws some closed contour Γ on S^2/Z_2 that might be of two types: (a) a contour that starts and terminates at the very same point (for example, a circle); (b) a contour that connects two diametrically opposite points of S^2/Z_2. Contours (a) can be continuously contracted into a single point. When Γ shrinks smoothly into a point, the corresponding director field in real space becomes uniform, **n(r)** = const, and the singularity disappears. Contours (b) cannot be contracted: under any continuous deformations, their ends remain the ends of a diameter of S^2/Z_2. The corresponding defect lines are topologically stable since they cannot be transformed into a uniform state (although they can be transformed one into another).

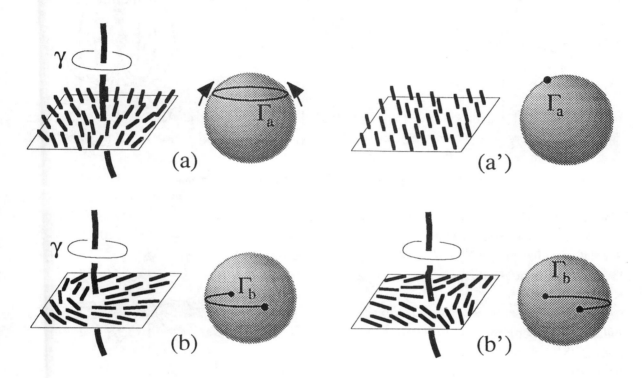

FIGURE 1 Topologically unstable (a) and stable (b) defect lines - disclinations in a uniaxial nematic.

We conclude that there are two classes of nematic line defects, called also disclinations: topologically stable and unstable. Transformation between these classes is possible only when the nematic order is destroyed at the whole half-plane ending at the line. The energy of such a singular wall is much larger than the energy of a singular line, which gives a physical interpretation of the topological stability of disclinations.

Point defects-hedgehogs are another type of topological defect in the bulk nematic [3]. The simplest is a radial hedgehog **n(r)** = ±**r**/|**r**|: a point with a radial director field around it, as shown in FIGURE 2. Generally, to elucidate the stability of a point defect, it is enclosed by a closed surface (e.g. a sphere) σ. The function **n(r)** produces a mapping of σ onto some surface Σ in the order parameter space. If Σ can be contracted to a single point, the point defect is topologically unstable. If Σ is wrapped N ≠ 0 times around the sphere S^2/Z_2, the point singularity is a stable defect with a topological charge N ≠ 0 (see FIGURE 2). Analytically,

$$N = \frac{1}{4\pi} \oiint \left(\frac{\partial \theta}{\partial u} \frac{\partial \varphi}{\partial v} - \frac{\partial \theta}{\partial v} \frac{\partial \varphi}{\partial u} \right) \sin\theta \, du \, dv = 0, \pm 1, \pm 2, \ldots \quad (1)$$

for the director parametrised as $\mathbf{n} = \{\sin\theta\cos\varphi; \sin\theta\sin\varphi; \cos\theta\}$, with the polar angle θ and the azimuthal angle φ being functions of the coordinates u and v on σ.

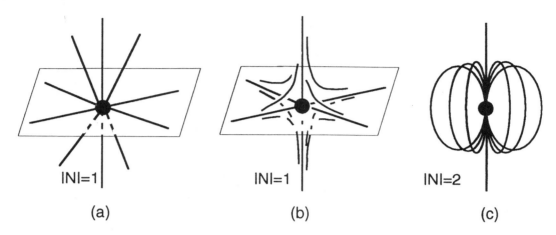

|N|=1 |N|=1 |N|=2

(a)   (b)   (c)

FIGURE 2 Topological point defects in a uniaxial nematic.

Since $\mathbf{n} \equiv -\mathbf{n}$, each point defect can be equally labelled by N and -N. The coalescence of two points N_1 and N_2 can result in a defect with a charge $|N_1 + N_2|$ or $|N_1 - N_2|$, depending on the presence of disclinations in the system and the path of coalescence [3].

Point defects-boojums (see FIGURE 3) are special point defects that, in contrast to hedgehogs, exist only at the boundary [4]. Any attempt to move a boojum from the surface into the bulk is accompanied by energetically costly additional deformations. In addition to the integer N, boojums are characterised by a two-dimensional topological charge k of the unit vector field $\mathbf{t}$ projected by the director onto the surface:

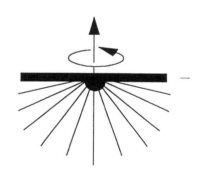

FIGURE 3 An axisymmetric boojum at a surface with tangential anchoring.

$$k = \frac{1}{2\pi} \oint \left(t_x \frac{dt_y}{dl} - t_y \frac{dt_x}{dl} \right) ds = 0, \pm 1, \pm 2, \ldots \quad (2)$$

Here s is the natural parameter defined along the loop at the surface enclosing the defect core; k shows how many times $\mathbf{t}$ rotates by the angle 2π when we move once around the defect.

Bounded nematic volumes. Usually, defects are considered as perturbations of the uniform state, caused, for example, by some mechanical admixtures. There are, however, many situations when topological defects correspond to the equilibrium state of a system. Nematic droplets suspended in an isotropic matrix (a fluid such as water or a polymer such as polyvinylalcohol) and inverted systems (water droplets in a nematic matrix) provide the most evident examples.

The balance of the elastic energy of director distortions and surface energy defines the equilibrium of a bounded nematic such as a droplet. Representative estimates are $\sigma_0 R^2$ for the isotropic part of the surface energy, $W_a R^2$ for the anisotropic surface energy, and, finally, KR for the elastic energy. Here R is the radius of the droplet, σ_0 is the surface tension coefficient and W_a is the surface anchoring coefficient, which measures the energy penalty for director deviations from some preferred surface orientation (e.g. molecular interactions might favour perpendicular orientation of **n** at the boundary). Usually, $\sigma_0 \gg W_a$, so that the droplets are practically spherical with the interior director field defined by the balance of KR and $W_a R^2$.

Small droplets with $R \ll K/W_a$ avoid spatial variations of **n** at the expense of violated boundary conditions. In contrast, large droplets, $R \gg K/W_a$, satisfy boundary conditions by aligning **n** along the preferred direction(s) at the surface. Since the boundary of the droplet is curved, this anchoring effect leads to a distorted director distribution in the bulk, for example a radial hedgehog in the case where the surface director orientation is normal. With typical values of $W_a \approx 10^{-5}$ J/m^2 and $K \approx 10^{-11}$ N, the characteristic radius R is of the order of 1 μm. Generally, bounded nematic volumes at scales $R \gg K/W_a$ contain defects with total topological charges satisfying the following two relationships that have their roots in the Poincaré and Gauss theorems of differential geometry:

$$\sum_i k_i = E \text{ and } \sum_j N_j = E/2 \tag{3}$$

Here E is the topological invariant of the bounding surface, called the Euler characteristic; for a sphere E = 2 and for a torus E = 0. FIGURE 4 shows nematic droplets freely suspended in a glycerin matrix; each droplet contains a pair of boojums at the poles, $k_1 = k_2 = 1$, in agreement with the first expression of EQN (3).

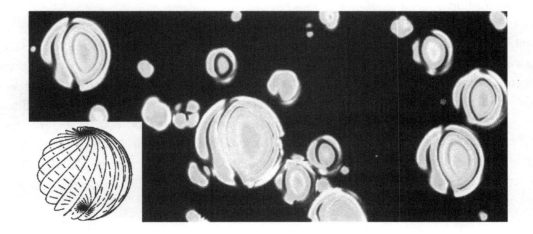

FIGURE 4 Bipolar nematic droplets with point defects-boojums at the poles. The droplets are suspended in a glycerol matrix and illuminated by polarised light. The inset shows the director configuration at the surface of the droplet.

C   ENERGETICS OF THE DEFECTS

Disclinations within the same topological class but of different configurations can be continuously transformed into one another. Their relative stability depends on the Frank elastic constants of splay (K_1), twist (K_2), bend (K_3) and saddle-splay (K_{24}) in the elastic free energy density functional:

$$F = \frac{1}{2} K_1 (\mathrm{div}\mathbf{n})^2 + \frac{1}{2} K_2 (\mathbf{n}.\mathrm{curl}\mathbf{n})^2 + \frac{1}{2} K_3 (\mathbf{n} \times \mathrm{curl}\mathbf{n})^2 - K_{24}\mathrm{div}(\mathbf{n}.\mathrm{div}\mathbf{n} + \mathbf{n} \times \mathrm{curl}\mathbf{n}) \qquad (4)$$

Here we assume the so-called K_{13} constant to be zero.

Frank [5] considered planar disclinations in which $\mathbf{n}$ is perpendicular to the line. For such disclinations, the K_{24}-term in EQN (4) is always zero. In the one-constant approximation $K_1 = K_2 = K_3 = K$, the equilibrium director field around the planar disclination reads

$$\mathbf{n} = \{\cos[k\varphi + c], \sin[k\varphi + c], 0\} \qquad (5)$$

where $\varphi = \arctan(y/x)$, x and y are Cartesian coordinates in the plane normal to the line, and c and k are constants; k is called the strength of the disclination that shows the number of 2π-rotations of the director around the line; it can be integer or half-integer. For the line in FIGURE 1(b), k = 1/2, while for the line in FIGURE 1(b'), k = -1/2.

The energy per unit length (line tension) of a planar disclination is

$$F_{11} = \pi K k^2 \ln \frac{R}{r_c} + F_c \qquad (6)$$

where R is the characteristic size of the system, and r_c and F_c are respectively the radius and the energy of the disclination's core, a region in which the distortions are too strong to be described by a phenomenological theory.

The Frank theory does not distinguish lines of integer and half-integer strength, except for the fact that the lines with |k| = 1 tend to split into pairs of lines with |k| = 1/2, which reduces the energy, according to EQN (6). However, the lines of integer strength are unstable in a more fundamental topological sense: they can be continuously transformed into a non-singular uniform state, as already discussed. Imagine a circular cylinder with normal orientation of molecules at the boundaries, as shown in FIGURE 5(a). The planar disclination would have a radial-like director field normal to the axis of the cylinder. However, the director can be reoriented along the axis, as indicated in FIGURE 5(b). The process, called 'escape in the third dimension', is energetically favourable, since the energy of the escaped configuration is $3\pi K$ [6,7].

Detailed calculations of the disclination energies have been performed by Anisimov and Dzyaloshinskii [8]. They showed that, in addition to planar lines, 'bulk' disclinations can exist, in which the director does not lie in a single plane. Planar lines are stable when $2K_2 > K_1 + K_3$; bulk lines are preferable when $2K_2 < K_1 + K_3$.

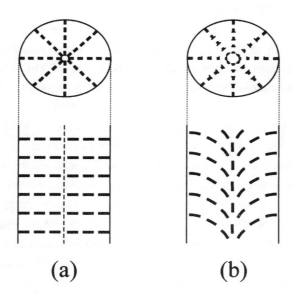

FIGURE 5 A cylinder with a disclination k = 1 (a) and an escaped configuration (b).

Point defects. Unlike point defects such as vacancies in solids, the topological point defects in nematics cause disturbances over the whole nematic volume. The energy of the point defect is proportional to the size R of the system. For example,

$$F_{rh} = 8\pi R(K_1 - K_{24}) + F_{cr} \tag{7}$$

$$F_{hh} = 8\pi R\left(\frac{K_1}{5} + \frac{2K_3}{15} + \frac{K_{24}}{3}\right) + F_{ch} \tag{8}$$

for the radial hedgehog (see FIGURE 2(a)) with

$$\mathbf{n} = (x, y, z)/\sqrt{x^2 + y^2 + z^2}$$

and the hyperbolic hedgehog (FIGURE 2(b)) with

$$\mathbf{n} = (-x, -y, z)/\sqrt{x^2 + y^2 + z^2}$$

respectively [9].

Interaction between defects. The interaction energy between two planar disclinations with strength k_1 and k_2 separated by a distance d is [10]

$$F = \pi K(k_1 + k_2)^2 \ln(R/r_c) - 2\pi K k_1 k_2 \ln(d/r_c) \tag{9}$$

The lines with opposite signs of k attract each other. Note that if $k_1 = -k_2$ the energy of the pair does not depend on the size R of the system.

D   EXPERIMENTAL OBSERVATIONS

When viewed under a polarising microscope, a nematic slab between two glass plates shows distinctive textures. The simplest one is the homeotropic texture that occurs when the director is oriented strictly perpendicular to the bounding plates. Between crossed polarisers, the entire area of the texture appears dark, since the optical axis of the nematic is oriented along the optical axis of the microscope. If the surface orientation of **n** is tangential (**n** is in the plane of the plates) or tilted conical, then a so-called Schlieren texture can form. The main feature of Schlieren textures (see FIGURE 6) is the presence of two types of centres from which two or four extinction bands emerge. The extinction bands (also called brushes) occur in areas where **n** is parallel to either polariser or analyser of the microscope. The centres with two bands have a sharp (singular) core, insofar as can be seen, of submicron dimensions and correspond to the ends of disclinations. The two ends of the disclinations can be located on the opposite plates or on the same plate. The centres with four brushes correspond to boojums, or, on rare occasions, to hedgehogs. On some occasions, points with higher numbers of brushes are encountered [11,12]. There is a simple relationship between the number of brushes B emerging from the point and the defect strength k: $|k| = B/4$. Note, however, that this relationship has limited validity: when the director field is distorted non-uniformly around the defect, the number of brushes fails to provide the information about k [12].

FIGURE 6 A typical Schlieren texture in a film of 4-pentyl-4′-cyanobiphenyl (thickness 23 μm) between two glass plates coated with thin layers of glycerol.

E   ENTANGLEMENT OF DISCLINATIONS

Disclinations can pass through each other. Experiments [13,14] show the process of reconnection: two initial lines exchange ends as they cross so that each of the two ensuing lines has segments of the two original disclinations (see FIGURE 7). The result of crossing depends on the local director field in the region of crossing. In the cells that favour a planar distribution of the director around the disclinations, the result of crossing depends on the total strength of the pair [14,15]: $k_1 + k_2 = 0$ favours scheme (a) and $k_1 + k_2 = 1$ favours scheme (b) shown in FIGURE 7.

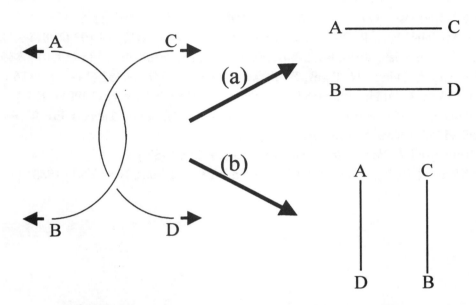

FIGURE 7 Two possible results of reconnection of line defects.

F   CONCLUSION

Although the topological classification of defects in nematics has been firmly established, there are still many open questions concerning the behaviour of defects in external fields, their dynamics and interaction. Note that the classification of defects in biaxial nematics is drastically different from that in the uniaxial nematics considered here: in biaxial nematics, there are no hedgehogs (although boojums are allowed), and there are five topological classes of disclinations. Some pairs of these disclinations cannot cross each other without a creation of a third disclination that joins the original pair [16]. A detailed discussion of defects in liquid crystals can be found in the book by Kléman [17].

REFERENCES

[1]   P.G. de Gennes, J. Prost [*The Physics of Liquid Crystals* 2nd edition (Oxford Science Publishers, Oxford, 1993) p.166-8]

[2]   G. Toulouse, M. Kléman [*J. Phys. Lett. (France)* vol.37 (1976) p.L-149]

[3]   G.E. Volovik, V.P. Mineyev [*Zh. Eksp. Teor. Fiz. (USSR)* vol.72 (1977) p.2256; *Sov. Phys.- JETP Lett. (USA)* vol.45 (1977) p.1186]

[4]   G.E. Volovik [*Pis'ma Zh. Eksp. Teor. Fiz. (USSR)* vol.28 (1978) p.65; *JETP Lett. (USA)* vol.28 (1978) p.59]

[5]   F.C. Frank [*Trans. Faraday Soc. (UK)* vol.25 (1958) p.19]

[6]   P.E. Cladis, M. Kléman [*J. Phys. (France)* vol.33 (1970) p.1443]

[7]   R.B. Meyer [*Philos. Mag. (UK)* vol.27 (1973) p.405]

[8]   S.I. Anisimov, I.E. Dzyaloshinskii [*Zh. Eksp. Teor. Fiz. (USSR)* vol.63 (1972) p.1460; *Sov. Phys.-JETP (USA)* vol.36 (1973) p.774]

[9]   M.V. Kurik, O.D. Lavrentovich [*Usp. Fiz. Nauk (USSR)* vol.154 (1988) p.381; *Sov.Phys.-JETP (USA)* vol.31 (1988) p.196]

[10]   C.M. Dafermos [*Quart. J. Mech. Appl. Math. (UK)* vol.23 (1970) p.S49]

[11]   N.V. Madhusudana, R. Pratibha [*Mol. Cryst. Liq. Cryst. (UK)* vol.103 (1983) p.31]

[12]   O.D. Lavrentovich, Yu.A. Nastishin [*Europhys. Lett. (France)* vol.12 (1990) p.135]

[13]   I. Chuang, R. Durrer, N. Turok, B. Yurke [*Science (USA)* vol.251 (1991) p.1336]

[14]   T. Ishikawa, O.D. Lavrentovich [*Europhys. Lett. (France)* vol.41 (1998) p.171]

[15]   K. Janich, H.-R. Trebin [in *Physique de Defauts/Physics of Defects* Ed. R. Balian (North Holland Publishers, Amsterdam, 1981) p.421-9]

[16]   G. Toulouse [*J. Phys. Lett. (France)* vol.38 (1977) p.L-67]

[17]   M. Kléman [*Points, Lines, and Walls* (John Wiley and Sons, New York, 1983)]

5.5   Measurements and interpretation of flexoelectricity

A.G. Petrov

September 1998

A   INTRODUCTION

Theory, experiment and critical evaluation of data on nematic flexoelectricity are presented. Direct and converse flexoeffects are discussed, in either homogeneous or inhomogeneous electric fields, including both static and dynamic responses. Photoflexoelectricity, a new phenomenon, is also mentioned.

B   SUBSTANCES AND ABBREVIATIONS

APAPA   -   4-methoxybenzylidene-4′-acetoxybenzene
Azpac   -   Pd-complex of 4,4′-dihexyloxyazoxybenzene
BBBA   -   4-butoxybenzilidene-4′-butylaniline
BMAOB   -   4-butyl-4′-methoxyazoxybenzene
CCH-7   -   trans-1-heptyl-4-(4′-cyanocyclohexyl)cyclohexane
DOBAMBC -   4-decyloxybenzylidene 4′-amino-(2-methylbutyl)-cinnamate
EBBA   -   4-ethoxybenzilidene-4′-butylaniline
EPPC   -   4-ethoxyphenyl trans-4-propylcyclohexylcarboxylate
E7   -   commercial material (Merck)
HCPB   -   4-heptyl-4′-cyanophenyl benzoate
HOAB   -   4,4′-dihexyloxyazoxybenzene
HOBA   -   4-heptyloxybenzoic acid
HOACS   -   4-hexyloxy-4′-amyl-α-cyanstilbene
HOT   -   4,4′-dihexyloxytolane
MBBA   -   4-methoxybenzilidene-4′-butylaniline
Mixture A   -   (4-butyl-4′-methoxyazoxybenzene + 4-butyl-4′heptanoyloxy-azoxybenzene, 2:1)
OOBA   -   4-octyloxybenzoic acid
ROCP-7037 -   5-heptyl-2-(4-cyanophenyl)-pyrimidine (Roche)
PAA   -   4,4′-dimethoxyazoxybenzene
PB   -   quaternary mixture of phenylbenzoates
5CB   -   4-pentyl-4′-cyano biphenyl
7CB   -   4-heptyl-4′-cyano biphenyl
8CB   -   4-octyl-4′-cyano biphenyl

C   THEORY

Flexoelectricity denotes curvature-induced electric polarisation in liquid crystals subjected to orientational deformations of the director field $\mathbf{n}(\mathbf{r})$ [1,2]; the term is due to de Gennes [2]. Originally, a term such as liquid crystal piezoelectricity was employed by Meyer [1]. This was used to underline the analogy of the phenomenon to the piezoelectricity of solid crystals.

Two of the three basic modes of deformation of a nematic, splay and bend, possess the symmetry of a polar vector:

$$\mathbf{S} = \mathbf{n}\nabla \cdot \mathbf{n}$$

$$\mathbf{B} = -\mathbf{n} \times \nabla \times \mathbf{n} \tag{1}$$

Thus, these deformations can polarise the initially centrosymmetric structure of the nematic, according to the Curie principle:

$$\mathbf{P}_f = e_{1z}\mathbf{S} + e_{3x}\mathbf{B} \tag{2}$$

where the coefficients e_{1z} and e_{3x} are flexoelectric coefficients of splay and bend in Meyer's notation. Flexocoefficients are measured in C m^{-1} and an order of magnitude estimation yields: $e \approx$ (electron charge)/(molecular length) $= 1.6 \times 10^{-19}/4 \times 10^{-9} = 4 \times 10^{-11}$ C m^{-1}.

EQN (2) is the constitutive equation of the direct flexoelectric effect. Experimental studies of this direct effect in nematics are scarce (see later). Experiments mostly deal with the converse flexoelectric effect: the appearance of torques and orientational deformations in the director field $\mathbf{n}(\mathbf{r})$ due to external electric fields $\mathbf{E}(\mathbf{r})$. In this case the bulk flexoelectric torque can be written in the form $\Gamma_f^b = \mathbf{n} \times \mathbf{h}_f$, where $\mathbf{h}_f$ is the bulk flexoelectric component of the molecular field [3,4]:

$$\mathbf{h}_f = e_{1z}\left[\mathbf{E}\nabla \cdot \mathbf{n} - \nabla(\mathbf{E} \cdot \mathbf{n})\right] + e_{3x}\left[\mathbf{E} \times \nabla \times \mathbf{n} - \nabla \times (\mathbf{E} \times \mathbf{n})\right] \tag{3}$$

This expression can be transformed into an equivalent form (whereby a term e_{3x} $\mathbf{n}\nabla \cdot \mathbf{E}$, parallel to the director $\mathbf{n}$, is neglected) [5]:

$$\mathbf{h}_f = (e_{1z} - e_{3x})\left[\mathbf{E}\nabla \cdot \mathbf{n} - \nabla \mathbf{n} \cdot \mathbf{E}\right] + (e_{1z} + e_{3x})\mathbf{n} \cdot \nabla \mathbf{E} \tag{4}$$

This equation shows that bulk flexodeformations can be obtained when there is a gradient in the distribution of either the nematic director $\mathbf{n}$ or an external electric field $\mathbf{E}$, or both. Effects of the first kind are known as linear (with respect to electric field) and they depend on the difference of splay and bend flexocoefficients. Those of the second kind are known as gradient effects, and they depend on the sum of both flexocoefficients (that is the total flexocoefficient).

Recently, a proposal was made [6] to redefine the flexocoefficients by inversion of the sign of the bend coefficient. This amounts to choosing a positive sign in the bend vector expression in EQN (1), i.e. defining it to point outward from the centre of curvature. With such a definition gradient effects would be proportional to the difference (i.e. the anisotropy) of flexocoefficients. However, all published data so far follow the original Meyer definition.

It can be easily shown that for planar problems, where the director field is confined to a plane, the first term in EQN (4) is identically zero. In such a case, with a homogeneous field, the only source of flexodeformation comes from the surface torque at the liquid crystal/substrate interface, $\Gamma_f^s = \mathbf{n} \times \mathbf{g}_f$, where $\mathbf{g}_f$ is the flexoelectric surface molecular field [4]:

$$\mathbf{g}_f = -e_{1z}(\mathbf{E} \cdot \mathbf{n})\mathbf{s} + e_{3x}[\mathbf{s} \times (\mathbf{n} \times \mathbf{E})] \tag{5}$$

where $\mathbf{s}$ is the surface normal. A complete solution of a static flexoelectric problem requires the inclusion of the other torques in the torque balance equation, such as elastic and dielectric ones in the bulk, and elastic, surface anchoring and surface polarisation ones at the surface [4,8]. Surface torque balance equations serve as boundary conditions to the bulk one, as shown by the Harvard and Sofia groups [67,4]. In the dynamic case viscous torques are also included, again bulk and surface ones [7].

A molecular interpretation of flexoelectricity was given first in terms of the combination of (dipole) electric and shape asymmetry of nematics by Helfrich and by the Sofia group [9-11], following qualitative ideas of Meyer. Subsequently, another molecular mechanism was put forward by the Bordeaux group [12], exploring the role of quadrupole electric asymmetry. The definition of a nematic as a quadrupolar ferroelectric (with non-zero bulk quadrupole density) which develops a polarisation under splay or bend [13] is interesting in this respect: the vector-gradient of a tensor (bulk quadrupole) gives rise directly to a vector (bulk polarisation). The complete dependence of both dipole and quadrupole mechanisms on temperature and the orientational order parameter S was discussed in [7] and in [14]. Conformational flexoelectricity was proposed [63] for the bend flexocoefficient of mesogens with a transverse dipole depending on their conformation (trans or cis, cf. [37]), when the cis conformation is favoured by the macroscopic bend of the nematic [63]. In principle, the same conformational mechanism can exist in the case of a longitudinal dipole and splay deformation, as in the flexoelectricity of lyotropic lipid bilayers.

Finally, we should mention that an ordoelectric polarisation was also predicted by the Orsay group when the director field is constant, but gradients of the order parameter S($\mathbf{r}$) are allowed [12,15,13]. In the quadrupolar approximation then:

$$\mathbf{P}_o = \frac{3}{2} e \nabla S \left(\mathbf{nn} - \frac{1}{3}\mathbf{I} \right) \tag{6}$$

where e is the total quadrupolar flexoelectric coefficient for perfect orientational order S = 1.

D   EXPERIMENTAL TECHNIQUES

D1   Direct Flexoeffect

In static situations the flexoelectric polarisation (EQN (2)) will be screened by ionic impurities and so the way to measure the direct flexoelectric effect is to time-modulate $\mathbf{P}_f$. Time-dependent changes of $\mathbf{P}_f$ should be faster than the Maxwell relaxation time of space charges τ_c, which is typically about 0.1 s for very pure nematics. Two ways for the modulation of $\mathbf{P}_f$ are known at present: by pressure pulses and by heat pulses. The first attempt to correlate electric signals in nematics subjected to oscillating shear with flexoelectricity was made in [16].

The first observation of the direct flexoeffect in homeotropic nematic layers of an azoxy compound (mixture A) and of 8CB subjected to pressure pulses from a loudspeaker with a repeating frequency of 50 Hz was reported by the Moscow group [17,18]. The authors measured the induced surface charge on one of the electrodes by a charge-sensitive amplifier and related it to the bend flexocoefficient e_{3x} by solving the corresponding nematodynamic problem. Making allowance for the contribution from the electrokinetic effect, which also exists in the isotropic phase, they estimated for mixture A: $|e_{3x}| \approx 3 \times 10^{-13}$ C m^{-1}. Subsequently, the Moscow group studied a similar effect in hybrid aligned nematic (HAN) layers (homeo-planar), where acoustically induced Poiseuille flow induced a voltage difference across the layer, proportional to $e_{1z} + e_{3x}$ [19,20]. At the same time, e_{1z} and e_{3x} of MBBA and four other nematics have been measured by a technique based on acoustic wave-induced oscillations in a nematic cell with one flexible wall [85,86], yielding values that are relatively higher, but in accord with the converse flexoeffect data listed in Section D3 ($|e_{1z}| \approx 1.5 \times 10^{-11}$ C m^{-1}; $|e_{3x}| \approx 3 \times 10^{-11}$ C m^{-1} for MBBA). The temperature dependence of the flexocoefficients was analysed in terms of quadrupole and dipole contributions ($e_{ij} \propto \alpha S + \beta S^2$, where for MBBA $\alpha \approx 0.2$, $\beta \approx 0.8$ [86]).

In another series of experiments by the same group flexopolarisation of a homeo-planar nematic was modulated by heat pulses from a YAG-laser [21,19] or from a He-Ne laser [22]. In the second case a small amount of anthraquinone dye was dissolved in the nematic to ensure absorption of the red light, and thus heating. This resulted in voltage pulses across the nematic layer, proportional to the pyroelectric coefficient $\gamma(T) = dP_z(T)/dT$ where $P_z = (e_{1z} + e_{3x})\sin\theta\cos\theta d\theta/dz$ where θ is the angle between the director and cell normal at point z in the cell. In the isotropic phase a small signal due to surface polarisation was also observed. By integration of the pyroelectric signal from T_{NI} to any given temperature in the nematic phase the temperature dependence of $e_{1z} + e_{3x}$ was obtained. Absolute calibration of the measurements in this case was achieved by comparing the pyroelectric coefficient of deformed nematic structures with the known pyroelectric coefficient of a ferroelectric liquid crystal, DOBAMBC [19,21,22]. The authors reported $e_{1z} + e_{3x} = -4.7 \times 10^{-12}$ C m^{-1} for 5CB at 30°C [19] and $e_{1z} + e_{3x} = -2.3 \times 10^{-12}$ C m^{-1} for MBBA at 36°C [19]. From the data in FIGURE 4 of [22] $|e_{1z} + e_{3x}| = 3 \times 10^{-12}$ C m^{-1} for dye-doped 5CB at 27°C can be calculated [87]. In a recent pyroelectric experiment with 5CB using an absolute internal calibration from a dielectrically polarised 5CB a still higher value of $e_{1z} + e_{3x}$ (close to the theoretical upper limit for flexocoefficients [76]) was determined and the negative sign was confirmed: $e_{1z} + e_{3x} = -4.7 \times 10^{-11}$ C m^{-1} for 5CB at 24°C [88]. However, the sign of $e_{1z} + e_{3x}$ for 5CB appears to be positive from measurements of the direct flexoeffect [75,47] (see later).

By studying polar surface bifurcations in dimerised nematics (HOBA and OOBA), and modelling them as due to ordoelectric surface polarisation, the total quadrupolar coefficient, e, (for S = 1) of OOBA was estimated to be $+3 \times 10^{-11}$ C m^{-1} [23].

D2 Converse Flexoeffect in Homogeneous Fields

According to the Maxwell relations [1] the coefficients of the direct and converse flexoelectric effects have to be equal in the thermodynamic limit. The validity of the Maxwell relations has not been specifically checked until now. The converse flexoeffects in homogeneous electric fields can be observed as due to surface torques, EQN (5), or to the bulk torques (see the first term in EQN (4)).

D2.1 Surface flexoelectric torques

There are at least four basic geometries where surface torques are of importance [4]. When the electric field is parallel to the nematic layer, the flexoelectric torques depend on one of the flexocoefficients only, while with the field normal to the layer the torques depend on the sum of the flexocoefficients. Surface-induced flexodeformations additionally require relatively weak anchoring over the layer substrates; this is achieved by different surface treatments. However, inherent to weak nematic anchoring are a number of complications, related to the unknown form of the anchoring potential, the presence of surface polarisation, surface electric fields, non-linear elasticity, ordoelectricity, etc. Due to the fact that some of these complications were recognised only recently, and were not properly taken into account in the earlier analyses, any numerical data on flexoelectric coefficients determined from surface-induced deformations should be considered with great caution.

Historically, the first flexoelectric experiment that gave numerical values of e_{3x} was the bending of a homeotropic MBBA layer under an electric field parallel to the layer but normal to the director [24,25,28]. The electric field was applied by metal foil electrodes that also serve as spacers. The reported values differed by an order of magnitude: 1×10^{-11} C m^{-1}, sign not clarified (according to the analysis in [26] of the data from [24]) against $+1.2 \times 10^{-12}$ C m^{-1} [25]. Subsequent theoretical analysis revealed the fact that the surface torque depends not just on e_{3x}, but also on $e_{3x} \pm m_p$, where m_p is the surface polarisation due to the surface breaking of the non-polar symmetry of the nematic [67,4]. Accordingly this difference was attributed to the longitudinal biphilic asymmetry of the MBBA molecule and to the opposite orientation of m_p in these two experiments [27]. The true e_{3x} value for MBBA should be between these two values, but the data from a single experiment may well be explained by surface polarisation only, without invoking any flexoelectricity.

The Moscow group performed extended studies of flexoelectric bending for a number of nematics (MBBA, BMAOB, azoxy-mixture A), some of them with smectic polymorphism at lower temperatures (BBBA, HOACS). Separate measurements of refractive indices and bend elastic coefficients were also carried out. The authors determined $e^*_{3x} = (e_{3x} + m_p)/(1 + W_S d/2K_{33})$, an effective bend flexocoefficient, that includes the Rapini-Papoular anchoring energy W_S contribution according to the authors [29] (here d is the layer thickness), but also the surface polarisation m_p contribution, according to us [4]. In all substances the order of magnitude for e^*_{3x} was $\approx 1 \times 10^{-12}$ C m^{-1}, which is at least one order of magnitude less than theoretical expectations. The temperature dependence for MBBA and mixture A was reported to be weaker than $e \propto S$; for BBBA and HOACS on decreasing temperature there is an initial growth of e as in ordinary nematics but this was later suppressed by short range smectic ordering. Four types of completely different homeotropic surface treatments resulted in only a small difference between the measured effective bend flexocoefficients, in the range $0.7 \rightarrow 1.3 \times 10^{-12}$ C m^{-1}. Eventually, these seemingly contradictory results were resolved by dedicated experiments to determine the anchoring energy of homeotropic layers (Freedericksz transition [30] and magnetically stabilised flexoelectric bending [31]) that yield large values for W_S ($10^{-5} \rightarrow 10^{-6}$ J m^{-1}, which is thickness dependent), assuming $|e_{3x} + m_p| = 1.3 \times 10^{-11}$ C m^{-1} for MBBA and surfactant-treated glass substrates. A negative value for the bare bend flexocoefficient of MBBA ($e_{3x} = -1.3 \times 10^{-11}$ C m^{-1}) was quoted in the interpretation of the same type of experiments, but with crystal substrates (Al_2O_3, $LiNbO_3$); the remaining experimental difference was attributed to m_p; the anchoring strength was lower in that case [32]. The quoted value comes from bulk flexoelectric measurements of French and Bulgarian groups [62,63,71,72].

The first and so far only measurement of the splay flexocoefficient e_{1z} of MBBA was made by the Sofia group, using an asymmetrically anchored (strong-weak, planar-tilted) layer and an electric field, parallel both to the layer and to the director [34]. The e_{1z}-dependent surface torque at the weakly anchored substrate, namely a glass substrate, rubbed with diamond paste just like the strong one, but with additional deposition of a soap layer, changes the nematic orientation much like the bending case: m_p was not negligible on the pre-tilted substrate. The result was: $e_{1z}(\pm m_p) = +(3.0 \pm 0.7) \times 10^{-12}$ C m^{-1} at room temperature (about 20°C, with $T_{NI} = 44$°C). A positive sign was inferred from the control experiments with pre-tilted layers, assuming that the rubbing direction pre-determined the sign of the tilt. A method for measuring e_{1z}/e_{3x} by using a homeotropic layer bending in a destabilising magnetic field has also been proposed [83].

Surface flexoeffects in weakly anchored HAN layers have been described [35,36,77,84]; here layer substrates serve also as electrodes. By comparing HAN (homeotropic: silane treatment; planar-formvar treatment) cell transmittance in AC and in DC fields, normal to the layer, in order to isolate the flexoelectric effect from the dielectric one, and with the hypothesis that the true inner electric field is due to the effective conduction anisotropy, an Italian group reported $e_{1z} + e_{3x} = -(1.5 \pm 0.2) \times 10^{-11}$ C m^{-1} for MBBA at room temperature [35]; this and other values [77, 84] are listed in TABLE 1. Still, some m_p contribution is probably contained in these values, judging from a recent result of the Orsay group [36]. There, m_p was explicitly included in the model: in a planar-tilted 5CB layer on SiO evaporated under different angles, and applying very low voltage pulses, below 150 mV (in order to avoid electrochemistry and the dielectric torque), the authors reported a value of $m_p + (e_{1z} + e_{3x})/2 = +8.7 \times 10^{-11}$ C m^{-1} for 5CB at room temperature, which is substantially larger than the $e_{1z} + e_{3x}$ measured by the Bangalore group [75].

TABLE 1 Total and difference flexocoefficients of some nematics. The first two rows report on the measurements of the direct flexoeffect, the rest on the converse one. When two references are given, the second one refers to a justification of the sign.

Substance	$(e_{1z} - e_{3x})/10^{-12}$ C m^{-1}	T/°C	Ref	$(e_{1z} + e_{3x})/10^{-12}$ C m^{-1}	T/°C	Ref
5CB				-4.7	30	[19]
MBBA				-2.3	36	[19]
MBBA				-25	Room	[77]
MBBA				-15 ± 2	Room	[35]
MBBA	$+3.3 \pm 0.7$	20	[62]	$-29.2 \pm 20\%$	20	[70,72]
MBBA	$+14 \pm 1$	30	[84]	-54 ± 10	30	[84]
8OCB				$+21 \pm 5$	73	[70,75]
BMAOB	-5.7	25	[55,56]			
ZLI-4792	-15 ± 1	30	[84]	$\leq |10|$	30	[84]

Further, polar flexoelectric surface instabilities should be mentioned [37-44]. The first polar instability, in an asymmetrically anchored planar MBBA layer at very low voltages, was observed by the Sofia group [37,4]. The theory for this was developed in a number of papers [38,81,4,39]. Experiments on asymmetrically anchored homeotropic layers of 5CB and 7CB soon followed [40-44]. The role of m_p in these instabilities is crucial. The reported values are $m_p \approx 1 \times 10^{-10}$ C m^{-1} [40-42]; $e_{1z} + e_{3x} + |m_p| = 1.3 \times 10^{-10}$ C m^{-1} for the 5CB-In$_2$O$_3$:Sn interface and 2.0×10^{-10} C m^{-1} for the 5CB-silicone elastomer interface [43]; $e_{1z} + e_{3x} - 2m_p \approx 1 \times 10^{-10}$ C m^{-1} for 5CB and a number of other interfaces [44]. Surface-induced flexoelectric domains were observed under a DC field in reversely-pre-tilted 5CB layers with a longitudinal gradient of lecithin [45].

Finally, while the methods described so far are either DC or pulsed, AC methods also exist, with the advantage of using phase-sensitive detection. Surface-driven flexoelectric oscillations were first observed by the Sofia group in symmetric homeotropic layers of MBBA under a horizontal AC + DC field [7]. These measurements evolved recently into a method of flexoelectric spectroscopy for the investigation of visco-elastic properties of nematics and resulted in some of the first data on MBBA surface viscosity [33,46]. Such oscillations in planar, homeotropic or pre-tilted layers of MBBA, 5CB and MBBA + HCPB mixtures were recently studied with the modulation ellipsometry technique by the Moscow and Orsay groups [47]. They reported a value for $e_{1z} + e_{3x} = 1 \times 10^{-11}$ C m^{-1} for a dielectrically compensated MBBA + HCPB mixture, which is about half of the bulk value for MBBA (see later). This was explained by a lower surface order at the rough SiO surface, but the m_p contribution should also be discussed for homeotropic and tilted geometry. The signs of the total flexocoefficients for pure MBBA and 5CB were found to be opposite, in agreement with earlier findings (see later). In planar, asymmetrically anchored nematic E7 layers on poly(1,4-butylene terephthalate) and nylon 6,6 with sufficient difference in anchoring strengths (10^{-5} J m^{-1}) a polar flexoelectric effect was observed under AC and pulsed excitation [48,49] and interpreted theoretically [50]. Further, the Bangalore group developed recently an AC electrooptic technique with numerical analysis for measuring both sign and magnitude of $e_{1z} + e_{3x}$ in HAN cells (rubbed polyimide - octadecyl triethoxy silane) [51]. $e_{1z} + e_{3x}$ entered in both surface and bulk torque balance equations, because of the inhomogeneity of the bulk electric field in a deformed nematic with a strong dielectric anisotropy. The m_p contribution was not considered. Experiments on CCH-7 resulted in a positive sign for the sum, and a good accuracy for the magnitude (5%). The magnitude was measured as a function of temperature (from 56°C to 83°C); however, only the ratio $(e_{1z} + e_{3x})/S$ was provided in the paper, and from this it follows that $e_{1z} + e_{3x} = 1.8 \times 10^{-11}$ C m^{-1}. The ratio showed a relative constancy up to 72°C (quadrupolar contribution) and a steep decay beyond this temperature due to a structural change in the nematic, for example unpairing of cyanogroups.

D2.2 Bulk flexoelectric torques

In homogeneous electric fields such torques depend on $e_{1z} - e_{3x}$ (the first term in EQN (4)). The first prediction of a modulated (domain) structure in unbound nematics caused by bulk flexoelectric torques is due to Meyer [1,3]. By slightly generalising that calculation, we can easily verify that the period of these domains depends on $e_{1z} - e_{3x}$; they seem to have finally been observed in Azpac and HOAB [33]. A threshold flexoelectric instability expressed in longitudinal domains parallel to the director in planar nematic layers with strong anchoring was predicted later [52,53]. These were eventually identified with second order Vistin's domains in low conductance nematics, whose period is inversely proportional to the voltage [54]. By comparing experiment and theory, the Moscow group then determined the absolute value of e_{1z} - e_{3x} at 25°C in BMAOB (5.7×10^{-12} C m^{-1}) and in PBS (6.0×10^{-12} C m^{-1}) [55]; by further experiments in twisted (nematic + chiral nematic) structures the sign of BMAOB was determined to be negative [56]. More recently, a value for $|e_{1z} - e_{3x}| = 3.3 \times 10^{-12}$ C m^{-1} has been reported from measurements on flexoelectric domains in Azpac [57]. A number of other longitudinal modulated structures in strongly-weakly and weakly-weakly anchored MBBA films were studied by Hinov [58-61]. These permit, in principle, evaluation of either e_{1z} - e_{3x}, or e_{1z} + e_{3x}, or both.

The Orsay group studied flexoelectrically controlled twist of the texture in HAN cells subject to transverse DC electric fields [62,63] and flexoelectric instabilities in HAN cells under an electric field in the plane of the initial splay-bend deformation [64]. The permanent flexopolarisation, which is proportional to e_{1z} - e_{3x}, is rotated by the field, causing measurable twist. Results obtained are: e_{1z} - e_{3x} = +(3.3 ± 0.7) × 10^{-12} C m^{-1} for MBBA at 20°C [62], $(e_{1z} - e_{3x})/K = 0.90$ C m^{-1} N^{-1} for 8CB at 36.3°C [63] and $(e_{1z} - e_{3x})/K = -4.6$ C m^{-1} N^{-1} for 8OCB at 77°C [63], where K is an average elastic constant close to K_2. The latter two substances differ only by the presence of an oxygen in 8OCB having a transverse dipole whose orientation depends on the conformation of the octyloxy group: cis or trans. The striking difference between 8CB and 8OCB was interpreted as a manifestation of conformational flexoelectricity: a bend will favour the cis-conformation of the alkoxy group, aligning the oxygen dipole towards the centre of curvature, thus strongly enhancing a positive e_{3x}. The ratio $(e_{1z} - e_{3x})/K$ for MBBA was found to be practically temperature independent, in accordance with the predominantly dipolar character ($\propto S^2$) of the difference [63]. In contrast, the same ratio for 8OCB was found to be rather S-like (far enough above the smectic A-nematic transition) and explained by the peculiarity of conformational flexoelectricity [63]. The next systematic study of $(e_{1z} - e_{3x})/K$ and its temperature dependence for sixteen nematogens by this method was reported by the Bangalore group [75]. All substances, except CCH-7, showed a positive $(e_{1z} - e_{3x})/K$ ranging from 0.28 C m^{-1} N^{-1} for EPPC at 50°C to 2.4 C m^{-1} N^{-1} for 5CB at 30°C and even up to 4.36 C m^{-1} N^{-1} for ROCP-7037 at 46°C. For CCH-7, $(e_{1z} - e_{3x})/K = -1.29(\pm10\%)$ C m^{-1} N^{-1} at 70°C. If the corresponding elastic constants of some of these substances are known then the different flexocoefficients proper can be calculated from the results in the Table in [75].

Recent results about both the sum and difference of flexocoefficients of two room temperature nematics (MBBA and ZLI-4792) were reported by a Japanese group [84]. These authors performed fits of the electrooptical characteristics of HAN cells with numerical simulations; DC fields were applied along the cell normal and along the cell surface. The results (see TABLE 1) for MBBA match in sign, but not in value, with those quoted before (especially the difference).

D3 Converse Flexoeffect in Inhomogeneous Fields

Bulk flexoelectric torques in inhomogeneous electric fields are described by the second term of EQN (4) and depend on e_{1z} + e_{3x}. This effect was first recognised as a gradient flexoelectric effect by the Sofia group [65] and eventually observed in the planar layers of four nematics (MBBA, EBBA, APAPA, PAA) at very low DC voltages (down to 100 mV). The field inhomogeneity was related to the space charge accumulation in the thin layers. Assuming a linear field dependence across the layer, both field gradient and e_{1z} + e_{3x} for MBBA could be determined from a Fourier analysis of (random-like) layer deformations [66]. A positive sign was also ascribed to the sum of the flexocoefficients under the assumption that the anode field is larger than that at the cathode. As this could not be claimed for certain, just the value should be quoted now: $|e_{1z} + e_{3x}| = 1 \times 10^{-11}$ C m^{-1} for MBBA at room temperature [66].

The gradient flexoelectric effect was subsequently observed by the Harvard and Bordeaux groups in homeotropic nematic (and smectic) layers (of BBMBA, CBOOA, MBBA, 3PBClB, HOT, 8OCB) in an inhomogeneous field created by interdigital electrodes deposited on one of the glass substrates [67-70]. In these experiments the field is a function of two coordinates and the tensor of the field gradient has four non-zero components. Measurements are essentially made under AC excitation, so that the charge screening effect can be avoided (very important) and the frequency dependence of flexoelectricity can be followed, up to 10^7 Hz [68]. Calculations of absolute values of $e_{1z} + e_{3x}$ from the interdigital method data are rather involved, and require a fully characterised nematic (elastic and dielectric constants, rotational viscosity, refractive indices). The anchoring conditions have also to be clarified: weak [67] versus strong [70]. Originally, it was claimed that the method could provide the sign of $e_{1z} + e_{3x}$ [67]; this was later rejected [72,73], cf. [71]. The most precise data reported are: $|e_{1z} + e_{3x}| = (2.4 \pm 0.5) \times 10^{-11}$ C m^{-1} for MBBA at 35°C [70] and $|e_{1z} + e_{3x}| = (2.1 \pm 0.5) \times 10^{-11}$ C m^{-1} for 8OCB at 73°C [70]. Other estimates are: $e_{1z} + e_{3x} \approx 8.3 \times 10^{-12}$ C m^{-1} for BBMBA at 35.6°C [67], the same order of magnitude for CBOOA [67], $(e_{1z} + e_{3x})/K \approx 2.6$ C m^{-1} N^{-1} for HOT at 99.3°C [70] and $(e_{1z} + e_{3x})/K \approx 0.9$ C m^{-1} N^{-1} for 3PBClB at 48.9°C [70]. Temperature and frequency dependencies revealed the relative contribution of dipole and quadrupole flexoelectricity in BBMBA [12], MBBA (see FIGURE 1) [70], 8OCB [70]; limiting cases of a purely quadrupolar nematic (symmetric tolane HOT) and a well developed dipolar contribution (3PBClB) were identified [68,70]. A pre-transitional flexoelectric signal in the isotropic phase of MBBA was even observed, due to the short range nematic ordering close to nematic-isotropic transition [69]. This experiment revealed the build-up of macroscopic quadrupolar density that condensed at the nematic-isotropic transition.

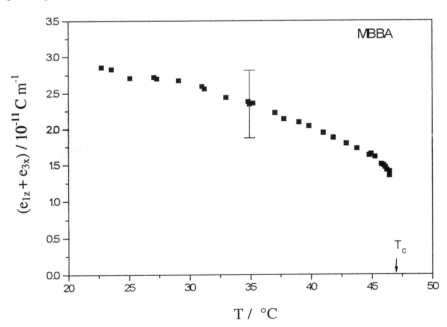

FIGURE 1 Absolute value of the total flexoelectric coefficient of high purity MBBA ($T_{NI} = 47$°C) as a function of temperature. Data are recalculated from [70] and scaled to the reported value at 35°C. According to [72,77,19,35,84] the sign of the total MBBA flexocoefficient is negative: $e_{1z} + e_{3x} < 0$.

Dozov et al [72,73] have suggested a very simple method with which to measure the sign of $e_{1z} + e_{3x}$ based on strongly anchored homeotropic layers of MBBA and 8OCB in a controlled field gradient created by a four-electrode cell structure, with two in-plane electrodes on each substrate, and a high resistivity SnO_2 layer between them. By conoscopy they observed a layer tilt very clearly related to the field gradient and this reversed when the latter reversed. However, the reported absolute values were rather low, probably due to an ion screening effect and the necessity to work at low fields to avoid hydrodynamic flow: $e_{1z} + e_{3x} = -3.3 \times 10^{-12}$ C m^{-1} for MBBA and $e_{1z} + e_{3x} = 4.6 \times 10^{-12}$ C m^{-1} for 8OCB. This was the first indication of a negative sign of the total flexocoefficient of MBBA, later confirmed by other groups [77,19,35]. Subsequently, the Bangalore group used this method to evaluate systematically $(e_{1z} + e_{3x})/K_3$ for nine substances [75]. The most important finding was that all these substances, except a three component mixture, display a positive total flexocoefficient. Their absolute values were probably also affected by the screening problem: compare $|e_{1z} + e_{3x}| = (2.1 \pm 0.5) \times 10^{-11}$ C m^{-1} for 8OCB at 73°C [70] with $e_{1z} + e_{3x} = +(0.6 \pm 0.2) \times 10^{-11}$ C m^{-1} [75]. For 5CB the result was $(e_{1z} + e_{3x})/K_3 \approx +1.08$ C m^{-1} N^{-1} at 30°C. However, this positive sign for 5CB is in disagreement with results of the Moscow group [19], as noted previously, and further studies are needed to resolve it. The first theoretical predictions of a bulk flexoelectric effect (allowing also for measurements of $e_{1z} + e_{3x}$ in DC or AC fields) in an initially deformed, wedge-like nematic layer were made by the Sofia group [74,82]. Such an experiment was later performed by the Bangalore group [75].

By combining the extrapolated value of the sum for MBBA at 20°C from FIGURE 1 and the sign from [72] with the difference at 20°C [62] (row 5 of TABLE 1), we can finally arrive (cf. [73]) at the only reasonably well-established values (with twice as high relative errors): $e_{1z} = -13 \times 10^{-12}$ ($\pm 40\%$) C m^{-1}; $e_{3x} = -16 \times 10^{-12}$ ($\pm 40\%$) C m^{-1}.

For 8OCB it was likewise established from [63,72,70] that: $e_{1z} \approx 0$; $e_{3x} = +17 \times 10^{-12}$ ($\pm 40\%$) C m^{-1} at 77°C.

D4   Photoflexoelectricity

Photoflexoelectricity is a novel effect that combines light, electric field and liquid crystal deformation [78]. It can be observed in photoactive liquid crystals, e.g. guest-host azo dye-nematic mixtures, capable of trans-cis isomerisation. Very briefly, the influence of UV illumination on the flexoelectric twist of the texture in HAN cells containing azobenzene derivatives was reported in [79] and the UV influence on the flexoelectric bend of homeotropic nematic cells with azo dyes was studied in [80]. In the last case a strong UV effect on the surface polarisation was also inferred. Very recently, UV influence on the direct flexoeffect was studied by a pyroelectric response technique, indicating that cis azo-isomers actually have a lower flexocoefficient than the trans-isomers (Blinov, private communication).

E   CONCLUSION

Despite 30 years' study of nematic flexoelectricity, there is still considerable controversy in the published data, not only in value, but also in sign (e.g. there is no agreement about the sign of total flexocoefficient for 5CB). Therefore, further efforts in this direction are necessary. It seems, though, that a consensus order of magnitude for flexocoefficients of a few 10^{-11} C m^{-1} is now established, which makes it indispensable to pay close attention to flexoelectricity in practically all electrostructural and electrooptic effects. The direct flexoeffect seems to be studied best by the pyroelectric method, which allows us, in principle, to determine both the sum and the difference of the flexocoefficients [19], although the last has not yet been demonstrated. Converse flexoeffects are studied best by the bulk effect with strong anchoring conditions. Despite its complexity, the interdigital electrode method seems to be most reliable in determining the sum of the flexocoefficients [70], while its sign should be separately determined by the four-electrode [72] or wedge cell [74,75] gradient methods. The flexoelectric twist of the HAN structure is most appropriate for measuring both the value and the sign of the difference [62,63]. The use of a HAN cell to measure both the sum and the difference [84] deserves further attention. Finally, the validity of the Maxwell relations between direct and converse flexocoefficients should also be confirmed.

REFERENCES

[1]   R. Meyer [*Phys. Rev. Lett. (USA)* vol.22 (1969) p.918]

[2]   P.G. de Gennes [*Physics of Liquid Crystals* (Clarendon Press, Oxford, 1974) p.97]

[3]   C. Fan [*Mol. Cryst. Liq. Cryst. (UK)* vol.13 (1971) p.9]

[4]   A. Derzhanski, A.G. Petrov, M.D. Mitov [*J. Phys. (France)* vol.39 (1978) p.273]

[5]   A. Derzhanski, A.G. Petrov [*Adv. Liq. Cryst. Res. Appls.* (Pergamon, Oxford - Kiadó, Budapest, 1980) vol.1 p.515]

[6]   P. Rudquist, S.T. Lagerwall [*Liq. Cryst. (UK)* vol.23 (1997) p.503]

[7]   A. Derzhanski, A.G. Petrov [*Acta Phys. Pol. A (Poland)* vol.55 (1979) p.747]

[8]   S.A. Pikin [*Structural Transformations in Liquid Crystals* (Gordon and Breach, New York, 1991)]

[9]   W. Helfrich [*Z. Naturforsch. (Germany)* vol.26c (1971) p.833]

[10]   A. Derzhanski, A.G. Petrov [*Phys. Lett. A (Netherlands)* vol.36 (1971) p.483]

[11]   A. Derzhanski, A.G. Petrov, I. Bivas [*Adv. Liq. Cryst. Res. Appls.* (Pergamon, Oxford - Kiadó, Budapest, 1980) vol.1 p.505]

[12]   J. Prost, J.P. Marcerou [*J. Phys. (France)* vol. 38 (1977) p.315]

[13]   G. Durand [*Physica A (Netherlands)* vol.163 (1990) p.94]

[14]   J.-P. Marcerou [Master Thesis, University of Bordeaux, France, 1978, p.111]

[15]   G. Barbero, I. Dozov, J.F. Palierne, G. Durand [*Phys. Rev. Lett. (USA)* vol.56 (1986) p.2056]

[16]   Y. Kagawa, T. Hatakeyama [*J. Sound Vib. (UK)* vol.53 (1977) p.585]

[17]   S.V. Yablonski, L.M. Blinov, S.A. Pikin [*Pis'ma Zh. Eksp. Teor. Fiz. (USSR)* vol.40 (1984) p.226]

[18]   S.V. Yablonski, L.M. Blinov, S.A. Pikin [*Mol. Cryst. Liq. Cryst. (UK)* vol.127 (1985) p.381]

[19]   L.M. Blinov, L.A. Beresnev, S.A. Davydyan, S.G. Kononov, S.V. Yablonski [*Ferroelectrics (UK)* vol.84 (1988) p.365]

[20]   L.M. Blinov, D.B. Subachyus, S.V. Yablonski [*J. Phys. II (France)* vol.1 (1991) p.459]

[21]   L.A. Beresnev, L.M. Blinov, S.A. Davydyan, S.G. Kononov, S.V. Yablonski [*Pis'ma Zh. Eksp. Teor. Fiz. (USSR)* vol.45 (1987) p.592]

[22]   L.M. Blinov, D.Z. Radzhabov, S.V. Yablonski, S.S. Yakovenko [*Nuovo Cimento D (Italy)* vol.12 (1990) p.1353]

[23]   M. Petrov, G. Durand [*J.Phys. II (France)* vol.6 (1996) p.1259]

[24]   W. Haas, J. Adams, J.B. Flannery [*Phys. Rev. Lett. (USA)* vol.25 (1970) p.1326]

[25]   D. Schmidt, M. Schadt, W. Helfrich [*Z. Naturforsch. (Germany)* vol.27a (1972) p.277]

[26]   W. Helfrich [*Phys. Lett. A (Netherlands)* vol.35 (1971) p.393]

[27]   A.G. Petrov, A. Derzhanski [*Mol. Cryst. Liq. Cryst. Lett. (UK)* vol.41 (1977) p.41]

[28]   A.S. Vasilevskaya, A.S. Sonin [*Fiz. Tverd. Tela (USSR)* vol.21 (1979) p.196]

[29]   B.A. Umanski, L.M. Blinov, M.I. Barnik [*Kristallografiya (USSR)* vol.27 (1982) p.729]

[30]   L.M. Blinov, A.A. Sonin, M.I. Barnik [*Kristallografiya (USSR)* vol.34 (1989) p.413]

[31]   L.M. Blinov, A.A. Sonin [*Poverkhn. (USSR)* vol.10 (1988) p.29]

[32]   L.M. Blinov, A.A. Sonin [*Zh. Tekh. Fiz. (USSR)* vol.87 (1984) p.476]

[33]   A.G. Petrov, A.Th. Ionescu, C. Versace, N. Scaramuzza [*Liq. Cryst. (UK)* vol.19 (1995) p.169]

[34]   H.P. Hinov, A.I. Derzhanski [in *Liquid Crystals and Ordered Fluids* vol.4, Eds. A.C. Griffin, J.E. Johnson (Plenum Press, New York - London, 1984) p.1103]

[35]   B. Valenti, C. Bertoni, G. Barbero, P. Taverna-Valabrega, R. Bartolino [*Mol. Cryst. Liq. Cryst. (UK)* vol.146 (1987) p.307]

[36]   S. Forget, I. Dozov, Ph. Martinot-Lagarde [*17th Int. L.C. Conf.* Strasbourg, 1998; submitted to *Mol. Cryst. Liq. Cryst. (Switzerland)*]

[37]   A.G. Petrov [PhD Thesis, Bulgarian Academy of Science, Sofia, 1974, p.81 and p.93]

[38]   W. Helfrich [*Appl. Phys. Lett. (USA)* vol.24 (1974) p.451]

[39]   G. Barbero, G. Durand [*Phys. Rev. A (USA)* vol.35 (1987) p.1294]

[40]   M. Monkade, Ph. Martinot-Lagarde, G. Durand [*Europhys. Lett. (Switzerland)* vol.2 (1986) p.299]

[41]   O. Lavrentovich, V. Pergamenshchik, V. Sergan [*Mol. Cryst. Liq. Cryst. (UK)* vol.192 (1990) p.239]

[42]   O. Lavrentovich, V.G. Nazarenko, V.M. Pergamenshchik, V.V. Sergan, V.M. Sorokin [*Sov. Phys.-JETP (USA)* vol.72 (1991) p.431]

[43]   O.D. Lavrentovich, V.G. Nazarenko, V.V. Sergan, G. Durand [*Phys. Rev. A (USA)* vol.45 (1992) p.R6969]

[44]   V.G. Nazarenko, R. Klouda, O.D. Lavrentovich [*Phys. Rev. E (USA)* vol.57 (1998) p.R36]

[45]   L. Komitov, H. Hinov [*Cryst. Res. Technol. (Germany)* vol.18 (1983) p.97]

[46]   Y. Marinov, N. Shonova, C. Versace, A.G. Petrov [*17th Int. L.C. Conf.* Strasbourg, 1998; submitted to *Mol. Cryst. Liq. Cryst. (Switzerland)*]

[47]   L.M. Blinov, G. Durand, S.V. Yablonsky [*J. Phys. II (France)* vol.2 (1992) p.1287]

[48]   S.-D. Lee, J.S. Patel [*Phys. Rev. Lett. (USA)* vol.65 (1990) p.56]

[49]   S.-D. Lee, B.K. Ree, Y.J. Jeon [*J. Appl. Phys. (USA)* vol.73 (1993) p.480]

[50]   J. Lee, S.-W. Suh, K. Lee, S.-D. Lee [*Phys. Rev. E (USA)* vol.49 (1994) p.923]

[51]   S.R. Warrier, N.V. Madhusudana [*J. Phys. II (France)* vol.7 (1997) p.1789]

[52]   Yu.P. Bobylev, S.A. Pikin [*Sov. Phys.-JETP (USA)* vol.45 (1977) p.195]

[53]   Yu.P. Bobylev, V.G. Chigrinov, S.A. Pikin [*J. Phys. (France)* vol.40 suppl. (1979) p.C3-331]

[54]   L.K. Vistin' [*Dokl. Akad. Nauk SSSR (USSR)* vol.194 (1970) p.1318]

[55] M.I. Barnik, L.M. Blinov, A.N. Trufanov, B.A. Umanski [*J. Phys. (France)* vol.39 (1978) p.417]

[56] B.A. Umanski, V.G. Chigrinov, L.M. Blinov, Yu.B. Pod'yachev [*Zh. Eksp. Teor. Fiz. (USSR)* vol.81 (1981) p.1307]

[57] N. Scaramuzza, M.C. Pagnotta [*Mol. Cryst. Liq. Cryst. (UK)* vol.239 (1994) p.263]

[58] H.P. Hinov, L.K. Vistin' [*J. Phys. (France)* vol.40 (1979) p.269]

[59] H.P. Hinov [*Z. Naturforsch. (Germany)* vol.37a (1982) p.334]

[60] H.P. Hinov [*Mol. Cryst. Liq. Cryst. (UK)* vol.74 (1981) p.1639]

[61] H.P. Hinov [*Mol. Cryst. Liq. Cryst. (UK)* vol.89 (1982) p.227]

[62] I. Dozov, Ph. Martinot-Lagarde, G. Durand [*J. Phys. Lett. (France)* vol.43 (1982) p.L-365]

[63] I. Dozov, Ph. Martinot-Lagarde, G. Durand [*J. Phys. Lett. (France)* vol.44 (1983) p.L-817]

[64] N.V. Madhusudana, J.F. Palierne, Ph. Martinot-Lagarde, G. Durand [*Phys. Rev. A (USA)* vol.30 (1984) p.2153]

[65] A.I. Derzhanski, A.G. Petrov, Chr.P. Khinov, B.L. Markovski [*Bulg. J. Phys. (Bulgaria)* vol.1 (1974) p.165]

[66] A.I. Derzhanski, M.D. Mitov [*C.R. Acad. Bulg. Sci. (Bulgaria)* vol.28 (1975) p.331]

[67] J. Prost, P.S. Pershan [*J. Appl. Phys. (USA)* vol.47 (1976) p.2298]

[68] J.-P. Marcerou, J. Prost [*Ann. Phys. (France)* vol.3 (1978) p.269]

[69] J.-P. Marcerou, J. Prost [*Phys. Lett. A (Netherlands)* vol.66 (1978) p.218]

[70] J.-P. Marcerou, J. Prost [*Mol. Cryst. Liq. Cryst. (UK)* vol.58 (1980) p.259]

[71] J.-P. Marcerou [PhD Thesis, University of Bordeaux, France, 1982, no.d'ordre 734]

[72] I. Dozov, I. Penchev, Ph. Martinot-Lagarde, G. Durand [*Ferroelectr. Lett. Sect. (UK)* vol.2 (1984) p.135]

[73] I. Dozov, G. Durand, Ph. Martinot-Lagarde, I. Penchev [*Abstracts, 5 LC Conf. Soc. Countries (Odessa, USSR)* vol.1 part II (1983) p.11]

[74] A. Derzhanski, H.P. Hinov, M.D. Mitov [*Acta Phys. Pol. A (Poland)* vol.55 (1979) p.567]

[75] P.R.M. Murthy, V.A. Raghunathan, N.V. Madhusudana [*Liq. Cryst. (UK)* vol.14 (1993) p.483]

[76] W. Helfrich [*Mol. Cryst. Liq. Cryst. (UK)* vol.26 (1974) p.1]

[77] N.V. Madhusudana, G. Durand [*J. Phys. Lett. (France)* vol.46 (1985) p.L-195]

[78] A.G. Petrov [*Europhys. News (Switzerland)* vol.27 (1996) p.92]

[79] D.S. Hermann, P. Rudquist, K. Ichimura, K. Kudo, L. Komitov, S.T. Lagerwall [*Phys. Rev. E (USA)* vol.55 (1997) p.2857]

[80] Y. Marinov, N. Shonova, L.M. Blinov, A.G. Petrov [*Europhys. Lett. (Switzerland)* vol.41 (1998) p.513]

[81] A.I. Derzhanski, H.P. Hinov [*J. Phys. (France)* vol.38 (1977) p.1013]

[82] H.P. Hinov, A.I. Derzhanski [*Adv. Liq. Cryst. Res. Appls.* (Pergamon, Oxford - Kiadó, Budapest, 1980) vol.1 p.523]

[83] A.I. Derzhanski, H.P. Hinov [*Phys. Lett. A (Netherlands)* vol.62 (1977) p.36]

[84] T. Takahashi, S. Hashidate, H. Nishijou, M. Usui, M. Kimura, T. Akahane [*Jpn. J. Appl. Phys. (Japan)* vol.37 (1998) p.1982]

[85] O.A. Skaldin, A.N. Lachinov, A.N. Chuvyrov [*Sov. Phys.-Solid State (USA)* vol.27 (1985) p.734]

[86] O.A. Skaldin, A.N. Chuvyrov [*Sov. Phys.-Crystallogr. (USA)* vol.35 (1990) p.505]

[87] L.M. Blinov [private communication]

[88]   L.M. Blinov, M. Ozaki, K. Yoshino [submitted for publication]

CHAPTER 6

ELECTRIC AND MAGNETIC PROPERTIES

6.1   Static dielectric properties of nematics

S. Urban

January 1999

A   INTRODUCTION

The static dielectric permittivity is an important parameter that characterises the response of a medium to the application of an electric field. Its value is determined by the distribution of the electric charges in molecules (polar and non-polar compounds) as well as by the intermolecular interactions (for example anisotropy of the medium and intermolecular correlations). In nematics the dielectric permittivity is a tensorial quantity. The value and the sign of the dielectric anisotropy play an important role in the application of nematics in display technologies.

B   THEORETICAL BACKGROUND

The external static electric field applied to a dielectric material induces the polarisation $\mathbf{P}$, that is the dipole moment per unit volume. For low fields $\mathbf{P}$ is proportional to the electric field $\mathbf{E}$ [1-3], $\mathbf{P} = \varepsilon_0 (\varepsilon_s - 1) \mathbf{E}$, where ε_s is the relative dielectric permittivity or dielectric constant and ε_0 is the dielectric permittivity of free space. All these quantities concern the macroscopic volume of the dielectric medium. In order to relate them to the relevant microscopic parameters (for example dipole moment and polarisability) the local electric field $\mathbf{E}_{loc}$ acting on a molecule must be known. The relation between $\mathbf{E}_{loc}$ and $\mathbf{E}$ is the crucial problem of the physics of dielectrics and has not been solved in general. For isotropic fluids the Onsager theory is commonly used [4].

The dielectric constant can be defined as $\varepsilon_s = C/C_0$, where C and C_0 are the capacitances of the filled and empty capacitor, respectively. Thus, the measurement of ε_s can be reduced to the measurement of the capacitance.

The polarisation $\mathbf{P}$ comes from two main sources. First, the electric field causes the relative displacement of atoms and the electronic charges. This contribution is relatively small (the atomic displacements in particular can be ignored) and the corresponding relative permittivity, ε_∞, may be approximated by the refractive index squared, n^2. The second contribution to the polarisation is given by the permanent dipole moment, μ, of the molecules when they possess a rotational freedom allowing the alignment along the external field. Usually this contribution dominates the dielectric properties of molecular systems.

For an anisotropic medium, like the nematic phase, the dielectric permittivity is a tensorial quantity. Due to the axial symmetry of the nematic phase we are dealing with two principal components of the permittivity, $\widetilde{\varepsilon}_\parallel$ and $\widetilde{\varepsilon}_\perp$, which define the dielectric anisotropy $\Delta\widetilde{\varepsilon} = \widetilde{\varepsilon}_\parallel - \widetilde{\varepsilon}_\perp$. $\widetilde{\varepsilon}_\parallel$ is measured if the electric field $\mathbf{E}$ is parallel to the director $\mathbf{n}$, $\mathbf{E}\|\mathbf{n}$, whereas $\widetilde{\varepsilon}_\perp$ corresponds to the perpendicular geometry $\mathbf{E}\perp\mathbf{n}$ (see FIGURE 1).

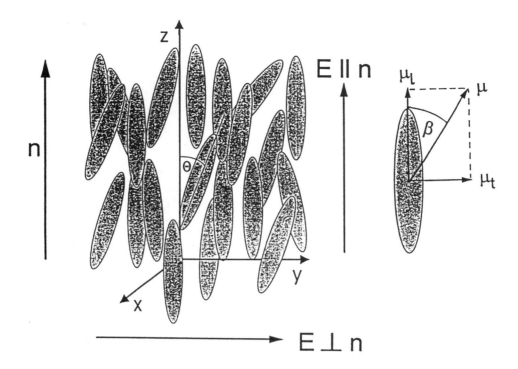

FIGURE 1 The geometry of dielectric experiments on the nematic phase. The orientations of the elongated molecules are statistically distributed around the director **n**. The molecule is assumed to possess a dipole μ at an angle β to its long axis giving the longitudinal μ_l and transverse μ_t components. The measuring electric field **E** can be either parallel (**E**‖**n**) or perpendicular (**E**⊥**n**) to the director which can be aligned by a strong magnetic field (**B** ≥ 0.4 T [6]), by a DC electric field (E ≥ 300 V/cm [6], see FIGURE 2(c)) or by a special treatment of the electrode surfaces [7-10].

A general interpretation of the dielectric anisotropy of the nematic phase as a function of the molecular parameters was given by Maier and Meier [5]. These authors extended the Onsager theory of isotropic liquids [4] to the nematic phase. For simplicity, the molecules are considered as spherical, but their polarisability is anisotropic with longitudinal α_l and transverse α_t components. The point dipole moment μ in the centre of the molecule makes an angle β with the molecular long axis (see FIGURE 1). The surroundings of the molecules are treated as a continuum which limits the validity of the model to $|\Delta\widetilde{\varepsilon}| \ll \overline{\varepsilon}$, where $\overline{\varepsilon} = (\widetilde{\varepsilon}_\parallel + 2\widetilde{\varepsilon}_\perp)/3$ is the mean permittivity. Moreover, the nematic orientational order defined by the order parameter $S = (3\cos^2\theta - 1)/2$ is not influenced by the external electric field. On this basis Maier and Meier obtained the following equations for the principal permittivity components and the dielectric anisotropy of the nematic phase:

$$\left(\widetilde{\varepsilon}_\parallel - 1\right) = \varepsilon_0^{-1} NFh \left\{ \overline{\alpha} + \frac{2}{3}\Delta\alpha S + F\frac{\mu^2}{3k_B T}\left[1 - \left(1 - 3\cos^2\beta\right)S\right] \right\} \tag{1}$$

$$\left(\widetilde{\varepsilon}_\perp - 1\right) = \varepsilon_0^{-1} NFh \left\{ \overline{\alpha} - \frac{1}{3}\Delta\alpha S + F\frac{\mu^2}{3k_B T}\left[1 + \frac{1}{2}\left(1 - 3\cos^2\beta\right)S\right] \right\} \tag{2}$$

$$\Delta\widetilde{\varepsilon} = \left(\widetilde{\varepsilon}_{\parallel} - \widetilde{\varepsilon}_{\perp}\right) = \varepsilon_0^{-1} NFh\left[\Delta\alpha - F\frac{\mu^2}{2k_BT}\left(1 - 3\cos^2\beta\right)\right]S \tag{3}$$

where N is the number density, $\Delta\alpha = \alpha_{\parallel} - \alpha_{\perp}$ and $\overline{\alpha} = (\alpha_l + 2\alpha_t)/3$. The local field parameters F and h are expressed by the mean polarisability $\overline{\alpha}$ and mean permittivity $\overline{\varepsilon}$: $F = 1/(1 - f\overline{\alpha})$ with $f = (N/3\varepsilon_0)[(2\overline{\varepsilon} - 2)/(2\overline{\varepsilon}_s + 1)]$, and $h = 3\overline{\varepsilon}_s/(2\overline{\varepsilon}_s + 1)$. It has been checked by Cummins et al [6] that the inclusion of the anisotropic F and h factors has little effect on the molecular parameters derived from EQNS (1) and (2), although they do have a significant effect on the orientational order parameter determined. More realistic molecular shapes and the anisotropy of the medium have also been considered in some theories (e.g. [11-14]). However, the usefulness of the equations derived there is doubtful due to restrictive assumptions about either the local order parameters or the molecular polarisability anisotropy.

C   DIPOLE-DIPOLE CORRELATIONS

In the Onsager theory of isotropic dielectrics as well as in its extension to the nematic phase given by Maier and Meier the short range dipole-dipole correlations were ignored. Therefore the dipole moment μ in EQNS (1) - (3) cannot be identified with its value measured in the gas state [16]. The dipole-dipole correlations were considered in the theory developed by Fröhlich [17] who generalised the former Kirkwood approach [18]. Fröhlich has introduced the dipole-dipole correlation factor (known as the Fröhlich-Kirkwood g-factor) in the form

$$g = 1 + \sum_{i,j}\frac{\langle\mu_i \cdot \mu_j\rangle}{\mu^2} \tag{4}$$

where the summation is over all neighbouring molecules. For the nematic phase we should consider separate g-factors for the directions parallel and perpendicular to the director. According to Dunmur et al [15,19,20] the rod-like molecules embedded in the anisotropic medium characterised by the orientational order parameter S exhibit anti-parallel correlation with longitudinal dipoles ($g_{\parallel} < 1$) and parallel correlation with transverse dipoles ($g_{\perp} > 1$). Usually the g-factors are calculated as

$$g_{\parallel} = (\mu_{\parallel})^2/\mu_l^2$$

$$g_{\perp} = (\mu_{\perp})^2/\mu_t^2 \tag{5}$$

where $\mu_{\parallel} = \mu^2[1 - (1 - 3\cos^2\beta)]S$ and $\mu_{\perp} = \mu^2[1 + (1 - 3\cos^2\beta)/2]S$ (compare EQNS (1) and (2)), and μ_l and μ_t are the longitudinal and transverse components of the dipole moment, respectively (see FIGURE 1). Thus, the correlation factors should be related to the actual order parameter. Dielectric studies of many strongly polar substances indicate that the dipole-dipole association plays an important role in determining the dielectric properties of substances in the nematic and isotropic phase as well as in solution [19-30]. The cyano-compounds, in particular, exhibit a considerable degree of the antiparallel associations leading to $g_{\parallel} < 1$. With increasing temperature the g-factors go to unity, although the antiparallel correlations persist in the isotropic phase as well [30-32]. The increase of hydrostatic pressure also destabilises such associates markedly [30].

D   TYPICAL EXAMPLES

According to EQN (3) when $\mu = 0$ (non-polar molecules) the dielectric anisotropy is determined by the polarisability anisotropy $\Delta\alpha$ which is always positive for rod-like molecules. For polar molecules ($\mu > 0$) the anisotropy $\Delta\tilde{\varepsilon}$ depends on the angle β. The dipole contribution to $\Delta\tilde{\varepsilon}$ is positive for $\beta < 54.7°$ and negative for $\beta > 54.7°$. In the latter case the sign of $\Delta\tilde{\varepsilon}$ depends on the relative magnitude of the two contributions. Typical dielectric permittivities measured for substances with different polarities are presented in FIGURE 2.

There is no polar group in the molecule 1-(4-propylphenyl)-2-(4-hexyl[2,2,2-bicyclooctyl]) ethane (6BAP3) (see FIGURE 2(a)) and therefore the permittivities are comparable with the corresponding refractive indices squared [33]. The mean permittivity, $\bar{\varepsilon}$, is larger than the value obtained by extrapolation of ε_{is} due to the increase in the density. For 1-(4-propyloxy-3-fluorophenyl)-2-(4-hexyl[2,2,2-bicyclooctyl])ethane (6BAP(F)) (see FIGURE 2(b)) the molecules possess two polar groups giving a net dipole moment of ~2.3 D (1D = 3.334×10^{-30} C m) inclined by ~68° with respect to the molecular long axis [34]. As a consequence, the substance exhibits a negative dielectric anisotropy with permittivities markedly exceeding the respective refractive indices squared. 4-hexyloxy-4′-cyanobiphenyl (6OCB) (see FIGURE 2(c)) is a substance representative of the large group of para-cyanophenyl compounds having a strong longitudinal dipole moment. Dielectric studies of such compounds [35] indicate that μ is inclined from the long axis by about 25° - 33° (at least in the isotropic phase) depending on the structure of the molecular cores and the lengths of the terminal group. The dielectric anisotropy is very large and the permittivities are considerably larger than n^2. Such compounds show antiparallel dipole-dipole correlations which are manifested as a small negative step in $\bar{\varepsilon}$ at the nematic-isotropic transition (e.g. [22-24]). It is worth noting that 6OCB can be well oriented by strong magnetic and electric fields (see FIGURE 2(c)).

E   ANALYSIS OF THE MAIER-MEIER EQUATION

EQNS (1), (2) and (3) are commonly used as the basis of the molecular interpretation of the static permittivities measured as a function of temperature (e.g. [6,12-15,20,21,28-30,36,37]) and/or pressure [30,38]. Recently, quite successful predictions of anisotropic optical and dielectric constants from molecular modelling calculations were achieved with the aid of the Maier-Meier theory [39,40]. It seems worthwhile, therefore, to analyse these equations in order to point out the weak and strong points of the theory. EQN (3) is the most convenient for this purpose (both components of the permittivity are discussed by Jadzyn et al in a recent paper [41]). The parameters N, F and h in the Maier-Meier equations vary little with temperature. Therefore, the contribution from the polarisability anisotropy $\Delta\alpha$ to $\Delta\varepsilon$ varies with temperature in the same way as the order parameter S, whereas that connected with the orientation polarisation varies like S/T. Especially interesting seems to be the case of constant temperature discussed in [38] where $\Delta\tilde{\varepsilon}$ was measured as a function of pressure, p. The discussion of the measured permittivities $\tilde{\varepsilon}_{\parallel}$, $\tilde{\varepsilon}_{\perp}$ and the anisotropy $\Delta\tilde{\varepsilon}$ as a function of the order parameter S obtained from the independent experiment seems to be the best way of verifying the assumptions on which the theory is based.

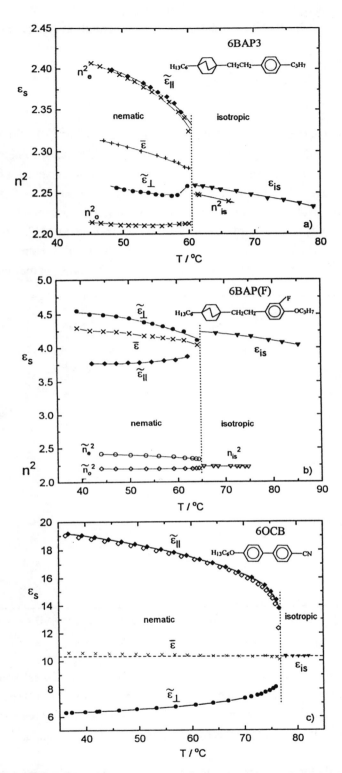

FIGURE 2 Dielectric permittivities and the refractive indices squared in the nematic and isotropic phases of typical nematogens: (a) non-polar [33], (b) polar with $\beta \approx 68°$ [34], and (c) strongly polar [33] with μ inclined by a small angle from the effective symmetry axis of the molecule ($\beta \approx 30°$) [35]. In case (c) the parallel component was measured for two kinds of orienting fields: B = 0.7 T (full points) and E = 2500 V/cm (open points). For 6OCB the values of n^2 are not shown as they are considerably smaller than the permittivities.

As an example, let us consider the data presented in FIGURE 2(c) in order to check the applicability of the Maier-Meier EQN (3) for the determination of S. This equation can be rewritten as

$$S^{MM} = \Delta\widetilde{\varepsilon}^{exp} \bigg/ \left\{ \varepsilon_0^{-1} NFh \left[\Delta\alpha - F\frac{\mu^2}{2k_BT}\left(1 - 3\cos^2\beta\right) \right] \right\} \tag{6}$$

FIGURE 3 shows a comparison of the calculated order parameter $S^{MM}(T)$ for 6OCB with the order parameter obtained from NMR [42] and optical anisotropy [43] measurements (the details of the calculations will be published elsewhere [33]). However, such consistency of the results could be achieved assuming in the calculations that the $g_\parallel$-factor changes linearly from 0.58 at $T_{NI} - T = 40$ K to 0.71 close to the nematic-isotropic temperature. This is in good agreement with the results of the analysis of both permittivity components performed by Jadzyn et al [41]. A crucial point in this analysis is the appropriate choice of the angle β. As seen in FIGURE 3 a small decrease in β leads to a marked reduction of S^{MM}. Similar results were also obtained for other nematics [33].

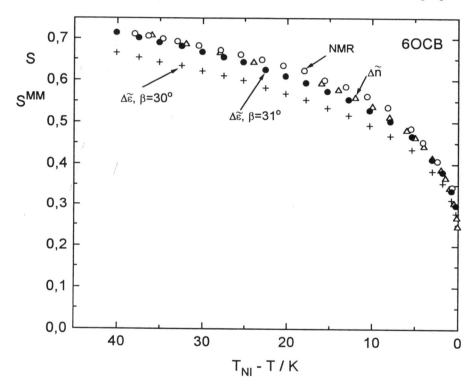

FIGURE 3 Comparison of the order parameters for the nematic phase of 6OCB calculated with the aid of the Maier-Meier equation (closed circles, plus sign) [33], and obtained from NMR [42] (open circles) and optical anisotropy [43] (open triangles) measurements. The angle β was chosen as: 31° (closed circles) [35] and 30° (plus sign). The $g_\parallel$-factor changes linearly between 0.58 and 0.71 within the nematic phase.

The dipolar part of the dielectric permittivity, $\delta\varepsilon = \varepsilon_s - \varepsilon_\infty$, decreases with lengthening of the end groups in a given homologous series [35,44]. FIGURE 4 shows this effect for three homologous series (nCB, nOCB and nPCH) in the nematic and isotropic phases [35]. This feature can be attributed to the effective dilution of the dipole moment by the longer alkyl chain.

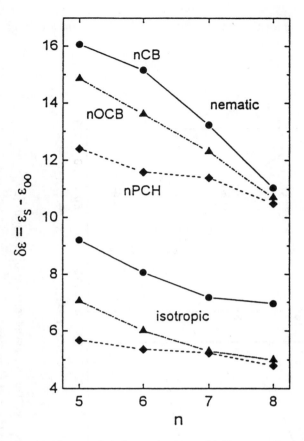

FIGURE 4 Dipolar part of the static permittivity at $T = T_{NI} \pm 10$ K versus the number n of carbon atoms in the alkyl or alkoxy chains in three homologous series (nCB, nOCB and nPCH) [35].

F   DIELECTRIC CONSTANT OF BINARY MIXTURES

For applications, mixtures of mesogenic compounds are used rather than pure substances. The dielectric permittivities of mixtures depend upon the polarity of the particular components and their concentrations. However, some specifications of the mutual influence of the components of a mixture can be expected for binary mixtures only. Interesting dielectric studies of binary mixtures of substances with different capabilities to form associates in the pure state have been performed by Raszewski et al [28,29]. FIGURE 5 presents examples of the concentration dependences of the principal permittivities. Usually marked deviations (positive or negative) from the straight lines predicted by the theory of ideal solutions were observed. This was interpreted as being caused either by dissociation of the cyanomesogen dimers or by a non-additive behaviour of the mixture density [29].

G   CONCLUSION

Static dielectric properties of the nematic phase are determined by the molecular parameters (μ, β, $\overline{\alpha}$, $\Delta\alpha$), intermolecular interactions (characterised by the Fröhlich-Kirkwood g-factors), the density, and

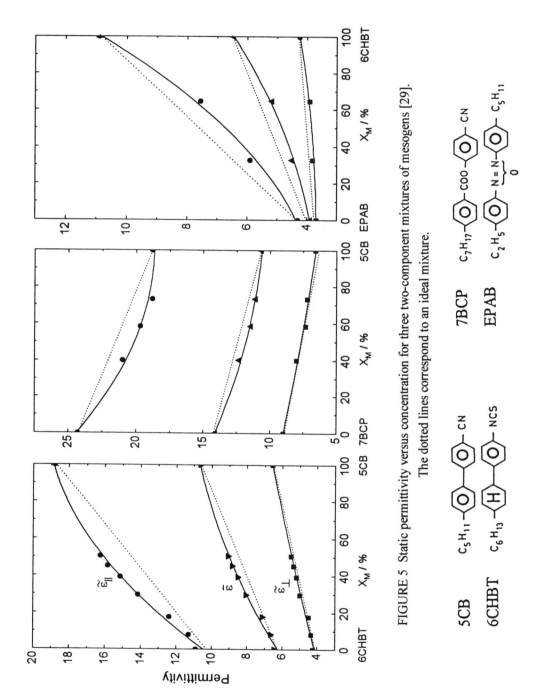

FIGURE 5 Static permittivity versus concentration for three two-component mixtures of mesogens [29]. The dotted lines correspond to an ideal mixture.

the orientational order parameter S. In spite of numerous assumptions and simplifications in consideration of the anisotropic medium the Maier-Meier theory can be very useful in understanding the relationship between the microscopic and macroscopic properties of the nematic phase. The role of the order parameter S in this discussion cannot be overestimated. Unfortunately, particular experimental techniques often give quite different results for S for the same substance (e.g. [41,45,46]).

REFERENCES

[1]   C.J.F. Böttcher, P. Bordewijk [*Theory of Dielectric Polarisation* vol.1 (Elsevier, Amsterdam, 1973)]

[2]   A. Chelkowski [*Dielectric Physics* (PWN, Warszawa, 1993)]

[3]   P. Debye [*Polar Molecules* (Chemical Catalogue Co., New York, 1929)]

[4]   L. Onsager [*J. Am. Chem. Soc. (USA)* vol.58 (1936) p.1486]

[5]   W. Maier, G. Meier [*Z. Nat.forsch. A (Germany)* vol.16 (1961) p.262 and p.470]

[6]   P.G. Cummins, D.A. Dunmur, D.A. Laidler [*Mol. Cryst. Liq. Cryst. (UK)* vol.30 (1975) p.109]

[7]   J.L. Jenning [*Appl. Phys. Lett. (USA)* vol.44 (1972) p.73]

[8]   W.J.A. Goosens [*Mol. Cryst. Liq. Cryst. (UK)* vol.124 (1985) p.305]

[9]   J.H. Kim, H.R. Ha, Y. Shi, S. Kumar [Poster P1-92 presented during *17th Int. L.C. Conf.* Strasbourg, 1998]

[10]   B. Jérôme [in *Handbook of Liquid Crystals, Vol.1 Fundamentals* Eds. D. Demus, J. Goodby, G.W. Gray, H.-W. Spiess, V. Vill (Wiley-VCH, Weinheim, 1998) ch.VI.10]

[11]   P. Bordewijk [*Physica (Netherlands)* vol.75 (1974) p.146]

[12]   W.H. de Jeu [in *Liquid Crystals* Ed. L. Liebert (Solid State Physics Suppl.14, Academic Press, New York, 1978) p.109]

[13]   W.H. de Jeu [*Physical Properties of Liquid Crystalline Materials* (Gordon and Breach, New York, 1980) ch.5]

[14]   K. Toriyama, D.A. Dunmur, S.E. Hunt [*Liq. Cryst. (UK)* vol.5 (1988) p.1001]

[15]   D. Dunmur, K. Toriyama [in *Handbook of Liquid Crystals, Vol.1 Fundamentals* Eds. D. Demus, J. Goodby, G.W. Gray, H.-W. Spiess, V. Vill (Wiley-VCH, Weinheim, 1998) ch.VI.4]

[16]   L.G.P. Dalmolen, S.J. Picken, A.F. de Jong, W.H. de Jeu [*J. Phys. (France)* vol.46 (1985) p.1443]

[17]   H. Fröhlich [*Theory of Dielectrics* (Clarendon Press, Oxford, 1949)]

[18]   J.G. Kirkwood [*J. Chem. Phys. (USA)* vol.7 (1939) p.911]

[19]   D.A. Dunmur, P. Palffy-Muhoray [*Mol. Phys. (UK)* vol.76 (1992) p.1015]

[20]   K. Toriyama, S. Sugimori, K. Moriya, D.A. Dunmur, R. Hanson [*J. Phys. Chem. (USA)* vol.100 (1996) p.307]

[21]   H. Kresse [in *Handbook of Liquid Crystals, Vol.2A Low Molecular Weight Liquid Crystals* Eds. D. Demus, J. Goodby, G.W. Gray, H.-W. Spiess, V. Vill (Wiley-VCH, Weinheim, 1998) ch.III.2.2]

[22]   J. Jadzyn, P. Kedziora [*Mol. Cryst. Liq. Cryst. (UK)* vol.145 (1987) p.17]

[23]   H. Herba, B. Zywucki, G. Czechowski, D. Bauman, A. Wasik, J. Jadzyn [*Acta Phys. Pol. A (Poland)* vol.87 (1995) p.985]

[24]   J. Jadzyn, G. Czechowski [*Liq. Cryst. (UK)* vol.4 (1989) p.157]

[25]   P. Kedziora, J. Jadzyn [*Liq. Cryst. (UK)* vol.8 (1990) p.445]

[26]   P. Kedziora, J. Jadzyn, P. Bonnet [*Ber. Bunsenges. Phys. Chem. (Germany)* vol.97 (1993) p.864]

[27]   A. Buka, L. Bata [*Mol. Cryst. Liq. Cryst. (UK)* vol.135 (1986) p.49]

[28]   Z. Raszewski, R. Dabrowski, Z. Stolarzowa, J. Zmija [*Cryst. Res. Technol. (Germany)* vol.22 (1987) p.835]

[29]   Z. Raszewski [*Liq. Cryst. (UK)* vol.3 (1988) p.307]

[30]   S. Urban, A. Würflinger [*Adv. Chem. Phys. (USA)* vol.98 (1997) p.147]

[31]   M.J. Bradshaw, E.P. Raynes [*Mol. Cryst. Liq. Cryst. Lett. (UK)* vol.72 (1981) p.73]

[32]   J. Thoen, G. Menu [*Mol. Cryst. Liq. Cryst. (UK)* vol.97 (1983) p.163]

[33]   S. Urban, J. Kedzierski, R. Dabrowski [in preparation]

[34]   S. Urban, B. Gestblom, R. Dabrowski, H. Kresse [*Z. Nat.forsch. A (Germany)* vol.53 (1998) p.134]

[35]   S. Urban, B. Gestblom, A. Würflinger [*Mol. Cryst. Liq. Cryst. (Switzerland)* in print]

[36]   D.A. Dunmur, M.R. Manterfield, W.H. Miller, J.K. Dunleavy [*Mol. Cryst. Liq. Cryst. (UK)* vol.45 (1978) p.127]

[37]   Z. Raszewski, J. Rutkowska, J. Kedzierski, J. Zielinski, J. Zmija, R. Dabrowski [*Mol. Cryst. Liq. Cryst. (UK)* vol.215 (1992) p.349]

[38]   S. Urban [*Z. Nat.forsch. A (Germany)* vol.50 (1995) p.826]

[39]   M. Bremer, A. Goetz, M. Klasen, A. Manabe, S. Naemura, K. Tarumi [*Jpn. J. Appl. Phys. (Japan)* vol.37 (1998) p.L945-8]

[40]   A. Fujita, M. Ushioda, H. Takeuchi, T. Inukai, D. Demus [Poster P2-71 presented during *17th Int. L.C. Conf.* Strasbourg, 1998, submitted to *Mol. Cryst. Liq. Cryst. (Switzerland)*]

[41]   J. Jadzyn, S. Czerkas, G. Czechowski, A. Burczyk, R. Dabrowski [*Liq. Cryst. (UK)* vol.26 (1999) p.437-42]

[42]   R.Y. Dong, G. Ravindranath [*Liq. Cryst. (UK)* vol.17 (1994) p.47]

[43]   M. Mitra [*Mol. Cryst. Liq. Cryst. (UK)* vol.241 (1994) p.17]

[44]   S. Urban, B. Gestblom, R. Dabrowski [*Liq. Cryst. (UK)* vol.24 (1998) p.681]

[45]   S. Urban, B. Gestblom, T. Brückert, A. Würflinger [*Z. Nat.forsch. A (Germany)* vol.50 (1995) p.984]

[46]   S. Urban, B. Gestblom, H. Kresse, R. Dabrowski [*Z. Nat.forsch. A (Germany)* vol.51 (1996) p.834]

6.2   Dynamic dielectric properties of nematics

H. Kresse

November 1998

A   INTRODUCTION

Every medium increases the capacitance of a capacitor from C_0 (the vacuum value) to C_M. The sample can be characterised by the (relative) dielectric constant

$$\varepsilon' = C_M/C_0 \tag{1}$$

According to

$$P = \varepsilon\,(\varepsilon' - 1)E \tag{2}$$

with $\varepsilon = 8.85 \times 10^{-12}$ F m^{-1} and E the electric field strength, the polarisation P can be calculated. On switching off the electric field P decays to zero. Three different time constants are possible for this decay in liquids and liquid crystals:

(i)   the very fast displacement of the electron density and atom positions back to the equilibrium position,

(ii)   the reorientation of the electric dipoles via rotational diffusion (P_μ) with the relaxation time $\tau > 10^{-12}$ s and

(iii)   a slow relaxation connected with charged particles in the sample.

Experimentally only the last two processes can be measured by analysing the decay function (time domain spectroscopy). Thus, the dipole relaxation time, τ, in which P_μ decreases to P_μ/e (where e is the base of natural logarithms), is available.

In the frequency domain it is necessary to measure the complex dielectric constant

$$\varepsilon^* = \varepsilon' - i\varepsilon'' \tag{3}$$

as a function of the frequency, f. The dielectric constant $\varepsilon'(f)$ as the real part and the loss $\varepsilon''(f)$ as the imaginary part in EQN (3) can be fitted to

$$\varepsilon^* = \varepsilon_\infty + \frac{\varepsilon_0 - \varepsilon_\infty}{\left(1 + \left(\mathrm{if}/f_R\right)^{1-\alpha}\right)^{1-\beta}} \tag{4}$$

Here ε_∞ is the high frequency limit of ε', ε_0 is the static dielectric constant (low frequency limit of ε'), $\varepsilon_0 - \varepsilon_\infty = \Delta$ is the dielectric increment, f_R is the relaxation frequency, α is the Cole-Cole distribution parameter, and β is the asymmetry parameter. The relaxation frequency is related to the relaxation time by $f_R = (2\pi\tau)^{-1}$. A simple exponential decay of P_μ ($\alpha,\beta = 0$) is characterised by a single relaxation time (Debye-process [1]), $\beta = 0$ and $1 < \alpha < 0$ describe a Cole-Cole-relaxation [2] with a symmetrical distribution function of τ whereas the Havriliak-Negami equation (EQN (4)) is used for an asymmetric distribution of τ [3]. The symmetry can be readily seen by plotting ε' versus ε'' as the so-called Cole-Cole plot [4-6].

B   DIELECTRIC RELAXATION IN NEMATICS

There have been many attempts to connect the phenomenological quantities in EQN (4) with molecular parameters like the polarisability and dipole moment. Indeed Debye [1] describes the relaxation as a rotational diffusion process characterised by a molecular rotational viscosity. Based on that and using the Onsager equations [7] as well as the Maier-Saupe model [8], Maier and Meier [9] and Martin et al [10] developed a theory for the dielectric behaviour of nematic liquid crystals. Here nematics are regarded as a uniaxial system. Consequently, dielectric measurements have to be carried out parallel and perpendicular to the director which in most cases is magnetically oriented. The central idea for the dynamics is that the nematic potential of mean torque [8], responsible for the formation of the nematic phase, is the reason for an additional hindrance of the reorientation of the rod-like molecules around their short axis. This model is sketched in FIGURE 1 for a molecule exhibiting only a longitudinal dipole moment (direction 1 in the molecular system).

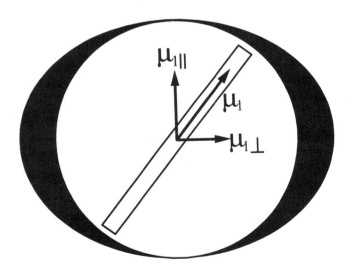

FIGURE 1 A rod-like molecule partially oriented by the nematic potential of mean torque. The dipole moment in the direction of the long axis is resolved into the two components of the laboratory system, parallel and perpendicular to the director.

The respective dielectric increments can be measured in the laboratory system parallel ($\Delta_{1\parallel}$) and perpendicular ($\Delta_{1\perp}$) to the director. They are given by

$$\Delta_{1\parallel} = \frac{NhF^2}{3k_BT}\mu_1^2\left(1+2S\right) \quad \text{and} \quad \Delta_{1\perp} = \frac{NhF^2}{3k_BT}\mu_1^2\left(1-S\right) \tag{5}$$

where N, h and F are parameters associated with the internal field, and S is the second rank orientational order parameter [9]. As an example the absorption data of Ratna and Shashidhar [11] on 4-pentyl-4'-cyanobiphenyl are presented in FIGURE 2. The increase of the absorption intensity of 27% from 31°C to 19.5°C results, according to EQN (5), from 4% associated with the decreasing temperature and 23% from the increasing degree of orientational order. FIGURE 3 shows the

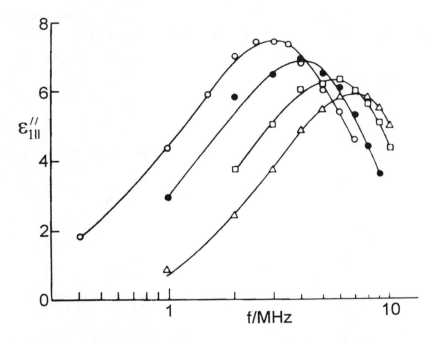

FIGURE 2 Dielectric absorption curves for 4-pentyl-4'-cyanobiphenyl related to $\mu_{1\parallel}$ from [11] at 19.5°C (open circles), 24°C (closed circles), 28°C (open squares) and 31°C (open triangles).

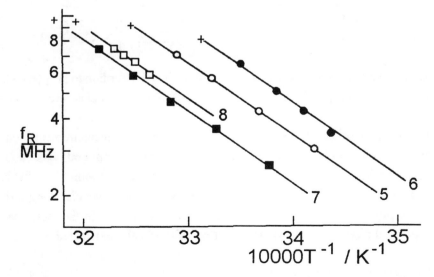

FIGURE 3 Relaxation frequencies for four 4-alkyl-4'-cyanobiphenyls. The numbers are related to the number of carbon atoms in the alkyl chain [11]. Crosses indicate the nematic-isotropic transitions.

relaxation frequencies of four different 4-alkyl-4′-cyanobiphenyls as an Arrhenius plot; a mean activation energy of 54 kJ mol^{-1} can be determined. There is a systematic increase of f_R with falling nematic-isotropic transition temperatures.

The change of relaxation times especially at the nematic-isotropic transition of 4-heptyl-4′-cyano-biphenyl measured by Davis [12], Buka [13] and Lippens [14] together with their respective co-workers is given in FIGURE 4. In accordance with the theoretical prediction [10], the relaxation time associated with $\mu_{1\parallel}$ denoted as $\tau_{1\parallel}$ increases stepwise by a factor four at the transition into the nematic phase. The component $\mu_{1\perp}$, measured in the perpendicular direction, is less intense and reorients, as predicted [10], faster in the nematic phase due to the higher orientational order. The interpretation of the additional absorption ranges of $\varepsilon''_\perp$ detected by the three authors differs.

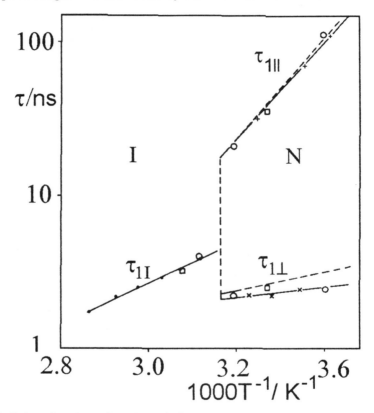

FIGURE 4 Relaxation times for 4-heptyl-4′-cyanophenyl according to [12] ($\bullet$,+,×), [13] (open circles), [15] (open squares), - - - - i.e. dashed line calculated according to [10].

Recently Urban and co-workers [15] have published dielectric measurements on 1-(4-n-hexyl-bicyclo[2,2,2]octyl)-2-(3-fluoro-4-methoxyphenyl)ethane. This compound has a much stronger perpendicular dipole moment μ_2 than μ_1. It is therefore a suitable compound with which to study the influence of the nematic phase on motion reorientational about the molecular long axis. FIGURE 5 shows the Cole-Cole plots measured in the isotropic and nematic state. In the mesophase only a small absorption in the parallel direction with an increment $\Delta_{1\parallel}$ of 0.38 can be detected.

According to EQN (5) we can estimate from the relation

$$\Delta_{1\parallel}/\Delta_{1\perp} = (1 + 2S)/(1 - S) \tag{6}$$

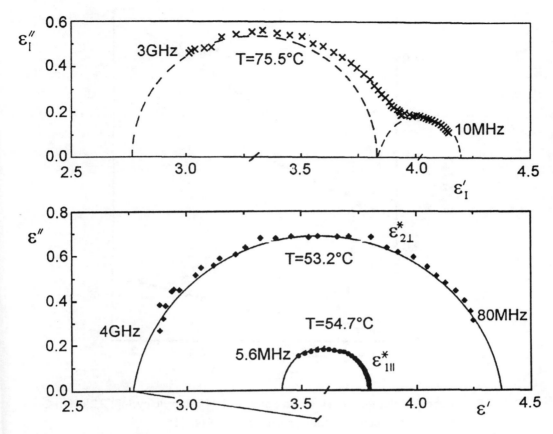

FIGURE 5   Cole-Cole plots for 1-(4-n-hexylbicyclo[2,2,2]octyl)-2-(3-fluoro-4-methoxyphenyl)ethane in the isotropic (75.5°C) and nematic phase (∥: 54.7°C; ⊥: 53.2°C) from [15].

the value expected for $\Delta_{1\perp}$. With S = 0.6 at 54°C [15] we have calculated the relation to be 1:0.18, or $\Delta_{1\perp}$ = 0.07. This small contribution from the parallel component of dipole moment cannot be separated from the intense one for $\varepsilon^*_{2\perp}$. It means that the dielectric absorption at high frequencies is dominated by reorientation about the molecular long axis characterised by $\tau_{2\perp}$. As seen in FIGURE 6 the reorientation about the molecular long axis is practically independent of the transition into the nematic phase in contrast to reorientation around the short axis.

The separation of the high frequency processes in the nematic phase indicated by the dielectric increments $\Delta_{1\perp}$ and $\Delta_{2\perp}$ and the respective reorientation about the two main molecular axes in the isotropic state (Δ_{1I} and Δ_{2I}) from each other is an experimental problem. Jadzyn and co-workers have investigated 1-(4-isothiocyanatophenyl)-2-(4-n-hexylbicyclo[2,2,2]octane)ethane which is a molecule with a strong dipole component parallel to the molecular long axis and a small value orthogonal to this [16]. The results in FIGURE 7 are a good example for the two observed components parallel to the nematic director and the two main reorientational processes in the isotropic phase. The high frequency data were obtained as the result of a fitting procedure assuming a Debye relaxation. Kresse and co-workers have used a sample with a high viscosity [17]. In this way the dielectric constants and dispersion ranges in both main directions could be measured in the MHz-range. Strong deviations of $\varepsilon^*_\perp$ from a simple Cole-Cole plot are discussed in [17] and attributed to separate motions of the polar end groups.

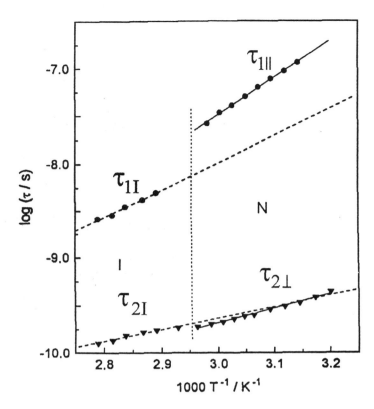

FIGURE 6 Arrhenius plot for the relaxation times for 1-(4-n-hexylbicyclo[2,2,2]octyl)-2-(3-fluoro-4-methoxyphenyl)ethane from [15].

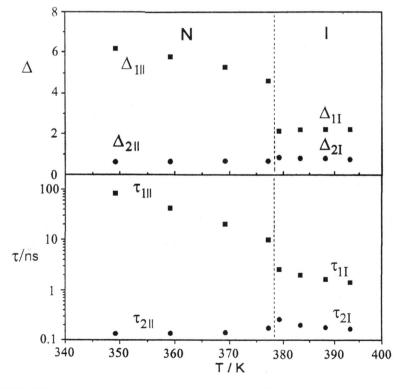

FIGURE 7 Increments and relaxation times for 1-(4-isothiocyanatophenyl)-2-(4-n-hexylbicyclo[2,2,2]octane)ethane from [16].

Due to scientific and technical interest there are a lot of data for $\varepsilon^*_{1\parallel}$ in the literature. The first dielectric relaxation measurements on 4,4'-(di-alkylazoxybenzenes) were made by Maier and Meier [18] and Weise together with Axmann [19]. Later Rondelez et al [20] (4-methoxy-benzylidine-4'-butylaniline (MBBA)) and Schadt [21] (4-subst.-benzylidene-4'-cyanoanilines) have published their data. Strongly polar compounds have also been investigated by Galerne et al (4-nitrobenzylidene-4'-octylaniline) [22], Tsvetkov and co-workers (substituted cyanostilbenes) [23] and Druon together with Wacrenier (4-octyl-4'-cyanobiphenyl) [24]. 4-heptyl-4'-cyanobiphenyl where the phenyl is stepwise substituted by cyclohexane was presented by Parneix and Chapoton [25]. Gupta and co-workers [26] have studied the less polar 4-n-octyloxybenzylidene-4'-toluidine.

A simple model to summarise some data based on the clearing temperature as reference was given by Kresse [27]. Using the formula $\lg \tau_R = 0.4342 \varepsilon_R (T_{NI} - T)/T$, with $\tau_R = \tau_{1\parallel}/\tau_{1\parallel}^{NI}$ and $\varepsilon_R = E_A/RT_{NI}$ (E_A is the activation energy), experimental values of different samples can be compared. Own data on derivatives of phenyl benzoates (1-9) [28] and results of de Jeu and Lathouwers on 4-butylphenyl 4-(4-alkyloxybenzoyloxy)benzoates (10,11) [29] are plotted in FIGURE 8. The three ring systems clearly differ from the two ring compounds. The method described shifts the problems under discussion only to the different relaxation times at the clearing temperature $\tau_{1\parallel}^{NI}$ [27] or, in other words, to the rotational viscosity in the isotropic phase.

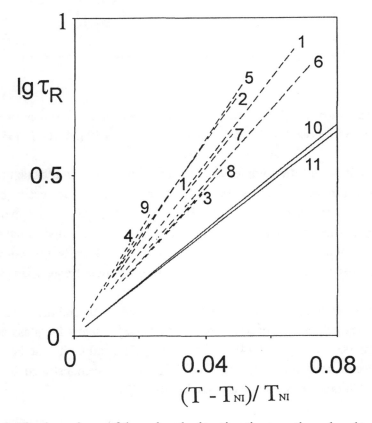

FIGURE 8 The dependence of the reduced relaxation times on the reduced temperature
for derivatives of phenyl benzoates according to [27].

A further question is related to the activation energies. Looking to FIGURE 3 and to the model [10] E_A should be practically given by the activation energy for the reorientation of μ_1 in the isotropic phase.

Some data are plotted in FIGURE 9 versus the nematic range. Below the nematic phase, in all cases, different smectic modifications exist. The relatively good correlation demonstrates that a molecular field approximation like that in [8] cannot be applied to samples with a short nematic range influenced by smectic clusters. Probably, the degree of order changes more strongly with temperature as predicted in [8] and this is reflected in higher activation energies. Thus, E_A contains the classical part from the rotational diffusion in the isotropic state, the nematic potential of mean torque according to Meier and Saupe and pre-transitional smectic effects.

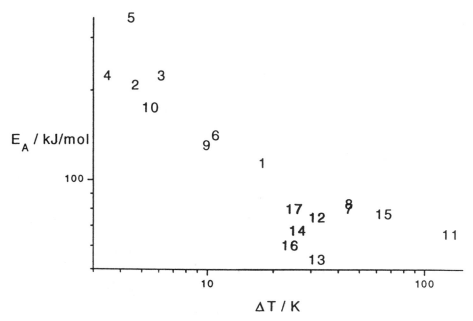

FIGURE 9   Activation energies as a function of the nematic range. The numbers are related to the references: 1-4 [30], 5,6 [31], 7 [32], 8 [33], 9,10 [34], 11-17 [35].

The individual behaviour of nematic samples can be seen in mixtures. In a basic mixture consisting of components with two aromatic systems the relaxation frequency decreases by one decade on lengthening the solute by one phenyl ring [36-38]. Such behaviour can be explained by different degrees of orientational order for solvent and solute. It should be noted that in the same way in which f_R decreases an increase of the activation energy is observed [38]. The low relaxation frequency can be used to design two-frequency addressing mixtures with low cross-over frequencies [39-42].

A freezing process of the main reorientational motions, the rotation around the molecular short and long axes, is seen in nematics with a high glass temperature [43-46]. Below the glass transition only local motions, as in polymers, are detected [46]. From this observation it can be concluded that the description of a molecule as a rod is too simple. This is also demonstrated by laterally branched molecules where an additional absorption has been detected [47].

Interesting effects can be observed if the molecules show stronger interactions. So, dipole correlations [48], associations [49], steric effects [50] or interactions due to hydrogen bridges [51] are observed. As an example the dielectric absorption curves for a swallow-tailed compound in the smectic A and nematic phases are presented in FIGURE 10. In contradiction to FIGURE 2 where a sample with a strong dipole correlation is given, the much stronger steric interaction results in a decrease of the absorption intensity with decreasing temperature especially in the smectic phase. By mixing molecules with a suitable shape such interactions can be systematically destroyed [50]. The smectic A phase disappears in the same way in which normal absorption intensities according to EQN (5) are observed. This experiment demonstrates that special interactions are directly related to the transition between the nematic and smectic A phase.

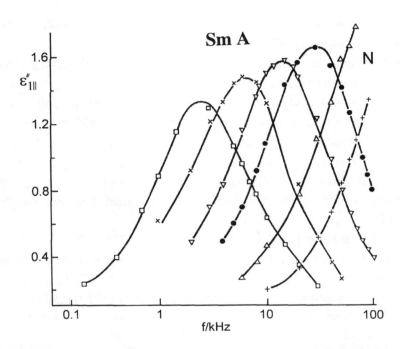

FIGURE 10 The dielectric absorption curves for a swallow-tailed compound from [52].

The increase of the relaxation time with increasing pressure gives further experimental possibilities which are also very interesting from a theoretical point of view. In this way, the validity of some general assumptions of the Maier-Saupe theory [8] was established by Urban and Würflinger [53-55].

REFERENCES

[1]   P. Debye [in *Polare Molekeln* (Hirzel Verlag, Leipzig, 1929) p.98-108]

[2]   K.S. Cole, R.H. Cole [*J. Chem. Phys. (USA)* vol.9 (1941) p.341]

[3]   S. Havriliak, S. Negami [*J. Polym. Sci. C (USA)* vol.14 (1966) p.89]

[4]   C.J.F. Böttcher, P. Bordewijk [in *Theory of Electric Polarization* vol.2, 2nd ed. (Elsevier, Amsterdam, 1992)]

[5]   N.E. Hill, W.E. Vaughan, A.H. Price, M. Davis [in *Dielectric Properties and Molecular Behaviour* (van Nostrand Reinhold Co., London, 1969)]

[6]   C.P. Smyth [in *Dielectric Behavior and Structure* (McGraw-Hill Book Co., 1955)]

[7]   L. Onsager [*J. Am. Chem. Soc. (USA)* vol.58 (1936) p.1486]

[8]   W. Maier, A. Saupe [*Z. Nat.forsch. A (Germany)* vol.15 (1960) p.287]

[9]   W. Maier, G. Meier [*Z. Nat.forsch. A (Germany)* vol.16 (1961) p.262]

[10]   A.J. Martin, G. Meier, A. Saupe [*Symp. Faraday Soc. (UK)* vol.5 (1971) p.119]

[11]   B.R. Ratna, R. Shashidhar [*Mol. Cryst. Liq. Cryst. (UK)* vol.42 (1977) p.185]

[12]   M. Davis, R. Moutran, A.H. Price, M.S. Beevers, G. Williams [*J. Chem. Soc. Faraday Trans. II (UK)* vol.72 (1976) p.1447]

[13]   A. Buka, P.G. Owen, A.H. Price [*Mol. Cryst. Liq. Cryst. (UK)* vol.51 (1979) p.273]

[14]   D. Lippens, J.P. Parneix, A. Chapoton [*J. Phys. (France)* vol.38 (1977) p.1465]

[15]   S. Urban, B. Gestblom, R. Dabrowski, H. Kresse [*Z. Nat.forsch. A (Germany)* vol.53 (1998) p.134]

[16]   J. Jadzyn, C. Legrand, P. Kedziora, B. Zywudski, G. Czechowski, D. Baumann [*Z. Nat.forsch. A (Germany)* vol.51 (1996) p.933]

[17]   H. Kresse, K. Worm, W. Schäfer, H. Stettin, D. Demus, W. Otowski [*Cryst. Res. Technol. (Germany)* vol.21 (1986) p.293]

[18]   W. Maier, G. Meier [*Z. Nat.forsch. A (Germany)* vol.16 (1961) p.1200]

[19]   H. Weise, A. Axmann [*Z. Nat.forsch. A (Germany)* vol.21 (1966) p.1316]

[20]   F. Rondelez, D. Diguets, C. Durand [*Mol. Cryst. Liq. Cryst. (UK)* vol.15 (1971) p.183]

[21]   M. Schadt [*J. Chem. Phys. (USA)* vol.56 (1972) p.1494]

[22]   Y. Galerne [*C.R. Acad. Sci. B (France)* vol.278 (1974) p.347]

[23]   V.N. Tsvetkov, E.I. Ryumtsev, A.P. Kovshik, I.P. Kolomiets [*Kristallografiya (USSR)* vol.20 (1975) p.885]

[24]   C. Druon, J.M. Wacrenier [*J. Phys. (France)* vol.38 (1977) p.47]

[25]   J.P. Parneix, A. Chapoton [in *Advances in Liquid Crystal Research and Applications* (Pergamon Press/Academiai Kiado, Budapest, 1980) vol.1 p.297-303]

[26]   G.K. Gupta, V.K. Argawal [*J. Chem. Phys. (USA)* vol.71 (1979) p.5290]

[27]   H. Kresse [in *Advances in Liquid Crystals* (Academic Press, 1983) p.133-6]

[28]   H. Kresse [*Z. Phys. Chem. (Germany)* vol.262 (1981) p.801]

[29]   W.H. de Jeu, T.W. Lathouwers [*Mol. Cryst. Liq. Cryst. (UK)* vol.26 (1974) p.225]

[30]   H. Kresse, A. Wiegeleben, D. Demus [*Krist. Tech. (Germany)* vol.15 (1980) p.341]

[31]   A. Buka, L. Bata, H. Kresse [*Proc. Hung. Acad. Sci. KFKI (Hungary)* vol.1980-4 (1980)]

[32]   J. Chrusciel, J.A. Janik, J.M. Janik, S. Wrobel, H. Kresse [*Acta Phys. Pol. A (Poland)* vol.59 (1981) p.431]

[33]   J. Chrusciel, S. Wrobel, J.A. Janik, J.M. Janik, H. Kresse [in *Advances in Liquid Crystal Research and Applications* (Pergamon Press/Academiai Kiado, Budapest, 1980) vol.1 p.279-86]

[34]   J. Chrusciel, H. Kresse, S. Urban [*Liq. Cryst. (UK)* vol.11 (1992) p.711]

[35]   S. Chandrasekhar [*Mol. Cryst. Liq. Cryst. (UK)* vol.124 (1985) p.1]

[36]   H. Kresse, H. Stettin, F. Gouda, G. Anderson [*Phys. Status Solidi A (Germany)* vol.111 (1989) p.265]

[37]   G. Heppke, J. Keyed, U. Müller [*Mol. Cryst. Liq. Cryst. (UK)* vol.98 (1983) p.309]

[38]   H. Kresse, P. Rabenstein, D. Demus [*Mol. Cryst. Liq. Cryst. (UK)* vol.154 (1987) p.1]

[39]   M. Schadt [*Mol. Cryst. Liq. Cryst. (UK)* vol.89 (1982) p.77]

[40]   T.S. Chang, E.E. Loebner [*Appl. Phys. Lett. (USA)* vol.25 (1974) p.1]

[41]   H.K. Buecher, R.T. Klingbiel. J.P. VanMeter [*Appl. Phys. Lett. (USA)* vol.25 (1974) p.186]

[42]   H. Stettin, H. Kresse, W. Schäfer [*Cryst. Res. Technol. (Germany)* vol.24 (1989) p.111]

[43]   D. Lippens, C. Druon, J.M. Wacrenier [*J. Phys. Colloq. (France)* vol.40 (1979) p.306]

[44]   C.P. Johari [*Philos. Mag. B (UK)* vol.46 (1982) p.549]

[45]   H.R. Zeller [*Phys. Rev. A (USA)* vol.26 (1982) p.1785]

[46]   H. Kresse, S. Ernst, W. Wedler, D. Demus, F. Kremer [*Ber. Bunsenges. Phys. Chem. (Germany)* vol.94 (1990) p.1478]

[47]   H. Kresse, S. Tschierske, A. Hohmuth, C. Stützer, W. Weissflog [*Liq. Cryst. (UK)* vol.20 (1996) p.715]

[48]   W.H. de Jeu, Th.W. Lathouwers [*Z. Nat.forsch. A (Germany)* vol.29 (1974) p.905]

[49]   K. Toriyama, D.A. Dunmur [*Mol. Cryst. Liq. Cryst. (UK)* vol.139 (1986) p.123]

[50]   H. Kresse, S. Heinemann, R. Paschke W. Weissflog [*Ber. Bunsenges. Phys. Chem. (Germany)* vol.97 (1993) p.1337]

[51]   H. Kresse, S. König, D. Demus [*Wiss. Z. Univ. Halle (Germany)* vol.27M (1978) p.47]

[52]   H. Kresse, P. Rabenstein, H. Stettin, S. Diele, D. Demus, W. Weissflog [*Cryst. Res. Technol. (Germany)* vol.23 (1988) p.135]

[53]   S. Urban, D. Büsing, A. Würflinger, B. Gestblom [*Liq. Cryst. (UK)* vol.25 (1998) p.253]

[54]   S. Urban, A. Würflinger [*Adv. Chem. Phys. (USA)* vol.98 (1997) p.143-216]

[55]   A. Würflinger [*Z. Nat.forsch. A (Germany)* vol.53 (1998) p.141]

6.3 Measurements of diamagnetic properties

W. Haase

June 2000

A   INTRODUCTION

Diamagnetism is one of the fundamental properties of all materials. If unpaired electrons are present, the compounds are paramagnetic, and sometimes collective magnetism appears. Because most organic liquid crystals are closed shell systems, diamagnetism or more importantly the diamagnetic anisotropy is a property of practical relevance. For instance the orientational order parameter $\overline{P}_2$ can be extracted from the diamagnetic susceptibility anisotropy in the liquid crystalline state $\Delta\widetilde{\chi}$ and because the sign of $\Delta\widetilde{\chi}$ can be positive or negative, control of the alignment of the liquid crystal director is possible.

B   SOME RELATIONS

In a homogeneous magnetic field, **H**, a material acquires an induced magnetisation **M**

$$\frac{\partial \mathbf{M}}{\partial \mathbf{H}} = \chi \tag{1}$$

where **H** is an axial vector, **M** is a vector and χ, the volume susceptibility, is a second rank tensor. In an isotropic liquid χ becomes a scalar. If the magnetic field is weak enough χ is independent of **H** and so

$$\mathbf{M} = \chi\,\mathbf{H} \tag{2}$$

For $\chi < 0$ the material is called diamagnetic which is valid for liquid crystals with closed shells. The dimensionless volume susceptibility, χ, is weakly temperature dependent because of the temperature dependence of the density. Usually either the mass diamagnetic susceptibility $\chi_g \equiv \chi\,\rho$ in $(m^3\ kg^{-1})$ with the density ρ in $(kg\ m^{-3})$ or the molar diamagnetic susceptibility $\chi_{mol} \equiv \chi_g\,M$ in $(m^3\ mol^{-1})$ with the molar mass M in $(kg\ mol^{-1})$ is used: both are temperature independent. In a vacuum the magnetic induction or magnetic flux density **B** is

$$\mathbf{B} = \mu_0\,\mathbf{H} \tag{3}$$

where μ_0 is the vacuum permeability. Because in magnetochemistry the CGS-system has been widely used, some relations will be presented in TABLE 1. In the literature the molar magnetic susceptibility is mostly given in $(cm^3\ mol^{-1})$.

TABLE 1 Relationships between CGS and SI systems.

Unit	SI	CGS	Multiply CGS for SI units
B: magnetic induction	T, Tesla	G, Gauss	10^{-4} T/G
H: magnetic field strength	A m^{-1}	Oe, Oersted	$(4\pi)^{-1} \times 10^{3}$ (A m^{-1}) (Oe)$^{-1}$
M: magnetisation	A m^{-1}	G, Gauss	10^{3} A m^{-1} G^{-1}
χ: volume susceptibility	1	1	4π
χ_g: mass susceptibility	m^{3} kg^{-1}	cm^{3} g^{-1}	$4\pi \, 10^{-3}$ (m^{3} kg^{-1})(cm^{-3} g)
χ_{mol}: molar susceptibility	m^{3} mol^{-1}	cm^{3} mol^{-1}	$4\pi \, 10^{-6}$ m^{3} cm^{-3}

For a uniaxial system in the laboratory fixed frame (x, y, z) with the z axis parallel to the director **n** we find $\widetilde{\chi}$ with principal components $\widetilde{\chi}_{xx} = \widetilde{\chi}_{yy} = \widetilde{\chi}_{\perp}$, $\widetilde{\chi}_{zz} = \widetilde{\chi}_{\parallel}$; in biaxial nematics $\widetilde{\chi}_{xx} \neq \widetilde{\chi}_{yy}$.

C   ALIGNMENT BY AN EXTERNAL MAGNETIC FIELD

The magnetic free energy density F_{mag} is

$$F_{mag} = -\frac{1}{2} \Delta \widetilde{\chi} \, \mu_0^{-1} (\mathbf{B} \cdot \mathbf{n})^2 \tag{4}$$

where the diamagnetic susceptibility anisotropy $\Delta \widetilde{\chi}$ is defined as

$$\Delta \widetilde{\chi} = \widetilde{\chi}_{\parallel} - \widetilde{\chi}_{\perp} \tag{5}$$

Both components $\widetilde{\chi}_{\parallel}$ and $\widetilde{\chi}_{\perp}$ are negative and small; $\widetilde{\chi}_{\parallel}$ is the component parallel to the director **n**, $\widetilde{\chi}_{\perp}$ is perpendicular to it. $\Delta \widetilde{\chi}$ can be either positive or negative.

(a)   For most calamitic liquid crystals containing one benzene ring or more the lowest energy, i.e. the minimum, is reached when the director is parallel to the magnetic field **H** (FIGURE 1(a)). In this case $\Delta \widetilde{\chi}$ is positive because $|\widetilde{\chi}_{\perp}| > |\widetilde{\chi}_{\parallel}|$. The ring current associated with the delocalised charges in the aromatic system tends to reduce the magnetic flux perpendicular to it and result in a large negative component of diamagnetic susceptibility, $\widetilde{\chi}_{\perp}$. In an ensemble of molecules as in a nematic liquid crystal thus the orientation is always favourable parallel to the benzene ring (see FIGURE 1(a)); the magnetic field avoids orientations perpendicular to the aromatic ring (see FIGURE 1(b)).

(b)   For calamitic liquid crystals not containing benzene units, the sign of $\Delta \widetilde{\chi}$ can also be negative, $|\widetilde{\chi}_{\perp}| < |\widetilde{\chi}_{\parallel}|$. For example the compound 4'-heptyl-bicyclohexyl-4-carbonitrile has a negative $\Delta \widetilde{\chi}$ [1].

(c)   For discotic nematic liquid crystals with aromatic units $\Delta \widetilde{\chi}$ is negative and so the director **n** is oriented perpendicular to the disc plane.

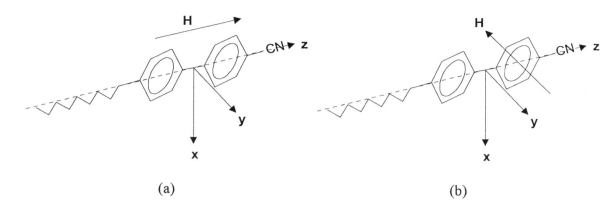

FIGURE 1 Laboratory fixed x, y, z system with magnetic field **H** parallel (a)
or perpendicular (b) to the benzene ring for 4′-alkyl-4-cyanobiphenyl.

It should be mentioned that the orientation of a liquid crystal by an external magnetic field is enhanced due to forces inside the nematic domains. For low molar mass nematics a magnetic field strength of 10^4 to 10^5 A/m (equivalent to a magnetic induction of 10^{-2} to 10^{-1} T) is enough to orient the bulk sample. For polymer nematic liquid crystals [2,3] a higher magnetic field strength is usually needed. In comparison with this, a single molecule, e.g. in the gaseous state, is in essence not aligned by a magnetic field.

D   RELATION BETWEEN THE DIAMAGNETIC SUSCEPTIBILITY ANISOTROPY AND THE ORDER PARAMETER

Many bulk properties reflect the balance between (nematic) order and thermal fluctuations. Whereas the averaged mass or molar diamagnetic susceptibilities $\overline{\chi}$ ($\overline{\chi}_g$ and $\overline{\chi}_{mol}$) are strictly temperature independent and, therefore, constant in the crystalline, liquid crystalline and isotropic states,

$$\overline{\chi} = \frac{1}{3}\sum_r \widetilde{\chi}_{rr} = \frac{1}{3}\left(\widetilde{\chi}_\parallel + 2\widetilde{\chi}_\perp\right) \tag{6}$$

where r denotes x, y, and z, the temperature dependence of $\widetilde{\chi}$ or $\Delta\widetilde{\chi}$ is controlled by the temperature dependent order parameter

$$\overline{P}_2 = \frac{1}{2}\left\langle 3\cos^2\beta - 1\right\rangle \tag{7}$$

Here β is the angle between the director **n** and the molecular long axis ς (see FIGURE 2).

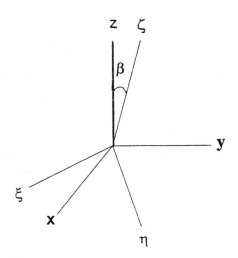

FIGURE 2 The molecular frame ξ, ς, η with respect to the
laboratory frame x, y, z; the director is parallel to z.

The diamagnetic susceptibility anisotropy, $\Delta\chi$, for the crystal is

$$\Delta\chi = N\left[\kappa_{\varsigma\varsigma} - \frac{1}{2}\left(\kappa_{\xi\xi} + \kappa_{\eta\eta}\right) \right] \tag{8}$$

where N is the number of molecules per unit volume and κ is the molecular diamagnetic susceptibility tensor; ς, ξ, and η are the molecule fixed axes. For a uniaxial system $\kappa_{\xi\xi} = \kappa_{\eta\eta}$ and the averaging imposed by the nematic gives

$$\Delta\widetilde{\chi} = \overline{P}_2 \, N\left(\kappa_{\varsigma\varsigma} - \kappa_{\xi\xi}\right) = \overline{P}_2\left(\chi_1 - \chi_t\right) = \overline{P}_2 \, \Delta\chi \tag{9}$$

where χ_1 is the longitudinal diamagnetic susceptibility (parallel to ς), χ_t the transversal component in the plane formed by the ξ,η-axes. In addition

$$\widetilde{\chi}_{\|} = \overline{\chi} + \frac{2}{3}\overline{P}_2 \, \Delta\chi \tag{10}$$

and

$$\widetilde{\chi}_{\perp} = \overline{\chi} - \frac{1}{3}\overline{P}_2 \, \Delta\chi \tag{11}$$

Knowing $\Delta\chi$ we can determine the order parameter $\overline{P}_2$ by measuring $\widetilde{\chi}_{\|}$ ($\Delta\widetilde{\chi} > 0$) or $\widetilde{\chi}_{\perp}$ ($\Delta\widetilde{\chi} < 0$) as a function of temperature.

E METHODS TO OBTAIN THE DIAMAGNETIC SUSCEPTIBILITY ANISOTROPY $\Delta\widetilde{\chi}$

$\Delta\widetilde{\chi}$ is the diamagnetic susceptibility anisotropy of a large number of non-interacting but completely ordered individual molecules. If, for instance, we are interested in numerical values of the orientational order parameter as a function of temperature it is necessary to measure $\widetilde{\chi}_{\parallel}$ and $\overline{\chi}$ or $\widetilde{\chi}_{\perp}$ and $\overline{\chi}$ by knowing $\Delta\chi$. There are several methods to obtain $\Delta\chi$ which include the following.

(a) Measuring the principal susceptibilities of a single crystal by knowing the crystal structure and transforming these components in the molecular frame. The method of calculating molecular susceptibilities from crystal susceptibilities is given in [4]. Unfortunately, only a few mesogenic compounds have been investigated in the solid state. The diamagnetic data for 4,4'-dimethoxyazoxybenzene (PAA) are $\chi_{\eta\eta} = -3.08$, $\chi_{\varsigma\varsigma} = -1.33$, $\chi_{\xi\xi} = -1.13 \; 10^{-9} \; m^3 \; mol^{-1}$ [5,6].

(b) Quantum mechanical calculations although here only a few attempts have been described [7-9].

(c) Calculating $\Delta\chi$ from $\Delta\widetilde{\chi}$ if $\overline{P}_2$ is obtained by a different technique, e.g. ^{1}H-NMR dipolar splitting, ^{13}C-NMR chemical shifts [10,11] or the widely used optical methods via birefringence measurements [6,12]. The last method suffers from the problem of the internal field which must be taken into account, while for the former some parametrisation relating to the molecular geometry is necessary.

(d) The order parameter can be predicted by some models, e.g. by the Maier-Saupe theory [13]; from that $\Delta\chi$ can be calculated from $\Delta\widetilde{\chi}$.

(e) One of the most recommended methods is to calculate the $\Delta\chi$ values by means of incremental systems. Diamagnetism can be calculated using values given by Pascal et al [14,15] for atoms or group susceptibilities. A detailed discussion and some data have been given by Haberditzl [16,17]. In addition, Flygare et al [18-22] have introduced a very efficient tensor incremental system for atoms and chemical groups. These data are reported by Stannarius [23]. Some additional incremental data for some groups are given by Ibrahim and Haase [24], Buka and de Jeu [25], and again given in [23].

(f) A method following Haller et al [26] is widely used, based on extrapolation procedures from the nematic state to $T = 0$ K where $\Delta\widetilde{\chi}$ tends to become $\Delta\chi$. Here, the order parameter is assumed to vary with temperature according to

$$\overline{P}_2 = \left(1 - \frac{T}{T_{NI}}\right)^{\gamma} \tag{12}$$

where T_{NI} is the nematic-isotropic transition temperature, and γ is a fitting parameter, usually $\gamma \sim 0.17$. Using this method, EQN (9) becomes

$$\Delta\widetilde{\chi} = \Delta\chi\left(1 - \frac{T}{T_{NI}}\right)^{\gamma} \tag{13}$$

Similar extrapolation procedures have been used by Ibrahim and Haase [24], Müller and Haase [27], and Leenhouts et al [28]. It must be mentioned that the extrapolation procedures are not theoretically founded, but because of their easy applicability they are often used, although the results must be taken with caution and should always be compared with other methods, e.g. the tensor increment methods of Flygare et al [18-22].

A collection of $\Delta\chi$ data mainly obtained by extrapolation methods is given by Stannarius [23]. Values obtained by different methods or given by different authors for comparable compounds differ by up to 10%.

F   EXPERIMENTAL METHODS TO MEASURE DIAMAGNETIC PROPERTIES

The diamagnetic susceptibility of the nematic state or of single crystals can be obtained in principle by applying any kind of susceptometer, sensitive enough to detect diamagnetism. For single crystals the experiments are more complicated because large crystals are needed, the crystal structure must be known and the principal axes must be evaluated [5]. For liquid crystals the susceptometer must be combined with an oven and temperature control for accuracy of ± 0.01 K is needed. The liquid crystal (ca. 100 mg) has to be transferred to a sample holder, and a first heating and cooling is advisable. To avoid the contribution of paramagnetic oxygen the system must be under vacuum. Among the many methods for magnetic measurements two susceptometers of practical importance will be introduced. In addition NMR techniques allow the determination of diamagnetic data. The diamagnetic susceptibility anisotropy $\Delta\widetilde{\chi}$ can also be obtained by magneto-electro-optical methods.

(a)   Faraday-balance method
    The method [29] is based on the measurement of the force K on a sample with susceptibility χ in an inhomogeneous magnetic field. The Faraday method is very flexible and the accuracy is better than 1%.

(b)   SQUID method (supraconducting quantum interference devices method)
    The advantage of this method [30] is varying the applied field (e.g. zero field) during the measurements, but this is more important for real magnets than for diamagnetic species. The accuracy is slightly better than for a good Faraday-balance apparatus. The high temperature equipment is not necessarily present in all commercial apparatus, and to install this is not easy. This may be the reason that only a few data for diamagnetic liquid crystals using SQUID have been published so far, although the SQUID technique is now used for most magnetic susceptibility measurements.

(c)   NMR techniques
    The method was first applied to liquid crystals by Rose [11]. The splitting $\delta\nu$ of the proton NMR line is used to calculate the susceptibility.

(d)   Magneto-electro-optical methods

Following EQN (4) for the magnetic free energy, a similar expression is valid for the electric case, namely

$$F_{diel} = -\frac{1}{2}\varepsilon_0 \Delta\widetilde{\varepsilon} \cdot (\mathbf{E} \cdot \mathbf{n})^2 \qquad (14)$$

Thus the magnetic and electric torques in a liquid crystal can be balanced via

$$\Delta\widetilde{\chi}\,\mu_0^{-1}\,\mathbf{B}^2 = \Delta\widetilde{\varepsilon}\,\varepsilon_0\mathbf{E}^2 \qquad (15)$$

Here $\Delta\widetilde{\varepsilon}$ is the dielectric susceptibility anisotropy for the nematic, ε_0 is the vacuum permittivity and $\mathbf{E}$ is the applied electric field strength. It follows that $\Delta\widetilde{\chi}$ can be extracted by knowing all of the other data. This is widely used by Freederickz transition experiments [31,32]. Moreover, by knowing the threshold fields the elastic constants K_i can be determined. A good overview of this topic is given in [33]. For lyotropic nematic liquid crystals the elastic constants and diamagnetic susceptibility anisotropy data are given in [34].

Methods (a), (b) and in part (c) give absolute diamagnetic susceptibilities. The diamagnetic susceptibility anisotropy $\Delta\widetilde{\chi}$ can be calculated via EQNS (5), (6), (10), and (11). In comparison to this, the magneto-electro-optical method gives only $\Delta\widetilde{\chi}$ values. $\widetilde{\chi}_{\parallel}$, $\widetilde{\chi}_{\perp}$ and $\Delta\widetilde{\chi}$ (EQNS (9), (10) and (11)) are temperature dependent because of their dependence on the orientational order parameter. Therefore, the data for different compounds or obtained by different methods can only be compared at the same reduced temperature, T/T_{NI}. Some data are given in [23] and [35]. For example, it is of interest to compare the diamagnetic data of the biphenyl, phenylcyclohexane and bicyclohexane compounds presented in TABLE 2.

TABLE 2 Diamagnetic susceptibilities for nematics.

Compound	Ref	$\Delta\chi$ $(10^{-9}\text{ m}^3\text{ kg}^{-1})$	χ $(10^{-9}\text{ m}^3\text{ kg}^{-1})$
4'-heptyl-4-cyanobiphenyl	[36]	1.37	8.66
4-heptyl-(4'cyanophenyl)-cyclohexane	[37]	0.42	9.32
4'-heptyl-bicyclohexyl-4-carbonitrile	[36]	-0.38	8.87

G   CONCLUSION

For calamitic nematics diamagnetic data or diamagnetic anisotropies are available with an accuracy of about 10%. Based on tensor incremental methods diamagnetic data for most new compounds synthesised can be calculated with sufficient accuracy, but there is still a need for accurate data for discotic, polymeric and lyotropic nematics.

ACKNOWLEDGEMENT

Support by the Fonds der Chemischen Industrie is gratefully acknowledged. I thank Dr. M.A. Athanassopoulou for valuable discussions.

REFERENCES

[1]   H. Schad, G. Baur, G. Meier [*J. Chem. Phys. (USA)* vol.71 (1979) p.3174]

[2]   W. Haase, H. Pranoto [*Progr. Colloid Polym. Sci. (Germany)* vol.69 (1984) p.139]

[3]   M.F. Achard, G. Sigaud, F. Hardouin, C. Weill, H. Finkelmann [*Mol. Cryst. Liq. Cryst. (UK)* vol.92 (1983) p.111]

[4]   J.W. Rohleder, R.W. Munn [*Magnetism and Optics of Molecular Crystals* (John Wiley, Chichester, 1992)]

[5]   G. Foëx [*Trans. Faraday Soc. (UK)* vol.29 (1933) p.958]

[6]   W.H. de Jeu [*Physical Properties of Liquid Crystalline Materials* (Gordon and Breach Science Publishers, New York, 1980)]

[7]   M. Schindler, W. Kutzelnigg [*J. Am. Chem. Soc. (USA)* vol.105 (1983) p.1360]

[8]   U. Fleischer, W. Kutzelnigg, P. Lazzeretti, V. Muhlenkamp [*J. Am. Chem. Soc. (USA)* vol.116 (1994) p.5298]

[9]   C. van Wullen, W. Kutzelnigg [*Chem. Phys. Lett. (Netherlands)* vol.205 (1993) p.563]

[10]   J.R. Zimmermann, M.R. Foster [*J. Phys. Chem. (USA)* vol.61 (1957) p.282]

[11]   P.I. Rose [*Mol. Cryst. Liq. Cryst. (UK)* vol.26 (1974) p.75]

[12]   I.H. Ibrahim, W. Haase [*Z. Nat.forsch. A (Germany)* vol.31 (1976) p.1644]

[13]   W. Maier, A. Saupe [*Z. Nat.forsch A (Germany)* vol.15 (1960) p.287]

[14]   A. Pacault, J. Hoarau, A. Marchand [*Adv. Chem. Phys. (USA)* vol.3 (1961) p.171]

[15]   P. Pascal, A. Pacault, J. Hoarau [*C.R. Acad. Sci. (France)* vol.233 (1951) p.1078]

[16]   W. Haberditzl [*Angew. Chem. (Germany)* vol.68 (1966) p.277]

[17]   W. Haberditzl [*Magnetochemie* (Akademie Verlag, Berlin, Germany, 1968)]

[18]   W.H. Flygare, R.C. Benson [*Mol. Phys. (UK)* vol.20 (1971) p.225]

[19]   W.H. Flygare [*Chem. Rev. (USA)* vol.74 (1974) p.685]

[20]   T.D. Gierke, H.L. Tigelaar, W.H. Flygare [*J. Am. Chem. Soc. (USA)* vol.94 (1972) p.330]

[21]   T.D. Gierke, W.H. Flygare [*J. Am. Chem. Soc. (USA)* vol.94 (1972) p.7277]

[22]   T.G. Schmalz, C.L. Norris, W.H. Flygare [*J. Am. Chem. Soc. (USA)* vol.95 (1973) p.7961]

[23]   R. Stannarius [in *Handbook of Liquid Crystals* vol.2A, Eds. D. Demus, J. Goodby, G.W. Gray, H.-W. Spiess, V. Vill (Wiley-VCH, Weinheim, 1998) p.113]

[24]   I.H. Ibrahim, W. Haase [*J. Phys. Colloq. (France)* vol.40 no.C-3 pt.4 (1979) p.164]

[25]   A. Buka, W.H. de Jeu [*J. Phys. (France)* vol.43 (1982) p.361]

[26]   I. Haller, H.A. Huggins, H.R. Lilienthal, T.R. McGuire [*J. Phys. Chem. (USA)* vol.77 (1973) p.950]

[27]   H.J. Müller, W. Haase [*J. Phys. (France)* vol.44 (1983) p.1209]

[28]   F. Leenhouts, W.H. de Jeu, A.J. Dekker [*J. Phys. (France)* vol.40 (1979) p.989]

[29]   A. Weiss, H. Witte [*Magnetochemie* (Verlag Chemie, Weinheim, 1973)]

[30]   J.S. Philo, W.M. Fairbank [*Rev. Sci. Instrum. (USA)* vol.48 (1977) p.1529]

[31]   P.G. de Gennes, J. Prost [*The Physics of Liquid Crystals* (Oxford University Press, 1993)]

[32]   L.M. Blinov, V.G. Chigrinov [*Electrooptic Effects in Liquid Crystal Materials* (Springer-Verlag, New York, 1994)]

[33]   J. Kedzierski et al [*Mol. Cryst. Liq. Cryst. (UK)* vol.249 (1994) p.29]

[34]   A.J. Palangana, A.M. Figueiredo Neto [*Phys. Rev. A (USA)* vol.41 (1990) p.7053]

[35]   D. Dunmur, K. Toriyama [in *Handbook of Liquid Crystals* vol.1, Eds. D. Demus, J. Goodby, G.W. Gray, H.-W. Spiess, V. Vill (Wiley-VCH, Weinheim, 1998) p.204-14]

[36]   J.D. Bunning, D.A. Crellin, T.E. Faber [*Liq. Cryst. (UK)* vol.1 (1986) p.37]

[37]   H. Schad, G. Baur, G. Meier [*J. Chem. Phys. (USA)* vol.70 (1979) p.2770]

6.4   Measurements of paramagnetic properties

P.J. Alonso and J.I. Martínez

February 1999

A   INTRODUCTION

For a long time the majority of liquid crystals known have been diamagnetic since they are built with purely organic molecules. With the appearance of the metallomesogens in the middle of the 1970s [1] the possibility of looking for paramagnetic liquid crystals emerged as a reality. Their interest matches that of magnetic molecular materials. The presence of molecules with an intrinsic magnetic moment combined with the structural order in the mesophase opens new possibilities of finding interesting materials [2]. The use of free radical entities with mesogenic properties is, in principle, another way to reach paramagnetic liquid crystals, but this strategy has scarcely been explored particularly for nematogenic materials [3]. We shall therefore restrict ourselves to the case of metallomesogens when the metal is paramagnetic (e.g. Cu, VO, Mn, Fe). Although several mesogenic lanthanide complexes with exotic magnetic properties have been reported in the last years none of them has a nematic phase.

In FIGURE 1 we show the principal molecular structures of the metallomesogens studied. It is considered classically that mesogenic molecules approximate to cylindrical symmetry and thermotropic

a) SCHIFF'S BASE DERIVATIVES

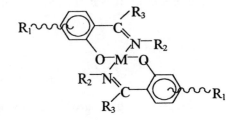

position of R$_1$: 4 or 5
2,4- or 2,5-substitution

b) SALICYLALDEHYDE DERIVATIVES

position of R$_1$: 4 or 5
2,4- or 2,5-substitution

c) β-DIKETONE DERIVATIVES

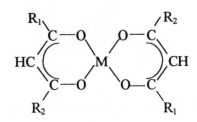

d) ENAMINOKETONE DERIVATIVES

FIGURE 1 Molecular structure of typical paramagnetic nematogenic (a) Schiff's base, (b) salicylaldehyde, (c) β-diketone and (d) enaminoketone derivatives. M is the metal atom.

liquid crystals are divided into two sets: calamitic (rod-like) that give the nematic phase and discotic (disk-like) that form discotic nematic and columnar nematic. Such a classification is based on the ratio between the molecular sizes along the symmetry axis and transverse to it [1] (e.g. all the Schiff's base derivatives are described as calamitic whereas the β-diketones assignment depends on the relative lengths of the R_1 and R_2 substituents). In any event it is worth noting that recently this classification scheme has been criticised and the need to include new molecular shapes is claimed [4].

In spite of its interest we have to note that in the present state of knowledge there are very few data about magnetic properties of the nematic phase. Moreover there is incomplete information on these compounds and so it is difficult to reach general conclusions about the magnetic behaviour of paramagnetic nematics. For that reason this Datareview will be structured according to the properties studied; these can be classified in three categories: (i) magnetic susceptibility, χ, (ii) magnetic field-induced orientation in the mesophase and (iii) EPR spectra.

B MAGNETIC SUSCEPTIBILITY, $\widetilde{\chi}$

A sample in a homogeneous magnetic field, **B**, exhibits a magnetisation, **M**, and the magnetic susceptibility, $\widetilde{\chi}$, is defined by:

$$\mathbf{M} = \widetilde{\chi}\,\mathbf{B} \tag{1}$$

In general $\widetilde{\chi}$ is a second rank symmetric tensor and it contains several additive contributions (diamagnetic, paramagnetic, etc.). In particular the temperature dependence of the paramagnetic contribution follows the Curie law and the components of the susceptibility tensor, $\chi_{\alpha\beta}$, are given by:

$$\chi_{\alpha\beta} = \frac{N\mu_o\mu_B^2\left(\mathbf{g}^2\right)_{\alpha\beta}s(s+1)}{3k_BT} \tag{2}$$

when N is the density of paramagnetic entities, μ_B is the Bohr magneton, $(\mathbf{g}^2)_{\alpha\beta}$ are the components of $\widetilde{g}\cdot\widetilde{g}$, $\widetilde{g}$ being the g-tensor, s is the effective spin, k_B is the Boltzmann constant and T is the temperature. In an isotropic sample (like a powder or a polydomain sample) the susceptibility measured is:

$$\chi = \frac{1}{3}\mathrm{tr}\left(\chi_{\alpha\beta}\right) = \frac{N\mu_o\mu_B^2 g^2 s(s+1)}{3k_BT} = \frac{C}{T} \tag{3}$$

$$g^2 = \frac{1}{3}\left(g_{xx}^2 + g_{yy}^2 + g_{zz}^2\right) \text{ and } C = \frac{N\mu_o\mu_B^2 g^2 s(s+1)}{3k_B} \tag{4}$$

Often the Curie constant, C, is expressed as a function of the effective magnetic moment, μ_{eff}:

$$\mu_{eff} = \mu_B\sqrt{\frac{3k_BC}{N}} = \mu_B g\sqrt{s(s+1)} \tag{5}$$

If an exchange interaction occurs between the paramagnetic entities the Curie-Weiss law replaces the Curie law, and:

$$\chi = \frac{C}{T - \theta} \tag{6}$$

where $\theta > 0$ for a ferromagnetic interaction and $\theta < 0$ for an antiferromagnetic one. In some other cases a temperature independent paramagnetic contribution to the magnetic susceptibility has to be included.

For magnetic molecular materials in general and for nematics in particular the magnetic susceptibility depends on the molecular arrangement since it gives a measurement of the spatially averaged molecular magnetic moment response to the magnetic field [5,6]. So, it is to be expected that the $\tilde{\chi}$-tensor provides information about the structure of the mesophase. For such a reason as well as for the potential applications of paramagnetic liquid crystals the study of their magnetic susceptibility tensor (and the evolution of the magnetic moments) has been claimed to be very interesting.

However, and restricting ourselves to nematogenic compounds, the data on macroscopic magnetic properties (the magnetic susceptibility tensor and magnetisation) are scarce and often misleading. An example of that is provided by the pioneer work by Chandrasekhar et al [7] that reports the thermal evolution of the magnetic susceptibility of a calamitic β-diketone with $R_1 = -\phi-\phi-C_{10}H_{21}$ and $R_2 = -\phi-O-CH_3$ which forms a monotropic nematic phase. The authors find a continuous evolution of the magnetic susceptibility for the isotropic and nematic phases that can be described by a Curie law but the values of the Curie constant in both phases are different. The value in the isotropic phase is twice that in the nematic. This is interpreted as due to the existence of antiparallel magnetic correlation in the nematic associated with the orientational ordering, but as discussed in [6] this interpretation is in conflict with their own EPR data and magnetic field induced orientation effects cannot be ignored.

As for Schiff's base derivatives there are a few reports describing magnetic measurements on some nickel and copper compounds; in TABLE 1 we summarise these results.

In spite of the difficulty of obtaining general conclusions due to the small number of data, it is observed that no difference is found between the chiral nematic and nematic phases. On the other hand a discontinuity in the phase transition is not generally observed. In fact the influence of the orientation effects on the magnetic susceptibility (or magnetic moment) cannot be discarded; this point was discussed in [6].

Galyametdinov et al [11] reported magnetic measurements on a mesogenic μ-oxo-bridged iron(III) complex that forms a nematic phase between 115°C and 159°C. The temperature dependence of the magnetic susceptibility does not show any discontinuity at the phase transition and it can be described by considering the two Fe(III) in the molecule magnetically coupled by an antiferromagnetic interaction with an isotropic coupling constant, J, of -91.8 cm^{-1} as independent magnetic entities.

There are also few reports of polymeric nematogenic compounds. A study of the magnetic susceptibility and of the magnetisation of a nematogenic polyester has been presented [12]. Some differences are found depending on the thermal history of the sample. In any case the behaviour can be described as a pure paramagnet introducing background susceptibility.

TABLE 1 Magnetic susceptibility measurements on
Schiff's base derivatives (R_1 in 4-position, R_3 = H).

M	R_1	R_2			Ref
Cu	$H_{13}C_6$-ϕ-COO-	-ϕ-C_4H_9	N	Discontinuity at the phase transition ($\chi_N > \chi_K$) Curie law (g = 2.01 ±0.02)	[8]
Ni	$H_{15}C_7$-O-ϕ-COO-	$CH_2C^*H(CH_3)C_2H_5$	N*	Jump of μ_{eff} on heating Hysteresis behaviour	[9]
Ni	$H_{25}C_{12}$-O-ϕ-COO-	$CH_2C^*H(CH_3)C_2H_5$	N*	Jump of μ_{eff}	[9]
Cu	$H_{13}C_6(CH_3)$ C*-O- $H_{21}C_{10}$-O-ϕ-COO-	-ϕ-ϕ-O-$C_{12}H_{25}$ -O-$C_{12}H_{25}$	N* N	No discontinuity at the transition Curie law (S = ½, g = 2.23 - 2.24) Temperature independent paramagnetism contribution	[10]

C   MAGNETIC FIELD INDUCED ORIENTATION OF NEMATIC MESOPHASES

In the presence of a magnetic field the sample tends to orient reaching a minimum of its magnetic energy, and the orientation of the director with respect to the magnetic field is determined by the sign of the susceptibility anisotropy. Then, qualitative information about the highest principal component of the susceptibility tensor can be obtained from this type of experiment.

Magnetic field induced orientation of the mesophase has been reported since the early studies of liquid crystals. This behaviour is described as a coupling between the partially averaged magnetic susceptibility and the magnetic field [5,13]. For purely organic diamagnetic nematics the director tends to orient in an applied magnetic field: parallel to the field for most calamitics. A richer behaviour is found for paramagnetic calamitic liquid crystals where parallel ($\Delta\chi > 0$) and perpendicular ($\Delta\chi < 0$) orientations are observed depending on the molecular structure and on the paramagnetic entity. Perpendicular orientation has been detected only in some nematogenic copper compounds. A summary of the results for Schiff's base derivatives of Cu(II) with R_3 = H (see FIGURE 1) is given in TABLE 2. For most of the cases only the sign of $\Delta\chi$ is reported; this information is obtained from the observation of the orientation in the mesophase and not by measuring the magnetic susceptibility tensor, apart from one compound listed in TABLE 2.

As has been noted in [16,20], a parallel or perpendicular orientation of the director results as the consequence of competition between the paramagnetic contribution due to the metal atom and the diamagnetic contribution to the magnetic susceptibility provided by the organic skeleton which is mainly due to the phenyl rings. Whereas the second contribution tends to orient the director parallel to the magnetic field, the first for Schiff's base derivatives of Cu(II) favours a perpendicular orientation of the director. That explains why the diamagnetic contribution wins in the compounds with six phenyl rings in the molecule ($\Delta\chi > 0$). When only four phenyl rings are present in the molecule the orientation behaviour depends on the position of the R_1 substituent. A perpendicular orientation occurs ($\Delta\chi < 0$) if R_1 is in the 4-position whereas if it is in the 5-position there is a parallel orientation.

TABLE 2 Magnetic field induced orientation behaviour of nematic
Schiff's bases derivatives of Cu(II) ($R_3 = H$).

Position of R_1	R_1	R_2	# ϕ	$\Delta\chi$	Ref
4	$H_{21}C_{10}O$-ϕ-COO-	-C_nH_{2n+1} $n = 1 - 10$ $n = 1, 10$	4	<0	[14] [15]
4	$H_{15}C_7O$-ϕ-COO-	-$C_{12}H_{25}$	4	<0	[16]
4	$C_7H_{15}O$-	-ϕ-CN	4	<0	[16]
4	$C_7H_{15}O$-ϕ-COO-	$(-)CH_2C^*H(CH_3)\ C_2H_5$	4	<0	[16]
4	$C_7H_{15}O$-ϕ-COO-	$(\pm)CH(CH_3)(CH_2)_4CH_3$	4	<0	[16]
4	$(+)C_2H_5C^*H(CH_3)CH_2O$	-ϕ-c-C_6H_{10}-C_6H_{13}	4	<0	[16]
5	$H_{21}C_{10}O$-ϕ-COO-	-C_nH_{2n+1} $n = 1 - 3, 6 - 9$ $n = 1$	4	>0	[17] [15]
4	H	-ϕ-COO-ϕ-O C_nH_{2n+1} $n = ?$ $n = 12$	6	>0	[18] [16]
4	$H_{13}C_6O$-ϕ-COO-	-ϕ- C_4H_9	6	>0	[8] [19]
4	$H_{2n+1}C_nO$-ϕ-COO- n=6,4	-ϕ- OCH_3	6	>0	[19]
4	$H_{21}C_{10}O$-ϕ-COO-	-ϕ-OC_nH_{2n+1} $n = 5$ $n = 10$	6	>0	[4] [14]
4	$C_7H_{15}O$-	-ϕ-COO-ϕ-$OC_{12}H_{25}$	6	>0	[16]
4	$(-)C_6H_{13}C^*H(CH_3)O$	-ϕ-ϕ-$OC_{12}H_{25}$	6	24.76×10^{-6} $cm^3\ mol^{-1}$	[16]

In [20] it is pointed out that there is the possibility of some magnetic field orientation effect in some chiral Schiff's base derivatives of Ni(II) compounds in the N phase on the basis of some modification of the effective magnetic moment, but the authors do not discard the existence of some molecular polymorphism.

The same behaviour happens for the Cu(II) salicylaldehyde derivative with $R_1 = H_{21}C_{10}O$-ϕ-COO- in the 5-position and $R_2 = H$ giving a molecular shape similar to the Schiff's base derivatives with four phenyl rings and R_1 in the 5-position.

Considering that the orientation behaviour is a consequence of the anisotropy of the macroscopic magnetic susceptibility, and taking into account all these facts, a discussion of the relationship between the orientation behaviour of the mesophases of the metallomesogens derived from Schiff's bases and the mesophase structure has been presented recently [20]. The model contrasts with that previously introduced in [16] where only the anisotropy of the molecular susceptibility is taken into account and the phase structure is neglected. The authors conclude [20] that, at least for the perpendicular orientation of Schiff's base derivatives, an additional ordering of the molecular short axes has to be included. This provides a strong indication of the existence of biaxial order in the mesophase of this derivative, which is associated with the shape of the 2,4-substituted molecules that can be better described as box shaped rather than calamitic. On this important point further experimental evidence needs to be obtained; at the present it remains an open question.

For the enaminoketone derivatives of Cu(II) there is limited information about the orientation by a magnetic field [16, 21, 22]; a summary of this is given in TABLE 3.

TABLE 3 Magnetic field induced orientation behaviour
of nematic enaminoketone derivatives of Cu(II).

R_1	R_2	$\Delta\chi$	Ref
$H_{19}C_9$-O-ϕ-	-ϕ-OCH_3	>0	[21]
$H_{19}C_9$-O-ϕ-	-ϕ-$OC_{12}H_{25}$	>0	[16]
$H_{19}C_9$-O-ϕ-	-ϕ-OC_8H_{17}	>0	[16]
$H_{13}C_6$-O-N=N-ϕ-	-$C_{18}H_{37}$	>0	[22]
$C_{10}H_{21}$-c-C_6H_{10}-ϕ-	-i- $C_{13}H_{27}$	<0	[16]

As can be observed the type of orientation, parallel or perpendicular, also depends in this case on the number of phenyl rings present in the molecule.

D   ELECTRON PARAMAGNETIC RESONANCE (EPR) STUDIES OF THE NEMATIC MESOPHASE

The EPR spectra are described using the general spin-hamiltonian:

$$H = H_{Ze} + H_{ZFS} + H_{hf} = \mu_B B\tilde{g}s + s\tilde{D}s + I\tilde{A}s \tag{7}$$

where the three terms correspond to the electronic Zeeman, zero-field splitting and hyperfine interaction respectively. **B** is the magnetic field, **s** and **I** the electronic and nuclear spin operators and $\tilde{g}$, $\tilde{D}$ and $\tilde{A}$ denote the tensors that account for the corresponding interactions. For axially symmetric systems the following notation is widely used:

$$g_{xx} = g_{yy} = g_\perp, \qquad g_{zz} = g_\parallel$$

$$A_{xx} = A_{yy} = A_\perp, \qquad A_{zz} = A_\parallel \tag{8}$$

$$D_{xx} = D_{yy} = -\frac{1}{3}D \qquad D_{zz} = \frac{2}{3}D$$

In a system with S = ½, such as Cu^{2+} or VO^{2+} complexes, the zero field contribution is necessarily absent.

When there is a long-range magnetic exchange interaction between the paramagnetic entities then an averaging process causes the collapse of the magnetic hyperfine structure.

A classical type of study in liquid crystals is the temperature dependence of the EPR spectra of paramagnetic probes dissolved in them. For paramagnetic liquid crystals the paramagnetic entity is the mesogenic molecule itself. In this case EPR studies can provide direct information about the structure and the dynamic behaviour of the mesophase [23]. A summary of the main results is given in TABLE 4.

TABLE 4 EPR data of calamitic paramagnetic metallomesogens in the nematic phase.

Type	M	R_1	R_2		Ref
βd	Cu	$H_{21}C_{10}$-φ-φ-	-φ-OCH_3 -φ-OC_2H_5	$g_{\parallel} = 2.261$ $g_{\perp} = 2.062$ hf resolved in the ∥ feature which is not resolved in the solid	[7]
BS	VO	$H_{21}C_{10}$O-φ-COO-	-C_5H_{11}	$g_{\parallel} = 1.945$ $g_{\perp} = 1.989$ $A_{\parallel} = 485$ MHz $A_{\perp} = 180$ MHz Difference if the nematic is reached from the isotropic phase (orientation effects)	[24]
BS	VO	$H_{21}C_{10}$O-φ-COO-	-φ-C_5H_{11}	Orientation effects Only ⊥ features are observed $g_{\perp} = 1.970$ $A_{\perp} = 190$ MHz	[24]
BS	Cu	$H_{21}C_{10}$O-φ-COO-	-C_5H_{11}	$g_{\parallel} = 2.21$ $A_{\parallel} = 500$ MHz No orientation effects The exchange interaction is destroyed in the nematic	[4]
BS	Cu	$H_{21}C_{10}$O-φ-COO-	-φ-C_5H_{11}	The nematic orients with the director ∥ to magnetic field $g_{\perp} = 2.09$	[4]
BS	FeCl	$H_{21}C_{10}$O-φ-COO-	-φ-OC_6H_{13}	Fe(III) in axial coordination $g \approx 2$, $D \approx 0.23$ cm^{-1} Some indications of decomposition	[25]
EN	Cu	$H_{13}C_6$-cC_6H_4-φ-	-C_6H_{13}	Broad EPR signal: $g \approx 2.09$ No analysis	[26]
EN	Cu	$H_{19}C_9$-O-φ-	-φ-OCH_3	The nematic orients with the director ∥ to magnetic field $g_{\perp} = 2.055$	[21]
EN	Cu	$H_{13}C_6$-O-N=N-φ-	-$C_{18}H_{37}$	The nematic orients with the director ∥ to magnetic field	[22]

βd: β-diketonate; BS: Schiff's base; EN: enaminoketone.

The results on EPR spectra of nematics are very poor. It is worth noting that in those copper compounds in which an exchange interaction is present in the more ordered phases (crystalline or smectic), the smeared out hyperfine structure emerges in the nematic [7,25]. It can be understood as a consequence of the lack of translational order in the nematic phase that induces a significant reduction of the exchange interaction. In addition because of the low viscosity of the nematic phase, orientation effects induced by the magnetic field of the spectrometer are observed. This complicates the analysis of the EPR spectra and only partial information is usually obtained.

E   CONCLUSION

The magnetic properties of the paramagnetic liquid crystals play a very special role in their characterisation as well as in determining the properties of the molecular materials that can make them useful for applications. However these studies are very limited and there is a lack of systematic investigation of the paramagnetic properties in selected families of nematogen compounds. Furthermore there is a need to deepen the understanding of the relationship between magnetic behaviour and structure of the mesophase.

REFERENCES

[1]   J.L. Serrano [in *Metallomesogens. Synthesis, Properties and Applications* Ed. J.L. Serrano (VCH, Weinheim, 1996) ch.1 and references therein]

[2]   W. Haase, B. Borchers [*NATO ASI Ser. E, Appl. Sci (Netherlands)* vol.198 (1991) p.245]

[3]   J. Allgaier, H. Finkelmann [*Macromol. Chem. Phys. (Switzerland)* vol.195 (1994) p.1017]

[4]   J.I. Martínez, M. Marcos, J.L. Serrano, V.M. Orera, P.J. Alonso [*Liq. Cryst. (UK)* vol.19 (1995) p.603]

[5]   P.G. de Gennes, J. Prost [*The Physics of Liquid Crystals* 2nd edition (Clarendon Press, Oxford 1993)]

[6]   P.J. Alonso [in *Metallomesogens. Synthesis, Properties and Applications* Ed. J.L. Serrano (VCH, Weinheim, 1996) ch.10]

[7]   S. Chandrasekhar, B.K. Sadashiva, B.S. Srikanta [*Mol. Cryst. Liq. Cryst. (UK)* vol.151 (1987) p.93]

[8]   W. Haase, S. Gehring, B. Borchers [*Mater. Res. Soc. Symp. Proc. (USA)* vol.175 (1990) p.249]

[9]   K. Griesar, Yu. Galyametdinov, M. Athanassopoulou, I.V. Ovchinnikov, W. Haase [*Adv. Mater. (Germany)* vol.6 (1995) p.381]

[10]   M.A. Athanassopoulou, S.Hiller, L.A. Beresnev, Yu.G. Galyametdinov, M. Schweissguth, W. Haase [*Mol. Cryst. Liq. Cryst. (Switzerland)* vol.261 (1995) p.29]

[11]   Yu. Galyametdinov, G. Ivanova, K. Griesar, A. Prosvirin, I.V. Ovchinnikov, W. Haase [*Adv. Mater. (Germany)* vol.4 (1992) p.739]

[12]   P.J. Alonso et al [*Macromolecules (USA)* vol.26 (1993) p.4304]

[13]   G. Vertogen, W.H. de Jeu [*Thermotropic Liquid Crystals. Fundamentals* (Springer Verlag, Berlin, 1987)]

[14]   M. Marcos, P. Romero, J.L. Serrano [*J. Chem. Soc., Chem. Commun. (UK)* (1989) p.1641]

[15]   M. Marcos, P. Romero, J.L. Serrano, J. Barberá, A.M. Levelut [*Liq. Cryst. (UK)* vol.7 (1990) p.251]

[16]   I. Bikchantaev, Yu. Galyametdinov, A. Prosvirin, K. Griesar, E.A. Soto-Bustamante, W. Haase [*Liq. Cryst. (UK)* vol.18 (1995) p.231]

[17]   E. Campillos, M. Marcos, J.L. Serrano [*J. Mater. Chem. (UK)* vol.3 (1993) p.1049]

[18]   I.V. Ovchinnikov, I. Bikchantaev, Yu. Galyametdinov, R.M. Galimov [*24th Congress Ampere* Poznan, 1988, p.567]

[19]   B. Borchers, W. Haase [*Mol. Cryst. Liq. Cryst. (UK)* vol.209 (1991) p.319]

[20]   P.J. Alonso, J.I. Martínez [*Liq. Cryst. (UK)* vol.21 (1996) p.597]

[21]   Yu. Galyametdinov, A.P. Polishchuk, I.G. Bikchantaev, I.V. Ovchinnikov [*J. Struct Chem. (USA)* vol.34 (1993) p.872]

[22]   J. Szydlowska, W. Pyzuk, A. Krówczynski, I. Bikchantaev [*J. Mater. Chem. (UK)* vol.6 (1996) p.733]

[23]   P.J. Alonso [in *Metallomesogens. Synthesis, Properties and Applications* Ed. J.L. Serrano (VCH, Weinheim, 1996) ch.9]

[24]   P.J. Alonso, M.L. Sanjuán, P. Romero, M. Marcos, J.L. Serrano [*J. Phys., Condens. Matter (UK)* vol.2 (1990) p.9173]

[25]   M. Marcos, J.L. Serrano, P.J. Alonso, J.I. Martínez [*Adv. Mater. (Germany)* vol.7 (1995) p.173]

[26]   W. Pyzuk, E. Górecka, A. Krówczynski, J. Przedmojski [*Liq. Cryst. (UK)* vol.14 (1993) p.773]

6.5　Measurements on ferromagnetic mesophases

A.M. Figueiredo Neto

August 1998

A　INTRODUCTION

Ferromagnetic nematic liquid crystals can be obtained by doping [1] nematics with magnetic fluids, usually called ferrofluids [2,3]. These magnetic fluids are colloidal suspensions of small magnetic grains (typical dimension 10 nm), coated with surfactant agents [3] (surfactant ferrofluids) or electrically charged [4] (ionic ferrofluids), dispersed in a polar or non-polar liquid carrier. The doping of thermotropic liquid crystals with ferrofluids is a very delicate task, due to their low solubility in thermotropics [5,6]. On the other hand, lyotropic ferronematics are easily obtained [7,8] by doping lyotropic liquid crystals with water-based ferrofluids. Due to the small anisotropy in the diamagnetic susceptibility of usual nematics ($\Delta\widetilde{\chi} \sim 10^{-6}$ ($4\pi \times 10^{-7}$ MKS)) [9,10], magnetic fields of the order of 0.2 T are needed to orient these materials. After doping with the ferrofluid for a concentration of magnetic grains larger than some critical value, C_m, the ferronematic responds collectively to small magnetic fields, typically of the order of 10^{-2} T. The mechanical coupling [1] between the nematic director and the magnetic grains is responsible for the response of the ferronematic to small magnetic fields. This theoretical prediction was verified experimentally for lyotropic ferronematic liquid crystals [11]. For ellipsoidal grains C_m can be written as [12]:

$$C_m = \frac{\log\left(y + \left(y^2 - 1\right)^{\frac{1}{2}}\right)}{4\pi D^2 b\left(y^2 - 1\right)^{\frac{1}{2}}} \tag{1}$$

where $y = (a/b)$, a and b are the largest and smallest semi-axes of the ellipsoid, respectively, and D is the sample thickness. When a >> b, EQN (1) reduces to $C_m \approx 1/(b\,D^2)$, proposed in [1]. A typical value of C_m [11,12] for lyotropic ferronematics (mixtures based on potassium laurate - decanol - water, placed in sample holders 200 µm thick) is 10^9 grains/cm^3.

A serious problem that experimentalists usually face when they dope lyotropics with a small quantity of ferrofluids (usually 1 µl of concentrated ferrofluid [13] per 1 ml of the liquid crystal) is to control the stability of this doping over time. It is well-known that a magnetic field gradient applied to a ferronematic [14] modifies the bulk concentration of magnetic grains, favouring their agglomeration in an irreversible way. In normal ferrofluids, even a uniform magnetic field favours the formation of small chains. This process of chaining is responsible for the optical birefringence observed [15] in these materials in the presence of a magnetic field. However, in normal ferrofluids this chaining process is reversible, and after the removal of the field the grains became isolated again. Despite the problem of magnetic field gradients, ferrofluid doping is a powerful tool with which to investigate the physical properties of nematic liquid crystals.

B   BIREFRINGENCE MEASUREMENTS

Ferrofluid doping can be used to obtain well-oriented [16] lyotropic nematic samples (both uniaxial and biaxial [17]) for birefringence measurements. FIGURE 1 shows the linear optical birefringence, obtained using laser conoscopy [18], of the lyotropic mixture potassium laurate (KL)/decylammonium chloride (DaCl)/water, doped with a water-based ferrofluid (grains of Fe_3O_4 coated with oleic acid). In this technique, a linearly polarised HeNe laser beam converges in a well-oriented ferronematic sample by using a wide-aperture microscope objective (half-angle of aperture ~50°). The transmitted beam is analysed by a linear polariser and interference fringes are obtained. The phase diagram of this mixture obtained from conoscopic measurements and optical microscopic observations of the textures is shown in FIGURE 2.

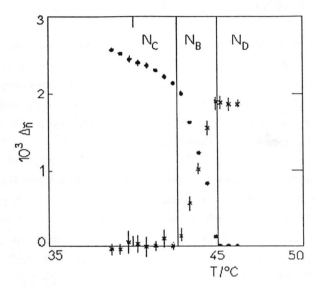

FIGURE 1   The temperature dependence of the birefringence $\Delta\tilde{n}$: (•), $\tilde{n}_1 - \tilde{n}_2$; (x), $\tilde{n}_2 - \tilde{n}_3$.
Extracted from [16]. 1, 2 and 3 are the laboratory frame axes.

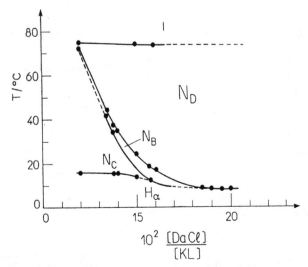

FIGURE 2   Part of the phase diagram for the mixture KL/DaCl/water. H_α, N_C, N_D, N_B and I denote the hexagonal, uniaxial calamitic, uniaxial discotic and biaxial nematic, and isotropic phases, respectively.
Extracted from [16].

The experimental procedure to orient ferronematics in small magnetic fields (H ~ 10^{-2} T) is the following: N_C (uniaxial calamitic phase) orients with the director **n** parallel to the magnetic field (**H**, oriented along the 1-axis of the laboratory frame [16]); the N_D (uniaxial discotic phase) orients with **n** perpendicular to **H**, and to break the degeneracy the sample is rotated (in the presence of **H**) around an axis perpendicular to **H** (3-axis of the laboratory frame [16]); the N_B phase orients with one of the directors parallel to **H** (this director is essentially parallel to the longest axis of the biaxial micelles [8,18]) and to orient the second director the sample oscillates (in the presence of **H**, by about ±20°) around an axis perpendicular to **H** (3-axis of the laboratory frame [16]). The maximum $\Delta\tilde{n}$ measured in ferronematic lyotropics is typically about 10^{-3}.

C   CRITICAL EXPONENT MEASUREMENTS

As the birefringence is proportional to the second rank orientational order parameter [9], birefringence measurements in lyotropic ferronematics have been used to investigate the critical properties of these complex fluids. In the particular case of the lyotropic ferronematic mixtures of KL/1-decanol/water (see FIGURE 3), the critical exponent β for the order parameter was measured [19] along the N_D - N_B transition line, by varying the relative concentration of KL and water, for a fixed concentration of decanol (7.10 weight %). The results are shown in FIGURE 4.

The birefringence was measured using an optical microscope with a Berek compensator. The temperature of the experiments ranged from about 0.01°C to about 0.1°C from the uniaxial-to-biaxial nematic transition temperature. Two distinct intervals of values of β can be distinguished. Values between 0.27 and 0.50 (region I) are measured for lower water concentrations and correspond to a narrow width of stability in temperature for the N_B phase. In region II, $0.52 \le \beta \le 0.7$ is obtained for larger water concentrations where the biaxial region is enlarged. This behaviour seems to be due to the particular characteristics of lyotropic nematics where the micelles change their anisotropic shape (keeping the intrinsically biaxial shape) with temperature and the relative concentrations of the components of the mixture [19-21].

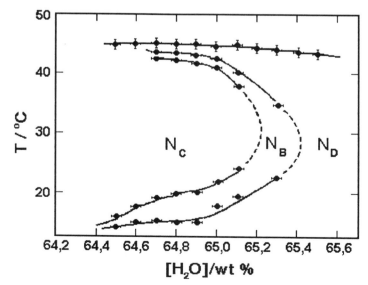

FIGURE 3 Phase diagram for the ferrolyotropic mixture KL/decanol/water. Extracted from [19].

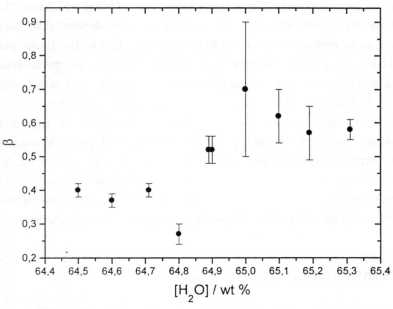

FIGURE 4 The critical exponent β as a function of the water concentration. Extracted from [19].

D   ELASTIC CONSTANT AND DIAMAGNETIC SUSCEPTIBILITY MEASUREMENTS

The bend elastic constant, K_3, and the anisotropic diamagnetic susceptibility, $\Delta\tilde{\chi}$, of the lyotropic nematic mixtures of KL/1-decanol/water and sodium decylsulphate (SdS)/1-decanol/water, doped with water-based surfactant ferrofluid (grains of Fe_3O_4 coated with oleic acid) in the N_C phase were measured [10] using small magnetic fields to produce bend periodic distortions in the sample. Initially the ferronematic was oriented in a planar geometry and then a magnetic field was applied in the direction perpendicular to **n**. For fields smaller than 3×10^{-3} T the wavelength, P, of the periodic bend distortion scales [10] with the magnetic flux density, B, as described by

$$P^{-2} = \mu_0^{-1}\frac{\Delta\tilde{\chi}}{\pi K_3}B^2 \tag{2}$$

where μ_0 is the magnetic permeability of free space.   The results obtained for both ferronematic mixtures are presented in TABLE 1.

TABLE 1 The anisotropic diamagnetic susceptibility, $\Delta\tilde{\chi}$, and bend elastic constant K_3.
Calamitic ferronematic lyotropic mixtures of KL and SdS. Extracted from [10].

Mixture	$\Delta\tilde{\chi} \times 10^{-8}$	$K_3/10^{-11}$ N
KL	0.7 ± 0.2	2 ± 1
SdS	1.2 ± 0.4	2 ± 1

The effective splay-bend elastic constant K_{13} [22] of the lyotropic mixture $KL/DaCl/NH_4Cl/water$ in the ferronematic N_D phase was measured [14] as a function of the ferrofluid concentration at the glass/liquid crystal interface. The order of magnitude of K_{13} is 10^{-11} N, and is positive.

The lyotropic mixture was doped with a water-based surfactant ferrofluid (grains of Fe_3O_4 coated with oleic acid), with different concentrations of the grains. An optical microscope with a Berek compensator was used to measure the optical path difference due to the ferronematic sample. This parameter is directly related to K_{13} by means of a phenomenological model described in [14]. A magnetic field gradient is applied to the sample to produce a migration of the magnetic grains from the bulk to the glass/liquid crystal interface. K_{13} is found to decrease with increasing amount of ferrofluid at the bounding surfaces. The intrinsic contribution to K_{13} is connected to the intermolecular interactions [23] depending on the relative molecular orientation. The extrinsic contribution to K_{13} depends on the amplitude and on the decay length of the surface field, which are expected to be increasing functions of the ferrofluid concentration, at least for low concentrations. Increasing the ferrofluid concentration by a factor of five (initially, the concentration was 0.96×10^8 grains/cm^3), K_{13} decreases by about 6% in the middle of the ferronematic domain.

E   NON-LINEAR OPTICAL MEASUREMENTS

The non-linear refractive indices ($\tilde{n}_2$) of the lyotropic ferronematic of KL/1-decanol/water doped with a surfactant water-based ferrofluid (grains of Fe_3O_4 coated with oleic acid) in the N_C phase were measured [24] using the Z-scan technique, with a timescale of milliseconds. This technique [25] consists of moving a sample along the propagation direction (called z) of the laser beam and detecting the transmittance as a function of z. As the sample moves along the beam focus, self-focusing and defocusing modify the wave front phase, thereby modifying the transmittance. TABLE 2 shows the values of $\tilde{n}_2$ with the laser beam polarised parallel ($\tilde{n}_2^{\parallel}$) and perpendicular ($\tilde{n}_2^{\perp}$) to the director.

TABLE 2 Pulse width in the Z-scan set-up and non-linear refractive indices $\tilde{n}_2^{\parallel}$ and $\tilde{n}_2^{\perp}$.
The sample contained 3 μl of ferrofluid per 1 ml of the lyotropic mixture.
Extracted from [24].

Pulse width/ms	$-\tilde{n}_2^{\parallel} \times 10^{-6}$	$-\tilde{n}_2^{\perp} \times 10^{-6}$
10	1.13 ± 0.08	0.78 ± 0.11
15	1.21 ± 0.04	0.91 ± 0.06
20	1.28 ± 0.04	0.98 ± 0.06
25	1.39 ± 0.05	1.03 ± 0.07
30	1.39 ± 0.05	1.09 ± 0.07
35	1.47 ± 0.04	1.09 ± 0.07
40	1.46 ± 0.04	1.09 ± 0.07
45	1.54 ± 0.04	1.12 ± 0.07
50	1.52 ± 0.04	1.13 ± 0.07

The values of $\tilde{n}_2$ are negative and their modulus increases with the pulse width, reaching a constant value for widths larger than about 35 ms. The non-linear optical birefringence ($\tilde{n}_2^{\parallel}$ - $\tilde{n}_2^{\perp}$) remains almost constant ($\sim -10^{-7}$), for all the pulse widths used. The values of $\tilde{n}_2$ obtained in ferronematics are about ten times larger than those obtained in normal nematics. A possible mechanism which seems to be present in ferronematics is the indirect heating of the lyotropic via the ferrofluid grains. The energy absorbed by the grain increases its temperature and, by heat conduction, heats the lyotropic mixture. The increase in temperature would modify the density of the lyotropic and increases its non-linear response. The value of $\tilde{n}_2$ is a function of the concentration of the magnetic grains in the sample. For the concentrations of 3 and 9 μl of ferrofluid per ml of lyotropic mixture the values of $\tilde{n}_2$ are $-(1.0 \pm 0.1) \times 10^{-6}$ and $-(7.4 \pm 0.1) \times 10^{-6}$, respectively.

F   THERMAL DIFFUSIVITY MEASUREMENTS

The collinear mirage technique [26] was used to measure [27] the thermal diffusivity, $\tilde{\alpha}$, of a lyotropic ferronematic in the N_C phase (KL/1-decanol/water, doped with an ionic water-based ferrofluid of γ-Fe_2O_3 grains). In this technique, a modulated exciting (pump) laser beam heats the sample generating a refractive index gradient within it. A probe laser beam reaches the sample in a direction anti-parallel to the pump beam and is deflected by the gradient in the refractive index due to the thermal gradients generated by the pump beam. The magnitude of the deflection is the mirage effect and is related to the thermo-optical properties of the medium. The initial concentration of grains in the ferrofluid used to dope the lyotropic mixture was 10^{16} grains/cm^3. The concentration in weight percent (wt%) of ferrofluid added to the lyotropic mixture varied from 0.5 to 1.0 wt% and the calamitic ferronematic was oriented by a small magnetic field (9×10^{-2} T). The thermal diffusivities parallel, $\tilde{\alpha}_{\parallel}$, and perpendicular, $\tilde{\alpha}_{\perp}$, to the director are presented in TABLE 3.

TABLE 3   Thermal diffusivities parallel and perpendicular to the director for ferronematic samples with different ferrofluid concentrations. Extracted from [27].

Ferrofluid concentration/wt%	$\tilde{\alpha}_{\parallel}$ /10^{-3} cm^2 s^{-1} ($\pm$5%)	$\tilde{\alpha}_{\perp}$ /10^{-3} cm^2 s^{-1} ($\pm$2%)
0.50	1.55	1.30
0.75	1.40	1.15
1.00	1.40	1.10

G   CONCLUSION

In conclusion, the doping of lyotropic nematics with ferrofluids provides a powerful tool with which to investigate the physical-chemical properties of these complex fluids. Some of the properties of lyotropic ferronematics are very different from the usual properties of undoped lyotropics and pure ferrofluids. In particular, the non-linear optical response of ferronematics is larger than the response of undoped nematics.

REFERENCES

[1]   F. Brochard, P.G. de Gennes [*J. Phys. (France)* vol.31 (1970) p.691]

[2]   S.W. Charles, J. Popplewell [in *Ferromagnetic Materials* vol.2, Ed. E.P. Wohfarth (North-Holland Publishing Co., 1980)]

[3]   A.R.V. Bertrand [*Rev. Inst. Fr. Pet. (France)* vol.25 (1970) p.1]

[4]   R. Massart [*IEEE Trans. Magn. (USA)* vol.17 (1981) p.1247]

[5]   S.H. Chen, N.M. Amer [*Phys. Rev. Lett. (USA)* vol.51 (1983) p.2298]

[6]   I. Potocova et al [*8th Int. Conf. on Magnetic Fluids* Tmisoara, Romania, 29 June-3 July 1998, abstract II-p-28, p.235]

[7]   L. Liébert, A. Martinet [*J. Phys. Lett. (France)* vol.40 (1979) p.L-363]

[8]   A.M. Figueiredo Neto, Y. Galerne, A.M. Levelut, L. Liébert [in *Physics of Complex and Supermolecular Fluids* EXXON Monograph Series, Eds. S. Safran, N.A. Clark (Wiley, New York, 1987) p.347]

[9]   P.G. de Gennes, J. Prost [*The Physics of Liquid Crystals* 2nd edition (Clarendon Press, Oxford, 1993)]

[10]   T. Kroin, A.J. Palangana, A.M. Figueiredo Neto [*Phys. Rev. A (USA)* vol.39 (1989) p.5373]

[11]   A.M. Figueiredo Neto, M.M.F. Saba [*Phys. Rev. A (USA)* vol.34 (1986) p.3483]

[12]   C.Y. Matuo, F.A. Tourinho, A.M. Figueiredo Neto [*J. Magn. Magn. Mater. (Netherlands)* vol.122 (1993) p.53]

[13]   Normal ferrofluids typically have 10^{16} grains/cm^3.

[14]   S. Fontanini, A.L. Alexe-Ionescu, G. Barbero, A.M. Figueiredo Neto [*J. Chem. Phys. (USA)* vol.106 (1997) p.6187]

[15]   J.C. Bacri, V. Cabuil, R. Massart, R. Perzynski, D. Salin [*J. Magn. Magn. Mater. (Netherlands)* vol.65 (1987) p.285]

[16]   E.A. Oliveira, L. Liébert, A.M. Figueiredo Neto [*Liq. Cryst. (UK)* vol.5 (1989) p.1669]

[17]   L.J. Yu, A. Saupe [*Phys. Rev. Lett. (USA)* vol.45 (1980) p.1000]

[18]   A.M. Figueiredo Neto, Y. Galerne, A.M. Levelut, L. Liébert [*J. Phys. Lett. (France)* vol.46 (1985) p.L499]

[19]   L.T. Thieghi, S.M. Shibli, A.M. Figueiredo Neto, V.P. Tolédano, V. Dmitriev [*Phys. Rev. Lett. (USA)* vol.80 (1998) p.3093]

[20]   P. Tolédano, A.M. Figueiredo Neto [*Phys. Rev. Lett. (USA)* vol.73 (1994) p.2216]

[21]   A.M. Figueiredo Neto [in *Phase Transitions in Complex Fluids* Eds. P. Tolédano, A.M. Figueiredo Neto (World Sci. Publ., Singapore, 1998) p.151]

[22]   J. Nehring, A. Saupe [*J. Chem. Phys. (USA)* vol.54 (1971) p.337]

[23]   S. Fontanini, G. Barbero, A.M. Figueiredo Neto [*Phys. Rev. E (USA)* vol.53 (1996) p.2454]

[24]   S.L. Gomez, F.L.S. Cuppo, A.M. Figueiredo Neto, T. Kosa, M. Muramatsu, R. Horowicz [submited for publication]

[25]   M. Sheik-Bahae, A.A. Said, E.W. van Stryland [*Opt. Lett. (USA)* vol.14 (1989) p.955]

[26]   A.C. Boccara, D. Fournier, W. Jackson, A.M. Amer [*Opt. Lett. (USA)* vol.5 (1980) p.377]

[27]   S.M. Shibli, A.L.L. Dantas, D. Walton [*Appl. Phys. Lett. (USA)* vol.72 (1998) p.674]

CHAPTER 7

LINEAR AND NON-LINEAR OPTICAL PROPERTIES

7.1 Refractive indices of nematics

D.A. Dunmur

January 2000

A INTRODUCTION

Apart from high strength materials formed from nematic polymer fibres, most applications of nematic liquid crystals depend on their anisotropic optical properties. As a consequence the refractive indices of nematics are of prime importance in the development of materials for applications. The refractive indices are determined by the molecular polarisability coupled to the orientational order of the mesogens in the liquid crystal phase, so refractive indices can provide a direct probe of the order parameter. Furthermore the optical properties of liquid crystal films are frequently used to determine phase behaviour, identify phase types through characteristic optical textures or explore the properties of defects, and such experiments rely on the anisotropy in the refractive index of the material. The first tool of a liquid crystal scientist is the polarising microscope, which emphasises the importance of optical properties in general and refractive indices in particular to the study of liquid crystals.

The electrooptical effects, which form the basis of many liquid crystal applications, depend on a change in optical properties on application of an electric field. This may be a realignment of the director (Freedericksz transition) or a change in the optical transmission of the film as in devices based on the scattering or absorption of light. These effects depend on the anisotropy of the real part of the refractive index (the birefringence), while effects related to optical absorption depend on the imaginary part of the refractive index. Recent reviews [1,2] deal with many aspects of the optical properties of liquid crystals. Here, theory and experimental techniques will be briefly reviewed, and data presented for common nematic liquid crystals, as well as for materials having some special optical features. Other Datareviews will deal with non-linear optical response [3] and the optical properties of materials for applications [4].

B THEORETICAL BACKGROUND

The refractive indices of anisotropic materials are conveniently represented in terms of the optical indicatrix, the surface of which maps the refractive indices of propagating waves as a function of angle. Solution of Maxwell's equations for an anisotropic medium leads to the result that for a particular wave-normal, two waves may propagate with orthogonal plane-polarisations and different refractive indices. An ellipsoid having as semi-axes the three principal refractive indices defines the optical indicatrix. In general for any wave-normal, the section of the indicatrix perpendicular to the wave-normal direction will be an ellipse, and the semi-axes of this ellipse are the refractive indices of the two propagating waves.

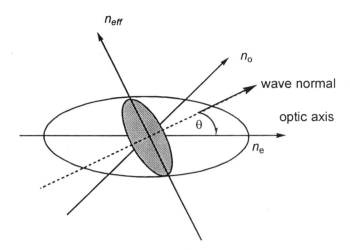

FIGURE 1 Optical indicatrix.

The overall phase symmetry of an anisotropic material will determine the number of independent refractive indices. A cubic or isotropic phase has only one refractive index, which is independent of direction, so the corresponding indicatrix is a sphere, and only one ray will propagate in any direction. Phases with a greater than two-fold symmetry axis have only two principal refractive indices, so the corresponding indicatrix is an ellipsoid of revolution. Thus for rays travelling along the symmetry axis, the perpendicular section is circular, and the corresponding refractive index is called the ordinary refractive index n_o. When viewed along this special direction, known as the optic axis, any material behaves optically like an isotropic material. Such materials are known as uniaxial, since there is a single optic axis. For wave-normals other than along the optic axis, two waves may propagate with different refractive indices. If the wave-normal is perpendicular to the optic axis, then one refractive index is n_o, and the other corresponds to the semi-axis along the symmetry axis, and is called the extraordinary refractive index n_e. In general for wave-normals at some arbitrary direction making an angle of θ to the optic axis, two waves can propagate; one is always an ordinary ray, refractive index n_o, while the other propagates with an effective refractive index given by:

$$n_{eff}^2(\theta) = \left(\frac{\cos^2(\theta)}{n_o^2} + \frac{\sin^2(\theta)}{n_e^2} \right)^{-1} \tag{1}$$

All low molecular weight thermotropic nematic liquid crystals so far characterised are uniaxial, so they have just two principal refractive indices, n_e and n_o, and the difference between these is the birefringence $\Delta n = (n_e - n_o)$. The optic axis is coincident with the director, and the principal refractive indices are often denoted as $ñ_{\parallel} \equiv n_e$ (parallel to the director), and $ñ_{\perp} \equiv n_o$ (perpendicular to the director).

If a phase is of a sufficiently low symmetry, then the corresponding indicatrix will be a general ellipsoid, the three semi-major axes of which will be three principal refractive indices. For such an ellipsoid there are two directions for which the normal section is circular, so such phases have two optic axes and are known as biaxial. The radius of the circular section corresponds to the intermediate principal refractive index. In 1980 [5] a biaxial phase was reported and characterised in a lyotropic nematic consisting of potassium laurate, 1-decanol and deuterated water. The biaxiality derives from asymmetric micelles, which are thought to be ellipsoidal in shape, having different dimensions in three perpendicular directions. In order to probe such biaxiality it is obviously necessary to align the three axes of the micelles. Some nematic polymers have been reported which may be optically biaxial, but the evidence for the claimed biaxial nematic phases of low molecular weight thermotropic mesogens is still controversial [6].

C   EXPERIMENTAL METHODS

Various techniques are available to measure the refractive indices of anisotropic materials. All require that the sample has some defined alignment of the director or optic axis, which ideally should be uniform through the sample. Direct methods, which give the individual refractive indices, rely on measuring the deviation of a ray, or rays, through the sample. Thus conventional refractometers can be used for liquid crystals, provided they are adapted to accommodate aligned samples. The Abbé refractometer is the most popular instrument used to measure the refractive indices of uniaxial nematics. It is customary to align the liquid crystal so that the optic axis is perpendicular to the refracting prism. With this configuration, two refracted beams can be detected corresponding to the ordinary and extraordinary refractive indices. By using polarisers on the incident light and on the exit eye-piece it is possible to select either of the two rays for ease of measurement. It is also possible to measure the refractive indices of liquid crystals from the deviation of a ray through a wedge cell containing the liquid crystal [7]. Once again the liquid crystal must be aligned, but this can be achieved by an external magnetic field. The advantage of this method is that it does not rely on treatments of surfaces to create the liquid crystal alignment, and in addition there is less restriction on the temperature range over which measurements can be made.

If the interest is in measuring just the birefringence, then there are many techniques which rely on optical interference in thin films. For example a wedge (angle ϕ) of liquid crystal, aligned so that the optic axis is perpendicular to the wedge angle, will display a series of equally-spaced fringes when viewed in monochromatic light (wavelength λ) between crossed polarisers aligned at 45° to the optic axis. The separation (Δx) of the fringes is given by [8]:

$$(\tilde{n}_e - \tilde{n}_o) = \frac{\lambda}{\Delta x \tan \phi} \qquad (2)$$

The birefringence having been measured, then an independent measurement of one of the other refractive indices will define the indicatrix.

A very accurate interference method has been reported using a rotating analyser [9,10]. If a birefringent film is illuminated by plane polarised light, with the plane of polarisation making an angle of 45° to the optic axis, then the light transmitted by the film is elliptically polarised with an ellipticity expressed as a phase retardation of δ given by:

$$\delta = \frac{2\pi \ell}{\lambda} (\tilde{n}_e - \tilde{n}_o) \qquad (3)$$

Passing through a quarter wave plate, the elliptically polarised light may be converted to plane polarised light rotated through half the ellipticity angle. This angle can be very accurately determined by using a rotating analyser and measuring the phase angle of the rotated light with respect to a reference beam.

The wavelength dependence of the refractive indices is of importance primarily because it can influence the operational behaviour of liquid crystal devices. A variable wavelength light source can be used to determine the dispersion of the refractive indices with wavelength. A technique has been described [11] which allows the wavelength dependence of the refractive indices to be determined in a single experiment. Insertion of a phase plate into the field of a spectrometer results in a series of interference bands giving a channelled spectrum. By using an aligned liquid crystal as the phase plate, and measuring the positions of the bands, the wavelength dispersion of the refractive indices can be determined. Another method utilising a spectrometer has been reported [12].

The transmissions of an oriented liquid crystal film between parallel polarisers ($T_{\parallel}$) or crossed polarisers ($T_{\perp}$) are given by

$$T_{\parallel} = T_{o}\left[1 - \sin^{2} 2\theta \sin^{2}\left(\delta/2\right)\right]$$

$$T_{\perp} = T_{o}\left[\sin^{2} 2\theta \sin^{2}\left(\delta/2\right)\right]$$

(4)

where θ is the angle between the polariser axis and the optic axis of the aligned liquid crystal film, δ is the phase retardation of the film, and T_{o} is the transmission of the liquid crystal cell without polarisers including reflection and absorption losses at the glass interfaces. Setting θ equal to 45° simplifies the equations, and allows the phase retardation angle to be determined from measurements of $T_{\parallel}$ and $T_{\perp}$:

$$\delta = m\pi + 2\tan^{-1}\sqrt{T_{\perp}/T_{\parallel}} \text{ for integer m even}$$

(5)

$$\delta = (m+1)\pi - 2\tan^{-1}\sqrt{T_{\perp}/T_{\parallel}} \text{ for integer m odd}$$

(6)

The birefringence is in turn related to the phase retardation by

$$\Delta\tilde{n} = \delta\lambda/2\pi\ell$$

(7)

where ℓ is the cell thickness. In order to obtain $\Delta\tilde{n}$ values precise to 0.001 over the wavelength range 400 nm to 900 nm, it is necessary to correct for scattering losses, which are assumed to be proportional to λ^{-4}.

Analysis of the optical properties of liquid crystals in more complex geometries, such as the twisted nematic state, can also be used to determine the refractive indices of nematics. Generalised transmission ellipsometry [13] has been used to measure the principal refractive indices and their dependence on wavelength over the spectral range 350 to 1700 nm for 5CB and mixtures of 5CB with its chiral isomer CB15 (4-cyano-4′-(2methyl)-butylbiphenyl).

318

D   DEPENDENCE OF REFRACTIVE INDICES ON TEMPERATURE AND WAVELENGTH

To understand how the refractive indices of a liquid crystal depend on temperature and wavelength, it is necessary to consider the molecular basis for optical refraction. For isotropic materials, the refractive index can be related to the molecular polarisability through the Lorenz-Lorentz expression:

$$\frac{n^2 - 1}{n^2 + 2} = \frac{N\alpha}{3\varepsilon_o} \tag{8}$$

where N is the number density and α is the polarisability. For anisotropic liquid crystals, this expression must be modified to take account of the polarisability anisotropy of the molecules and their orientational order. The resulting expression is:

$$\frac{\tilde{n}_i^2 - 1}{\tilde{n}^2 + 2} = \frac{N\langle\alpha_{ii}\rangle}{3\varepsilon_o} \tag{9}$$

where $\langle\alpha_{ii}\rangle$ is the component of the polarisability along the direction of the principal refractive index $\tilde{n}_i$, averaged over the orientational distribution of the liquid crystal molecules, and

$$\tilde{n}^2 = \frac{1}{3}\left(\tilde{n}_1^2 + \tilde{n}_2^2 + \tilde{n}_3^2\right) \tag{10}$$

is the mean square refractive index. The use of EQN (9) assumes that the local field [14] is isotropic, which has been shown [15] to be a satisfactory approximation. For uniaxial liquid crystal phases, the average polarisability components can be expressed in terms of the order parameters for the long (z) molecular axis $S_{zz} = S$, and the difference between the order parameters for the short molecular axes (S_{xx}, S_{yy}) expressed as $D = (S_{xx} - S_{yy})$. Hence

$$\langle\alpha_\parallel\rangle = \alpha + \frac{2}{3}\left\{S\left[\alpha_{zz} - \frac{1}{2}(\alpha_{xx} + \alpha_{yy})\right] + \frac{D}{2}(\alpha_{xx} - \alpha_{yy})\right\}$$

$$\langle\alpha_\perp\rangle = \alpha - \frac{1}{3}\left\{S\left[\alpha_{zz} - \frac{1}{2}(\alpha_{xx} + \alpha_{yy})\right] + \frac{D}{2}(\alpha_{xx} - \alpha_{yy})\right\} \tag{11}$$

where $\parallel$ and $\perp$ refer to directions parallel and perpendicular to the extraordinary refractive index. It is now clear that the primary dependence of the refractive indices on temperature comes from the order parameters S and D, with a small dependence on temperature through the number density. Effects of thermal expansion or compression will be reflected in changes of the mean square refractive index.

The variation with wavelength of the refractive indices derives solely from the dependence of the polarisability on wavelength. The polarisability determines the extent to which electromagnetic radiation interacts with the charge distribution within the molecule, and the in-phase and out-of-phase response can be conveniently represented in terms of real and imaginary contributions to the polarisability tensor:

$$\alpha_{\alpha\beta} = \text{Re}\alpha_{\alpha\beta} + i\text{Im}\alpha_{\alpha\beta} \tag{12}$$

The real part determines the refractive index, while the imaginary part measures the extent of absorption of the radiation by the molecule, i.e. the optical absorption coefficient. For visible and UV radiation, the wavelength or frequency (ω) dependence of the polarisability is given by:

$$\text{Re}\,\alpha_{\alpha\beta} = \frac{2}{\hbar}\sum_{n\neq o} f(\omega)\omega_{on}\langle o|\mu_\alpha|n\rangle\langle n|\mu_\beta|o\rangle \tag{13}$$

$$\text{Im}\,\alpha_{\alpha\beta} = -\frac{2}{\hbar}\sum_{n\neq o} g(\omega)\omega_{on}\langle o|\mu_\alpha|n\rangle\langle n|\mu_\beta|o\rangle \tag{14}$$

where $f(\omega)$ and $g(\omega)$ are lineshape functions defined by:

$$f(\omega) = \frac{\omega_{on}^2 - \omega^2}{\left(\omega_{on}^2 - \omega^2\right)^2 + \omega^2\Gamma_{on}^2} \tag{15}$$

$$g(\omega) = \frac{\omega\Gamma_{on}}{\left(\omega_{on}^2 - \omega^2\right)^2 + \omega^2\Gamma_{on}^2} \tag{16}$$

$\hbar\omega_{on}$ is the energy difference between electronic states $|o\rangle$ and $|n\rangle$, and Γ_{on} is the linewidth of the transition $o \rightarrow n$. In the region of an absorption there is an anomaly in both the real and imaginary parts of the polarisability, which is reflected in the wavelength dependence of both the absorption coefficient and the refractive index.

For wavelengths λ far removed from an absorption, EQN (15) can be approximated to

$$f(\lambda) = \text{constant} \times \left(\frac{\lambda_{on}^2\lambda^2}{\lambda^2 - \lambda_{on}^2}\right) \tag{17}$$

where $\lambda_{on} \equiv \lambda_j$ is the wavelength corresponding to a particular absorption labelled (j). Combining EQNS (9), (13) and (17) results in parametrised expressions for the wavelength dependence of the principal refractive indices as:

$$\tilde{n}_i(\lambda, T) \approx 1 + \sum_j G_j^i(T)\left(\frac{\lambda_j^2\lambda^2}{\lambda^2 - \lambda_j^2}\right) \tag{18}$$

$$= \tilde{n}_{oi} + \sum_{j=1,2} G_j^i(T)\left(\frac{\lambda_j^2\lambda^2}{\lambda^2 - \lambda_j^2}\right) \tag{19}$$

where i = o,e refers to the ordinary and extraordinary refractive indices. These expressions [16] are used to fit the variation of refractive indices with wavelength for non-absorbing liquid crystals in the wavelength regions of visible and near infrared. Depending on the material, it may be sufficient to include a single term in the summation in EQN (18), the 'single band model', or two terms in EQN (19), the so-called 'three band model'.

E   RESULTS FOR SELECTED MESOGENS

Various applications of liquid crystals require specific optical properties such as high or low optical anisotropy, small or large temperature dependence, and a weak dependence on wavelength. These optical properties can be manipulated through the mesogenic molecular structure, or by using mixtures of different mesogens. The magnitudes of the principal refractive indices are determined by the molecular electronic polarisability and the appropriate order parameters. Thus increasing the electronic conjugation along the major axis of a rod-like molecule will tend to increase the refractive index parallel to the director. Lateral substitution of electron-withdrawing groups, such as F or CN, will reduce the major axis polarisability, and reduce the refractive index anisotropy. Similarly for disc-like molecules, the nematic discotic phases of which have a negative birefringence, manipulation of the molecular structure can be used to adjust the values of the mesophase refractive indices. However, it is not just the molecular structure that determines the anisotropic optical properties: the orientational order is also an important factor. Control of this by molecular engineering is much more difficult.

In this section results will be given for the refractive indices of some specimen liquid crystals. First results for standard materials will be given to illustrate the temperature and wavelength dependence of the refractive indices of nematic liquid crystals. Additionally selected results are given for a variety of other mesogens, to illustrate the effect of molecular structure modifications on the refractive indices. Further results for materials of importance in display applications will be given in Chapter 11.

E1   Standard Materials

E1.1   4-alkyl-4′-cyanobiphenyls

The most widely studied nematic mesogens are the 4-alkyl-4′-cyanobiphenyls, and there are a number of measurements of refractive indices of various homologues reported in the literature. Of these the standard is 5CB, for which many data are available. Those selected for this review are from a variety of sources, which are mostly in good agreement. While the refractive indices of liquid crystals can be measured to very high precision <0.0001, results from different methods or laboratories may differ in the third decimal place due to the use of different samples, or slight uncertainties in temperature. Thus the results quoted here should be accurate at worst to ±0.1%.

TABLE 1 and FIGURE 2 give measurements for the ordinary ($\tilde{n}_o$) and extraordinary ($\tilde{n}_e$) refractive indices as a function of temperature for 5CB at a wavelength of 589 nm. These results are taken from [17].

TABLE 1 Refractive indices of 5CB as a function of shifted temperature
(λ = 589 nm) (*values are for the isotropic phase).

Temperature (T - T_{NI})/°C	$\tilde{n}_o$	$\tilde{n}_e$	(T - T_{NI})/°C	$\tilde{n}_o$	$\tilde{n}_e$
-34.2	1.5304	1.762	-8.93	1.5355	1.711
-32.64	1.5305	1.759	-8.44	1.5359	1.71
-30.8	1.5307	1.757	-7.48	1.5367	1.706
-29.57	1.5308	1.755	-6.52	1.5376	1.703
-28.02	1.531	1.752	-5.88	1.5383	1.7
-26.18	1.5311	1.749	-5.61	1.5386	1.699
-24.33	1.5312	1.746	-4.28	1.5404	1.693
-22.79	1.5314	1.743	-3.79	1.5414	1.691
-20.9	1.5317	1.74	-2.59	1.544	1.687
-19.1	1.5319	1.736	-2	1.5456	1.684
-17.03	1.5323	1.732	-1.1	1.5494	1.676
-15.28	1.5326	1.729	-0.58	1.5532	1.669
-13.58	1.5331	1.724	1	1.583*	
-12.1	1.5336	1.721	5	1.582*	
-10.5	1.5345	1.715	9	1.581*	

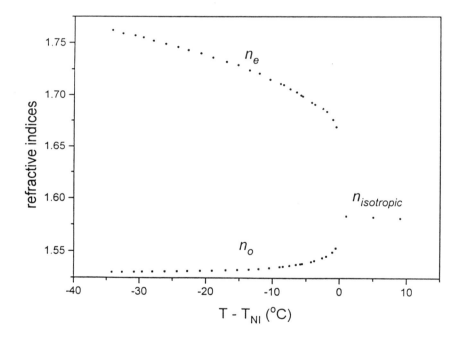

FIGURE 2 Refractive indices of 5CB as a function of temperature (λ = 589 nm).

In FIGURE 3 is given the variation with wavelength of the refractive indices of 5CB (T_{NI} = 35.2°C) at a shifted temperature of (T - T_{NI}) = -10°C; these data are taken from [18]. The data are accurately fitted by EQN (19) including two terms in the wavelength expansion, λ_1 = 210 nm and λ_2 = 282 nm for both the ordinary and extraordinary refractive indices. The parameters of the fitting are given in the first data row of TABLE 2: parameters for other temperatures are included for completeness.

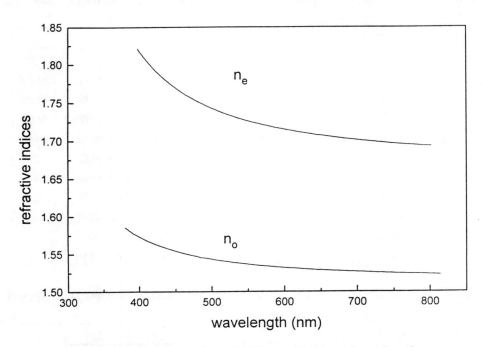

FIGURE 3 Refractive indices of 5CB as a function of wavelength.

TABLE 2 Wavelength dependence of the refractive
indices of 5CB: fitting parameters for EQN (19).

Temperature	$\tilde{n}_e$			$\tilde{n}_o$		
$T/°C$	$\tilde{n}_{0e}$	$G_1{}^e/10^{-6}\ nm^{-2}$	$G_2{}^e/10^{-6}\ nm^{-2}$	$\tilde{n}_{0o}$	$G_1{}^o/10^{-6}\ nm^{-2}$	$G_2{}^o/10^{-6}\ nm^{-2}$
25.1	1.4552	2.3250	1.3970	1.4136	1.3515	0.4699
27.2	1.4588	2.2051	1.3866	1.4137	1.3631	0.4789
29.9	1.4603	2.0874	1.3544	1.4138	1.3842	0.4982
32.6	1.4588	1.9206	1.3248	1.4157	1.4135	0.5272
34.8	1.4680	1.7205	1.1406	1.4165	1.4574	0.5824

Refractive indices for the nematic phases of other homologues of the alkylcyanobiphenyls are given in TABLES 3 - 5 for a wavelength of 633 nm [19].

TABLE 3 Refractive indices of 6CB as a function of shifted temperature (λ = 633 nm).

Temperature (T - T_{NI})/°C	$\tilde{n}_o$	$\tilde{n}_e$	(T - T_{NI})/°C	$\tilde{n}_o$	$\tilde{n}_e$
-13.35	1.5301	1.6978	-3.6	1.5373	1.6677
-12.75	1.5305	1.6954	-3.15	1.5381	1.6655
-11.8	1.5309	1.6930	-2.7	1.5389	1.6629
-10.75	1.5313	1.6909	-2.25	1.5398	1.6604
-9.8	1.5318	1.6883	-1.55	1.5419	1.6548
-8.8	1.5324	1.6859	-1.2	1.5432	1.6516
-7.8	1.5331	1.6832	-0.6	1.5468	1.6432
-6.8	1.5339	1.6801	-0.15	1.5499	1.6355
-5.8	1.5347	1.6768	0.9	1.5773*	
-4.9	1.5355	1.6734	1.65	1.5769*	
-4.05	1.5366	1.6698	2.7	1.5765*	

TABLE 4 Refractive indices of 7CB as a function of shifted temperature (λ = 633 nm).

Temperature (T - T_{NI})/°C	$\tilde{n}_o$	$\tilde{n}_e$	(T - T_{NI})/°C	$\tilde{n}_o$	$\tilde{n}_e$
-18.6	1.5157	1.6991	-5.05	1.5205	1.6697
-17.4	1.5158	1.6974	-3.95	1.5215	1.6657
-15.95	1.5160	1.6949	-3.2	1.5230	1.6604
-14.5	1.5162	1.6925	-1.7	1.5242	1.6553
-13.35	1.5165	1.6903	-1.05	1.5267	1.6503
-11.7	1.5169	1.687	-0.65	1.5278	1.6473
-10.4	1.5172	1.6842	-0.2	1.5297	1.6427
-8.9	1.5179	1.6808	0.85	1.5652*	
-7.65	1.5186	1.6773	1.65	1.5648*	
-6.2	1.5194	1.6736	2.6	1.5642*	

TABLE 5 Refractive indices of 8CB as a function of shifted temperature (λ = 633 nm).

Temperature (T - T_{NI})/°C	$\tilde{n}_o$	$\tilde{n}_e$	(T - T_{NI})/°C	$\tilde{n}_o$	$\tilde{n}_e$
-7.1	1.514	1.6676	-2.5	1.5212	1.6467
-6.55	1.5153	1.6644	-1.8	1.5228	1.6423
-6.0	1.5161	1.6623	-1.1	1.5252	1.6364
-5.5	1.5167	1.6603	-0.55	1.5281	1.6295
-4.95	1.5174	1.6583	0.3	1.5606*	
-4.35	1.5183	1.6558	1.25	1.5602*	
-3.65	1.5193	1.6525	2.3	1.5596*	
-3.45	1.5204	1.6489	3.4	1.5591*	

E1.2 Cyanophenylcyclohexanes and cyanobicyclohexanes

A group of widely studied nematic materials are the analogues of the cyanobiphenyls with one or both rings replaced by a cyclohexane ring. The effect of this structural modification is to reduce the refractive indices and birefringence, and also to reduce the wavelength dependence of the refractive indices in the visible region. This is illustrated in TABLE 6, where results are given for the refractive indices of 7CB, PCH7 and CCH7 [20].

TABLE 6 Refractive indices of PCH7, CCH7 and 7CB for comparison,
as a function of shifted temperature for a wavelength of 589 nm.

Temperature	7CB	$T_{NI} = 42°C$	PCH7	$T_{NI} = 57°C$	CCH7	$T_{NI} = 83°C$
$(T - T_{NI})/°C$	$ñ_o$	$ñ_e$	$ñ_o$	$ñ_e$	$ñ_o$	$ñ_e$
-33					1.457	1.515
-28					1.455	1.512
-27			1.484	1.596		
-23					1.454	1.510
-22			1.483	1.590		
-18					1.453	1.506
-17			1.482	1.584		
-13					1.453	1.503
-12	1.523	1.697	1.482	1.577		
-8					1.452	1.497
-7	1.526	1.683	1.482	1.568		
-3					1.452	1.491
-2	1.534	1.658				

The dispersion of refractive indices has been measured for one of the cyanophenylalkylcyclohexanes, PCH5 (cyanophenylpentylcyclohexane) [18], and fitting parameters according to EQN (19) are given in TABLE 7.

TABLE 7.

Temperature	$ñ_e$			$ñ_o$		
$T/°C$	$ñ_{0e}$	$G_1^e/10^{-6}$ nm^{-2}	$G_2^e/10^{-6}$ nm^{-2}	$ñ_{0o}$	$G_1^o/10^{-6}$ nm^{-2}	$G_2^o/10^{-6}$ nm^{-2}
25.0	1.4560	1.7501	1.1423	1.3869	1.4874	0.5303
30.4	1.4566	1.7355	1.1251	1.3869	1.4866	0.5293
34.8	1.4551	1.7211	1.1109	1.3866	1.4900	0.5300
40.1	1.4531	1.7034	1.0830	1.3862	1.4937	0.5327
45.4	1.4499	1.6850	1.0548	1.3849	1.5133	0.5432
50.8	1.4454	1.6469	1.0166	1.3838	1.5408	0.5562
53.4	1.4429	1.6101	0.9606	1.3844	1.5603	0.5602

E1.3   Schiff's bases

Some of the first room temperature nematic liquid crystals discovered were Schiff's bases, and these were subjected to a number of physical measurements. However, as standard materials they were very unsatisfactory, since they readily underwent thermal decomposition and hydrolysis, and their high electrical conductivity often resulted in electrolytic decomposition. These materials also tended to decompose over time, or under exposure to light. The material 4-methoxybenzylidene-4'-butylaniline (MBBA) was widely used for a variety of physical experiments, and refractive indices are given in TABLE 8 for a wavelength of 589 nm [21].

TABLE 8 Refractive indices of MBBA as a function of shifted temperature.

Temperature $(T - T_{NI})/°C$	$ñ_o$	$ñ_e$	$(T - T_{NI})/°C$	$ñ_o$	$ñ_e$
- 20.8	1.5485	1.7805	-7.2	1.5561	1.7411
-14.8	1.5507	1.7647	-5.2	1.5584	1.7334
-12.8	1.5518	1.7598	-3.2	1.5616	1.7241
-10.8	1.5531	1.7531	-2.2	1.5638	1.7188
-9.0	1.5544	1.7474	-1.0	1.5676	1.7096

E1.4   4,4'-dimethoxyazoxybenzene (PAA)

One of the first nematic liquid crystals to be studied in any detail was 4,4'-dimethoxyazoxybenzene (PAA). This material is only nematic at elevated temperatures (T_{NI} = 137°C), but is relatively stable, and is a rare example of a nematogen for which the shape is almost independent of internal motions (i.e. there is no attached flexible alkyl chain), and for which measurements have been made. Results are given in TABLE 9 for refractive indices as a function of temperature and wavelength [22].

TABLE 9 Refractive indices of PAA as a function of shifted temperature and wavelength.

Temperature $(T - T_{NI})/°C$	λ = 644 nm		λ = 589 nm		λ = 509 nm		λ = 480 nm	
	$ñ_o$	$ñ_e$	$ñ_o$	$ñ_e$	$ñ_o$	$ñ_e$	$ñ_o$	$ñ_e$
-42	1.550	1.864	1.558	1.894	1.581	1.974	1.593	2.029
-37	1.550	1.857	1.558	1.885	1.581	1.964	1.594	2.018
-27	1.550	1.839	1.559	1.866	1.588	1.942	1.595	1.995
-17	1.552	1.819	1.562	1.843	1.585	1.917	1.600	1.967
-12	1.554	1.807	1.565	1.829	1.589	1.902	1.604	1.949
-7	1.558	1.791	1.570	1.813	1.593	1.882	1.611	1.928
-4	1.562	1.778	1.574	1.800	1.598	1.865	1.617	1.910
-1	1.568	1.762	1.581	1.779	1.605	1.843	1.626	1.885
1	1.629*		1.642*		1.682*		1.707*	

E1.5   Discotic materials

There are as yet no accepted standard materials for discotic nematic liquid crystals. It is expected that the birefringence of these materials will be negative, and they have already found application as optical compensation films for liquid crystal displays [23]. In order to illustrate an example of the optical properties of discotic nematic liquid crystals, we give in TABLE 10 measurements of the ordinary and extraordinary refractive indices of hexakis[(4-octylphenyl)ethinyl]benzene (T_{NI} = 100°C). These are taken from [24] and were measured for a wavelength of 589 nm.

TABLE 10 Refractive indices of hexakis[(4-octylphenyl)ethinyl]benzene
as a function of shifted temperature.

Temperature $(T - T_{NI})/°C$	$ñ_o$	$ñ_e$	$(T - T_{NI})/°C$	$ñ_o$	$ñ_e$
-35	1.686	1.493	-10	1.662	1.520
-30	1.681	1.498	-5	1.650	1.531
-25	1.677	1.500	0	1.610*	
-20	1.671	1.507	5	1.608*	
-15	1.667	1.511			

E2 Materials for High and Low Birefringences

For display applications it is often necessary to tune the birefringence to suit the particular display type, and this can be achieved by mixing materials of different refractive indices. Also some display applications require materials of high birefringence, while others utilise materials of low birefringence. However, optimisation of the optical properties is rarely sufficient to produce a material suitable for a display application, and the desired refractive indices have to be reached in mixtures which also have the correct dielectric, elastic and viscous behaviour. Thus the refractive indices of very many compounds have been measured to provide a library of materials of different structures with particular optical properties. A large number of these are listed in Datareview 11.1 in this volume. The refractive indices and birefringences of these compounds have mostly been determined by extrapolation of measurements on mixtures.

In this section, the refractive indices of a few compounds of high and low birefringence will be reviewed to illustrate the molecular design principles that can be used to achieve particular birefringences.

Since the order parameters of nematics tend to be almost the same at the same reduced temperature T/T_{NI}, the predicted result of the Maier-Saupe molecular field theory, there is little scope for increasing the birefringence through the order parameter. In order to produce nematic materials of high birefringence it is necessary to design mesogens with a high polarisability anisotropy. This can be achieved by extending the length of the electron conjugation parallel to the major axis of rod-like molecules. Thus mesogens with three conjugated phenyl rings (terphenyls) will have higher birefringences than biphenyls, but at the expense of greatly increased nematic-isotropic transition temperatures. Increased electron conjugation can also be introduced into molecules through unsaturated double or triple carbon-carbon bonds. The refractive indices of some recently reported materials of high birefringence are given in TABLE 11. These are taken from [25] and measurements were made at a wavelength of 514 nm.

TABLE 11 Refractive indices of high birefringence materials.

Temperature $(T - T_{NI})$/°C	I $T_{NI} = 112$°C		II $T_{NI} = 160$°C	
	$\tilde{n}_o$	$\tilde{n}_e$	$\tilde{n}_o$	$\tilde{n}_e$
-20	1.52	1.83	1.60	1.97
-15	1.53	1.82	1.61	1.94
-10	1.54	1.81	1.61	1.92
-5	1.54	1.78	1.62	1.90
0	1.55	1.75	1.63	1.87

Compound I =

Compound II =

Low birefringence materials can be created by lateral substitution of the phenyl rings of rod-like molecules, usually using fluorine atoms, or by the replacement of aromatic rings by heterocycles containing nitrogen [4] or oxygen [4,26]. An interesting new approach to the design of low birefringence nematic liquid crystals is using the carborane ring as part of the mesogen [27]. Measurements at a wavelength of 589 nm of refractive indices for a carboranephenyl ester liquid crystal (Compound III) are given in TABLE 12; also included in TABLE 12 are values for a carbon analogue bicyclooctanephenyl ester (Compound IV).

TABLE 12 Refractive indices of low birefringence materials.

Temperature $(T - T_{NI})$/°C	Compound III $T_{NI} = 36.1$°C		Compound IV $T_{NI} = 93.5$°C	
	$\tilde{n}_o$	$\tilde{n}_e$	$\tilde{n}_o$	$\tilde{n}_e$
-35			1.476	1.550
-30.2			1.475	1.547
-26.9			1.474	1.545
-22.9			1.473	1.542
-19			1.472	1.538
-15			1.472	1.535
-14			1.472	1.534
-13.3	1.514	1.571		
-13			1.471	1.533
-12.2			1.471	1.532
-11			1.471	1.531
-10			1.471	1.530
-9.9	1.514	1.570		
-8.4	1.514	1.568		
-7.4	1.513	1.566		
-5.2	1.513	1.564		
-2.3	1.513	1.562		
-1.4	1.514	1.559		
-0.4	1.514	1.556		
0.6	1.527*			
1.6	1.527*			
3.3	1.526*			

Compound III =

C_5H_{11}—C-B ... C—C(=O)—O—⬡—OC_5H_{11}

Compound IV =

C_5H_{11}—⬡—C(=O)—O—⬡—OC_5H_{11}

E3 Other Materials

Almost any new mesogenic material will be assessed from the point of view of its optical properties, and so it is impossible to review all the data in the literature in a short article. The important issues of high and low birefringence have been considered, but equally of importance is the achievement of particular optical properties and other physical properties in the same materials. Low viscosity is an important requirement for most displays, and an important range of materials having very low rotational viscosities and also near zero dielectric anisotropies have been reported. These are based on a flexible cyclohexylbiphenyl alkane core. Results for the refractive indices of a single example (I52: pentylcyclohexylethylfluorobiphenylethane, T_{NI} = 103°C) are given in TABLE 13: these are taken from [7], and are for a wavelength of 633 nm.

TABLE 13 Refractive indices of pentylcyclohexylethylfluorobiphenylethane (T_{NI} = 103°C) as a function of shifted temperature for a wavelength of 633 nm.

Temperature ($T - T_{NI}$)/°C	$ñ_o$	$ñ_e$	$T - T_{NI}$/°C	$ñ_o$	$ñ_e$
-63	1.495	1.635	-23	1.487	1.605
-58	1.494	1.632	-18	1.486	1.602
-53	1.493	1.628	-13	1.485	1.597
-48	1.492	1.624	-8	1.486	1.591
-43	1.491	1.621	-3	1.487	1.581
-38	1.490	1.617	0	1.497	1.560
-33	1.489	1.613	7	1.510*	
-28	1.488	1.609			

E4   Other Wavelengths

All of the data given so far in this Datareview refer to wavelengths in the visible region of the spectrum, which are of course most relevant to displays. However, there are a number of applications of liquid crystals that use infrared or longer wavelength radiation, so the optical properties of liquid crystals in these regions of the electromagnetic spectrum are also of interest. Unfortunately there have been few systematic studies of the optical properties of nematics at infrared and far infrared wavelengths. The transmissivity of a thin film of 5CB has been measured [28] as a function of the angle of incidence, for three sub-millimetre wavelengths, and fitting these measurements to the optical properties of the film yields the principal refractive indices given in TABLE 14.

TABLE 14 Submillimetre refractive indices of 5CB at room temperature.

λ = 435 µm		λ = 215 µm		λ = 118 µm	
$ñ_o$	$ñ_e$	$ñ_o$	$ñ_e$	$ñ_o$	$ñ_e$
1.819	2.017	1.935	2.035	2.119	2.277

There have been measurements of refractive indices of nematic mixtures of mesogens at longer wavelengths [29], and the birefringences at 30 GHz (λ = 10 mm) are in the region of 50 - 70% smaller than those measured for visible light (589 nm). All liquid crystals absorb in the infrared region of the spectrum, although transmission windows can be identified for specific materials, which have been utilised in a variety of applications. The materials mostly used are mixtures, and there are few comprehensive data available for the refractive indices of pure mesogens in the infrared region of the spectrum. Measurements have been made on nematic mixtures for wavelengths in the range 0.7 µm to 1.0 µm [30]; at room temperature for biphenyl mixtures $\Delta n \sim 0.2$, while for a mixture of fluorodiphenyldiacetylenes values of $\Delta n \sim 0.3$ have been reported.

F CONCLUSION

The optical properties of liquid crystals are of great importance both from a fundamental point of view, and because they are of direct relevance to liquid crystal optical devices. The most important aspect of the optical properties of liquid crystals is their optical anisotropy or birefringence. Values for the birefringences and refractive indices of nematic liquid crystals are comparable to corresponding values for solid crystals. However the optical response of liquid crystals to changes of temperature or applied electric and magnetic fields is dramatically different from the behaviour of solid crystals. The responses of liquid crystals to external forces reflect their particular structural features, which combine anisotropy with fluidity and molecular mobility. It is these features which have provided the stimulus for research, and also resulted in many real and potential applications.

REFERENCES

[1] R. Sambles, S. Elston (Eds.) [*The Optics of Liquid Crystals* (Taylor and Francis, London, 1997)]

[2] I.C. Khoo, S.-T. Wu [*Optics and Nonlinear Optics of Liquid Crystals* (World Scientific, RiverEdge, New Jersey, 1993)]

[3] I.C. Khoo [Datareview in this book: *7.4 Non-linear optical properties of nematic liquid crystals*]; Y.R. Shen [Datareview in this book: *7.5 Giant non-linear optical effects in nematic mesophases*]

[4] T. Ohtsuka, H. Ohnishi, H. Takatsu [Datareview in this book: *11.1 Optical properties of nematics for applications*]

[5] L.J. Yu, A. Saupe [*Phys. Rev. Lett. (USA)* vol.45 (1980) p.1000]

[6] D.W. Bruce, G.R. Luckhurst, D.J. Photinos (Eds.) [*Mol. Cryst. Liq. Cryst. (Switzerland)* vol.323 (1998) p.154]

[7] D.A. Dunmur, D.A. Hitchen, Hong Xi-Jun [*Mol. Cryst. Liq. Cryst. (UK)* vol.140 (1986) p.303]

[8] I. Haller, H.A. Huggins, M.J. Freiser [*Mol. Cryst. Liq. Cryst. (UK)* vol.16 (1972) p.53]

[9] K.-C. Lim, J.T. Ho [*Mol. Cryst. Liq. Cryst. (UK)* vol.47 (1978) p.173]

[10] D.W. Bruce, D.A. Dunmur, P.M. Maitlis, M.M. Manterfield, R. Orr [*J. Mater. Chem. (UK)* vol.1 (1991) p.255]

[11] M. Warenghem, G. Joly [*Mol. Cryst. Liq. Cryst. (UK)* vol.207 (1991) p.205]

[12] S.T. Wu, U. Efron, L.D. Hess [*Appl. Opt. (USA)* vol.23 (1984) p.3911]

[13] M. Schubert et al [*J. Opt. Soc. Am. (USA)* vol.13 (1996) p.1930]

[14] A.A. Minko, V.S. Rachevich, S.Ye. Yakovenko [*Liq. Cryst. (UK)* vol.4 (1989) p.1]

[15] D.A. Dunmur, R.W. Munn [*Chem. Phys. (Netherlands)* vol.76 (1983) p.249]

[16] S.T. Wu [*J. Appl. Phys. (USA)* vol.69 (1991) p.2080]

[17] J.W. Baran, F. Borowski, J. Kedzierski, Z. Raszewski, J. Zmija, K. Sadowska [*Bull. Pol. Acad. Sci., Series des sciences chimiques (Poland)* vol.26 (1978) p.117]

[18] S.-T. Wu, C.-S. Wu, M. Warenghem, M. Ismaili [*Opt. Eng. (USA)* vol.32 (1993) p.1775]

[19] H.J. Coles [in *Optics of Thermotropic Liquid Crystals* Eds. S. Elston, R. Sambles (Taylor and Francis, London 1998) p.82]

[20]   L. Pohl, R. Eidenschink, J. Krause, G. Weber [*Phys. Lett. A (Netherlands)* vol.65 (1978) p.169]

[21]   I. Haller, H.A. Huggins, M.J. Freiser [*Mol. Cryst. Liq. Cryst. (UK)* vol.16 (1972) p.53]

[22]   P. Chatelain, M. Germain [*C.R. Acad. Sci. (France)* vol.259 (1964) p.127]

[23]   H. Mori, Y. Itoh, Y. Nishiura, T. Nakamura, Y. Shinagawa [*Jpn. J. Appl. Phys. (Japan)* vol.39 (1997) p.143]

[24]   G. Heppke, H. Kitzerow, F. Oestreicher, S. Quentel, A. Ranft [*Mol. Cryst. Liq. Cryst. Lett. (UK)* vol.6 (1988) p.71]

[25]   Y. Takanishi, M. Yoshimoto, K. Ishikawa, H. Takezoe [*Mol. Cryst. Liq. Cryst. (Switzerland)* vol.331 (1999) p.619]

[26]   P. Kirsch, E. Poetsch [*Adv. Mater. (Germany)* vol.10 (1998) p.602]

[27]   A.G. Douglass, K. Czuprynski, M. Mierzwa, P. Kaszynski [*J. Mater. Chem. (UK)* vol.8 (1999) p.2391]

[28]   T. Nose, S. Sato, K. Mizuno, J. Bae, T. Nozokido [*Appl. Opt. (USA)* vol.36 (1997) p.6363]

[29]   K.C. Lim, J.D. Margerum, A.M. Lackner, L.J. Miller, E. Sherman, W.H. Smith [*Liq. Cryst. (UK)* vol.14 (1993) p.327]

[30]   S.-T. Wu, J.D. Margerum, H.B. Meng, C.S. Hsu, L.R. Dalton [*Appl. Phys. Lett. (USA)* vol.64 (1994) p.1204]

7.2   Chiral nematics

H.-G. Kuball and E. Dorr

September 1999

A   INTRODUCTION

Thermotropic chiral nematics, also called thermotropic cholesterics, are liquid crystals formed by chiral compounds or by mixtures of chiral and achiral compounds with long range chiral orientational order but without long range positional order. Their properties have been extensively discussed and reviewed [1-3] and here the materials will be presented from the points of view of chirality and structure-property relations.

B   THE LONG RANGE ORIENTATIONAL ORDER

The chirality of a chiral nematic, neglecting the chirality of atoms, is based on the chirality of the molecules building the phase and the chiral long range orientational order of the molecules, that is the suprastructural chirality [4,5]. Within the length scale of ultraviolet, visible, infrared and microwave light the chiral nematic phase is inhomogeneous because of the helical twist of the material about the helical axis $\mathbf{P}$ (optical axis in crystal optics). The macroscopic appearance is uniaxial because of the periodicity and the rotational symmetry C_∞ about $\mathbf{P}$. With the $\infty C_2'$ axes perpendicular to the helical axis, the symmetry group (symmetry class) D_∞ results. The local symmetry in a slice perpendicular to C_∞ (see FIGURE 1) is determined by three mutually perpendicular C_2 axes which correspond to the symmetry class D_2 (see FIGURE 1). The three two-fold rotation axes are the helical axis $\mathbf{P}$ (chosen here as the z direction), the local director $\mathbf{n}$ within the slice and $\mathbf{P} \times \mathbf{n}$. The local orientational distribution function $f(\alpha,\beta,\gamma,z)$ in a slice at position z determines the local orientational distribution coefficients as a function of z:

$$g_{ijkl}(z) = \frac{1}{8\pi^2} \int f(\alpha,\beta,\gamma,z) a_{ik} a_{jl} \sin\beta \, d\alpha \, d\beta \, d\gamma \tag{1}$$

where a_{ij} are the elements of the transformation matrix from the space-fixed x_i' to the molecule-fixed x_i coordinate system. α, β, and γ are Eulerian angles as defined in [6].

The orientational distribution function varies along the helical axis. The condition

$$f(\alpha,\beta,\gamma,0) = f(\alpha,\beta,\gamma,p/2) = f(\alpha,\beta,\gamma,p) \tag{2}$$

defines the pitch p as an additional order parameter. The 81 orientational distribution coefficients $g_{ijkl}(z)$ (i,j,k,l = 1,2,3) can be grouped by the nine pairs of indices kl into nine order tensors with respect to the molecule-fixed coordinate system (pair of indices ij). The tensor $g_{ij33}(z)$ (kl = 33) is equivalent to the Saupe ordering matrix [7]

333

$$S_{ij}(z) = (3g_{ij33}(z) - \delta_{ij})/2 \qquad (3)$$

Diagonalisation of g_{ijkl} for each pair kl, with respect to the first pair of indices ij, gives the molecule-fixed principal axes of the order tensors (x^*_i) and the corresponding order parameters. In general, the same principal axes are not obtained for all pairs of indices kl. If the largest eigenvalue of g_{ij33} is associated with the axis x^*_3, i.e. $g_{3333}(z)$, this x^*_3 axis is defined as the orientation axis of the molecule which is on average the best ordered axis in the slice of a chiral nematic as given in FIGURE 1. The principal axes x^*_i are determined by symmetry for molecules which do not possess the point symmetry group C_1, C_2, C_i, C_s and C_{2h}. Whereas for C_2, C_s and C_{2h} only one axis is fixed by symmetry, no axes are fixed for C_1 and C_i. Conformers of a single compound may or may not have their own sets of order parameters, whereas different compounds always have their own sets of order parameters.

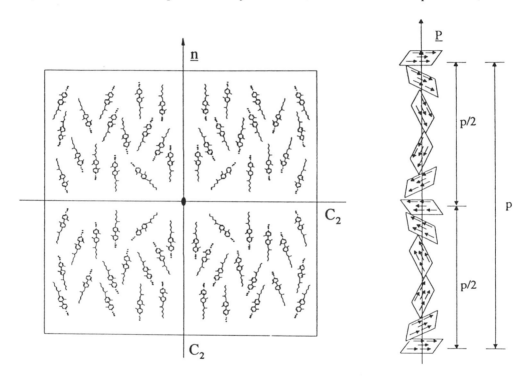

FIGURE 1 Left: A slice of the chiral nematic phase perpendicular to **P** with its local D_2 (E, C_2, C'_2, C''_2) symmetry. Right: Continuous rotation of these slices about the helical axis builds up the uniaxial chiral nematic phase of symmetry class D_∞ (E, C_∞, $\infty C'_2$). The periodicity is given by the pitch p.

For a one-component chiral nematic of D_∞ symmetry, a local D_2 symmetry, and an effective molecular symmetry of D_2, the four local order parameters S*, D*, A* and B* of the slices of the chiral nematic phase need to be considered [8]. Because the numerical factors in the definitions are different in the literature, the functions of the Eulerian angles are given also in EQNS (4) to (7) where now the orientational distribution function is normalised:

$$S* = \frac{1}{2}\left(3g*_{3333} - 1\right) = \frac{1}{2}\left\langle 3\cos^2\beta - 1\right\rangle \qquad (4)$$

$$D* = \frac{\sqrt{3}}{2}\left(g*_{2233} - g*_{1133}\right) = -\frac{\sqrt{3}}{2}\left\langle \sin^2\beta\cos 2\gamma\right\rangle \qquad (5)$$

334

$$A^* = \frac{\sqrt{3}}{2}\left(g^*_{3322} - g^*_{3311}\right) = -\frac{\sqrt{3}}{2}\left\langle \sin^2 \beta \cos 2\alpha \right\rangle \tag{6}$$

$$B^* = \frac{1}{2}\left(g^*_{1111} - g^*_{1122} - g^*_{2211} + g^*_{2222}\right) \tag{7}$$

$$= \left\langle \frac{1}{2}\left(1 + \cos^2 \beta\right)\cos 2\alpha \cos 2\gamma - \cos \beta \sin 2\alpha \sin 2\gamma \right\rangle$$

The stars indicate that the order parameters are given with respect to their principal axes [9]. S^* describes the distribution of the molecule-fixed x^*_3 axis with respect to the local director $\mathbf{n}$; D^* is the order parameter which describes the distribution of the molecule-fixed x^*_1 and x^*_2 axes around the molecule fixed x^*_3 axis. With $D^* \neq 0$ the molecule is biaxial, whereas the local phase biaxiality is expressed by A^* and B^*. The model of de Vries for chiral nematics [10] neglects the biaxiality of the slices and thus assumes the order parameters A^* and B^* to be zero. Then the slices are uniaxial with a director $\mathbf{n} \perp \mathbf{P}$ of the so-called quasi-nematic slice. Only S^* and D^* can be determined experimentally [11]. Some selected examples are given in TABLE 1. However, there is a lack of knowledge about the number of order parameters needed in addition to S^* and D^* to describe the properties of chiral nematics [5]. It has been found experimentally that the local director and the orientation axis of the molecules are always perpendicular to the helical axis. Is there a hidden symmetry in these findings? There are no experimental indications for Pleiner and Brand's postulate [12] that chiral nematics possess a local C_1 instead of a local D_2 symmetry.

C   CHIRALITY OF CHIRAL NEMATICS

C1   Chirality from Elasticity Theory

The Oseen-Zöcher-Frank equation [13] for the free energy density F allows a good understanding of chiral nematic phases [1,2]:

$$F = k_2\left(\mathbf{n} \cdot \nabla \times \mathbf{n}\right) + \frac{1}{2}K_1\left(\nabla \cdot \mathbf{n}\right)^2 + \frac{1}{2}K_2\left(\mathbf{n} \cdot \nabla \times \mathbf{n}\right)^2 + \frac{1}{2}K_3\left(\mathbf{n} \times \nabla \times \mathbf{n}\right)^2 \tag{8}$$

Here K_1, K_2 and K_3 are the splay, the twist, and the bend elasticity constants, respectively, while $\mathbf{n}$ is a local director of the phase. The term k_2 introduces the chirality and forces the phase to have a left-handed ($k_2 < 0$) or right-handed ($k_2 > 0$) twist. This twist reaches an equilibrium because of the counteraction of the torques connected with the first (k_2-term) and the third term (K_2-term). De Gennes [13] has shown that the reciprocal pitch is proportional to the number of chiral 'points', i.e. the concentration c of the chiral molecules. The pseudoscalar proportionality constant $\beta(k_2)$ determines size and sense of the helix:

$$p^{-1} = \beta c \tag{9}$$

TABLE 1 Phase and molecular properties of some selected compounds. All data are measured under atmospheric pressure and at the given reduced temperature T/T_{N*I}, unless a different temperature is given.

Comp.	Single components					Mixtures			
	4	5	6	7	8 [26]	R-1 [27]	R-9 [24]	R-10 [24]	R-11 [27]
$T/T_{N*I}^{(1)}$	0.980 [28]	0.987 [28]	0.990 [28]	0.991 [28]	-	0.908	0.907	0.900	0.907
$T_{N*I}/°C^{(1)}$	92 [29]	115 [31]	92,5 [30]	83 [29]	-	71.7	72.2	74.7	72.3
$p^{-1}/\mu m^{-1}$	4.00 T=85°C [25]	3.45 T=106°C [25]	2.7 T=80°C [25]	4.0 T=84°C [26]	2.52	0.22 $x = 3 \times 10^{-3(2)}$	-0.17 $x = 3 \times 10^{-3(2)}$	-0.26 $x = 2 \times 10^{-3(2)}$	0.092 $x = 0.5 \times 10^{-3(2)}$
$HTP/\mu m^{-1}$	-	-	-	-	-	73.8	-55.1	-130.0	183.9
$k_2/10^{-3}\,N\,m^{-1}$	-	-	-	-	0.09	-$^{(3)}$	-$^{(3)}$	-$^{(3)}$	-$^{(3)}$
K_1/pN					-	8$^{(3)}$	8$^{(3)}$	8$^{(3)}$	8$^{(3)}$
K_2/pN	2.2$^{(4)}$ ±0.7	-			5.7	-$^{(3)}$	-$^{(3)}$	-$^{(3)}$	-$^{(3)}$
K_3/pN	7.45$^{(4)}$ ±1.1				-	10$^{(3)}$	10$^{(3)}$	10$^{(3)}$	10$^{(3)}$
$S*_{guest}^{(4)}$	-	-	-	-	0.32	0.50	0.29	0.30	(0.41)
$D*_{guest}^{(4)}$	-	-	-	-	-	0.07	0.25	0.22	(0.10)
$S*_{host}^{(6)}$	0.24$^{(6)}$ [28]	0.24$^{(6)}$ [28]	0.33$^{(6)}$ [28]	0.25$^{(6)}$ [28]	-	0.65$^{(5)}$	0.65$^{(5)}$	0.68$^{(5)}$	-
$W*_{11}^{(7)}$							-1	-63	
$W*_{22}^{(7)}$							-202	20	
$W*_{33}^{(7)}$							41	-246	

(1) 2°C below the nematic-isotropic transition temperature of the host 4-hexyl-4'-cyanobiphenyl [32]; (2) approximate mole fraction of measurements; measured with a modified Cano method [32]; the accuracy of the pitch or HTP measurement is about 3%; (3) for the component with n = 7 of the nematic host phase (0.5 < K_3/K_1 > 3.0 and 0.5 < K_2/K_1 < 0.8 [33]); (4) for 4 in 4-methoxybenzylidene-4'-butylaniline (MBBA) (0.15 - 0.2 % per weight) for T = 22°C; (5) the accuracy for S* from ^{2}H-NMR is about 2% and for D* 10%; (6) from polarisability anisotropy; (7) estimated.

In a similar approach [14] the pseudoscalar constant $\lambda(k_2)$ in

$$p^{-1} = \frac{\lambda}{K_2} \qquad (10)$$

has been expressed via perturbation theory. Taking into account the form chirality of a molecule with its anisotropic chiral structure, a different approach for the helical twisting power (HTP) has been obtained for guest/host systems which has been applied successfully for inherently dissymmetric molecules [15]:

$$HTP = \frac{p^{-1}}{x_2} = \frac{\varepsilon_{an} V_m}{K_2} \sum_{i,j} S_{ij} Q_{ij} \qquad (11)$$

Here V_m is the molar volume of the phase and x_2 is the mole fraction of the chiral guest molecule. In principle, EQN (11) is also valid for pure phases ($x_2 = 1$). The anchoring energy ε_{an} characterises the interaction of the molecular surface with the surrounding host phase and allows, to some extent, a scaling for different phases and classes of compounds. Q_{ij} is the helicity tensor determined from the form of the surface of the chiral molecules. The use of the traceless helicity tensor Q_{ij} in EQN (11) takes into account the anisotropy of the molecule but leads to a neglect of its pseudoscalar contribution to the HTP. The pseudoscalar contribution is described by EQNS (9) and (10) because β and λ are pseudoscalars; but here the anisotropy of chirality phenomena is neglected. Within the approximation of EQN (11) the HTP of a phase built up by a chiral compound distributed isotropically in a nematic phase is identically zero. This is, in principle, not possible because chiral phenomena cannot disappear for isotropically distributed chiral molecules because of their lack of symmetry of the second kind (rotation reflection axes).

C2 Molecular Mechanism for Chiral Phenomena in Chiral Liquid Crystal Phases

Molecules can be characterised by their absolute configuration R and S or their helicity P and M according to Cahn, Ingold, Prelog (CIP [16]). Besides configurational chirality, given by asymmetric carbon atoms (chiral centres) or by chiral axes or chiral planes, a conformational chirality produced, for example, by hindered rotation has to be taken into account. Examples are atropisomers like 1,1'-binaphthols or even the gauche conformers of 1,2-dichlorethane [5,16]. Under suitable thermodynamic conditions these chiral conformers can also be separated into stable enantiomers.

The phase chirality is given by the molecular chirality and the suprastructural chirality of the orientational or positional order. Thus, a phase has to be specified by symbols for the molecular and the suprastructural chirality. Chiral nematics as given in FIGURE 1 are then described by R or S or P or M for the molecules (first letter) and by P or M for the helicity of the phase (second letter): RP or SP.

From the four levels of chirality which can be introduced [4,5] only the second level, the molecules, and the third level, the phase, are of interest here. Because there is a need to know about the correlation between the chirality of the molecule and the phase, the concept of different levels of chirality implies a concept for the transfer of the information chirality from a molecule to the phase and vice versa

[4,17,18], the so-called intermolecular chirality transfer. Besides the intermolecular chirality transfer, a transfer of the information chirality from the molecular region around a chiral element, the so-called chiral part of the molecule, to parts away from the chiral element [16], the intramolecular chirality transfer, has to be analysed in order to take into account that chirality can be induced in an achiral part of the molecule [17,18]. The achiral part is defined as the molecular region which does not possess a chiral element. Having defined the intra- and intermolecular chirality transfer as a transfer of information, the question arises how chirality effects measured in one level can be compared to those measured in an adjacent level. An example is the comparison of the circular dichroism (CD) and the optical rotatory dispersion (ORD) measurement for single molecules in the gas phase and for chiral nematic phases. The latter is, in general, orders of magnitude larger because the effect has a different origin. This is not an enhancement of molecular chirality as has often been claimed in the literature, but only an enhancement of a chirality measurement [17,18] due to the interaction of light with the ordered system of molecules in the phase, the suprastructural chirality (chirality of the third level) which is determined by the intermolecular chirality transfer from the molecules to the phase [4,17,19].

C3   Molecular Structure-Property Relation for the Reciprocal Pitch and the Helical Twisting Power

For a structure-property relation, structural features have to be taken into account which cannot be covered with EQNS (9) and (10). Taking all structural features into consideration by an atom-atom Lennard-Jones potential in Monte Carlo or molecular dynamic calculations would allow the calculation of the chirality transfer [3]. Unfortunately at present these techniques allow only calculations to a sufficiently good approximation for larger ensembles by the use of the Gay-Berne model potential with the simplest possible chiral term [3]. Thus, it is of interest to find a description which introduces molecular parameters as a bridge between the structure and the measurable reciprocal pitch or the HTP. Experimentally the HTP for a chiral molecule in achiral or chiral liquid crystal phase, in which a helical structure is induced, can be given by [18]

$$(HTP)_e = \frac{1}{2} \left\{ \left(\frac{\partial p^{-1}}{\partial x_e} \right)_{x_e} - \left(\frac{\partial p^{-1}}{\partial x_{e^+}} \right)_{x_{e^+}} \right\} \tag{12}$$

where e^+ denotes the enantiomer of e, and x_e and x_{e^+} are the mole fractions $(0 \leq x \leq 1)$ of e and e^+, respectively. If only one chiral compound is present in the phase the relation

$$\left(\frac{\partial p^{-1}}{\partial x_e} \right)_{x_e} = - \left(\frac{\partial p^{-1}}{\partial x_{e^+}} \right)_{x_{e^+}} \tag{13}$$

holds. For $x \ll 1$ it follows that

$$HTP = \sum_i x_i (HTP)_i \tag{14}$$

With respect to EQNS (12) to (14) it is unimportant whether a mixed phase of chiral and achiral compounds or a dilute solution of a chiral guest in an achiral or chiral host or a phase of only a chiral compound (x = 1) is discussed because the chiral - chiral interactions between the molecules in the phase are small. But in principle one should be aware that for the extrapolation from a mixed phase (solution) to the pure phase the HTP in EQNS (12) to (14) should be described as a mixed phase quantity, i.e. it depends on guest-host interactions.

Equation (11) of Nordio et al [20] can be generalised to develop a theory with a non-traceless pseudotensor of second rank W_{ij}, the chirality interaction tensor. Because the contribution of the interaction energy from the chiral/chiral potentials of the molecules is small in comparison to the chiral/achiral contribution, the helical twisting power is then given for a guest/host system as well as for a single component chiral nematic phase by:

$$HTP = \sum_{i,j} g_{ij33} W_{ij} = \left\{ \frac{1}{3} W + \left(W^*_{33} - \frac{1}{3} W \right) S^* + \frac{1}{\sqrt{3}} \left(W^*_{22} - W^*_{11} \right) D^* \right\} \tag{15}$$

where g_{ij33} are the coordinates of the order tensor and S^* and D^* are the Saupe order parameters of the chiral molecules.

$$W = \sum_i W^*_{ii} \neq 0 \tag{16}$$

is the pseudoscalar contribution of a chiral molecule to the HTP given by EQNS (9) and (10). According to EQN (15) the trace W is the order-independent, i.e. S^*, D^*-independent, contribution of a dopant to the HTP of a guest/host phase or to the pitch of a pure chiral nematic. It is interesting that EQN (15) and the well-known equation for CD of oriented molecules have an identical structure [19]. The chirality interaction tensor can be rewritten as

$$W_{ij} = \sum_k C_{ik} L_{kj} = \frac{A \left(S_{host/guest} \right)}{K_2 \left(S_{host/guest} \right) V_m} \sum_k C_{ik} \overline{L}_{kj} \tag{17}$$

The coordinates of the pseudotensor of second rank W_{ij} are chirality parameters for the interaction of the chiral molecule with its anisotropic surroundings; the relevant properties of the host phase are approximated by a second rank tensor L_{ij}. $S_{host/guest}$ is the order parameter of the host in the guest/host phase. For a convenient comparison to EQN (11) the proportionality constant $A(S_{host/guest})$, the twist constant K_2 and the molar volume V_m of the guest/host phase are separated out of L_{ij} which yields $\overline{L}_{ij}$. $\overline{L}_{ij}$ is given for a fixed host order and, therefore, depends on $S_{host/guest}$. In a single component phase, host and guest are identical. The coordinates of the chirality interaction tensor W^*_{ij} can be determined from experimental HTP values and experimentally determined coordinates of the order tensor g^*_{ij33} as a function of temperature by a multiple regression procedure assuming a constant order for the host as shown for 1,1'-binaphthyls [24]. It is necessary to take into account the fact that S_{guest} is also a function of the host order, and so the HTP is a function of a quotient $S_{guest}/S_{host/guest}$.

From EQN (15) the helix inversion, that is when the reciprocal pitch p^{-1} passes through zero and changes sign as a function of an achiral variable such as the temperature without a change of the chiral structure, is easily explained. In general, sign inversion fits into the general behaviour of chirality measurements and no special explanation must be found. Altogether, four mechanisms can be discussed [18]: (1) equilibrium of conformers [21], (2) action of independent chiral elements in a chiral molecule [22], (3) anisotropy of a chiral property ($W_{ij} \neq 0$) of a molecule (EQN (15)) [24] and (4) interaction of the molecule with a molecular field potential with more than one pseudoscalar parameter [23].

C4 Typical Data for Chirality Measurements of Chiral Nematics

To relate macroscopic properties, especially the results of chirality measurements, to mesoscopic and further to pseudoscalar molecular properties, experimental data should be available in order to develop and check structure-property relations, mechanisms, and models. A set of usable data for the chiral nematic phase consists of the composition of the phase, p^{-1}, HTP, k_2, K_1, K_2, and K_3. On the microscopic scale the (HTP)$_i$ as well as the microscopic order parameters S*, D*, A*, B*, the helicity tensor Q_{ij} or the chirality interaction tensor W^*_{ij}, and a chirality tensor C_{ij} or an equivalent quantity should be known for every component. At the moment, the coordinates W^*_{ij} can only be estimated because a variation of the order of the chiral dopant for a constant host order needs to be known for their measurement. Some data are collected in TABLE 1 in order to give a feeling for the size and sign of available quantities [25-33] but, as can be seen, no complete sets of data for a system have yet been given in the literature.

In TABLE 1 the data for cholesteryl nonanoate (**4**), cholesteryl propanoate (**5**), cholesteryl decanoate (**6**), and cholesteryl myristate (**7**) are given for the single component chiral nematic phase.

The chiral compounds

dinaphtho-[2,1-d:1′,2′-f][1,3]-dioxepin] (**9**), spiro-[cyclohexanon-ethylenacetal-dinaphtho[2,1-1,2′-d:1′,2′-f][1,3]-dioxepin] (**10**), and the TADDOLs

R,R-**11**   R,R-**12**

were measured in a mixed phase. For **8** the host is 4-hexyl-4′-cyanobiphenyl (about 20% per weight). For R-**1**, R-**9**, R-**10**, R-**11** and the TADDOLs R,R-**11** and R,R-**12** the nematic host ZLI-1695 (Merck) has been chosen (ZLI-1695 is a mixture of four 4-alkyl-4′-cyanobicyclohexanes; the alkyls (R) are:

(R = C_2H_5, x = 0.28), (R = C_3H_7, x = 0.19), (R = C_4H_9, x = 0.23), (R = C_7H_{15}, x = 0.30), the chosen mole fractions of the mixture are given in brackets; (T_{N*I} = 72.9°C)). 5CB is 4-pentyl-4'-cyanobiphenyl (T_{NI} = 35.1°C). For **4**, K_2 and K_3 were measured in 4-methoxy-benzylidene-4-butylaniline (MBBA).

The data in TABLE 1 were selected on the one hand in order to demonstrate that, in principle, no differences with respect to the pitch or HTP between one-component, guest-host phases and real mixtures of any composition exist. The reciprocal pitches of the guest/host systems chosen are smaller because of the measurement conditions in comparison to those for the single component and systems. Increasing the mole fraction to 0.01 or to 0.2, the single component and mixtures have an equal pitch in spite of the fact that in the guest/host system there are 80% - 99% achiral molecules. The order parameters S* for **4** to **7** are relatively small for the steroids. Because of the method used to determine them, we suggest that these are not the order parameters with respect to the principal axes of the order tensor [9]. S* is always larger than S, and for steroids about 0.5 to 0.6 [9].

At the moment, the TADDOL R,R-**12** possesses the largest known HTP value with 341/μm^{-1} in ZLI-1695 (Merck) and with 534/μm^{-1} in MBBA [34]. This seems to support the often cited conclusion that the chiral induction is the largest when the inducing guest and the host have similar structures. This means that the largest induction of a compound should be in its pure state because then the neighbouring molecules are not only similar but identical. A systematic analysis from which such a general conclusion can be drawn does not seem to exist, as yet.

D   CHIRALITY AND PHASE TRANSITIONS

A phase transition from the uniaxial chiral nematic to the isotropic blue phase [1] appears when the helicity increases. The critical length for the periodicity of the helix structure by which the phase transition appears is about 0.5 μm [35]. The constant k_2 is then about 8×10^{-5} N m^{-1} and K_2 is about 5×10^{-12} N [26]. Chiral nematic to blue phase transitions have been found for single components as well as for mixtures [36] under similar conditions.

Phase transitions from chiral nematics to chiral smectics, or to blue or even to isotropic phases can be connected to chirality-induced pretransitional effects which often exist at more than 10°C above or below the transition temperature. Thus measurements of ORD have been performed, where an enormous enhancement of the rotation [37] has been found. This effect can be related [38] to chiral fluctuations within the phase. The handedness of the effect induced is determined by the sign of $(\mathbf{n} \cdot \nabla \times \mathbf{n})$, the term which introduces the chiral interaction in continuum theory.

E   PHASE PROPERTIES AND EXTERNAL PARAMETERS

Dependences of properties of chiral nematic phases on achiral variables such as pressure, temperature, electric and magnetic fields, flow and others are not considered in this Datareview. For this information we refer to [1,2]. Also optical, electro-optical, and magneto-optical properties like optical rotation, CD, and selective reflection, in the region where the wavelength is or is not in the length-scale of the periodicity, are not taken into account here [1,35,39]. The internal reflection as a consequence of the internal periodicity can be used as a phenomenon for measuring a parameter for the suprastructural chirality directly.

REFERENCES

[1]   D. Demus, J. Goodby, G.W. Gray, H.-W. Spiess, V. Vill (Eds.) [*Handbook of Liquid Crystals* vol.I-III (Wiley-VCH, Weinheim, 1998)]

[2]   P.J. Collings, J.S. Patel (Eds.) [*Handbook of Liquid Crystal Research* (University Press, Oxford, 1997)]

[3]   M.A. Bates, G.R. Luckhurst [*Struct. Bond. (Germany)* vol.94 (1999) p.65]

[4]   H.-G. Kuball [*Liquid Crystals Today (UK)* vol.9 (1999) p.1]

[5]   H.-G. Kuball, R. Memmer, O. Türk [in *Ferroelectrics* Eds. S. Lagerwall, L. Komitov (World Scientific Press, in press 2000)]

[6]   H.-G. Kuball, T. Karstens, A. Schönhofer [*Chem. Phys. (Netherlands)* vol.12 (1976) p.1]

[7]   A. Saupe [*Z. Naturforsch. A (Germany)* vol.19 (1964) p.161]

[8]   W.J.A. Goossens [*Mol. Cryst. Liq. Cryst. (UK)* vol.12 (1971) p.237]

[9]   H.-G. Kuball, M. Junge, B. Schultheis, A. Schönhofer [*Ber. Bunsenges. Phys. Chem. (Germany)* vol.95 (1991) p.1219]; H.-G. Kuball, B. Schultheis, M. Klasen, J. Frelek, A. Schönhofer [*Tetrahedron Asymmetry (UK)* vol.4 (1993) p.517]

[10]   H. de Vries [*Acta Crystallogr. (Denmark)* vol.4 (1951) p.219]

[11]   G.R. Luckhurst, G.W. Gray (Eds.) [*The Molecular Physics of Liquid Crystals* (Academic Press, 1979)]

[12]   H. Pleiner, H.R. Brand [*J. Phys. II (France)* vol.3 (1993) p.1397]

[13]   P.G. de Gennes, J. Prost [*The Physics of Liquid Crystals* (Clarendon Press, Oxford, 1995)]

[14]   M.A. Osipov [in *Liquid Crystalline and Mesomorphic Polymers* Eds. V.P. Shibaev, L.P. Lam (Springer-Verlag, 1994) ch.1 p.1]

[15]   A. Ferrarini, P.L. Nordio [*J. Chem. Soc., Perkin Trans. 2 (UK)* (1998) p.455]

[16]   E.L. Eliel, S.H. Wilen [*Stereochemistry of Organic Compounds* (John Wiley & Sons, New York, 1994)]

[17]   H.-G. Kuball, O. Türk [*Pol. J. Chem. (Poland)* vol.73 (1999) p.209]

[18]   H.-G. Kuball, H. Brünning [*Chirality (USA)* vol.9 (1997) p.407]

[19]   H.-G. Kuball, A. Schönhofer, T. Höfer [in *Circular Dichroism: Principles and Applications* Eds. K. Nakanishi, N. Berova, R.W. Woody (VCH/John Wiley & Sons, New York, 1994/second edition 2000)]

[20]   A. Ferrarini, G.J. Moro, P.L. Nordio [*Mol. Phys. (UK)* vol.87 (1996) p.485]

[21]   G. Gottarelli, M. Hibert, B. Samori, G. Solladié, G.P. Spada, R. Zimmermann [*J. Am. Chem. Soc. (USA)* vol.105 (1983) p.7318]

[22]   I. Dierking, F. Gießelmann, P. Zugenmaier [*Mol. Cryst. Liq. Cryst. (Switzerland)* vol.281 (1996) p.79]

[23]   A.N. Zakhlevnykh, M.I. Shliomis [*Sov. Phys.-JETP (USA)* vol.59 (1984) p.764]

[24]   H.-G. Kuball, O. Türk, I. Kiesewalter, E. Dorr [*Liq. Cryst. (UK)* (in press 2000)]

[25]   Shu-Hsia Chen, L.S. Chou [*Mol. Cryst. Liq. Cryst. (UK)* vol.67 (1981) p.221]

[26]   R.J. Miller, H.F. Gleeson [*Liq. Cryst. (UK)* vol.14 (1993) p.2001]

[27] H.-G. Kuball, R. Kolling, H. Brüning, B. Weiß [*Proc. SPIE (USA)* vol.3318 (1998) p.1229-34]

[28] P. Adamski, L.A. Dylik-Gromiec, M. Wojciechowski [*Mol. Cryst. Liq. Cryst. (UK)* vol.75 (1981) p.33-8]

[29] Z. Luz, R. Poupko, E.T. Samulski [*J. Chem. Phys. (USA)* vol.74 (1981) p.5825-37]

[30] G.W. Gray [*J. Chem. Soc. (UK)* (1956) p.3733-9]

[31] E.M. Barrall, J.F. Johnson, R.S. Porter [*Mol. Cryst. Liq. Cryst. (UK)* vol.8 (1969) p.27-44]

[32] G. Heppke, F. Oestreicher [*Mol. Cryst. Liq. Cryst. Lett. (UK)* vol.41 (1978) p.245-9]

[33] G. Vertogen, W.H. De Jeu [*Thermotropic Liquid Crystals, Fundamentals* (Springer Verlag, Berlin, 1988)]

[34] H.-G. Kuball, B. Weiß, A.K. Beck, D. Seebach [*Helv. Chim. Acta (Switzerland)* vol.80 (1997) p.2507-14]

[35] S. Chandrasekhar [*Liquid Crystals* (Cambridge University Press, 1992)]

[36] H. Stegemeyer, T.H. Blümel, K. Hiltrop, H. Onusseit, F. Porsch [*Liq. Cryst. (UK)* vol.1 (1986) p.3-28]

[37] P.J. Collings [*Mod. Phys. Lett. B (Singapore)* vol.6 (1992) p.425-46]

[38] T.C. Lubensky, R.D. Kamien, H. Stark [*Mol. Cryst. Liq. Cryst. (Switzerland)* vol.288 (1996) p.15-23]

[39] L.M. Blinov, V.G. Chigrinov [*Electrooptical Effects in Liquid Crystal Material* (Springer Verlag, New York, 1994)]

7.3 Measurements of the physical properties of nematic liquid crystals by guided wave techniques

Fuzi Yang and J.R. Sambles

March 1999

A INTRODUCTION

The primary low power devices for display applications are liquid crystal cells. In these the liquid crystal is in the form of a thin layer, with a thickness of the order of a few microns, sandwiched between two glass plates (the two cell walls) with some transparent conductive coatings (e.g. ITO) and alignment layers on them. Knowledge of the form of the director alignment in such cells and subsequently the ability to predict their optical response to changing external conditions are essential requirements for the basis for display device development. Thus the investigation of the alignment of liquid crystals in such thin films is essential.

A range of guided wave techniques has been developed to study the index profiles in thin film optical guiding structures [1,2]. These techniques utilise the fact that a series of discrete (quantised waveguide momentum) modes may be excited in thin films sandwiched between two other materials. Their mode spectrum is dependent upon the refractive index distribution within, and the dimensions of, the structure. The guided mode momenta are sensitive to the optical tensor distribution in the guiding layer. Each different order mode has a different optical field profile and so will be sensitive to different parts of the waveguide. In addition, for an anisotropic waveguide when the optical axis of the guiding film is twisted out of the incidence plane and/or tilted out of the plane of the cell walls the mode (mixed p-s or s-p eigenmode) spectrum will be very complex and very sensitive to the optical axis distribution through the guiding layer [3,4]. This sensitivity is only in one dimension, through the thickness of the guide, but since a monodomain should be invariant in the plane of the guide this is exactly the sensitivity required. The guided wave technique is about the only one which is going to yield the required spatial selectivity across the thickness of the guiding film.

An experimental procedure is required to couple radiation into the liquid crystal film to allow detailed probing of its optical structure. For a conventional cell, prism or grating coupling [5,6] is by far the most appropriate. However, for flat panel display-like cells the prism-coupling technique is more often used. Over recent years three somewhat different geometries have been exploited using prism coupling of light to guided modes in liquid crystal layers [7].

For all three of the geometries, discussed in more detail in the following sections, the essential technique is that of monitoring the angle dependent reflectivity of a plane parallel, monochromatic linearly polarised optical beam, incident through the coupling prism at the glass/liquid crystal layer boundary. At certain angles of incidence the momentum of the incident radiation along the surface will match that of one of the guiding modes in the layered structure incorporating the liquid crystal, and there will then be a reduction in reflectivity at these angles. Hence by simply monitoring the reflectivity as a function of angle of incidence we find all of the mode momenta. More generally, instead of just evaluating the mode momenta, we monitor accurately the angle dependent reflectivity over a wide range of angles and then fit this to predictions from a multilayer optics model of the system. From this the optical tensor distribution of the liquid crystal layers can be deduced.

B   EXPERIMENTAL METHODS

B1   Fully Guided Geometry

For exciting a series of fully guided modes a low refractive index cladding, a high index film and a low index substrate are required. In this geometry metal-clad waveguides have been the focus of attention: early work [8] investigated surface plasmons, located at the interface between a metal and the liquid crystal, to explore the director profile near the aligning surface. In order to provide a well aligned liquid crystal and to allow the application of external electric fields as well as to allow the study of fully guiding modes special cell geometries are required. This includes a high index prism with a thin metal coating, the optical tunnel barrier, which is overcoated with an alignment film, the aligned liquid crystal layer and a glass substrate also coated with a metal film (generally optically opaque) and an alignment layer. Because the refractive index of some common metals (Ag, Au and Al) for the visible region of the spectrum is less than unity then the metal layer acts as a tunnel barrier. If the metal layers are of the correct thickness then the complete structure may be a nearly symmetric fully guiding waveguide. Hence a series of sharp fully guided modes can be excited by high index prism coupling as well as the broader surface plasmon resonance. A significant amount of work [9-15] has been done using this novel technique. Unfortunately, there are several limitations that inhibit the usefulness of the metal-clad fully guided technique for practical devices. First, there are no metal layers in most real devices, and certainly no top, thin, metal film. Secondly, these fragile metal layers limit the use of strongly rubbed aligning layers found in many commercial cells. Thirdly, the thin metal layer tends to reflect strongly s-polarised radiation, thereby limiting the technique's sensitivity to polarisation conversion; that is, the polarisation-conversion signals, R_{sp} and R_{ps}, are rather weak, yet it is just these signals that are particularly sensitive to director twist and tilt. Because of the limitations caused by using the thin metal layers other guided mode techniques are required which avoid the use of these layers and are more closely akin to real device structures.

B2   Fully Leaky Geometry

If the metallic layers are removed from the previous geometry than all three limitations will be avoided. However this will inevitably mean losing the surface plasmon, and also all of the guided modes will now become leaky. Thus a technique based upon such a geometry may be called a fully leaky guided mode technique [16]. Now the liquid crystal waveguide geometry is that of a thin liquid crystal layer sandwiched between two high index prisms, both of which are coated, as in conventional cells, with indium tin oxide (ITO) layers for field application and rubbed polyimide or equivalent layers for alignment of the liquid crystal. The use of high index prisms means that there are still critical edges available to help in quantifying the refractive index tensor of the liquid crystal.

Because the sharp-reflectivity features recorded for the metal-clad waveguide are now completely absent, the sensitivity of data fitting to the specifics of the director profile is reduced. This limits the precision with which the technique may be used to determine the director twist/tilt structure through the cell. Nevertheless, the technique has value in that it allows the study of cells that are much nearer in structure to those fabricated commercially.

B3   Half-Leaky Guided Mode Geometry

To improve upon the fully leaky geometry a new technique, the half-leaky guided mode technique, has been developed [3]. In this arrangement the bottom, high index, prism is replaced by a low index glass plate. This means that there is now one real critical angle in the geometry, corresponding to that between the high index entrance prism and this low index glass plate, and one pseudo-critical angle between the high index prism and the liquid crystal. For the angle range between these two the interface between the liquid crystal layer and the low index glass plate acts as a perfect mirror. Thus over this angle range the structure is only half leaky, and the reflectivity features recorded are much sharper. The clear advantages of this technique are that not only can it be used with cell structures that are similar to standard device cells, like the fully leaky procedure, but now the reflectivity features are much sharper, giving improved resolution of the director profile. In addition, as for the fully leaky procedure, the p-to-s conversion signal may also be strong containing quite sharp features.

However, this powerful half-leaky guided-mode procedure still has two limitations. First, it continues the use of the high index entrance prism; secondly; the two glasses from which the cell is comprised are very different from each other and from commercial cell glass. So if the director profile of a real commercial device cell, which has low index, normally 1.52, glass plates, is to be explored by a guided mode technique, some further improvement over the geometries mentioned previously is still required.

B4   Improved Fully Leaky Guided Mode Technique

If a standard commercial liquid crystal cell with low index glass plates is to be investigated by a guided mode technique to unravel the director profile through the cell, then only the low index fully leaky guided mode technique may be chosen. As we have mentioned, because all the guided modes will now be leaky and will give correspondingly broad features in the reflectivity data, this may severely limit the precision with which the director twist/tilt structure through the cell will be determined. In addition, the use of low index glass means that no longer will there be any critical angle available to help determine the refractive indices of the liquid crystal. However, recently some improvements [4] have been introduced, in which two refinements to the fully leaky technique have been made to make it much more useful. First, the full set of both transmission, T, and reflection, R, data may be utilised, including all the polarisation-conversion signals R_{sp}, R_{ps}, T_{sp} and T_{ps} as well as the polarisation-conserving signals R_{pp}, R_{ss}, T_{pp} and T_{ss}. The polarisation-conversion signals are particularly sensitive to the director twist and tilt. Secondly, two matching prisms with matching fluid have been used to allow rotation of the cell to a position that allows optimisation of sensitivity to director twist and tilt. In addition, this allows the acquisition of data for a set of different azimuthal angle settings that, by fitting all of the data sets, further removes ambiguity in determining the director profile.

All these guided wave techniques are very powerful tools for exploring the director profile and measuring the optical properties of thin liquid crystal films. Which technique is chosen is dependent on the sample and the information required. However it seems likely that the advantages of being able to use off-the-shelf cells will see progressively more use of the fully leaky technique. We now go on to discuss some applications of guided mode techniques to the study of nematic liquid crystals.

C   PROPERTIES OF NEMATIC LIQUID CRYSTALS DETERMINED BY OPTICAL WAVEGUIDING

C1   The Influence of an Electric Field upon the Optical Permittivity

The optical properties of uniaxial nematics are described by $\widetilde{\varepsilon}_\parallel(\omega)$ and $\widetilde{\varepsilon}_\perp(\omega)$ which are respectively the relative optical permittivities (here ω is the frequency) for polarisation parallel to the director and orthogonal to it. These parameters are not only functions of frequency, but through changes in the orientational order parameters, they are also temperature dependent and influenced by an applied electric field. This influence may be through two possible effects. First, on a microscopic level the electric field interacts with the molecules through both permanent and induced moments to change the order parameter. Secondly, director fluctuations may be progressively quenched. Some experiments [17,18], which involve measuring the change of a combination of $\widetilde{\varepsilon}_\parallel$ and $\widetilde{\varepsilon}_\perp$ using transmission birefringence changes in relatively thick cells of homeotropically aligned liquid crystal, have been reported.

Ruan et al [12] used a fully guiding technique to excite a series of optical guided modes in a thin homeotropically aligned nematic (E7-Merck UK: a mixture of cyanobiphenyls) layer to determine accurately and independently the changes in $\widetilde{\varepsilon}_\parallel$ and $\widetilde{\varepsilon}_\perp$ by an applied electric field. Such changes are very small and so the very sharp guided modes of the fully guided geometry were felt to be the most sensitive procedure. A number of s (transverse electric, TE) and p (transverse magnetic, TM) modes were excited and their changes measured upon application of the field to the cell. The experimental geometry comprises a high index prism - thin silver film - lecithin aligning layer - E7 layer - lecithin aligning layer - thick silver film - glass substrate. A pulsed AC(20 kHz) voltage was applied to the cell. Some typical reflectivity data recorded without and with the applied voltage at 21°C are shown in FIGURE 1. In FIGURE 1(a) reflectivity data for TE modes as a function of the incident angle, θ, are shown with several guided modes in a cell of thickness 5.99 μm with no applied field. In FIGURES 1(b) and 1(c) the reflectivity data using a pulsed voltage of 25 V obtained for fixed angles of 41.41°, the leading edge, and 41.48°, the trailing edge, of the middle TE guided mode in FIGURE 1(a) are shown, respectively. It is very clear that the mode is moved to lower momentum under application of a field, even though this movement is very small. Heating, B $\to$ C, and cooling, D $\to$ E, effects are also very clear (see FIGURES 1(b) and 1(c)). Thus the effects of thermal and possible cell dimension changes can be eliminated by using the pulsing technique. Carefully fitting the shifts of these modes to Fresnel theory yields electric field effects alone.

The results of the variation of $\widetilde{\varepsilon}_\parallel$, $\widetilde{\varepsilon}_\perp$ and $\delta\Delta\tilde{n}(= (\tilde{n}_\parallel - \tilde{n}_\perp)_V - (\tilde{n}_\parallel - \tilde{n}_\perp)_0)$ with voltage are shown in FIGURES 2(a), 2(b) and 2(c) and TABLE 1, respectively. From FIGURE 2(c) it is clear that the change in the refractive index anisotropy, $\delta\Delta\tilde{n}$, varies linearly with voltage at higher fields. This result agrees with the prediction from the theory developed by Faber [19,20] for the suppression of director fluctuations by the applied field. Thus it appears that the experimental data obtained strongly support the director fluctuation suppression model for field induced changes in the system.

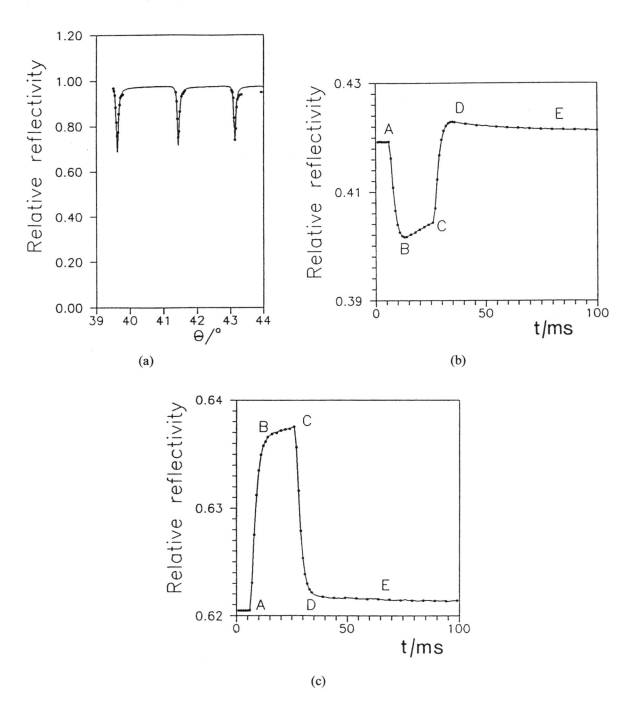

FIGURE 1 (a) Reflectivity data for TE modes as a function of the incident angle, θ, showing several guided modes in a cell, with a thickness of 5.99 μm, under no applied potential. (b) Reflectivity data with a pulsed voltage of 25 V obtained for a fixed angle of 41.41°, the leading edge of a TE guided mode (from [12]). (c) Reflectivity data with a pulsed voltage of 25 V obtained for a fixed angle of 41.48°, the trailing edge of the same TE mode (from [12]).

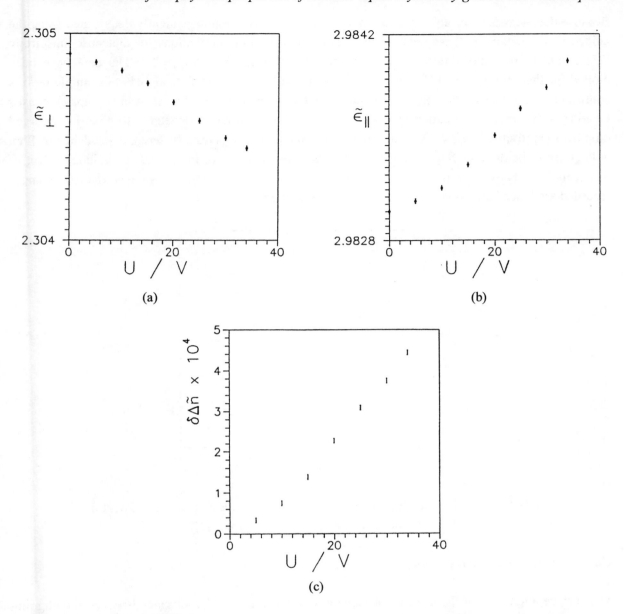

FIGURE 2 (a) The voltage dependence of $\widetilde{\varepsilon}_\perp$ for E7. (b) The voltage dependence of $\widetilde{\varepsilon}_\parallel$ for E7. (c) The variation in the refractive index anisotropy, $\delta\Delta\tilde{n}$, with the applied voltage, U, for E7 (from [12]).

TABLE 1 Electric field effects on refractive index of nematic liquid crystals.

U/V	$\widetilde{\varepsilon}_\parallel$	$\widetilde{\varepsilon}_\perp$	$\tilde{n}_\parallel$	$\tilde{n}_\perp$	$\Delta\tilde{n}$	$\delta\Delta\tilde{n}$ ($\times 10^{-4}$)
0	2.98300	2.30490	1.72714	1.571819	0.20895	0
5	2.98307	2.30486	1.72716	1.51818	0.20898	0.33
10	2.98316	2.30482	1.72718	1.51816	0.20902	0.74
15	2.98332	2.30476	1.72723	1.51814	0.20909	1.40
20	2.98352	2.30467	1.72729	1.51811	0.20918	2.30
25	2.98370	2.30458	1.72734	1.51808	0.20926	3.10
30	2.98384	2.30450	1.72737	1.51806	0.20932	3.60
34	2.98402	2.30445	1.72743	1.51804	0.20939	4.40

Because the changes in $\widetilde{\varepsilon}_{\parallel}$ and $\widetilde{\varepsilon}_{\perp}$ under applied voltages may be independently determined using the guided mode technique, these results can be compared with the predictions for dielectric anisotropy [21] obtained from the molecular field theory, which gives $|\delta\widetilde{\varepsilon}_{\parallel}/\delta\widetilde{\varepsilon}_{\perp}|=2$. The data have been plotted for this ratio in FIGURE 3 where it is clear that to within better than 10% this simple ratio is confirmed. It indicates that the molecular field theory expressions for $\widetilde{\varepsilon}_{\parallel}$ and $\widetilde{\varepsilon}_{\perp}$ may be used together with director fluctuation theory for the orientational order parameter S to give the observed ratio of permittivity changes. Although the Maier-Saupe theory gives a different dependence of S on voltage from the director fluctuation theory the ratio of the changes in $\widetilde{\varepsilon}_{\parallel}$ and $\widetilde{\varepsilon}_{\perp}$ is independent of this variation. Thus the experimental data provide very strong supporting evidence that the changes recorded are indeed due to voltage induced changes in the order parameter.

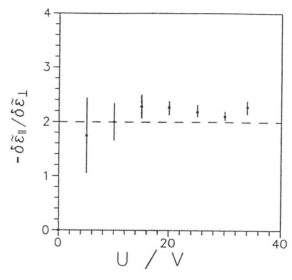

FIGURE 3 The ratio of the observed changes in the relative permittivities $\widetilde{\varepsilon}_{\parallel}$ and $\widetilde{\varepsilon}_{\perp}$ as a function of voltage, U, for E7 (from [12]).

C2   Surface Anchoring Energy

The surface alignment of the director of nematic liquid crystals at the interface between the aligning layer and the liquid crystal is determined by the competition between surface and bulk interactions. If there are no external torques on the liquid crystal, the director should be aligned along those directions (easy axes) which minimise the anchoring energy. This is a function of both the surface polar angle θ_s and the surface azimuthal angle ϕ_s, and represents the work which is needed to rotate the director from the easy axis towards the actual surface orientation. If θ_s or ϕ_s is held fixed and equal to the easy polar or azimuthal angle, the surface anchoring energy becomes a function of θ_s or ϕ_s only. From the continuum theory the equilibrium director distribution through the cell is deduced by minimising the total free energy which is given by the sum of the bulk free energy and of the surface anchoring energy. To determine the surface anchoring energy it is necessary to monitor changes in the director profile under some external torque.

There are broadly two techniques for measuring surface anchoring energies, in which the external torque can be generated either by exploiting the competition between different surface orientations (hybrid or twisted cell), or by applying external fields (magnetic, electric). In all these cases an external torque is applied to the director and the consequent change of the director at the surface is measured. In the first case, i.e. the geometrical technique, since no orientation field is present in the bulk, the mathematics involved is rather simple; however special cell geometries are needed. In the second case, namely the external field technique, the mathematics is rather more complex, especially under an electric field because of the high dielectric susceptibility and its anisotropy, even though no special cell is needed. This complexity is reduced for small electric fields but generally the distortions induced by weak fields are rather too small to measure. However, because the optical guided wave technique is very sensitive to details of the director profile through the liquid crystal cell it is an ideal technique for these measurements. Possibly a more important reason for choosing to use a guided mode technique is that often the measurements comprise the determination of the polarisation state of either transmitted or reflected light using an interpretative analysis which ignores the surface layers (ITO, aligning layer, etc.). Ignoring the optical influence of these layers and also the multiple reflections within the multilayer system may cause quite serious errors in the final results [22]. Because multilayer optics theory is used to fit the experimental data obtain using the guided wave technique, these possible errors are avoided.

Several studies of surface anchoring have been undertaken using the guided wave technique. Welford et al [9] have used the metal-clad fully guiding technique to determine the surface director reorientation in a cell containing E7. The sample geometry comprises a high index prism - silver coating - SiO_x thin film - E7 - SiO_x thin film - silver coating - high index prism. From their experimental results the surface polar anchoring energy between E7 and SiO_x is deduced as $(2.6 \pm 0.1) \times 10^{-4}$ J m^{-2} at 17°C.

Yang et al [23] have used the half-leaky guided mode technique to determine the azimuthal anchoring energy and twist elastic constant of a homogeneously aligned E7 liquid crystal. The sample geometry used is that of a high index (n = 1.732 at λ = 632.8 nm) glass plate and a low index (n = 1.458 at λ = 632.8 nm) glass substrate separated by 12 μm mylar spacers. The cell was optically matched to a high index (n = 1.732 at λ = 632.8 nm) prism by use of a suitable matching fluid (CH_2I_2) to give the complete experimental geometry. Before assembly of the cell the low index glass plate was coated with 200 nm thick gold electrodes having a 3 mm, parallel sided, gap. Also both of the inner glass surfaces were coated with obliquely (60° from normal) evaporated thin SiO_x layers to provide homogeneous aligning surfaces. The gold coated surface had the aligning director arranged closely parallel to the electrode gap edge. At a temperature of 75°C the cell was capillary-filled with the nematic E7. The experimental results show that even with very small fields ($\sim 2 \times 10^{-2}$ V cm^{-1}) there are easily recorded signal changes showing the enormous sensitivity to the director profile in the HLGW window. Using such low fields allows much simpler analytical treatment. From their experimental results the surface azimuthal anchoring energy between E7 and SiO_x and the twist elastic constant of E7 were determined as $(2.6 \pm 0.3) \times 10^{-6}$ J m^{-2} and $(6.5 \pm 0.5) \times 10^{-12}$ N, respectively, at 23.7°C.

More recently the improved fully leaky guided mode technique has been used to investigate the director profile of a commercial, conventional, liquid crystal cell constructed from normal glass plates with a refractive index of about 1.52 at λ = 632.8 nm. Hallam et al [24] have used the fully leaky technique to quantify the azimuthal anchoring energy of a homogeneously aligned nematic E7 on rubbed polyimide. Their sample geometry was almost the same as that mentioned previously for the half leaky technique, except that the high and low index glass plates were replaced by normal low index glass plates and the SiO_x aligning layers were replaced by rubbed polyimide. From their experimental results the surface azimuthal anchoring energy between E7 and rubbed polyimide and the twist elastic constant of E7 were determined to be $(2.9 \pm 0.2) \times 10^{-5}$ J m^{-2} and $(6.5 \pm 0.05) \times 10^{-12}$ N, respectively, at 23.5°C.

Yang et al [23] have also used the fully leaky guided mode technique to determine the surface polar anchoring energy of a homogeneously aligned nematic E7 on rubbed polyimide. The sample geometry for this investigation was that of a low index prism - matching fluid - normal glass plate - ITO coating - rubbed polyimide film - E7 - rubbed polyimide film - ITO coating - normal glass plate - matching fluid - low index prism. From the experimental results the surface polar anchoring energy between E7 and rubbed polyimide is $(1.2 \pm 0.1) \times 10^{-4}$ J m^{-2} at 26.8°C.

A summary of the above studies is shown in TABLE 2. It is worth noting that the experimental results are strongly dependent on the type of aligning material and the sample preparation.

TABLE 2 The surface anchoring energy between E7 and aligning layers.

	SiO$_x$ alignment	Polyimide alignment	Temperature/°C	Ref
Polar anchoring energy/J m^{-2}	$(2.6 \pm 0.1) \times 10^{-4}$		17.0	[9]
		$(1.2 \pm 0.1) \times 10^{-4}$	26.8	[23]
Azimuthal anchoring energy/J m^{-2}	$(2.6 \pm 0.3) \times 10^{-6}$		23.7	[23]
		$(2.9 \pm 0.2) \times 10^{-5}$	23.5	[24]

C3 Influence of an Electric Field on the Optical Permittivity of 4-Octyl-4′-Cyanobiphenyl (8CB)

As we saw in Section C1 Ruan et al [25] have also used the metal-clad fully guided wave technique with a pulsed voltage to study the influence of an electric field on the optical permittivity of both the nematic and smectic A phases of 8CB. The sample geometry was the same as that described in Section C1 with an 11.84 μm thick 8CB layer homeotropically aligned by thin lecithin films. For the nematic phase at 36.0°C the changes with voltage of the optical permittivities, $\widetilde{\varepsilon}_{\parallel}$ and $\widetilde{\varepsilon}_{\perp}$, and the optical refractive index anisotropy, $\delta\Delta\tilde{n} = \delta(\tilde{n}_{\parallel} - \tilde{n}_{\perp})$, were deduced from the experimental data and are shown in FIGURES 4(a) and 4(b). These results are very similar in magnitude to those obtained for E7 (see Section C1 and [12]) although the linear dependence found in that case at high fields now appears dominated by the lower field quadratic dependence. The changes of $\widetilde{\varepsilon}_{\parallel}$ and $\widetilde{\varepsilon}_{\perp}$ caused by an applied field should arise primarily from the suppression of director fluctuations by the field [12]. This is also true for the smectic A phase even though the layering within that phase gives much increased and strongly anisotropic elastic constants resulting in a much reduced quadratic dependence of the voltage induced changes.

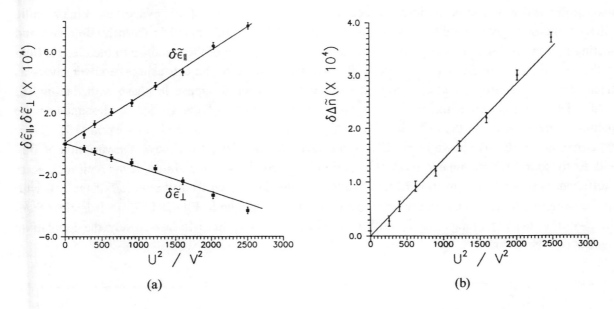

FIGURE 4 (a) Variation of $\delta\tilde{\epsilon}_{\parallel}$ and $\delta\tilde{\epsilon}_{\perp}$ with U^2 for the nematic
phase of 8CB. (b) Variation in the refractive index
anisotropy, $\delta\Delta\tilde{n}$, with U^2 in the nematic
phase of 8CB (from [25]).

C4 Pre-Transitional Electro-Optic Effect in 4-Hexyl-4′-Cyanobiphenyl (6CB)

In the pre-transitional effect, which occurs in the isotropic phase of liquid crystal materials at temperatures close to the nematic-isotropic phase transition, the field induced birefringence in the liquid crystal response can be orders of magnitude larger than for other standard liquids. These large effects are explained by the proximity of the nematic phase. On a microscopic level, the liquid has an inhomogeneous distribution of free energy density, and in regions favoured by a low free energy density transient nematic organisation occurs with a characteristic correlation length. The correlation length of these nematic regions is determined by the proximity of the nematic-isotropic phase transition. Under an applied field the local directors of the nematic regions will tend to align along the field, and hence the comparatively large pre-transitional Kerr effect is orientational in origin. The induced birefringence, δn (or $\delta\tilde{\epsilon}_r$), varies with the square of the applied electric field E and the Kerr coefficient is defined as

$$K = \delta n / \lambda E^2$$

where λ is the wavelength of incident radiation. The de Gennes' extended Landau theory [26,27] predicts that δn should be quadratic in the applied field, and in addition, that the reciprocal Kerr coefficient varies linearly with temperature. Both these results have been shown to be true for a number of materials [28,29].

To determine the very small field-induced birefringence in the pre-transitional region, Wood et al [13] have used a differential version of the metal-clad fully guided mode technique to investigate 6CB. Their sample geometry consisted of a ~4 μm layer of 6CB confined between two high refractive index (n = 1.80 at λ = 632.8 nm) 60° pyramids. The square pyramid faces were coated with ~50 nm of silver, after which half of each silver surface was coated with 20 nm of 60° evaporated silicon oxide (SiO_x). The SiO_x gives good homogeneous alignment of the liquid crystal in the nematic phase and coating half the cell with this aligning agent enabled a comparison of the response to the electric field in the two halves. An AC electric field of 900 Hz was applied to the cell using the silver layers as electrodes. The isotropic phase responded with an induced birefringence in phase with the applied field. This induced birefringence affects the angular positions of the guided modes and can be accurately determined by the differential procedure developed by Wood and co-workers [13]. Their experimental results are shown in FIGURES 5(a) and 5(b). FIGURE 5(a) shows the variation of $\delta\tilde{\varepsilon}_r$ with E^2 for four different temperatures: the relation is clearly linear. In addition, if the reciprocal Kerr coefficient is plotted as a function of temperature for the data taken from the region of the cell with surface aligning layers then a linear relationship is found as shown in FIGURE 5(b). It is clear from these results that the changes in $\tilde{\varepsilon}_r$ with both field and temperature follow closely the Landau-de Gennes type behaviour.

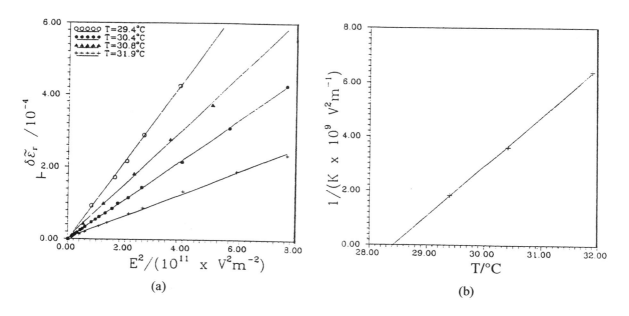

FIGURE 5 (a) Variation of $\delta\varepsilon_r$ with E^2 for 6CB. Note that the solid triangles are results obtained in the region of the cell with no surface aligning layers.
(b) Variation of the reciprocal Kerr coefficient with temperature for the region of the cell with surface aligning layers (from [13]).

Because the differential version of the metal-clad fully guided mode technique is very sensitive to changes of the optical properties of the liquid crystals, not only the induced birefringence, but also smaller changes in the light scattering properties of the liquid crystal under applied field are determined. These induced changes in light scattering affect the sharpness and depth of the reflectivity minima. This microscopic scattering effect is well modelled by changing the macroscopic value of the imaginary part of $\tilde{\varepsilon}_i$ for the material and adding this into the differential fitting procedure to compare theoretical prediction with the experimental data. The changes observed in $\tilde{\varepsilon}_i$ with electric field and temperature are more complex than the changes in the real part $\tilde{\varepsilon}_r$ and have been explained by considering both the surface and bulk free energies of the pre-transitional nematic in response to an applied field. A region of the cell with no surface aligning layers also gave a larger induced birefringence, at a given temperature, than that within the SiO_x coated region of the cell; this observation was explained by the same idea [13].

D CONCLUSION

Because of the nature of optical guided modes the waveguided mode techniques provide powerful tools for exploring the director distribution within liquid crystal cells. Thus the characterisation of those physical properties of liquid crystals which are related to the director profile and its response to external conditions (electric and magnetic fields, temperature, etc.) may be accurately accomplished by the guided mode techniques. By using a wide range of incident angles and multilayer optical theory to fit the obtained data some of the uncertainties and erroneous conclusions associated with integrated techniques (e.g. simple crossed polariser microscopy), which often ignore the surface layers, are avoided. However, it should be added that, dependent on the sample geometry and the measurements taken, the guided mode technique may require elaborate data analysis based on the Berreman transmission/reflection matrix and a complex fitting procedure to obtain the required results. It is essential that sufficient consideration is given to the sample geometry suitable for obtaining the information required before any one particular optical guided mode technique is chosen.

REFERENCES

[1]   P.K. Tien [*Appl. Opt. (USA)* vol.10 (1971) p.2395]

[2]   D. Marcuse (Ed.) [*Integrated Optics* (IEEE Press, New York, 1973)]

[3]   Fuzi Yang, J.R. Sambles [*J. Opt. Soc. Am. B (USA)* vol.10 (1993) p.858]

[4]   Fuzi Yang, J.R. Sambles [*J. Opt. Soc. Am. B (USA)* vol.16 (1999) p.1]

[5]   Y. Kawata, K. Takatoh, M. Hasegawa, M. Sakamoto [*Liq. Cryst. (UK)* vol.16 (1994) p.1027]

[6]   G.P. Bryan-Brown, J.R. Sambles, K.R. Welford [*J. Appl. Phys. (USA)* vol.73 (1993) p.3603]

[7]   Fuzi Yang, J.R. Sambles, G.W. Bradberry [*The Optics of Thermotropic Liquid Crystals* Eds. S.J. Elston, J.R. Sambles (Taylor & Francis, London, 1998) ch.5]

[8]   G.J. Sprokel, R. Santo, J.D. Swalen [*Mol. Cryst. Liq. Cryst. (UK)* vol.68 (1981) p.29]

[9]   K.R. Welford, J.R. Sambles [*Appl. Phys. Lett. (USA)* vol.50 (1987) p.871]

[10]   S.J. Elston, J.R. Sambles [*Appl. Phys. Lett. (USA)* vol.55 (1989) p.1621]

[11]   Lizhen Ruan, S.J. Elston, J.R. Sambles [*Liq. Cryst. (UK)* vol.10 (1991) p.369]

[12]   Lizhen Ruan, G.W. Bradberry, J.R. Sambles [*Liq. Cryst. (UK)* vol.11 (1992) p.655]

[13]   E.L. Wood, J.R. Sambles, P.S. Cann [*Liq. Cryst. (UK)* vol.16 (1994) p.983]

[14]   S.J. Elston, J.R. Sambles, M.G. Clark [*J. Mod. Opt. (UK)* vol.36 (1989) p.1019]

[15]   S. Ito, F. Kremer, E. Aust, W. Knoll [*J. Appl. Phys. (USA)* vol.75 (1994) p.1862]

[16]   C.R. Lavers, J.R. Sambles [*Ferroelectrics (UK)* vol.113 (1991) p.339]

[17]   D.A. Dunmur, P. Palffy-Muhoray [*J. Phys. Chem. (USA)* vol.92 (1988) p.1406]

[18]   A.R. MacGregor [*J. Phys. D (UK)* vol.21 (1988) p.1438]

[19]   T.F. Faber [*Proc. R. Soc. Lond. A (UK)* vol.24 (1977) p.493]

[20]   T.F. Faber [*Liq. Cryst. (UK)* vol.9 (1991) p.95]

[21]   S. Chandrasekhar [*Liquid Crystals* (Cambridge University Press, 1977) p.54]

[22]   Fuzi Yang, J.R. Sambles [*Jpn. J. Appl. Phys. (Japan)* vol.37 (1998) p.3998]

[23]   Fuzi Yang, J.R. Sambles, G.W. Bradberry [*J. Appl. Phys. (USA)* vol.85 (1999) p.728]

[24]   B.T. Hallam, Fuzi Yang, J.R. Sambles [*Liq. Cryst. (UK)* vol.26 (1999) p.657]

[25]   Lizhen Ruan, G.W. Bradberry, J.R. Sambles [*Liq. Cryst. (UK)* vol.12 (1992) p.799]

[26]   P.G. de Gennes [*The Physics of Liquid Crystals* (Clarendon Press, 1974) p.46]

[27]   G. Vertogen, W.H. de Jeu [*Thermotropic Liquid Crystals. Fundamentals* (Springer-Verlag, 1988) p.225]

[28]   B.R. Ratna, M.S. Vigava, R. Shashider, B.K. Sadashiva [*Pramana (India)* suppl.1 (1973) p.69]

[29]   J. Philip, T.A. Prasada Rao [*J. Phys. D (UK)* vol.25 (1992) p.1231]

7.4   Non-linear optical properties of nematic liquid crystals

I.C. Khoo

November 1998

A   INTRODUCTION

To date, almost all conceivable non-linear optical phenomena have been observed in nematic liquid crystals [1]. We present a review of their electronic and non-electronic optical non-linearities and some exemplary non-linear optical processes.

B   ELECTRONIC NON-LINEAR OPTICAL PROCESSES

FIGURE 1 depicts schematically the interaction of light with a liquid crystal molecule. Electronic interaction involves the perturbation of the electronic wave functions by the impinging optical fields [1]. The induced dipole moment of a molecule is given by:

$$\mu = \alpha{:}\mathbf{E} + \beta{:}\mathbf{EE} + \gamma{:}\mathbf{EEE} + \ldots\ldots \tag{1}$$

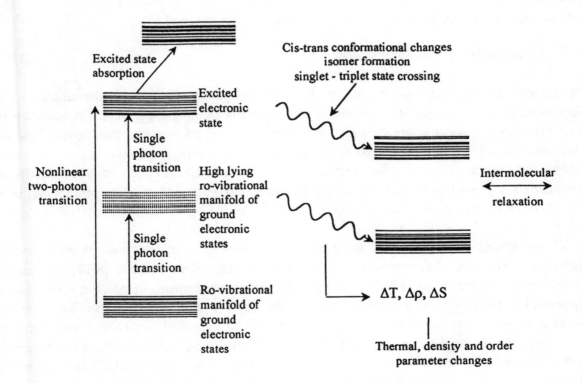

FIGURE 1 Schematic depiction of single and two-photon processes occurring in a liquid crystal molecule and various relaxation processes leading to temperature, density and order parameter changes [1,9].

where α, β, and γ are the linear, second order and third order non-linear polarisability tensors, **E** is the optical electric field and : denotes tensorial operation. The macroscopic polarisation **P** (dipole moment per unit volume) is

$$\mathbf{P} = \varepsilon_O \chi^{(1)} : \mathbf{E} + \chi^{(2)} : \mathbf{EE} + \chi^{(3)} : \mathbf{EEE} + \dots \dots \tag{2}$$

where $\chi^{(1)}$, $\chi^{(2)}$, $\chi^{(3)}$ are linear, second order and third order non-linear susceptibilities, respectively. The macroscopic parameters (χ's) are related to α, β and γ by the local field correction factors. For example, the third order non-linearities $\chi^{(3)}_{\parallel}$ and $\chi^{(3)}_{\perp}$ involved in DC field-induced second harmonic generation are related to γ by $\chi_{\parallel,\perp} = N f_o f_\omega^2 f_{2\omega} \gamma_{zzzz,xxxx}$, where N is the molecular number density, the f's are the local field correction factors, and ω is the frequency of the light [2]. We shall consider here typical effects associated with the second and third order non-linear susceptibilities.

B1   Second Harmonic Generation

In second harmonic generation [2-4], two photons from the incident laser photon at frequency ω excite the molecule to a virtual state, which emits a photon at 2ω. This process requires that the medium be non-centrosymmetric. In nematic liquid crystals which are centrosymmetric, an applied DC field, discontinuity at a surface or interface, or director orientational curvature could provide the required symmetry breaking mechanism [4,5]. In DC field-induced measurement, the non-linear polarisation involved is of the third order given by $P(2\omega) = \chi^{(3)} E(0) E(\omega) E(\omega)$. Typically measured values of $\chi^{(3)}$ are ~10^{-12} esu (see TABLE 1), expressed in the commonly used cgs unit. The conversion factor between cgs and SI unit is given by [1]:

$$\chi^{(3)} \text{ (esu)} = 80.77 \times 10^{17} \chi^{(3)} \text{ (SI unit)} \tag{3}$$

In ferroelectric or smectic C* liquid crystals, the molecules possess permanent polarisation, i.e. the centrosymmetry is broken, and it is possible to observe second harmonic generation associated with the second order non-linear polarisation [6]. Typically, the observed non-linear coefficients β_{2HG} are on the order of ~10^{-14} m/V. In contrast to inorganic crystals of much larger dimensions, these thin liquid crystal films are not useful for practical devices. Nonetheless, because of the sensitivity of second harmonic generation to surface and interface conditions, second harmonic generation has been proven to be an effective spectroscopic tool [7].

Two-photon absorption occurs where the molecule is excited to a real electronic excited state by simultaneous absorption of two photons from the incident light. Recent studies [8-10] have shown that this process could be an efficient optical self-limiting mechanism. Since the absorption is proportional to the square of the optical intensity, the incident light will experience an increasing absorption effect, and its transmission at high intensity will be 'quenched' to an almost constant level [9]. Two-photon absorptions are usually characterised by the two-photon absorption coefficient β, which is related to the imaginary part of $\chi^{(3)}$ [11]. For nematic liquid crystals [10,12], they are on the order of 3×10^{-12} m/W (equivalent to 0.5 cm/GW in the more commonly used unit).

B2 Third Order Non-Linear Susceptibilities and Intensity Dependent Refractive Index Change

There is relatively very little work done on measuring or studying the third order non-linear electronic polarisabilities of nematic liquid crystals. Most of the studies are centred on the molecular correlation effects associated with the nematic/isotropic phase transition, as exemplified by the work of Saha and Wong [2]. The typical magnitude of electronic $\chi^{(3)}$ is 10^{-12} esu.

A particular process associated with the third order non-linear polarisation is degenerate four wave mixing where three photons of the same frequency from the incident lasers are mixed to produce a fourth photon at the same frequency. Usually, this is performed with two coherent laser beams intersecting on the nematic liquid crystal film, with the generated laser beam emerging in the phase matched directions (FIGURE 2). The non-linear polarisation is of the form:

$$P(\omega) = \chi^{(3)} E(\omega) E^*(-\omega) E(\omega) \tag{4}$$

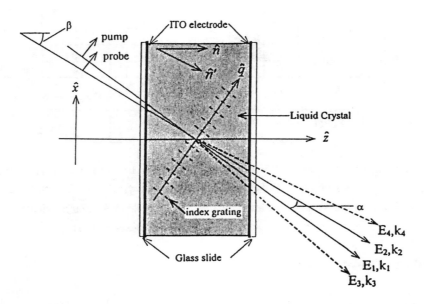

FIGURE 2 Schematic depiction of the interaction of polarised laser beams with a homeotropically aligned nematic liquid crystal film in a degenerate four wave mixing geometry [1].

This may be expressed in terms of an induced dielectric constant change $\Delta\varepsilon = \chi^{(3)} E(\omega) E^*(-\omega)$, or an intensity dependent refractive index change $\Delta n = n_2(\omega) [E(\omega)]^2 = n_2 I$. The conversion relationship between $\chi^{(3)}$ and the non-linear index coefficient n_2 is given by [1]:

$$\chi^{(3)} \text{ (esu)} = 9.54 \, n_o^2 n_2 \tag{5}$$

where n_o is the refractive index. As an example, for $\chi^{(3)} = 10^{-12}$ esu, and $n_o = 1.5$, we have $n_2 = 4.6 \times 10^{-14}$ cm^2/W, if the optical intensity I is in units of cm^2/W.

C   NON-ELECTRONIC NON-LINEAR OPTICAL PROCESSES

Following photo-absorption, the energy deposited in the molecules shows up in the form of temperature, density and order parameter changes. The dipolar interaction of the optical electric field with the birefringent liquid crystal also gives rise to collective reorientation of the director axis. These processes are characterised by much longer response and relaxation times (see FIGURE 3), but the resultant non-linearities such as n_2 are many orders of magnitude larger than their electronic counterparts [1,24].

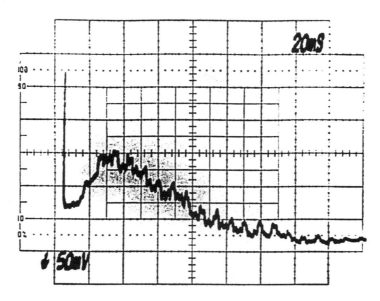

FIGURE 3 Oscilloscope trace showing the diffraction of a CW probe laser from gratings induced by a picosecond laser pulse in a nematic liquid crystal film. The first narrow spike is due to the temperature and density effects, while molecular reorientation and flow effects occur at much longer timescales [14].

C1   Thermal and Density Effects

Laser induced temperature and density changes are described by time dependent coupled hydrodynamics equations with three-dimensional boundary conditions [1]. In grating diffraction experiments (see FIGURE 2), the problem can be simplified to a one-dimensional thermal diffusion process along the grating wave-vector, if the grating spacing is smaller than the nematic film thickness and the beam size [13,14]. In that case, the thermal and Brillouin (sound wave) decay time constants are given by, respectively:

$$\tau_T = \rho_o \, C_v / \lambda_T q^2 \tag{6a}$$

$$\tau_B = 2\rho_o / \eta q^2 \tag{6b}$$

where $q = 2\pi/\lambda_q$ is the grating wave-vector, ρ_o the unperturbed density, C_v the specific heat, λ_T the thermal conductivity, and η the viscosity. The values of these parameters depend on the director axis alignment relative to the optical fields, thermal diffusion and sound propagation directions. For typical values of $\eta = 7 \times 10^{-2}$ kg m^{-1} s^{-1}, $\rho_o = 10^3$ kg m^{-3}, $D = \lambda_T/\rho_o \, C_v \sim 1.6 \times 10^{-7}$ m^2 s^{-1} and a grating spacing $\lambda_q = 20$ μm, we have $\tau_T \sim 50$ μs, whereas $\tau_B \sim 200$ ns. Such a vast difference between the relaxation times allows one to separate the thermal and density components in transient grating diffraction experiments [13,14].

In the steady state, i.e. the laser-nematic interaction time is much longer than τ_T and τ_B, the non-linear index coefficient is given by [1]:

$$n_2^{ss}(\Delta T) = (dn/dT) \, \alpha \, \tau_r/\rho_o \, C_v \tag{7}$$

where (dn/dT) is the appropriate thermal index coefficient. The absorption constant α of liquid crystals varies widely in the visible-infrared spectrum; near the phase transition temperature $dn_{\|,\perp}/dT$ also changes dramatically [1]. The magnitude of $n_2^{ss}(\Delta T)$ could therefore vary by several orders of magnitude. For $\alpha = 100$ cm^{-1}, and $dn_{\|,\perp}/dT \sim 10^{-3}$ K^{-1}, and $\lambda_q = 20$ μm, we have $n_2^{ss}(\Delta T) \sim 10^{-6}$ cm^2/W. This is the typical value observed in wave mixing experiments with infrared (CO_2 laser) in pure 4,4$'$pentylcyanobiphenyl or E7 (Merck mixture) or visible lasers in doped nematics.

In the transient regime ($\tau_p \ll \tau_r, \tau_B$), the non-linear index coefficient is proportional to the laser pulse duration [1]:

$$n_2(\tau_p) = n_2^{ss}(\Delta T) \, \tau_p/\tau_r \tag{8}$$

With $\tau_r \approx 0.5 \times 10^{-4}$ s, and $\tau_p = 1$ ns, $n_2(\tau_p) \sim 5 \times 10^{-11}$ cm^2/W. Such non-linearities are indeed observed in transient dynamic grating diffraction studies [14] involving nanosecond laser pulses. In these studies, it is found that the contribution from the electrostrictive effect (movement of the nematic due to the optical intensity gradient) to the refractive index change can be as large as the absorptive thermal component.

C2   Molecular Reorientation

By analogy to reorientation by AC or static fields, an impinging optical field can also realign the director axis of the nematics. Consider the interaction of a linearly polarised laser beam with a homeotropically aligned nematic liquid crystal film (see FIGURE 2). The equation describing the angular acceleration $d^2\theta/dt^2$ of the director axis is given by [1,13,14]:

$$I_m d^2\theta/dt^2 + \gamma_1 d\theta/dt + T_{el} + T_{op} + T_{vis} = 0 \tag{9}$$

where I_m is the moment of inertia of the liquid crystal, γ_1 is the viscosity coefficient, T_{el} is the elastic restoring torque, T_{op} the optical torque and T_{vis} the flow-rotational viscous torque.

The magnitude of the induced reorientation angle θ, and therefore the extraordinary index change $\Delta n = n_e (\beta + \theta) - n_e (\beta)$, depend on whether the interaction is transient or steady state, just as in the case of the thermal heating effect, as well as on other boundary conditions [1]. In the steady state case, to the first order of approximation, Δn is proportional to the optical intensity, i.e. $\Delta n = n_2^{ss}(\theta) I$. Under some experimental conditions, one can also show [1] that the transient non-linear index coefficient $n_2(\tau_p)$ is given by $n_2(\tau_p) \sim n_2^{ss}(\theta) \tau_p/\tau_r$, where $\tau_r = \gamma d^2/K\pi^2$ is the orientational relaxation time constant. For $d = 100$ μm, $\gamma = 0.1$ poise and $K = 10^{-6}$ dyn, we have $\tau_r \sim 1$ s. Typical observed values of $n_2^{ss}(\theta)$ are in the range of 10^{-5} - 10^{-4} cm^2/W. In the transient case, if $\tau_p = 1$ ns, then $n_2(\tau_p) \sim 10^{-12}$ cm/W ($\chi^{(3)} \sim 2 \times 10^{-11}$ esu). Experimental observations using nanosecond laser pulses [1,13,14] indeed give non-linearities close to this value.

The steady state non-linear index coefficient $n_2^{ss}(\theta)$ may be enhanced by about an order of magnitude with the application of a bias magnetic or electric field [15,16]. In dye-doped nematic films, studies have shown that the optically excited dye dopant molecules could exert an intermolecular torque T_m that could also enhance the optical reorientation process. The non-linear index coefficients $n_2^{ss}(\theta)$ obtained in grating diffraction or self-phase modulation experiments [17,18] in dye-doped nematic films are in the range of 10^{-3} - 10^{-2} cm^2/W.

C3   Photorefractive Effect

Several groups [19,20] have reported the observation of 'photorefractive' effects in nematic films containing photocharge producing dopants such as R6G dye and fullerene C_{60}. The basic mechanisms are illustrated in FIGURE 4. An incident optical intensity grating creates various DC space charge fields, which in combination with the applied DC field cause director axis reorientation. One of these DC space charge fields is similar to those usually encountered in inorganic crystals such as lithium niobate or semiconductors. For an incident optical intensity grating function of the form $I_{op} = I_o (1 + m \cos(q\xi))$, it is given by [20] that:

$$E_{ph} = q\nu \, mk_BT \, [(\sigma - \sigma_d)/(2e\,\sigma)] \cos(q\xi - \pi/2) \tag{10a}$$

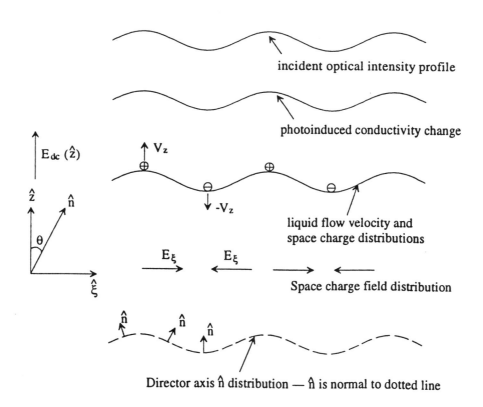

FIGURE 4 Schematic depiction of photorefractive effect, in which an incident optical intensity grating creates DC space charge fields, which in combination with the applied DC field, cause director axis reorientation.

where k_B = Boltzmann constant, σ = illuminated conductivity, σ_d = dark state conductivity, $\nu = (D+ - D-)/(D+ + D-)$, where $D+$ and $D-$ are the diffusion constants for the positively and negatively charged ions, respectively, m is the optical intensity modulation factor, $q = 2\pi/\lambda$ is the magnitude of the grating wave-vector, and ξ is the coordinate along q.

Note that there is a $\pi/2$ phase shift between the incident optical intensity function and the photorefractive space charge field. This phase shift is responsible for the two-beam coupling effect, a gain in energy of one incident beam from the other (see FIGURE 2). Gains as high as 3000 cm^{-1} were observed [21] in C_{60}-doped nematic film.

The other two sources of space charge fields arise from the conductivity and dielectric anisotropies of the nematic. For a spatially varying director axis reorientation angle θ, the space charge fields are given by [19]:

$$E_{\Delta\sigma} = E_{DC} \, [(\sigma_\| - \sigma_\perp) \sin\theta \cos\theta]/[\sigma_\| \sin 2\theta + \sigma_\perp \cos 2\theta] \tag{10b}$$

$$E_{\Delta\varepsilon} = E_{DC} \, [(\varepsilon_\| - \varepsilon_\perp) \sin\theta \cos\theta]/[\varepsilon_\| \sin 2\theta + \varepsilon_\perp \cos 2\theta] \tag{10c}$$

where $(\sigma_\| - \sigma_\perp)$ is the conductivity anisotropy and $(\varepsilon_\| - \varepsilon_\perp)$ the dielectric anisotropy, and E_{DC} is the applied DC field. In general, for a 25 μm thick homeotropically aligned doped film under a DC bias voltage of 1.5 V, the non-linear index coefficients observed are in the range of 10^{-3} - 10^{-2} cm^2/W. To date, fullerene C_{60} is found to be the best dopant.

D   THE MOST NON-LINEAR OPTICAL EFFECT OBSERVED TO DATE

Recent studies [22,23] in methyl-red doped nematic liquid crystal films have revealed an even larger non-linear effect. The responsible mechanism is the nematic director axis reorientation by the optically induced space charge fields. The much larger effect is possibly due to higher photo-charge and space charge field production efficiency in the methyl-red dye doped nematics, as evidenced by the generation of detectable DC photo-voltages across the ITO-electrode coated windows. More importantly, the effect does not require a DC bias voltage at all, and may be further enhanced by a very small AC bias field. This is a useful feature for practical application, as it avoids DC bias field-induced instabilities and dynamic scattering. In grating diffraction experiments [22,23], non-linear index coefficients n_2 of 6 cm^2/W ($\chi^{(3)} \sim 150$ esu) have been obtained.

This is arguably the most non-linear optical material known to date. Such extraordinarily large non-linearity enables the performance of several all-optical switching and limiting, image modulation and sensing processes at unprecedented low threshold power. In particular, self-defocusing and optical limiting effects with nanowatt power lasers, and incoherent/coherent image conversion at μW/cm^2 optical intensity have been demonstrated [23]. Such sensitivity rivals that obtainable from liquid crystal spatial light modulators [1].

E   CONCLUSION

In conclusion, nematic liquid crystals in their various pure and doped forms possess many interesting non-linear optical responses, with non-linearities ranging over sixteen orders of magnitude.   In TABLE 1, we summarise the non-linear electronic and non-electronic properties of nematic liquid crystals. Because of widely varying temporal, optical and liquid crystal parameters, only the typical magnitudes for these non-linearities could be quoted. Current active research is likely to result in even more interesting and useful non-linear effects in the near future.

TABLE 1 Electronic and non-electronic non-linearities of nematic liquid crystals.

Non-linear optical phenomena and non-linearities	Magnitude	Timescale	Liquid crystal
2nd harmonic generation (DC field induced) [3] $\chi^{(3)}_{\parallel}$	2.7×10^{-12} esu	ns	5CB
$\chi^{(3)}_{\perp}$	0.55×10^{-12} esu	ns	5CB
2nd harmonic generation β_{2HG} [6]	10^{-12} m/V	ps	(ferroelectric)
Two-photon absorption, β [12]	3×10^{-12} m/W	80 ps	5CB
Degenerate four wave mixing [3] electronic $\chi^{(3)}$ (ω; ω, $-\omega$, ω)	10^{-12} esu	ns	5CB
Orientational non-linearities [1] purely optically induced n_2	10^{-4} cm^2/W	ms	pure 5CB
excited dopant assisted n_2	10^{-3} cm^2/W	ms	dyed 5CB
photorefractive	10^{-3} cm^2/W	ms	5CB
photorefractive +	6 cm^2/W	ms	methyl-red doped 5CB
Thermal non-linearities [1] n_2	10^{-6} cm^2/W	ms	E7
n_2	10^{-11} cm^2/W	ns	E7
Density effect [1] electrostrictive n_2	10^{-11} cm^2/W	ns	E7

REFERENCES

[1]   I.C. Khoo [*Liquid Crystals: Physical Properties and Nonlinear Optical Phenomena* (Wiley Interscience, New York, 1984)]; I.C. Khoo, S.T. Wu [*Optics and Nonlinear Optics of Liquid Crystals* (World Scientific, River Edge, New Jersey, USA, 1993)]

[2]   S.K. Saha, G.K. Wong [*Appl. Phys. Lett. (USA)* vol.34 (1979) p.423]; S.K. Saha, G.K. Wong [*Opt. Commun. (Netherlands)* vol.30 (1979) p.119]

[3]   M.I. Barnik, L.M. Blinov, A.M. Dorozhkin, N.M. Shtykov [*Sov. Phys.-JETP (USA)* vol.54 (1981) p.935]

[4]   S.J. Gu, S.K. Saha, G.K.L. Wong [*Mol. Cryst. Liq. Cryst. (UK)* vol.69 (1981) p.287]

[5]   A.V. Sukhov, R.V. Timashev [*Pis'ma Zh. Eksp. Teor. Fiz. (Russia)* vol.51 (1990) p.364]

[6]   J.Y. Liu, M.G. Robinson, K.M. Johnson, D. Doroski [*Opt. Lett. (USA)* vol.15 (1990) p.267]

[7]   Y.R. Shen [*Liq. Cryst. (UK)* vol.5 (1989) p.635]

[8]   R.J. McEwan, R.C. Hollins [*J. Nonlinear Opt. Phys. Mater. (Singapore)* vol.4 (1995) p.245]

[9]   I.C. Khoo, M.V. Wood, B.D. Guenther, Min-Yi Shih, P.H. Chen [*J. Opt. Soc. Am. B (USA)* vol.15 (1998) p.1533]

[10]   A. Hochbaum, J.L. Ferguson, J.D. Buck [*SPIE (USA)* vol.1692 (1992) p.97]

[11]   F.W. Deeg, M.D. Feyer [*J. Chem. Phys. (USA)* vol.91 (1989) p.2269]

[12]   R. Macdonald, J. Schwartz, H.J. Eichler [*Int. J. Nonlinear Opt. Phys. (Singapore)* vol.1 (1992) p.103]; H.J. Eichler, R. Macdonald, B. Trosken [*Mol. Cryst. Liq. Cryst. (UK)* vol.231 (1993) p.1]

[13]   H.J. Eichler, R. Macdonald [*Phys. Rev. Lett. (USA)* vol.67 (1991) p.2666]

[14]   I.C. Khoo, R.G. Lindquist, R.R. Michael, R.J. Mansfield, P.G. LoPresti [*J. Appl. Phys. (USA)* vol.69 (1991) p.3853]

[15]   I.C. Khoo, S.L. Zhuang [*Appl. Phys. Lett. (USA)* vol.37 (1980) p.3]

[16]   C.L. Pan, S.H. Chen, H.H. Liao [*Phys. Rev. A (USA)* vol.33 (1986) p.4312]

[17]   I. Janossy, L. Csillag, A.D. Lloyd [*Phys. Rev. A (USA)* vol.44 (1991) p.8410]

[18]   W.M. Gibbons, P.J. Shannon, S.T. Sun, B.J. Swetlin [*Nature (UK)* vol.351 (1991) p.4]

[19]   I.C. Khoo [*IEEE J. Quantum Electron. (USA)* vol.32 (1996) p.525; *Opt. Lett. (USA)* vol.20 (1995) p.2137]; I.C. Khoo, H. Li, Yu Liang [*Opt. Lett. (USA)* vol.19 (1994) p.1723]

[20]   E.V. Rodenko, A.V. Sukhov [*JETP (Russia)* vol.78 (1994) p.875]

[21]   I.C. Khoo, B.D. Guenther, M. Wood, P. Chen, Min-Yi Shih [*Opt. Lett. (USA)* vol.22 (1997) p.1229]

[22]   I.C. Khoo, S. Slussarenko, B.D. Guenther, Min-Yi Shih, P.H. Chen, M.V. Wood [*Opt. Lett. (USA)* vol.23 (1998) p.253]

[23]   I.C. Khoo, M.V. Wood, P. Chen, Min-Yi Shih, B.D. Guenther [*SPIE (USA)* vol.3475 (1998) p.143]

[24]   Y.R. Shen [Datareview in this book: *7.5 Giant non-linear optical effects in nematic mesophases*]

7.5   Giant non-linear optical effects in nematic mesophases

Y.R. Shen

July 1998

A   INTRODUCTION

Nematic liquid crystals are composed of highly anisotropic organic molecules. The orientational correlations between molecules are extraordinarily strong so that they behave like anisotropic fluids. They are optically birefringent with the optical axis defined by the average molecular orientation known as the director. The latter can be easily oriented by an external field, such as a DC electric, a magnetic or an optical field. Optical-field-induced director reorientation gives rise to an extraordinarily large refractive index change, and hence a giant optical non-linearity in the nematic [1]. It leads to a number of very unusual non-linear optical effects in nematics [2-7].

B   GIANT OPTICAL NON-LINEARITY

That the director of a nematic can be oriented by a DC field was known at the beginning of the century, but the same effect with an optical field was only recognised in 1979 by Herman and Serinko [1]. Optical-field-induced director reorientation works against resistance to distortion of the original director alignment imposed on the medium by the boundary conditions. The optical torque per unit volume on the director **n** by an optical field **E** is given by [2-7]

$$T_{opt} = (\Delta\tilde{\varepsilon}/2\pi)[(\mathbf{n} \bullet E)(\mathbf{n} \times E)^*] \tag{1}$$

where $\Delta\tilde{\varepsilon} = \tilde{\varepsilon}_\parallel - \tilde{\varepsilon}_\perp$, and $\tilde{\varepsilon}_\parallel$ and $\tilde{\varepsilon}_\perp$ are the optical dielectric constants parallel and perpendicular to **n**, respectively. The elastic torque per unit volume against orientational deformation is obtained from $T_{el} = \delta F/\delta\theta$, where θ is the angle of rotation of **n**, and F_{el} is the deformation energy density given by [8]

$$F_{el} = \left(\frac{1}{2}\right)\left\{K_1(\nabla \bullet \mathbf{n})^2 + K_2(\mathbf{n} \bullet \nabla \times \mathbf{n})^2 + K_3[\mathbf{n} \times (\nabla \times \mathbf{n})]^2\right\} \tag{2}$$

Here K_1, K_2 and K_3 are the splay, twist, and bend elastic constants, respectively. Balance of the two torques $T_{opt} + T_{el} = 0$ yields the equilibrium configuration of the director described by $\mathbf{n}(r)$. In the time-dependent case, there is in addition a viscous torque on the director acting against its rotation. Strictly speaking, a complete analysis of director rotation in a nematic requires five independent viscosity coefficients [9,10]. However, for simplicity, a single phenomenological viscosity coefficient (γ) is often used and the viscous torque takes the form [2,3]

$$T_{vis} = -\gamma \, \mathbf{n} \times (\partial\mathbf{n}/\partial t) \tag{3}$$

366

The torque balance $T_{opt} + T_{el} + T_{vis} = 0$ then yields $\mathbf{n}(\mathbf{r},t)$. The local refractive index tensor is uniaxial, having values $\tilde{n}_{\parallel}$ and $\tilde{n}_{\perp}$ parallel and perpendicular to $\mathbf{n}$, respectively, with $|\tilde{n}_{\parallel} - \tilde{n}_{\perp}| \sim 0.15$ for many nematics. Director reorientation by an angle $\Delta\theta$ leads to a refractive index change $\Delta\tilde{n} = (\partial\tilde{n}/\partial\theta)\,\Delta\theta$.

For an order-of-magnitude estimate, we note that appreciable reorientation of $\mathbf{n}(\mathbf{r})$ is expected when the optical field energy density is close to the elastic energy density, i.e. $\varepsilon_{o}\Delta\tilde{\varepsilon}\,|E|^{2} \sim K/L^{2}$, where L is the nematic film thickness. Taking typical values of $K \sim 10^{-11}$ N, $\Delta\varepsilon \sim 0.1$, and $L \sim 100$ μm, we find $|E| \sim 300$ V/cm. Such an intensity can be easily obtained from a focused CW laser beam. It can be further reduced by a DC bias field [2,7] or in a doped nematic [11,12].

The refractive indices of a nematic can also be changed appreciably by laser heating, but the effect is much less significant than that of director reorientation and will not be discussed further here.

C   GIANT NON-LINEAR OPTICAL EFFECTS

We describe non-linear optical effects resulting from strong refractive index changes induced by a laser via director reorientation in a nematic.

C1   Degenerate Wave Mixing

Interference of laser beams can lead to the formation of a pattern of refractive index changes $\Delta\tilde{n}$ in a nematic film, which in turn diffracts the incoming laser beams. Diffraction can be of multiple orders because of the very large $\Delta\tilde{n}$; an example is shown in FIGURE 1 [13]. Phase conjugation is a special case of degenerate four-wave mixing. In a nematic film, because of the large $\Delta\tilde{n}$, phase conjugation can be self-started with a laser power of only about 100 mW [14].

C2   Self-Phase Modulation

The intensity profile of a laser beam traversing a nematic film naturally produces a $\Delta\tilde{n}$ depending on the transverse coordinate ρ, and hence a phase modulation or wavefront distortion $\Delta\psi(\rho)$ (known as self-phase modulation) [15]. The phase modulation is given by

$$\Delta\Psi(\rho) = (2\pi/\lambda)\int_{0}^{L}\Delta n(\rho, z)\,dz \tag{4}$$

and the maximum $\Delta\Psi$ can reach a value of 100π or more. This strong spatial phase modulation gives rise to strong self-focusing and a spectacular diffraction effect in the form of multiple rings; an example is shown in FIGURE 2(a).

The diffraction effect can be understood physically. For simplicity, consider a cylindrically symmetric laser beam. The induced $\Delta\tilde{n}$, and hence $\Delta\Psi$, should have a bell-like shape with respect to the radial coordinate ρ. As displayed in FIGURE 2(b), there are always two points on the curve of $\Delta\Psi(\rho)$ with the same slope. Since $\partial\Delta\Psi/\partial\rho$ corresponds to the transverse wavevector of a propagating wave, the waves originating from two points with the same $\delta\Delta\Psi/\delta\rho$ must propagate in the same

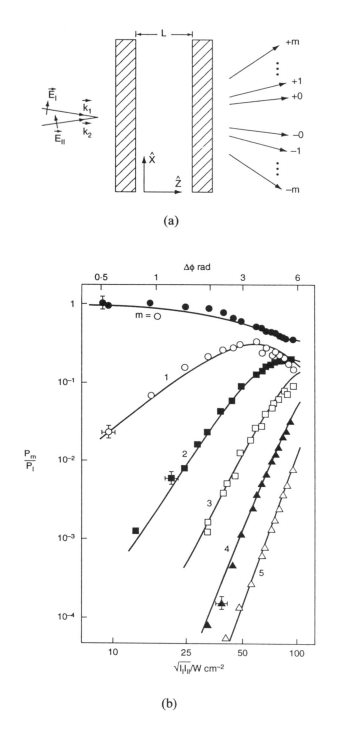

(a)

(b)

FIGURE 1 (a) Experimental geometry for degenerate wave mixing in a homeotropic nematic film 250 μm thick. (b) Diffracted beam powers of various orders from a laser-induced 71 μm grating. The solid lines are theoretical curves (after [13]).

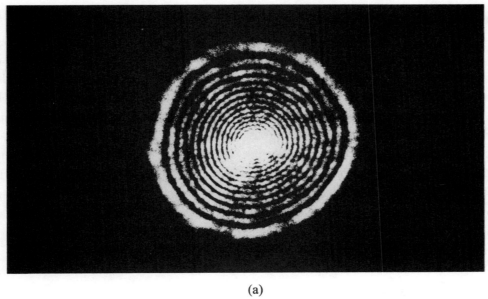

(a)

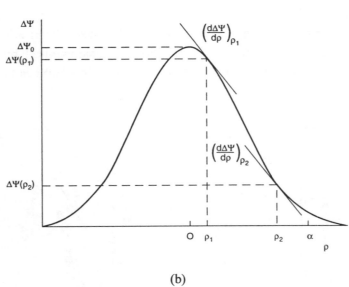

(b)

FIGURE 2 (a) Picture of typical far-field diffraction ring pattern due to self-phase modulation in a nematic film. (b) Profile of laser-induced phase shift. Light diffracted at ρ_1 and ρ_2 has the same wavevector and interferes in the far field to produce rings (after [15]).

direction, and should, therefore, interfere in the far field. The interference is constructive or destructive, depending on the phase difference between the two waves, namely $2n\pi$ for maximum constructive interference and $(2n + 1)\pi$ for maximum destructive interference, with n being an integer. This then leads to the ring pattern in diffraction. The total number of rings in the pattern is given by $N \sim \Delta\Psi_0/2\pi$.

In general, there can be many other non-linear optical effects that may result from laser-induced refractive index change, for example optical bistability [16], optical self-switching [17], optical transistor action [18] and transverse pattern formation [19]. They are not unique to nematics but here they can be extraordinarily strong and readily observed.

C3   Optical Fréedericksz Transition

The optical Fréedericksz transition is analogous to the well-known Fréedericksz transition induced by a DC field. Similar to the DC-field-induced alignment, the optical field tends to align the director **n** parallel to **E**. If initially **n** and **E** are orthogonal, the optical torque on **n** must vanish. However, fluctuations of **n** away from equilibrium result in a finite optical torque on **n** which would amplify the deviation from equilibrium. If the optical torque is smaller than the elastic restoring torque, then the fluctuations are damped and the original equilibrium state remains stable. Otherwise, reorientation of **n** would occur. The transition from absence to presence of reorientation happens when the laser intensity reaches a threshold value I_{th}. This transition is known as the optical Fréedericksz transition and is second order in nature [20-22]. One example is the case of a linearly polarised laser beam normally incident on a homeotropically aligned nematic film. If the laser beam cross-section is much larger than the film thickness, the threshold intensity is given by

$$I_{th} = \pi^2 n_e^2 cK /(\Delta\widetilde{\epsilon})n_o L^2 \tag{5}$$

where K is an effective elastic constant and L is the film thickness. Typically, I_{th} is of the order of 100 to 1000 W/cm^2 with L ~ 100 μm. The transition is characterised by the onset of optical birefringence due to director reorientation seen by the normally incident beam; an example is shown in FIGURE 3.

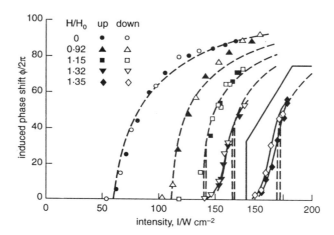

FIGURE 3 Phase retardation $\phi/2\pi$ versus laser intensity for fixed magnetic field strengths. Open symbols were measured with increasing laser intensity; solid symbols with decreasing intensity. Solid lines are guides to the eye. Broken lines are theoretical fits. H_o is the magnetic field at which the second order optical Fréedericksz transition is expected to become first order (after [24]).

Because the orienting actions of DC and optical fields are similar, the optical Fréedericksz transition threshold can be lowered by biasing the nematic film with a destabilising DC electric or magnetic field. Conversely, if a stabilising DC bias field is used, I_{th} would increase. When the bias field is sufficiently strong, the transition could change from second order to first order. This is also illustrated in FIGURE 3, where the first order transition is characterised by a hysteresis loop in the curve [23,24].

There are other geometries for the observation of optical Fréedericksz transitions using linearly polarised laser beams [2]; some are of first order, and others second order. Unlike the DC case, optical Fréedericksz transitions can also be induced by circularly or elliptically polarised laser beams [2,7]. For the case of a normally incident beam on a homeotropic nematic film the threshold intensity for the transition with circular polarisation is two times higher than with linear polarisation, and that with elliptical polarisation is in between.

C4   Laser-Induced Non-Linear Dynamics

The laser field can orient the director above the Fréedericksz transition threshold, but the orientation depends on the laser polarisation. We then expect, in general, that as reorientation alters the local optical birefringence in the nematic, the latter in turn can modify the polarisation of the beam propagating in the medium. The interplay of the two can lead to interesting non-linear dynamic phenomena that have been experimentally observed and are physically understood [25-27].

Consider a circularly polarised laser beam propagating normally into a homeotropic nematic film. Director reorientation makes the film locally birefringent and converts the circular polarisation into elliptical at the output. Hence a net angular momentum $(e_{in} - e_{out})I/\omega$ per unit time is deposited in the film, where e describes the ellipticity of the polarisation and ω is the optical frequency. This produces a torque on the director and sets the director into rotational motion. The latter in turn causes the elliptical output polarisation to rotate. In the steady state, the viscous torque, which is proportional to the angular velocity of the rotation, Ω, must be in balance with the optical torque, and Ω can be determined from the result. To set the director into motion, energy must be transferred from the optical field to the nematic. The loss of laser energy in a transparent medium can happen only if part of the transmitted laser beam is red-shifted. In fact, the rotating elliptical polarisation can be decomposed into two circular polarisation components, with one at a frequency 2Ω lower than the other. The process can then be described as a stimulated Raman-like scattering process in which the Stokes component generated is red-shifted by 2Ω and has an opposite circular polarisation from the pump laser beam. Since Ω is not a characteristic constant of the medium, but depends on the laser intensity, the process can be called self-induced stimulated light scattering [26].

With elliptical input polarisation, the situation is more complicated although the underlying basic physics is the same: the interplay between director orientation and beam polarisation change causes angular momentum exchange and energy transfer between the optical field and the medium. Depending on the input laser intensity and ellipticity, the induced non-linear dynamics fall into different regimes characterised by undistorted configuration, distorted equilibrium configuration, persistent oscillation, and precession-nutation of the director [27]. The phase diagram describing the various dynamic regimes is shown in FIGURE 4. With increasing laser intensity, the non-linear dynamics may go through a sequence of transitions and enter a deterministic chaotic regime [28]. The route to chaos may be a mixing of two scenarios: (a) period oscillation $\Rightarrow$ quasiperiodic oscillation $\Rightarrow$ chaos, and (b) periodic oscillation $\Rightarrow$ period-doubling cascade $\Rightarrow$ chaos.

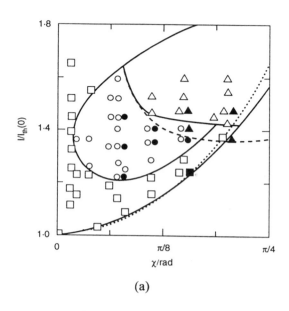

(a)

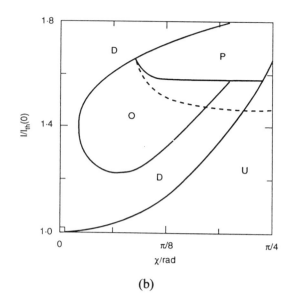

(b)

FIGURE 4 Phase diagrams of various dynamic regimes in the (I, χ) plot, where χ is related to the polarisation ellipticity e by $e = \sin(2\chi)$. (a) Experimental observations: squares, circles, and triangles denote distorted-equilibrium, persistent-oscillation, and precession-nutation states, respectively. Open symbols refer to states obtained with increasing laser intensity, and solid symbols to states with decreasing intensity. (b) Theoretical simulation; U, D, O, and P refer to undistorted, distorted-equilibrium, persistent-oscillation, and precession-nutation regimes, respectively. Solid lines describe boundaries between different dynamic regimes. The dashed lines describe boundaries at which P switches back to U, D, or O as the pump intensity increases (after [27]).

D   CONCLUSION

Director reorientation by a laser field can render extraordinarily large refractive index changes in nematics and lead to giant non-linear optical effects, many of which are not observable in other media.

E   ACKNOWLEDGEMENT

This work was supported by NSF Grant no. DMR-9704384.

REFERENCES

[1]   R.M. Herman, R.J. Serinko [*Phys. Rev. (USA)* vol.19 (1979) p.1757]
[2]   N.V. Tabiryan, A.V. Sukhov, B.Y. Zel'dovich [*Mol. Cryst. Liq. Cryst. (UK)* vol.136 (1986) p.1]
[3]   Y.R. Shen [*Philos. Trans. R. Soc. Lond. A (UK)* vol.313 (1984) p.327]
[4]   I.C. Khoo, Y.R. Shen [*Opt. Eng. (USA)* vol.24 (1985) p.579]

[5]   I.C. Khoo [*Prog. Opt. (Netherlands)* vol.26 (1988) p.107]

[6]   F. Simoni [in *Physics of Liquid Crystalline Materials* Eds. I.C. Khoo, F. Simoni (Gordon and Breach, New York, 1991) p.365]

[7]   E. Santamento, Y.R. Shen [in *Handbook of Liquid Crystal Research* Eds. P.J. Collings, J.S. Patel (Oxford University Press, New York, 1997) p.539]

[8]   F.C. Frank [*Discuss. Faraday Soc. (UK)* vol.25 (1958) p.19]

[9]   J.L. Ericksen [*Phys. Fluids (USA)* vol.9 (1966) p.1205]

[10]   F.M. Leslie [*Arch. Ration. Mech. Anal. (Germany)* vol.28 (1968) p.265]

[11]   I. Janossy, A.D. Lloyd, B.S. Wherret [*Mol. Cryst. Liq. Cryst. (UK)* vol.179 (1990) p.1]

[12]   I.C. Khoo, S. Slussarenko, B.D. Guentier, M.Y. Shih, P. Chen, W.V. Wood [*Opt. Lett. (USA)* vol.23 (1998) p.253]

[13]   S.D. Durbin, S.M. Arakelian, Y.R. Shen [*Opt. Lett. (USA)* vol.7 (1982) p.145]

[14]   I.C. Khoo, P.Y. Yan, G.M. Finn, T.H. Liu, R.R. Michael [*J. Opt. Soc. Am. B (USA)* vol.5 (1988) p.202]

[15]   S.D. Durbin, S.M. Arakelian, Y.R. Shen [*Opt. Lett. (USA)* vol.6 (1981) p.411]

[16]   M.M. Cheung, S.D. Durbin, Y.R. Shen [*Opt. Lett. (USA)* vol.6 (1983) p.39]

[17]   I.C. Khoo [*Appl. Phys. Lett. (USA)* vol.40 (1982) p.645]

[18]   E. Santamato, A. Sasso, R. Bruzzese, Y.R. Shen [*Opt. Lett. (USA)* vol.11 (1986) p.452]

[19]   M. Tamburrini, M. Bonavita, S. Wabnitz, Y.R. Shen [*Opt. Lett. (USA)* vol.18 (1993) p.855]

[20]   A.S. Zolot'ko, V.F. Kitaeva, N. Kroo, N.N. Sobolev, L. Csillag [*JETP Lett. (USA)* vol.32 (1980) p.158]

[21]   B.Y. Zel'dovich, N.F. Pilipetskyi, A.V. Sukhov, N.V. Tabiryan [*JETP Lett. (USA)* vol.32 (1980) p.263]

[22]   S.D. Durbin, S.M. Arakelian, Y.R. Shen [*Phys. Rev. Lett. (USA)* vol.47 (1981) p.1411]

[23]   H.L. Ong [*Phys. Rev. A (USA)* vol.28 (1983) p.2393]

[24]   A.J. Karn, S.M. Arakelian, Y.R. Shen [*Phys. Rev. Lett. (USA)* vol.57 (1986) p.448]

[25]   E. Santamato, B. Saino, M. Romagnoli, M. Settembre, Y.R. Shen [*Phys. Rev. Lett. (USA)* vol.57 (1986) p.2423]

[26]   E. Santamato, B. Saino, M. Romagnoli, M. Settembre, Y.R. Shen [*Phys. Rev. Lett. (USA)* vol.61 (1988) p.113]

[27]   E. Santamato, G. Abbete, P. Maddalena, L. Marrucci, Y.R. Shen [*Phys. Rev. Lett. (USA)* vol.64 (1990) p.1377]

[28]   G. Cipparrone, V. Carbone, G. Versace, C. Umeton, R. Bartolino, F. Simoni [*Phys. Rev. Lett. (USA)* vol.47 (1993) p.3741]

CHAPTER 8

FLOW PROPERTIES

8.1 Introduction to nematodynamics

F.M. Leslie

April 1999

A   INTRODUCTION

Continuum theory for nematic liquid crystals has its origins in the 1920s in the work of Oseen [1] and Zocher [2], who largely developed the static theory. The first to attempt the formulation of a dynamic theory was Anzelius [3], who was a student of Oseen, but an acceptable version had to await developments in non-linear continuum mechanics many years later, as well as further experimental studies by Zwetkoff [4] and Miesowicz [5]. A full account of the early development of dynamic theory for nematics can be found in a paper by Carlsson and Leslie [6].

Ericksen [7] was the first to attack this problem from a sound mechanics viewpoint, deriving the balance laws of nematodynamics as a generalisation of his formulation of equilibrium theory based on virtual work, and Leslie [8] essentially completed the theory by proposing expressions for the various dynamic contributions. Their derivation employs generalised forces that arise naturally in their formulation, but in the following sections we give a somewhat more straightforward presentation of their successful theory [9].

B   BALANCE LAWS

The basis of any continuum theory is the balance or conservation laws for mass, linear and angular momentum, and energy. For the present purposes our attention is limited to incompressible, isothermal processes, and therefore balance laws for mass and energy are somewhat superfluous, the former simply reducing to the obvious constant density plus a constraint upon the velocity, while the latter is essentially replaced by a balance of rate of work. Our two balance laws are therefore those for linear and angular momentum, the former basically that for an isotropic continuum, except that the stress need no longer be symmetric, but that for angular momentum is augmented by the inclusion of explicit external body and surface moments [9]. For any volume V of a nematic bounded by a surface S, we assume for linear momentum that

$$\frac{d}{dt} \int_V \rho \mathbf{v} dv = \int_V \mathbf{F} dv + \int_S \mathbf{t} ds \tag{1}$$

where ρ denotes the density, $\mathbf{v}$ is the velocity, $\mathbf{F}$ is the external body force per unit volume, and $\mathbf{t}$ is the surface force per unit area, and also for angular momentum that

$$\frac{d}{dt} \int_V \rho \mathbf{x} \times \mathbf{v} dv = \int_V (\mathbf{x} \times \mathbf{F} + \mathbf{K}) dv + \int_S (\mathbf{x} \times \mathbf{t} + \boldsymbol{\ell}) ds \tag{2}$$

Here **x** represents the position vector, **K** is the external body moment per unit volume, and ℓ is any surface moment per unit area. In this equation, the velocity **v** is subject to the constraint

$$\text{div } \mathbf{v} = 0 \tag{3}$$

and the time derivative is the material time derivative, while in the latter any contribution to the angular momentum arising from the local microstructure is ignored given that it must be negligible.

The force and moment per unit area on a surface with unit normal $\boldsymbol{\nu}$ can be shown to take the forms

$$\mathbf{t} = -p\nu + \mathbf{T}\nu, \quad \boldsymbol{\ell} = \mathbf{L}\nu \tag{4}$$

where p is an arbitrary pressure arising from incompressibility, and **T** and **L** are the stress and couple stress matrices or tensors, respectively. As a result we can rewrite EQNS (1) and (2) in point form (i.e. true at all points in the fluid) as

$$\rho \frac{d\mathbf{v}}{dt} = \mathbf{F} - \text{grad } p + \text{div } \mathbf{T}, \quad \mathbf{K} + \hat{\mathbf{T}} + \text{div } \mathbf{L} = 0 \tag{5}$$

where $\hat{\mathbf{T}}$ is the axial vector associated with the asymmetric stress, and, denoting the elements or components of **T** by t_{ij}, then

$$\hat{T}_x = t_{zy} - t_{yz}, \quad \hat{T}_y = t_{xz} - t_{zx}, \quad \hat{T}_z = t_{yx} - t_{xy} \tag{6}$$

In EQNS (5) the divergence is with respect to the second index of both **T** and **L**. Clearly, if external moments are absent, the stress must be symmetric as for isotropic liquids.

In addition to these equations, we employ a rate of work hypothesis, which is effectively a balance of energy with thermal effects assumed to be negligible. Thus we consider

$$\int_V (\mathbf{F} \cdot \mathbf{v} + \mathbf{K} \cdot \mathbf{w}) dv + \int_S (\mathbf{t} \cdot \mathbf{v} + \boldsymbol{\ell} \cdot \mathbf{w}) ds = \frac{d}{dt} \int_V \left(\frac{1}{2} \rho \mathbf{v} \cdot \mathbf{v} + W \right) dv + \int_V D dv \tag{7}$$

where **w** denotes the angular velocity of the fluid element, W is the Frank-Oseen energy of static theory, so that

$$W = W(\mathbf{n}, \nabla \mathbf{n}), \quad \mathbf{n} \cdot \mathbf{n} = 1 \tag{8}$$

n again is a unit vector, and D is the rate of viscous dissipation per unit volume. Once again using the representations (4), and EQNS (3) and (5), EQN (7) can be expressed in point form as

$$\text{tr}\left\{ \mathbf{T}(\nabla \mathbf{v})^T \right\} + \text{tr}\left\{ \mathbf{L}(\nabla \mathbf{w})^T \right\} - \mathbf{w} \cdot \hat{\mathbf{T}} = \frac{dW}{dt} + D \tag{9}$$

which we exploit later to derive expressions for the stress and couple stress; the superscript T indicates the transpose of the matrix.

C   STRESS AND COUPLE STRESS

To proceed we must make certain assumptions concerning the kinematic variables to include in the equations for the stress and couple stress, but first it is possible to deduce the equilibrium values for these quantities by appeal to our rate of working assumption EQN (7) or EQN (9). This requires that

$$\frac{dW}{dt}$$

is calculated by using

$$\frac{d\mathbf{n}}{dt} = \mathbf{w} \times \mathbf{n}, \quad \frac{d}{dt}(\nabla \mathbf{n}) = \nabla\left(\frac{d\mathbf{n}}{dt}\right) - \nabla \mathbf{n} \nabla \mathbf{v} \tag{10}$$

and an identity satisfied by the Frank-Oseen energy

$$\mathbf{n} \otimes \frac{\partial W}{\partial \mathbf{n}} + \nabla \mathbf{n}\left(\frac{\partial W}{\partial \nabla \mathbf{n}}\right)^{T} + (\nabla \mathbf{n})^{T} \frac{\partial W}{\partial \nabla \mathbf{n}} = \frac{\partial W}{\partial \mathbf{n}} \otimes \mathbf{n} + \frac{\partial W}{\partial \nabla \mathbf{n}} (\nabla \mathbf{n})^{T} + \left(\frac{\partial W}{\partial \nabla \mathbf{n}}\right)^{T} \nabla \mathbf{n} \tag{11}$$

where the notation $\mathbf{a} \otimes \mathbf{b}$ represents the 3×3 matrix with (i,j)th element $a_i b_j$. The first of EQNS (10) of course uses the fact that the director is a unit vector. The identity follows from the invariance requirement

$$W(\mathbf{n}, \nabla \mathbf{n}) = W(\mathbf{Pn}, \mathbf{P}\nabla \mathbf{n}\mathbf{P}^{T}) \tag{12}$$

by choosing the proper orthogonal matrix $\mathbf{P}$ in the form

$$\mathbf{P} = \mathbf{I} + \varepsilon\mathbf{R}, \quad \mathbf{R}^{T} = -\mathbf{R} \tag{13}$$

$\mathbf{I}$ is the unit matrix and $\mathbf{R}$ is an arbitrary skew symmetric matrix; the result follows by expanding the right hand side of EQN (12) to first order in the small parameter ε.

Exploiting EQNS (10) and (11) we find that

$$\frac{dW}{dt} = \text{tr}\left\{\mathbf{T}^{S}(\nabla \mathbf{v})^{T}\right\} + \text{tr}\left\{\mathbf{L}^{S}(\nabla \mathbf{w})^{T}\right\} - \mathbf{w} \cdot \hat{\mathbf{T}}^{S} \tag{14}$$

where $\hat{\mathbf{T}}^{S}$ is the axial vector associated with the matrix $\mathbf{T}^{S}$ and

$$\mathbf{T}^{S} = -(\nabla \mathbf{n})^{T} \frac{\partial W}{\partial \nabla \mathbf{n}}, \quad \mathbf{L}^{S} = \mathbf{n} \times \frac{\partial W}{\partial \nabla \mathbf{n}} \tag{15}$$

379

the vector product in the latter with respect to the first index of the derivative. In this way we can rewrite EQN (9) as

$$\mathrm{tr}\left\{\left(\mathbf{T}-\mathbf{T}^{S}\right)\left(\nabla\mathbf{v}\right)^{T}\right\}+\mathrm{tr}\left\{\left(\mathbf{L}-\mathbf{L}^{S}\right)\left(\nabla\mathbf{w}\right)^{T}\right\}-\mathbf{w}\cdot\left(\hat{\mathbf{T}}-\hat{\mathbf{T}}^{S}\right)=D \tag{16}$$

However, the rate of viscous dissipation D must of course be positive, and this at once requires that the coefficients of terms linear in the gradients of velocity and angular velocity and in the angular velocity be zero, allowing us to deduce that stress and couple stress must take the forms

$$\mathbf{T}=-\left(\nabla\mathbf{n}\right)^{T}\frac{\partial W}{\partial\nabla\mathbf{n}}+\mathbf{T}^{d}, \quad \mathbf{L}=\mathbf{n}\times\frac{\partial W}{\partial\nabla\mathbf{n}}+\mathbf{L}^{d} \tag{17}$$

$\mathbf{T}^{d}$ and $\mathbf{L}^{d}$ denote dynamic terms, and the rate of work assumption (9) reduces to

$$\mathrm{tr}\left\{\mathbf{T}^{d}\left(\nabla\mathbf{v}\right)^{T}\right\}+\mathrm{tr}\left\{\mathbf{L}^{d}\left(\nabla\mathbf{w}\right)^{T}\right\}-\mathbf{w}\cdot\hat{\mathbf{T}}^{d}=D\geq0 \tag{18}$$

These expressions (17) are in agreement with the original results obtained by Ericksen [7].

To derive expressions for the dynamic contribution we must make some assumptions as to the relevant variables to include in these terms. It seems reasonable given the experiments of Zwetkoff [4] and Miesowicz [5] to assume that at any material point $\mathbf{T}^{d}$ and $\mathbf{L}^{d}$ are functions of $\mathbf{n}$, $\nabla\mathbf{v}$ and $\mathbf{w}$, evaluated at that point at that instant. However, since the gradient of the angular velocity is not included in our list of kinematic variables, it immediately follows from EQN (18) that

$$\mathbf{L}^{d}=0 \tag{19}$$

a conclusion reached by Leslie [8] as a result of a thermodynamic argument. Our inequality (18) now reduces to

$$\mathrm{tr}\left\{\mathbf{T}^{d}\left(\nabla\mathbf{v}\right)^{T}\right\}-\mathbf{w}\cdot\hat{\mathbf{T}}^{d}=D\geq0 \tag{20}$$

which restricts the viscous stress.

Invariance to superposed rigid body motions requires instead that $\mathbf{T}^{d}$ is a function of $\mathbf{n}$, $\mathbf{A}$ and ω, where $\mathbf{A}$ is the familiar rate of strain matrix and ω is the angular velocity relative to the background rotation of the continuum; thus

$$2\mathbf{A}=\nabla\mathbf{v}+\left(\nabla\mathbf{v}\right)^{T}, \quad \omega=\mathbf{w}-\frac{1}{2}\,\mathrm{curl}\,\mathbf{v} \tag{21}$$

This leads us to assume that $\mathbf{T}^{d}$ is a function of $\mathbf{n}$, $\mathbf{A}$ and $\mathbf{N}$, where

$$\mathbf{N}=\omega\times\mathbf{n}=\frac{d\mathbf{n}}{dt}-\frac{1}{2}\left(\nabla\mathbf{v}-\left(\nabla\mathbf{v}\right)^{T}\right)\mathbf{n} \tag{22}$$

essentially discounting a rotation parallel to the director, which can be shown to be zero in any event [9]. Nematic symmetry requires that $\mathbf{T}^d$ be isotropic, and independent of a change of sign in the director $\mathbf{n}$. The experiments by Zwetkoff [4] and Miesowicz [5] imply a linear dependence upon $\mathbf{A}$ and $\mathbf{N}$, and accepting this we can show that

$$\mathbf{T}^d = \alpha_1 \mathbf{n} \cdot \mathbf{Ann} \otimes \mathbf{n} + \alpha_2 \mathbf{N} \otimes \mathbf{n} + \alpha_3 \mathbf{n} \otimes \mathbf{N} + \alpha_4 \mathbf{A}$$

$$+ \alpha_5 \mathbf{An} \otimes \mathbf{n} + \alpha_6 \mathbf{n} \otimes \mathbf{An} \tag{23}$$

where again $\mathbf{a} \otimes \mathbf{b}$ denotes the 3×3 matrix with (i,j)th element $a_i b_j$, as the α's are constant coefficients.

It follows from this that we may write the axial vector

$$\hat{\mathbf{T}}^d = \mathbf{n} \times \mathbf{g} \tag{24}$$

where the vector $\mathbf{g}$ is given by

$$\mathbf{g} = -\gamma_1 \mathbf{N} - \gamma_2 \mathbf{An}, \quad \gamma_1 = \alpha_3 - \alpha_2, \quad \gamma_2 = \alpha_6 - \alpha_5 \tag{25}$$

and this allows us to recast the inequality (20) as

$$D = \mathrm{tr}\left(\mathbf{T}^d \mathbf{A}\right) - \boldsymbol{\omega} \cdot \hat{\mathbf{T}}^d = \mathrm{tr}\left(\mathbf{T}^d \mathbf{A}\right) - \mathbf{g} \cdot \mathbf{N} \geq 0 \tag{26}$$

This inequality, of course, restricts the possible range of values for the viscosity coefficients, and by choosing axes parallel to $\mathbf{n}$ and $\mathbf{N}$ it quickly follows that we must have

$$\alpha_4 > 0, \quad 2\alpha_4 + \alpha_5 + \alpha_6 > 0, \quad 3\alpha_4 + 2\alpha_5 + 2\alpha_6 + 2\alpha_1 > 0$$

$$\gamma_1 > 0, \quad (\alpha_2 + \alpha_3 + \gamma_2)^2 < 4\gamma_1 (2\alpha_4 + \alpha_5 + \alpha_6) \tag{27}$$

By appeal to Onsager relations, Parodi [10] proposed that the viscosity coefficients are restricted by

$$\gamma_2 = \alpha_3 + \alpha_2 = \alpha_6 - \alpha_5 \tag{28}$$

and subsequently Currie [11] obtained this relationship from a stability argument. As a consequence this reduction in the number of independent viscous coefficients is now generally accepted, and leads to some simplification in the theory. When EQN (28) applies, Ericksen [12] shows that the viscous stress and the vector $\mathbf{g}$ follow directly from the dissipation function D through

$$\mathbf{T}^d = \frac{1}{2} \frac{\partial D}{\partial \nabla \mathbf{v}}, \quad \mathbf{g} = -\frac{1}{2} \frac{\partial D}{\partial \dot{\mathbf{n}}}, \quad \dot{\mathbf{n}} = \frac{d\mathbf{n}}{dt} \tag{29}$$

These results are required in the following section.

D   BODY FORCES AND MOMENTS

Relatively simple considerations lead to the conclusion that the application of magnetic or electric fields to a nematic liquid crystal can give rise to body torques and forces. To fix our ideas consider a magnetic field which induces a magnetisation **M**, and straightforwardly this suggests a body force **F** and a body moment **K**, given by

$$\mathbf{F} = (\mathbf{M} \cdot \mathrm{grad})\mathbf{B}, \quad \mathbf{K} = \mathbf{M} \times \mathbf{B} \tag{30}$$

where **B** is the magnetic flux density.

For a nematic the magnetisation can take the form

$$\mathbf{M} = \widetilde{\chi}_\perp \mathbf{B} + \Delta\widetilde{\chi} \mathbf{n} \cdot \mathbf{Bn}, \quad \Delta\widetilde{\chi} = \widetilde{\chi}_\| - \widetilde{\chi}_\perp \tag{31}$$

where $\widetilde{\chi}_\|$ and $\widetilde{\chi}_\perp$ denote the components of diamagnetic susceptibility tensor parallel and perpendicular to the director, respectively. As a result it is clearly possible to have a body moment

$$\mathbf{K} = \Delta\widetilde{\chi} \mathbf{n} \cdot \mathbf{Bn} \times \mathbf{B} \tag{32}$$

and in this event we can write

$$\mathbf{K} = \mathbf{n} \times \mathbf{G}, \quad \mathbf{G} = \Delta\widetilde{\chi} \mathbf{n} \cdot \mathbf{BB} \tag{33}$$

any contribution parallel to **n** proving to be of no consequence.

The associated magnetic energy is

$$u = \frac{1}{2}\mathbf{M} \cdot \mathbf{B} = \frac{1}{2}\left(\widetilde{\chi}_\perp \mathbf{B}^2 + \Delta\widetilde{\chi}(\mathbf{n} \cdot \mathbf{B})^2\right) \tag{34}$$

and, if we regard u as a function of **x** and **n**, it follows that

$$\mathbf{G} = \frac{\partial u}{\partial \mathbf{n}}, \quad \mathbf{F} = \frac{\partial u}{\partial \mathbf{x}} \tag{35}$$

the latter assuming that the magnetic field is irrotational. In general the nematic diamagnetic susceptibilities are very small, and therefore it is not necessary to consider the influence of the presence of the nematic upon the field.

Similar expressions to these follow for an electric field **E**, which lead to an electric displacement **D** of the form

$$\mathbf{D} = \widetilde{\varepsilon}_\perp \mathbf{E} + \Delta\widetilde{\varepsilon} \mathbf{n} \cdot \mathbf{En}, \quad \Delta\widetilde{\varepsilon} = \widetilde{\varepsilon}_\| - \widetilde{\varepsilon}_\perp \tag{36}$$

382

$\tilde{\epsilon}_{\parallel}$ and $\tilde{\epsilon}_{\perp}$ denote the corresponding dielectric permittivities, which give rise to analogous body forces and moments. However, in the case of an electric field the dielectric permittivities can be sufficiently large for the liquid crystal to influence the applied field, leading to the need to consider an appropriate reduced version of Maxwell's equations.

E   EQUATIONS OF MOTION

The equations derived in the previous sections can be recast in a somewhat more convenient and familiar form as follows. If we employ EQNS (17), (19), (24) and (33), it is possible to show with the aid of the identity (11) that the balance law for angular momentum, the second of EQNS (5), reduces to

$$\mathbf{n} \times \left(\operatorname{div} \frac{\partial W}{\partial \nabla \mathbf{n}} - \frac{\partial W}{\partial \mathbf{n}} + \mathbf{G} + \mathbf{g} \right) = 0 \tag{37}$$

or equivalently to

$$\operatorname{div} \frac{\partial W}{\partial \nabla \mathbf{n}} - \frac{\partial W}{\partial \mathbf{n}} + \mathbf{G} + \mathbf{g} + \gamma \mathbf{n} = 0 \tag{38}$$

The divergence is again taken with respect to the second index and γ is an arbitrary scalar which absorbs all contributions parallel to the director $\mathbf{n}$. Noting the first of EQNS (35) this is effectively the Euler-Lagrange equation of static theory with a dynamic term, $\mathbf{g}$, added. We can also rewrite the first of EQNS (5) representing balance of linear momentum by substituting expression (17) for the stress tensor, and adding an inner product of EQN (38) with $\nabla \mathbf{n}$ to obtain

$$\rho \frac{d\mathbf{v}}{dt} = \mathbf{F} + (\nabla \mathbf{n})^{\mathrm{T}} \mathbf{G} - \operatorname{grad}(p + W) + (\nabla \mathbf{n})^{\mathrm{T}} \mathbf{g} + \operatorname{div} \mathbf{T}^{\mathrm{d}} \tag{39}$$

noting that because $\mathbf{n}$ is a unit vector

$$(\nabla \mathbf{n})^{\mathrm{T}} \mathbf{n} = 0 \tag{40}$$

Should the relationships (35) hold, EQN (39) reduces further to

$$\rho \frac{d\mathbf{v}}{dt} = -\operatorname{grad} \tilde{p} + (\nabla \mathbf{n})^{\mathrm{T}} \mathbf{g} + \operatorname{div} \mathbf{T}^{\mathrm{d}}, \quad \tilde{p} = p + W - u \tag{41}$$

Clearly EQNS (38) and (41) are more straightforward to use that the original versions.

However, in many problems it is convenient to select initially a particular form for the vector $\mathbf{n}$ referred to Cartesian axes so that it remains a unit vector; this invariably involves two angles θ and ϕ with

$$\mathbf{n} = \mathbf{f}(\theta, \phi), \quad \text{where } \frac{\partial \mathbf{f}}{\partial \theta} \times \frac{\partial \mathbf{f}}{\partial \phi} \neq 0, \quad \text{and } \mathbf{n} \cdot \frac{\partial \mathbf{f}}{\partial \theta} = \mathbf{n} \cdot \frac{\partial \mathbf{f}}{\partial \phi} = 0 \tag{42}$$

Straightforwardly this gives

$$\nabla \mathbf{n} = \frac{\partial \mathbf{f}}{\partial \theta} \otimes \nabla \theta + \frac{\partial \mathbf{f}}{\partial \phi} \otimes \nabla \phi, \quad \frac{d\mathbf{n}}{dt} = \frac{\partial \mathbf{f}}{\partial \theta} \frac{\partial \theta}{dt} + \frac{\partial \mathbf{f}}{\partial \phi} \frac{d\phi}{dt} \tag{43}$$

and consequently the Frank-Oseen energy W (EQN (8)) and that associated with an external field (EQN (34)) take the forms

$$W = W(\theta, \phi, \nabla\theta, \nabla\phi), \quad u = \chi(\theta, \phi, \mathbf{x}) \tag{44}$$

while the rate of dissipation D (EQN (26)) can be written as

$$D = 2\Delta\left(\theta, \phi, \frac{d\theta}{dt}, \frac{d\phi}{dt}, \nabla\mathbf{v}\right) \tag{45}$$

Relatively straightforwardly we can show that

$$\frac{\partial W}{\partial \theta} = \frac{\partial W}{\partial \mathbf{n}} \cdot \frac{\partial \mathbf{f}}{\partial \theta} + \text{tr}\left\{ \frac{\partial W}{\partial \nabla\mathbf{n}} \left(\frac{\partial^2 \mathbf{f}}{\partial \theta^2} \otimes \nabla\theta + \frac{\partial^2 \mathbf{f}}{\partial\theta\partial\phi} \otimes \nabla\phi \right)^T \right\}$$

$$\frac{\partial W}{\partial \phi} = \frac{\partial W}{\partial \mathbf{n}} \cdot \frac{\partial \mathbf{f}}{\partial \phi} + \text{tr}\left\{ \frac{\partial W}{\partial \nabla\mathbf{n}} \left(\frac{\partial^2 \mathbf{f}}{\partial\phi\partial\theta} \otimes \nabla\theta + \frac{\partial^2 \mathbf{f}}{\partial \phi^2} \otimes \nabla\phi \right)^T \right\}$$

$$\tag{46}$$

$$\frac{\partial W}{\partial \nabla\theta} = \left(\frac{\partial W}{\partial \nabla\mathbf{n}} \right)^T \frac{\partial \mathbf{f}}{\partial \theta}, \quad \frac{\partial W}{\partial \nabla\phi} = \left(\frac{\partial W}{\partial \nabla\mathbf{n}} \right)^T \frac{\partial \mathbf{f}}{\partial \phi}$$

$$\frac{\partial \chi}{\partial \theta} = \frac{\partial u}{\partial \mathbf{n}} \cdot \frac{\partial \mathbf{f}}{\partial \theta}, \quad \frac{\partial \chi}{\partial \phi} = \frac{\partial u}{\partial \mathbf{n}} \cdot \frac{\partial \mathbf{f}}{\partial \phi}, \quad \frac{\partial \Delta}{\partial \dot\theta} = \frac{1}{2} \frac{\partial D}{\partial \dot{\mathbf{n}}} \cdot \frac{\partial \mathbf{f}}{\partial \theta}, \quad \frac{\partial \Delta}{\partial \dot\phi} = \frac{1}{2} \frac{\partial D}{\partial \dot{\mathbf{n}}} \cdot \frac{\partial \mathbf{f}}{\partial \phi}$$

where in the latter the superposed dot again denotes a material time derivative. Combining the appropriate members of these equations, it follows that

$$\text{div}\left(\frac{\partial W}{\partial \nabla\theta} \right) - \frac{\partial W}{\partial \theta} + \frac{\partial \chi}{\partial \theta} - \frac{\partial \Delta}{\partial \dot\theta} = \left(\text{div} \frac{\partial W}{\partial \nabla\mathbf{n}} - \frac{\partial W}{\partial \mathbf{n}} + \frac{\partial u}{\partial \mathbf{n}} - \frac{1}{2} \frac{\partial D}{\partial \dot{\mathbf{n}}} \right) \cdot \frac{\partial \mathbf{f}}{\partial \theta} \tag{47}$$

$$\text{div}\left(\frac{\partial W}{\partial \nabla\phi} \right) - \frac{\partial W}{\partial \phi} + \frac{\partial \chi}{\partial \phi} - \frac{\partial \Delta}{\partial \dot\phi} = \left(\text{div} \frac{\partial W}{\partial \nabla\mathbf{n}} - \frac{\partial W}{\partial \mathbf{n}} + \frac{\partial u}{\partial \mathbf{n}} - \frac{1}{2} \frac{\partial D}{\partial \dot{\mathbf{n}}} \right) \cdot \frac{\partial \mathbf{f}}{\partial \phi} \tag{48}$$

With the aid of EQNS (29), (35) and (42) it is immediately deduced that EQNS (38) can be replaced by

$$\text{div}\left(\frac{\partial W}{\partial \nabla \theta}\right) - \frac{\partial W}{\partial \theta} + \frac{\partial \chi}{\partial \theta} - \frac{\partial \Delta}{\partial \dot{\theta}} = 0, \quad \text{div}\left(\frac{\partial W}{\partial \nabla \phi}\right) - \frac{\partial W}{\partial \phi} + \frac{\partial \chi}{\partial \phi} - \frac{\partial \Delta}{\partial \dot{\phi}} = 0 \tag{49}$$

and EQN (41) can be recast as

$$\rho \frac{d\mathbf{v}}{dt} = -\text{grad}\,\tilde{p} - \frac{\partial \Delta}{\partial \dot{\theta}} \nabla \theta - \frac{\partial \Delta}{\partial \dot{\phi}} \nabla \phi + \text{div}\left(\frac{\partial \Delta}{\partial \nabla \mathbf{v}}\right) \tag{50}$$

Again, these forms are more readily obtainable that EQNS (38) and (41). A similar reformulation is possible should we choose to introduce curvilinear coordinates; the details can be found in the paper by Ericksen [12].

F   BOUNDARY CONDITIONS

To complete the theory we require initial conditions as well as boundary conditions for the resulting partial differential equations for the velocity $\mathbf{v}$ and the director $\mathbf{n}$. If appropriate, the former are largely dictated by the particular problem under consideration, but the latter are in some respects more stereotyped. The most common boundary conditions are the assumption that both the velocity and director are prescribed at a bounding solid surface, the former to satisfy the familiar no-slip condition, while the particular alignment of the director is determined by the prior treatment of the solid surface, uniformly parallel or perpendicular being the most common. However, surface alignments with prescribed uniform tilt at a given angle to the surface normal are also possible. Such prescribed fixed surface alignments are known as strong anchoring conditions.

More recently, there is an increasing interest in situations where the anchoring is weak and so can vary when subject to competing torques. In this case it is usual to assume a surface energy per unit surface area, w, where

$$w = w(\mathbf{n}, \nu, \tau) \tag{51}$$

ν is the unit surface normal, and the unit vector τ describes a fixed direction in the surface. Here the boundary condition is a balance between the couple stress arising from the Frank-Oseen energy and the torque from this surface energy [13], so that

$$\mathbf{n} \times \frac{\partial W}{\partial \nabla \mathbf{n}} \nu + \mathbf{n} \times \frac{\partial w}{\partial \mathbf{n}} = 0 \tag{52}$$

at each point of the surface. A boundary condition of this type is also appropriate at a free surface, or an isotropic liquid - liquid crystal interface, although a vector τ is unlikely to appear in the surface energy (EQN (51)) in such cases.

REFERENCES

[1]   C.W. Oseen [*Trans. Faraday Soc. (UK)* vol.29 (1933) p.883]

[2]   H. Zocher [*Trans. Faraday Soc. (UK)* vol.29 (1933) p.945]

[3]   A. Anzelius [*Uppsala Univ. Arsskr. Mat. Naturv. (Sweden)* (1931) p.1]

[4]   W. Zwetkoff [*Acta Physicochim. URSS (USSR)* vol.10 (1939) p.555]

[5]   M. Miesowicz [*Bull. Int. Acad. Pol. Sci. Lett., Cl. Sci. Math. Nat., Ser. A (Poland)* (1936) p.228]

[6]   T. Carlsson, F.M. Leslie [*Liq. Cryst. (UK)* to appear]

[7]   J.L. Ericksen [*Trans. Soc. Rheol. (USA)* vol.5 (1961) p.23]

[8]   F.M. Leslie [*Arch. Ration. Mech. Anal. (Germany)* vol.28 (1968) p.265]

[9]   F.M. Leslie [*Contin. Mech. Thermodyn. (Germany)* vol.4 (1992) p.167]

[10]   O. Parodi [*J. Phys. (France)* vol.31 (1970) p.581]

[11]   P.K. Currie [*Mol. Cryst. Liq. Cryst. (UK)* vol.28 (1974) p.335]

[12]   J.L. Ericksen [*Q. J. Mech. Appl. Math. (UK)* vol.29 (1976) p.203]

[13]   J.T. Jenkins, P.J. Barratt [*Q. J. Mech. Appl. Math. (UK)* vol.27 (1974) p.111]

8.2   Measurements of viscosities in nematics[*]

J.K. Moscicki

April 2000

A   INTRODUCTION

Contemporary applications of liquid crystals [1,2] exploit the unique properties of these materials arising from their anisotropic response to external fields and forces. For example, the anisotropy in the dielectric properties makes it possible to construct electro-optical displays, and the characteristic response time of such devices is determined by the anisotropic viscoelastic properties of the liquid crystal [3]. In turn, these viscoelastic properties are related to various kinds of flows and deformations of the material in question. The exact number and nature of viscoelastic constants required to characterise fully the properties of the phase are determined by careful consideration of both static and dynamic behaviour [4]. The specific focus of this Datareview is the description of experimental techniques for measuring the various types of viscosity coefficients allowed in nematic phases.

B   VISCOSITY OF THE NEMATIC PHASE

Viscosity is related to the dissipation of kinetic energy in a flow, and, in general, the rate of kinetic energy dissipation by viscous forces can be viewed as the sum of the rate at which body forces, internal stresses, and surface stresses do work on the volume of interest. Therefore, we may study viscosity experimentally by observing the response to body forces derived from externally applied fields, by means of externally applied surface stresses, or by observing the damping of spontaneous fluctuations in internal stresses. Miesowicz was probably the first to realise that there exist geometrically distinct classes of initial conditions, boundary conditions, and driving terms on these quantities which make it possible to measure all of the physically unique viscosity coefficients in nematic phases. In particular, Miesowicz distinguished three base cases of the relative orientations of the director, $\mathbf{n}$, flow velocity, $\mathbf{v}$, and the flow velocity gradient, $\nabla \mathbf{v}$ (the so-called flow geometries) [5-7] (see FIGURE 1):

(i)   director parallel to the flow velocity, $\mathbf{n} \| \mathbf{v}$,
(ii)   director parallel to the flow velocity gradient, $\mathbf{n} \| \nabla \mathbf{v}$, and
(iii)   director orthogonal to both the flow and velocity gradient, $\mathbf{n} \| (\mathbf{v} \times \nabla \mathbf{v})$.

Viscosity coefficients measured in these geometries when $\mathbf{n}$ is immobilised by body forces, say η_1, η_2, and η_3, respectively, are known as the Miesowicz viscosities. (Note, that in the literature a variety of alternative notations are common; in particular the definitions of η_1 and η_2 are frequently interchanged.) If the orientation of $\mathbf{n}$ is fixed in an arbitrary direction with respect to $\mathbf{v}$ and $\nabla \mathbf{v}$, then the effective viscosity coefficient is given by a linear combination of the Miesowicz viscosities, and another viscosity constant η_{12}, which cannot be visualised in a pure shear-flow:

---

[*] In memoriam of my mentor, Marian Miesowicz.

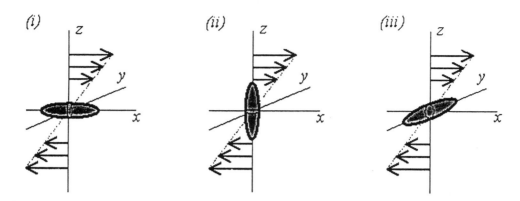

FIGURE 1 Schematic representation of three basic cases of mutual orientations of **n**, **v**, and ∇v; ellipsoids represent **n**, shear is in the x-z plane, **n** along x, z, and y, respectively.

$$\eta(\theta,\phi) = \eta_1 \cos^2\theta + (\eta_2 + \eta_{12} \cos^2\theta) \sin^2\theta \cos^2\phi + \eta_3 \sin^2\theta \sin^2\phi \qquad (1)$$

where θ and ϕ are the angles **n** makes with **v** and ∇v, respectively. The maximum contribution of η_{12} to the effective viscosity coefficient is when **n** is immobilised in the shear plane at $\pi/4$ with respect to **v**:

$$\eta_{12} = 4\eta(\pi/4,0) - 2(\eta_1 + \eta_2) \qquad (2)$$

If **n** is mobile (no body forces present), then in cases (i) and (ii) above a viscous torque rotates **n**, and conversely rotation of **n** by body forces induces a flow. Two additional viscosity coefficients, γ_1 and γ_2, which have no counterpart in isotropic liquids, are necessary to describe this situation. The first coefficient, γ_1, characterises the torque associated with rotation of **n**. The latter coefficient, γ_2, gives the contribution to the torque due to a shear velocity gradient in the nematic. The two coefficients also define the flow alignment of the director under stationary shear flow:

$$\cos 2\theta_o = -\gamma_1/\gamma_2, \quad |\gamma_1/\gamma_2| < 1 \qquad (3)$$

Straightforward evaluation of γ_1 from measurements of the torque exerted by a rotating director on the sample holder walls was pioneered by Tsvetkov in the late 1930s [8,9]. The alignment can be measured by optical methods, but it can also be found indirectly from measurements of the viscosity coefficient, η_o, in the simple shear flow experiment in the absence of a locking external field, when the director is aligned (and immobilised) by the flow,

$$\eta_o = \eta_1 \cos^2\theta_o + \eta_2 \sin^2\theta_o + \eta_{12} \sin^2\theta_o \cos^2\theta_o \qquad (4)$$

provided that η_1, η_2, and η_{12} are already known; see EQN (1). Since, typically, θ_o is of the order of a few degrees, values of η_o are very close to those of η_1.

To describe the hydrodynamics of the nematic phase five independent viscosity parameters (dynamic behaviour) and three elastic constants (static behaviour) are necessary [10-17]. These viscosity parameters cannot be identified with the experimental coefficients directly, but with certain linear combinations of them. Following the most widely accepted nematodynamics of Leslie [4] we have:

388

$$\eta_1 = \frac{1}{2}\left(\alpha_2 + 2\alpha_3 + \alpha_4 + \alpha_5\right) \tag{5}$$

$$\eta_2 = \frac{1}{2}\left(-\alpha_2 + \alpha_4 + \alpha_5\right) \tag{6}$$

$$\eta_3 = \frac{1}{2}\alpha_4 \tag{7}$$

$$\eta_{12} = \frac{1}{2}\alpha_1 \tag{8}$$

$$\gamma_1 = \alpha_3 - \alpha_2 \tag{9}$$

$$\gamma_2 = \alpha_3 + \alpha_2 \tag{10}$$

where $\{\alpha_i\}$, $i = 1, 2, \ldots 5$, is the complete set of independent viscosity parameters (Leslie viscosity parameters). Hence, four shear viscosity coefficients, η_1, η_2, η_3, and η_{12}, and the rotational (or twist) viscosity γ_1 also form an alternative canonical set of (experimental) viscosity coefficients describing dynamics of a nematic liquid crystal. Since γ_1 defines the timescale of the director relaxation process (i.e. the switching behaviour) [18], the convention of de Gennes for the sign of γ_1 ($\gamma_1 \geq 0$) [19] rather than the original one of Leslie [11] ($\gamma_1 \leq 0$) is used in EQN (9). The complete set of coefficients has been measured for only a few liquid crystals [20-24]. Results for 4-methoxybenzylidine-4'-butylaniline (MBBA) shown in FIGURE 2 illustrate very well typical properties of those liquid crystals which exhibit only the nematic phase.

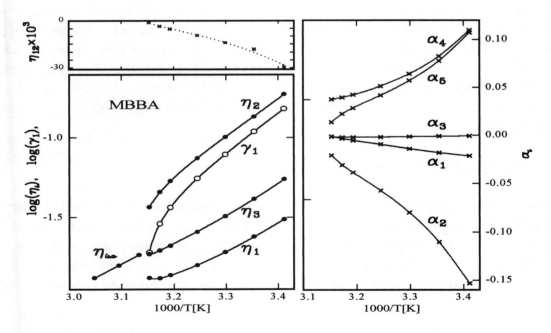

FIGURE 2 Canonical sets of (left) viscosity coefficients and (right) Leslie viscosity parameters of MBBA in units of Pa s [23,25].

Apart from basic viscosity experiments, there is an increasing number of contemporary methods monitoring the director field modulations, i.e. the splay, twist and bend deformations [19]. Flows related to these deformations are characterised by viscosity coefficients η_{splay}, η_{twist}, and η_{bend}, respectively. These viscosities can be conveniently expressed via combinations of the Miesowicz viscosity coefficients and/or the Leslie parameters [20,21,26]:

$$\eta_{splay} = \gamma_1 - \alpha_3^2 / \eta_1 = \gamma_1 - (\eta_1 - \eta_2 + \gamma_1)^2 / 4\eta_1$$

$$\eta_{twist} = \gamma_1 \tag{11}$$

$$\eta_{bend} = \gamma_1 - \alpha_2^2 / \eta_2 = \gamma_1 - (\eta_1 - \eta_2 - \gamma_1)^2 / 4\eta_2$$

C   EXPERIMENTAL TECHNIQUES

Nematoviscosimetric techniques can be roughly divided into two distinct categories. The first, which encompasses direct methods, determines the viscosity coefficients in a classical, from first principles way. These techniques can provide precise and highly accurate data; however, they require substantial amounts of sample. The second set of methods includes a variety of indirect methods. They explore the diversity of physical phenomena that are influenced by the viscoelastic behaviour of liquid crystals, and the viscosity coefficients are then extracted through theoretical considerations of the particular phenomenon and subsequent data manipulation. These techniques usually require much less sample, and can be used to study properties of liquid crystals in heterogeneous systems such as polymer-dispersed liquid crystals or composite optical cells. However, frequently these techniques yield the viscosity coefficients only when other quantities such as the elastic constants or other viscosity coefficients [26-36], the anisotropy of electric [34,37-39] or magnetic [40-50] susceptibility, or refractive index [34,37,51-53] are known from independent measurements. Variable purity or chemical instability of different samples of the same substance studied by different authors also adds to the proliferation of inconsistent results in the literature. Consequently, there are substantial discrepancies among results from the same substance obtained by different techniques, and with a few notable exceptions, the available data on the viscous properties of nematogenic substances are still far from satisfactory.

C1   First Principles Methods

The first viscosity measurements of nematic liquid crystals were performed with classical shear flow viscometers; flow induces a change in orientation of the director, see EQN (4), so the effective viscosity measured is η_o [54,55]. Since η_o can be considered a relatively good approximation to the Miesowicz η_1, conventional viscometers are still in use in conjunction with γ_1 measurements [44,56,82,86]. However, for the purpose of measuring the viscosity anisotropy older instruments had to be modified and new ones developed.

C1.1   Shear flow capillary method

The most mature direct technique, which was pioneered by Gahwiller [57-59] and explored over the years by others [21,24,25,52,60-66], most recently and exhaustively by Schneider and his colleagues [25,63,64,66], is an adaptation of a classical method for measuring the velocity of a laminar flow in a capillary induced by a slight pressure difference between the capillary ends, Δp [67]

$$dV/dt \propto \Delta p/\eta \qquad (12)$$

where dV/dt is the volume flow rate of a liquid. Adaptation of the method for liquid crystals requires the use of a rectangular capillary (FIGURE 3). Precise and reliable estimation of the viscosity coefficients of interest requires a number of corrections to be taken into account, including correction for a finite capillary cross-section [63,68], flow-alignment effects [25,69], boundary ordering effects [25,60,61,70], and corrections in the flow velocity and the pressure difference measurements [66]. With all the corrections properly accounted for, the typical accuracy of the capillary method is about 1 - 2% [25].

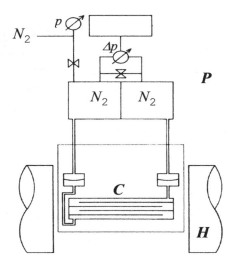

FIGURE 3 A schematic of a capillary viscometer. C - multilayer capillary, H - electromagnet, P - pressure difference generator and controller.

C1.2   Oscillating plate method

A technique pioneered by Miesowicz is based on the observation of damping of very small oscillations of a thin glass plate completely immersed in the investigated liquid [5-7]. The oscillation amplitude, A, decreases with time due to viscous drag on the plate, and the rate of decrease can be related to the viscosity of the liquid. The damping coefficient has the form

$$\delta = n^{-1} \ln(A_0/A_n) = k_1(f)(\eta\rho)^{1/2} + k_2 \qquad (13)$$

where A_0 is the initial amplitude, ρ is the density, $k_1(f)$ and $k_2 \ll k_1(f)$ are apparatus parameters, and f is the driving frequency [72]. The plate is suspended from one end of a balanced beam. The system undergoes extremely slow (f < 1 Hz) underdamped harmonic motion, which is monitored by optical methods [5,71-76]. Use of an analytical balance limited the original Miesowicz measurements to η_2, η_3 and η_0. To remove this limitation a dedicated beam balance was employed [72]. Improved precision for detecting the plate position was realised by using lasers to measure the displacements accurately, which allowed the method to be extended to the overdamped mode, enabling measurements of high viscosity nematics [76]. A DC version of the method, i.e. a direct measurement of the viscous drag on the plate slowly pulled from the liquid, has also been designed [77]. The accuracy of the oscillating plate method is comparable to that of the capillary viscometers [72].

C1.3　Rotational viscosimetry

The rotational viscosity coefficient γ_1 is the most frequently determined viscosity coefficient of liquid crystals. By a straightforward adaptation of the shear flow technique, Tsvetkov originated the idea of measuring a specific torque, T/V, exerted on the sample by **n** rotating with a constant angular velocity ω [8]:

$$T/V = \gamma_1 \omega \qquad (14)$$

For this purpose **n** is usually locked to an external rotating magnetic field. Tsvetkov's method has been subsequently explored by several groups [23,78-88]. In order to obtain precise viscosity data a number of corrections due to the presence of walls, including wall alignment effects, generation of inversion walls and backflow effects, have to be considered [89]. By careful design of the experimental setup and procedure these corrections can be properly accounted for [85]. The maximum accuracy of the method (0.3%) is achieved if a sample is rotated in a stationary magnetic field [85].

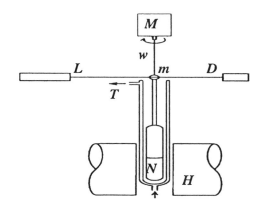

FIGURE 4　A schematic of the rotational viscosimeter. N - sample, H - electro-magnet,
T - temperature control, L - laser, D - detector, w - wire, M - wire mounting,
m - mirror. Either M or H is made to rotate about the wire axis.

C2　Mechanical Wave Method

Determination of the viscosity coefficients from the mechanical wave propagation and attenuation in the ordered nematic phase is probably the closest to the first principles methods. The shear impedance technique is based on measuring the reflection and attenuation of ultrasonic shear waves [90-92]. The complex shear impedance of the nematic sample, $Z_n^* = R_n + iX_n$, is determined from the complex reflection coefficient r* for a transverse shear wave at the solid(quartz)-nematic interface r* = $(Z_q^* - Z_n^*)/(Z_q^* + Z_n^*)$, where Z_q^* is the known complex impedance of the quartz substrate. Z_n^* is related to the dynamic viscosity,

$$\eta = RX(\pi\rho f)^{-1} \tag{15}$$

where ρ is the density and f the shear wave frequency. The ordered nematic fluid, see FIGURE 1, behaves like an ordinary liquid with a characteristic viscosity coefficient for a given geometry of experiment. Three geometries of the displacement were considered: **n** planar at the solid-nematic interface and (A) parallel or (B) perpendicular to the displacement, and (C) **n** perpendicular to the interface. The relevant viscosity coefficients are

$$\eta_A = \eta_1 - (\alpha_3/2)(1 + \gamma_2/\gamma_1), \quad \eta_B = \eta_2 - (\alpha_2/2)(1 + \gamma_2/\gamma_1), \quad \eta_C = \eta_3 \tag{16}$$

A thin nematic layer is formed on top of a fused quartz rod with obliquely cut ends to which a quartz generator and receiver are glued. The transducer generates waves at its fundamental frequency and its odd harmonics. The wave transmitted to the nematic layer is rapidly damped, while the reflected wave is detected by the receiver. Multiple reflection at the interface and both ends leads to a pulse echo pattern at the detector, from which the reflection coefficient can be determined [90,91]. The accuracy of the determination of the viscosity coefficients is about 6% to 15%. Use of a mechanical wave method was also reported in [93].

C3   Transient Director Pattern Methods

The nematic liquid crystal is orientationally soft, since restoring forces associated with deformation in the director field are very weak. This softness makes alignment of **n** in bulk samples occur even in very weak external magnetic or electric fields, **F** $\equiv$ **H** or **E**, or by interaction with boundary surfaces and flows in the liquid. This softness also allows for long wavelength thermal fluctuations in the director field. The Leslie viscosity parameters rather than the viscosity coefficients are the more natural quantities of interest for those methods that monitor the viscoelastic response of the nematic to director field modulations. Modulation of **n** in space and time manifests itself in variations of many bulk properties, e.g. the refractive index [27-37,41-44,48,51,94-106], electric susceptibility [38,39,107-110], or magnetic resonance spectra [40,45-47,111-113]. However, only a limited number of the viscosity parameters/coefficients can be precisely determined by these methods.

C3.1   Dynamic Freedericksz effect

For a thin nematic film (a few to several tens of μm thick) with **n** strongly anchored at the boundaries, the reorienting field (electric or magnetic, **F** $\equiv$ **E** or **H**) must be sufficiently strong, F > F_c, to overcome the elastic restoring force of the material and deform the equilibrium pattern of **n**; $\Delta\widetilde{\chi}_F F_c^2 d^2 = \pi^2 K$, where $\Delta\widetilde{\chi}_F$ is the anisotropy of the field relevant susceptibility, d is the film thickness and K is some effective elastic constant characteristic for a given distortion geometry. The magnitude of the deformation in this case is also a function of the field value and of the material elastic constants. Of particular importance are simple deformation geometries, where the distorting field is perpendicular to **n** of the thin uniformly oriented nematic film. The undistorted **n** field is such that the monodomain layer is usually either planar (also called homogeneous; **n** parallel to the film walls and in the same direction at both walls), or twisted (**n** parallel to the walls but the directions at both walls mutually orthogonal), or homeotropic (**n** normal to the walls). The transition between the undistorted (F < F_c) and distorted (F ≥ F_c) patterns in these films is known as the Freedericksz transition [114].

When the deforming field is rapidly switched on and off, the transient behaviour of **n** that follows is determined by the viscoelastic properties of the sample, the boundary conditions, and the initial and final states of the director pattern. Such experiments typically provide the most reliable information on the rotational viscosity coefficient. In order to model transient behaviour in a particular geometry a set of the Leslie equations of motion is solved. This solution gives the time evolution of the azimuthal, $\varphi(t,\mathbf{r})$, and polar, $\vartheta(t,\mathbf{r})$, angles describing the orientation of **n** with respect to some reference frame at any given arbitrary position **r** in the sample. These functions are parametrised by the Leslie viscosity parameters and the elasticity constants.

Basic Freedericksz effect geometries involve pure splay, twist, and bend deformations, FIGURE 5. In each case the deformation is not uniform across the slab and can be expressed as:

$$\vartheta(t,z) \approx \vartheta_m(t)\cos(\pi z/d) \tag{17}$$

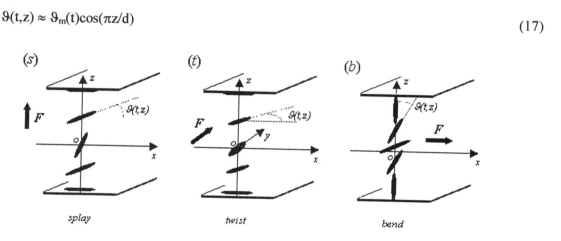

(s)　　　　　　(t)　　　　　　(b)

splay　　　　　　twist　　　　　　bend

FIGURE 5 Three basic geometries of distortion for the Freedericksz effect.

When the field is instantly removed, $\vartheta_m(t)$ decays exponentially with a relaxation time, $\tau = \eta^+ d^2/(\pi^2 K)$, where η^+ is a more or less complicated function of the canonical viscosity coefficients of the material, and is characteristic for a particular geometry and magnitude of the distorting field [95]. For F >> F_c, η^+ reduces, depending on the case, to η_{splay}, η_{twist}, or η_{bend} given in EQN (11). In more general cases, nematodynamic equations for particular experimental conditions are solved and numerical results fitted to the observed transient behaviour [34,35,51]. Such a procedure enables, in principle, estimation of all the Leslie parameters, except for α_1 [34,35]. However, often the fits are fairly insensitive to many of these parameters, thus making it difficult to determine these quantities reliably. A typical claimed accuracy is 5 - 10% for some parameters, but values of other viscosity parameters cannot be estimated, and are taken from the literature as known constants.

Any anisotropic property, from optical birefringence, electric susceptibility, electric or thermal conductivity, or orientation dependent spectra, can be used to probe the average state of alignment. Quantities measured in such experiments are most often proportional to the square of the distortion angle; thus the transient behaviour observed is that of $\vartheta_m^2(t)$. The distortion is commonly produced by either a magnetic [41-43] or an electric [34,35,51] field.

Optical detection. The distortion in thin films is most accurately detected optically [37,41-44,52]. Consider the splay geometry in FIGURE 5. For light polarised along the y axis the refractive index is n_o irrespective of the director pattern. For light polarised along the x axis the refractive index is n_e at $\pm d/2$ boundaries, but elsewhere is

$$n(t,z) = n_o n_e \left[n_e^2 \sin^2 \vartheta(t,z) + n_o^2 \cos^2 \vartheta(t,z) \right]^{-1/2} \tag{18}$$

The difference in optical path length, Δl, of the ordinary and extraordinary rays of a polarised light beam of wavelength λ at any arbitrary direction on passing through the film is

$$2\Delta l(t)/\lambda = (4/\lambda) \int_0^{d/2} \left[n(t,z) - n_o \right] dz = N(t) \tag{19}$$

For small distortions $\Delta l(t)/\lambda \propto \vartheta_m^2(t)$. Interference figures are usually observed with the aid of an analyser. Constructive or destructive interference of transmitted light is then observed depending on whether N(t) is an odd or even integer. Similar transient optical behaviour is also characteristic for the other two dynamic Freedericksz effect geometries [42,43,95]. A typical optical set-up to measure the birefringence of the liquid crystal film is schematically shown in FIGURE 6 [44,95].

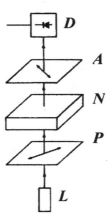

FIGURE 6 A scheme of optical observation of the Freedericksz effect; N - nematic film
between glass plates, L - light source, P -polariser, A - analyser,
and D - photodetection system.

A photodetector follows the variation of the light intensity between maxima and minima following transient behaviour of the director pattern, and $N(t) = G\vartheta_m^2(t)$ is recorded.

The average alignment of the initially planar nematic layer twisted by an in-layer magnetic field, FIGURE 5(t), can also be monitored by following the rotation of a conoscopic figure [41,48,115]. The conoscopic figure associated with a uniform planar nematic film consists of two hyperbolas. The twist deformation of **n** leads to a rotation of the pattern by an angle δ, $\tan 2\delta(t) \approx \langle \sin 2\vartheta(t,z) \rangle_r / \langle \cos 2\vartheta(t,z) \rangle_r$, where $<...>_r$ denotes averaging over the sample volume. Since the rotation can take place in either direction, to avoid experimental problems the direction of $\mathbf{H}$ should deviate slightly from being normal to $\mathbf{n}$ [48]. The transient $\delta(t)$ is then determined from negatives of the conoscopic figure photographed at fixed time intervals [48,115].

An optical set-up similar to that in FIGURE 6 was used in the torsional shear flow study of the director field distortion by the combined action of flow and an AC electric field [37,52,53,116-119] which were mutually perpendicular. A photographic film was used instead of a photoelement as the detector. In the absence of flow, the nematic sample was homeotropically aligned between circular glass plates, as in FIGURE 5(b). The lower boundary glass disk was made to rotate at a sufficiently slow rate to secure a disclination-free flow everywhere in the sample. Simultaneously, away from the rotation axis the flow rates were sufficiently high to produce a nearly uniform flow alignment throughout the layer (along z) except at the boundaries. Numerical analysis of the data yields γ_1 and γ_2 (or, alternatively α_2 and α_3: see EQNS (9) and (10)), provided that the refractive indices, dielectric anisotropy and Miesowicz viscosity coefficient η_2 are known [52].

Contemporary stimulated director pattern deformation methods offer information on a wider range of viscosity coefficients. In [51] a slowly varying deforming electric field was employed. For sufficiently slow changes of this field the director pattern is in equilibrium, and follows the field adiabatically. However, once the rate of change becomes comparable with τ, the director response is delayed and the amplitude of the director reorientation decreases. This is reflected in the phase shift between the optical transmission and a driving voltage, and a decrease in the transmission amplitude. A low frequency modulated electric field produces a stationary cyclic behaviour of the director field. The optical detection set-up is shown schematically in FIGURE 6, with the initial planar orientation of the sample. The frequency dependence of amplitude and phase of the transmitted light contain information on α_1, γ_1, γ_2 and η_1. Extraction of these values requires solving the nematodynamic equations numerically and fitting these results to the experimental curves [51]. Similar complexity and information is involved when a layer with a twisted director pattern is subjected to electric field deformation [34,35]. The transmission time curve features in this case the so-called optical bounce, which can be reproduced numerically on the basis of the nematic equations of motion. Numerical integration requires knowledge of the elastic constants, the dielectric and the optical anisotropies, and γ_1. The method is insensitive to α_1 and by fitting transient calculated responses to experimental curves it is possible to determine α_3, α_4, and α_5.

Electrical (capacitive) detection. Electric methods provide an alternative way of monitoring the director realignment. They are based on following the time dependence of the effective electric susceptibility of the nematic layer. The effective susceptibility (in general a complex quantity) across the initially planar nematic sample of thickness d is

$$\overline{\chi}_E^{*-1}(f;t) = d^{-1} \int_{-d/2}^{d/2} \chi_E^{*-1}[f; \vartheta(t,z)] \, dz \qquad (20)$$

where $\chi_E^*[f; \vartheta(t,z)] = \chi_{E\perp}^*(f) + \Delta\chi_E^*(f) \sin^2 \vartheta(t,z)$ is the electric susceptibility at the probing frequency f, at time t and at position z across the sample, and $\Delta\chi_E^*(f) = \chi_{E\|}^*(f) - \chi_{E\perp}^*(f)$ is the susceptibility anisotropy, $\chi_{Ei}^*(f)$, $i \equiv \|$ and $\perp$ being the electric susceptibilities across the planar and homeotropic layers, respectively. In practice, either the real or imaginary part of the complex susceptibility can be used for the purpose of monitoring the director distribution. For small distortions (when the deforming (electric) field is just above the Freedericksz transition threshold value) the transient change in capacitance is analogous to that of the optical birefringence:

$$C_R(f;t) = G_C(f)\vartheta_m^2(t) = C_o(f)\exp(-2t/\tau).$$

The deformation is caused by a low-frequency (≈ 100 Hz, square wave) electric field. The capacitance was measured at ≈ 10 kHz. In [94] the method was used with a planar layer while in [110] it was applied to a twisted nematic film. An alternative electric detection method is to record the transient displacement current, I [38,39]. The nearly planar film is strongly anchored at capacitor plates with a very small but non-negligible pretilt angle ϑ_0. At t = 0 a DC voltage is applied stepwise to the plates. Due to the rotation of **n**, the transient displacement current exhibits a characteristic maximum, I_p, which is shown to be a function of d, ϑ_0, γ_1, applied voltage V, and $\Delta\chi_E(f)$: $V/(d \cdot I_p^{1/3}) = [2\gamma_1/(A \cdot \Delta\chi_E^2)]^{1/3}[1 + \beta/V]$, where A and β are the electrode area and a positive constant related to the contribution of elastic deformation to the current, respectively [38]. Plotting $V/(d \cdot I_p^{1/3})$ versus $1/V$ gives $[2\gamma_1/(A \cdot \Delta\chi_E^2)]^{1/3}$; hence if $\Delta\chi_E$ is known, γ_1 can be determined. The dependence on $\Delta\chi_E$ can be eliminated if the time of occurrence of I_p, t_p, and the total charge Q_s are also known [39].

Thermal detection. Pieranski et al [95] found that monitoring the transient behaviour of the thermal conductivity is also well suited for studying the dynamic Freedericksz effect, since the relaxation time of the transition, τ, is large compared to the thermal relaxation time, τ_{th}: $\tau/\tau_{th} \approx 100$. The effective thermal conductivity $k_e(t)$ across an initially planar layer of thickness d is given by

$$k_e^{-1}(t) = d^{-1}\int_{-d/2}^{d/2} k^{-1}[\vartheta(t,z)]\,dz \tag{21}$$

where $k[\vartheta(t,z)] = k_\perp + \Delta k \cdot \sin^2[\vartheta(t,z)]$, $k_\perp$ is the heat conductivity across the planar layer, and Δk is the thermal conductivity anisotropy. The experimental set-up was, in principle, similar to that used in electro-optical studies (FIGURE 6), but in addition the cell accommodated thin-film metallic temperature sensors on both glass plates, and a gold heater film on the lower plate for measuring small changes in the thermal properties of a nematic layer (d $\approx$ few hundreds μm).

C3.2   Director dynamics in bulk samples

Optical detection becomes inconvenient for thick (bulk) layers. The transient director behaviour is more easily monitored with the aid of NMR [49,111-113,120] and dielectric [108,109] spectroscopies. These experiments were performed on sufficiently large samples that effects associated with alignment at the boundaries could be safely neglected. The stepwise reorientation of the ordering magnetic field with respect to **n** was realised either over a relatively small angle, $\vartheta^o < \pi/4$, or by $\vartheta^o \cong \pi/2$. In the former case the homogeneous pattern of **n** remains undisturbed during the reorientation, and relaxation of **n** towards the new equilibrium is exponential, with a relaxation time

$$\tau = \gamma_1 / \Delta\tilde{\chi}_F F^2 \tag{22}$$

where $F \equiv H$ or E, and $\Delta\chi_F$ denotes relevant susceptibility. For $\vartheta^\circ \cong \pi/2$ the director pattern undergoes a spatially periodic, transient distortion coupled with backflow effects. The relaxation process begins as a pure bend deformation gradually transforming into splay deformation at later times.

Magnetic resonance methods. In the NMR experiment, step rotation of the aligning magnetic field was done either by mechanical rotation of the sample [111,112,120], or by fast reorientation of the field realised via electronic circuitry [49,108]. The progress of the reorientation of the director pattern towards the new equilibrium position is followed by monitoring the separation of the doublet lines, $\Delta v(t) = \Delta v_o <3\cos^2[\vartheta^\circ - \vartheta(t)] - 1>/2$ in the (proton or deuteron) NMR spectrum, with the transient behaviour of $\vartheta(t)$ given by $\vartheta^\circ - \vartheta(t) = \arctan[\tan\vartheta^\circ \exp(t/\tau)]$. For details of the NMR methods see Datareview 8.3 in this book.

Electric methods. Gerber used a capacitance bridge to follow the transient behaviour of the director field under small stepwise spatial oscillations of an ordering magnetic field [108]. The nematic inside a measuring capacitor ($d \approx 1$ mm) was under the influence of two superimposed, mutually orthogonal magnetic fields: the main field $\mathbf{H_0}$, and a small transverse switchable field of $H_1 = H_0/1000$. Sudden reversal of the direction of $\mathbf{H_1}$ caused a small change in the direction of $\mathbf{H_0} + \mathbf{H_1}$. The transient change of the director alignment led to an exponential change in the sample capacitance, with a relaxation time given by EQN (22).

The $\vartheta^\circ \cong \pi/2$ reorientation of $\mathbf{n}$ from a planar to a homeotropic geometry under the electric field has been studied by dielectric spectroscopy by Kozak et al [109]. The mean director pattern deformation, $<\sin^2\vartheta(t)>$, was determined from the effective loss factor transient behaviour, $\overline{\chi}_E''(f;t)$ (see EQN (20)):

$$<\sin^2\vartheta(t)> = [\overline{\chi}_E''(f;t) - \chi_\perp''(f)]/[\chi_\parallel''(f) - \chi_\perp''(f)] \tag{23}$$

$\overline{\chi}_E''(f;t)$ is directly related to the director pattern variations in space and time, in an analogous way to that of NMR [113]. Extraction of the Leslie viscosity parameters requires numerical solution of the nematic equations of motion, calculating $\chi_E''(f;t,z)$, integrating over the sample volume to get $\overline{\chi}_E''$, and fitting the latter to the experimentally observed transient behaviour. Fitted parameters include the set of the Leslie parameters [109].

C3.3   Light scattering methods

Volume scattering. Even in a perfectly ordered nematic liquid crystal there are present thermally excited long wavelength fluctuations of $\mathbf{n}$ which subsequently relax to zero: $\delta\mathbf{n}(t,\mathbf{r}) = \mathbf{n}(t,\mathbf{r}) - \mathbf{n_o}$, where $\mathbf{n_o}$ is the time-averaged director. This dynamic process can be described by the nematodynamic equations. Such time-dependent fluctuations can be observed because they give rise to a large depolarised light scattering, FIGURE 7 [96,121]. The light scattering can be described by fluctuations of the electric susceptibility tensor, $\chi_E(t,\mathbf{r})$. In the nematic liquid crystal the off-diagonal elements of the tensor depend linearly on $\delta\mathbf{n}$. The scattered amplitude, for a given scattering wavevector $\mathbf{q}$, depends on two independent Fourier components of $\delta\mathbf{n}$; $n_1(\mathbf{q})$, which describes a deformation involving splay and bend, and $n_2(\mathbf{q})$, which involves twist and bend deformations. Introducing the reference frame defined by the base vectors $\{\hat{e}\}$:

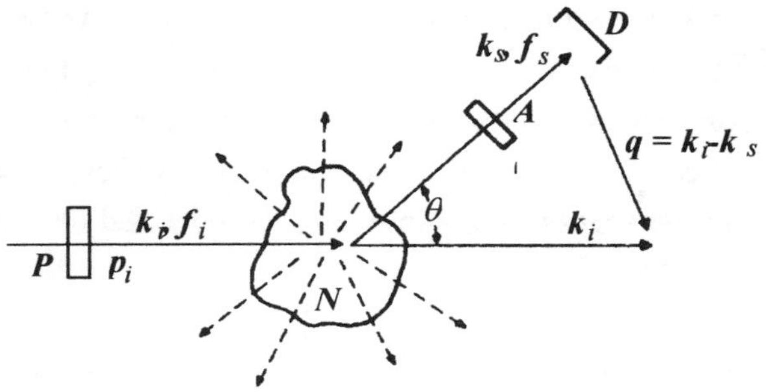

FIGURE 7 Light scattering experiment principles. P - polariser, A - analyser, D - detector, N - scattering volume. Incident beam has polarisation $\mathbf{p}_i$ and wavevector $\mathbf{k}_i$. The scattered beam at A has polarisation $\mathbf{p}_s$ and wave vector $\mathbf{k}_s$. The scattering vector $\mathbf{q}$ is defined by the geometry.

$$\hat{\mathbf{e}}_3 \equiv \mathbf{n}_o, \quad \hat{\mathbf{e}}_2 \equiv (\mathbf{n}_o \times \mathbf{q})/|\mathbf{n}_o \times \mathbf{q}|, \quad \hat{\mathbf{e}}_1 \equiv \hat{\mathbf{e}}_2 \times \hat{\mathbf{e}}_3$$

the equations governing relaxation of the director fluctuation modes are

$$\partial \delta n_j(\mathbf{q})/\partial t = -\tau_j^{-1} \delta n_i(\mathbf{q}) \tag{24}$$

with $\tau_j^{-1} = K_j(q)/\eta_j^{\text{eff}}(\{\alpha_i\}, q)$, $j = 1, 2$, where K_j and η_j^{eff} are some effective elastic constant and viscosity coefficient characteristics for a particular mode [26]. Each mode gives rise to an experimentally observed single lorentzian line with a width τ_i^{-1}. The scattering geometries are generally specified by the scattering angle, orientation of the director, $\{\vartheta, \varphi\}$, together with the states of polarisation of the incident and scattered light. In general, different geometries correspond to different relative contributions of different types of distortions to the scattered light. In the simplest cases, i.e. when $\mathbf{q} \| \mathbf{n}$, both modes reduce to a pure bend, so $\eta_i^{\text{eff}} = \eta_{\text{bend}}$. If $\mathbf{q} \perp \mathbf{n}$, modes 1 and 2 become pure splay and twist deformations, respectively; $\eta_1^{\text{eff}} = \eta_{\text{splay}}$, and $\eta_2^{\text{eff}} = \eta_{\text{twist}}$ (EQN (11)). By selecting a sufficient number of suitably varied geometries, the whole set of Leslie parameters, $\{\alpha_i\}$, normalised by some elastic constant can be determined [26,30]. Therefore, each of the viscosity coefficients can be found if the elastic constant is known from another method. A typical light scattering cell consists of two parallel glass plates separated by a spacer with thickness of a few to several tens of μm. This is thick enough to permit light scattering but thin enough to prevent multiple scattering. The nematic film is strongly anchored at the plates with uniform homeotropic or planar director pattern, as desired. The cell is mounted on a goniometric head to enable its orientation with respect to the light beam to be varied. The scattered light was detected using either homodyne [26,30,97] or heterodyne techniques [100,101,122]. A signal from the photomultiplier was processed either by the photon-correlation technique [123] or by noise intensity spectrum analysis [97]. The reported standard deviation of the viscosity coefficients estimates is about 20%, but that of the Leslie parameters can be as low as, for example, 4% for α_2, or as high as 50% for α_5 [104].

Free surface scattering. Viscosity coefficients have also been determined by light scattering from mechanical waves on the free surface of the nematic layer [105,106]. The waves either originated from thermal fluctuations [105], or were stimulated mechanically [106]. Of special interest for the first method are fluctuations at (sufficiently high ~10 kHz) frequencies where the contribution from elastic forces to hydrodynamic motion is negligible. The power spectrum of the electric field scattered by the surface has a contribution that is well-correlated with these fluctuations, that can be detected optically. The spectrum of vertical fluctuations of points on the free surface depends on the relative orientation of **n** with respect to **k**, the wavevector. Three effective viscosity coefficients were determined from the dependence of the spectrum on the scattering angle, using well-defined experimental geometries [105], i.e. **n** planar at the surface and **n∥k**

$$\eta_{\|1} = (\alpha_4 + \alpha_6 - \alpha_3\gamma_2/\gamma_1)/2 \text{ and } \eta_{\|2} = \alpha_4/2 \tag{25}$$

or **n⊥k**

$$\eta_{\perp3} = \alpha_1 + \gamma_2^2/\gamma_1 \tag{26}$$

For **n** perpendicular to the interface, the effective viscosity coefficient measured is some combination of $\eta_{\|1}$, $\eta_{\|2}$ and $\eta_{\perp3}$.

The second method explores low frequency (<1 kHz) waves generated on the surface mechanically. The waves were generated by a sharp metal wedge positioned very close to the free nematic surface. An AC voltage was applied between the wedge and the surface. The wave propagated a distance away from the wedge where it caused an incident light beam to be deflected. This deflection was detected by a position sensing photodiode [106]. Dispersion relations for waves propagating at an oblique angle with respect to **n** depend in a more or less complicated manner on the viscosity coefficients [106]. For well defined geometries of **k** with respect to **n**, this dependency becomes computationally tractable, and the Leslie parameters can be determined numerically [106]. The estimated standard deviation of these values ranges between 6% and 15%.

D   CONCLUSION

In conclusion we observe that despite the rather well-understood interplay between body forces, internal and surface stresses in defining nematodynamics, the design and development of reliable, routine experiments yielding the complete, canonical set of viscosity coefficients encounters a number of limitations and obstacles which are yet to be satisfactorily solved.

REFERENCES

[1]   B. Bahadur (Ed.) [*Liquid Crystals Applications and Uses* vol.1 (World Scientific, Singapore, 1990)]

[2]   [Datareviews in this book: *Chapter 11 Material properties for applications*]

[3]   E. Jakeman, E.P. Raynes [*Phys. Lett. A (Netherlands)* vol.39 (1972) p.69]

[4]   F.M. Leslie [Datareview in this book: *8.1 Introduction to nematodynamics*]

[5]   M. Miesowicz [*Nature (UK)* vol.17 (1935) p.261]

[6]   M. Miesowicz [*Bull. Int. Acad. Pol. Sci. Lett. Ser. A (Poland)* (1936) p.228]

[7]   M. Miesowicz [*Nature (UK)* vol.158 (1946) p.27]

[8]   W. Zwetkoff (V.N. Tsvetkov) [*Zh. Eksp. Teor. Fiz. (USSR)* vol.9 (1935) p.603]

[9]   W. Zwetkoff (V.N. Tsvetkov) [*Acta Physicochim. (USSR)* vol.10 (1939) p.555]

[10]   J.L. Ericksen [*Arch. Ration. Mech. Anal. (Germany)* vol.4 (1960) p.231]

[11]   F.M. Leslie [*Arch. Ration. Mech. Anal. (Germany)* vol.28 (1968) p.265]

[12]   D. Forster, T.C. Lubensky, P.C. Martin, J. Swift, P.S. Pershan [*Phys. Rev. Lett. (USA)* vol.26 (1971) p.1016]

[13]   S. Hess [*Z. Nat.forsch. A (Germany)* vol.30 (1975) p.728]

[14]   S. Hess [*Z. Nat.forsch. A (Germany)* vol.30 (1975) p.1224]

[15]   G. Vertogen [*Z. Nat.forsch. A (Germany)* vol.38 (1983) p.1273]

[16]   Yu.S. Chilingaryan, R.S. Kakopian, N.V. Tabirian, B.Ya. Zel'dovich [*J. Phys. (France)* vol.45 (1984) p.413]

[17]   O. Parodi [*J. Phys. (France)* vol.31 (1970) p.581]

[18]   J. Kelly, S. Jamal, M. Cui [*J. Appl. Phys. (USA)* vol.86 (1999) p.4091]

[19]   P.G. de Gennes [*The Physics of Liquid Crystals* (Clarendon, Oxford, 1975) p.171]

[20]   W.W. Beens, W.H. de Jeu [*J. Phys. (France)* vol.44 (1983) p.129]

[21]   W.H. de Jeu [*Phys. Lett. A (Netherlands)* vol.69 (1978) p.122]

[22]   W.H. de Jeu [*Physical Properties of Liquid Crystalline Materials* (Gordon and Breach, London, 1980) ch.7]

[23]   H. Kneppe, F. Schneider, N.K. Sharma [*J. Chem. Phys. (USA)* vol.77 (1982) p.3203]

[24]   M.G. Kim, S. Park, Sr. M. Cooper, S.V. Letcher [*Mol. Cryst. Liq. Cryst. (UK)* vol.36 (1976) p.143]

[25]   H. Kneppe, F. Schneider [*Mol. Cryst. Liq. Cryst. (UK)* vol.65 (1981) p.23]

[26]   J.P. van der Meulen, R.J.J. Zijlstra [*J. Phys. (France)* vol.45 (1984) p.1347]

[27]   H. Fellner, W. Franklin, S. Christensen [*Phys. Rev. A (USA)* vol.11 (1975) p.1440]

[28]   J.P. van der Meulen, R.J.J. Zijlstra [*J. Phys. (France)* vol.43 (1982) p.411]

[29]   J.P. van der Meulen, R.J.J. Zijlstra [*J. Phys. (France)* vol.45 (1984) p.1627]

[30]   R. Akiyama, S. Abe, A. Fukuda, E. Kuze [*Jpn. J. Appl. Phys. (Japan)* vol.21 (1982) p.L266]

[31]   Y. Saito et al [*Proc. Soc. Inf. Disp. (USA)* vol.32 (1991) p.213]

[32]   R. Akiyama, K. Tomida, A. Fukuda, E. Kuze [*Jpn. J. Appl. Phys. (Japan)* vol.25 (1986) p.769]

[33]   J.-I. Hirakata, G.-P. Chen, T. Toyooka, S. Kawamoto, H. Takezoe, A. Fukuda [*Jpn. J. Appl. Phys. (Japan)* vol.25 (1986) p.L607]

[34]   O. Cossalter, B. Cramer, D.A. Mlynski [*J. Phys. II (France)* vol.6 (1996) p.1663]

[35]   O. Cossalter, D.A. Mlynski [*Liq. Cryst. (UK)* vol.19 (1995) p.545]

[36]   G.-P. Chen, H. Takezoe, A. Fukuda [*Liq. Cryst. (UK)* vol.5 (1989) p.341]

[37]   K. Skarp, T. Carlsson [*Mol. Cryst. Liq. Cryst. Lett. (UK)* vol.49 (1978) p.75]

[38]   M. Imai, H. Naito, M. Okuda, A. Sugimura [*Jpn. J. Appl. Phys. (Japan)* vol.33 (1994) p.3482]

[39]   O. Nakagawa, M. Imai, H. Naito, A. Sugimura [*Jpn. J. Appl. Phys. (Japan)* vol.35 (1996) p.2762]

[40]   F.M. Leslie, G.R. Luckhurst, H.J. Smith [*Chem. Phys. Lett. (Netherlands)* vol.13 (1972) p.368]

[41]   P.E. Cladis [*Phys. Rev. Lett. (USA)* vol.28 (1972) p.1629]

[42]   F. Brochard, P. Pieranski, E. Guyon [*Phys. Rev. Lett. (USA)* vol.28 (1972) p.1681]

[43]   P. Pieranski, F. Brochard, E. Guyon [*J. Phys. (France)* vol.34 (1973) p.35]

[44]   R. Stannarius, W. Gunther, M. Grigutsch, A. Scharkowski, W. Wedler, D. Demus [*Liq. Cryst. (UK)* vol.9 (1991) p.285]

[45]   S.G. Carr, G.R. Luckhurst, R. Poupko, H.J. Smith [*Chem. Phys. (Netherlands)* vol.7 (1975) p.278]

[46]   J.W. Emsley, J.C. Lindon, G.R. Luckhurst, D. Shaw [*Chem. Phys. Lett. (Netherlands)* vol.19 (1973) p.345]

[47]   J.W. Emsley, S.K. Khoo, J.C. Lindon, G.R. Luckhurst [*Chem. Phys. Lett. (Netherlands)* vol.77 (1981) p.609]

[48]   J.W. van Dijk, W.W. Beens, W.H. de Jeu [*J. Chem. Phys. (USA)* vol.79 (1983) p.3888]

[49]   H. Gotzig, S. Grunenberg-Hassanein, F. Noack [*Z. Nat.forsch. (Germany)* vol.49 (1994) p.1179]

[50]   E. Ciampi, J.W. Emsley, G.R. Luckhurst, B.A. Timimi [*J. Chem. Phys. (USA)* vol.107 (1997) p.5907]

[51]   H. Schmiedel, R. Stannarius, M. Grigutsch, R. Hirning, J. Stelzer, H.-H. Trebin [*J. Appl. Phys. (USA)* vol.74 (1993) p.6053]

[52]   K. Skarp, S.T. Lagerwall, B. Stebler [*Mol. Cryst. Liq. Cryst. (UK)* vol.60 (1980) p.215]

[53]   K. Skarp, T. Carlsson, S.T. Lagerwall, B. Stebler [*Mol. Cryst. Liq. Cryst. (UK)* vol.66 (1981) p.199]

[54]   R.S. Porter, E.M. Barrall II, J.F. Johnson [*J. Chem. Phys. (USA)* vol.45 (1966) p.1452]

[55]   B.C. Benicewicz, J.F. Johnson, M.T. Shaw [*Mol. Cryst. Liq. Cryst. (UK)* vol.65 (1981) p.111]

[56]   H. Kuss [*Mol. Cryst. Liq. Cryst. (UK)* vol.91 (1983) p.59]

[57]   Ch. Gahwiller [*Phys. Lett. A (Netherlands)* vol.36 (1971) p.311]

[58]   Ch. Gahwiller [*Phys. Rev. Lett. (USA)* vol.28 (1972) p.1554]

[59]   Ch. Gahwiller [*Mol. Cryst. Liq. Cryst. (UK)* vol.20 (1973) p.301]

[60]   L. Leger, A. Martinet [*J. Phys. (France)* vol.37 (1976) p.C3-89]

[61]   P.K. Currie [*J. Phys. (France)* vol.40 (1979) p.501]

[62]   V.A. Tsvetkov [in *Advances in Liquid Crystal Research and Applications* Ed. L. Bata (Pergamon Press, Oxford, 1980) p.567]

[63]   F. Schneider [*Z. Nat.forsch. A (Germany)* vol.35 (1980) p.1426]

[64]   H. Kneppe, F. Schneider, N.K. Sharma [*Ber. Bunsenges. Phys. Chem. (Germany)* vol.85 (1981) p.784]

[65]   A.G. Chmielewski [*Mol. Cryst. Liq. Cryst. (UK)* vol.132 (1986) p.329]

[66]   H.-H. Graf, H. Kneppe, F. Schneider [*Mol. Phys. (UK)* vol.77 (1992) p.521]

[67]   J.R. Van Wazer, J.W. Lyons, K.Y. Kim, R.E. Colwell [*Viscosity and Flow Measurement* (Wiley-Interscience, 1963) ch.4]

[68]   C. Oldano [*Nuovo Cimento D (Italy)* vol.11 (1989) p.1101]

[69]   G.J. O'Neill [*Liq. Cryst. (UK)* vol.3 (1986) p.271]

[70]   J. Fisher, A.G. Fredrickson [*Mol. Cryst. Liq. Cryst. (UK)* vol.8 (1969) p.267]

[71]   L.T. Siedler, A.J. Hyde [in *Advances in Liquid Crystal Research and Applications* Ed. L. Bata (Pergamon Press, Oxford, 1980) p.561]

[72]   H.-C. Tseng, B.A. Finlayson [*Mol. Cryst. Liq. Cryst. (UK)* vol.116 (1985) p.265]

[73]   F. Hennel, J. Janik, J.K. Moscicki, R. Dabrowski [*Mol. Cryst. Liq. Cryst. (UK)* vol.191 (1990) p.401]

[74]   K. Czuprynski, J. Janik, J.K. Moscicki [*Liq. Cryst. (UK)* vol.14 (1993) p.1371]

[75] J. Janik, J.K. Moscicki, K. Czuprynski, R. Dabrowski [*Phys. Rev. E (USA)* vol.58 (1998) p.3251]

[76] P. Wasowicz [Thesis, Jagiellonian University, Krakow, 1997]

[77] J.W. Summerford, J.R. Boyd, B.A. Lowry [*J. Appl. Phys. (USA)* vol.46 (1975) p.970]

[78] S. Meiboom, R.C. Hewitt [*Phys. Rev. Lett. (USA)* vol.30 (1973) p.261]

[79] J. Prost, H. Gasparoux [*Phys. Lett. A (Netherlands)* vol.36 (1971) p.245]

[80] H. Gasparoux, J. Prost [*J. Phys. (France)* vol.32 (1971) p.953]

[81] A.C. Diogo, A.F. Martins [*Mol. Cryst. Liq. Cryst. (UK)* vol.66 (1981) p.133]

[82] P.R. Gerber, M. Schadt [*Z. Nat.forsch. A (Germany)* vol.37 (1982) p.179]

[83] L.T.S. Siedler, A.J. Hyde, R.A. Pethrick, F.M. Leslie [*Mol. Cryst. Liq. Cryst. (UK)* vol.90 (1983) p.255]

[84] H. Kneppe, F. Schneider [*Mol. Cryst. Liq. Cryst. (UK)* vol.97 (1983) p.219]

[85] H. Kneppe, F. Schneider [*J. Phys. E, Sci. Instrum. (UK)* vol.16 (1983) p.512]

[86] M. Schadt, M. Petrzilka, P.R. Gerbar, A. Villiger [*Mol. Cryst. Liq. Cryst. (UK)* vol.122 (1985) p.241]

[87] H. Dorrer, H. Kneppe, E. Kuss, F. Schneider [*Liq. Cryst. (UK)* vol.1 (1986) p.573]

[88] H.L. Dorrer, H. Kneppe, F. Schneider [*Liq. Cryst. (UK)* vol.11 (1992) p.905]

[89] P.G. de Gennes [*The Physics of Liquid Crystals* (Clarendon, Oxford, 1975) p.179-81]

[90] P. Martinoty, S. Candau [*Mol. Cryst. Liq. Cryst. (UK)* vol.14 (1971) p.243]

[91] F. Kiry, P. Martinoty [*J. Phys. (France)* vol.38 (1977) p.153]

[92] J.C. Bacri [*J. Phys. Lett. (France)* vol.35 (1974) p.L-141]

[93] A.S. Lagunov, A.N. Larionov [*Russ. J. Phys. Chem. (UK)* vol.57 (1983) p.1005]

[94] Hp. Schad [*J. Appl. Phys. (USA)* vol.54 (1983) p.4994]

[95] P. Pieranski, F. Brochard, E. Guyon [*J. Phys. (France)* vol.7 (1972) p.681]

[96] Orsay Liquid Crystal Group [*Mol. Cryst. Liq. Cryst. (UK)* vol.13 (1971) p.187]

[97] D.C. van Eck, W. Westera [*Mol. Cryst. Liq. Cryst. (UK)* vol.38 (1977) p.319]

[98] D.C. van Eck, M. Perdeck [*Mol. Cryst. Liq. Cryst. Lett. (UK)* vol.49 (1978) p.39]

[99] G.A. DiLisi, Ch. Rosenblatt, A.C. Griffin, U. Hari [*Phys. Rev. A (USA)* vol.45 (1992) p.5738]

[100] H.J. Coles, M.S. Sefton [*Mol. Cryst. Liq. Cryst. Lett. (UK)* vol.3 (1986) p.63]

[101] H.J. Coles, M.S. Sefton [*Mol. Cryst. Liq. Cryst. Lett. (UK)* vol.4 (1987) p.131]

[102] E. Ciampi, J.W. Emsley [*Liq. Cryst. (UK)* vol.22 (1997) p.543]

[103] E. Miraldi, L. Trossi, P. Taverna Valabrega, C. Oldano [*Nuovo Cimento B (Italy)* vol.60 (1980) p.165]

[104] M. Hasegawa, K. Miyachi, A. Fukuda [*Jpn. J. Appl. Phys. (Japan)* vol.34 (1995) p.5694]

[105] D. Langevin [*J. Phys. (France)* vol.33 (1972) p.249]

[106] C.H. Sohl, K. Miyano, J.B. Ketterson, G. Wong [*Phys. Rev. A (USA)* vol.22 (1980) p.1256]

[107] G. Heppke, F. Schneider [*Z. Nat.forsch. A (Germany)* vol.29 (1974) p.1356]

[108] P.R. Gerber [*Appl. Phys. A (Germany)* vol.26 (1981) p.139]

[109] A. Kozak, G.P. Simon, J.K. Moscicki, G. Williams [*Mol. Cryst. Liq. Cryst. (UK)* vol.193 (1990) p.155]

[110] F. Leenhouts [*J. Appl. Phys. (USA)* vol.58 (1985) p.2180]

[111] J.K. Moscicki, B. Robin-Lherbier, D. Canet, S.M. Aharoni [*J. Phys. Lett. (France)* vol.45 (1984) p.L379]

[112] A.F. Martins, P. Esnault, F. Volino [*Phys. Rev. Lett. (USA)* vol.57 (1986) p.1745]

[113]   A.F. Martins [Datareview in this book: *8.3 Measurement of viscoelastic coefficients for nematic mesophases using magnetic resonance*]

[114]   V. Freedericksz, V. Zolina [*Trans. Faraday Soc. (UK)* vol.29 (1933) p.919]

[115]   F. Leenhouts, A.J. Dekker [*J. Chem. Phys. (USA)* vol.74 (1981) p.1956]

[116]   J. Wahl, F. Fischer [*Mol. Cryst. Liq. Cryst. (UK)* vol.22 (1973) p.359]

[117]   T. Waltermann, F. Fischer [*Z. Nat.forsch. A (Germany)* vol.30 (1975) p.519]

[118]   F. Fischer, J. Wahl, T. Waltermann [*Ber. Bunsenges. Phys. Chem. (Germany)* vol.78 (1974) p.891]

[119]   S. Holmstrom, S.T. Lagerwall [*Mol. Cryst. Liq. Cryst. (UK)* vol.38 (1977) p.141]

[120]   N. Schwenk, H.W. Spiess [*J. Phys. II (France)* vol.3 (1993) p.65]

[121]   B.J. Berne, R. Pecora [*Dynamic Light Scattering* (Wiley, New York, 1976)]

[122]   S.A. Shaya, H. Yu [*J. Chem. Phys. (USA)* vol.63 (1975) p.221]

[123]   G. Matsumoto, H. Shimizu, J. Shimada [*Rev. Sci. Instrum. (USA)* vol.47 (1976) p.861]

8.3 Measurement of viscoelastic coefficients for nematic mesophases using magnetic resonance

A.F. Martins

July 1999

A INTRODUCTION

The bulk mechanical properties of (homogeneous, incompressible) nematic liquid crystals are characterised by nine viscoelastic coefficients, namely three curvature (Frank) elastic constants K_i ($i = 1, 2, 3$), for splay, twist and bend deformations, respectively, and six (Leslie) viscosities α_i ($i = 1, ..., 6$). The number of independent viscosity coefficients is five, because of the Onsager-Parodi relation [1,2]: $\alpha_2 + \alpha_3 = \alpha_6 - \alpha_5$. The effective viscosity measured in a given experiment is always some combination of the α_i's.

The measurement of the bulk elastic constants in low molecular mass liquid crystals is usually based on the observation of an electric or magnetic field-induced reorientation transition (Freedericksz transition). The measurement of the Leslie viscosities has been achieved in a number of ways including the Miesowicz oscillating plate technique, light scattering, Poiseille flow with magnetic field controlled director orientation in tubes of rectangular cross section, etc. The rotational viscosity, $\gamma_1 = \alpha_3 - \alpha_2$, has been more frequently obtained from experiments similar to Tsvetkov's, where the shift of the director relative to a magnetic field applied perpendicular to the axis of a slow spinning cylindrical sample is observed. All such experiments give up to four viscosity coefficients only; therefore at least two different experiments are required to obtain the whole set of nematic viscosities. These techniques and the corresponding data are reviewed elsewhere in this book and so will not be further considered here.

The techniques developed to measure the viscoelastic properties of low molecular mass nematics can hardly be applied to high molecular mass and high viscosity materials such as nematic polymers. In this context, nuclear magnetic resonance techniques (NMR) or appropriate combinations of traditional rheometry with NMR (Rheo-NMR) have shown significant comparative advantages [3-8]. NMR spectroscopy [9] has the potential to measure molecular structure, orientational order and dynamics in almost every kind of material. It has been widely used to study molecular dynamics, conformation and order in liquids and liquid crystals. Rheology, i.e. the study of macroscopic deformation and flow of materials, has been approached through NMR as well, and the corresponding Rheo-NMR techniques are currently being rapidly developed [10].

B NMR METHODS FOR THE EVALUATION OF THE VISCOELASTIC PROPERTIES OF LIQUID CRYSTALLINE POLYMERS

The line shape of a typical proton or deuterium NMR spectrum is highly sensitive to the director orientation. By recording the time evolution of the line shape of the NMR spectrum of a flowing nematic, for example, we can accurately probe the rotational dynamics of the director. Then, by fitting the experimental data with an appropriate model derived from continuum theory [11], we can obtain the viscoelastic parameters involved in the dynamic process under study.

The basic relationship between a deuteron (or isolated proton-pair) NMR spectrum in a nematic and the director orientation θ(t) with respect to the magnetic field **H** of the spectrometer is given by

$$\Delta\tilde{v}(t) = \Delta\tilde{v}_o \frac{1}{2}\left(3\cos^2\theta(t) - 1\right) \tag{1}$$

where $\Delta\tilde{v}_o$ is the doublet splitting measured for the equilibrium spectrum (that is when the nematic director **n** is aligned with the magnetic field) and $\Delta\tilde{v}(t)$ is the corresponding splitting measured during flow, at time t. If the director orientation is spatially non-homogeneous, i.e. **n** ≡ **n(r)**, therefore θ ≡ θ(**r**,t), what is observed is a frequency distribution $\Delta\tilde{v}(\mathbf{r}, t)$ rather than a doublet splitting. See for example [12] for more details.

The NMR spectroscopy techniques reported so far to measure the viscoelastic properties of nematics differ in the way chosen to generate the required non-homogeneous flow in the sample, and may be classified as follows.

B1   Steady-State Rheo-NMR Techniques

Two cases may be distinguished: (i) continuous rotation of the sample container about an axis normal to the magnetic field **H** of the spectrometer; (ii) integration of miniature Couette or cone-and-plate rheometers into the probe head of the NMR spectrometer.

In the continuous, steady state, rotation experiment [13,14], the angle θ in EQN (1) is stationary and related to the angular velocity Ω of the container by

$$\theta = \frac{1}{2}\arcsin(\Omega / \Omega_c) \tag{2}$$

provided Ω is below a critical value Ω_c, given by

$$\Omega_c = \left(\Delta\tilde{\chi}B^2 / 2\mu_o\gamma_1\right) \tag{3}$$

where B is the magnetic flux density, $\mu_o = 4\pi \times 10^{-7}$ H m^{-1} is the magnetic constant, and $\Delta\tilde{\chi}$ is the anisotropy of the magnetic susceptibility. If $\Omega > \Omega_c$ the director is predicted [14] to rotate with a mean angular velocity

$$\overline{\omega} = \left(\Omega^2 - \Omega_c^2\right)^{1/2} \tag{4}$$

and the angle θ(t) in EQN (1) is given by

$$\theta(t) = \arctan\left\{\left(\overline{\omega}/\Omega\right)\tan\left[\overline{\omega}(t - t_o)\right] - \Omega_c/\Omega\right\} \tag{5}$$

where t_o is an arbitrary time constant. Prediction (4) of a uniform rotating director is observed only in the initial stage of the experiment; in fact, the director evolves with time to a stationary distribution that is non-uniform in space [6,14-16]. In spite of this, EQNS (3) - (5), combined with EQN (1), have been used to evaluate, with good accuracy, the rotational viscosity γ_1 of a side-chain nematic LC polymer [6]. Early NMR experiments of this kind [14-16] have been performed on low molecular mass nematics in order to test the predictions of the hydrodynamic theory [11] concerning the existence of two qualitatively different behaviours of the nematic director below and above Ω_c, rather than to measure γ_1. However, it has been recognised [14,17,18] that the spinning experiment (on low molecular mass nematics) provides a reliable route to the determination of the ratio of material constants $\Delta\tilde{\chi}/\gamma_1$ (and therefore of γ_1 if $\Delta\tilde{\chi}$ is known; in fact this quantity may also be measured by NMR [18]).

Experiments where a miniature cone-and-plate viscometer was integrated into the probe head of an NMR spectrometer have been reported [7,8,19]. The set-up allowed for the simultaneous acquisition of the NMR spectrum and the measurement of an effective viscosity of the sample. In this type of experiment there is a competition between the orienting effects of the magnetic field of the spectrometer and of the imposed shear. When the cone axis of the viscometer is collinear with the magnetic field, the director orientation angle θ in EQN (1) is predicted [8] to be a stationary function of the shear rate $\dot{\gamma}$ given by

$$\theta(\dot{\gamma}) = \arctan \left\{ \mp \frac{\Delta\tilde{\chi}B^2}{2\mu_o |\alpha_3|\dot{\gamma}} \pm \sqrt{\left(\frac{\Delta\tilde{\chi}B^2}{2\mu_o |\alpha_3|\dot{\gamma}}\right)^2 \pm \left|\frac{\alpha_2}{\alpha_3}\right|} \right\} \qquad (6)$$

where the upper signs (- + +) are to be used when $\alpha_2/\alpha_3 > 0$ (flow-aligning nematics) and the lower signs (+ - -) when $\alpha_2/\alpha_3 < 0$ (tumbling nematics). Notice that, due to the orienting influence of the magnetic field, tumbling is suppressed even for nematics of the tumbling type when the shear rate $\dot{\gamma}$ is lower than a critical value

$$\dot{\gamma}_c = \Delta\tilde{\chi}B^2 / 2\mu_o (|\alpha_2\alpha_3|)^{1/2} \qquad (7)$$

For $\dot{\gamma} < \dot{\gamma}_c$, the steady-state effective viscosity simultaneously measured in the viscometer is also a function of the shear-rate dependent director orientation $\theta(\dot{\gamma})$:

$$\eta_{ef}(\dot{\gamma}) = \alpha_1 \sin^2\theta(\dot{\gamma}) \cos^2\theta(\dot{\gamma}) + \eta_b \sin^2\theta(\dot{\gamma}) + \eta_c \cos^2\theta(\dot{\gamma}) \qquad (8)$$

where $\eta_b = (\alpha_3 + \alpha_4 + \alpha_6)/2$ and $\eta_c = (\alpha_4 + \alpha_5 - \alpha_2)/2$ are Miesowicz viscosities [1,2]. Therefore, fitting EQNS (6) and (8) to the values $\theta(\dot{\gamma})$ derived from the NMR spectra through EQN (1) and $\eta_{ef}(\dot{\gamma})$ measured with the viscometer integrated into the probe head of the NMR spectrometer, four out of the five Leslie viscosities can be evaluated, namely α_1, α_2, α_3 and η_b (note that $\eta_c = \eta_b - \alpha_2 - \alpha_3$). Measurements have been reported on selectively deuterated side-chain nematic liquid crystal polymers, and on mixtures of these with low molecular mass nematics, at several shear rates and temperatures [7,8,19].

B2   Transient-State Rheo-NMR Techniques

This case usually involves an impulse rotation of the sample about an axis normal to the magnetic field [3], although a variant has been proposed which uses a field-cycling spectrometer allowing for the rotation of the magnetic field instead of the sample [24].

In practice, the nematic polymer sample is left at rest in the strong magnetic field of the NMR spectrometer for a time long enough to reach the equilibrium state where the director aligns parallel to the magnetic field direction (z-axis), and $\mathbf{n(r)}$ becomes homogeneous in space.   Then, the tube containing the sample is instantly (but carefully) rotated through some angle ϑ about an axis normal to the magnetic field direction and the relaxation of the director field $\mathbf{n(r)}$ back to equilibrium is observed by recording the NMR spectrum as a function of time.   This relaxation appears to be qualitatively different for initial rotations of $\vartheta < \pi/4$ and $\vartheta \cong \pi/2$ (or, more generally, $\vartheta > \pi/4$, but here the interesting case is $\vartheta \cong \pi/2$).

When $\vartheta < \pi/4$ the spatially homogeneous configuration of the director field is not significantly disturbed and the observed reorientation of $\mathbf{n}$ towards equilibrium ($\mathbf{n}//\mathbf{H}$) follows the well-known equation

$$\gamma_1 d\theta / dt + \left(\Delta\tilde{\chi}B^2 / 2\mu_o \right) \sin 2\theta = 0 \tag{9}$$

where θ still represents the orientation of the director with respect to the magnetic field.   The solution to this equation is

$$\theta(t) = \arctan\left\{ \tan\vartheta . \exp\left[-\left(\Delta\tilde{\chi}B^2 / \mu_o \gamma_1 \right) t \right] \right\} \tag{10}$$

When $\vartheta \cong \pi/2$ an instability occurs which breaks down the uniform alignment and induces some flow (back-flow) in the magnetic field direction so that the relaxation to equilibrium ($\mathbf{n}//\mathbf{H}$) proceeds via a transient distortion of the director field $\mathbf{n(r,t)}$, which forms a spatially periodic pattern of oriented regions separated by inversion walls.   This pattern originates in a thermal fluctuation of the director, with wave vector q_x, which is amplified by the magnetic field.   The reorientation of the director back to equilibrium then follows the more complex equation [3]:

$$\left[\gamma_1 - \frac{j(\varphi)^2}{g(\varphi)} \right] \frac{\partial\varphi}{\partial t} - \left(\Delta\tilde{\chi}B^2 / 2\mu_o \right) \sin 2\varphi - K(\varphi) = 0 \tag{11}$$

where

$$\varphi = \frac{\pi}{2} - \theta$$

$$K(\varphi) = f(\varphi)\frac{\partial^2\varphi}{\partial x^2} + \frac{1}{2}\frac{\partial f(\varphi)}{\partial\varphi}\left(\frac{\partial\varphi}{\partial x} \right)^2 \tag{12}$$

$$f(\varphi) = K_1 - (K_1 - K_3)\cos^2\varphi \tag{13}$$

$$j(\varphi) = \alpha_2 - \gamma_2\sin^2\varphi \tag{14}$$

$$g(\varphi) = (\alpha_1\cos^2\varphi + \gamma_2)\sin^2\varphi + \eta_c \tag{15}$$

and α_1, α_2, $\gamma_2 = \gamma_1 + 2\alpha_2$, and η_c are nematic viscosities (described previously).

It is important to note at this point that the effective viscosity for the (magnetic field driven) director reorientation is γ_1 in the homogeneous process, EQN (9), and is a function of φ (or $\theta = \pi/2 - \varphi$) in the non-homogeneous process, EQN (11): $\gamma_{eff} = \gamma_1 - j(\varphi)^2/g(\varphi)$. In this case, solving EQN (11) for the dominant Fourier component of $\varphi(x,t) = \varphi_0(t)\sin q_x x$ gives the relaxation times

$$\tau_i = \left[\gamma_1 - \left(\alpha_2^2/\eta_c\right)\right]/\left(\mu_o^{-1}\Delta\widetilde{\chi}B^2 - K_3 q_x^2\right) \tag{16}$$

and

$$\tau_i = \left[\gamma_1 - \left(\alpha_3^2/\eta_b\right)\right]/\left(\mu_o^{-1}\Delta\widetilde{\chi}B^2 - K_1 q_x^2\right) \tag{17}$$

in the limits $\varphi \to 0$ ($\theta \to \pi/2$) and $\varphi \to \pi/2$ ($\theta \to 0$), respectively [4]. The solution (EQN (10)) to EQN (9) gives the characteristic time for the homogeneous reorientation: $\tau_o = \mu_o\gamma_1/\Delta\widetilde{\chi}B^2$. In practice, it is found that $\tau_i \ll \tau_o < \tau_f$, i.e. when $\vartheta \cong \pi/2$ the observed reorientation of the director is much faster than predicted by EQN (9) at the beginning of the relaxation process and slower than predicted by EQN (9) at the end.

Experimentally, the NMR spectrum is recorded as a function of time, during the relaxation process. In the case of experiments with $\vartheta = \pi/2$, an approximate solution to EQN (11) can easily be obtained numerically [3,4] and fitted to the experimental data through EQN (1). In practice, both the doublet splitting $\Delta\widetilde{v}(t)$ [3] and the line shape [4,20-23] of the (deuteron or proton) NMR spectra can be used. The numerical results are found to fit very accurately to the experimental data and give four out of the five independent nematic viscosities, namely γ_1, η_c, α_1, and α_2, as well as the ratio K_3/K_1 of the bend to splay Frank elastic constants and the product $K_1 q_x^2$ (or both elastic constants if the wavelength $\lambda_x = 2\pi/q_x$ is measured, e.g. by optical microscopy). In the homogeneous case ($\vartheta < \pi/4$), the experimental data are equally well fitted by EQN (10), and give an independent value for γ_1 that may be used to check the result obtained from the non-homogeneous case. Measurements have been made so far for thermotropic main-chain [3,4], combined main-chain/side-chain [20], and lyotropic [21-23] nematic polymers.

It will be shown in the next section that the fifth Leslie viscosity can be estimated from those given by the transient state technique just described.

C REPRESENTATIVE EXPERIMENTAL DATA

In all cases it has been assumed that the anisotropy of the magnetic susceptibility is known (otherwise, this quantity may also be measured by NMR [18]), and $\Delta\tilde{\chi} > 0$. The same concepts can be applied, mutatis mutandis, to nematic materials with $\Delta\tilde{\chi} < 0$, in which the director aligns, in equilibrium, perpendicular to the magnetic field. In this case the steady-state techniques based on the cone-and-plate viscometer may not be applicable and the use of the Couette geometry may be advantageous; the transient techniques may require another, different, external field (e.g. electric) to prepare the initial state of the sample. Both techniques require that the spatial configuration of the velocity field be known for the geometry considered, or can be derived from the Leslie equations, which is not a serious limitation in general but may cause problems or uncertainties in some cases. The combined Rheo-NMR (steady-state) techniques are technically more difficult to implement, and may not work properly with high molecular weight polymers, a domain where the transient techniques (eventually including shearing devices to prepare the initial state) are particularly easy to apply.

Both NMR methods described allow for the evaluation of four out of the five independent nematic viscosities. These are extracted from the NMR data alone when the method based on transient-state flows is used; employing methods based on steady-state flows requires, in addition to the NMR data, an independent, simultaneous measurement of the effective viscosity (given by EQN (8) in the case of cone-and-plate geometry) to obtain the same number of independent viscosities.

In order to determine the fifth independent viscosity, and so be able to derive the whole set of Leslie coefficients, we may either use complementary data from a different rheological measurement [21,23] or use theoretical arguments [25] to estimate it. In this case it can be shown that

$$\alpha_5 = -\frac{1}{7}\left[2\alpha_1 + \frac{1}{10}(3P + 7)(10\alpha_2 + 5\gamma_1)\right] \tag{18}$$

where P is a shape factor related to the polymer conformation; P is less than but close to unity for long molecules and can also be estimated from the measured viscosities [25]. Using the data derived from the Rheo-NMR experiments (γ_1, η_c, α_1, and α_2) and EQN (18) we finally obtain the whole set of the Leslie viscosities: α_1, α_2, $\alpha_3 = \gamma_1 + \alpha_2$, $\alpha_4 = 2\eta_c + \alpha_2 - \alpha_5$, α_5, and $\alpha_6 = \gamma_1 + 2\alpha_2 + \alpha_5$.

The experimental data so far published on the Leslie viscosities for nematic polymers are very scarce. The first claim for the evaluation of a complete set of viscosities has only recently been made and refers to a chiral nematic solution of poly(γ-benzyl-L-glutamate) in m-cresol [21]. TABLE 1 gives a number of new complete sets of data for several representative systems, which have been established from the published experimental data and estimates of α_5 according to EQN (18) [25]. The values of $\Delta\tilde{\chi}$ and P used are also given in TABLE 1.

PSi4 is a side-chain thermotropic nematic polymer; DDA9, AZA9 and TPB10 are main-chain thermotropic nematic polymers; PBLG and PPTA are lyotropic nematic polymers. It is interesting to note that the polymers DDA9, AZA9 and PSi4 have $\alpha_2/\alpha_3 > 0$ and therefore are of the flow-aligning type, and the remaining four systems (high molecular weight polymers) are of the tumbling type, i.e. have $\alpha_2/\alpha_3 < 0$.

TABLE 1 Leslie viscosities α_i for several nematic polymer systems.

	DDA9	AZA9	PBLG 12% in m-cresol	PBLG 17% in m-cresol	PPTA 8.8% in SO$_4$D$_2$	TPB10	PSi4
Mn=	4300	4300	280000	280000	35000	16000	27000
T=	394 K	393 K	302 K	302 K	300 K	360 K	348 K
α_1/Pa s	-1.620×10^2	-1.320×10^3	-1.930×10^2	-1.212×10^3	-1.177×10^6	-1.793×10^5	-9.600×10^2
α_2/Pa s	-1.700×10^2	-1.595×10^3	-3.900×10^2	-1.328×10^3	-2.136×10^6	-1.810×10^5	-2.000×10^3
α_3/Pa s	-2.000	-2.500×10^1	2.900×10^1	4.000	6.972×10^4	0.000	-1.270×10^2
α_4/Pa s	1.601×10^1	2.089×10^2	5.983×10^1	1.137×10^2	2.415×10^5	7.838×10^3	3.066×10^3
α_5/Pa s	1.620×10^2	1.470×10^3	2.916×10^2	1.208×10^3	1.767×10^6	1.752×10^5	1.534×10^3
α_6/Pa s	-1.001×10^1	-1.499×10^2	-6.943×10^1	-1.157×10^2	-2.992×10^5	-5.838×10^3	-5.932×10^2
$\Delta\tilde{\chi}$	1.57×10^{-6}	1.57×10^{-6}	0.96×10^{-7}	1.62×10^{-7}	1.57×10^{-6}	?	1×10^{-6}
P	0.806	0.815	0.723	0.705	0.898	0.862	0.430
Ref	[3]	[4]	[23]	[21]	[22]	[19]	[7]

DDA9 = poly(4,4'-dioxy-2,2'-dimethylazoxybenzene-dodecanediyl)

AZA9 = poly(4,4'-dioxy-2,2'-dimethylazoxybenzene-nonanediyl)

PBLG = poly(γ-benzyl-L-glutamate)

PPTA = poly(p-phenylene-terephthalamide)

TPB10 = poly[1,10-decylene-1-(4-hydroxy-4'-biphenylyl)-2-(4-hydroxyphenyl)butane]

PSi4 = poly[(2,3,5,6-tetradeuterio-4-methoxyphenyl-4'-butanoxybenzoate)-methylsiloxane]

The errors in the NMR evaluation of the viscosity coefficients are usually low (5 - 10% can be easily achieved) for γ_1, α_2 and η_c, and about twice as large for α_1. These viscosities are of the same order of magnitude and may be very high in nematic polymers. On the other hand, α_3 is always much smaller (for materials with rod-like molecules); therefore when we try to evaluate this viscosity from the previous ones ($\alpha_3 = \gamma_1 + \alpha_2$, where $\gamma_1 > 0$ and $\alpha_2 < 0$) the error incurred may be rather significant. In spite of this, experience suggests that the sign of α_3 is, in general, correctly given by the transient-state technique, i.e. the error is reflected only on the absolute value of this viscosity.

D   CONCLUSION

The main complication that severely limits the utility of conventional rheometry in the field of liquid crystals is the existence of at least five independent viscosity parameters and three elastic constants that play a role in their behaviour but cannot be discriminated by conventional experiments. The special techniques developed to measure these parameters in low molecular mass nematics can hardly be applied to polymer systems. Optical techniques, for example, demand the use of thin (transparent) samples and a careful control of molecular orientation at the sample surfaces, which is very difficult to achieve with nematic polymers, in most practical cases. In contrast, the Rheo-NMR technique works better with thick samples and therefore the influence of the boundaries on the observed dynamics is negligible in most practical cases. Moreover, this technique may easily be applied to characterise the viscous properties of nematic polymers up to very high degrees of polymerisation. The transient recovery Rheo-NMR techniques offer an enormous potential for the study of the director reorganisation processes and to elucidate in depth the structure-properties relationships in soft materials and complex fluids, therefore substantially complementing the more traditional optical and mechanical techniques. As compared to conventional rheometry, the NMR technique requires minute amounts of sample and can be used over an extremely wide range of temperatures. Another major advantage of NMR is its non-invasive nature. This is relevant in many laboratory and industrial applications, and may be of paramount importance for in-vivo bio-rheology.

ACKNOWLEDGEMENT

This work was partly supported by the European Union under research contract FMRX-CT97-0121 and by PRAXIS XXI Program (Portugal) under contract Nr. 2/2.1/MAT/380/94.

REFERENCES

[1]   P.G. de Gennes, J. Prost [*The Physics of Liquid Crystals* (Clarendon Press, Oxford, 1993)]
[2]   S. Chandrasekhar [*Liquid Crystals* (Cambridge University Press, Cambridge, 1992)]
[3]   A.F. Martins, P. Esnault, F. Volino [*Phys. Rev. Lett. (USA)* vol.57 (1986) p.1745]
[4]   P. Esnault, F. Volino, A.F. Martins, S. Kumar, A. Blumstein [*Mol. Cryst. Liq. Cryst. (UK)* vol.153 (1987) p.145]
[5]   P. Esnault, J.P. Casquilho, F. Volino, A.F. Martins, A. Blumstein [*Liq. Cryst. (UK)* vol.7 (1990) p.607]
[6]   D. van der Putten, N. Schwenk, H.W. Spiess [*Liq. Cryst. (UK)* vol.4 (1989) p.341]

[7] D.A. Grabowski, C. Schmidt [*Macromolecules (USA)* vol.27 (1994) p.2632]

[8] H. Siebert, D.A. Grabowski, C. Schmidt [*Rheol. Acta (Germany)* vol.36 (1997) p.618]

[9] A. Abragam [*Principles of Nuclear Magnetism* (Clarendon Press, Oxford, 1961)]

[10] P.T. Callaghan [*Rep. Prog. Phys. (UK)* vol.62 (1999) p.599]

[11] F.M. Leslie [*Arch. Ration. Mech. Anal. (Germany)* vol.28 (1968) p.265]

[12] J.W. Emsley [*NMR of Liquid Crystals* (D. Reidel Publ. Co., Dordrecht, 1985)]

[13] V.N. Tsvetkov [*Acta Physicochimica (USSR)* vol.11 (1939) p.537]

[14] F.M. Leslie, G.R. Luckhurst, H.J. Smith [*Chem. Phys. Lett. (Netherlands)* vol.13 (1972) p.368]

[15] S.G. Carr, G.R. Luckhurst, R. Poupko, H.J. Smith [*Chem. Phys. Lett. (Netherlands)* vol.7 (1975) p.278]

[16] J.W. Emsley, S.K. Khoo, J.C. Lindon, G.R. Luckhurst [*Chem. Phys. Lett. (Netherlands)* vol.77 (1981) p.609]

[17] S.K. Khoo, G.R. Luckhurst [*Liq. Cryst. (UK)* vol.15 (1993) p.729]

[18] E. Ciampi, J.W. Emsley [*Liq. Cryst. (UK)* vol.22 (1997) p.543]

[19] H. Siebert [PhD Thesis, University of Freiburg, Germany, 1998]

[20] M. Tittelbach, G. Kothe, C.R. Leal, J.B. Ferreira, A.F. Martins [to be published]

[21] A. Veron, A.E. Gomes, C.R. Leal, J. Van Der Klink, A.F. Martins [*Mol. Cryst. Liq. Cryst. (Switzerland)* vol.331 (1999) p.499]

[22] J.B. Ferreira, J.R. Hughes, C.R. Leal, G.R. Luckhurst, A.F. Martins [to be published]

[23] A. Veron, A. Gomes, J. van der Klink, A.F. Martins [presented at *Eurorheo99-1* Sophia-Antipolis, France, 3-7 May 1999]

[24] H. Gotzig, S. Grunenberg-Hassanein, F. Noack [*Z. Naturforsch. A (Germany)* vol.49 (1994) p.1179]

[25] A.F. Martins [to be published]

8.4 Relationship between nematic viscosities and molecular structure

V.V. Belyaev

September 1998

A INTRODUCTION

Early results on viscosities of nematics and their temperature dependences were discussed systematically in [1] and [2]. More detailed information on the effect of molecular structure on viscosity was considered in [3-6]. Here the results are discussed to show the dependence of the different nematic viscosities on thermodynamic parameters (temperature and pressure), molecular dimensions and shape and the electronic density distribution in the molecules or their fragments. The possibility of calculating the nematic viscosities according to different molecular models and theories is also discussed.

B NEMATIC VISCOSITIES

The viscosities considered and methods for their measurement are listed in TABLE 1.

The best accuracy (0.1 - 1%) was achieved in the experiments performed by Kneppe and Schneider [7,8] using a large quantity (~30 - 70 g) of nematic. Usually the accuracy in the Fréederiksz transition and light scattering methods is 7 - 10% of the viscosity and about 0.01 eV of an activation energy but the quantity of nematic required is significantly less (~50 mg).

Much of the data has been obtained by extrapolation with respect to temperature or concentration; for temperature the data are shown in round brackets and for concentration in square brackets (e.g. (23.3 cP) or [1.70 P]). The nematic-isotropic transition temperatures given in round brackets indicate a monotropic transition. The viscosities η, γ_1 are given in Pa s or poise (P = 10^{-1} Pa s) and kinematic viscosities ν in stokes (St = 10^{-4} m^2 s^{-1}). To reduce the length of this review references to data published in [6] are omitted. Here we give some frequently used abbreviations: CB - cyanobiphenyl, PCH - phenylcyclohexane, BCO - bicyclo[2.2.2]octane, T - tolane, Ph - phenyl ring, Pyr - pyridine ring, Cy - cyclohexane moiety. A number associated with the abbreviation corresponds to the number of carbon atoms in an alkyl substituent.

TABLE 1 The nematic viscosities and their measurement.

Designation	Relation to other viscosities	Name of parameter	Method of measurement	
α_i (i = 1, ..., 6)		Leslie viscosities		
η	$\approx \eta_2 - \alpha_3 \approx \eta_2$ $(-\alpha_2 \gamma_2/\gamma_1 + \alpha_4 + \alpha_5)/2$	Dynamic viscosity	Capillary or rotation viscometer Ultrasonic shear	
ν	η/ρ (ρ - density)	Kinematic viscosity	Capillary viscometer	
η_1* η_2* η_3 η_{12}	$(-\alpha_2 + \alpha_4 + \alpha_5)/2$ $(\alpha_3 + \alpha_4 + \alpha_6)/2$ $\alpha_4/2$ α_1	Miesowicz viscosities	Poiseille or couette flow	$\mathbf{v} \perp \mathbf{n}, \mathbf{n} \mid\mid \nabla\mathbf{v}$ $\mathbf{v} \mid\mid \mathbf{n}, \mathbf{n} \perp \nabla\mathbf{v}$ $\mathbf{v} \perp \mathbf{n}, \mathbf{n} \perp \nabla\mathbf{v}$ $\angle(\mathbf{v},\mathbf{n}) = 45°$
γ_1	$\alpha_3 - \alpha_2$	Rotational (Zwetkoff) viscosity	Rotating magnetic field Dynamics of Fréedericksz transition Light scattering on thermal director fluctuations	
γ_2	$\alpha_3 + \alpha_2$	Torsion coefficient in velocity gradient	Couette flow	
η_S	$\gamma_1 - \alpha_3^2/\eta_1$	Splay viscosity	Dynamics of Fréedericksz transition Light scattering on thermal director fluctuations	
η_B	$\gamma_1 - \alpha_2^2/\eta_2$	Bend viscosity	Dynamics of Fréedericksz transition Light scattering on thermal director fluctuations	

*Editors note - n_1 and n_2 are interchanged with respect to definitions in 8.2.

C   VISCOSITIES: TEMPERATURE AND PRESSURE DEPENDENCES

The temperature dependences of different viscosities measured by various techniques have been reported [9-13]. The Miesowicz viscosities η_i are collected in TABLE 2 for 4-methoxybenzylidine-4'-butylaniline (MBBA) [7]. All the data can be reduced to an activation or a free volume formula, e.g. for the rotational viscosity the relationships

$$\gamma_1 = g \, S^x \exp (E \, S^y/k_B \, T) \tag{1}$$

$$\gamma_1 = b \, S^x \exp [\, 1/v_{ff}] = b \, S^x \exp [B/(T - T_0)] \tag{2}$$

are valid [5], where g and b are constants, x = 0, 1 or 2, y = 0 or 1, E is the activation energy (y = 0) or the height of the Maier-Saupe potential (y = 1), v_{ff} is the fluctuation free volume, B is an interaction coefficient which is equal to 1225 K (x = 0) [14] or 949 K (x = 2) [15] for all substances, and T_0 is the temperature at which the director motion freezes or the temperature at which the free volume is zero. In [14] the coefficient b = 6.03×10^{-5} Pa s for all the nematic materials studied.

TABLE 2 Numerical values of the Miesowicz viscosity coefficients
of MBBA in the nematic and isotropic phases (T_{NI} = 45.1°C) [7].

T/°C	η_1/Pa s	η_2/Pa s	η_3/Pa s	η_{12}/Pa s
20	0.1891	0.0308	0.0548	-0.0297
25	0.1361	0.0240	0.0413	-0.0181
30	0.1007	0.01927	0.0322	-0.0141
35	0.0741	0.01596	0.0257	-0.0095
40	0.0534	0.01377	0.0211	-0.0054
42	0.0456	0.01320	0.01972	-0.0036
44	0.0367	0.01323	0.01871	-0.0012
46	0.01852			
50	0.01585			
55	0.01323			

The validity of both EQNS (1) and (2) was considered in [5,6]. The activation mechanism fits the temperature dependence of γ_1 for all nematics with a narrow mesophase range (up to 40°C) at T > 290 K. The free volume model fits better for nematics with a low activation energy (E < 0.45 eV = 4950 K at x = y = 0) and broad nematic range (>60°C) achieving low temperatures (<0°C). The dependence on the second rank orientational order parameter S is determined by the type of intermolecular interaction [5]. For bicyclic polar substances with both aromatic moieties (benzene, heterocycle with nitrogen atoms) and an elongated conjugation chain the proportionality $\gamma_1 \sim \Delta ñ \sim S$ is valid; here $\Delta ñ$ is the birefringence. For polar substances containing a moiety with saturated bonds (cyclohexane, bicyclooctane) and weakly polar compounds with a dielectric constant $< \varepsilon > \sim$ 3 - 5 the fitting with $\gamma_1 \sim (\Delta ñ)^2 \sim S^2$ is valid. The activation energy for substances with the first type of intermolecular interaction is typically as high as 5000 K or more, and for the second class of substances as high as 2500 - 4500 K. In fact the dependence of viscosity on the order parameter is an expansion in a power series in S or Legendre polynomials $<P_L>$, L = 1, 2, 4, ...

Parameters obtained from some fittings for some common nematics are listed in TABLE 3; the formulae used for the rotational viscosity and the orientational order parameter were

$$\ln(\gamma_1/S) = A + (x - 1)S + B/T$$

$$S = S_0 (1 - T/T^*)^\beta$$

(3)

In [3] a free volume approximation

$$\gamma_1/Pa \; s = 3.482 \times 10^{-4} (1 - T/319.83K)^{0.2942} \exp[841.35/(T/K - 170.0)]$$

(4)

was obtained for MBBA. The activation energy is of the same order of magnitude as other parameters which are determined by the orientational part of the intermolecular potential energy, namely the molar melting enthalpy ΔH_m [15], the anisotropic part of the van der Waals interaction energy calculated from density data [16], and the height of the Maier-Saupe potential εS_{NI} = (3/2) 4.54 k_B T_{NI} S_{NI}. The last

TABLE 3 Fitting parameters for MBBA, 4-butyl-4′-methoxyazoxybenzene (N4 Merck), 4,4′-dipentylazoxybenzene (5AB) and 4-pentyl-4′-cyanobiphenyl (5CB). R-correlation coefficient at confidence level $p = 0.95$.

Substance	x	-A	B/K	R	S_0	T*/K	T_{NI}/K	β
MBBA	2	12.66 ± 0.88	4108 ± 270	0.9984	0.677	318.71	318.2	0.1737
	2.183 ± 0.477	11.52 ± 3.13	3795 ± 866	0.9994				
N4	2	10.68 ± 0.55	3511 ± 177	0.9981	1.08 ± 0.10	347.4 ± 0.2	346.4	0.215 ± 0.015
	2.604 ± 0.218	12.18 ± 0.88	3922 ± 248	0.9997				
5AB	2	9.12 ± 0.98	2951 ± 318	0.9914	1.065 ± 0.05	341.2 ± 0.2	340.7	0.17 ± 0.01
	2.296 ± 0.636	7.77 ± 3.05	2566 ± 891	0.9965				
5CB	1	18.48	5617	0.9943	1.13	309.6	308.7	0.19

relation is widely used in theory and experiment as a measure of the intermolecular interaction energy [17-20].

Viscosities for flow and reorientation processes are related by the following identity [21]:

$$\eta_1/\eta_2 \equiv \eta_S/\eta_B \tag{5}$$

The order parameter dependence of the ratios for 4-pentyl-4'-methoxytolane (5TO1) is described by the following experimental relation [13] which corresponds to theoretical equations for η_1 and η_2 [22]:

$$\eta_S/\eta_B = 1 + 3.51 \ p \ S$$

$$= 1 + 10.5S \tag{6}$$

where p is a shape factor of the occupied volume [15] (see Section D).

In [3] the following universal approximation with an accuracy better than 20% is proposed for the ratio of rotational and dynamic viscosities of a variety of nematics:

$$\gamma_1/\eta = [1 - T/K/(T_{NI}/K + 0.35)]^{0.35}/(0.1076 - 5.048 \times 10^{-4} \ T_{NI}/K) \tag{7}$$

For various nematics the relationship

$$\Sigma \ A_i \ \eta_i = 0 \ , \ i = 1, 2, 3 \tag{8}$$

is valid over the whole temperature range. $A_1:A_2:A_3 = 0.08:1.1:(-1)$ for N4, MBBA, and 5CB [11] or $0.114:0.608:(-1)$ for anisilidene-N-aminophenylacetate (APAPA) [21]. Each Miesowicz coefficient of 5CB has thermal activation energies E_i which are equal to $E_1 = 0.49 \pm 0.06$ eV, $E_2 = 0.21 \pm 0.04$ eV, $E_3 = 0.27 \pm 0.04$ eV [23]. The dependence of η_i on the orientational order is described for reduced η_{ir} [11] or absolute [21] viscosity values by

$$\eta_{ir} = \eta_i/\eta_a = 1 + (\eta_i^{(0)} / \eta_a - 1)S$$

$$\eta_a = (\eta_1 + \eta_2 + \eta_3)/3 \tag{9}$$

$$\eta_i = \eta_i^{(0)} + \beta_i \ S \tag{10}$$

where $\eta_i^{(0)}$ is the viscosity of a completely ordered nematic (S = 1). In EQN (10) the fitting parameters for the APAPA data are: $\eta_1^{(0)} = -0.068$ Pa s, $\beta_1 = 0.214$ Pa s; $\eta_2^{(0)} = 0.007$ Pa s, $\beta_2 = -0.029$ Pa s; $\eta_3^{(0)} = -0.012$ Pa s, $\beta_3 = 0.042$ Pa s. EQN (9), however, fits the viscosity data of other substances described here better.

The pressure dependence of the dynamic and rotational viscosities of Schiff's bases are described in [24,25]. The activation energy for viscous flow increases with both rising temperature and pressure: up to 5 kbar it is as high as 35.6 - 41.8 kJ/mol for MBBA and 28.1 - 28.0 kJ/mol for EBBA [24]. A fitting has been proposed [25] for $\gamma_1(T, p)$ data for pressures between 1 and 2500 bar:

$$\gamma_1 = A(1 - T/T^*)^{2\beta} \exp[C(T^* - T_0)/(T - T_0)]$$

$$T^* = T_0^*(1 + p/a)^c$$

(11)

For MBBA the fitting parameters $A = 4.5160 \times 10^{-4}$ Pa s, $\beta = 0.1520$, $C = 5.4247$, $T_0 = 168.3$ K, $T_0^* = 320.0$ K, $a = 3827$ bar and $c = 0.4486$ have been obtained.

D   INFLUENCE OF MOLECULAR STRUCTURE ON NEMATIC VISCOSITIES

The viscosity is affected mainly by two parameters describing the mesogenic microscopic structure: the molecular packing and the molecular friction or rotational diffusion coefficients. If we know the first then the free volume parameters can be calculated. The second is determined by molecular or cluster rotation and translation.

Molecular packing in nematics has been studied in [16,26,27]. The usual value of the Kitaigorodsky packing coefficient, k_p, is in the range from 0.6 to 0.7. $k_p = N_a v_0/V = N_a v_0 \rho/M = n v_0$ where v_0 is the van der Waals volume of a single molecule, V is the molar volume, M is the molar mass, and n is the concentration of molecules in a cubic centimetre. The larger is the geometrical free volume $v_{fg} = 1 - k_p$ the lower is nematic viscosity (see FIGURE 1) [6].

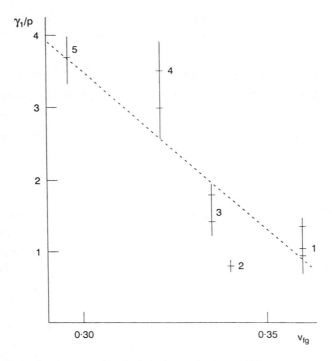

FIGURE 1   Relation between the rotational viscosity γ_1 (Poise) of bicyclic polar compounds and the geometrical free volume v_{fg} at 25°C.   1 - C_nH_{2n+1}-Cy-Ph-CN, 2 - C_nH_{2n+1}-Ph-Ph-CN, 3 - C_nH_{2n+1}-Pyr-Ph-CN, 4 - C_nH_{2n+1}-BCO-Ph-CN, 5 - C_nH_{2n+1}-O-Ph-Ph-CN.

In the theory proposed by Diogo and Martins [17] the free volume in the nematic phase is the difference between the molar volume V^a extrapolated from the isotropic phase and the active molar volume V_N. When density data are available the value of the parameter $G = T^{1/2} T_{NI} S^2 / [V_N (V^a - V_N)] \sim \gamma_1$ can be calculated. In FIGURE 2 values of γ_1 and G are compared. A similar variation of both γ_1 and G is observed.

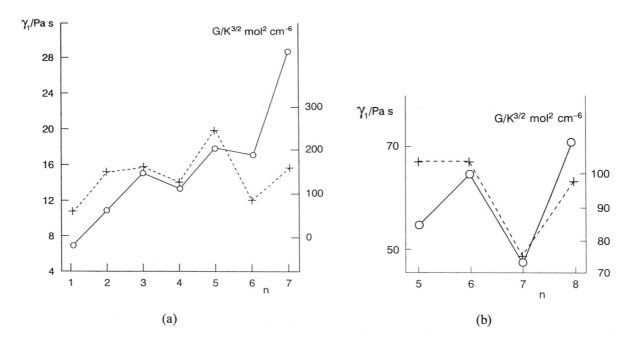

FIGURE 2   Relation between the rotational viscosity γ_1 and the free volume parameter G in the homologous series of (a) 4,4'-dialkyloxyazoxybenzenes (nOAB), $\Delta T = T_{NI} - T = 10$ K, (b) 4-alkyl-4'-cyanobiphenyls (nCB), $\Delta T = 5$ K.

Let us consider the influence of the structure of various molecular fragments on nematic viscosities and their temperature dependence. Each molecule consists of a rigid core (aromatic ring or moieties with saturated bonds which are connected by different linkages), and polar or weakly-polar terminal and lateral substituents. These parts and the molecule as a whole are characterised by geometrical factors and the electron density distribution. Various intermolecular interactions in mesogens are described in [28].

The polarity of a compound can be described by the anisotropy of its polarisability, $\Delta\alpha$, the electric dipole moment, μ, the average dielectric constant $<\varepsilon> = (\widetilde{\varepsilon}_\parallel + 2\widetilde{\varepsilon}_\perp)/3$ or its anisotropy $\Delta\widetilde{\varepsilon}$. It has been shown [6] that for nematic materials of Hoffmann-La-Roche (now ROLIC, Switzerland) in the range of $\Delta\widetilde{\varepsilon} = 3 - 20$ and viscosities $\eta = 15 - 50$ cP the following linear relationship is valid for the majority of materials:

$$\eta / cP = A/cP + 0.02\Delta\widetilde{\varepsilon} \tag{12}$$

where A is 0.105 ± 0.05 cP. The proportionality of the viscosity coefficients or their logarithms to the polarity parameters ($\Delta\alpha$, polarisability and dipole moment of a terminal substituent $\Delta\alpha_x$ and μ_x) has been demonstrated [6,29]. In TABLE 4 the influence of the dipole moment of a heterocyclic moiety on

γ_1 and the dielectric anisotropy of substances with these fragments is presented. All the heterocycles have approximately the same value of dipole moment but their conjugation with cyanophenyl fragment results in consequent differences in $\Delta\varepsilon$ values and rotational viscosities.

TABLE 4 Influence of structure of a heterocycle moiety with an electric dipole moment μ on the dielectric anisotropy $\Delta\tilde{\varepsilon}$ and rotational viscosity γ_1.

Substance	γ_1/P	μ/D	$\Delta\varepsilon$ ($T_{NI} - T = 10$ K)
C_nH_{2n+1} —◯—◯— CN	1.02	~0	11.73
C_nH_{2n+1} —◯—◯— CN	1.34	2.14 - 2.55	13.30
C_nH_{2n+1} —◯—◯— CN	1.84	2.11 - 2.20	16.60
C_nH_{2n+1} —◯—◯— CN	2.01	2.00 - 2.42	19.72

The influence of the linkage on γ_1 and its temperature dependence is displayed in TABLE 5. The increase in the dipole moment of the linkages results in a significant increase in the rotational viscosity and the activation energy.

TABLE 5 Influence of the dipole moment, μ, of the molecular core with a linkage group on the rotational viscosity, γ_1, and its activation energy E.

Substance	γ_1/P (25°C)	$E/k_B K$	μ/D
C_4H_9—◯—C≡C—◯—CC_2H_5	1.2	3600	0 - 0.3
CH_3O—◯—CH=N—◯—C_4H_9	1.09	5820	1.57
CH_3O—◯—N=N—◯—C_4H_9 (O)	1.57	4550	1.70
C_4H_9—◯—COO—◯—CC_2H_5	2.20	5110	1.90
CH_3O—◯—CH O-H N—◯—C_4H_9	(6)	7350	2.39

To quantify the molecular structure a set of models has been proposed [30]. The simplest is an ellipsoid of revolution or a spherocylinder. However, the best agreement with the experimental data is achieved when the actual position of atoms and bonds in a molecule and the internal molecular rotations are taken into account [15,30-35]. A model of the occupied volume or the space which a molecule occupies has been proposed to explain and predict the change of various physical properties in different homologous series of mesogens. The properties include transition temperatures [12,35], viscosity coefficients and elastic constants [13], refractive indices [36], anisotropy of the local field [15], and the molecular friction [20]. In this model values of the length L, width W and shape of the space which a molecule occupies (occupied volume) p = L/W are calculated in a coordinate system set in the molecular core instead of the long axis of a molecule. The use of exact values for the bond lengths and valence angles allows us to explain by geometric factors alone a set of results for steric isomers with different configurations of the substituents.

In FIGURE 3 the dependence of the visco-elastic ratio η_S/K_1 on the total number of carbon atoms in both alkyl and alkoxy chains, n+m, for nTOm are presented [13]. The odd-even alteration for pentyl homologues is obvious as well as the strong dependence on the configuration of alkyl and alkoxy chains. The $\eta_S/K_1(p^{-2})$ dependence is linear when the shape factor is calculated in the rigid core coordinates. The same is also true for the series of nAB, nOAB and the 4-alkoxy-2,3-dicyanophenyl ester of 4'-alkylcyclohexanecarboxylic acid. The last series is of special interest since an unusual reduction in the viscosity with an increase in the molecular length is observed [37].

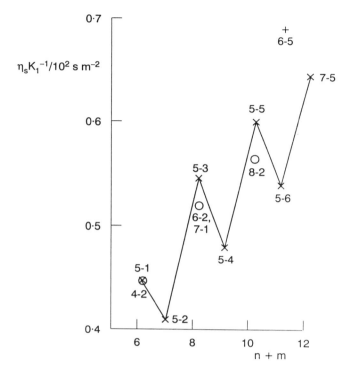

FIGURE 3 Dependence of the visco-elastic ratio η_S/K_1 of nTOm (n-m) on the total number of carbon atoms in both substituents at 25°C.

For other viscosity coefficients the expression

$$\eta/\gamma_1 \cong \eta_B/\eta_S = 1.341\kappa^2 + 0.031 \tag{13}$$

holds for the change in the viscosities along the homologous series of nTOm [13], where $\kappa^2 = (p^2 - 1)/(p^2 + 1)$ and p was calculated in the coordinate system of the molecular core.

The (n+m) dependence of the activation energy E for the visco-elastic ratio for the splay deformation illustrates the influence of internal molecular rotations on a macroscopic parameter (see FIGURE 4) [13]. The energy does not change when the addition of a methylene group results in appearance of the end C-C bond parallel to the core. In contrast its change is equal to 200 K when the end C-C group is inclined to the core. In the first case the alkyl or alkoxy group cannot form new gauche-configurations ($\Delta E_{t-t} = 0$); in the second new rotational isomers can be formed with a conformation energy change $\Delta E_{t-g} = 1.7$ kJ/mol $\equiv 208$ K which coincides exactly with the change in the activation energy.

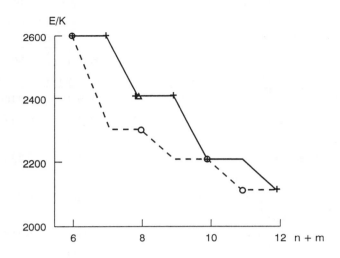

FIGURE 4 Activation energy for the visco-elastic ratio η_S/K_1 of nTOm.
Dashed line is for even alkyl and odd alkyloxy substituents.

In TABLES 6 and 7 the influence of the position of the double bond in the alkenyl substituent is illustrated. When this is connected to the benzene ring the viscosity increases in comparison with that of the alkyl subsituent. The viscosity reduces when the alkenyl chain replaces the alkyl chain at the cyclohexane moiety. For PCH cyano-derivatives the maximal viscosity is obtained when a vinyl group is separated from the core by an ethane linkage -CH_2-CH_2- and the minimal viscosity is achieved when the vinyl group is directly bonded to the cyclohexane moiety.

The effect of successive substitution of alkyl group, polar CN group and a linkage on the viscosity t_{off} is illustrated in TABLE 8. The rotational viscosity reduces when an isothiocyanate group (NCS) or ethane linkage or alkenyl substituent is introduced into a molecule. The cooperative action of all these moieties offers to minimise all the parameters controlled. Kinematic viscosity reduction from 50 cSt to 18 cSt and increase in T_{NI} from 43.5°C to 98.5°C was described also for substitution of CN group for NCS group in 5-pentyl-2-(4'-cyano- or 4'-isothianatephenyl)pyridines [38].

The effect of the cyclohexane and bicyclooctane moieties has been displayed and discussed previously (see TABLE 8 and FIGURE 1). In cyclohexene derivatives the viscosity does not depend on the position of the double -C=C- bond (see TABLE 9). When the molecular core consists only of cyclohexane moieties the viscosity increases in cyano derivatives due to more dense packing. However bicyclohexane derivatives with alkyl and alkoxy substituents and ether chains (oxygen atom is between

TABLE 6 Viscosity of phenylcyclohexanes R_1-Cy-Ph-R_2 with alkenyl substituents.

R_1	R_2	η/mPa s (20°C)	T_{NI}/K
C_3H_7	C_3H_7	[7]	[217]
C_3H_7	OC_2H_5	[13.5]	
C_3H_7	CH=CH-CH_3	[17]	358
C_3H_7-CH=CH-	OC_2H_5	11.8	329.7

TABLE 7 Physical properties of 4-alkenyl-(4'-cyanophenyl)cyclohexanes at 22°C C_nH_{2n+1}-CH=CH-C_mH_{2m}-Cy-Ph-CN. Fitting $\ln \gamma_1 = \ln \gamma_0 + E/T$.

n, m	γ_1/P	η/P	T_{NI}/K	$-\ln \gamma_0$	E/K	$\Delta \tilde{n}$ ($T_{NI} - T = 10$ K)
n = 3, m = 0	1.34	0.221	331.7	16.62	4990	0.116
n = 3, m = 0 (60%)	1.21	0.228	325.4	16.32	4870	0.117
n = 2, m = 0 (40%)						
n = 3, m = 0 (60%)	1.19	0.209	324.1	16.33	4870	0.115
n = 0, m = 2 (40%)						
n = 1, m = 2 (60%)	1.50	0.190	338.4	15.73	4760	0.127
n = 0, m = 2 (40%)						
n = 3, m = 0 (60%)	0.97	0.278	308.3	20.10	5920	0.105
n = 0, m = 3 (40%)						
n = 0, m = 5	0.99		305.4	14.65	4320	0.086
PCH-5	1.28	0.215	328.1	16.67	4990	0.099
5CB	1.01	0.263	308.1	21.24	6270	0.178

TABLE 8 Rotational viscosity and activation energy E for bicyclic polar compounds.

Compound	γ_1/P	E/K
C$_5$H$_{11}$⟨⟩⟨O⟩CN	1.28	4870
C$_5$H$_{11}$⟨⟩⟨O⟩NCS	0.83	4300
CH$_3$-CH=CH-C$_2$H$_4$⟨⟩⟨O⟩CN	1.50	4760
CH$_2$=CH-C$_2$H$_4$⟨⟩⟨O⟩CN C$_2$H$_5$-CH=CH-C$_2$H$_4$⟨⟩⟨O⟩NCS	1.00	4820
C$_5$H$_{11}$⟨⟩CH$_2$-CH$_2$⟨O⟩CN	1.80	4870
CH$_3$-CH=CH⟨⟩⟨O⟩CN CH$_2$=CH-C$_2$H$_4$⟨⟩⟨O⟩CN	2.15	4990
C$_3$H$_7$/C$_5$H$_{11}$⟨⟩CH$_2$-CH$_2$⟨O⟩NCS	1.05	4820
C$_5$H$_{11}$/C$_7$H$_{15}$⟨⟩COO⟨O⟩CN	2.69	4430
C$_5$H$_{11}$⟨O⟩⟨O⟩CN	1.01	6270
C$_5$H$_{11}$/C$_7$H$_{15}$⟨O⟩CH=N⟨O⟩CN	3.60	5360
C$_5$H$_{11}$/C$_7$H$_{15}$⟨O⟩COO⟨O⟩CN	4.10	6480

carbons) have very low viscosities (see TABLE 9). The substitution of the cyano group in the equatorial position produces high negative $\Delta \tilde{\varepsilon}$ with a reduced viscosity in comparison with 2,3-dicyanophenyl ethers of alkylcyclohexanecarboxylic ethers. The data presented in TABLE 10 illustrate the influence of other cyclic moieties. Alkyl substitution in the axial position of a cyclobutane fragment reduces the viscosity in comparison with substitution in the equatorial position. Introduction of the spyroheptane moiety instead of the cyclobutane increases the viscosity significantly. Substances with the cyclopentane moiety have a reduced viscosity and nematic-isotropic transition temperature in comparison with similar cyclohexane derivatives.

TABLE 9 Viscosities of cyclohexylcyclohexene and bicyclohexyl derivatives
R_1-Cy-X-R_2 (trans), R_3 (cis) at 20°C.

X	R_1	R_2	R_3	Viscosity, η/P ν/St$^+$ γ_1*/P	T_{NI}/K
⬡	C_3H_7	C_5H_{11}	-	[0.079]	300
⬡	C_3H_7	C_5H_{11}	-	[0.079]	295
Cy	C_3H_7	C_5H_{11}	-	[0.065]	370
Cy	C_3H_7	OCH_3	-	(0.07)$^+$	290
Cy	C_3H_7	CH_2OCH_3	-	[0.05]	324.2
Cy	C_5H_{11}	$COOCH_3$ (50%) $COOC_3H_7$ (50%)	-	[0.14]	351
Cy	C_5H_{11}	CN	-	[0.65]$^+$	358
Cy	C_nH_{2n+1}	C_5H_{11}	CN	0.70	338
Cy	C_5H_{11}	C_2H_4-CH=CH$_2$	CN	[0.50]$^+$ (22°C) [1.58]* (22°C) E = [6620 K]	[325.8]
CH$_3$-Cy-COO--O C$_5$H$_{11}$				[1.5]	(329.4)

TABLE 10 Viscosity at 20°C of cyclobutane, spyro[3.3]heptane and cyclopentane [39] derivatives; C_nH_{2n+1}-X -(COO)$_m$-Ph-(COO)$_p$-Ph-CN.

n	X	Alkyl position	m	p	η/P ν/St$^+$	T_{NI}/K
3	◇	cis	1	0	[1.33]	336.2
3	◇	trans	1	0	[0.74]	414.7
3	◇◇		1	0	[1.70]	428.0
5	⬠		0	0	[0.62]$^+$	334
5	⬡		0	0	[0.62]$^+$	495
3	⬠		0	1	[0.80]$^+$	363
3	⬡		0	1	[1.00]$^+$	500

In TABLE 11 the viscosities of complex esters with direct and inverse succession of atoms in the ester link are compared. The compounds with the C_nH_{2n+1}-Ph-O- fragment have lower viscosities in comparison with those containing the C_nH_{2n+1}-Cy-O- moiety. For alkoxy compounds with ester linkages in the core the viscosity decreases when a benzene ring is positioned between two oxygen atoms. The 4-methylphenyl ester of 4'-propylbicyclohexanecarboxylic acid has the smallest value of η and a high nematic-isotropic transition temperature amongst the tricyclic weakly-polar compounds. Amongst the polar compounds the 1-(4'-butylcyclohexyl)-2-(4''-(3''',4''',5'''-trifluorophenyl)phenyl)ethane has η = [16.1 mPa s], $\Delta\tilde{\varepsilon}$ = [11.8] and T_{NI} of 51.8°C [40].

TABLE 11 Viscosity of some complex esters R_1-A-(COO)$_n$-B-(COO)$_m$-D-R_2 at 20°C.

R_1	A	n	B	m	D	R_2	η/P	T_{NI}/K
C_5H_{11}	Cy	1	-	0		F	[0.194]	(258)
C_5H_{11}	Cy	1	-	0		F	[0.17]	(282)
F		1	-	0	Cy	C_5H_{11}	[0.70]	(267.2)
F		1	-	0	Cy	C_5H_{11}	[0.62]	(289.9)
C_5H_{11}	Cy	1	Ph	0	Cy	C_3H_7	(0.40)	466
C_3H_7	Cy	0	Ph	1	Cy	C_3H_7	[1.30]	433
C_3H_7	Cy	0	Cy	1	Ph	CH_3	[0.16]	477
C_3H_7	Cy	0	Cy	1	Cy	CH_3	[0.19]	406

Fluorine derivatives are now the main components of materials for active matrix liquid crystal displays. The influence of position of the substituted fluorine atoms is displayed in TABLE 12 [41]. The main trend is the increase in γ_1 with that of the polarisability and the dimension of the core. Minimal viscosity values are obtained for 3,4-difluorophenyl- and 3-fluor-4-perfluoromethoxyphenyl derivatives. The second substitution in the ortho-position reduces the viscosity of nematics with a terminal fluorine atom and increases its value for compounds with the -OCF$_3$ group.

TABLE 12 Physical properties of tricyclic substances with fluorine atoms at 20°C [41].

C_3H_7-Cy-A-B-R; Ph(F) = ; Ph(F,F) =

A	B	R	[γ_1]/mPa s	T_{NI}/°C	[$\Delta\tilde{n}$]	U_{th}/V
Cy	Ph(F)	F	138	121	0.079	1.00
Ph	Ph(F)	F	138	81	0.142	0.95
Ph(F)	Ph(F)	F	165	50	0.139	0.88
Ph(F,F)	Ph(F)	F	220	48	0.135	0.83
Ph(F)	Ph(F,F)	F	153	50	0.135	0.81
Ph(F)	Ph(F,F)	OCHF$_2$	225		0.147	0.89
Ph(F)	Ph(F,F)	OCF$_3$	198		0.123	0.85
Ph(F)	Ph(F,F)	OCHFCF$_3$	232		0.138	0.82
Ph(F)	Ph(F,F)	OC$_2$F$_5$	290		0.135	0.88
Ph(F)	Ph(F,F)	OCH$_2$C$_2$F$_5$	220		0.138	0.85
Ph(F)	Ph(F,F)	OCH$_2$CF$_2$CHFCF$_3$	252		0.121	0.80
Ph(F,F)	Ph(F,F)	F	143			0.93

E   CONCLUSION

Numerous data on the different nematic viscosities measured in experiments with flow and director reorientation are considered. The temperature dependence of the viscosity is determined by the nature of the intermolecular interactions. The viscosity is dependent mainly on molecular packing and molecular friction. The effect of polarity, dimensions and shape of molecules and their moieties has been discussed. As a rule viscosities and their activation energy increase with the polarisability anisotropy or the electric dipole moment of the molecule or a constituent group. A non-compensated steric dipole moment can reduce these parameters. The main geometrical factor which determines the molecular packing and reorientation is the occupied or excluded volume. The influence of the position, dimension and shape of various atoms and groups has been presented in detail in tables and figures.

REFERENCES

[1]   B.C. Benicewicz, J.F. Johnson, M.T. Shaw [*Mol. Cryst. Liq. Cryst. (UK)* vol.65 (1981) p.111]

[2]   W.H. de Jeu [*Physical Properties of Liquid Crystalline Materials* (Gordon and Breach Science Publishers, New York, 1980)]

[3]   F.-J. Bock, H. Kneppe, F. Schneider [*Liq. Cryst. (UK)* vol.1 (1986) p.239]

[4]   M.F. Grebyonkin, A.V. Ivashchenko [*Liquid Crystal Materials* (Khimiya Publishers, Moscow, 1983) in Russian]

[5]   V.V. Belyaev, S.A. Ivanov, M.F. Grebenkin [*Sov. Phys.-Crystallogr. (USA)* vol.30 (1985) p.674]

[6]   V.V. Belyaev [*Russian Chem. Rev. (UK)* vol.58 (1989) p.917]

[7]   H. Kneppe, F. Schneider [*Mol. Cryst. Liq. Cryst. (UK)* vol.65 (1981) p.23]

[8]   H. Kneppe, F. Schneider [*J. Phys. E (UK)* vol.16 (1983) p.512]

[9]   H. Kneppe, F. Schneider, N.K. Sharma [*J. Chem. Phys. (USA)* vol.77 (1982) p.3203]

[10]   O.Y. Shmelev, S.V. Pasechnik, V.A. Balandin, V.A. Tsvetkov [*Zh. Fiz. Khim. (Russia) (Russ. J. Phys. Chem. (UK))* vol.59 (1985) p.2036]

[11]   H. Kneppe, F. Schneider, N.K. Sharma [*Berichte Bunsenges. Phys. Chem. (Germany)* vol.85 (1981) p.784]

[12]   V.V. Belyaev [DSc Thesis, Russia, 1996, in Russian]

[13]   V.V. Belyaev, S.A. Ivanov [*Sov. Phys.-Crystallogr. (USA)* vol.37 (1992) p.123]; V.V. Belyaev, A.B. Kuznetsov, S.A. Ivanov [in *Proc. 15th Int. Display Research Conference* Hamamatsu, Japan, 1995, p.935]

[14]   Hp. Schad, H.R. Zeller [*Phys. Rev. A (USA)* vol.26 (1982) p.2940]

[15]   V.V. Belyaev [*Bull. Russ. Acad. Sci. (USA)* vol.60 (1996) p.512]

[16]   V.V. Belyaev, M.F. Grebenkin, V.F. Petrov [*Russ. J. Phys. Chem. (UK)* vol.64 (1990) p.512]; V.V. Belyaev, T.P. Antonyan, L.N. Lisetski, M.F. Grebenkin, G.G. Slashcheva, V.F. Petrov [*Mol. Cryst. Liq. Cryst. (UK)* vol.129 (1985) p.221]

[17]   A.C. Diogo, A.F. Martins [*Mol. Cryst. Liq. Cryst. (UK)* vol.66 (1981) p.133]

[18]   M.A. Osipov, E.M. Terentjev [*Z. Nat.forsch. A (Germany)* vol.30 (1989) p.785]

[19]   V.B. Nemtsov [*Mol. Cryst. Liq. Cryst. (UK)* vol.192 (1990) p.137]

[20]   V.V. Belyaev, V.B. Nemtsov [*Russ. J. Phys. Chem. (UK)* vol.66 (1992) p.1471]

[21]   J.P. van der Meulen, R.J.J. Zijlstra [*Physica B (Netherlands)* vol.132 (1985) p.153]

[22]   A.A. Tskhai [in *Up-to-Date Problems of Thermal Physics* (Nauka, Novosibirsk, Russia, 1987) p.258]

[23]   E. Szwajczak, A. Szymanski [*Mol. Cryst. Liq. Cryst. (UK)* vol.139 (1986) p.253]

[24]   E. Kuss [*Mol. Cryst. Liq. Cryst. (UK)* vol.47 (1978) p.71]

[25]   H. Doerrer, H. Kneppe, E. Kuss, F. Schneider [*Liq. Cryst. (UK)* vol.1 (1986) p.573]

[26]   D.A. Dunmur, W.H. Miller [*J. Phys. Colloq. (France)* vol.40 (1979) p.C3]

[27]   W.H. de Jeu, W.A.P. Claassen [*J. Chem. Phys. (USA)* vol.68 (1978) p.102]; F. Leenhouts, W.H. de Jeu, A.J. Dekker [*J. Phys. (France)* vol.40 (1979) p.989]

[28]   G.R. Luckhurst, G.W. Gray (Eds.) [*The Molecular Physics of Liquid Crystals* (Academic Press, 1979)]

[29]   H. Takatsu, K. Takeuchi, H. Sato [*Mol. Cryst. Liq. Cryst. (UK)* vol.108 (1984) p.157]

[30]   M.A. Osipov [*Bull. Russ. Acad. Sci. (USA)* vol.53 (1989) p.1915 in the Russian edition]

[31]   E.T. Samulski, H. Toriumi [*J. Chem. Phys. (USA)* vol.79 (1983) p.5194]

[32]   D.J. Photinos, E.T. Samulski, H. Toriumi [*Mol. Cryst. Liq. Cryst. (UK)* vol.204 (1991) p.161]

[33]   A. Hauser, Ch. Selbmann, R. Rettig, D. Demus [*Cryst. Res. Technol. (Germany)* vol.21 (1986) p.685]

[34]   G.R. Luckhurst [in *Recent Adv. Liq. Cryst. Polym., Proc. Eur. Sci. Found. Polym. Workshop Liq. Cryst. Polym. Syst. 1983* Eds. Chapoy, L. Lawrence (Elsevier Appl. Sci., London, UK, 1983) p.105]

[35]   V.V. Belyaev [*Mol. Cryst. Liq. Cryst. (UK)* vol.265 (1995) p.675]

[36]   V.V. Belyaev, A.B. Kuznetsov [*Sov. J. Opt. Technol. (USA)* vol.60 (1993) p.456]

[37]   T. Inukai, K. Furukawa, H. Inoue, K. Terashima [*Mol. Cryst. Liq. Cryst. (UK)* vol.94 (1983) p.109]

[38]   V.F. Petrov, M.F. Grebyonkin [*Zh. Fiz. Khim. (Russia) (Russ. J. Phys. Chem. (UK))* vol.65 (1991) p.1356]

[39]   V.F. Petrov, M.F. Grebyonkin, L.A. Karamysheva, R.C. Geivandov [*Zh. Fiz. Khim. (Russia) (Russ. J. Phys. Chem. (UK))* vol.65 (1991) p.1359]

[40]   D. Demus, Y. Goto, S. Sawada, E. Nakagawa, H. Saito, R. Tarao [*Mol. Cryst. Liq. Cryst. (UK)* vol.260 (1995) p.1]

[41]   K. Tarumi, E. Bartmann, T. Geelhaar, B. Schuler, H. Ichinose, H. Numata [in *Proc. 15th Int. Display Research Conference* Hamamatsu, Japan, 16-18 Oct. 1995 (Inst. Telev. Eng. Japan & SID, Tokyo, Japan & Santa Ana, CA, USA, 1995) p.559]

8.5 Flow and backflow in nematics

R.J. Atkin and F.M. Leslie

August 1999

A INTRODUCTION

Our understanding of flow phenomena has improved considerably over recent years through the stimulus of technological application coupled with the emergence of the viable continuum theory described in Datareview 8.1 and summarised in Section B. This theory supports the view, first suggested by Porter, Barrall and Johnson [1], that the non-Newtonian behaviour of nematic liquid crystals stems from the competition between flow and solid boundaries to dictate alignment. The theory also predicts an unusual relationship for the apparent viscosity in that it is not simply a function of the relative shear rate (see Section C). Fisher and Fredrickson's [2] experimental verification of the corresponding prediction for capillary flow provided important confirmation of several assumptions behind the continuum theory and established its relevance.

To illustrate the influence of flow upon alignment it is useful initially to neglect boundary and other effects. With these simplifications, solutions discussed in Section D demonstrate the crucial role of the signs of the viscosities α_2 and α_3. The solution of the continuum equations outlined in Section E demonstrates the non-Newtonian behaviour that arises from the competition between flow and the viscometer surfaces. Throughout this Datareview thermal effects are neglected and incompressibility is assumed. When discussing flow phenomena attention is restricted to the simplest geometry of two parallel flat plates. On the boundaries strong anchoring is adopted for the director and the no-slip condition imposed on the velocity. The flow may arise from the motion of the plates relative to each other with a uniform speed along a straight line in their plane (simple shear flow). Alternatively it may occur with both plates stationary when a pressure gradient applied in a given direction to the plates produces the motion (plane Poiseuille flow). Even in the simplest cases a variety of solutions is possible. A wider range of flow instabilities occur than in isotropic liquids, and some of these are discussed in Section F. As well as flow influencing the director, changes in alignment of the anisotropic axis can also induce flow, which in turn can influence the transient behaviour of the director. This induced flow, commonly referred to as backflow, can lead to some unexpected behaviour which is particularly relevant to electrooptic devices. The theoretical description of this phenomenon is discussed in Section G.

B BASIC EQUATIONS

In the absence of body forces and external fields the balance laws of linear and angular momentum reduce to (see Datareview 8.1 EQNS (41) and (38)):

$$\rho \frac{d\boldsymbol{\nu}}{dt} = -\operatorname{grad}\widetilde{p} + (\nabla\mathbf{n})^{\mathrm{T}}\mathbf{g} + \operatorname{div}\mathbf{T}^{\mathrm{d}}, \quad \widetilde{p} = p + W \tag{1}$$

$$\mathrm{div}\left(\frac{\partial W}{\partial \nabla \mathbf{n}}\right) - \frac{\partial W}{\partial \mathbf{n}} + \mathbf{g} + \gamma \mathbf{n} = 0 \tag{2}$$

Here ρ denotes the density, the velocity, such that div $\mathbf{v} = 0$, p is the pressure, $\mathbf{n}$ is the director, W is the Frank-Oseen free energy and γ is an arbitrary scalar arising from the constraint $\mathbf{n}.\mathbf{n} = 1$. The dynamic part of the stress tensor, $\mathbf{T}^d$, and the vector $\mathbf{g}$ are given by (Datareview 8.1 EQNS (23), (25) and (28)):

$$\mathbf{T}^d = \alpha_1 \, \mathbf{n}.\mathbf{A}\mathbf{n} \otimes \mathbf{n} + \alpha_2 \mathbf{N} \otimes \mathbf{n} + \alpha_3 \mathbf{n} \otimes \mathbf{N} + \alpha_4 \mathbf{A} + \alpha_5 \mathbf{A}\mathbf{n} \otimes \mathbf{n} + \alpha_6 \mathbf{n} \otimes \mathbf{A}\mathbf{n} \tag{3}$$

$$\mathbf{g} = -\gamma_1 \mathbf{N} - \gamma_2 \mathbf{A}\mathbf{n}, \quad \gamma_1 = \alpha_3 - \alpha_2, \quad \gamma_2 = \alpha_6 - \alpha_5 = \alpha_3 + \alpha_2 \tag{4}$$

where

$$2\mathbf{A} = \nabla \mathbf{v} + (\nabla \mathbf{v})^T, \quad \mathbf{N} = \frac{d\mathbf{n}}{dt} - \frac{1}{2}\left\{ \nabla \mathbf{v} - (\nabla \mathbf{v})^T \right\}\mathbf{n} \tag{5}$$

$\mathbf{a} \otimes \mathbf{b}$ denotes the 3×3 matrix with the (ij)th element $a_i b_j$, and the ith components of div $\mathbf{T}^d$ and $\mathrm{div}(\partial W/\partial \nabla \mathbf{n})$ are $T^d_{ij,j}$ and $(\partial W/\partial n_{i,j})_j$ respectively, summation being over the repeated suffixes. The range of values of the five viscosities α_i is restricted, in particular $\alpha_4 > 0$ and $\gamma_1 > 0$.

C   SCALING ANALYSIS

Ericksen [3] noticed that EQNS (1) and (2) may be scaled in a way that preserves the form of the equations. If we scale space and time by the transformation

$$\mathbf{x} = d\mathbf{x}^*, \quad t = d^2 t^* \tag{6}$$

where d is some constant, the new equations are identical to the original provided that

$$p = d^{-2} p^*, \quad \gamma = d^{-2} \gamma^* \tag{7}$$

are chosen.

In this scaling the velocity and stress tensor $\mathbf{T}$ (see Datareview 8.1 EQN (17)) transform according to

$$\mathbf{v}^* = d\mathbf{v}, \quad \mathbf{T}^* = d^2 \mathbf{T} \tag{8}$$

but the director being a unit vector is unaltered. In the context of simple shear flow it is natural to choose the constant d as the gap width. As a result, if V denotes the relative shearing speed, the original problem transforms into one with unit gap width and relative speed V* where

$$V^* = Vd \tag{9}$$

with the strong anchoring boundary condition remaining unchanged.

Given a unique solution to this problem, it follows therefore that

$$\mathbf{n} = \mathbf{n}^* = \mathbf{n}^*(\mathbf{x}^*, V^*) = \mathbf{n}^*(d^{-1}\mathbf{x}, Vd) \tag{10}$$

and so in simple shear flow an optical property must depend upon the speed V through the product Vd. Wahl and Fischer [4] find such scaling in optical measurements with a flow of 4-methoxybenzylidine-4'-butylaniline (MBBA) between parallel discs under conditions closely approximating simple shear. Defining an apparent viscosity, η, by

$$\eta = \sigma d/V \tag{11}$$

where σ is the shear stress applied per unit area of the plate in the direction of shear, using EQNS (8), (9) and (11) it follows that

$$\eta = \frac{\sigma d^2}{Vd} = \sigma^*/V^* = \eta^* = F(V^*) = F(Vd) \tag{12}$$

the function F being unknown. The same simple scaling must occur in viscosity measurements if the theory applies. This result contrasts with that for an isotropic viscoelastic fluid for which η is a function of the shear rate V/d (see Rivlin [5], Coleman and Noll [6]). Unfortunately no experimental evidence appears to exist with which to check this prediction.

Adapting this analysis to Poiseuille flow through a circular capillary of radius R under a constant pressure gradient P, Atkin [7] predicts that the flux per unit time Q and the apparent viscosity $\eta (= \pi PR^4/8Q)$ satisfy the functional relationships

$$Q = RG(PR^3), \quad \eta = H(Q/R) \tag{13}$$

where the functions G and H are unknown. These results are also in contrast with the corresponding results for isotropic viscoelastic fluids where the dependence upon radius differs. The prediction EQN (13) led to the first experimental confirmation of the continuum theory. Learning of this result, Fisher and Fredrickson [2] plotted η against Q/R using their data from capillary flow experiments with the nematic 4,4'-dimethoxyazoxybenzene (PAA). Their results for perpendicular (homeotropic) surface alignment fit a single curve in good agreement with the prediction. For parallel surface orientation the results are not in such good agreement with this type of scaling, but as the authors acknowledge the rubbing technique used may not have been satisfactory. This confirmation of the novel scaling is important since it vindicates a number of special assumptions made in formulating the constitutive relations in the continuum theory. These are discussed in detail by Leslie [8]. It also justifies the strong anchoring assumption on the surface. In view of the relative importance of this topic, it is surprising that essentially only two experimental studies have investigated it.

D FLOW ALIGNMENT

Ignoring the effect of surfaces and external fields the influence of flow upon the alignment in simple shear can be seen by considering velocity and director fields with Cartesian components

$$\mathbf{v} = (\kappa z, 0, 0), \quad \mathbf{n} = (\cos\theta\cos\phi,\ \cos\theta\sin\phi,\ \sin\theta) \tag{14}$$

where κ is a positive constant and θ and ϕ are functions solely of time. For this spatially uniform director field EQN (2) reduces to

$$\gamma_1\dot{\theta} + \kappa m(\theta)\cos\phi = 0, \quad \gamma_1\cos\theta\dot{\phi} - \kappa\alpha_2\sin\theta\sin\phi = 0 \tag{15}$$

where the dots indicate differentiation with respect to time and

$$m(\theta) = \alpha_3\cos^2\theta - \alpha_2\sin^2\theta \tag{16}$$

Whatever the values of the viscosities, EQNS (15) always have a steady solution of the form $\theta = 0$, $\phi = \pi/2$ corresponding to uniform alignment normal to the plane of shear. Such an orientation pattern has been demonstrated experimentally by Pieranski and Guyon [9]. If

$$\alpha_2\alpha_3 > 0 \tag{17}$$

there are solutions with the director in the plane of shear given by

$$\theta = \pm\theta_0, \quad \phi = 0 \tag{18}$$

where the acute angle θ_0 is defined by

$$\tan^2\theta_0 = \alpha_3/\alpha_2 \tag{19}$$

Since $\gamma_1 > 0$ there are two possibilities, either

$$\alpha_2 < \alpha_3 < 0, \quad 0 < \theta_0 < \pi/4 \tag{20}$$

or

$$\alpha_3 > \alpha_2 > 0, \quad \pi/4 < \theta_0 < \pi/2 \tag{21}$$

For consistency with the observations of Gähwiller [10] for MBBA and Wahl and Fischer [4], which indicate flow alignment at relatively small angles to the streamlines, the former option must be selected.

When restriction (17) holds, a detailed examination of EQNS (15) reveals that all time dependent solutions must tend to one of the in-plane solutions, the other steady solution being unstable. When α_2 and α_3 satisfy (20), $\theta = \theta_0$ is the stable solution, but if they satisfy (21) the stable solution is

$\theta = -\theta_0$. However, when restriction (17) does not hold, time dependent solutions are periodic (see Ericksen [11]).

Whilst solution (14) with θ and ϕ functions only of time is perhaps a little too simplistic it does indicate the behaviour away from the boundaries for high shear rates. It also indicates the importance of the viscosities α_2 and α_3 and the restrictions necessary to ensure agreement with observed flow alignment. Materials whose viscosity coefficients α_2 and α_3 satisfy (17) (and (20)) are known as flow-aligning materials whilst those for which (17) does not hold, particularly those for which $\alpha_3 > 0$, $\alpha_2 < 0$, are non-flow-aligning materials. This latter class of materials is considered in Section F.

E   SIMPLE SHEAR FLOW

The simplest solution of the continuum equations for shear flow between parallel flat plates is that with the director uniformly aligned normal to the plane of shear. Taking solutions in which the velocity and the director have Cartesian components

$$\mathbf{v} = (u(z),0,0), \quad \mathbf{n} = (0,1,0) \tag{22}$$

EQN (2) reduces to an identity and EQNS (1) give

$$\frac{\partial p}{\partial x} = \eta_a \frac{d^2 u}{dz^2}, \quad \frac{\partial p}{\partial y} = \frac{\partial p}{\partial z} = 0 \tag{23}$$

where

$$2\eta_a = \alpha_4 \tag{24}$$

For parallel boundary conditions for the director, these equations predict Newtonian behaviour for both simple shear and plane Poiseuille flow in which the director remains constant and unaffected by the flow.

The non-Newtonian behaviour that arises from competition between flow and the boundaries to dictate the director orientation is illustrated by taking the velocity and the director with Cartesian components

$$\mathbf{v} = (u(z),0,0), \quad \mathbf{n} = (\cos\theta(z), 0, \sin\theta(z)) \tag{25}$$

For simple shear EQNS (1) and (2) yield

$$T_{xz} = g(\theta)u' = c \tag{26}$$

where a prime denotes differentiation with respect to z,

$$g(\theta) = \eta_b \cos^2 \theta + \eta_c \sin^2 \theta + \alpha_1 \sin^2 \theta \cos^2 \theta$$

(27)

$$2\eta_b = \alpha_4 + \alpha_3 + \alpha_6 = 2g(0), \quad 2\eta_c = \alpha_4 + \alpha_5 - \alpha_2 = 2g(\pi/2)$$

and

$$2f(\theta)\theta'' + \frac{df(\theta)}{d\theta}(\theta')^2 = 2m(\theta)u'$$

(28)

with

$$f(\theta) = K_1 \cos^2 \theta + K_3 \sin^2 \theta$$

(29)

In EQN (26) T_{xz} denotes the component of the stress tensor, the static contribution being zero, and c is a positive constant equal to the total shear force per unit area applied to the plates. Comparing this result with the corresponding equation for a Newtonian fluid shows that $g(\theta)$ can be interpreted as an orientation-dependent viscosity which can be shown to be positive. EQN (28) illustrates the competition between flow and the boundaries. For flow-aligning materials the viscous torques are trying to orient the material at the angle $\pm\theta_0$, whereas the elastic torques are resisting this tendency and are accommodating the boundary orientation. Eliminating u' EQN (28) may be rewritten and integrated to yield

$$f(\theta)(\theta')^2 = 2c \int_{\theta_m}^{\theta} m(\Psi)/g(\Psi)d\Psi$$

(30)

where θ_m is an arbitrary constant. On account of the inequalities $K_1 > 0$, $K_3 > 0$ proposed by Ericksen [12], the left hand side of this equation is always positive and this places restrictions on possible solutions.

When the alignment at the boundaries is prescribed by suitable prior treatment of the boundary surfaces the boundary conditions are of the form

$$u(h) = V, \quad u(-h) = 0, \quad \theta(h) = \theta_1, \quad \theta(-h) = \theta_2 + n\pi$$

(31)

where θ_1 and θ_2 are given constant values, n is an integer and h denotes half of the gap-width. The two most frequent cases are planar alignment in which θ_1 and θ_2 are both zero, and homeotropic alignment in which both are $\pi/2$.

When (17) applies it can be shown from the integral in EQN (30) (see Leslie [13]) that there are simple symmetric solutions in which for planar alignment $0 \leq \theta \leq \theta_m < \theta_0$, and for homeotropic alignment $\pi/2 \geq \theta \geq \theta_m > \theta_0$, where θ_m now denotes the value of θ at the turning point in the centre of the gap. For these solutions the value θ_m tends to θ_0 as the product ch^2 becomes large. On the other hand, if there is no flow alignment, the symmetric solutions are such that $\theta_m \leq \theta \leq \theta_w$, where θ_w denotes the value at the boundary. In this case θ_m increases without bound as ch^2 becomes large.

Further it can be shown that $\theta_m = \Theta(Vh)$ and the apparent viscosity $\eta = F(Vh)$ agreeing with the earlier scaling.

As Currie [14] points out, an examination of the phase plane for EQN (30) shows that the symmetric solution is not unique. Restricting their attention to flow-aligning nematics in the special case when $K_1 = K_3$ and $\alpha_1 = 0$, Currie and MacSithigh [15] show that it is possible to distinguish between the various solutions of EQN (30). However, even with these simplifications there are still seven stable solutions. A solution with the least viscous dissipation can be selected as the most likely to occur in practice but this depends upon the prescribed surface alignment. McIntosh et al [16] also discuss this topic employing numerical methods, and considering symmetric solutions claim that one of the above is in fact unstable. They also find that results can be sensitive to the particular surface alignment chosen.

For plane Poiseuille flow EQNS (1) give

$$T_{xz} = g(\theta)u' = -Pz, \quad \widetilde{p} = \widetilde{p}(z) - Px \qquad (32)$$

where P is a constant. EQN (28) still applies. The explicit appearance of z in T_{xz} and in the subsequent equation for θ prevents progress analytically. Zúñiga and Leslie [17] discuss this problem numerically for MBBA taking boundary conditions for θ of the form of EQN (31) with $\theta_1 = \theta_2$ and zero velocity on the plates. Since there is no unique solution three cases are considered, one for planar boundary alignment and two for homeotropic, with conditions imposed on the form of $\theta(z)$. Two values of the pressure gradient are taken. Assuming that $\theta(z)$ is odd about $z = 0$, for the higher pressure gradient the solution for planar alignment exhibits flow alignment with either $\theta = \theta_0$ or $\theta = -\theta_0$ across most of the gap except near the boundaries and in a thin central layer. In the solution for $\theta_1 = -\pi/2$, n = 1 and $\theta(0) = 0$ again with $\theta(-z) = -\theta(z)$ the director rotates monotonically from the lower plate to the upper plate with θ changing from $\pi/2$ to $-\pi/2$. However, if $\theta_1 = \pi/2$, n = 0 and $\theta(0) = \pi/2$ with $\theta - \pi/2$ odd in z, for $0 < z < h$, $\theta - \pi/2$ increases to a maximum before returning to zero on the central plane. In the lower half $\theta - \pi/2$ decreases to a minimum finally increasing to satisfy the boundary condition on the lower plate. For homeotropic alignment the flow profiles are very nearly the same as the corresponding profile for a Newtonian fluid.

Exact solutions for other viscometric flows of flow-aligning materials have been considered. For Couette flow between concentric circular cylinders Atkin and Leslie [18] give a symmetric solution, non-uniqueness being discussed by Currie [14]. Poiseuille flow through circular cylinders presents difficulties. Atkin [7] makes some progress particularly for flow through concentric circular cylinders but less so for a capillary. In the absence of an exact solution for capillary flow Tseng and co-workers [19], Finlayson [20] and Kini and Ranganath [21] compute numerical solutions for particular compounds. Currie [22] discusses approximate solutions. Details of these developments are given in the review articles by Jenkins [23] and Leslie [8]. More recently Zúñiga [24] gives a numerical solution for Couette flow of non-flow-aligning materials. For 4-hexyloxybenzylidene-4'-aminobenzonitrile (HBAB) at 85°C he finds that the velocity profile is very nearly Newtonian in qualitative agreement with the observations of Cladis and Torza [25]. Flow of PAA and MBBA in converging and diverging channels is discussed by Rey and Denn [26]. Neither torsional flow nor flow in the cone and plate Weissenberg rheometer appear to have been considered in any great detail using the current theory, although some aspects of torsional flow and radial Poiseuille flow are discussed by Lam and Shu [27]. Hiltrop and Fischer [28] have investigated the radial Poiseuille flow of MBBA experimentally.

F   FLOW INSTABILITIES

As seen in Section E in one steady flow pattern in shear the director is perpendicular to the plane of shear. This configuration is possible for all values of the viscosities, but is it stable for all shear rates? In experiments with MBBA Pieranski and Guyon [9] find that this orientation is stable for low shear rates, but as the speed of the moving plate exceeds a critical value the director in the interior of the gap begins to turn into the plane of shear, and tilts relative to the plane of the plates, as might be anticipated for such a flow-aligning material. The elastic couples successfully maintain the uniform director orientation consistent with the boundary conditions until the viscous couples are strong enough to force the director in the interior of the gap into the plane of shear. The distortion is homogeneous being uniform in the plane of the plates, and no domain or roll structures occur. The critical speed varies inversely with the gap width giving further experimental verification of the scaling analysis given in Section C. A review of related experimental and theoretical work on similar flows with relevant references up to the mid 1970s is given by Dubois-Violette et al [29,30], the former article concentrating on effects observed in MBBA. Pieranski and Guyon's simple analysis gives values of the critical speed that agree qualitatively with the experimental results. However, Manneville and Dubois Violette [31] and Leslie [32] realised that associated with the deflection of the director axis the continuum theory requires that a transverse secondary flow, later observed experimentally, must accompany such an instability. They gave a more detailed analysis, the predictions of which again agree with experiment. In particular Leslie predicts a critical speed, V_c, given by

$$V_c d = 4(K_1 K_2 \eta_b)^{1/2} \xi^2 / (\alpha_2 \alpha_3 \eta_a)^{1/2} \tag{33}$$

where d is the gap width and ξ is the smallest non-trivial root of the equation

$$2\eta_b \xi + (\eta_a - \eta_b)(\tan \xi + \tanh \xi) = 0 \tag{34}$$

More recently the behaviour in shear of non-flow-aligning compounds for which $\alpha_3 > 0$, $\alpha_2 < 0$ has received attention. With a uniform director field parallel to the velocity of the plate, using HBAB, Pieranski and Guyon [33] observe that as the speed of the plate increases the parallel orientation gradually distorts, and at a critical value of shear rate turns out of the plane eventually becoming perpendicular to the plane of shear. If the shear rate is increased further at a high critical shear rate convective rolls develop. This is in contrast to observations on HBAB and 4-cyanobenzylidene-4'-octyloxyaniline (CBOOA) by Cladis and Torza [25] for Couette flow between concentric circular cylinders. With the director initially orientated radially, they observe an instability called tumbling when the alignment changes abruptly to a new configuration, in which the director remains in the plane of shear except in boundary layers near each cylinder where the axial component is non-zero. At some temperatures at a second critical value convective rolls develop. These conflicting observations have prompted a number of analyses for simple shear between parallel plates.

Taking the earlier symmetric solution with $\alpha_1 = 0$, Carlsson [34] integrates the equations numerically using values of K_1, K_3 and α_i for 4-octyl-4'-cyanobiphenyl (8CB) at different temperatures. He finds that the angle θ_m is not a single-valued function of V for certain ranges of the parameter $\varepsilon = \alpha_3/|\alpha_2|$. Earlier Manneville [35] reached a similar conclusion by a different approach. As both authors discussed, this many-valued property can model tumbling of the alignment at certain thresholds. Both these studies appear to support the observation of tumbling rather than the behaviour noted by Pieranski and Guyon. However, they assume that the director remains in the plane of shear and tend not to pursue the possibility of an instability involving the director turning out of the plane. Pieranski and co-workers [36] did discuss this latter possibility but their approximate analysis is limited and does not resolve the matter.

More recently Zúñiga and Leslie [37,38], Luskin and Pan [39] and Han and Rey [40] have discussed this problem and, using data for 8CB at different temperatures, they confirm the importance of the parameter ε. For an initial planar alignment Zúñiga and Leslie show that the director comes out of the plane following the instability for all their data. However, with an initial homeotropic orientation, for two cases a tumbling instability occurs producing a tilt configuration that is immediately unstable to perturbations out of the plane. The critical speed at which the instabilities occur also depends upon ε. Using the same six data sets Han and Rey find that the director comes out of the plane for both initial alignments. Extending the range of values of ε Luskin and Pan find that the first instability can be either in the plane of the shear (tumbling) or out of the plane depending upon the size of ε. For $0.5 \le \varepsilon < 0.75$ the non-planar instability occurs first, whilst for $0.75 < \varepsilon < 1.1$ the planar instability occurs at a lower speed. For $\varepsilon = 0.20$ with planar boundary orientation for speeds greater than 0.00025 cm s^{-1} the director is nearly normal to the plane of shear. All of these analyses support the observations of Pieranski and Guyon. The Cladis-Torza instabilities are possibly peculiar to the Couette geometry, although Zúñiga's results [24] suggest that the rotation and cylindrical geometry have only a small effect on the first threshold. In contrast to simple shear, he finds that beyond the first instability there is a stable solution with both the velocity and the director in the plane of shear. Assuming that the director comes back into the plane of shear this latter solution could be that observed in the experiments. However, his analysis is only valid for small perturbations of the basic solution in the plane of shear, and so before more definitive conclusions can be drawn consideration of the non-linear terms in the full set of equations is needed.

Plane Poiseuille flow of flow-aligning nematics has also been considered experimentally by Guyon and Pieranski [41] and Janossy and co-workers [42] when the director is parallel to the boundaries and perpendicular to the plane of shear. Their behaviour resembles that observed in simple shear, the critical speed now being replaced by a critical pressure gradient. Analysis of the linearised equations for homogeneous perturbations by Manneville and Dubois-Violette [43] and Manneville [44] give predictions in agreement with experiment. When the alignment lies in the plane of shear the corresponding equations, which are more complex, have been solved numerically by Zúñiga and Leslie [17]. For 8CB their results are similar to those for simple shear, the planar alignment becoming unstable above a critical value of the pressure gradient, when the director moves out of the plane of shear. On the other hand, for MBBA the stability is dependent upon the particular wall alignment. When the alignment is parallel to the velocity the behaviour is again similar to that for simple shear with an apparently stable alignment at a small angle to the streamlines. However, with an initial homeotropic alignment the corresponding solution becomes unstable and the director moves out of the plane of shear. At higher pressure gradients there is a second planar solution which is stable to both types of perturbation. Hiltrop and Fischer [28] observe similar behaviour in radial Poiseuille flow, but no experimental results have appeared for plane Poiseuille flow.

G   BACKFLOW

A problem which demonstrates the relevance of transient flow effects in display devices is that of optical bounce in a twisted nematic cell. Gerritsma et al [45] observe that upon the removal of a strong electric field applied across a twisted nematic cell the optical transmission does not decrease monotonically to zero as might be expected; instead, following an initial decrease, it increases to a value slightly less than the initial level before decaying to zero. Van Doorn [46] explains this phenomenon as a consequence of flow induced close to the cell walls, commonly called backflow, by the rapidly relaxing alignment following the removal of the field. With the field on, the director is approximately parallel to the applied field in most of the sample, the elastic torque, which is strongest near the cell walls, balancing the electric torque. When the field is switched off, the unbalanced elastic torque causes a rapidly changing director orientation near the cell walls which induces fluid motion. This creates a shear flow in the central part of the cell causing the director to tilt over to an angle greater than $\pi/2$ in the bulk of the sample before relaxing back to zero.

The simplest approach to this problem is to consider a layer of nematic confined between parallel plates with its field-free uniform parallel orientation distorted by an external field perpendicular to the plates. The transient behaviour on the removal of the field is illustrated by taking the velocity and director with Cartesian components given by EQN (25) but allowing u and θ to depend upon z and t. EQNS (1) and (2) reduce to

$$\rho \frac{\partial u}{\partial t} = \frac{\partial}{\partial z}\left[g(\theta)\frac{\partial u}{\partial z} + m(\theta)\frac{\partial \theta}{\partial t}\right] \tag{35}$$

$$2f(\theta)\frac{\partial^2 \theta}{dz^2} + \frac{df}{d\theta}\left(\frac{\partial \theta}{\partial z}\right)^2 - 2\gamma_1 \frac{\partial \theta}{\partial t} - 2m(\theta)\frac{\partial u}{\partial z} = 0 \tag{36}$$

where the functions $f(\theta)$, $g(\theta)$ and $m(\theta)$ are given by EQNS (16), (27) and (29). The need for a flow component when θ depends upon both z and t is now clear, since if u is zero the problem is overdetermined. Appropriate boundary conditions are

$$\theta(\pm h, t) = 0, \quad u(\pm h, t) = 0, \quad t > 0 \tag{37}$$

with initial conditions

$$\theta(z, 0) = \theta(z), \quad u(z, 0) = 0, \quad |z| < h \tag{38}$$

the function $\theta(z)$ being known from the corresponding static Freedericksz problem.

The main obstacle to solving EQNS (35) and (36) analytically is the fact that most of the coefficients depend upon θ, making the system non-linear. Van Doorn [46] has integrated the equations numerically for MBBA neglecting the inertia term $\rho \partial u / \partial t$. Clark and Leslie [47] also neglect inertia but make some analytical progress by making some approximations. It seems reasonable to assume that during the first stages of reorientation a good approximation to solutions is given by replacing the variable coefficients by their initial values. Since the strength of the field is well above threshold, and so the distorted alignment is approximately normal to the plates in the greater part of the sample, they put $\theta = \pi/2$ in the coefficients in EQNS (35) and (36) which then reduce to

$$K_3 \frac{\partial^2 \theta}{\partial z^2} - \gamma_1 \frac{\partial \theta}{\partial t} + \alpha_2 \frac{\partial u}{\partial z} = 0, \quad \eta_c \frac{\partial^2 u}{\partial z^2} - \alpha_2 \frac{\partial^2 \theta}{\partial z \partial t} = 0 \tag{39}$$

A solution of EQNS (39) in the form of an infinite series now follows and this reduces the computational effort required to obtain detailed predictions.

For MBBA, at a given z, θ initially increases with time and attains a maximum value in a very short time relative to the overall decay before decreasing monotonically to zero. This flow-induced kickback is most pronounced and occurs over the longest time at the centre of the cell, where, in the present approximation, the maximum kickback angle is about 25°. As we move towards the edge of the cell the maximum decreases and θ decays to zero at even shorter times. The graphs of θ at the centre of the cell are qualitatively similar to those obtained by Van Doorn with the maximum occurring after approximately the same time interval. Clark and Leslie appear to obtain significantly larger values for the maximum value of θ during kickback. The velocity is an odd function of z with the centre of the gap and the walls being planes of zero velocity. After an initial, very large backflow, the velocity quickly diminishes and changes sign, finally approaching zero rather slowly through small positive values. The velocity graphs given by Van Doorn have the form that Clark and Leslie associate with a later stage of the relaxation process. The cause of this discrepancy appears uncertain.

The treatment of the twisted cell is more complex. A three-dimensional director field is needed with θ and ϕ dependent upon z and t, as well as a transverse velocity component in the y direction. This problem has been treated numerically by Van Doorn [48] and Berreman [49] again neglecting inertia.

REFERENCES

[1]   R.S. Porter, E.M. Barrall, J.F. Johnson [*J. Chem. Phys. (USA)* vol.45 (1966) p.1452]

[2]   J. Fisher, A.G. Fredrickson [*Mol. Cryst. Liq. Cryst. (UK)* vol.8 (1969) p.267]

[3]   J.L. Ericksen [*Trans. Soc. Rheol. (USA)* vol.13 (1969) p.9]

[4]   J. Wahl, F. Fischer [*Mol. Cryst. Liq. Cryst. (UK)* vol.22 (1973) p.359]

[5]   R.S. Rivlin [*J. Ration. Mech. Anal. (Germany)* vol.5 (1956) p.179]

[6]   B.D. Coleman, W. Noll [*Arch. Ration. Mech. Anal. (Germany)* vol.3 (1959) p.289]

[7]   R.J. Atkin [*Arch. Ration. Mech. Anal. (Germany)* vol.38 (1970) p.224]

[8]   F.M. Leslie [*Adv. Liq. Cryst. (USA)* vol. 4 (1979) p.1]

[9]   P. Pieranski, E. Guyon [*Solid State Commun. (USA)* vol.13 (1973) p.435]

[10]   C. Gähwiller [*Phys. Rev. Lett. (USA)* vol.28 (1972) p.1554]

[11]   J.L. Ericksen [*Kolloidn. Zh. (USSR)* vol.173 (1960) p.117]

[12]   J.L. Ericksen [*Phys. Fluids (USA)* vol.9 (1966) p.1205]

[13] F.M. Leslie [*Mol. Cryst. Liq. Cryst. (UK)* vol.63 (1981) p.111]

[14] P.K. Currie [*Arch. Ration. Mech. Anal. (Germany)* vol.37 (1970) p.222]

[15] P.K. Currie, G.P. MacSithigh [*Q. J. Mech. Appl. Math. (UK)* vol.32 (1979) p.499]

[16] J.G. McIntosh, F.M. Leslie, D.M. Sloan [*Contin. Mech. Thermodyn. (Germany)* vol.9 (1997) p.293]

[17] I. Zúñiga, F.M. Leslie [*J. Non-Newton. Fluid Mech. (Netherlands)* vol.33 (1989) p.123]

[18] R.J. Atkin, F.M. Leslie [*Q. J. Mech. Appl. Math. (UK)* vol.23 (1970) p.S3]

[19] H.C. Tseng, D.L. Silver, B.A. Finlayson [*Phys. Fluids (USA)* vol.15 (1972) p.1213]

[20] B.A. Finlayson [*Liquid Crystals and Ordered Fluids* vol.2, Eds. J.F. Johnson, R.S. Porter (Plenum Publishing Corp., New York, 1974) p.211-23]

[21] U.D. Kini, G.S. Ranganath [*Pramãna (India)* vol.4 (1975) p.19]

[22] P.K. Currie [*Rheol. Acta (Germany)* vol.14 (1975) p.688]

[23] J.T. Jenkins [*Annu. Rev. Fluid Mech. (USA)* vol.10 (1978) p.197]

[24] I. Zúñiga [*Phys. Rev. A (USA)* vol.41 (1990) p.2050]

[25] P.E. Cladis, S. Torza [*Phys. Rev. Lett. (USA)* vol.35 (1975) p.1283]

[26] A.D. Rey, M.M. Denn [*J. Non-Newton. Fluid Mech. (Netherlands)* vol.27 (1988) p.375]

[27] L. Lam, C.Q. Shu [*Solitons in Liquid Crystals* Eds. L. Lam, J. Prost (Springer-Verlag, New York, 1992) p.51-109]

[28] H. Hiltrop, F. Fischer [*Z. Naturforsch. (Germany)* vol.31 (1976) p.800]

[29] E. Dubois-Violette, E. Guyon, I. Janossy, P. Pieranski, P. Manneville [*J. Méc. (France)* vol.16 (1977) p.733]

[30] E. Dubois-Violette, G. Durand, E. Guyon, P. Manneville, P. Pieranski [*Solid State Physics Supplement (Japan)* vol.14 (1978) p.147]

[31] P. Manneville, E. Dubois-Violette [*J. Phys. (France)* vol.37 (1976) p.285]

[32] F.M. Leslie [*J. Phys. D (UK)* vol.9 (1976) p.925]

[33] P. Pieranski, E. Guyon [*Phys. Rev. Lett. (USA)* vol.32 (1974) p.924]

[34] T. Carlsson [*Mol. Cryst. Liq. Cryst. (UK)* vol.104 (1984) p.307]

[35] P. Manneville [*Mol. Cryst. Liq. Cryst. (UK)* vol.70 (1981) p.223]

[36] P. Pieranski, E. Guyon, S.A. Pikin [*J. Phys. (France)* vol.37C1 (1976) p.3]

[37] I. Zúñiga, F.M. Leslie [*Europhys. Lett. (Switzerland)* vol.9 (1989) p.689]

[38] I. Zúñiga, F.M. Leslie [*Liq. Cryst. (UK)* vol.5 (1989) p.725]

[39] M. Luskin, T.W. Pan [*J. Non-Newton. Fluid Mech. (Netherlands)* vol.42 (1992) p.369]

[40] W.H. Han, A.D. Rey [*J. Non-Newton. Fluid Mech. (Netherlands)* vol.48 (1993) p.181]

[41] E. Guyon, P. Pieranski [*J. Phys. (France)* vol.36C1 (1975) p.203]

[42] I. Janossy, P. Pieranski, E. Guyon [*J. Phys. (France)* vol.37 (1976) p.1105]

[43] P. Manneville, E. Dubois-Violette [*J. Phys. (France)* vol.37 (1976) p.1115]

[44] P. Manneville [*J. Phys. (France)* vol.40 (1979) p.713]

[45] C.J. Gerritsma, C.Z. Van Doorn, P. Van Zanten [*Phys. Lett. A (Netherlands)* vol.48 (1974) p.263]

[46] C.Z. Van Doorn [*J. Phys. (France)* vol.36C1 (1975) p.261]

[47] M.G. Clark, F.M. Leslie [*Proc. R. Soc. Lond. A (UK)* vol.361 (1978) p.463]

[48] C.Z. Van Doorn [*J. Appl. Phys. (USA)* vol.46 (1975) p.3738]

[49] D.W. Berreman [*J. Appl. Phys. (USA)* vol.46 (1975) p.3746]

8.6 Electrohydrodynamic convection in nematics

L. Kramer and W. Pesch

July 2000

A   INTRODUCTION

Electrohydrodynamic convection (or, briefly, electroconvection (EC)) occurs when a voltage above a critical threshold strength is applied across a thin layer of a nematic liquid crystal with non-zero conductivity [1-7]. At onset typically periodic patterns of convection rolls are observed (see FIGURE 1). With increasing voltage, transitions take place either to rather complex spatio-temporal states, which are heavily influenced by defects or, under appropriate conditions, to more complicated (quasi-)stationary patterns, typically periodic in two directions. Eventually turbulent states, which are characterised by strong light scattering (dynamic scattering mode, DSM [8]) occur.

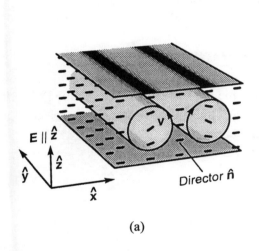
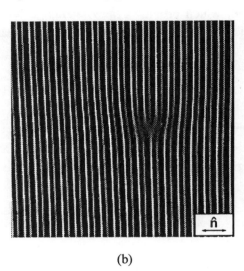

(a)   (b)

FIGURE 1 (a) Cell geometry with a section of a roll pattern for EC (planar configuration). **E** = electric field, **v** = velocity. (b) Normal roll pattern for EC with a dislocation (courtesy I. Rehberg).

After the first systematic experimental characterisation of EC in 1963 by Williams [9] as well as by Kapustin and Larinova [10] the phenomenon was intensively investigated in the early 1970s. In the last fifteen years EC in nematics has developed into an important model system for pattern formation in hydrodynamic instabilities [11]. The specific anisotropy of nematics brings out totally new phenomena compared to the canonical example of Rayleigh-Bénard convection in simple fluids [12].

From an experimental point of view EC is, up to a point, a convenient system: simple, with multiple controls (e.g. amplitude and frequency of voltage, and additional magnetic fields), convenient timescales, and easy visualisation of the gross features. Limitations result from limited accuracy, lack of reproducibility and optical resolution, as well as from the lack of three-dimensional visualisation of the director field inside the layer. From a theoretical point of view the picture is also two-fold. The full hydrodynamic description (see Section B) is mostly well-established and transparent, whereas extracting the consequences of its various competing mechanisms is highly demanding. A complete description of the onset behaviour is now feasible (Section C). On the non-linear level there is also considerable progress in understanding the secondary instabilities of the roll pattern and more complex states (Section D). The problem has been treated by purely numerical methods, and also by a description in terms of appropriate order parameters. In comparison to full hydrodynamics the states satisfy much simpler equations, which can even be approached phenomenologically. The reduced level of order parameter equations allows valuable analytical insight, and by numerical simulations the description of defects and other spatio-temporal disordered states.

A problem that touches both experiment and theory, is connected with the large number of material parameters involved in the hydrodynamic description. There are only two room temperature nematics with negative dielectric anisotropy where all of the material parameters have been measured: 4-methoxybenzylidene-4'-butylaniline (MBBA), see [13], and Merck Phase 5, see [14] (and references therein). Since these nematics are similar in their properties, the investigation of other classes is highly desirable and promising. A recent successful example is the very stable material I52 doped with iodine [15]. Even at onset qualitatively new localised structures ('worms') [16] have been detected, which provide a new challenge for the theory. This applies also to materials where the nematic changes to the smectic phase on decreasing the temperature [17].

B   BASIC EQUATIONS AND INSTABILITY MECHANISMS

The dynamics of nematics is described by a set of macroscopic equations (see for example [2,18,19]), which couple the director **n**, the velocity field **v** and the (static or low-frequency) electric and magnetic fields **E**, **B**. The rate equation for the director can be written as (we use $\partial_j = \partial x_j$)

$$\frac{dn_i}{dt} = \Omega_{ki} n_k + \lambda(\delta_{ij} - n_i n_j) A_{jk} n_k + \frac{h_i}{\gamma_1} \tag{1}$$

where $\dfrac{d}{dt} = \partial_t + \mathbf{v} \cdot \nabla$

$$\Omega_{ik} = \frac{1}{2}(\partial_i v_k - \partial_k v_i), \quad A_{ik} = \frac{1}{2}(\partial_i v_k + \partial_k v_i)$$

$$\tag{2}$$

$$h_i = (\delta_{ij} - n_i n_j)\left(\partial_k \pi_{kj} - \frac{dF_d}{dn_j}\right), \quad \pi_{kj} = \frac{\partial F_d}{\partial(\partial_k n_j)}$$

$\lambda = -\gamma_2/\gamma_1$ is the flow-alignment parameter (≈ 1 for ordinary nematics). The orientational free energy density is given by

$$F_d = K_1(\nabla \cdot \mathbf{n})^2 + K_2[\mathbf{n} \cdot (\nabla \times \mathbf{n})]^2 + K_3[\mathbf{n} \times (\nabla \times \mathbf{n})]^2 -$$

$$\tag{3}$$

$$\frac{1}{2}\Delta\widetilde{\chi}(\mathbf{n} \cdot \mathbf{B})^2 - \frac{1}{2}\Delta\widetilde{\varepsilon}(\mathbf{n} \cdot \mathbf{E})^2 - \mathbf{P}^{\text{flexo}} \cdot \mathbf{E}$$

$$\mathbf{P}^{\text{flexo}} = e_1 \mathbf{n}(\nabla \cdot \mathbf{n}) + e_3 (\mathbf{n} \cdot \nabla)\mathbf{n} \tag{4}$$

with the orientational elastic constants K_1, K_2, K_3 describing the three basic deformations splay, twist and bend of the director field ($K_i \sim 10^{-11}$ N). $\Delta\widetilde{\chi}$ and $\Delta\widetilde{\varepsilon}$ are the anisotropic parts of the magnetic and electric susceptibilities. A distortion of the director field leads to an electric polarisation $\mathbf{P}^{\text{flexo}}$ ('flexoeffect') with the flexoelectric coefficients e_1, e_3, which are hard to measure. In EQN (1) only the last contribution is dissipative (irreversible).

The momentum balance equation (generalised Navier-Stokes equation) and (approximate) incompressibility are given by

$$\rho \frac{dv_i}{dt} = -\partial_i p + \partial_k (s_{ik} + E_i D_k), \quad s_{ik} = s_{ik}^R + s_{ik}^v, \quad \partial_j v_j = 0 \tag{5}$$

where

$$s_{ik}^R = -\pi_{kj}\partial_i n_j + \frac{\alpha_2}{\gamma_1} n_k h_i + \frac{\alpha_3}{\gamma_1} n_i h_k, \quad D_i = \widetilde{\varepsilon}_{ik} E_k + P_i^{\text{flexo}}, \quad \widetilde{\varepsilon}_{ik} = \widetilde{\varepsilon}_\perp \delta_{ik} + \Delta\widetilde{\varepsilon} n_i n_k \tag{6}$$

The viscous part of the stress tensor s_{ik} can be written as

$$s_{ik}^v = \alpha_4 A_{ik} + (\alpha_6 + \lambda\alpha_3)(n_i n_j A_{jk} + n_k n_j A_{ji}) + (\alpha_1 - \lambda(\alpha_2 + \alpha_3))n_i n_k n_j n_l A_{jl} \tag{7}$$

Thus there are three independent shear viscosities and the two parameters which characterise the orientational behaviour ($\gamma_1 = \alpha_3 - \alpha_2$, $\gamma_2 = \alpha_3 + \alpha_2$). We note that in the usual Leslie-Ericksen formulation the last two terms in s_{ik}^R appear together with the viscous terms, which from a systematic point of view may be less appealing. s_{ik} can be symmetrised by adding terms that do not change the body force $\partial_k s_{ik}$ [19-21]. For the occurrence of EC a non-zero conductivity is essential. It is possible or sometimes essential to add an ionisable dopant to the nematic in order to obtain sufficient and/or well-controlled conductivity. Then the following quasi-static Maxwell equations determine the current $\mathbf{J}$ and the charge density ρ_{el}

$$\frac{d\rho_{\text{el}}}{dt} = -\nabla \cdot \mathbf{J}, \quad \nabla \cdot \mathbf{D} = \rho_{\text{el}} \tag{8}$$

The condition $\nabla \times \mathbf{E} = 0$ is used to introduce an electric potential. In the 'standard model' (SM) of EC Ohm's law is assumed to hold,

$$J_i = \widetilde{\sigma}_{ik} E_k, \quad \widetilde{\sigma}_{ik} = \widetilde{\sigma}_\perp \widetilde{\sigma}'_{ik}, \quad \widetilde{\sigma}'_{ik} = (\delta_{ik} + \Delta\widetilde{\sigma}' n_i n_k) \tag{9}$$

with fixed, anisotropic conductivity. Diffusion currents are usually negligible. From EQNS (8) and (9) it can be seen that, as long as $\Delta\widetilde{\sigma}/\widetilde{\sigma}_\perp \neq \Delta\widetilde{\varepsilon}/\widetilde{\varepsilon}_\perp$ holds, any spatial variation of the director in the presence of an electric field (i.e. when a current is flowing) leads to the generation of non-zero ρ_{el}. This is an intrinsic property of any anisotropic and inhomogeneous conductor.

Now we can appreciate the main driving mechanism for EC. The important point is that in almost all nematics $\Delta\widetilde{\sigma}'$ is substantially positive. Choosing materials with negative or only slightly positive dielectric anisotropy $\Delta\widetilde{\varepsilon}$ (here the materials show great diversity) it is easily seen that charges are generated (or focused) at locations where the director bends. The Coulomb force $\rho_{el}\mathbf{E}$ inherent in the term $\partial_k(E_iD_k)$ in EQN (5) will then drive a velocity field $\mathbf{v}$. Via flow-alignment coupling this enhances the spatial variation of the director and thus generates a positive feedback. Above some threshold V_c this will overcome the stabilising elastic and viscous forces. For typical materials at low frequencies (usually AC driving is used) the threshold voltage is of the order

$$V_0 = \sqrt{\pi^2 K_1 / \widetilde{\varepsilon}_\perp}$$

The introduction of the reduced control parameter $R = V^2 / V_0^2$ is often useful. When $\Delta\widetilde{\varepsilon}$ becomes too positive a bend Fréedericksz transition will preempt EC [13].

The SM has three typical (linear) timescales associated with the three dynamical equations shown previously: the director relaxation time $\tau_d = \gamma_1 d^2/(K_1\pi^2) \sim 1$ s, the viscous diffusion time $\tau_v = \rho d^2/\alpha_4 \sim 10^{-5}$ s (d = thickness of the layer ~10 μm - 100 μm), and the charge relaxation time

$$\tau_q = \widetilde{\varepsilon}_\perp / \widetilde{\sigma}_\perp \sim 10^{-2} \text{ s}$$

In the frequency ranges usually covered, the effect of viscous relaxation is a very fast process, so the velocity field can be treated adiabatically, which means dropping the left-hand side in the Navier-Stokes equation (i.e. inertial terms). The flexoeffect is of minor relevance for AC driving, but under DC conditions it is certainly important (see later).

There is a very distinct effect which is not captured by the standard model, namely the travelling rolls arising via a Hopf bifurcation, often observed at threshold, in particular in thin and clean cells [22]. A rather natural generalisation, where the conduction mechanism via two types of mobile ions (generated by a (slow) dissociation-recombination reaction) is included, has been shown to describe all the experimental results. Then the ion densities n^+ and n^- with $\rho_{el} = e(n^+ - n^-)$ become dynamic variables (e is the elementary charge). This may be accommodated in the simplest manner by letting the conductivity $\widetilde{\sigma}_\perp$ (EQN (9)) become a dynamic variable

$$\widetilde{\sigma}_\perp = e(\mu_\perp^+ n^+ + \mu_\perp^- n^-)$$

satisfying the balance equation

$$\partial_t\widetilde{\sigma}_\perp + \partial_i(v_i\widetilde{\sigma}_\perp + \mu_\perp^+\mu_\perp^-\widetilde{\sigma}'_{ij}E_j\rho) = -\tau_{rec}^{-1}(\widetilde{\sigma}_\perp - \widetilde{\sigma}_\perp^{eq}) \tag{10}$$

This description has been called the weak electrolyte model (WEM) [23,24]. The quantities $\mu_\perp^\pm$ denote the mobilities of the positive/negative charges. In particular the recombination timescale τ_{rec} is essential for the WEM effects.

444

C   BEHAVIOUR NEAR THRESHOLD

The theoretical task is to solve the equations presented in the previous section with boundary conditions appropriate for the usual slab geometry (see FIGURE 1(a)): the director and the electric potentials are fixed at the confining plates and the velocity vanishes there. The large lateral extension allows the use of periodic boundary conditions, and thus a transformation from position to Fourier space ($\mathbf{x} = (x,y) \rightarrow \mathbf{q} = (q,p)$) in the horizontal directions is advantageous. Also, the time-periodic driving (AC-frequency ω) suggests an expansion in a Fourier series in ωt. The remaining transverse direction (z) can be treated by (truncated) expansions in terms of suitable test functions which satisfy the boundary conditions (Galerkin method). In general highly non-linear algebraic equations for the expansion coefficients are then obtained. The resulting solutions have to be tested for stability with respect to general fluctuations [25].

This program is simplified considerably when only the onset of the convection instability is to be determined. At onset certain linear perturbations of the basic (primary) state start to grow exponentially (linear stability analysis). It is necessary to analyse an eigenvalue problem, where the eigenvalue $\Lambda(\mathbf{q},R) = \Sigma(\mathbf{q},R) \pm i\Omega(\mathbf{q},R)$ with the largest real part, determines the growth rate Σ and the frequency Ω. The condition $\Sigma(\mathbf{q},R) = 0$ defines the neutral surface $R = R_0(\mathbf{q})$. Minimising $R_0(\mathbf{q})$ gives the threshold $R_c = R_0(\mathbf{q}_c)$ with the critical wavevector $\mathbf{q}_c = (q_c,p_c)$ and the critical frequency $\Omega_c = \Omega(\mathbf{q}_c)$, which vanishes for a stationary bifurcation (the more common case in EC) but differs from zero for a Hopf (oscillatory) bifurcation.

The linear stability analysis within the SM for planar aligned samples ($\mathbf{n} \parallel \hat{\mathbf{x}}$) has a long history starting with Helfrich [26], who considered the case of DC driving. He chose the wavevector $\mathbf{q} = (q,0)$ parallel to $\hat{\mathbf{x}}$ (normal rolls, NRs as in FIGURE 1) and discarded any z dependence (i.e. a one-dimensional model). The resulting growth rate has its maximum at $q = 0$. By setting $q = 2\pi/d$ a reasonable low-frequency threshold was obtained. The analysis was generalised to AC driving by the Orsay group [27], who thereby established the increase of the threshold with increasing frequency and found a crossover from the low-frequency conduction mode (conduction regime in FIGURE 2(a)) to a destabilising mode with opposite time-reflection symmetries above a frequency ω_c (the scale is set by τ_q^{-1}) in the dielectric regime (FIGURE 2(a)). This dielectric mode can usually be described well within the one-dimensional model, since its wave number is not determined by the width d of the layer but rather by the (intrinsic) orientational diffusion length

$$\sqrt{K_3/(|\alpha_2|\,\omega)}$$

In the dielectric range this length is typically small compared to d and then variations in the z direction are indeed negligible. The predicted threshold voltage $\sim\omega^{1/2}$ is in agreement with many experiments, but conflicts with some others, in particular at very high frequency [4,28,29]. Note that for materials with positive $\Delta\tilde{\varepsilon}$ there exists a competing homogeneous ($q = 0$) splay Fréedericksz destabilisation, so here the threshold curve remains finite at high frequencies [13]. Recently there was renewed interest in the interaction between EC and the Fréedericksz transition, which can also be influenced by an additional vertical magnetic field [30].

Inclusion of the variations in the z direction (two-dimensional model), thereby allowing for a complete description of the normal-roll threshold, was initiated by Pikin (DC case with some approximations) [31] and Penz and Ford (DC, rigorous) [32]. The two-dimensional theory gives a good account of the threshold behaviour in the NR regime (i.e. the transition line normal rolls in FIGURE 2(a)).

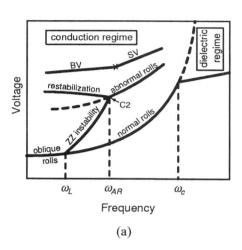

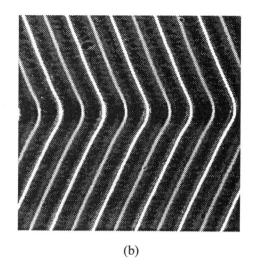

(a)   (b)

FIGURE 2 (a) Schematic stability diagram for planar EC. Convection sets in above the lowest solid curves. For the secondary bifurcations, see text. (b) Zig-zag (ZZ) pattern after increasing the voltage in FIGURE 1(b).

The apparent divergence of the conduction mode threshold at the cut-off frequency for negative $\Delta\tilde{\varepsilon}$ is a result of the lowest-order time-Fourier expansion. Including higher Fourier modes [33,13] or considering a square-wave voltage, where the problem can be handled essentially analytically [34], it was found that there is in fact restabilisation of the conduction mode at high voltage.

Fifteen years ago, the two most important tasks left for linear theory were understanding the oblique rolls that typically occur at low frequencies and the travelling rolls that appear predominantly in thin and clean cells. The first task proved comparatively simple: the SM, even without inclusion of the flexoeffect, when evaluated properly in three dimensions, describes the oblique rolls [35,13,36]. The crossover from NRs to oblique rolls occurs at a characteristic co-dimension-2 point called the Lifshitz point at frequency ω_L. As first shown by Madhusudana et al [37] in the DC case the flexoeffect provides an additional mechanism for oblique rolls. In the AC case the flexoeffect becomes important only for not too thin and clean cells [36,38]. For a discussion of the dielectric regime, see [39]. Meanwhile there exist computer programs to calculate the threshold curves and the critical wavevector $\mathbf{q}_c$ from the SM to any desired accuracy (with flexoeffect and higher Fourier modes) [40]. In addition there exist approximate closed expressions based on one-mode Galerkin expansions which describe satisfactorily the threshold behaviour of the SM over a large parameter range [41].

Understanding the Hopf bifurcation [22] proved more difficult. After all possibilities within the SM were exhausted the WEM was constructed and shown to provide for the Hopf bifurcation. The threshold behaviour in MBBA [42,43], I52 [23] and Phase 5 [14] can be described.

For homeotropic alignment ($\mathbf{n} \parallel \hat{\mathbf{z}}$) there are two very different cases: for negative dielectric anisotropy $\Delta\widetilde{\varepsilon}$ (not too near to zero) there is first a bend Fréedericksz transition, where the director gains a planar component (planar director $\mathbf{c}$). Increasing the voltage there is eventually a transition to convection, which is, on the linear level, in many ways similar to that in planar cells [44]. The tendency to oblique rolls is enhanced. In Merck Phase 5 there are two Lifshitz points, so that at very low frequency NRs are recovered [45]. For dielectric anisotropy around zero and negative α_3 there is a direct transition to EC with a very small wavelength [44].

Very near to a symmetry-breaking bifurcation thermal fluctuations become important. The first clear identification in EC (in fact in any pattern forming instability to our knowledge) was presented in [47] (for recent measurements, see [48]). For their description one has to generalise the hydrodynamic framework: see for example [49].

D   NON-LINEAR REGIME

The first task of the non-linear theory is to describe the saturation of the pattern evolving from the linear modes in the spirit of a (time-dependent) Landau theory. This was achieved by Bodenschatz et al [13], thereby establishing that the SM leads to a supercritical (continuous) bifurcation. In the oblique-roll regime a superposition of zig and zag rolls leading to rectangular structures is in principle possible, but for small angles of obliqueness this can be excluded by general phenomenological arguments [50].

Next, slow spatial modulations of the ideal periodic pattern with wavevector $\mathbf{q}_c$ can be included in the spirit of a Ginzburg-Landau theory by introducing a complex amplitude A such that the pattern is described by the real part of $A(\mathbf{x})\exp(i\mathbf{q}_c\mathbf{x})$. The generic amplitude equation for anisotropic systems in the range of static normal rolls is the real Ginzburg-Landau equation [25]

$$\tau\partial_t A = \left(\varepsilon - g\,|\,A\,|^2 + \xi_1^2\partial_x^2 + \xi_2^2\partial_y^2\right)A \qquad (11)$$

where

$$\varepsilon = (V^2 - V_c^2)/V_c^2$$

with the critical voltage V_c, is the reduced control parameter. $g > 0$ describes the non-linear saturation. $\xi_{1,2}$ represent the coherence lengths. Generalisation to the neighbourhood of the Lifshitz point is possible [50]. Deep inside the oblique-roll range one has to use two coupled equations. The coefficients of the amplitude equations were calculated from the SM in [13] (without flexoeffect). The Ginzburg-Landau equations can be used in particular to study the structure and dynamics of dislocations [51] (see FIGURE 1(b)) in good agreement with experiments [52]. In this framework the motion of defects provides a mechanism for the selection of the preferred wavevector $\mathbf{q}_p$. The non-linear velocity versus wavevector relation at small mismatch $\delta q = |\mathbf{q} - \mathbf{q}_p|$ (in fact, there is a logarithmic singularity for $\delta q \to 0$) has recently been verified for motion along the rolls [53] and (in a homeotropic system with a planar magnetic field, see below) perpendicular to the rolls [54].

In the framework of the WEM it was established that the Hopf bifurcation to travelling rolls is supercritical and then one has the complex Ginzburg-Landau equation (actually two or four coupled such equations for the counter-propagating, and possibly oblique, roll systems [55]) to describe the weakly non-linear behaviour [24]. It turns out that the stationary bifurcation near the crossover to travelling rolls is typically subcritical. The resulting small hysteresis has been measured in various materials [47,56,23]. Particularly interesting scenarios involving extended spatio-temporal chaos at onset [57] and subcritically arising localised structures ('worms') [16] have been found in I52. The former can be understood on the basis of two coupled complex Ginzburg-Landau equations describing zig and zag rolls travelling in the same direction [58]. A phenomenological model has been proposed to describe the worms [59].

To extend the range of validity of the above description finite-ε corrections have to be taken into account. Then various higher-order terms appear. In particular the curvature of rolls is known to induce a so-called mean flow. In the presence of two-dimensional lateral spatial variations (three-dimensional on the hydrodynamic level) the mean flow cannot be fully eliminated due to the singular structure of its spatial dependence. Thus, one is left with an additional (static) equation [60]. Meanwhile there exist efficient computer programs to calculate the coefficients [61,40]. The analysis showed that, at least near to the transition to oblique rolls, normal rolls are destabilised with increasing ε by a zig-zag (or undulatory) instability (see the ZZ instability line in FIGURE 2(a)) as found in experiments [62] (see FIGURE 2(b) for a ZZ pattern).

A full numerical Galerkin calculation confirmed this result and extended it to larger ε and frequencies [63,61]. Surprisingly, at frequencies above some value $\omega_{AR} > \omega_L$ (to the right of C2 in FIGURE 2(a)) destabilisation of normal rolls occurs at $\varepsilon = \varepsilon_{AR}$ via a spatially homogeneous (in the plane of the layer) mode involving a twist of the director. The mode is the analog of that which destabilises the basic state in a twist Fréedericksz transition (with the magnetic field in the y direction). The instability signals a (continuous) pitchfork bifurcation from normal to abnormal rolls (ARs) where the director attains such a twist deformation, either to the left or to the right. Another interesting effect is the restabilisation of ARs for $\omega < \omega_{AR}$ above ε_{ARstab} (see the line marked restabilisation in FIGURE 2(a)). At larger ε the ARs destabilise either via a long-wave skewed-varicose instability (here the modulation wavevector of the destabilising mode is at an oblique angle), or, at lower frequency, via a short-wave skewed-varicose instability. This is also called a bimodal varicose instability [64,61], because it indicates the transition to a bimodal state composed of the superposition of two roll systems with different orientations.

When this diagram is extended to oblique rolls (wavevector $\mathbf{q} = (q,p)$ with non-zero p, as in FIGURE 2(b)), the AR bifurcation becomes imperfect (smooth), since in oblique rolls the left-right symmetry is already broken. Also, the destabilisation is shifted upward and restabilisation downward, so that the curves meet at some value $P_m(\omega)$ (with vertical slope) [63]. Thus there is an unstable bubble in the ε-P plane which is bounded from below by ε_{ZZ}, and from above by ε_{ARstab} [41,67]. Thus there is a very interesting co-dimension-2 (C2) point at ω_{AR}, ε_{AR}.

The reason why ARs had escaped the notice of experimentalists is that in planar aligned cells and with the ordinary visualisation (at most one polariser in the x direction) they cannot be distinguished from NRs. For homeotropic alignment this is different, and there the signature of ARs had indeed been observed before (see later). Indirect evidence comes from the observation of domain walls between the two variants of ARs. They are observable because inside the wall the amplitude of the periodic n_z deformation is larger than in ARs [63]. Direct evidence has been obtained from measurements of the ellipticity of the light induced by ARs [65,66]. In these measurements the ZZ instability, together with the restabilisation line, could also be identified. Apparently the rather small width of the system in the x direction, which allowed for twelve roll pairs, stabilised the ZZ structures, which appear above the instability. Interestingly, another line ε_{HB} was found, which lies above ε_{ARstab} and also goes through the C2 point: whereas the angle of the ZZ structures first increases with ε when ε_{ZZ} is crossed, it subsequently decreases again, becoming zero at ε_{HB}. Then there remain ARs with domain walls, because the different orientations in the ZZs induce different variants of ARs. The domain walls quickly annihilate and a single AR domain is left. When ε decreases a massive hysteresis occurs: the ARs persist down to ε_{ARstab}. Then there is a discontinuous transition to the ZZ branch, which is followed down.

These features can be understood in terms of a simple, phenomenological description of the AR bifurcation [65,67]. The two active modes involved are the twist mode, characterised by an angle ϕ, and the phase of the roll pattern θ. The equations are

$$\partial_t \phi = (\mu - g\phi^2)\phi + (K_1 \partial_x^2 + K_2 \partial_y^2)\phi - \gamma \partial_y \theta$$

$$\tag{12}$$

$$\partial_t \theta = (D_1 \partial_x^2 + D_2 \partial_y^2)\theta - (\nu + h\phi^2)\partial_y \phi$$

The control parameters μ and ν are to be associated with $\varepsilon - \varepsilon_{AR}$ and $\omega - \omega_{AR}$. The coupling terms are obtained from symmetry considerations. The term proportional to h is included because ν goes through zero. (A similar term in the ϕ equation would allow the destabilisation of ARs at larger values of μ to be included.) This model contains all the features described above. The slopes of the different lines are easily expressed in terms of the parameters of the model. In particular, for $\nu < 0$, the ZZ instability of NRs at $\mu = (\gamma/D_2)\nu$ preempts the AR instability and ARs exhibit the observed restabilisation. At the HB line there is a heteroclinic connection between ZZ solutions and NRs. Above the HB line domain walls perpendicular to the rolls move spontaneously [67].

This description is well-founded only when modulations along y occur, as in the ZZ instability and in ZZ solutions. The coefficients can, in principle, be deduced from hydrodynamics, including (regular) mean flow. In order to describe y and x variations, additional terms that include in particular the singular mean flow have to be included. This changes the restabilisation into a skewed-varicose instability and moves it upward [68,40].

Before going to a more general description let us discuss homeotropically aligned systems in materials with manifestly negative dielectric anisotropy, where there is first a bend Fréedericksz transition through which the director acquires a planar component (planar director **c**). The transition to convection occurring at a higher voltage is in many ways similar to that in planar cells, except that the preferred axis (the **c** director) is not fixed externally. Consequently, in a weakly non-linear description, the Goldstone mode related to rotation of the **c** director has to be included from the beginning. In generalising EQN (11) for small angles ϕ, which now denotes the angle between **c** and the x axis [69,70], the equations

$$\tau \partial_t A = [\varepsilon - g \, | \, A \, |^2 + \xi_1^2 \partial_x^2 + \xi_2^2 (\partial_y^2 - 2iC_1 \phi \partial_y - C_2 \phi^2 - iv \partial_y \phi)] A$$

$$\partial_t \phi = G(iq_c A * (\partial_y - iq_c \phi) A + c.c) - T\phi + (K_1 \partial_x^2 + K_2 \partial_y^2) \phi$$

(13)

are obtained, with $C_1 = C_2 = 1$ (we introduce the coefficients for a later purpose). Then the first three terms in the bracket proportional to ξ_2^2 can be combined to a full square, which expresses (local) invariance with respect to rotation of the rolls together with **c**. If this were an equilibrium system derivable from a potential, then $v = 1$ would be needed. Here, however, v is an independent coefficient that can (and will, for $\omega < \omega_{AR}$) even become negative. The first term in the ϕ-equation expresses the abnormal torque on the **c** director ($G > 0$), which arises at second order in the convection amplitude. We have included a (small) linear damping, which appears only in the presence of an additional planar magnetic field (then $T \approx \Delta\tilde\chi B^2$). For $T = 0$ all roll solutions are unstable. In simulations ϕ grows without bound (for $C_1 = C_2 = 1$), and then a globally invariant generalisation of these equations must be used, which exhibits dynamic disorder (defect turbulence) [69,70], which is essentially what is found experimentally [71-73].

For $T > 0$ (non-zero planar field) EQNS (13) describe NRs at the band centre

$$(| \, A \, | = \sqrt{\varepsilon / g}, \quad \phi = 0)$$

which are stable against homogeneous ϕ perturbations for $\varepsilon < \varepsilon_{AR} = T/(2G)$ and against ZZ fluctuations for $\varepsilon < \varepsilon_{ZZ} = \varepsilon_{AR}/(1 - v)$. Thus, the situation is similar to that in planar cells with the near-Goldstone mode in homeotropic systems corresponding to the twist mode in planar cells. Here, for $C_1 = C_2 = 1$, the ARs

$$A = \sqrt{\varepsilon_{AR} / g}, \quad \phi = \sqrt{(\varepsilon - \varepsilon_{AR})} / (r_1 q_c)$$

destabilise for $v > 0$ at $\varepsilon = 3/2\varepsilon_{AR}$. There is no restabilisation for $v < 0$ because the short-wave instability sets in right at the restabilisation line (also at $\varepsilon = 3/2\varepsilon_{AR}$). For larger values of ε there is dynamic disorder (defect turbulence). In this system ARs are easily identified by birefringence measurements, and the symmetry breaking was indeed first detected in such a system (with $B = 0$) in the disordered state [71]. Meanwhile a quantitative comparison of the pitchfork bifurcation to ARs ($B \neq 0$) with theory has been made [45].

For $\varepsilon \gg \varepsilon_{AR}$ (this can be achieved for any positive value of ε by choosing the planar field sufficiently small) there is a spontaneous ordering of defects along periodically arranged lines, which appears to explain the most common types of chevrons observed in the dielectric range of planar aligned cells [74]. The theory is indeed applicable to the dielectric range because here the orienting effects of the boundaries can be considered as small perturbations [39], which is consistent with experiments [75,76]. The prediction that chevrons should occur in homeotropic systems also in the conduction range has been verified [72,73].

Let us now come back to the case of planar alignment. The scenario found there can be described qualitatively by EQNS (13) with $1 \geq C_2^2 > C_1$. The surface anchoring now leads to $T \approx \Delta\tilde\chi B_F^2$

(B_F = twist Fréedericksz field). The destabilisation of NRs is independent of C_1 and C_2. For negative ν there is a restabilisation curve for ARs which passes through the C2 point with slope $d\varepsilon/d\nu = -1/(2(1 - C_1))$ and saturates for $\nu \to -\infty$ at $\varepsilon_\infty = C_2/(2C_1 + C_2)$. Destabilisation of ARs at large ε is also captured. For negative values of ν there is a short-wave instability merging with a long-wave instability curve at some positive ν (for C_1, $C_2 \to 1$ this point becomes $\nu = 0$, $\varepsilon = 3/2\varepsilon_{AR}$). For $\nu \to \infty$ the instability curve tends to ε_∞ from above. In the range of stable ARs the equations describe interesting defect scenarios. For $\varepsilon \gg \varepsilon_{AR}$ they describe the dynamic chevrons mentioned before and also a new type of static chevron [68] (at larger ε). The equations can only be taken as a quantitative description when the AR bifurcation occurs sufficiently near to the primary instability. This can be achieved in planar systems by applying an additional (destabilising) magnetic field in the y direction. Then $T \approx \Delta\widetilde{\chi}(B_F^2 - B^2)$ so that T, and therefore also ε_{AR}, tends to zero for $B \to B_F$. In simulations ϕ remains bounded even for $T = 0$, in contrast to the rotationally invariant case.

When ε_{AR} is not sufficiently small, as is the case for planar systems without an additional magnetic field, corrections have to be included, which in particular involve mean flow. This has been carried out for the case with modulations only in the y direction [61] (as mentioned before, mean flow can then be eliminated leading to important higher-order terms in the A equation). Otherwise the equations become rather complicated [68,40].

Some of these studies have been carried over to the oblique-roll regime [69,70,63,61]. The signature of the twist mode (as in ARs) has also been observed in travelling rolls [77].

E CONCLUSION

Although EC in nematics has come a long way since the early days there remains much to do. There are unsolved problems concerning the threshold behaviour of materials near a nematic-smectic transition. It appears that in a region where $\Delta\widetilde{\sigma}'$ has become negative there can be EC in the form of localised structures (worms) [17,78]. In highly doped MBBA [28,79,73] and other materials [29,80] at high frequency a stationary periodic domain structure (period ~ cell thickness) is found, often without detectable convection rolls. The structure appears to persist at least in some cases with increasing temperature up to the nematic-isotropic phase transition. In the swallow tailed compounds used in [29] a treatment of the bounding plates by surfactants led to a considerable increase of the frequency range where the domains appeared. The periodic behaviour of the in-plane director is reminiscent of chevrons, but the origin is at present not clear (see also [39]).

There are various complex structures which are as yet only partially understood; see e.g. [81,82]. Because of space limitations we have concentrated on the basic phenomena in standard EC. There exist various other interesting possibilities, which have been mentioned in previous reviews [41,7]. Of particular interest is the use of a periodic modulation of the driving AC-voltage [42,83,77] with the superposition of noise [84]. It should also be mentioned, that EC in small aspect-ratio systems (very few short rolls) yields interesting bifurcation scenarios [85]. Finally, it would certainly be rewarding to understand better electroconvectively driven turbulence, i.e. the so-called dynamic scattering modes [8].

REFERENCES

[1]   L.M. Blinov [*Electrooptical and Magnetooptical Properties of Liquid Crystals* (John Wiley, New York, 1983)]

[2]   P.G. de Gennes, J. Prost [*The Physics of Liquid Crystals* (Clarendon Press, Oxford, 1993)]

[3]   S. Chandrasekhar [*Liquid Crystals* (University Press, Cambridge, 1992)]

[4]   S.A. Pikin [*Structural Transformations in Liquid Crystals* (Gordon & Breach Science Publishers, New York, 1991)]

[5]   L. Kramer, W. Pesch [*Annu. Rev. Fluid Mech. (USA)* vol.27 (1995) p.515]

[6]   For a recent review see A. Buka, L. Kramer [*Pattern Formation in Liquid Crystals* (Springer-Verlag, New York, 1996)]

[7]   W. Pesch, U. Behn [in *Evolution of Spontaneous Structures in Dissipative Continuous Systems* Eds. F.H. Busse, S.C. Müller (Springer, 1998)]

[8]   S. Kai, W. Zimmermann [*Prog. Theor. Phys. Suppl. (Japan)* vol.99 (1989) p.458]

[9]   R. Williams [*J. Chem. Phys. (USA)* vol.39 (1963) p.384]

[10]   A. Kapustin, L. Larinova [*Kristallografiya (USSR)* vol.9 (1963) p.297]

[11]   M.C. Cross, P.C. Hohenberg [*Rev. Mod. Phys. (USA)* vol.65 (1993) p.851]

[12]   L. Bodenschatz, W. Pesch, G. Ahlers [*Annu. Rev. Fluid Mech. (USA)* vol.32 (2000) p.709]

[13]   E. Bodenschatz, W. Zimmermann, L. Kramer [*J. Phys. (France)* vol.49 (1988) p.1875]

[14]   M. Treiber, N. Eber, Á. Buka, L. Kramer [*J. Phys. II (France)* vol.7 (1997) p.649-61]

[15]   M. Dennin, G. Ahlers, D. Cannell [in *Spatio-Temporal Patterns in Nonequilibrium Complex Systems* Sante Fe Institute Studies in the Sciences of Complexity XXI, Eds. P. Cladis, P. Palffy-Muhoray (Addison-Wesley, New York, 1994)]; M. Dennin, D. Cannell, G. Ahlers [*Mol. Cryst. Liq. Cryst. (Switzerland)* vol.261 (1995) p.337]

[16]   M. Dennin, G. Ahlers, D.S. Cannell [*Phys. Rev. Lett. (USA)* vol.77 (1996) p.2475; *Science (USA)* vol.272 (1997) p.388]; U. Bisang, G. Ahlers [*Phys. Rev. Lett. (USA)* vol.80 (1998) p.3061; *Phys. Rev. E (USA)* vol.60 (1999) p.3910]

[17]   H.R. Brand, C. Fradin, P.L. Finn, W. Pesch, P.E. Cladis [*Phys. Lett. A (Netherlands)* vol.235 (1997) p.508]

[18]   M.J. Stephen, J.P. Straley [*Rev. Mod. Phys. (USA)* vol.46 (1974) p.617]

[19]   H. Pleiner, H. Brandt [in *Pattern Formation in Liquid Crystals* (Springer-Verlag, New York, 1996)]

[20]   L.D. Landau, E.M. Lifshitz [*Lehrbuch der Theoretischen Physik* vol.7 (Akademie Verlag, Berlin, 1989)]

[21]   D. Forster, T. Lubensky, P.C. Martin, J. Smith, P.J. Pershan [*Phys. Rev. Lett. (USA)* vol.26 (1971) p.1016]; P.C. Martin, O. Parodi, P.J. Pershan [*Phys. Rev. A (USA)* vol.6 (1972) p.2401]

[22]   S. Kai, K. Hirakawa [*Prog. Theor. Phys. Suppl. (Japan)* vol.14 (1978) p.212]; A. Joets, R. Ribotta [*Phys. Rev. Lett. (USA)* vol.60 (1988) p.2164]; I. Rehberg, S. Rasenat, V. Steinberg [*Phys. Rev. Lett. (USA)* vol.62 (1989) p.756]

[23]   M. Dennin, M. Treiber, L. Kramer, G. Ahlers, D. Cannell [*Phys. Rev. Lett. (USA)* vol.76 (1996) p.319]

[24]   M. Treiber, L. Kramer [*Phys. Rev. E (USA)* vol.58 (1998) p.1973]

[25]   W. Pesch, L. Kramer [in *Pattern Formation in Liquid Crystals* (Springer-Verlag, New York, 1996)]

[26]   W. Helfrich [*J. Chem. Phys. (USA)* vol.51 (1969) p.4092]

[27]   E. Dubois-Violette, P.G. de Gennes, O.J. Parodi [*J. Phys. (France)* vol.32 (1971) p.305]; E. Dubois-Violette, G. Durand, E. Guyon, P. Manneville, P. Pieranski [in *Solid State Phys. Suppl. (Japan)* vol.14 (1978)]

[28]   A.N. Trufanov, L.M. Blinov, M.I. Barnik [in *Advances in Liquid Crystal Research and Applications* Ed. L. Bata (Pergamon Press, Oxford-Budapest, 1980) p.549]

[29]   W. Weissflog, G. Pelzl, H. Kresse, D. Demus [*Cryst. Res. Technol. (Germany)* vol.23 (1988) p.1259]

[30]   A. Hertrich, W. Pesch, J.T. Gleeson [*Europhys. Lett. (Switzerland)* vol.44 (1996) p.417]

[31]   S.A. Pikin [*Zh. Eksp. Teor. Fiz. (USSR)* vol.60 (1971) p.1185; *Sov. Phys.-JETP (USA)* vol.33 (1971) p.641]

[32]   P.A. Penz, G.W. Ford [*Phys. Rev. A (USA)* vol.6 (1972) p.414]

[33]   P. Sengupta, A. Saupe [*Phys. Rev. A (USA)* vol.9 (1974) p.2698]

[34]   E. Dubois-Violette [*J. Phys. (France)* vol.33 (1972) p.95]; R.A. Rigopoulos, H.M. Zenginoglou [*Mol. Cryst. Liq. Cryst. (UK)* vol.35 (1976) p.307]

[35]   W. Zimmermann, L. Kramer [*Phys. Rev. Lett. (USA)* vol.55 (1985) p.402]

[36]   L. Kramer, E. Bodenschatz, W. Pesch, W. Thom, W. Zimmermann [*Liq. Cryst. (UK)* vol.5 no.2 (1989) p.699]

[37]   N.V. Madhusudana, V.A. Raghunathan, K.R. Sumathy [*Pramana J. Phys. (India)* vol.28 (1987) p.L311]

[38]   The importance of the flexoeffect for the appearance of oblique rolls is overrated in [3], p.210 and in L.M. Blinov, V.G. Chigrinov [*Electrooptical and Magnetooptical Properties of Liquid Crystal Materials* (Springer, New York, 1996) p.263]

[39]   A.G. Rossberg [*Three-Dimensional Pattern Formation, Multiple Homogeneous Soft Modes, and Nonlinear Dielectric Electroconvection* preprint, http://arXiv.org/abs/nlin/0001065 (2000)]

[40]   B. Dressel, W. Pesch [unpublished]

[41]   L. Kramer, W. Pesch [in *Pattern Formation in Liquid Crystals* (Springer-Verlag, New York, 1996)]

[42]   I. Rehberg, S. Rasenat, J. Fineberg, M. del la Torre Juarez, V. Steinberg [*Phys. Rev. Lett. (USA)* vol.61 (1988) p.2449]

[43]   M. Treiber [PhD Thesis, Universität Bayreuth, 1996]

[44]   A. Hertrich, W. Decker, W. Pesch, L. Kramer [*J. Phys. II (France)* vol.2 (1992) p.1915]; L. Kramer, A. Hertrich, W. Pesch [in *Pattern Formation in Complex Dissipative Systems* Ed. S. Kai (World Scientific, Singapore, 1992) p.238]

[45]   A.R. Rossberg, N. Eber, A. Buka, L. Kramer [*Phys. Rev. E (USA)* vol.61 (2000) p.R25]

[46]   L. Kramer, A. Hertrich, W. Pesch [in *Pattern Formation in Complex Dissipative Systems* Ed. S. Kai (World Scientific, Singapore, 1992) p.238]

[47]   I. Rehberg et al [*Phys. Rev. Lett. (USA)* vol.67 (1991) p.596]

[48]   M.A. Scherer, G. Ahlers, F. Hörner, I. Rehberg [*Deviations from Linear Theory for Fluctuations Below the Supercritical Primary Bifurcation to Electroconvection* preprint (2000)]

[49]   M. Treiber [in *Pattern Formation in Liquid Crystals* (Springer-Verlag, New York, 1996)]

[50]   W. Pesch, L. Kramer [*Z. Phys. B (Germany)* vol.63 (1986) p.121]

[51]   E. Bodenschatz, W. Pesch, L. Kramer [*Physica D (Netherlands)* vol.32 (1988) p.135]; E. Bodenschatz, A. Weber, L. Kramer [*J. Stat. Phys. (USA)* vol.64 (1991) p.1007]

[52]   L. Kramer, E. Bodenschatz, W. Pesch [*Phys. Rev. Lett. (USA)* vol.64 (1990) p.2588]

[53]   V. Steinberg [private communication]

[54]   P. Toth, N. Eber, L. Kramer, A. Buka [to be published]

[55]   M. Silber, H. Riecke, L. Kramer [*Physica D (Netherlands)* vol.61 (1992) p.260]

[56]   I. Rehberg, F. Hörner, H. Hartung [*J. Stat. Phys. (USA)* vol.64 (1991) p.1017]

[57]   M. Dennin, D.S. Cannell, G. Ahlers [*Phys. Rev. E (USA)* vol.57 (1997) p.638]

[58]   H. Riecke, L. Kramer [*Physica D (Netherlands)* vol.137 (2000) p.124]

[59]   H. Riecke, G.D. Granzow [*Phys. Rev. Lett. (USA)* vol.81 (1998) p.333]

[60]   M. Kaiser, W. Pesch [*Phys. Rev. E (USA)* vol.48 (1993) p.4510]

[61]   E. Plaut, W. Pesch [*Phys. Rev. E (USA)* vol.59 (1999) p.1247]

[62]   A. Joets, R. Ribotta [*J. Phys. (France)* vol.47 (1986) p.595]; E. Braun, S. Rasenat, V. Steinberg [*Europhys. Lett. (Switzerland)* vol.15 (1991) p.597]; S. Nasuno, S. Kai [*Europhys. Lett. (Switzerland)* vol.14 (1991) p.779]; S. Nasuno, O. Sasaki, S. Kai, W. Zimmermann [*Phys. Rev. A (USA)* vol.46 (1992) p.4954]

[63]   E. Plaut et al [*Phys. Rev. Lett. (USA)* vol.79 (1997) p.2376]

[64]   E. Plaut, R. Ribotta [*Europhys. Lett. (Switzerland)* vol.38 (1997) p.441]

[65]   H. Zhao, L. Kramer, I. Rehberg, A. Rudroff [*Phys. Rev. Lett. (USA)* vol.81 (1998) p.4144]

[66]   S. Rudroff, V. Frette, I. Rehberg [*Phys. Rev. E (USA)* vol.59 (1999) p.1814]

[67]   H. Zhao, L. Kramer [*Phys. Rev. E (USA)* in press]

[68]   H. Zhao [PhD Thesis, Bayreuth, 2000]

[69]   A.G. Rossberg, A. Hertrich, L. Kramer, W. Pesch [*Phys. Rev. Lett. (USA)* vol.76 (1996) p.4729]

[70]   A.G. Rossberg [PhD Thesis, Bayreuth, 1998]; A.G. Rossberg, L. Kramer [*Phys. Scr. Vol. T (Sweden)* vol.67 (1996) p.121]

[71]   H. Richter, A. Buka, I. Rehberg [in *Spatio-Temporal Patterns in Nonequilibrium Complex Systems* Sante Fe Institute Studies in the Sciences of Complexity XXI, Eds. P. Cladis, P. Palffy-Muhoray (Addison-Wesley, New York, 1994)]

[72]   P. Toth, A. Buka, J. Peinke, L. Kramer [*Phys. Rev. E (USA)* vol.58 (1998) p.1983]

[73]   J.H. Hu, Y. Hidaka, A.G. Rossberg, S. Kai [*Phys. Rev. E (USA)* vol.61 (2000) p.2769]

[74]   A.G. Rossberg, L. Kramer [*Physica D (Netherlands)* vol.115 (1998) p.19]

[75]   M. Scheuring, L. Kramer, J. Peinke [*Phys. Rev. E (USA)* vol.58 (1998) p.2018]

[76]   H. Amm, R. Stannarius, A.G. Rossberg [*Physica D (Netherlands)* vol.126 (1999) p.171]

[77]   M. Dennin [*Direct Observation of Twist Mode in Electroconvection in I52* preprint, http://arXiv.org/abs/cond-mat/0003515 (2000)]

[78]   C.F. Fradin, P.L. Finn, P. Cladis, H.R. Brand [*Phys. Rev. Lett. (USA)* vol.81 (1998) p.2902]

[79]   L. Nasta, L. Lupu, M. Giurgea [*Mol. Cryst. Liq. Cryst. (UK)* vol.71 (1981) p.65]

[80]   N. Eber, T. Tóth Katona, A. Buka [private communication]

[81]   S. Nasuno, O. Sasaki, S. Kai, W. Zimmermann [*Phys. Rev. A (USA)* vol.46 (1992) p.4954]

[82]   A. Belaidi, R. Ribotta [*Restabilisation of Topological Disorder into an Ordered Lattice of Defects* preprint (1997)]

[83]   A. Belaidi [PhD Thesis, University of Orsay, Paris XII, 1998]

[84]   Th. John, R. Stannarius, U. Behn [*Phys. Rev. Lett. (USA)* vol.83 (1999) p.749]

[85]   T. Peacock, D.J. Binks, T. Mullin [*Phys. Rev. Lett. (USA)* vol.82 (1999) p.1446]

CHAPTER 9

DIFFUSION PROPERTIES

9.1   Measurement of translational diffusion in nematics

S. Miyajima

March 1999

A   INTRODUCTION

Translational self-diffusion of constituent particles in the condensed phase is described by the diffusion equation [1,2]

$$\frac{\partial c}{\partial t} = D \frac{\partial^2 c}{\partial \xi^2} \tag{1}$$

where $c = c(\xi, t)$ is the particle concentration per unit volume, ξ is the spatial coordinate, t is the time, and D is the self-diffusion coefficient. EQN (1) assumes that diffusion is one dimensional, i.e. the concentration gradient exists only along the ξ direction. In a real three dimensional system, D is a second rank symmetric tensor with six independent components. EQN (1) is now extended to

$$\frac{\partial c}{\partial t} = D_{\xi\xi} \frac{\partial^2 c}{\partial \xi^2} + D_{\eta\eta} \frac{\partial^2 c}{\partial \eta^2} + D_{\varsigma\varsigma} \frac{\partial^2 c}{\partial \varsigma^2}$$

$$+ \left(D_{\xi\eta} + D_{\eta\xi} \right) \frac{\partial^2 c}{\partial \xi \partial \eta} + \left(D_{\eta\varsigma} + D_{\varsigma\eta} \right) \frac{\partial^2 c}{\partial \eta \partial \varsigma} + \left(D_{\varsigma\xi} + D_{\xi\varsigma} \right) \frac{\partial^2 c}{\partial \varsigma \partial \xi} \tag{2}$$

where ξ, η and ς are the axes of the Cartesian coordinate system taken arbitrarily in space. If the three axes of D are taken so that the tensor D is diagonalised, namely, the principal axis system of diffusion is used, then only three elements, D_{xx}, D_{yy}, and D_{zz} remain. The other three degrees of freedom are then the three Eulerian angles which transform from the initial axis (ξ, η, ς) system to the principal axis (x, y, z) system.

For uniaxial nematic liquid crystals, the z axis for diffusion is defined to be the symmetry axis of the phase, i.e. the director. The number of independent components is now reduced to two: $D_\parallel = D_{zz}$ and $D_\perp = D_{xx} = D_{yy}$, and the diffusion equation becomes

$$\frac{\partial c}{\partial t} = D_\parallel \frac{\partial^2 c}{\partial z^2} + D_\perp \left(\frac{\partial^2 c}{\partial x^2} + \frac{\partial^2 c}{\partial y^2} \right) \tag{3}$$

For biaxial nematic liquid crystals, the three principal components of the diffusion tensor are independent. However, these three components have not been observed yet in biaxial nematic liquid crystals. Therefore, this brief account of the experimental studies of translational diffusion deals only with uniaxial nematics in which the two components are observed experimentally.

B   CLASSICAL EXPERIMENTAL METHODS

For uniaxial nematic liquid crystals, the two components are obtained by solving EQN (3). If $D_\parallel$ and $D_\perp$ are independent of time or coordinates, the solutions give

$$D_\parallel = \frac{\left\langle \Delta z^2 \right\rangle}{2\tau} \tag{4}$$

and

$$D_\perp = \frac{\left\langle \Delta x^2 \right\rangle}{2\tau} = \frac{\left\langle \Delta y^2 \right\rangle}{2\tau} \tag{5}$$

Here $\langle \Delta z^2 \rangle$ is the mean square displacement of the particle along the director within the time interval τ, and $\langle \Delta x^2 \rangle$ and $\langle \Delta y^2 \rangle$ are the corresponding quantities perpendicular to the director.

The components of **D** are obtained experimentally by measuring the mean square displacements along the specific directions in space. It is necessary, therefore, to label (or encode) the particles and to detect their displacements. Radioactive particles, dyes, and radicals with unpaired electrons have been used in radioactive [3,4], optical transient grating [5-9], and EPR (electron paramagnetic resonance) [10,11] measurements, respectively. Small organic molecules such as tetramethylsilane (TMS) have been used in NMR (nuclear magnetic resonance) measurements [12,13]. The review article by Krueger [13] summarises the data on anisotropic self-diffusion in liquid crystals until 1981. Diffusion of chiral dopants in an achiral nematic host has been detected by optical microscopy [14-16]. However, all of these methods suffer a common problem, namely that the tracer particles are different from the particles of the nematic medium. Therefore, the true self-diffusion coefficients cannot be measured with these methods. Among them, the radiotracer method can be most reliable because the difference between the tracer and the host molecule may be only in the small difference in masses. In fact, the classical work by Yun and Fredrickson [3] in 1970 shows the most reliable data on the anisotropic self-diffusion in 4,4'-dimethoxyazoxybenzene (PAA).

C   PULSED FIELD GRADIENT SPIN ECHO NMR METHOD APPLIED TO NEMATIC LIQUID CRYSTALS

The pulsed field gradient spin echo (PGSE, or PFG) NMR method [17-19] is preferable to the former methods because labelling is made for all the molecules in the sample, without changing the molecular structure, mass, or any of the molecular properties. FIGURE 1 shows the pulse sequence for a standard PGSE NMR experiment. Every molecule is labelled according to its position at t = 0 (initial time) by the first field gradient (FG) pulse (encoding FG pulse). The second FG pulse (decoding FG pulse) with exactly the same integrated intensity cancels out the effect of the first FG if the radio frequency (RF) π-pulse, which inverts the spin states, is interposed between the two FG pulses. If, on the other hand, the molecule changes its position within the time interval between the two FG pulses, the effect of the FGs does not cancel, and is detected as the decay of the spin-echo amplitudes. The spin echo decay is related to D by

$$\frac{I_{FG}}{I_0} = \exp\left[-D(\gamma G\delta)^2\left(\Delta - \frac{\delta}{3}\right)\right] \tag{6}$$

Here γ is the gyromagnetic ratio of the nuclei under observation, and G, δ, and Δ are the FG intensity, the width of the individual FG pulse, and the time interval between the two RF pulses ($\pi/2$-π), respectively. The diffusion-induced echo decay is accelerated by the integrated intensity of the FG pulses.

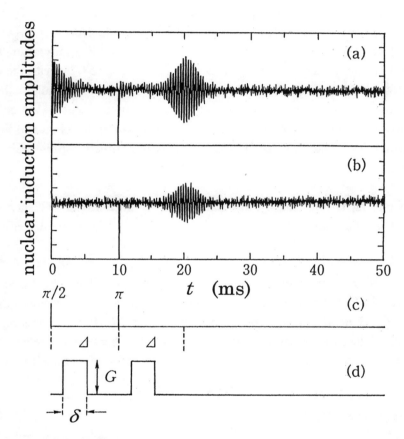

FIGURE 1 Pulsed field gradient spin echo experiment. (a) Free induction decay and spin echo, observed with the pulse sequence (c), that is without field gradients. (b) Spin echo observed with the pulse sequences (c) and (d). Reduction of the echo amplitude is seen compared with (a). (c) Radio frequency pulses for spin echo observation. (d) FG pulse sequence for PGSE study.

Application of PGSE NMR to nematic liquid crystals has been difficult because the short spin-spin relaxation time, T_2, makes the observation of the spin echo difficult, and also because vectorised FG, namely,

$$G_x = \frac{\partial B_{z0}}{\partial x} \tag{7}$$

$$G_y = \frac{\partial B_{z0}}{\partial y} \tag{8}$$

and

$$G_z = \frac{\partial B_{z0}}{\partial z} \tag{9}$$

are necessary. Here B_{z0} is the magnetic field component along the main external field. The first successful experiment in lengthening T_2 in ^{1}H NMR was made by Blinc et al [20] using multiple pulses and selective deuteriation. The major homonuclear dipolar interactions were eliminated by ring-deuteriation, and the rest were averaged by a train of RF pulses. A long T_2 was thus realised, though ring-deuteriation was a perturbation to the original material. Noack et al [21-23] employed a method involving fast switching of the external field directions. After equilibration, the external field is changed suddenly to the direction which makes the magic angle (54.7°) with the director. The ^{1}H NMR line is narrowed significantly by this procedure because the origin of line-broadening in the nematic is mainly the homonuclear dipolar interactions along the molecular long axis which tends to align along the director. The first success in ^{13}C NMR was obtained by Zhou and Frydman [24] who combined high power ^{1}H decoupling and the design of the FG directions; two pairs of Maxwell and saddle coils were used.

In these experiments, $D_\parallel$ is obtained by designing the two-dimensional FG such that $G_z \neq 0$ and $G_x = G_y = 0$. On the other hand, $D_\perp$ is obtained by setting $G_z = 0$ and making either G_x or G_y non-zero.

Apart from these studies, a number of studies have been reported on the PGSE NMR of small molecules dissolved in nematic liquid crystals. Among them are the multiple quantum spin echo experiments [25,26]. Generally these experiments give faster diffusion and less anisotropy than the non-probe methods. These results are not discussed any further because of the ambiguity inherent in the method. The next section describes the experimental results based on the non-probe methods.

D   EXPERIMENTAL RESULTS

The principal components of the self-diffusion tensor have been reported for nematic 4-methoxybenzylidene-4′-butylaniline (MBBA) [20] and (4-pentylphenyl)-2-chloro-4-(4-pentylbenzoyloxy)benzoate (PCPB) [24]. A systematic study was made by Noack et al for typical low molar mass nematics, such as PAA, MBBA, 4-alkyl-4′-cyanobiphenyl (nCB) and 4-alkoxy-4′-cyanobiphenyl (nOCB) [21-23]. Some of the results are summarised in TABLE 1. References [21-23] contain interesting comparisons of the different experimental methods.

The reported data on $D_\parallel$ and $D_\perp$ for a number of nematics and the scalar diffusion coefficients D_{iso} in their isotropic phases show the following common features.

(i)   Absolute values of $D_\parallel$ and $D_\perp$ range from 10^{-11} to 10^{-9} m^2 s^{-1} in the nematic phase, depending on the compound and the temperature. These values may be compared with the D_{iso} values of water (2.2×10^{-9} m^2 s^{-1}) and glycerol (1.8×10^{-11} m^2 s^{-1}) at 25°C.

TABLE 1 Translational self-diffusion coefficient tensors for some typical nematic liquid crystals, measured with PGSE NMR without using spin probes.

Compounds	$t/°C$	$D_\parallel/10^{-11}$ m^2 s^{-1}	$D_\perp/10^{-11}$ m^2 s^{-1}	$D_\parallel/D_\perp$	$E_{a\parallel}/$kJ mol^{-1}	$E_{a\perp}/$kJ mol^{-1}	$E_{a\,iso}/$kJ mol^{-1}	Ref
PAA	119	89	55	1.6	-	-	-	[22]
	-	-	-	-	33.9	36.2	41.5	[21]
MBBA	5	6.9	4.6	1.4	-	-	-	[20]
	28	-	2.9 - 7.1	1.8	-	-	-	[22]
5CB	30	6.4	2.8	2.3	-	-	-	[22]
	31.2	5.99	3.43	1.7	69.7	60.4	29.6	[23]
5OCB	35	4.1	1.4	2.9	-	-	-	[22]
	63.9	14.1	6.71	2.1	34.8	46.4	38.3	[23]
6OCB	72.5	19.4	7.89	2.5	36.1	24.8	30.9	[23]
7OCB	71.1	18.5	6.47	2.9	29.0	26.6	35.7	[23]
8OCB	75.6	11.6	7.02	1.7	43.1	36.3	36.4	[23]

(ii) All the diffusion components, $D_{\parallel}$ and $D_{\perp}$ in the nematic phase and D_{iso} in the isotropic liquid phase, exhibit an Arrhenius-like temperature dependence, namely

$$D_i = D_i^0 \exp\left(-\frac{E_{ai}}{RT}\right), \quad i = \parallel, \perp, iso \tag{10}$$

within the experimental error. The three activation energies $E_{a\parallel}$, $E_{a\perp}$ and $E_{a\,iso}$ are similar; they are typically 30 - 50 kJ mol^{-1}, depending on the compound.

(iii) The anisotropy in the diffusion, which may be defined by the ratio $D_{\parallel}/D_{\perp}$, is always more than unity for calamitic nematics. Molecules tend to diffuse along the director rather than in the direction perpendicular to it. The values of $D_{\parallel}/D_{\perp}$ are typically about two.

(iv) A discontinuity is observed for the components of the diffusion tensor at the nematic to isotropic transition temperature, T_{NI}. The relationship $D_{iso} = (D_{\parallel} + 2D_{\perp})/3$ does not hold if the value of D_{iso} is taken immediately above T_{NI}, and the values of $D_{\parallel}$ and $D_{\perp}$ are taken immediately below T_{NI}.

(v) As for the dependence on the molecular structure, changes in the core affect the diffusional properties more than a change in the terminal chain length does.

REFERENCES

[1] J. Crank [*The Mathematics of Diffusion* (Clarendon Press, Oxford, 1975)]
[2] J.P. Stark [*Solid State Diffusion* (John Wiley & Sons, New York, 1976)]
[3] C.K. Yun, A.G. Fredrickson [*Mol. Cryst. Liq. Cryst. (UK)* vol.12 (1970) p.73]
[4] A.G. Chmielewski [*Mol. Cryst. Liq.Cryst. (UK)* vol.212 (1992) p.205]
[5] F. Rondelez [*Solid State Commun. (USA)* vol.14 (1974) p.815]
[6] M. Hara, S. Ichikawa, H. Takezoe, A. Fukuda [*Jpn. J. Appl. Phys. (Japan)* vol.23 (1984) p.1420]
[7] K. Ohta, T. Terajima, N. Horota [*Bull. Chem. Soc. Jpn. (Japan)* vol.68 (1995) p.2809]
[8] H.W. Heuer, H. Kneppe, F. Schneider [*Ber. Bunsenges. Phys. Chem. (Germany)* vol.100 (1996) p.1818]
[9] B. Yoon, S.H. Kim, I. Lee, S.K. Kim, M. Cho, H. Kim [*J. Phys. Chem. B (USA)* vol.102 (1998) p.7705]
[10] D.A. Cleary, Y.K. Shin, D.J. Schneider, J.H. Freed [*J. Magn. Reson. (USA)* vol.79 (1988) p.474]
[11] Y.K. Shin, U. Ewert, D.E. Budil, J.H. Freed [*Biophys. J. (USA)* vol.59 (1991) p.950]
[12] G.J. Krueger, H. Spiesecke [*Z. Naturforsch. A (Germany)* vol.28 (1973) p.964]
[13] G.J. Krueger [*Phys. Rep. (Netherlands)* vol.82 (1982) p.229]
[14] H. Hakemi, M.M. Labes [*J. Chem. Phys. (USA)* vol.61 (1974) p.4020]
[15] H. Hakemi, M.M. Labes [*J. Chem. Phys. (USA)* vol.63 (1975) p.3708]
[16] H. Hakemi [*Liq. Cryst. (UK)* vol.3 (1988) p.453]
[17] P. Callaghan, A. Coy [in *Nuclear Magnetic Resonance Probes of Molecular Dynamics* Ed. R.

Tycko (Kluwer Academic, 1994) p.489]

[18]   J. Kaerger, H. Pfeifer, W. Heink [in *Advances in Magnetic Resonance* Ed. J.H. Waugh (Academic Press, 1988)]

[19]   P. Stilbs [*Prog. Nucl. Magn. Reson. Spectrosc. (UK)* vol.19 (1987) p.1]

[20]   I. Zupancic, J. Pirs, M. Luzar, R. Blinc, J.W. Doane [*Solid State Commun. (USA)* vol.15 (1974) p.227]

[21]   F. Noack [*Mol. Cryst. Liq. Cryst. (UK)* vol.113 (1984) p.247]

[22]   F. Noack [in *Handbook of Liquid Crystals* vol.1, Eds. D. Demus, J. Goodby, G.W. Gray, H.W. Spiess, V. Vill (Wiley-VCH, Weinheim, 1998) p.582]

[23]   F. Noack, St. Becker, J. Struppe [*Annu. Rep. NMR Spectrosc. (UK)* vol.33 (1997) p.1]

[24]   M. Zhou, L. Frydman [*Solid State Nucl. Magn. Reson. (Netherlands)* vol.4 (1995) p.301]

[25]   J. Martin, L.S. Selwyn, R.R. Vold, R.L. Vold [*J. Chem. Phys. (USA)* vol.76 (1982) p.2632]

[26]   D. Zax, A. Pines [*J. Chem. Phys. (USA)* vol.78 (1983) p.6333]

9.2 Rotational diffusion of liquid crystals in the nematic phase

R. Righini

February 2000

A INTRODUCTION

The orientational dynamics of liquid crystals in the nematic phase has been widely investigated using NMR and dielectric relaxation spectroscopy. Here, we briefly discuss the applications of the two techniques and present the experimental data concerning a few liquid crystals.

B NMR SPECTROSCOPY

The rotational relaxation of molecules in liquid crystals can be described on the basis of the rotational diffusion model, proposed by Nordio et al [1]. According to the model molecular reorientation is a Markov process in which the molecules are involved in a series of small angular steps, as a consequence of the collisions with the surrounding molecules. In an orientationally ordered phase, like the nematic, this motion takes place under the effect of an anisotropic potential, the nematic potential $U(\Omega)$, where Ω represents the set of angular variables defining the molecular orientation. The molecular angular diffusion is thus characterised by a diffusion tensor, $\mathbf{D}$, with diagonal elements D_{xx}, D_{yy} and D_{zz}. If cylindrical symmetry can be assumed for the molecules (an approximation that has been shown to be reasonable for many liquid crystals), only two elements are retained, i.e. $D_{xx} = D_{yy} = D_\perp$ and $D_{zz} = D_\parallel$. $D_\parallel$ and $D_\perp$ then describe the rotation about the molecular long axis, and the rotation about an axis orthogonal to it, respectively.

The time evolution of the angular variables Ω can be obtained by solving the diffusion equation

$$\frac{1}{\rho}\frac{\partial \hat{P}(\Omega_0 \mid \Omega_t)}{\partial t} = \hat{\Gamma}\,\hat{P}(\Omega_0 \mid \Omega_t) \tag{1}$$

where $\hat{P}(\Omega_0 \mid \Omega_t)$ is the conditional probability of finding the molecule with orientation Ω_t at time t, if its orientation was Ω_0 at $t = 0$. The rotational diffusion operator $\hat{\Gamma}$ can be expressed in terms of the three independent elements of the rotational diffusion tensor, D_{xx}, D_{yy} and D_{zz}.

Deuteron NMR spectroscopy has been widely employed [2-6] to determine experimental values for the diffusion tensor elements. The connection of NMR spectroscopy to the rotational diffusion model comes through the pertinent correlation functions. In fact, the orientational correlation functions for a tensor of rank L are written, in general, as

$$G_{mn}^{L}(t) = \int\int d\Omega_0\,d\Omega\,D_{mn}^{L}(\Omega_0)\,D_{mn}^{L*}(\Omega)\sqrt{P(\Omega_0)\,P(\Omega)}\;\hat{P}(\Omega_0 \mid \Omega_t) \tag{2}$$

where $P(\Omega)$ is the equilibrium probability of finding a molecule with orientation Ω, and D_{mn}^{L} are elements of the Wigner orientation matrices; m and n are projection indices of a tensor of rank L in the laboratory and molecular frames, respectively. In NMR spectroscopy L = 2, and the $G_{mn}^{2}(t)$ correlation function is simply related to the NMR spectral densities $J_1(\omega_0)$ and $J_2(2\omega_0)$ by a Fourier transformation. On the other hand, the spectral densities are directly obtained by measuring the spin-lattice longitudinal and quadrupolar relaxation times:

$$1/T_{1Z} = J_1(\omega_0) + 4J_2(2\omega_0)$$

(3)

$$1/T_{1Q} = 3J_1(\omega_0)$$

$\omega_0/2\pi$ being the Larmor frequency.

The solution of the diffusion equation (EQN (1)), obtained with an appropriate choice of the orienting potential, can then be compared to the experimentally determined correlation functions, and the elements of the diffusion tensor D_{xx}, D_{yy} and D_{zz} (or, more frequently, $D_\parallel$ and $D_\perp$) can thus be extracted by means of an optimisation procedure.

In FIGURES 1, 2 and 3 we have collected the components of the diffusion tensor obtained for 4-pentyl-4′-cyanobiphenyl (5CB) [6], 4-pentyloxybenzylidene-4′-heptylaniline (5O.7) [7] and 4-hexyloxy-4′-cyanobiphenyl (6OCB) [8] as a function of temperature. In the figures a third diffusion constant D_R is reported: in fact the deuterium NMR technique allows us to measure the relaxation times for the different lines corresponding to deuterium atoms at different locations in the molecule. In solving the diffusion equation then a third diffusion term D_R is included, describing the internal

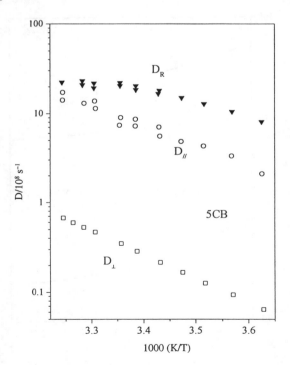

FIGURE 1 Temperature dependence of the components of the diffusion tensor for 5CB in the nematic phase. D_R describes the intramolecular rotation of the phenyl ring.

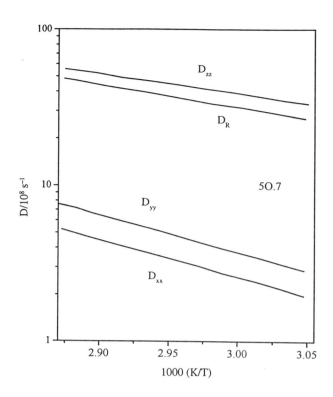

FIGURE 2 Temperature dependence of the components of the diffusion tensor for 5O.7 in the nematic phase. The diffusion equation was solved for an asymmetric rotator [9]; D_{zz} corresponds to the rotation about the molecular long axis. D_R describes the intramolecular rotation of the phenyl ring.

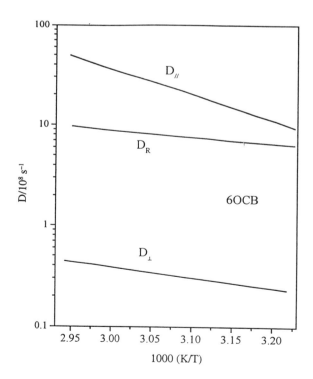

FIGURE 3 Temperature dependence of the components of the diffusion tensor for 6OCB in the nematic phase. D_R describes the intramolecular rotation of the phenyl ring.

466

rotation of the phenyl ring. Note that for 5O.7 in FIGURE 2 three independent elements of the diffusion tensor have been determined, using the treatment for an asymmetric diffuser [9]. Actually deuterium NMR, as a site-sensitive technique, has also been used to investigate the contribution of the alkyl chain motion to the overall relaxation [6]. The Arrhenius behaviour shown by the results of FIGURES 2 and 3 was directly imposed in the optimisation procedure; however no such assumption was made for the 5CB values reported in FIGURE 1.

The diffusion tensor for a nematic liquid is evidently a model-dependent quantity, and the numerical values obtained for its elements depend on the assumptions made in the calculation procedure. Actually the relation of $D_\parallel$ and $D_\perp$ to rotational relaxation times is not obvious. In fact, the correlation functions determined from the NMR experiments correspond to a weighted sum of decaying exponentials of the type [10]

$$G^2_{mn}(t) = \sum_k (\beta_{mn})_k \exp[\alpha_{mn}\tau_k t] \tag{4}$$

and the general relation of the time constants τ_k to the elements of the diffusion tensor is, for the first three terms:

$$\tau_0 = \frac{1}{6D_\perp}$$

$$\tau_1 = \frac{1}{5D_\perp + D_\parallel} \tag{5}$$

$$\tau_2 = \frac{1}{2D_\perp + 4D_\parallel}$$

The β and α coefficients in EQN (4) and the elements of the rotational diffusion tensor in EQN (5) are the results of the fitting procedure of the NMR data sketched previously, and their values depend on the details of the rotational diffusion model adopted. Expressions (5) are often written in terms of $\tau_\perp$ and $\tau_\parallel$ relaxation times, namely

$$\tau_\perp = \frac{1}{6D_\perp}, \quad \tau_\parallel = \frac{1}{6D_\parallel}$$

C   DIELECTRIC RELAXATION

In contrast to the NMR technique, dielectric relaxation measures the correlation function of a first rank tensor, the molecular permanent dipole moment, and the dipole rotational relaxation times can be directly extracted from the experiment. The technique has been widely used for investigating the anisotropic phases of liquid crystals [11]. Many liquid crystal molecules in fact possess a large dipole moment and have a high degree of rigidity of the molecular skeleton; if the dipole moment coincides with the molecular long axis, the time evolution of the dipole correlation function is a direct measure of the orientational dynamics of this axis.

Liquid crystals in the nematic phase are dielectrically anisotropic, and the real ($\tilde{\varepsilon}'$) and imaginary ($\tilde{\varepsilon}''$) parts of the dielectric permittivity have two independent components, corresponding to the parallel and perpendicular orientations, respectively, of the applied electric field with respect to the nematic director. The imaginary part of the permittivity [12] is given by the phenomenological Havriliak-Negami equation

$$\varepsilon'' = \text{Im} \frac{\delta\tilde{\varepsilon}}{\left[1 + (i\omega\tau)^\alpha\right]^\beta}$$

where $\delta\tilde{\varepsilon}$ is the dielectric strength and τ is the relaxation time; the parameters α and β describe the symmetric and asymmetric broadening of a distribution of Debye relaxation frequencies. If more than one relaxation process is active, the total dielectric loss is written as

$$\varepsilon'' = \frac{\sigma_o}{\varepsilon_o} \frac{1}{\omega^s} - \sum_j \text{Im}\left[\frac{\delta\tilde{\varepsilon}_j}{\left[1 + (i\omega\tau_j)^{\alpha_j}\right]^{\beta_j}}\right]$$

where ε_o is the vacuum permittivity; in the case of ohmic behaviour of the conductivity, $s = 1$ and σ_o is the DC conductivity. Dielectric spectroscopy gives access to a very broad frequency window, typically from 10^{-2} to 10^9 Hz.

In the following we consider the experimental data obtained for two liquid crystals belonging to the same family, i.e. 4-heptyl-4'-cyanobiphenyl (7CB) and 4-octyl-4'-cyanobiphenyl (8CB) in the bulk phase and in confined samples.

C1   Rotational Dynamics in the Bulk Phase

The nematic-isotropic transition temperature [13] for 7CB is 315.7 K, and for 8CB T_{NI} is 314.0 K. 8CB also has a smectic A phase, with a nematic-smectic transition of 294.3 K; crystallisation takes place at 302.7 K for 7CB and at 302.5 K for 8CB. Dielectric experiments have been performed by different authors for both compounds using frequency domain [14-16] and time domain [17,18] techniques.

In FIGURES 4 and 5 the available experimental data for the dielectric relaxation times are collected for 7CB and 8CB, respectively. The general behaviour is very similar in the two cases. In the isotropic phase a single relaxation time is measured, corresponding to the rotational dynamics of the molecules about their short axes. However, the relaxation becomes bimodal for temperatures lower than T_{NI}. In the nematic phase, dielectric data sets have been collected with the electric field parallel ($\tilde{\varepsilon}_\parallel$) and perpendicular ($\tilde{\varepsilon}_\perp$) to the nematic director. In the parallel geometry a single relaxation process is observed, characterised by a relaxation time which increases rapidly at low temperatures. A second relaxation process appears in the perpendicular geometry: its characteristic time is only slightly temperature dependent, and it is lower than the relaxation time measured in the isotropic phase.

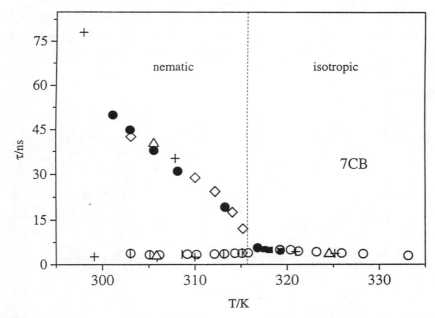

FIGURE 4 Temperature dependence of the rotational relaxation times for 7CB.

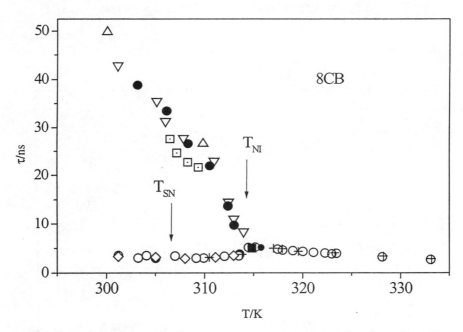

FIGURE 5 Temperature dependence of the rotational relaxation times for 8CB.

The semi-log plots in FIGURES 6 and 7 show the temperature dependence of the relaxation frequencies for the two compounds. The Debye behaviour of the isotropic phase with the expected Arrhenius temperature dependence is quite evident; in the oriented phases instead the temperature dependence of the two relaxation times is clearly non-linear on a log scale, thus preventing the definition of an activation energy for the relaxation processes involved. In TABLE 1 we have

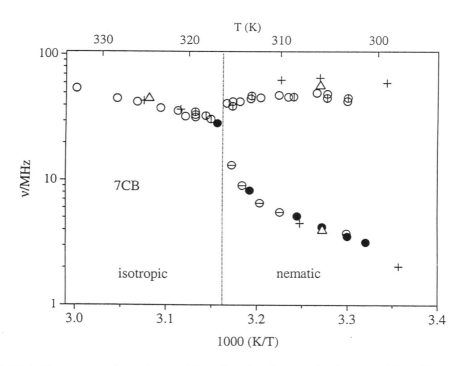

FIGURE 6 Temperature dependence of the relaxation frequencies (on a semi-logarithmic scale) for 7CB. The different symbols refer to different experiments [15,18,22].

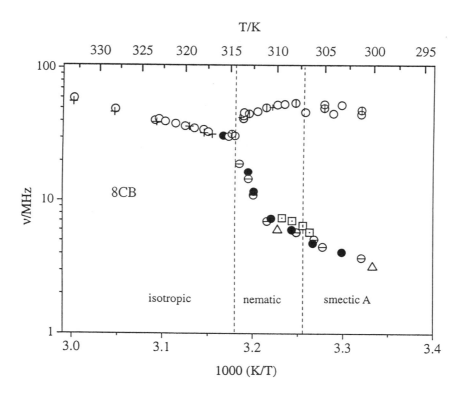

FIGURE 7 Temperature dependence of the relaxation frequencies (on a semi-logarithmic scale) for 8CB. The different symbols refer to different experiments [13-15,17,18,23].

TABLE 1 Rotational relaxation times for 7CB and 8CB.

7CB	Isotropic		Nematic		
T/K	333	317	315	310	303
$\tau/10^{-9}$ s	2.9 ± 0.2	5.6 ± 0.2	12.0 ± 0.5 4.0 ± 0.2	28.8 ± 0.8 3.4 ± 0.2	43.7 ± 1.0 3.5 ± 0.2

8CB	Isotropic		Nematic		Smectic A
T/K	333	315	313.5	310	303
$\tau/10^{-9}$ s	2.9 ± 0.2	5.6 ± 0.2	9.1 ± 0.5 3.8 ± 0.2	22.2 ± 1.0 3.0 ± 0.2	38.8 ± 1.0 3.0 ± 0.3

collected the values of the measured relaxation times at different temperatures; the estimated errors have been obtained from a comparison of the results of different experiments. It is worth noticing that the smectic A-nematic transition in 8CB has an almost negligible effect on the temperature dependence of the two relaxation times.

The interpretation of the experimental results of FIGURES 4 to 7 is now quite generally accepted [12-14,18,19]. In the aligned phases the dielectric spectra obtained with the electric field parallel to the director are peaked in the frequency region 3 - 20 MHz, depending on the temperature. The corresponding relaxation process is interpreted as due to the restricted rotation of the molecules about their short axes; the large increase of the rotational time and its non-Arrhenius behaviour are a direct consequence of the progressive establishment of the nematic potential which impedes the molecular rotation. The high frequency spectral feature observed in the perpendicular geometry is much less temperature dependent, being located around 50 - 60 MHz over all the temperature range for the two liquid crystals. The corresponding relaxation process involves the tumbling motion of the molecular long axis about the nematic director.

C2   Rotational Dynamics in Confined Samples

In recent years liquid crystals confined in micrometric and nanometric porous media have been widely investigated [19-21], in order to understand the changes induced by the confinement on the order parameters, phase transitions and dynamical properties of the material. Here we present the results obtained for 8CB in silica porous glass, with pore sizes of 100 and 1000 Å. In this case also the experimental data concerning the rotational dynamics of the mesogenic molecules were obtained by dielectric spectroscopy. The main difference with respect to the bulk phase is the appearance of a very low frequency component, corresponding to relaxation times at 295 K of 7.1×10^{-2} and 4.6×10^{-1} s for 8CB in 1000 Å and 100 Å pores, respectively. This relaxation is probably due to the orientation of the molecular long axis for the molecules in the wall-liquid interface. At higher frequencies, in the nematic and smectic temperature regions two main relaxation rates are observed. These appear similar to those characteristic of the bulk samples; the corresponding relaxation times at different temperatures are collected in TABLE 2.

TABLE 2 Bulk-like rotational relaxation times for 8CB in porous silica.

T/K	$\tau_1/10^{-9}$ s	$\tau_2/10^{-9}$ s
	8CB in 1000 Å pores	
300	46.3	1.81
310	23.5	1.75
	8CB in 100 Å pores	
300	55.6	1.91
310	19.5	0.80

The interpretation of the relaxation processes involved is the same as for the bulk material. The effect of the pore size is evident in the slowing down of the relaxation in the 100 Å pore samples. The temperature dependence of the slower relaxation time, τ_1, corresponding to the molecular rotation about the short axes, is shown in FIGURE 8.

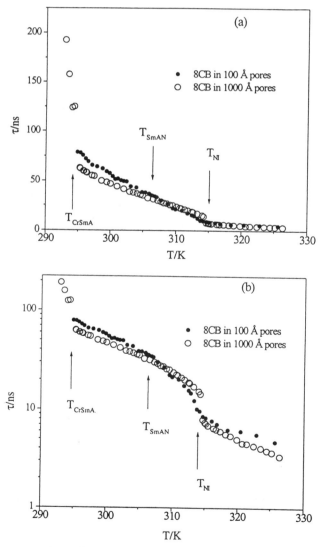

FIGURE 8 Temperature dependence of the rotational relaxation time for 8CB on a linear (a) and a logarithmic scale (b). Solid circles: in 100 Å pores; open circles: in 1000 Å pores.

The behaviour at the nematic-isotropic transition is much smoother for the smaller pore size, showing that the wall effect influences greatly the creation of the nematic order. The same effect is responsible for the incomplete crystallisation below T_{CrSmA}, resulting in the rotational dynamics observed (see FIGURE 8) well below the bulk crystallisation temperature.

REFERENCES

[1]   P.L. Nordio, P. Busolin [*J. Chem. Phys. (USA)* vol.55 (1971) p.5485]; P.L. Nordio, G. Rigatti, U. Segre [*J. Chem. Phys. (USA)* vol.56 (1972) p.2117]

[2]   R.R. Vold, R.L. Vold, N.M. Szeverenyi [*J. Phys. Chem. (USA)* vol.85 (1981) p.1934]; W.H. Dickerson, R.R. Vold, R.L. Vold [*J. Phys. Chem. (USA)* vol.87 (1983) p.166]

[3]   P.A. Beckman, J.W. Emsley, G.R. Luckhurst, D.L. Turner [*Mol. Phys. (UK)* vol.50 (1983) p.699]; P.A. Beckman, J.W. Emsley, G.R. Luckhurst, D.L. Turner [*Mol. Phys. (UK)* vol.59 (1986) p.97]

[4]   R.Y. Dong, G.M. Richards [*Mol. Cryst. Liq. Cryst. (UK)* vol.141 (1987) p.337]

[5]   R.R. Vold [in *NMR in Liquid Crystals* Ed. J.W. Emsley (Reidel, Dordrecht, 1985) p.253]

[6]   R.Y. Dong, G.M. Richards [*J. Chem. Soc. Faraday Trans. (UK)* vol.88 (1992) p.1885]

[7]   R.Y. Dong, X. Shen [*J. Chem. Phys. (USA)* vol.105 (1996) p.2106]

[8]   X. Shen, R.Y. Dong [*J. Chem. Phys. (USA)* vol.108 (1998) p.9177]

[9]   R. Tarroni, C. Zannoni [*J. Chem. Phys. (USA)* vol.95 (1991) p.4550]

[10]   R.R. Vold, R.L. Vold [*J. Chem. Phys. (USA)* vol.88 (1987) p.1443]

[11]   P.G. de Gennes [*The Physics of Liquid Crystals* (Clarendon Press, Oxford, 1974)]

[12]   S. Havriliak, S. Negami [*Polymer (UK)* vol.8 (1967) p.161]

[13]   J.M. Wacrenier, C. Druon, D. Lippens [*Mol. Phys. (UK)* vol.43 (1981) p.97]

[14]   C. Druon, J.M. Wacrenier [*J. Phys. (France)* vol.38 (1977) p.47]

[15]   D. Lippens, J.P. Parneix, A. Chapoton [*J. Phys. (France)* vol.38 (1977) p.1465]

[16]   C. Druon, J.M. Wacrenier [*Ann. Phys.* vol.3 (1978) p.199]

[17]   T.K. Bose, R. Chahine, M. Merabet, J. Thoen [*J. Phys. (France)* vol.44 (1984) p.1329]

[18]   T.K. Bose, R.B. Campbell, S. Yagihara, J. Thoen [*Phys. Rev. A (USA)* vol.36 (1987) p.5767]

[19]   Ch. Cramer, Th. Cramer, F. Kremer, R. Stannarius [*J. Chem. Phys. (USA)* vol.106 (1997) p.3730]

[20]   S. Rozanski, T. Stannarius, H. Grootens, F. Kremer [*Liq. Cryst. (UK)* vol.20 (1996) p.59]

[21]   G.P. Sinha, F.M. Aliev [*Phys. Rev. E (USA)* vol.58 (1998) p.2001]

[22]   M. Davies, R. Moutran, A.H. Price, M.S. Beevers, J. Williams [*J. Chem. Soc. Faraday Trans. (UK)* vol.72 (1976) p.1447]

[23]   B.R. Ratna, R. Shashidhar [*Mol. Cryst. Liq. Cryst. (UK)* vol.42 (1977) p.185]

9.3 Neutron scattering studies of dynamics in nematics

R.M. Richardson

November 1998

A INTRODUCTION

Incoherent quasi-elastic neutron scattering (IQENS) has been used to investigate the dynamics of nematic phases. When a monochromatic beam of neutrons is scattered from a nematic liquid crystal the shape of the elastic line in the energy-transfer spectra gives information on the molecular dynamics. A quasi-elastic neutron scattering measurement may be used to investigate the diffusive motions of the molecules in the nematic phase. The translational diffusion coefficients may be determined and localised motions such as molecular rotation and internal rotations may also be characterised.

B THEORETICAL BACKGROUND

When a neutron is scattered from a sample, it may exchange energy with the sample. If the scattered neutrons are analysed for energy transfer, spectra are obtained. For a normal crystalline sample of rigid molecules the spectra consist of a sharp elastic peak at zero energy transfer and inelastic peaks which arise from exchange of energy with phonon modes. For a liquid or liquid crystal sample, the elastic peak is generally broadened because of exchange of energy between the neutrons and the diffusive modes of the sample. In principle all molecular motions will cause movement of the scattering atoms and hence contribute to the shape of the quasi-elastic line in the spectra.

The incoherent scattered intensity is measured as a function of the scattering vector, $\mathbf{Q}$, and the energy transfer $\Delta E \equiv \hbar\omega$; it is known as the incoherent scattering law, $S_{inc}(\mathbf{Q},\omega)$, and is directly related to the motion of the scattering atoms. The motion may be described by a self-correlation function $G_S(\mathbf{r},t)$. If the scattering atom is at an arbitrary origin at $t = 0$, then $G_S(\mathbf{r},t)$ is the probability of finding the same atom at a displacement $\mathbf{r}$ from the origin at a later time, t, averaged over all choices of origin. The incoherent scattering law is the spatial and temporal Fourier transform of this correlation function [1]:

$$S_{inc}(\mathbf{Q},\omega) = \frac{\pi}{2} \int\int G_S(\mathbf{r},t)\exp\{i(\mathbf{Q}.\mathbf{r} - \omega t)\} \; d\mathbf{r} \; dt \qquad (1)$$

Normally the right hand side of this equation would be averaged over the different types of atom in a sample to obtain the scattering law. The usual way of interpreting the measured scattering law is to compare it with scattering laws that have been calculated for various models of molecular motion which have been discussed extensively elsewhere [2-4]. These comparisons may become complex because several motions contribute and because the $\mathbf{Q}$ and ω ranges as well as the resolution influence the observations. In the following sections we discuss briefly how the contributions of the different molecular motions to the observed scattering law may be identified. It is often possible to separate the contributions that the different motions make to the shape of the elastic line in the spectra.

(i)   Provided interference peaks that result from the structure of the phase are avoided, the neutron scattering from a nematic liquid crystal is incoherent. In organic molecules, the incoherent scattering is dominated by hydrogen atoms which have a much higher incoherent scattering cross-section ($\sigma_H = 80$ b) than any other atom. For instance, parts of a molecule may be rendered invisible to IQENS by substituting deuterium which has a much more typical cross-section ($\sigma_D = 2$ b). The spectra may be interpreted in terms of the motion of hydrogen atoms only.

(ii)   If the IQENS is measured from a monodomain sample, the spectra with scattering vector, $\mathbf{Q}$, parallel and perpendicular to the director give different information. Spectra for which $\mathbf{Q}$ is parallel to the director will be insensitive to motion perpendicular to it and vice versa. Although the perfection of molecular alignment is never very high for a nematic (i.e. $0.4 < <P_2> < 0.6$) this is useful in identifying the origin of broadening of the elastic line. For instance, rotation about the molecular long axes would contribute most broadening to spectra for which $\mathbf{Q}$ was perpendicular to the director.

(iii)   The shape of the broadened elastic line and its dependence on the magnitude of Q is also a powerful indicator of the type of motion. A major distinction is between localised motion such as rotation and those that explore larger distances such as translational diffusion. Translational diffusion gives a Lorentzian lineshape and the width of the line increases with Q^2. In fact, the full width at half maximum in energy transfer, ΔE, can be used to determine the translational diffusion coefficient, D_T:

$$\Delta E = 2\, \hbar\, D_T\, Q^2 \tag{2}$$

Rotational motion or other localised motions give rise to an unbroadened elastic component and a broadened quasi-elastic component that may be comprised of several Lorentzian terms. The width of the broadened term is inversely related to the characteristic time of the motion. The ratio of elastic to elastic plus quasi-elastic intensity varies with $\mathbf{Q}$ and is determined by the geometry of the motion. This ratio

$$EISF = \frac{I_e}{I_e + I_q} \tag{3}$$

is known as the elastic incoherent structure factor (EISF) and will decrease from unity over a Q range determined by the amplitude of the localised molecular motion. For instance rotational diffusion on a circle of radius a gives the EISF

$$EISF = J_0^2 \, (Qa \sin \theta_Q) \tag{4}$$

where J_0 is the zeroth order Bessel function of the first kind and θ_Q is the angle between the scattering vector and the axis of rotation. When applied to rotational diffusion of a molecule, the right hand side of this formula would be averaged over the different radii of the hydrogen atoms. If the rotation axes were perfectly aligned parallel to the director the EISF would be 1.0 for $\theta_Q = 0°$. For $\theta_Q = 90°$ it would be unity at $Q = 0$ and decreases to zero when the argument is 2.40. Since the second rank orientational order parameter in a nematic is typically $\langle P_2 \rangle \approx 0.5$ the observed EISFs for $\mathbf{Q} \| \mathbf{n}$ and $\mathbf{Q} \perp \mathbf{n}$ would be closer together than predicted by such an idealised model. At low Q, the full width at half maximum of the broadened component is related to the rotational diffusion coefficient

$$\Delta E = 2 \, \hbar \, D_R \tag{5}$$

C TECHNIQUES AND RESULTS FOR DIFFUSION TENSORS

The important instrumental factors for IQENS measurements are the $\mathbf{Q}$ range covered and the energy transfer resolution. For instance, to observe broadening due to translational diffusion alone it is necessary to make measurements at low Q. Low Q implies that the experiment is probing large distances and translational diffusion is the only diffusive motion that moves atoms over distances large compared with the size of a molecule. Therefore to measure translational diffusion, $Q \ll 2\pi/(\text{molecular size})$ is required. However, EQN (2) indicates that the broadening will be small at low Q and so an instrument with very high resolution is also required. The first instrument to be capable of this was the backscattering spectrometer at Jülich, Germany. This relied on backscattering from silicon single crystals to obtain the required energy transfer resolution of ~1 μeV. The method was later implemented on the IN10 spectrometer at the Institut Laue Langevin, Grenoble, France, and adapted to a pulsed neutron source at ISIS, Rutherford Appleton Laboratory, Oxfordshire, England. The backscattering results that are available for the nematic phase are summarised in TABLE 1. Measurements on monodomain samples give the two principal values of the diffusion tensor which correspond to diffusion parallel and perpendicular to the director, $D_\|$ and $D_\perp$. It should be noted that the real anisotropy of the coefficients is probably slightly greater than that observed because

TABLE 1 Typical results for the translational diffusion tensor of nematics.

Material	$T/°C$	$T_{NI}/°C$	$D_\|/10^{-10}$ m^2 s^{-1}	$D_\perp/10^{-10}$ m^2 s^{-1}	$\overline{D}/10^{-10}$ m^2 s^{-1}
PAA [6]	122	135		3.4 ± 0.2	4.1 ± 0.3
D-5CB [7]	23.5	35	0.53 ± 0.03	0.41 ± 0.03	
TBBA [8]	218	235			7.4
D-EABAC [5]	130	156	1.9	1.4	
D-IBPBAC [9]	208	214	3.6	4.4	
EBBA [10]	50	80		2.7	

PAA: 4,4′-dimethoxyazoxybenzene; D-5CB: deuteriated 4-pentyl-4′-cyanobiphenyl;

TBBA: terephthal-bis-butylaniline; D-EABAC: deuteriated 4-(4′-acetoxybenzylidene)aminocinnamate;

D-IBPBAC: deuteriated 4-isobutyl(4′-phenylbenzylidene)aminocinnamate;

EBBA: 4-ethoxybenzylidene-4′-butylaniline.

resolution effects tend to decrease the anisotropy. It has been shown that measurements on polydomain samples give the average:

$$\overline{D} \approx \left(D_{\parallel} + 2D_{\perp}\right)/3 \tag{6}$$

For measurements made at low Q, molecules with their terminal alkyl chains deuteriated give similar values for translational diffusion coefficients to the normal hydrogenous versions. It has generally been found that the values of the diffusion tensor measured by the high resolution neutron backscattering technique agree quite well with those determined by pulsed field gradient NMR [5] and tracer methods. Earlier measurements were not at sufficiently low Q and reported diffusion tensors were one or two orders of magnitude too large.

D TECHNIQUES AND RESULTS FOR LOCALISED MOLECULAR DYNAMICS

Incoherent quasi-elastic neutron scattering in the Q range 3 to 20 nm^{-1} reflects the geometry and dynamics of localised molecular motions. At $9 < Q/\text{nm}^{-1} < 18$ the incoherent scattering is contaminated by coherent scattering from the diffuse peak resulting from the short range order in a nematic phase which introduced some uncertainty in the interpretation. The measurements are made with energy transfer resolution of a few tens of μeV in order to resolve the elastic peak (which is generally broadened slightly by transitional diffusion) from the quasi-elastic peak (which results from the localised molecular motions). Unfortunately there is often some difficulty in making a detailed interpretation because there are several types of motion that contribute to the quasi-elastic broadening in this Q range. Separation of the components by lineshape analysis using an unbiased Bayesian approach has not been very promising [11]. Deuteriation of the terminal flexible chains and measurements of the EISF for $\mathbf{Q}\perp\mathbf{n}$ have been used to establish that rotational diffusion about the long molecular axes takes place in the smectic A and nematic phases. However, the spectra are remarkable in the similarity between those with $\mathbf{Q}\perp\mathbf{n}$ and those with $\mathbf{Q}\|\mathbf{n}$ and this suggests that there is a motion parallel to the director which has similar amplitude and timescale to the rotation. The experimental EISF for chain deuteriated 4-pentyl-4′-cyanobiphenyl (5CB) confirm the similarity of the scattering for the two directions of $\mathbf{Q}$.

It is also likely that the direction of the long molecular axes fluctuates on a similar scale since such motion has been observed in the underlying smectic phases in several materials [8,12]. The motion of hydrogen atoms in the rigid molecular cores will also be influenced by internal rotations of the terminal alkyl chains and the phenyl rings in the cores. TABLE 2 summarises the rotational diffusion coefficients that have been measured in the nematic phase by assuming that there is a coupling between the rotation about the long axis and the translation parallel to it so that a proton diffuses on an ellipse. The values are therefore model dependent to some extent but will be of the correct order of magnitude.

TABLE 2 The rotational diffusion coefficients for typical nematics.

Chain deuteriated	T/°C	$D_R/10^{-9}$ s^{-1}
EABAC [13]	130	25
5CB [14]	25	6
5CB	40 (isotropic)	10

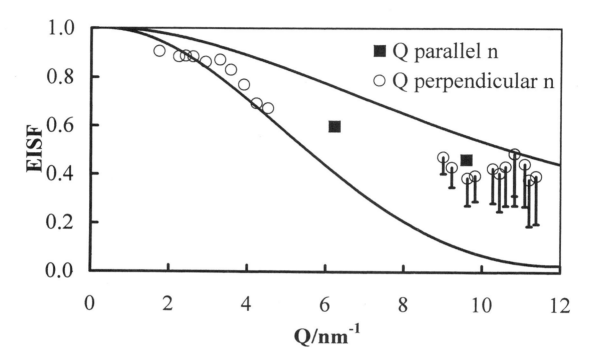

FIGURE 1 The experimental EISF for perdeuteriated 5CB with the values expected from a model of simple uniaxial rotational diffusion with the second rank orientational order parameter of 0.55. The error bars indicate the uncertainty introduced by the coherent scattering from the diffuse peak.

Quasi-elastic broadening has also been observed with coarser resolution (i.e. over 100 μeV) for 4,4′-dimethoxyazoxybenzene and its homologues. This is not able to resolve clearly the quasi-elastic scattering from rotational diffusion. The observed broadening has been interpreted in terms of internal rotations [15] that have short correlation times of a few picoseconds rather than the few tens to one hundred picoseconds normally associated with rotational diffusion about the molecular long axes.

E   CONCLUSION

Incoherent quasi-elastic neutron scattering measurements made at low Q and with very high energy transfer resolution give values of translational diffusion coefficients that are in good agreement with those determined by other methods. IQENS data from higher Q give information on the localised diffusive molecular motions. Interpretation is complex because of the many different motions that contribute to the measured scattering. Detailed analysis has established that rotational diffusion about the molecular long axis takes place on a timescale of about 100 ps and that motion parallel to these takes place on a similar timescale. Internal rotational relaxation of the molecules has a correlation time about one order of magnitude shorter.

REFERENCES

[1]   L. Van Hove [*Phys. Rev. (USA)* vol.95 (1954) p.249]

[2]   R.M. Richardson [in *The Molecular Dynamics of Liquid Crystals* Eds. G.R. Luckhurst, C.A. Veracini (Kluwer Academic Publishers, Dordrecht, 1994) ch.18]

[3]   F. Volino, A.J. Dianoux, H. Hervet [*Mol. Cryst. Liq. Cryst. (UK)* vol.38 (1977) p.125]

[4]   R.M. Richardson [in *Handbook of Liquid Crystals: Fundamentals* Eds. D. Demus, J. Goodby, G.W. Gray, H.W. Spiess, V. Vill (Wiley-VCH, Weinheim, 1998) p.680]

[5]   R.M. Richardson, A.J. Leadbetter, D.H. Bonsor, G.J. Kruger [*Mol. Phys. (UK)* vol.40 (1980) p.741]

[6]   J. Toepler, B. Alefeld, T. Springer [*Mol. Cryst. Liq. Cryst. (UK)* vol.26 (1974) p.297]

[7]   A.J. Leadbetter, F.P. Temme, A. Heidemann, W.S. Howells [*Chem. Phys. Lett. (Netherlands)* vol.34 (1975) p.363]

[8]   F. Volino, A.J. Dianoux, A. Heidemann [*J. Phys. (France)* vol.40 (1979) p.L583]

[9]   R.M. Richardson, A.J. Leadbetter, J.C. Frost [*Mol. Phys. (UK)* vol.45 (1982) p.1163]

[10]   M.P. Fontana, B. Rosi, M. Ricco [*Physica B (Netherlands)* vol.156-157 (1989) p.363]

[11]   J. Chrusciel, W. Zajac, C.J. Carlile [*Mol. Cryst. Liq. Cryst. (Switzerland)* vol.262 (1995) p.1649]

[12]   A.J. Dianoux, F. Volino [*J. Phys. (France)* vol.40 (1979) p.181]

[13]   A.J. Leadbetter, R.M. Richardson [*Mol. Phys. (UK)* vol.35 (1978) p.1191]

[14]   D.H. Bonsor, A.J. Leadbetter, F.P. Temme [*Mol. Phys. (UK)* vol.36 (1978) p.1805]

[15]   J.A. Janik, J.M. Janik, K. Otnes, T. Stanek [*Liq. Cryst. (UK)* vol.5 (1989) p.1045]

CHAPTER 10

SURFACES AND INTERFACES

10.1 Interfacial tension measurements for nematic liquid crystals

J. Springer and G.-H. Chen

March 1999

A   INTRODUCTION

The interface is a region of finite thickness (usually less than 0.1 μm) in which the composition and energy vary continuously from one bulk phase to the other. The types of interfaces can be summarised in a formal way in terms of the three states of matter - solid, liquid, and gas: gas-liquid, gas-solid, liquid-liquid, liquid-solid, and solid-solid. Thermodynamically, the interfacial tension γ is defined as the Gibbs free energy G of the system, which is required to create a unit interfacial area A at constant temperature T, pressure p, and total number of moles n in the system:

$$\gamma = \left(\frac{\partial G}{\partial A} \right)_{T,p,n} \tag{1}$$

For liquid crystalline substances, γ depends not only on the molecular interactions but also on the orientational order of the mesogenic molecules. Here, we restrict our discussion to the gas-liquid interface, on which interfacial tension measurements have mostly been performed. The corresponding interfacial tension is normally called the surface tension.

In principle, all the methods developed for the determination of surface tension of common liquids [1-5] are applicable to liquid crystals, especially for the nematic and isotropic phase states, with their low viscosities. That is why there have been a variety of methods used: e.g. the maximum bubble pressure method [6], the capillary height method [7,8], the DuNouy ring method [9,10], the Wilhelmy plate method [11-15], the sessile drop method [16], and the pendant drop method [17-27]. However, the rate of equilibrium is a limiting factor, since not all of the methods allow sufficient time for equilibration of the molecular arrangement at the surface, e.g. the maximum bubble pressure method. Another limiting factor is the liquid crystal-solid interactions. Due to their high sensitivity, the molecular order and orientation at liquid crystal surfaces will be seriously influenced by the solid surfaces with which they are in contact. For the capillary height, DuNouy ring, Wilhelmy plate and sessile drop methods, these solid surfaces are the glass capillary, the ring, the plate and the substrate, respectively. It has been found that the surface tension values obtained by using these methods are influenced by different aligning conditions [28-31].

In the case of the pendant drop method, however, the contact area between the liquid crystal drop and the syringe needle is very small. This results in a maximal elimination of the solid-liquid crystal interactions [17-28]. This small contact is also advantageous for a hanging drop to reach hydrostatic equilibrium in a relatively short time, which is especially important for viscous fluids, e.g. polymeric liquid crystals [2,21,22]. Moreover, this method has proved to be convenient for the investigation of the kinetics of the establishment of equilibrium of the surface tension [23,25,26]. For these reasons, the pendant drop method is considered to be the most successful for the surface/interfacial tension measurements of nematics. FIGURE 1 represents schematically a computer-aided measuring setup based on the pendant drop method. The surface/interfacial tensions are determined by fitting the experimental profiles recorded by on-line video camera to the Laplace-Young equation, which governs the profile of a liquid drop at its hydromechanical equilibrium:

$$\Delta p = \gamma \left(\frac{1}{R_1} + \frac{1}{R_2} \right) \tag{2}$$

Here Δp is the pressure difference across the interface, while R_1 and R_2 represent the two principle radii of curvature. The principle and the procedures for data analysis are given in detail elsewhere [1-5,32-38].

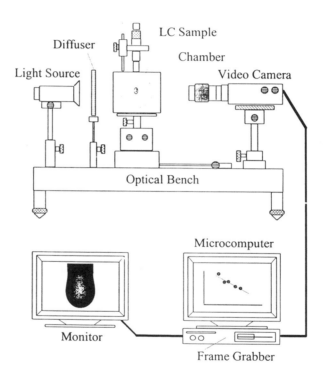

FIGURE 1   Schematic diagram of an experimental
setup based on the pendant drop method.

B   RESULTS

Since the pioneering work conducted by Schenck [39] and Jaeger [6], numerous surface tension measurements have been performed on low molar mass nematics by using various experimental techniques [7-28,40-46] (short reviews on this subject are available in [15] and [23]). Attention has been mainly paid to the change of surface tension at the nematic-isotropic phase transition. In contrast to common liquids, most of the temperature-scanning measurements ($\gamma(T)$-curves) show, with increasing temperature, positive values or discontinuities in the temperature coefficient $d\gamma/dT$. This unusual character is often referred to as the so-called anomaly. Employing the well-known Maxwell relation for a one-component system

$$d\gamma/dT = -[S^{surface} - S^{bulk}] = -S^{\sigma} \tag{3}$$

where $S^{surface}$ and S^{bulk} are the surface and the bulk entropy per unit area, respectively; it can be concluded that the surface excess entropy S^{σ} is negative. This implies that the molecular arrangement must be more ordered near the free surface than the bulk and so suggests there is surface-enhanced order [13,47-53]. Based on this principle, temperature-dependent measurements on surface tension provide the possibility of gaining an insight into pre-transitional phenomena at surfaces [13,23,24,26,54-63].

For a few frequently investigated nematogens, namely 4,4′-dimethoxyazoxybenzene (PAA), p-anisaldazine and 4,4′-diethoxyazoxybenzene (PAP), comparisons between the available γ(T)-results, obtained by different authors with different experimental methods, are shown in FIGURES 2 - 4.

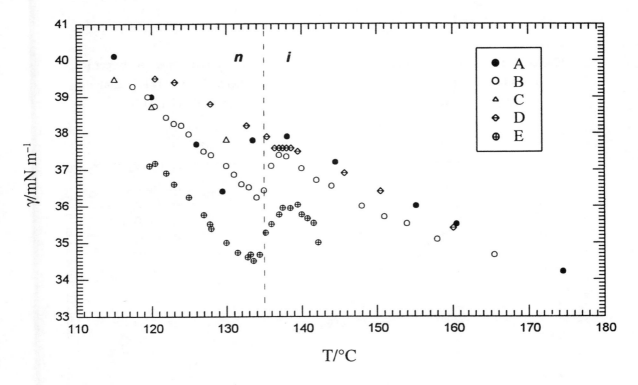

FIGURE 2 Temperature dependence of the surface tension of PAA (T_{NI} approx. 135°C). A: Jaeger [6], maximum bubble pressure method; B: Ferguson and Kennedy [7], capillary height method; C: Naggiar [44], radii of curvature (drop profile method); D: Schwarz and Moseley [9], DuNouy ring method; E: Krishnaswamy and Shashidhar [17], pendant drop method.

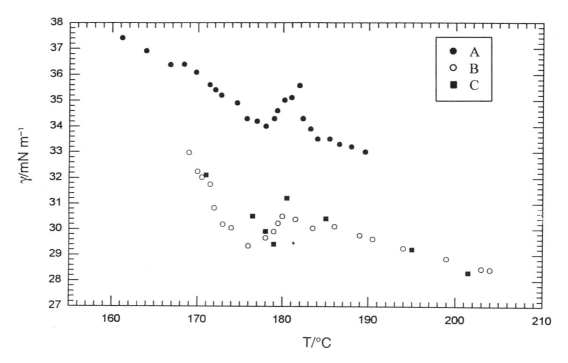

FIGURE 3 Temperature dependence of the surface tension of p-anisaldazine. A: Jaeger [6], maximum bubble pressure method (T_{NI} = 180°C); B: Ferguson and Kennedy [7], capillary height method (T_{NI} = 180.5°C); C: Krishnaswamy and Shashidhar [17], pendant drop method (T_{NI} = 180°C).

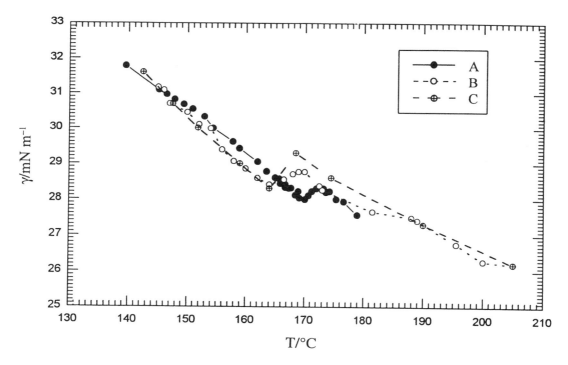

FIGURE 4 Temperature dependence of the surface tension of PAP. A: Krishnaswamy and Shashidhar [17], pendant drop method (T_{NI} = 166.6°C); B: Ferguson and Kennedy [7], capillary height method (T_{NI} = 162°C); C: Jaeger [6], maximum bubble pressure method (T_{NI} = 168°C).

486

As seen, the results are inconsistent not only in respect of the surface tension-temperature-behaviour but also from the aspect of the absolute values at a given temperature. TABLE 1 also summarises the surface tension values of the most investigated nematogen 4-methoxybenzylidene-4′-butylaniline (MBBA), published by different authors. Although nearly all the authors have given for their methods an absolute error of ~1 - 1.5%, the difference in surface tension amounts to about 30%. Further data for other nematics are listed in TABLE 2.

The inconsistency of the experimental data is supposed to be attributed mainly to the high sensitivity of the molecular arrangement at the surface, especially when the system is relatively highly ordered. Since, for nematics, there is an anisotropy of the surface tension, the different order and orientation of the surface molecules will cause a change in the surface tension [28]. For example, for MBBA, the surface with molecules parallel to it has a higher surface tension value ($\gamma_\parallel$) than that with perpendicularly aligned molecules ($\gamma_\perp$), i.e. $\gamma_\parallel > \gamma_\perp$ [8,65,66]. Accordingly, various factors such as the experimental method, measuring conditions, and sample purity can considerably affect the surface-molecular order and orientation, and consequently the surface tension.

TABLE 1 Surface tension values of MBBA (T_{NI} = 43 - 47°C) obtained by different authors by using various measuring methods.

T/°C/phase	γ/mN m^{-1}	Method*	Remarks[+]	Ref
23/N	38	?		[41]
23/N	35.8	PD	Effect of impurities is slight	[42]
?/N	-	Wilhelmy	Results are not reproducible	[13]
25/N	32 - 34	Ring		[10]
?/N	28.8	?	$\gamma^d = 20$ mN m^{-1}, $\gamma^p = 8.8$ mN m^{-1}	[43]
<T_{NI}/N	20/32.5	Capillary	$\gamma_\perp = 20$ mN m^{-1}, $\gamma_\parallel = 32$ mN m^{-1}	[8]
?/N	32	Wilhelmy		[15]
23/N	38	LS	Director aligns at an angle of 75° to the free surface, $\gamma_\parallel > \gamma_\perp$	[40]
25/N	34	SD		[16]
23/N	34	Ring	$\gamma^d = 33$ mN m^{-1}, $\gamma^p = 1$ mN m^{-1}	[64]
25/N	35.5	PD		[18]
50/I	32.5			
25/N	31.8	PD	In 1 bar N$_2$	[55,56]
50/I	31.4			

*? = unknown method; PD = pendant drop method; Wilhelmy = Wilhelmy plate method; Ring = DuNouy ring method; Capillary = capillary height method; LS = light scattering spectrum; SD = sessile drop method.
[+]γ^d = dispersive contribution of γ; γ^p = polar contribution of γ; $\gamma_\perp = \gamma$ for the surface with perpendicular ordered molecules; $\gamma_\parallel = \gamma$ for the surface with parallel ordered molecules.

487

TABLE 2 Surface tensions of some nematic liquid crystals.

LC/T_{NI}	T/°C/phase	γ/mN m^{-1}	Method*	Ref
4,4′-dimethoxyazoxybenzene	120/N	39	Bubble	[6]
(PAA) (T_{NI} = 135°C)	160/I	35.5		
	120/N	39	Capillary	[7]
	160/I	35		
	120/N	39.5	Ring	[9]
	160/I	35.5		
	120/N	38.7	RC	[44]
	160/I	37.8		
	120/N	37.2	PD	[17]
	142/I	34.8		
4,4′-diethoxyazoxybenzene	160/N	28.8	Bubble	[6]
(PAP) (T_{NI} = 162 - 168°C)	175/I	28.6		
	160/N	28.8	Capillary	[7]
	175/I	28.4		
	158/N	29.8	Ring	[9]
	160/N	29.5	PD	[18]
	175/I	28.4		
p-anisaldazine	170/N	32.2	Bubble	[6]
(T_{NI} = 180 - 182°C)	190/I	29.8		
	170/N	32	Capillary	[7]
	190/I	29.8		
	170/N	36.4	PD	[17]
	190/I	33.2		
4-pentyl-4′-cyanobiphenyl	25/N	27.9	Wilhelmy	[13]
(5CB) (T_{NI} = 35.5°C)	45/I	28.6		
	25/N	32.6	PD	[24]
	45/I	32.2		
4-octyl-4′-cyanobiphenyl	35/N	25.5	Wilhelmy	[13]
(8CB) (T_{NI} = 39.5°C)	45/I	26.4		
4-cyanobenzylidene-4′-octyloxyaniline	100/N	26	PD	[19]
(CBOOA) (T_{SmAN} = 83.2°C, T_{NI} = 107°C)	110/I	26.4		
4-ethoxybenzylidene-4′-butylaniline	50/N	21.7	PD	[24]
(EBBA) (T_{NI} = 78°C)	90/I	20.8		

*Bubble = maximum bubble pressure method; Capillary = capillary height method;
Ring = DuNouy ring method; RC = radii of curvature (drop profile method);
PD = pendant drop method; Wilhelmy = Wilhelmy plate method.

Employing different methods always means different solid-nematic interactions which are particularly dominant for the methods of capillary height, DuNouy ring, Wilhelmy plate and sessile drop [28]. In fact, even under identical experimental conditions, the different history of the nematic samples can lead to a considerable dispersion of the data [57,58]. The effect of impurities is the most frequently discussed factor influencing the experimental results. Some authors found that the presence of impurities will seriously affect the surface tension of nematics. For example, Neumann et al [11,12] have found that the temperature dependence of the surface tension for PAA is very sensitive to the purity of the sample. For a highly purified sample the variation of surface tension with temperature exhibits a similar anomaly to that observed by Jaeger, Ferguson and Kennedy, while that obtained from a relatively less pure PAA sample has a rather normal form. Moreover, they noticed that the surface tension of the less pure sample changed with time over a period of several hours at a given temperature; this was supposed to be caused by the strong adsorption of impurities on the liquid surface. In opposition to Neumann et al, Krishnaswamy and Shashidhar [17-20] observed that the sample purity does not critically affect the temperature-dependent characteristics of the nematics which they investigated. Recently, the influence of the surrounding gases has been recognised by Springer et al [23,25-27,55,56,67-70]. The underlying reason is the difference in gas solubility which has been verified to be a function of temperature, liquid crystal phase, gas type and gas pressure. FIGURE 5 shows, as an example, the remarkable time dependence of the surface tension of a freshly formed MBBA drop, caused by the gas sorption. Such time dependence is supposed to have confused the early measurements of the surface tension of nematics and to have contributed to some of the inconsistencies in the experimental results so far. The gas sorption also leads to fluctuations in the bulk properties of liquid crystals, e.g. the phase transition temperatures.

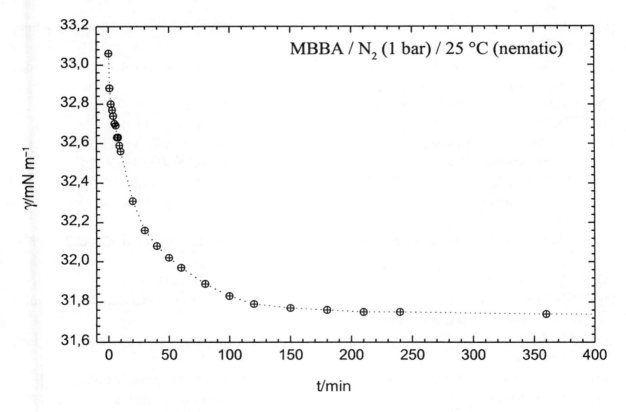

FIGURE 5 Time dependence of the surface tension of MBBA.

Surface tension measurements in polymeric liquid crystalline systems have rarely been reported possibly because such experiments are often limited by the high viscosity of the samples. Compared with low molar mass liquid crystals, much more time and care are required for the measurements on polymeric materials. The reports so far are restricted to several liquid crystalline side-group polymers, e.g. side-group polyacrylates [21,24,27,70] and polymethacrylates [22]. The γ(T)-characteristics of the polymeric samples studied are largely comparable with those of low molar mass nematics.

C   CONCLUSION

Surface tension measurements provide important information on the surfaces of liquid crystalline materials. The results are strongly dependent on the employed technique, the experimental condition, the degree of purification of the nematic samples and the sorption of surrounding gases. The dispersion of the data reported so far is considerable.

REFERENCES

[1]   N.K. Adam [in *The Physics and Chemistry of Surfaces* (Dover Publications, Inc., New York, 1968)]

[2]   S. Wu [in *Polymer Interface and Adhesion* (Marcel Dekker Inc., Basel, New York, 1982)]

[3]   A.W. Adamson [in *Physical Chemistry of Surfaces* 5th ed. (Wiley Interscience, New York, 1990)]

[4]   R.J. Hunter [in *Introduction to Modern Colloid Science* (Oxford Science Publication, Oxford, New York, Melbourne, 1993)]

[5]   A.W. Neumann, J.K. Spelt (Eds.) [in *Applied Surface Thermodynamics* (Marcel Dekker, Inc., New York, Basel, Hongkong, 1996)]

[6]   F.M. Jaeger [*Z. Anorg. Allg. Chem. (Germany)* vol.101 (1917) p.1]

[7]   A. Ferguson, S.J. Kennedy [*Philos. Mag. (UK)* vol.26 (1938) p.41]

[8]   B. Stryla, W. Kuczynski, J. Malecki [*Mol. Cryst. Liq. Cryst. Lett. (UK)* vol.1 (1985) p.33]

[9]   W.M. Schwarz, H.W. Moseley [*J. Phys. Coll. Chem. (USA)* vol.51 (1947) p.826]

[10]   L.T. Creagh, A.R. Kmetz [*Mol. Cryst. Liq. Cryst. (UK)* vol.24 (1973) p.59]

[11]   A.W. Neumann, P.J. Shell [in *5th Int. Congress Surface Activity* Barcelona, vol.2 (1968) p.125]

[12]   A.W. Neumann, R.W. Springer, R.T. Bruce [*Mol. Cryst. Liq. Cryst. (UK)* vol.27 (1974) p.23]

[13]   M.G.J. Gannon, T.E. Faber [*Philos. Mag. A (UK)* vol.37 (1978) p.117]

[14]   B. Tamamushi [in *Proc. VIth Int. Congress on Surface Active Substances* Zürich, 1972, vol.II (Carl Hanser Verlag, München, 1973) p.431]

[15]   B. Tamamushi [in *Colloid and Interface Science* Ed. M. Kerker, vol.5 (Academic Press Inc., New York, 1976) p.453]

[16]   D. Pabitra, K. Grzegorz, W.L. Aaron [*J. Colloid Interface Sci. (USA)* vol.82 (1981) p.1]

[17]   S. Krishnaswamy, R. Shashidhar [in *Proc. Int. Liquid Crystal Conf.* Bangalore, India, 1973, vol.1 (Pramana Supplement, India, 1975) p.247]

[18]   S. Krishnaswamy, R. Shashidhar [*Mol. Cryst. Liq. Cryst. (UK)* vol.35 (1976) p.253]

[19]   S. Krishnaswamy, R. Shashidhar [*Mol. Cryst. Liq. Cryst. (UK)* vol.38 (1977) p.353]

[20]   S. Krishnaswamy [in *Liquid Crystals* Ed. S. Chandrasekhar (Heyden & Son Ltd., London, Philadelphia, Rheine, 1980) p.487]

[21] M. Uzman, B. Song, T. Ruhnke, H. Cackovic, J. Springer [*Makromol. Chem. (Germany)* vol.192 (1991) p.1129]

[22] T. Runke, B. Song, J. Springer [*Ber. Bunsenges. Phys. Chem. (Germany)* vol.89 (1994) p.508]

[23] B. Song, J. Springer [*Mol. Cryst. Liq. Cryst. (Switzerland)* vol.293 (1997) p.39]

[24] B. Song, J. Springer [*Mol. Cryst. Liq. Cryst. (Switzerland)* vol.307 (1997) p.69]

[25] G.-H. Chen, J. Springer [*Mol. Cryst. Liq. Cryst. (Switzerland)* vol.307 (1997) p.89]

[26] G.-H. Chen, J. Springer [*Mol. Cryst. Liq. Cryst. (Switzerland)* vol.312 (1998) p.203]

[27] G.-H. Chen, J. Springer [*Macromol. Rapid. Commun. (Germany)* vol.19 (1998) p.625]

[28] J. Cognard [*Mol. Cryst. Liq. Cryst. Suppl. Ser. (UK)* vol.A5 (1982) p.1]

[29] A.A. Sonin [in *The Surface Physics of Liquid Crystals* (Gordon and Breach Publishers, Luxembourg, 1995)]

[30] J.D. Parsons [*Mol. Cryst. Liq. Cryst. (UK)* vol.31 (1975) p.79]

[31] J.D. Parsons [*J. Phys. C (UK)* vol.37 (1976) p.1187]

[32] S. Hartland, R. Hartley [in *Axisymmetric Fluid-Liquid Interfaces* (Elsevier Scientific Publishing Company, Amsterdam, Oxford, New York, 1979)]

[33] J.M. Andreas, E.A. Hauser, W.B. Tucke [*J. Phys. Chem. (USA)* vol.42 (1938) p.1001]

[34] H.H. Girault, J. Schriffrin, B.D.V. Smith [*J. Colloid Interface Sci. (USA)* vol.101 (1984) p.257]

[35] Y. Rotenberg, L. Boruvka, A.W. Neumann [*J. Colloid Interface Sci. (USA)* vol.93 (1983) p.169]

[36] S.H. Anatasiadis, J.-K. Chen, J.T. Koberstein, A.F. Siegel [*J. Colloid Interface Sci. (USA)* vol.119 (1987) p.55]

[37] B. Song, J. Springer [*J. Colloid Interface Sci. (USA)* vol.184 (1996) p.64]

[38] B. Song, J. Springer [*J. Colloid Interface Sci. (USA)* vol.184 (1996) p.77]

[39] R. Schenck [*Z. Phys. Chem. (Germany)* vol.25 (1898) p.337]

[40] M.A. Bouchiat, D. Langevin-Cruchon [*Phys. Lett. A (Netherlands)* vol.34 (1971) p.331]

[41] E. Proust, L. Ter-Minassian-Saraga [*J. Phys. (France)* vol.36 (1975) p.77]

[42] I. Haller [*Appl. Phys. Lett. (USA)* vol.24 (1974) p.394]

[43] F.J. Kahn, G.N. Taylor, H. Schonhorn [*Proc. IEEE (USA)* vol.61 (1973) p.823]

[44] V. Naggiar [*Ann. Phys. (France)* vol.18 (1943) p.5]

[45] K. Hatsuo, N. Huzio [*J. Phys. Soc. Jpn. (Japan)* vol.55-12 (1986) p.4186]

[46] D. Langevin [*J. Phys. (France)* vol.33 (1972) p.249]

[47] M.M. Telo Da Gama [*Mol. Phys. (UK)* vol.52 (1984) p.585]

[48] C.A. Croxton, S. Chandrasekhar [in *Proc. Int. Liquid Crystal Conf.* Bangalore, India, 1973, vol.1 (Pramana Supplement, India, 1975) p.237]

[49] C.A. Croxton [in *Liquid State Physics* (Cambridge University Press, UK, 1974) p.168]

[50] C.A. Croxton [*Phys. Lett. A (Netherlands)* vol.72 (1979) p.136]

[51] C.A. Croxton [in *Statistical Mechanics of the Liquid Surface* (John Wiley & Sons, Chichester, 1980)]

[52] L. Dufour, R. Defay [in *Thermodynamics of Clouds* (Academic Press, New York, London, 1963)]

[53] J.S. Rowlinson, B. Widom [in *Molecular Theory of Capillarity* (Clarendon Press, Oxford, 1982)]

[54] J.C. Earnshaw, C.J. Hughes [*Phys. Rev. A (USA)* vol.46 (1992) p.4494]

[55]   G.-H. Chen, J. Springer [*Mol. Cryst. Liq. Cryst. (Switzerland)* vol.325 (1998) p.185]

[56]   G.-H. Chen, J. Springer [*Mol. Cryst. Liq. Cryst. (Switzerland)* vol.326 (1998) p.1]

[57]   B. Song, G.-H. Chen, J. Springer, W. Thyen, P. Zugenmaier [*Mol. Cryst. Liq. Cryst. (Switzerland)* vol.323 (1998) p.89]

[58]   G.-H. Chen, J. Springer, W. Thyen, P. Zugenmaier [*Mol. Cryst. Liq. Cryst. (Switzerland)* vol.325 (1998) p.185]

[59]   D. Beaglehole [*Mol. Cryst. Liq. Cryst. (UK)* vol.89 (1982) p.319]

[60]   S. Immerschitt, T. Koch, W. Stille, G. Strobl [*J. Chem. Phys. (USA)* vol.96 (1992) p.6249]

[61]   H. Elben, G. Strobl [*Macromolecules (USA)* vol.26 (1993) p.1013]

[62]   P.S. Pershan, A. Braslau, A.H. Weiss, J. Als-Nielsen [*Phys. Rev. A (USA)* vol.35 (1987) p.4800]

[63]   B.M. Ocko, A. Braslau, P.S. Pershan, J. Als-Nielsen, M. Deutsch [*Phys. Rev. Lett. (USA)* vol.57 (1986) p.94]

[64]   P. Datta, A.W. Levine, G. Kaganowicz [in *150th Electrochem. Soc. Meeting* Las Vegas, 1976, no.215]

[65]   S. Naemura [*Appl. Phys. Lett. (USA)* vol.33 (1978) p.1]

[66]   S. Naemura [*J. Phys. Lett. Appl. (USA)* vol.40 (1979) p.514]

[67]   D.-S. Chen, G.-H. Hsiue, J.D. Schultze, B. Song, J. Springer [*Mol. Cryst. Liq. Cryst. (UK)* vol.7 (1993) p.85]

[68]   D.-S. Chen, G.-H. Hsiue [*Polymer (UK)* vol.35 (1994) p.2808]

[69]   G.-H. Chen, J. Springer [*Mol. Cryst. Liq. Cryst. (Switzerland)* vol.339 (2000) p.31]

[70]   G.-H. Chen, J. Springer [*Macromol. Chem. Phys. (Germany)* in press (2000)]

10.2   Anchoring energies for nematics

A. Sugimura

September 1998

A   INTRODUCTION

The structure of liquid crystalline phases in close proximity to an interface is different from that in the bulk, and this induced surface structure changes the boundary conditions and influences the behaviour of the liquid crystal in the bulk. The nematic phase is especially sensitive to external agents, in particular to surface forces [1]. Macroscopically, the surface effects are manifested by the director orientation in the bulk. There are two cases of particular interest: first the strong anchoring case, in which the director near the surface adopts a fixed orientation **e**, which is called the anchoring direction or the easy direction as termed by de Gennes [2], and secondly the weak anchoring case where the surface forces are not strong enough to impose a well-defined director orientation **n** at the surface, which is the situation for the majority of systems. When there are other fields (electric, magnetic, and flow) the director at the interface obviously deviates from the easy direction. The anisotropic surface anchoring energy reflects the ability of the surface director to deviate from the easy direction. To describe a weak anchoring surface for an untwisted nematic sample, Rapini and Papoular have introduced a simple phenomenological expression for the interfacial energy per unit area for a one-dimensional deformation [3]:

$$g_s = \frac{A}{2}\sin^2 \delta\theta \tag{1}$$

Here $\delta\theta$ is the angle between the easy direction **e** and the director **n** at the nematic/wall interface. The anchoring strength parameter A determines the ability of the surface director to deviate from the easy direction.

B   RAPINI-PAPOULAR SURFACE ANCHORING ENERGY

We consider a nematic cell located between the two planes located at $X_3 = 0$ and $X_3 = \ell$ as illustrated schematically in FIGURE 1. The easy directions at the top and bottom substrate surfaces are denoted by the unit vectors $\mathbf{e}^+$ and $\mathbf{e}^-$, respectively. Generally the Rapini-Papoular energy has been written as a linear combination of a polar angle anchoring energy

$$g_{s\theta}^- = \frac{A_\theta^-}{2}\sin^2\left(\theta^{o-} - \theta_o^-\right) \tag{2}$$

and an azimuthal angle anchoring energy

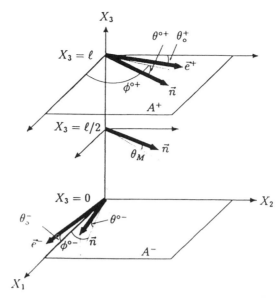

FIGURE 1 The geometry of the twisted chiral nematic cell
located between two planes $X_3 = 0$ and $X_3 = \ell$ [4].

$$g_{s\phi}^- = \frac{A_\phi^-}{2} \sin^2\left(\phi^{o-} - \phi_o^-\right) \tag{3}$$

where A_θ^- and A_ϕ^- are the anchoring strengths for the polar and azimuthal angles respectively at the bottom substrate surface. The anchoring energies $g_{s\theta}^+$ and $g_{s\phi}^+$ for the polar and azimuthal angles at the top substrate surface are expressed in the same manner. The total excess free energy density in the bulk is expressed as [2]

$$g_b = \frac{1}{2}\left[K_1(\nabla \cdot \mathbf{n})^2 + K_2\left(\mathbf{n} \cdot \nabla \times \mathbf{n} + \frac{2\pi}{p_o}\right)^2 + K_3(\mathbf{n} \times \nabla \times \mathbf{n})^2 \right] + g_f(\mathbf{n}) \tag{4}$$

where K_1, K_2 and K_3 are the splay, twist, and bend elastic constants of the nematic respectively, p_o denotes the pitch of the material induced by a chiral dopant, and $g_f(\mathbf{n})$ represents the interaction energy between the director and an external field which depends on $\mathbf{n}$ but not on $\nabla \mathbf{n}$. The total free energy F is the sum of the bulk energy and the surface free energy

$$F = \int g_b dV + \int g_{s\theta}^- dS^- + \int g_{s\phi}^- dS^- + \int g_{s\theta}^+ dS^+ + \int g_{s\phi}^+ dS^+ \tag{5}$$

where dV is the volume element of the bulk and dS is the surface area element. Minimisation of the total free energy yields the stable director configuration. The equilibrium condition is then determined by the variation equation $\delta F = 0$, that is by solutions of the Euler-Lagrange equations.

We consider a simple nematic cell located between the two planes $X_3 = 0$ and $X_3 = \ell$ with a symmetry with respect to the middle plane $X_3 = \ell/2$. The surface tilt and azimuthal angles are taken to be the same at both surfaces, $\theta^o = \theta^{o+} = \theta^{o-}$, $\theta_o = \theta_o^+ = \theta_o^-$, $\phi^o = \phi^{o+} = \pi - \phi^{o-}$, and $\phi_o = \phi_o^+ = \pi - \phi_o^-$. It is also assumed that $A_\theta = A_\theta^+ = A_\theta^-$ and $A_\phi = A_\phi^+ = A_\phi^-$.

494

When an electric field **E** is applied to the twisted chiral nematic cell, i.e. $g_f = -(\varepsilon_o \Delta \tilde{\varepsilon} / 2)(\mathbf{n} \cdot \mathbf{E})^2$ in EQN (4), the free energy density in the bulk may be expressed as

$$g_b = \frac{1}{2} \left[f(\theta) \left(\frac{d\theta}{dX_3} \right)^2 + h(\theta) \left(\frac{d\phi}{dX_3} \right)^2 \right] + k_2 \cos^2 \theta \frac{d\phi}{dX_3} + \frac{k_2^2}{2K_2} - \frac{\varepsilon_o \Delta \tilde{\varepsilon}}{2} E^2 \sin^2 \theta \qquad (6)$$

where ε_o is the dielectric constant in vacuum and $\Delta \tilde{\varepsilon}$ is the dielectric anisotropy. The minimisation of EQN (5) including EQNS (2), (3) and (6) gives the torque balance equations in the bulk as

$$f(\theta) \frac{d^2\theta}{dX_3^2} + \frac{1}{2} \frac{df(\theta)}{d\theta} \left(\frac{d\theta}{dX_3} \right)^2 - \frac{1}{2} \frac{dh(\theta)}{d\theta} \left(\frac{d\phi}{dX_3} \right)^2 + k_2 \sin 2\theta \frac{d\phi}{dX_3} + \frac{\varepsilon_o \Delta \tilde{\varepsilon}}{2} E^2 \sin 2\theta = 0 \quad (7)$$

$$\frac{d\phi}{dX_3} = \frac{1}{h(\theta)} \left(C_1 - k_2 \cos^2 \theta \right) \qquad (8)$$

where C_1 is a constant of integration, $k_2 = -2\pi K_2 / p_o$, where positive and negative signs of k_2 correspond to left- and right-handed helixes, $f(\theta) = K_1 \cos^2 \theta + K_3 \sin^2 \theta$ and $h(\phi) = \cos^2 \theta (K_2 \cos^2 \theta + K_3 \sin^2 \theta)$. The surface torque-balance equations at $X_3 = 0$ are expressed as

$$f(\theta) \frac{d\theta}{dX_3} \bigg|_{X_3 = 0} = \frac{A_\theta}{2} \sin 2(\theta - \theta_o) \qquad (9)$$

$$K_2 \frac{d\phi}{dX_3} \bigg|_{X_3 = 0} = -k_2 + \frac{A_\phi}{2} \sin 2\phi \qquad (10)$$

Using the Rapini-Papoular energy, the interfacial effect on the bulk orientation of the director has been investigated [1] and in this way an attempt was made to determine the values of A_θ and A_ϕ.

C   UNIFIED SURFACE ANCHORING ENERGY

For a twisted nematic sample the Rapini-Papoular energy for the director orientation must be extended to the more general form [1]

$$g_s = -\frac{A}{2} (\mathbf{n} \cdot \mathbf{e})^2 \qquad (11)$$

which is a non-linear combination of the azimuthal and polar angles. On the basis of the general expression for the anchoring energy given in EQN (11), the general expressions for the surface anchoring energy [4] can be used to calculate the field-controlled director orientation in a twisted nematic slab with weak anchoring. Using EQN (11) the corresponding surface energy densities are given by

$$g_s^+ = -\frac{A^+}{2}(\mathbf{n} \cdot \mathbf{e}^+)^2 \qquad (X_3 = \ell) \qquad (12)$$

$$g_s^- = -\frac{A^-}{2}(\mathbf{n} \cdot \mathbf{e}^-)^2 \qquad (X_3 = 0) \qquad (13)$$

where A^+ and A^- are the anchoring strengths at the top and bottom substrate surfaces, respectively. We consider the same conditions as used in the previous section (FIGURE 1) and $A = A^+ = A^-$. The minimisation of the total excess free energy including EQNS (6), (12) and (13) gives the torque-balance equations in the bulk in the same way as for EQNS (8) and (9). The surface torque-balance equations at $X_3 = 0$ can be solved generally as

$$f(\theta)\frac{d\theta}{dX_3}\bigg|_{X_3=0} = A(\sin\theta_o \sin\theta + \cos\theta_o \cos\theta \cos\phi)(\cos\theta_o \sin\theta \cos\phi - \sin\theta_o \cos\theta) \qquad (14)$$

$$h(\theta)\frac{d\phi}{dX_3}\bigg|_{X_3=0} = -k_2 \cos^2\theta + A(\cos\theta_o \cos\theta \cos\phi + \sin\theta_o \sin\theta)\cos\theta_o \sin\phi \cos\theta \qquad (15)$$

The surface torque-balance equations at $X_3 = \ell$ can be obtained simply by reversing the signs of the right-hand sides of EQNS (14) and (15). It is clear that EQN (14) is reduced to EQN (9) by setting $\phi = 0$ in EQN (14): that is, for an infinite value of A_ϕ. Similarly EQN (15) is reduced to EQN (10) by setting $\theta = \theta_o = 0$ in EQN (15): that is, for an infinite value of A_θ. Further manipulations of EQNS (8), (14) and (15) give the integration constant C_1 in EQN (8) as

$$C_1 = A(\cos\theta_o \cos\theta° \cos\phi° + \sin\theta_o \sin\theta°)\cos\theta_o \cos\theta° \sin\phi° \qquad (16)$$

D   MEASUREMENT TECHNIQUES OF THE SURFACE ANCHORING ENERGY

D1   External Field Off Method

The temperature dependence of the azimuthal anchoring energy defined in EQN (3) at the interface between the 4-pentyl-4'-cyanobiphenyl (5CB) and 60° obliquely evaporated silicon oxide (SiO) film has been measured [5] by using the relationship $d_e = 2\phi° \ell /(\phi_t - \phi°)$, where ϕ_t is the twist angle and d_e is the extrapolation length as $d_e = 2K_2/A_\phi$ derived from EQN (10). The azimuthal anchoring energy A_ϕ was estimated from the actual twist angle, $\phi_t - 2\phi°$, of the director, which was obtained by measuring the birefringence of the nematic and setting $\phi_t = 85°$ experimentally. The results are shown in FIGURE 2, and compared with those reported for the same interface in the literature [6].

The unified surface anchoring model leads to the following useful relationship between the surface anchoring strength and the twist angle for a twisted nematic slab by setting $\theta_o = 0$ in EQNS (7), (8) and (14) - (16):

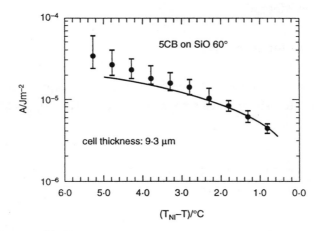

FIGURE 2 The temperature dependence of the azimuthal anchoring at a 5CB/SiO interface [5].
The solid line is the data obtained by Faetti et al [6].

$$\phi_t - 2\phi^\circ - \frac{2\pi\ell}{p_o} = \frac{A\ell}{2K_2}\sin 2\phi^\circ \tag{17}$$

which could be used to estimate the anchoring energy A. In [7] EQN (17) is applied to measure the surface anchoring energy for pure twisted nematic at the interface of 5CB and polyimide Langmuir-Blodgett layers with thickness less than 10 nm. The anchoring energy is found to increase and saturate with the number of Langmuir-Blodgett layers as shown in FIGURE 3. The open circles show the anchoring strength A deduced from the measured values ϕ° and EQN (17) (setting $\phi_t = 90^\circ$ and $p_o \to \infty$) as a function of polyimide Langmuir-Blodgett film thickness ℓ. The value of ϕ° was obtained by measuring the optical retardation of the nematic slab. The dashed line is just a guide to the eye for the experimental results and the solid line gives the theoretical curve based on a simple model of the generalised non-retarded van der Waals energy for the chain-chain interaction.

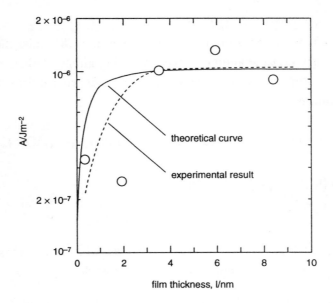

FIGURE 3 Polyimide-Langmuir-Blodgett film thickness
dependence of the anchoring strength A [7].

D2   **External Field On Method**

D2.1   **Fréedericksz transition technique**

The unified surface anchoring model leads to the following useful expression for the threshold electric field E_F of the Fréedericksz transition when setting $\theta_o = 0$ in EQNS (7), (8) and (14) - (16) as given in [4]:

$$E_F = \left[\frac{\ell^2 R + (\phi_t - 2\phi^\circ)[(K_3 - 2K_2)(\phi_t - 2\phi^\circ) + 4\pi\ell K_2 / p_o]}{\varepsilon_o \Delta\widetilde{\varepsilon}\ell^2} \right]^{1/2} \qquad (18)$$

where

$$A\cos^2\phi^\circ = \sqrt{K_1 R} \, \tan\left[\frac{\ell}{2}\left(\frac{R}{K_1}\right)^{1/2} \right] \qquad (19)$$

On the other hand the threshold electric field for an untwisted nematic slab with rigid boundary coupling $(A \rightarrow \infty)$ is expressed as

$$E_c = \frac{\pi}{\ell}\sqrt{K_1 / \varepsilon_o \Delta\widetilde{\varepsilon}} \qquad (20)$$

In the limiting case of $A \rightarrow \infty$ ($\phi^\circ \rightarrow 0$), EQN (18) reduces to

$$E_F \ell = \left[\frac{\pi^2 K_1 + (K_3 - 2K_2)\phi_t^2 + 4\pi\ell K_2 \phi_t / p_o}{\varepsilon_o \Delta\widetilde{\varepsilon}} \right]^{1/2} \qquad (21)$$

This reduces to EQN (20) and the result obtained by Becker et al [8] under the assumption of strong azimuthal anchoring and the consideration of polar anchoring alone. This is also the result reported by Hirning et al [9] in treating the tilt anchoring and twist anchoring independently and taking both anchoring strengths as infinite.

In the case of a twisted nematic cell with strong anchoring and $p_o \rightarrow \infty$, EQN (21) reduces to

$$E_F \ell = \left[\frac{\pi^2 K_1 + (K_3 - 2K_2)\phi_t^2}{\varepsilon_o \Delta\widetilde{\varepsilon}} \right]^{1/2} \qquad (22)$$

This is the same result as that derived by Leslie [10] as well as by Schadt and Helfrich [11]. Furthermore, for the homogeneous nematic slab with weak anchoring, $\ell / p_o = 0$ and $\phi_t = 0$, EQNS (17) to (19) lead to

$$A = \sqrt{K_1 \varepsilon_o \Delta \widetilde{\varepsilon}} E_F \tan\left[\frac{\ell}{2}\sqrt{\frac{\varepsilon_o \Delta \widetilde{\varepsilon}}{K_1}} E_F\right] \qquad (23)$$

which is the same result as that obtained by Rapini and Papoular [3].

Rosenblatt [12] measured the anchoring strength as a function of temperature using a Fréedericksz technique (EQNS (20) and (23)) in a very thin ($\ell < 2.5$ µm) surfactant-treated cell of 4-methoxybenzylidene-4'-butylaniline (MBBA) and reported that in the nematic phase A varies from approximately 6.5 to 1.3×10^{-5} J/m^2 for $T_{NI} - 20°C < T < T_{NI}$. The measurement method developed in which the electric Fréedericksz transition threshold is measured in a wedged capacitance cell is applied to the interface between MBBA and dodecyltrimethyl ammonium chloride to promote homeotropic alignment, and the value obtained is $(5.2 \pm 1.2) \times 10^{-5}$ J/m^2 at $T_{NI} - T = 2.5°C$ [13].

D2.2   High electric field technique

The high electric field technique could also be used to measure the polar anchoring energy [14]. In this technique the extrapolation length d_e is estimated by using the relationship between the measured values of the electric capacitance (C) and the optical retardation (R)

$$\frac{R}{R_o} = \frac{I_o}{CV} - \frac{2d_e}{\ell} \qquad V > 6V_{th} \qquad (24)$$

where I_o is a proportionality constant dependent on the nematic material; V stands for the applied voltage. The polar anchoring strength A_θ is obtained from the relation

$$A_\theta = f(\theta°)/d_e \qquad (25)$$

which is derived from EQN (9). Using this method Yokoyama et al reported that $A = (3.7 \pm 0.2) \times 10^{-5}$ J/m^2 at $0.230 \pm 0.001°C$ below the clearing point (35.30°C) for the interface 5CB/SiO [14].

The temperature dependence of the polar anchoring energy defined in EQN (2) at the interface between 5CB and rubbed polystyrene films has been measured [15] by using the high electric field technique. The results are shown in FIGURE 4. The same technique has been applied to determine the polar anchoring energy of 5CB aligned on rubbed polyimide films with various rubbing strengths [16]. Under the weakest rubbing condition, a polar anchoring energy of 8×10^{-4} J/m^2 was obtained at a cell temperature of $T = 30°C$. The results are shown in FIGURE 5.

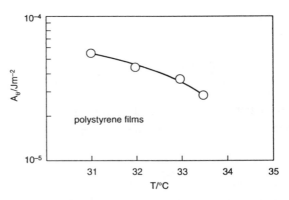

FIGURE 4 The temperature dependence of the polar anchoring energy for 5CB and strongly rubbed polystyrene films [14].

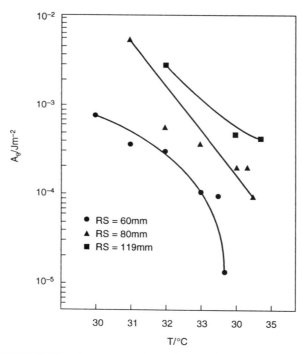

FIGURE 5 The temperature dependence of the polar anchoring
energy of 5CB on rubbed polyimide films [15].

D2.3   Saturation voltage method

The rigorous relationship between the anchoring strength A and the saturation voltage U_S, in which
the director becomes completely homeotropic, i.e. the entire nematic including the boundary layers is
oriented along the field direction at the saturation voltage, can be derived [17]:

$$\frac{\pi K_3}{A\ell} = \frac{\tanh(\pi Y/2)}{Y}\left[1 + \frac{\cos^2 Z}{\sinh^2(\pi Y/2)}\right] \tag{26}$$

where

$$Y = \sqrt{\frac{\varepsilon_o \Delta\widetilde{\varepsilon} U_S^2}{\pi^2 K_3} - \left(\frac{2\ell K_2}{p_o K_3}\right)^2} \quad \text{and} \quad Z = \frac{\phi_t}{2} - \frac{\pi\ell K_2}{p_o K_3} \tag{27}$$

For a nematic slab with a homogeneous director orientation, EQNS (26) and (27) give a rigorous
relationship between U_S and A, namely

$$A = \frac{U_S\sqrt{\varepsilon_o \Delta\widetilde{\varepsilon} K_3}}{\ell}\tanh\left(\frac{U_S}{2}\sqrt{\frac{\varepsilon_o \Delta\widetilde{\varepsilon}}{K_3}}\right) \tag{28}$$

and so a value of A can be determined by measuring U_S. As is expected from the definition of the
saturation transition, U_S can be determined precisely for the condition of zero optical retardation, and
then the unified surface anchoring energy can be estimated from EQN (28). Because the optical

retardation is inversely proportional to the applied voltage in the high voltage regime, the intersection of the extrapolated line of the values of the measured optical retardation and the horizontal inverse voltage axis gives the saturation voltage U_S. The saturation method is applicable for both homogeneous and twisted nematic sandwich slabs.

The saturation voltage method has been used to measure the surface anchoring strength at the interface of 5CB and a photo-alignment layer of polyvinylcinnamate (PVCi) [17]. This is an attractive technique free from a conventional rubbing method [18] of making a uniform director orientation at the substrate surface. TABLE 1 shows the anchoring strength dependence on the irradiation energy.

TABLE 1 The dependence of the anchoring strength of 5CB on PVCi films on the irradiation energy at 298 K. In the calculation based on EQN (28), $K_1 = 7.0 \times 10^{-12}$ N, $K_2 = 3.9 \times 10^{-12}$ N, $K_3 = 8.4 \times 10^{-12}$ N, and $\Delta\tilde{\varepsilon} = 10$ have been used for 5CB at 299 K [17].

Irradiation energy/J cm^{-2}	Thickness/μm	U_S/V	A/J m^{-2}
0.3	13.8	82.6	1.63×10^{-4}
0.6	10.1	128.6	3.46×10^{-4}
0.9	13.5	184.7	3.73×10^{-4}

D3 Other Data on the Surface Anchoring Energy Measured for Nematics

There are a number of reports in the literature concerning the measurement of the surface anchoring energies mainly, though, for MBBA and 5CB. Almost all of the data are for the polar and azimuthal anchoring energies based on the surface torque-balance equations (EQNS (9) and (10)) and according to the definitions given by EQNS (2) and (3). The data, which are summarised in the books by Blinov and Chigrinov [19] and by Sonin [20], show that the azimuthal energy is one to two orders of magnitude smaller than the polar anchoring energy for planar oriented nematics.

E THICKNESS DEPENDENCES OF THE ANCHORING ENERGY

Barbero and Durand [21] reported that the surface anchoring energy decreases with increasing cell thickness. The interpretation of this observation is based on the idea that the surface density of ions contained in the nematic increases with increasing sample thickness below a certain, critical, thickness. The thickness dependence of the anchoring is explained by the coupling of the surface field, created by the adsorbed surface charges, with the anisotropic properties of the nematic.

F CONCLUSION

The surface anchoring energy is generally expressed as a sum of independent terms $g_{s\theta}(\theta)$ and $g_{s\phi}(\phi)$ for the polar and azimuthal angles. A new approach in which the anchoring energy is written as a two-dimensional function $g_s(\theta,\phi)$ has been proposed theoretically and investigated. The coupling existing between the two angles θ and ϕ has recently been demonstrated unambiguously in experiments in which the electric field is applied parallel to the glass surface [22].

The expression for the surface energy can also be predicted for other shapes of the surface energy which are different from the Rapini-Papoular energy type, for example the elliptic type, Legendre expansion [14], and so on. There is presently a widespread interest in the inclusion of the surface elastic moduli K_{13} and K_{24} in the theory to allow for novel contributions to the surface anchoring energy to be taken into account [23] (see Datareview 5.3).

In conclusion, many of the available data and theoretical investigations are related so far to two liquid crystals only, MBBA and 5CB. These are not sufficient to understand fully the surface-induced effects produced by the anisotropic surface anchoring energy.

REFERENCES

[1]   B. Jérôme [*Rep. Prog. Phys. (UK)* vol.54 (1991) p.391]

[2]   P.G. de Gennes [in *The Physics of Liquid Crystals* (Oxford University Press, London, 1974)]

[3]   A. Rapini, M. Papoular [*J. Phys. Colloq. (France)* vol.30 (1969) p.C4-54]

[4]   A. Sugimura, G.R. Luckhurst, Z. Ou-Yang [*Phys. Rev. E (USA)* vol.52 (1995) p.681]

[5]   Y. Iimura, N. Kobayashi, S. Kobayashi [*Jpn. J. Appl. Phys. (Japan)* vol.34 (1995) p.1935]

[6]   S. Faetti, M. Gatti, V. Palleschi, T.J. Sluckin [*Phys. Rev. Lett. (USA)* vol.55 (1985) p.1681]

[7]   A. Sugimura, K. Matsumoto, Z. Ou-Yang, M. Iwamoto [*Phys. Rev. E (USA)* vol.54 (1996) p.5217]

[8]   M.E. Becker, J. Nehring, T.J. Scheffer [*J. Appl. Phys. (USA)* vol.57 (1985) p.4539]

[9]   R. Hirning, W. Funk, H.-R. Trebin, M. Schmidt, H. Schmiedel [*J. Appl. Phys. (USA)* vol.70 (1991) p.4211]

[10]   F.M. Leslie [*Mol. Cryst. Liq. Cryst. (UK)* vol.12 (1970) p.57]

[11]   M. Schadt, W. Helfrich [*Appl. Phys. Lett. (USA)* vol.18 (1971) p.127]

[12]   C. Rosenblatt [*J. Phys. (France)* vol.45 (1984) p.1087]

[13]   D.-F. Gu, S. Uran, C. Rosenblatt [*Liq. Cryst. (UK)* vol.19 (1995) p.427]

[14]   H. Yokoyama, H.A. van Sprang [*J. Appl. Phys. (USA)* vol.57 (1985) p.4520]

[15]   D.-S. Seo, K. Muroi, T. Isogami, H. Matsuda, S. Kobayashi [*Jpn. J. Appl. Phys. (Japan)* vol.31 (1992) p.2165]

[16]   D.-S. Seo, Y. Iimura, S. Kobayashi [*Appl. Phys. Lett. (USA)* vol.61 (1992) p.234]

[17]   A. Sugimura, T. Miyamoto, M. Tsuji, M. Kuze [*Appl. Phys. Lett. (USA)* vol.72 (1998) p.329]

[18]   M. Schadt, A. Schuster [*Nature (UK)* vol.381 (1996) p.212]

[19]   L.M. Blinov, V.G. Chigrinov [in *Electrooptic Effects in Liquid Crystal Materials* (Springer-Verlag, New York, 1994) p.118]

[20]   A.A. Sonin [in *The Surface Physics of Liquid Crystals* (Gordon and Breach Publishers, New York, 1995) p.42]

[21]   G. Barbero, G. Durand [*J. Appl. Phys. (USA)* vol.67 (1990) p.2678]

[22]   P. Jägemalm, L. Komitov, G. Barbero [*Appl. Phys. Lett. (USA)* vol.73 (1998) p.1]; P. Jägemalm, L. Komitov [*Liq. Cryst. (UK)* vol.23 (1997) p.1]

[23]   M. Faetti, S. Faetti [*Phys. Rev. E (USA)* vol.57 (1998) p.6741]

10.3 STM studies of anchoring phase transitions at nematic interfaces

M. Hara

June 2000

A INTRODUCTION

Molecular alignment on solid substrates is an important fundamental issue in both physics and chemistry as well as in practical applications such as optoelectronic devices. It is well-known that the ordering of adsorbed molecules in the anchoring region near the substrate, which we call the anchoring phase, is strongly affected by the substrate surface structure and the balance of molecule-molecule and molecule-substrate interactions. This is particularly evident for liquid crystals, where such an alignment technique with boundary effects at the substrate surface has been widely used to produce director configurations in monodomain cells such as flat-panel displays and optical shutters. However, the actual mechanism of their orientation at the interface has long been unclear, because there have been technical difficulties in the analysis of the anchoring structures at a molecular level.

Since its invention, scanning tunnelling microscopy (STM) has been successfully applied to direct visualisation of the interfacial structure with molecular resolution, and used to study positional and orientational order in organic monolayers on solid substrates, especially liquid-crystal monolayers [1]. The STM images of such monolayers have been reported for smectic [2-5], nematic [6,7], and antiferroelectric liquid crystals [8]. In fact, high-resolution STM images now allow us to carry out the real-space analysis of the molecular alignment at the interface on the individual molecular scale. In this Datareview, we discuss the anchoring phase on the basis of the highly reproducible STM images of a homologous series of 4-alkyl-4′-cyanobiphenyls (mCBs: where m (= 7 ~ 12) is the number of carbon atoms in the alkyl chains) and various bulk compositions of 8CB-12CB binary mixtures. We then consider the correlation between these anchoring structures and the bulk phase diagrams from the viewpoint of anchoring phase formation.

B EXPERIMENTAL

In a previous paper [4], we introduced the use of a molybdenum disulphide (MoS_2) single crystal as a substrate for the STM imaging of organic adsorbates for the first time and proposed that the anchoring structures on MoS_2 should well reflect the corresponding bulk phase sequences compared with those on graphite because of the higher degree of freedom allowed for the alignment of rod-like molecules on MoS_2 [7]. We will return to this important aspect in our discussion of substrate dependence.

The STM imaging samples were prepared by placing a drop of the pure and mixed mCBs on the surface of freshly cleaved MoS_2 and highly oriented pyrolytic graphite (HOPG) substrate and heating it to the isotropic phase. After slowly cooling the sample to room temperature, a sharp Pt/Ir tip was positioned at the interface and scanned over it in air at room temperature. All images were obtained in the constant-current mode. Typical operating conditions were 0.09 to 0.4 nA and 0.8 to 2.0 V (tip negative). The STM system used in this study was a commercially available NanoScope II (Digital Instruments, Inc., Santa Barbara, CA, USA). The STM images were digitally filtered to remove just the high-frequency noise.

C   STM IMAGING

FIGURE 1 shows typical STM images of a homologous series of pure mCB (m = 7 ~ 12) molecules on MoS$_2$. Individually distinguishable patterns of rods and regular alignment are observed. As we have proposed previously [6], each pattern shows individual mesogenic molecules. For example, the STM images of 10CB and 12CB (FIGURES 1(d) and 1(f)) exhibit a double-row structure similar to the smectic ordering in the bulk, as deduced from X-ray and neutron-scattering measurements. On the other hand, a completely different anchoring structure from 10CB and 12CB has been observed by STM for 8CB on MoS$_2$ as shown in FIGURE 1(b). In numerous studies on MoS$_2$, the anchoring structures are clearly divided into two categories: a single-row type (7CB, 8CB, 9CB and 11CB) in which the cyanobiphenyl head groups and the alkyl tails alternate in each row, and a double-row type (10CB and 12CB) in which the cyano groups face one another. For the binary mixtures, which will be discussed separately in Section F, it has been realised that there exists an inhomogeneous (mixed) double-row type consisting of 8CB and 12CB (see FIGURE 2).

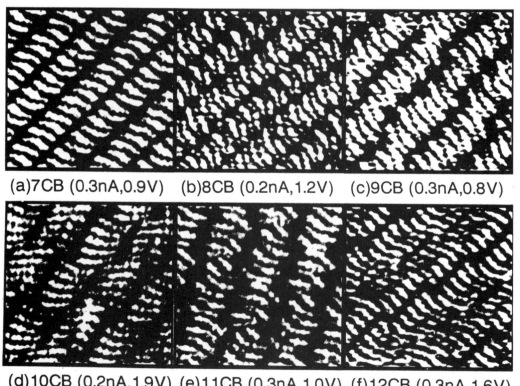

(a)7CB (0.3nA,0.9V)   (b)8CB (0.2nA,1.2V)   (c)9CB (0.3nA,0.8V)

(d)10CB (0.2nA,1.9V)   (e)11CB (0.3nA,1.0V)   (f)12CB (0.3nA,1.6V)

FIGURE 1 STM images of a homologous series of pure mCB (m = 7 ~ 12) molecules on MoS$_2$, from [7]. The numbers in parentheses are the tunnel current and bias voltage (tip negative). Each image area is 10 nm × 10 nm.

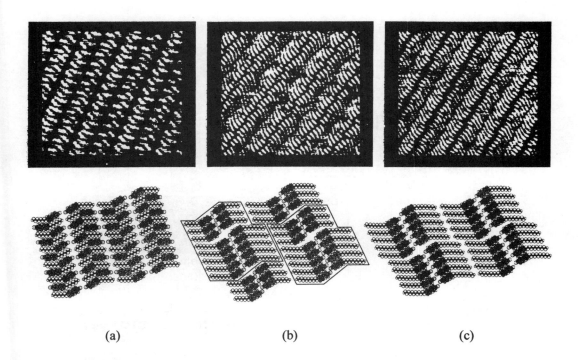

FIGURE 2 STM images and models showing the anchoring structures of mCBs on MoS$_2$. (a) Homogeneous 8CB single-row (15 × 15 nm). (b) Inhomogeneous (mixed) double-row formed by 8CB-12CB binary mixtures (20 × 20 nm). (c) Homogeneous 12CB double-row (20 × 20 nm). From [14].

D   ORIGIN OF SINGLE-ROW AND DOUBLE-ROW PHASES

While various interactions exist for the formation of the anchoring phases, one possible model for the origin of the two phases from the viewpoint of their phase sequences in the bulk has been proposed; they are shown schematically in FIGURE 3. First, while 10CB and 12CB only have a smectic liquid-crystal phase between isotropic and solid crystal phases, STM images of both molecules exhibit the same double-row structure equivalent to the bulk smectic phase ordering. Thus, it is natural to assume that the double-row is formed at the interface when the phase transition from the I to the ordered smectic phase takes place in the bulk. This double-row can then be attributed to the anchoring phase of the smectic phase (FIGURE 3(b)). Second, while 7CB only has a nematic liquid-crystal phase, the STM image exhibits a single-row structure on MoS$_2$. Comparing the results for 7CB (phase sequence Cr-N-I) with those for 10CB and 12CB (Cr-SmA-I), the single-row can be attributed to the anchoring phase of the nematic phase (FIGURE 3(a)). In the same manner, if the liquid crystals have the phase sequence Cr-SmA-N-I, the molecules should be ordered in the nematic phase which actually appears first when the temperature is decreased during the sample preparation. In fact, the STM images of pure 8CB samples, which, for example, possess the nematic phase in a higher temperature region than the smectic A phase in the bulk (Cr-SmA-N-I, FIGURE 4), exhibit a single-row phase on MoS$_2$. From this point of view, we proposed that the single-row and double-row are attributable to the formation of anchoring phases from nematic and smectic phases, respectively.

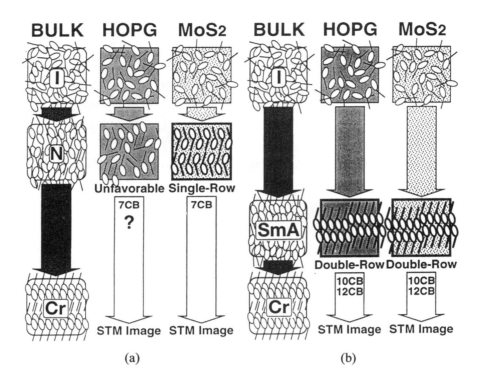

FIGURE 3 Phase sequences in bulk and anchoring region on MoS$_2$ and graphite (HOPG). (a) 7CB with the phase sequence Cr-N-I. The energetically favourable anchoring phase does not form on graphite because there is no smectic phase, while the single-row forms during the N-I bulk transition on MoS$_2$.
(b) 10 and 12CBs with the phase sequence Cr-SmA-I. The same double-row forms on both MoS$_2$ and graphite during the SmA-I bulk transition.

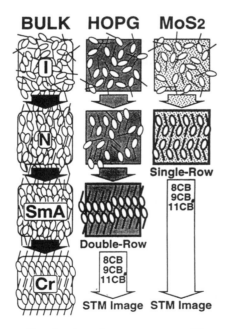

FIGURE 4 Substrate dependence of anchoring phase sequences. The anchoring phase transition takes place during the N-I bulk transition on MoS$_2$, while it takes place during the SmA-N bulk transition on HOPG. It has turned out that the SmA-N transition of 11CB was caused by impurity [10].

E   SUBSTRATE DEPENDENCE: MoS₂ AND GRAPHITE (HOPG)

Before we introduced MoS_2 as a substrate, the STM imaging of organic molecules was mainly carried out on graphite (HOPG). In fact, it has been well-accepted that 8CB, 10CB and 12CB exhibit the smectic-like double-row on graphite, and that these molecules form smectic A mesophases [3]. From the viewpoint of our interpretation of the anchoring phases, however, this result on graphite suggests that 8CB did not form an energetically favourable anchoring phase on graphite in the nematic phase (single-row), but formed it instead in the smectic A phase (double-row). We attributed this difference to the different surface structures of the substrates.

For graphite, the cleaved surface consists of a condensed ring of six carbon atoms. In addition, it is well-known that the hydrogen orbitals of n-alkanes can fit into the holes of the carbon rings on graphite under a nearly commensurate condition, resulting in only three degrees of freedom for the stable anchoring of alkyl chains. The alkyl chains are then most stable when the angle between chains is 0° or 60°. On the other hand, the cleaved surface of MoS_2 consists of a hexagonal lattice of sulphur atoms and there are at least six degrees of freedom for the anchoring of alkyl chains with multiples of 30°. The angle between alkyl chains in the single-row is 0° for every second molecule and approximately 30° for neighbouring molecules. For this reason, since the 30°-required single-row is unstable on graphite, 8CB is not adsorbed in the nematic phase which appears first with decreasing temperature, but is adsorbed in the subsequent smectic A phase (see FIGURE 4). In fact, STM images of 7CB on graphite exhibit neither single-row nor double-row structures (see FIGURE 3(a)). This result can be explained as 7CB forming only the nematic phase on graphite, resulting in the instability of the single-row on graphite.

F   ANCHORING PHASE OF BINARY MIXTURES

Although some questions and counter examples to our interpretation have been raised [9-12], we have proceeded to the miscibility test using binary mixtures of mCBs, which has long been accepted as one of the most useful and simple methods of studying aspects of phase sequences in the liquid crystal field. In the case of mixtures, bulk phase sequences can be continuously changed according to the composition. In 8CB-12CB binary mixtures, for example, there are nematic and smectic A phases (Cr-SmA-N-I) below 40 mol% of 12CB, but above it, only the smectic A phase exists between the isotropic and crystal phases (Cr-SmA-I).

As shown in FIGURE 5, a complete set of STM images for various compositions of the 8CB-12CB binary mixtures has revealed that the anchoring phases are drastically changed from the single-row to the double-row at 40 mol% of 12CB, while there are two types of double-rows: inhomogeneous (mixed) and homogeneous double-rows (see FIGURES 2(b) and 2(c)) [13,14]. This finding strongly supports our interpretation for the origin of the anchoring phases based on the phase sequences in the bulk. Namely, the single-row and double-row are determined by the N-I and SmA-I phase transitions in the bulk, respectively.

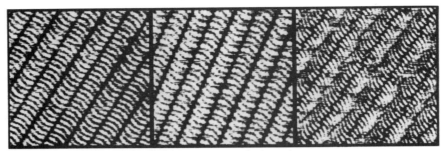

(a)80:20 (0.3nA,1.6V,15x15nm) (b)70:30 (0.3nA,1.0V,15x15nm)(c)60:40 (0.2nA,2.0V,20x20nm)

8CB Single Row (Cr-SmA-N-I)　8CB Single Row (Cr-SmA-N-I)　Mixed Double Row (Cr-SmA-I)

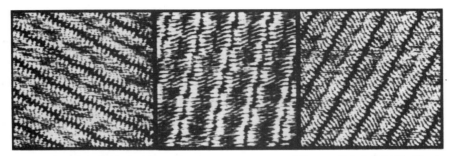

(d)30:70 (0.2nA,1.9V,20x20nm) (e)20:80 (0.3nA,1.6V,20x20nm) (f)10:90 (0.2nA,1.1V,20x20nm)

Mixed Double Row (Cr-SmA-I)　12CB Double Row (Cr-SmA-I)　12CB Double Row (Cr-SmA-I)

FIGURE 5　STM images of the mixtures of 8CB and 12CB, from (a) 8CB:12CB = 80:20 (mol%) to (f) 10:90 (mol%) on MoS_2, from [14]. The numbers in parentheses are the tunnel current, bias voltage (tip negative), and image area. Cr-SmA-N-I and Cr-SmA-I are the phase sequences in the bulk.

As schematically shown in FIGURE 6, another important finding is that the ratio of 8CB and 12CB adsorbed on MoS_2 in all mixtures is different from that in the bulk: a homogeneous 8CB single-row consisting of only 8CB (100% 8CB) below 40 mol% 12CB (Cr-SmA-N-I), an inhomogeneous (mixed) double-row consisting of three pairs of 8CB and five pairs of 12CB (8CB:12CB = 3:5) from 40 to 70 mol% 12CB (Cr-SmA-I), and a homogeneous 12CB double-row consisting of only 12CB (100% 12CB) above 80 mol% 12CB (Cr-SmA-I). In addition, it has been confirmed that there exists an additional phase transition at the boundary from inhomogeneous (mixed) to homogeneous double-row phases, while the bulk phase sequences remain the same as that of Cr-SmA-I.

Since these mixtures exhibit a continuous miscibility without significant phase separation, it has been considered that 12CB (8CB) should be continuously mixed with the 8CB single-row (12CB double-row) ordering. It is surprising, however, that 12CB (8CB) molecules are absent in the anchoring phase, in spite of the presence of 12CB (8CB) in the bulk below 40 mol% (above 80 mol%) 12CB. We assume that this is because such a homogeneous single-row (double-row) consisting of only 8CB (12CB) is energetically favourable in the anchoring phase in this mol% region showing the phase sequence Cr-SmA-N-I (Cr-SmA-I). If 12CB (8CB) molecules migrate into the 8CB single-row (12CB double-row), the substrate area covered with the molecules would decrease and the structure would lead to energetically unfavourable vacancies. For the inhomogeneous (mixed) double-row, we can, in the same manner, attribute the behaviour to the anchoring phase consisting of the 8CB-12CB unit cell (three pairs of 8CB and five pairs of 12CB) being energetically favourable at the boundary.

508

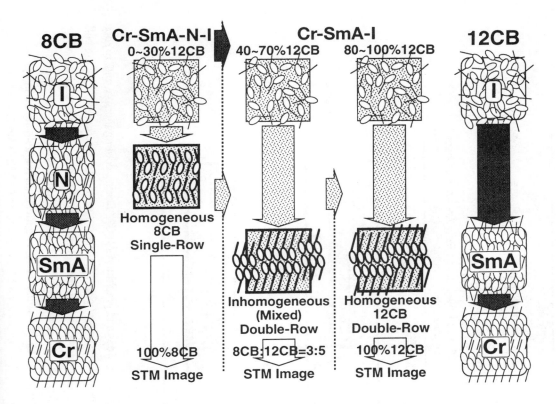

FIGURE 6 Phase sequences in the bulk and the anchoring region of 8CB-12CB binary mixtures on MoS$_2$. Below 40 mol% of 12CB, I-homogeneous 8CB single-row, while the bulk phase sequence is Cr-SmA-N-I. From 40 to 70 mol%, I-inhomogeneous (mixed) double-row, and above 80 mol%, I-homogeneous 12CB double-row, while that in the bulk is Cr-SmA-I. In this region, the anchoring phase transition takes place at the boundary.

Thus, the energetically favourable mixed double-row repeating units (see FIGURE 2(b)) determine the ratio of 8CB and 12CB at the boundary, resulting in the inconsistency with that of 8CB and 12CB in the bulk.

G   CONCLUSION

The anchoring structures of a homologous series of mCBs and various bulk compositions from 0 to 100 mol% 8CB-12CB binary mixtures are directly observed on MoS$_2$ by STM in order to investigate the boundary condition of adsorbed molecules in the anchoring region.   On the basis of highly reproducible STM images, the anchoring structures are clearly divided into two categories: a single-row type and a double-row type.   We have proposed that the single-row is the anchoring structure of a nematic phase, while the double-row is that of a smectic phase. Furthermore, a new phase transition, only possible at the boundary, has been confirmed for the first time in the binary mixtures.   The correlation between these anchoring structures and the bulk phase diagrams is discussed from the viewpoint of anchoring phase formation at a molecular level.

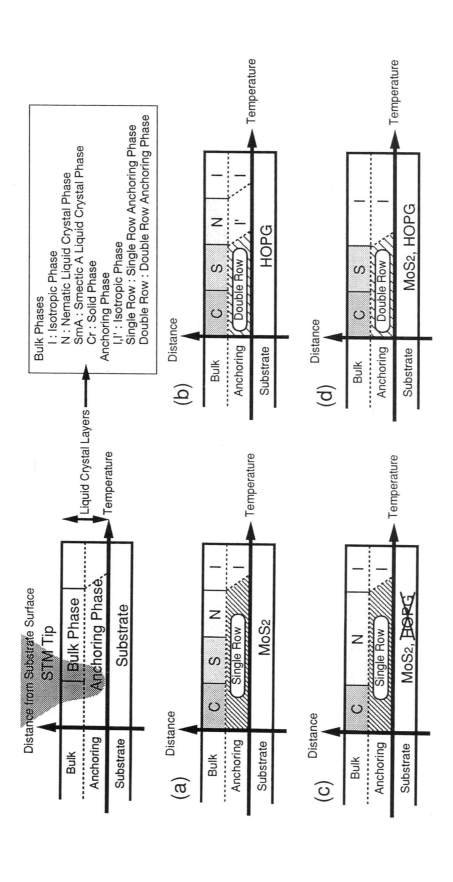

FIGURE 7 Schematic diagram of the phase sequences in the bulk and the anchoring region on MoS₂ and graphite (HOPG). During cooling the substrate to room temperature, phase transitions take place from liquid crystal phases, nematic (N) and/or smectic (SmA), in the bulk. In the anchoring region, on the other hand, the anchoring phase forms during the bulk phase transition from the I phase to the liquid crystal phase, N or SmA. (a) 8, 9 and 11CBs with bulk phase sequence, Cr-SmA-N-I. Anchoring phase transition takes place during the N-I bulk transition on MoS₂. (b) Anchoring phase transition takes place during the SmA-N bulk transition on graphite. (c) 7CB with the Cr-N-I sequence. Energetically favourable anchoring phase does not form on graphite because of no SmA phase, while it forms during the N-I bulk transition on MoS₂. (d) 10 and 12CBs with Cr-SmA-I sequence. The same anchoring phase forms on both MoS₂ and graphite during the SmA-I bulk transition.

Although it is not necessary for the anchoring phase transitions to occur at the same temperature at which the bulk N-I or SmA-I phase transition occurs, it is evident that such anchoring phase transitions are dominantly induced by the N-I or SmA-I phase transitions and that there is a strong correlation between the anchoring phase formation and the bulk phase diagram (see FIGURE 7). In addition, these STM images suggest that the energetically favourable repeating units at the boundary and phase separations also play an important role in the anchoring phase formation of organic molecular systems.

REFERENCES

[1] J.E. Frommer [*Angew. Chem., Int. Ed. Engl. (Germany)* vol.31 (1992) p.1265]

[2] J.S. Foster, J.E. Frommer [*Nature (UK)* vol.333 (1988) p.542]

[3] D.P.E. Smith, H. Hörber, Ch. Gerber, G. Binnig [*Science (USA)* vol.245 (1989) p.43]

[4] M. Hara, Y. Iwakabe, K. Tochigi, H. Sasabe, A.F. Garito, A. Yamada [*Nature (UK)* vol.344 (1990) p.228]

[5] D.P.E. Smith, J.K.H. Hörber, G. Binnig, H. Nejoh [*Nature (UK)* vol.344 (1990) p.641]

[6] Y. Iwakabe et al [*Jpn. J. Appl. Phys. (Japan)* vol.29 (1990) p.L2243]

[7] Y. Iwakabe et al [*Jpn. J. Appl. Phys. (Japan)* vol.30 (1991) p.2542]

[8] M. Hara, T. Umemoto, H. Takezoe, A.F. Garito, H. Sasabe [*Jpn. J. Appl. Phys. (Japan)* vol.30 (1991) p.L2052]

[9] D.P.E. Smith [private communication]

[10] Data Sheet distributed by BDH Ltd. (now Merck UK), Poole, UK

[11] J.S. Foster, J.E. Frommer, J.K. Spong [*Proc. SPIE, Liquid Crystal Chemistry, Physics and Applications (USA)* vol.1080 (1989) p.200]

[12] W.M. Heckl [*Thin Solid Films (Switzerland)* vol.210/211 (1992) p.640]

[13] Y. Iwakabe, M. Hara, K. Kondo, S. Oh-hara, A. Mukoh, H. Sasabe [*Jpn. J. Appl. Phys. (Japan)* vol.31 (1992) p.L1771]

[14] Y. Iwakabe, K. Kondo, S. Oh-hara, A. Mukoh, M. Hara, H. Sasabe [*Langmuir (USA)* vol.10 (1994) p.3201]

CHAPTER 11

MATERIAL PROPERTIES FOR APPLICATIONS

11.1 Optical properties of nematics for applications

T. Ohtsuka, H. Ohnishi and H. Takatsu

January 2000

A   INTRODUCTION

The optical anisotropy (i.e. the birefringence) of a nematic liquid crystal is a very important property for the contrast ratio and viewing angle in liquid crystal displays. The optical anisotropy of the nematic is caused by the anisotropy of the molecular polarisability of the constituent anisometric molecules (rod-like molecules) and their long range orientational order.

For a quantitative interpretation of the principal refractive indices ($\tilde{n}_e$, $\tilde{n}_o$) of a nematic it should be sufficient to use the empirical equations of Vuks:

$$\frac{\tilde{n}_e^2 - 1}{\tilde{n}^2 + 2} = \frac{\rho N_A}{3\varepsilon_0 M} \tilde{\alpha}_\parallel \tag{1}$$

$$\frac{\tilde{n}_o^2 - 1}{\tilde{n}^2 + 2} = \frac{\rho N_A}{3\varepsilon_0 M} \tilde{\alpha}_\perp \tag{2}$$

where

$$\tilde{n}^2 = 1/3\,(\tilde{n}_e^2 + 2\tilde{n}_o^2)$$
$$\tilde{\alpha}_\parallel = \alpha + 2/3\,(\alpha_1 - \alpha_t)S$$
$$\tilde{\alpha}_\perp = \alpha - 1/3\,(\alpha_1 - \alpha_t)S$$

with

$$\alpha = 1/3\,(\alpha_l + 2\alpha_t)$$

Here ρ is the density, M is the molar mass, ε_0 is the permittivity of a vacuum, $\tilde{n}$ is the average refractive index, α is the average molecular polarisability, $\tilde{\alpha}_\parallel$ and $\tilde{\alpha}_\perp$ are the average polarisabilities parallel and perpendicular to the director, α_l and α_t are the longitudinal and transverse molecular polarisabilities respectively and S is the second rank orientational order parameter. From EQNS (1) and (2), we obtain

$$\frac{\tilde{n}_e^2 - \tilde{n}_o^2}{\tilde{n}^2 + 2} = \frac{\rho N_A}{3\varepsilon_0 M}\left(\alpha_1 - \alpha_t\right)S \tag{3}$$

From EQN (3), we can conclude that

(i)   the magnitude of the birefringence ($\tilde{n}_e - \tilde{n}_o > 0$) is mainly determined by the anisotropy of the molecular polarisability ($\alpha_l - \alpha_t$), the orientational order parameter S and the molar volume M/ρ, and

(ii)   the temperature dependence of the birefringence is given by that of the orientational order parameter S and of the density ρ.

The dispersion of the principal refractive indices over a wide wavelength region has been studied for several nematogens and the refractive indices are found to decrease with increasing wavelength. The dispersion of the birefringence in the visible spectral region for a three state model is given by [1]:

$$\tilde{n}_e - \tilde{n}_o = G_o \lambda_o^2 + G_1 \frac{\lambda^2 \lambda_1^2}{\lambda^2 - \lambda_1^2} + G_2 \frac{\lambda^2 \lambda_2^2}{\lambda^2 - \lambda_2^2}$$

where λ is the wavelength in the visible region, λ_o is the $\sigma \rightarrow \sigma^*$ resonance wavelength, λ_1 and λ_2 are two $\pi \rightarrow \pi^*$ resonance wavelengths, and G_o, G_1 and G_2 are coefficients that depend on the magnitude and the anisotropy of oscillator strengths.

B   MEASUREMENT TECHNIQUES

B1   Measuring Equipment

The extraordinary and ordinary refractive indices $\tilde{n}_e$ and $\tilde{n}_o$ are most conveniently determined by an Abbé refractometer. As the refractive index is temperature and wavelength dependent, the prism of the Abbé refractometer is equipped with a water jacket to control the measuring temperature at a designated temperature and a sodium lamp of wavelength 589.3 nm is used as a monochromatic light source.

As a uniform director alignment is required to measure the refractive indices $\tilde{n}_e$ and $\tilde{n}_o$, the surface of the main prism is pre-treated to align the director homeotropically by using lecithin solution.

B2   Measurement Procedure

(1)   A solution of lecithin in methanol is coated over the main prism to align the director perpendicular to the prism surface.

(2) $\tilde{n}_e$, the refractive index parallel to the director, is measured with an eye-piece polariser parallel to the director.

(3) $\tilde{n}_o$, the refractive index perpendicular to the director, is measured with the polariser in a perpendicular position.

(4)   In order to estimate the birefringences for materials used for practical applications, it is common practice to make measurements of materials of interest as solutes in standard nematic hosts. 10 - 20% of each liquid crystal material (see TABLE 1) is dissolved in a host

liquid crystal mixture. The Δn value of the substance is obtained by extrapolation to 100% based on measurements taken usually at 20°C. This technique is valuable for comparing many different materials, but the birefringence obtained by extrapolation may not be an accurate measure of Δn for the pure material. Errors can arise if the orientational order of the dissolved material differs significantly from that of the host liquid crystal, and specific interactions between the solute and solvent may cause small changes in the polarisability components (see EQN (3)).

C   SPECIMEN RESULTS AND DATA

TABLE 1 shows $\Delta \tilde{n}$ values of standard materials and typical liquid crystals used for practical applications.

TABLE 1 Standard materials with positive dielectric anisotropy.

Liquid crystal material	$\Delta \tilde{n}$
	0.18
	0.23
	0.21
	0.12
	0.15
	0.06
	0.12
	0.16

517

TABLE 1 continued

Liquid crystal material	$\Delta\tilde{n}$
C5H11—⬡—COO—◯—CN	0.13
C5H11—◯—COO—◯(F)—CN	0.15
∿—⬡—◯—CN	0.14
CH3OCH2—⬡—◯—CN	0.14
C5H11—⬡—◯(F)—CN	0.12
C5H11—⬡—⬡—◯—CN	0.21
C5H11—⬡—◯—◯—CN	0.21
C5H11—⬡—◯—COO—◯(F)—CN	0.16
C5H11—◯—◯—◯—CN	0.37

TABLE 2 Standard materials with weak dielectric anisotropy.

Liquid crystal material	$\Delta \tilde{n}$
CH_3O—〈 〉—N=N—〈 〉—OCH_3 ↓	0.26
CH_3O—〈 〉—CH=N—〈 〉—C_4H_9	0.27
CH_3O—〈 〉—COO—〈 〉—C_5H_{11}	0.15
C_5H_{11}—〈 〉—COO—〈 〉—OCH_3	0.08
C_5H_{11}—〈 〉—C≡C—〈 〉—OCH_3	0.26
C_5H_{11}—〈 〉—〈 〉—CH_3	0.17
C_3H_7—〈 〉—〈 〉—OC_2H_5	0.10
C_3H_7—〈 〉—〈 〉—C_4H_9	0.04
C_3H_7—〈N 〉—〈 〉—C_3H_7	0.19
C_3H_7—〈 〉—CH=N—N=CH—〈 〉—CH_3	0.36
C_3H_7—〈 〉—〈 〉—C≡C—〈 〉—C_2H_5	0.25

TABLE 2 continued

Liquid crystal material	$\Delta\tilde{n}$
C_5H_{11}—[cyclohexyl]—[cyclohexyl]—[phenyl]—C_2H_5	0.08
CH$_2$=CH-CH$_2$—[cyclohexyl]—[cyclohexyl]—[phenyl]—CH_3	0.09
C_5H_{11}—[cyclohexyl]—[phenyl]—[phenyl]—C_2H_5	0.17
C_3H_7—[cyclohexyl]—[phenyl]—COO—[phenyl]—C_3H_7	0.16

TABLE 3 Fluorinated materials.

Liquid crystal material	$\Delta\tilde{n}$
C_3H_7—[cyclohexyl]—CH_2CH_2—[phenyl(F, F)]	0.02
C_3H_7—[cyclohexyl]—[cyclohexyl]—[phenyl]—F	0.09
C_5H_{11}—[cyclohexyl]—[cyclohexyl]—[phenyl]—OCF_3	0.10
C_5H_{11}—[cyclohexyl]—[cyclohexyl]—[phenyl]—$OCHF_2$	0.11
C_5H_{11}—[cyclohexyl]—[phenyl]—[phenyl]—OCF_3	0.14
C_3H_7—[cyclohexyl]—[cyclohexyl]—[phenyl(F, F)]	**0.10**

TABLE 3 continued

Liquid crystal material	$\Delta\widetilde{n}$
C₃H₇ ... F, F	0.12
C₃H₇ ... F, F, F	0.08
C₃H₇ ... F, F, F	0.09
C₃H₇ —CH₂CH₂— ... F, F, F	0.08
C₃H₇ —CH₂CH₂— ... F, F, F	0.08
C₃H₇ —CH₂CH₂— ... F, F, F	0.09
... F, F, F	0.08
C₃H₇ —C≡C— ... F, F, F	0.22
C₃H₇ —OCHF₂ ... F, F	0.08
C₅H₁₁ ... F, F, F, F	0.12

The optical anisotropy of a nematic increases with the increase of the π-electron conjugation in the molecule. Thus the replacement of a cyclohexane ring by an aromatic ring makes the birefringence larger. As a linkage a double bond or a triple bond affords a higher birefringence than a single bond or an alkylene bond.

In an actual display, the birefringence of a liquid crystal mixture is adjusted to fix the product of d, the cell gap, and Δñ to a certain value, depending on the mode and application. Therefore, materials with a wide variety of Δñ values are required to design an actual liquid crystal mixture. Liquid crystal materials with high birefringence are useful especially for improving the switching time by reducing the cell gap. Tolanes containing a triple bond linking group or biphenyl derivatives are representative materials as this type of component.

Liquid crystal materials with low birefringence are also required to design a liquid crystal mixture for a display with a wide viewing angle performance by setting dΔñ to a smaller value. Phenylbicyclohexanes are typical materials in this category.

REFERENCES

[1]   S.-T. Wu, C.-S. Wu [*J. Appl. Phys. (USA)* vol.66 (1989) p.5297]

[2]   H. Saito et al [*IEICE Trans. Electron. (Japan)* vol.E79-C no.8 (1996) p.1027]

[3]   Y. Goto [*Chemistry of Liquid Crystals, Kikan Kagaku Sosetsu (Japan)* no.22 (1994)]

[4]   V. Reiffenrath, U. Finkenzeller, E. Petsh, B. Rieger, D. Coates [*Proc. SPIE (USA)* (1990) p.84]

[5]   H. Takatsu, K. Takeuchi, Y. Tanaka, M. Sasaki [*Mol. Cryst. Liq. Cryst. (UK)* vol.141 (1986) p.279]

[6]   D. Demus, J. Goodby, G.W. Gray, H.-W. Spiess, V. Vill [*Handbook of Liquid Crystals* vol.2A (VCH, Germany, 1998) p.199]

[7]   K. Tarumi, E. Bartmann, T. Geelhaar, B. Schuler, H. Ichinose, H. Numata [Asia Display 1995 (1995) p.559]

11.2   Dielectric properties of nematics for applications

S. Naemura

July 1999

A   INTRODUCTION

We focus on the static dielectric permittivities. The correlation with molecular properties and structures will be reviewed and state-of-the-art data will be presented. The frequency dependence of the dielectric permittivities has been reviewed in [1].

B   DIELECTRIC ANISOTROPY AND MOLECULAR DESIGN

Due to the uniaxial or $D_{\infty h}$ symmetry of a nematic phase, the dielectric permittivity of a nematic is represented by a second rank tensor with two principal elements, $\widetilde{\varepsilon}_{\parallel}$ and $\widetilde{\varepsilon}_{\perp}$. The component $\widetilde{\varepsilon}_{\parallel}$ is parallel to the macroscopic symmetry axis, which is along the director, and $\widetilde{\varepsilon}_{\perp}$ is perpendicular to this. According to a molecular field theory, they are approximated by

$$\widetilde{\varepsilon}_{\gamma} = 1 + \left(NhF/\varepsilon_0\right)\left[\left\langle\alpha_{\gamma}\right\rangle + F\left\{\left\langle\mu_{\gamma}^2\right\rangle/k_BT\right\}\right]\ \left(\gamma = \parallel, \perp\right) \tag{1}$$

Here N is the number density, F is the reaction field factor, h is the cavity field factor, T is the absolute temperature, and ε_0 is the permittivity of the vacuum. The brackets $<\ >$ denote an average over the molecular orientations and $<\alpha_{\gamma}>$ $(\gamma = \parallel, \perp)$ is the average of the electronic polarisability and $<\mu_{\gamma}^2>$ $(\gamma = \parallel, \perp)$ the average of the square of the dipole moment.

The dielectric anisotropy, $\Delta\widetilde{\varepsilon} \equiv \widetilde{\varepsilon}_{\parallel} - \widetilde{\varepsilon}_{\perp}$, is given by

$$\Delta\widetilde{\varepsilon} = \left(NhF/\varepsilon_0\right)\left[\left(\alpha_l - \alpha_t\right) - F\left(\mu^2/2k_BT\right)\ \left(1 - 3\cos^2\beta\right)\right]S \tag{2}$$

where α_l and α_t are the principal elements of the molecular electronic polarisability, μ is the magnitude of the total permanent dipole moment, and β is the angle between the dipole and the molecular long or symmetry axis.

The temperature dependence of the dielectric properties of a nematic phase, for instance, can be well-understood by EQNS (1) and (2). Moreover, at a fixed temperature, or at a constant value of the order parameter S, EQN (2) gives a good estimation of the dielectric anisotropy using molecular dipole values calculated using MO methods [2].

The disagreement between calculated and measured dielectric anisotropy values, from a theoretical stand point, can be partly due to the change of the molecular conformation in a nematic phase. Investigations have been made on the effect of dipole-dipole association, and EQN (2) has been

523

modified by introducing the Kirkwood factor [3]; this provides a better correlation for some classes of substances [4].

From the viewpoint of display applications, liquid crystal mixtures are formulated to exhibit a large dielectric anisotropy, either positive or negative. Accordingly, most of the practical liquid crystal molecules are designed to possess a strong molecular dipole either parallel $(\beta \to 0°$ for $\Delta\tilde{\varepsilon} > 0)$ or perpendicular $(\beta \to 90°$ for $\Delta\tilde{\varepsilon} < 0)$ to the molecular long axis. It is also necessary to avoid the anti-parallel dipole-dipole association, which reduces the contribution of the molecular dipole to the macroscopic dielectric anisotropy.

In a general sense, however, molecules with a large dipole tend to dissolve relatively large amounts of ionic materials. Ions contained in liquid crystal materials reduce the picture quality and also the reliability of display devices, and so advanced liquid crystal materials are being designed to have the features of both large dielectric anisotropy and strong resistance against ionic contamination [5]. These features can be achieved by designing the molecule to minimise the interaction with charged particles [6].

Some classes of fluoro-substituted compounds have those features and are the main components of practical liquid crystal mixtures for use in sophisticated displays like active-matrix liquid crystal displays (LCDs).

The other trend for materials is an increasing interest in strongly polar compounds with a negative dielectric anisotropy. This class of substance is especially useful in such displays as IPS (in-plane switching), VA (vertically aligned), and so on, which give wide-viewing-angle properties. As such practical displays are also active-matrix LCDs, the main components of the liquid crystal mixtures used are fluorinated substances.

C   DIELECTRIC PROPERTIES OF TYPICAL COMPOUNDS

Many data are available for the dielectric anisotropy, $\Delta\tilde{\varepsilon}$, of liquid crystal substances and some are for the components $\tilde{\varepsilon}_{\parallel}$ and $\tilde{\varepsilon}_{\perp}$. In TABLE 1 measured values of those dielectric properties are listed mainly for modern substances with fluoro substitutions. For comparison a limited number of data are also included for conventional substances.

Most of the data are for practical applications in the design of mixtures and, therefore, the values are at room temperature. The measured temperatures, which are not mentioned clearly in the original report, are taken to be most probably 20°C.

The mesophases and the transition temperatures are also listed in TABLE 1. As the transition temperature is more sensitive to the purity of the materials, a wider dispersion among the literature values can be seen compared to the dielectric properties, and the dispersion is represented by (xx - yy) in the mesophase column, and mesophase transitions given in parentheses represent a monotropic phase.

TABLE 1 Dielectric properties of typical liquid crystals. Cr: crystal; SmA, SmB etc.: smectic subphases; N: nematic; I: isotropic liquid, T_{Cr}: melting point; T_{NI}: clearing point, Tr = T/T_{NI}: reduced temperature.

Group 1

Structure		Mesophases	Temp/°C (Tr.K)	$\tilde\varepsilon_\parallel$	$\tilde\varepsilon_\perp$	$\Delta\tilde\varepsilon$	Ref	Meas. conditions
R	X							
C_5H_{11}	CN	Cr(22.5–24°C)N(34.8–35.2°C)I	20 [1 kHz]			17	[7]	ex. ZLI-1132
			20		5.3	26.8	[8]	ex. 10 wt % I-Mix. ex.
			22		6.46	21.6	[9]	
			(22.9), Tr = 0.96			13.82	[10]	
				18.5		12.5	[11]	
			(24.8), T_{NI}-10		6.63	13.33	[10]	
			25		7	11.5	[12]	
				17	6	11	[13]	
			(28.8), Tr = 0.98	16.5	8.2	8.3	[14]	
			29.0 [1592 Hz]	17.1	7.2	9.9	[15]	
			Temp. dependence				[10,15]	
C_7H_{15}	CN	Cr(28.5–30°C)N(42.5–42.8°C)I	23	17.3	5.42	11.9	[16]	
			(26.8), Tr = 0.950	16.35	5.23	11.12	[17]	
			(35.0), Tr = 0.976	15.73	5.75	9.98	[17]	
			36	16.25	5.92	10.3	[16]	
			36.0 [1592 Hz]	15.7	6	9.7	[15]	
			(36.5), Tr = 0.98			8.7	[18]	
			40	15.8	6.2	9.6	[16]	
			(40.1), Tr = 0.992	14.87	6.4	8.47	[17]	
			(42.0), Tr = 0.998	13.9	7.1	6.8	[17]	
			Temp. dependence				[15,19]	

(Structure column: R–[biphenyl]–X)

TABLE 1 continued

Group 1

Structure: R–⟨ring⟩–⟨ring⟩–X

R	X	Mesophases	Temp/°C (Tr.K)	$\tilde{\varepsilon}_\parallel$	$\tilde{\varepsilon}_\perp$	$\Delta\tilde{\varepsilon}$	Ref	Meas. conditions
C_3H_7	SO_2F	Cr(94°C)I	20 [1 kHz]			27.9	[20]	ex.
C_3H_7	$OCHF_2$	Cr(84°C)I				11.9	[21]	ex. ZLI-1132
C_4H_9O	$OCHF_2$	Cr(122°C)I				11.4	[21]	ex. ZLI-1132
C_3H_7	$SCHF_2$	Cr(58°C)I	20 [1 kHz]			11.3	[20]	ex.
C_7H_{15}	$SOCHF_2$	Cr(72°C)I	20 [1 kHz]			14.8	[20]	ex.
C_7H_{15}	SO_2CHF_2	Cr(50°C)I	20 [1 kHz]			11.2	[20]	ex.
C_3H_7	CF_3	Cr(97°C)I				14.2	[21]	ex. ZLI-1132
C_4H_9O	CF_3	Cr(134°C)I				9.4	[21]	ex. ZLI-1132
C_3H_7	OCF_3	Cr(92°C)I				9.4	[21]	ex. ZLI-1132
C_5H_{11}	OCF_3					8.4	[9]	ex.
C_4H_9O	OCF_3	Cr(126°C)I				9.7	[21]	ex. ZLI-1132
C_5H_{11}	OCH_2CF_3	Cr(107°C)I	20 [1 kHz]			8.7	[20]	ex.
C_5H_{11}	$COCF_3$	Cr(13°C)I	20 [1 kHz]			17.5	[20]	ex.
C_5H_{11}	SCF_3	Cr(31°C)I	20 [1 kHz]			8.4	[20]	ex.
C_4H_9O	SCF_3	Cr(82°C)I	20 [1 kHz]			9.9	[20]	ex.

Group 2

Structure: R–⟨ring (F)⟩–⟨ring⟩–X

R	X	Mesophases	Temp/°C (Tr.K)	$\tilde{\varepsilon}_\parallel$	$\tilde{\varepsilon}_\perp$	$\Delta\tilde{\varepsilon}$	Ref	Meas. conditions
C_5H_{11}	CN	N(0°C)I	20		7.3	29.6	[8]	ex. 10 wt % I-Mix.
C_4H_9O	CN	N(35°C)I	20		7.9	33	[8]	ex. 10 wt % I-Mix.
C_5H_{11}	OCF_3					9.1	[9]	ex.

TABLE 1 continued

Group 3

Structure		Mesophases	Temp/°C (Tr.K)	$\tilde{\varepsilon}_{\parallel}$	$\tilde{\varepsilon}_{\perp}$	$\Delta\tilde{\varepsilon}$	Ref	Meas. conditions
R	X							
C$_5$H$_{11}$	OCF$_3$					12.8	[9]	ex.

Group 4

Structure		Mesophases	Temp/°C (Tr.K)	$\tilde{\varepsilon}_{\parallel}$	$\tilde{\varepsilon}_{\perp}$	$\Delta\tilde{\varepsilon}$	Ref	Meas. conditions
R	X							
C$_5$H$_{11}$	CH$_3$	Cr(25°C)(N-4°C)I	20 [1 kHz]			0.3	[22]	ex.
C$_5$H$_{11}$	OCH$_3$	Cr(41°C)(N31°C)I				-0.5	[7]	ex. ZLI-1132
C$_3$H$_7$–CH=CH	OC$_2$H$_5$	Cr(32°C)N(56.5°C)I	(45.5), T$_{NI}$-10		3.18	-0.17	[23]	
C$_3$H$_7$	CN	Cr(36-43.0°C)N(45.0-46°C)I	20			21.1	[9]	ex.
				17.1		13.5	[24]	ex. ZLI-1132
						11.6	[25]	ex. 10%
C$_5$H$_{11}$	CN	Cr(30-31°C)N(54.8-55.0°C)I	20 [1 kHz]		5.5	18	[2]	ex. ZLI-1132
			20 [1 kHz]			13	[7]	
			22		4.85	12.22	[10]	
			(41.9), Tr = 0.96			11	[11]	
			(53.8), T$_{NI}$-10		5.59	10.28	[10]	
C$_7$H$_{15}$	CN	Cr(30°C)N(59°C)I	(24.9), Tr = 0.900	14.83	4.32	10.51	[17]	
		Cr(54°C)(N52°C)I	(30.2), Tr = 0.916	14.69	4.44	10.25	[17]	
			15	15	5.1	9.9	[25]	
			(41.5), Tr = 0.950	14.25	4.75	9.5	[17]	
			(45.1), Tr = 0.961	14.06	4.88	9.18	[17]	

TABLE 1 continued

Group 4

Structure: R–(cyclohexane)–(phenyl)–X

R	X	Mesophases	Temp/°C (Tr.K)	$\tilde{\varepsilon}_\parallel$	$\tilde{\varepsilon}_\perp$	$\Delta\tilde{\varepsilon}$	Ref	Meas. conditions
CH₂:CHC₅H₁₁	CN	Cr(19°C)N(32.2°C)I	(45.5), Tr = 0.98	12.9	4.5	8.4	[25]	
C₅H₁₁	NCS	Cr(37°C)N(51°C)I	(50.1), Tr = 0.976	13.67	5.14	8.4	[18]	ex.
C₃H₇	Cl		(55.0), Tr = 0.991	12.98	5.6	8.53	[17]	ex.
CH₂:CHC₂H₅	F	Cr(31.3°C)(N-36°C)I	(22.2), T$_{NI}$-10		5.9	7.38	[17]	ex. binary mix.
C₅H₁₁	F	Cr(34°C)I				8.41	[26]	ex. 10%
			20 [1 kHz]			11	[22]	ex. ZLI-1132
			20 [1 kHz]			5.3	[9]	ex.
						1.6	[27]	ex.
						4	[2]	ex. ZLI-1132
						3.2	[7]	ex.
						3	[22]	ex.
C₃H₇	SO₂F	Cr(72°C)I	20 [1 kHz]			22.4	[20]	ex.
C₃H₇	OCHF₂	Cr(-14.9°C)I	20			8.3	[24]	ex. ZLI-1132
C₅H₁₁	OCHF₂	Cr(15°C)I				8.3	[21]	ex. ZLI-1132
C₇H₁₅	OCHF₂	Cr(1°C)(N-17°C)I	20			7.6	[21]	ex. ZLI-1132
C₃H₇	SCHF₂	Cr(7.0°C)I	20			5.6	[24]	ex. ZLI-1132
C₅H₁₁	SCHF₂	Cr(12.5°C)I				10.6	[24]	ex. ZLI-1132
		Cr(7°C)I	20 [1 kHz]			7.2	[20]	ex.
C₅H₁₁	COCHF₂	Cr(39°C)I	20 [1 kHz]			11.1	[20]	ex.
C₃H₇	CF₃	Cr(22°C)I				8.6	[21]	ex. ZLI-1132

TABLE 1 continued

Group 4

Structure R	Structure X	Mesophases	Temp/°C (Tr.K)	$\tilde{\varepsilon}_{\parallel}$	$\tilde{\varepsilon}_{\perp}$	$\Delta\tilde{\varepsilon}$	Ref	Meas. conditions
C$_5$H$_{11}$	CF$_3$	Cr(10°C)I	20 [1 kHz]			11	[7]	ex. ZLI-1132
		Cr(21°C)I				10.9	[21]	ex. ZLI-1132
C$_3$H$_7$	OCH$_2$CF$_3$					5.6	[9]	ex.
C$_3$H$_7$	OCF$_3$	Cr(14°C)I				7.1	[9]	ex.
						5.3	[21]	ex. ZLI-1132
C$_5$H$_{11}$	OCF$_3$	Cr(12-14°C)I				7.1	[21]	ex. ZLI-1132
						5.6	[2]	ex. 10%
			20 [1 kHz]			5	[7]	ex. ZLI-1132
C$_5$H$_{11}$	COCF$_3$	Cr(7°C)(N-24°C)I	20 [1 kHz]			15.5	[20]	ex.
C$_7$H$_{15}$	COCF$_3$	Cr(19°C)(N-5.1°C)I	20 [1 kHz]			15.5	[20]	ex.
C$_3$H$_7$	COOCH$_2$CF$_3$	Cr(43°C)I	20 [1 kHz]			6.8	[20]	ex.
C$_5$H$_{11}$	COOCH$_2$CF$_3$	Cr(49°C)I	20 [1 kHz]			5.9	[20]	ex.

Group 5

Structure R	Structure X	Mesophases	Temp/°C (Tr.K)	$\tilde{\varepsilon}_{\parallel}$	$\tilde{\varepsilon}_{\perp}$	$\Delta\tilde{\varepsilon}$	Ref	Meas. conditions
C$_3$H$_7$	CN	Cr(39°C)(N-11°C)I	20 [1 kHz]			26	[28]	ex.
C$_5$H$_{11}$	CN	Cr(31°C)(N5°C)I	20 [1 kHz]			22	[2]	ex. 10%
						18	[7]	ex. ZLI-1132
C$_7$H$_{15}$	F	Cr(7.9°C)I	20			3.4	[29]	ex. 20% FB-01
			25			4.5	[30]	ex. 20% FB-01

TABLE 1 continued

Group 6

Structure		Mesophases	Temp/°C (Tr.K)	$\tilde{\varepsilon}_{\parallel}$	$\tilde{\varepsilon}_{\perp}$	$\Delta\tilde{\varepsilon}$	Ref	Meas. conditions
R	X							
C_3H_7	CN	Cr(57°C)I	20 [1 kHz]			32.6	[28]	extr.
C_5H_{11}	CN					30	[2]	10% extr.
C_7H_{15}	F	Cr(25.6°C)I	25			6.8	[29,31]	20% FB-01
						5.2	[30]	20% FB-01

Group 7

Structure		Mesophases	Temp/°C (Tr.K)	$\tilde{\varepsilon}_{\parallel}$	$\tilde{\varepsilon}_{\perp}$	$\Delta\tilde{\varepsilon}$	Ref	Meas. conditions
R	X							
C_5H_{11}	CN	Cr(62.0°C)N(101.0°C)I	(86.0), Tr = 0.96			10	[11]	ex. ZLI-1132
C_7H_{15}	$OCHF_2$	Cr(10.8°C)N(31.6°C)I	20			3.8	[24]	

Group 8

Structure		Mesophases	Temp/°C (Tr.K)	$\tilde{\varepsilon}_{\parallel}$	$\tilde{\varepsilon}_{\perp}$	$\Delta\tilde{\varepsilon}$	Ref	Meas. conditions
R	X							
C_5H_{11}	C_2H_5	Cr(34°C)Sm(146°C)N(164.1°C)I	20 [1 kHz]			0.4	[7]	ex. ZLI-1132
C_5H_{11}	CN	Cr(96.0°C)N(219–222.0°C)I	20			12	[7,24]	ex. ZLI-1132
						11	[32]	ex. 10% ZLI-1132

TABLE 1 continued

Group 8

Structure: R—(cyclohexyl)—(phenyl)—(phenyl)—X

R	X	Mesophases	Temp/°C (Tr.K)	$\tilde{\varepsilon}_\parallel$	$\tilde{\varepsilon}_\perp$	$\Delta\tilde{\varepsilon}$	Ref	Meas. conditions
C_5H_{11}	NCS	Cr(123°C)SmA(134°C)N(234.3°C)I	20 [1 kHz]			10.9	[7]	ex. ZLI-1132
C_3H_7	F	Cr(98.3°C)N(153.4°C)I	20			4.3	[29]	ex. 20% FB-01
			25			4.7	[30]	ex. 20% FB-01
C_5H_{11}	F	Cr(100.1°C)N(153.9°C)I	20			3.8	[29]	ex. 20% FB-01
			25			4.6	[30]	ex. 20% FB-01
C_3H_7	SO_2F	Cr(156°C)I	20 [1 kHz]			27.1	[20]	ex.
C_5H_{11}	$COOCH_2CH_2F$	Cr(90°C)SmB(128°C)SmA(165°C)N(170.5°C)I	20 [1 kHz]			7.6	[20]	ex.
C_3H_7	$OCHF_2$	Cr(82.0°C)Sm(121.1°C)N(169.4°C)I				10.3	[21]	ex. ZLI-1132
		Cr(82°C)SmB(116°C)SmA(121°C)N(169.4°C)I	20			10.2	[24]	ex. ZLI-1132
C_5H_{11}	$OCHF_2$	Cr(67-69.5°C)SmB(119.6-120°C)N(161.8-167.5°C)I	20			7.9	[9]	ex.
						9.7	[24]	ex. ZLI-1132
						9.3	[21]	ex. ZLI-1132
			(109.6), Tr = 0.88	6.8	3.3	3.5	[24]	ex.
C_5H_{11}	$CO.CHF_2$	Cr(80°C)N(158°C)I	20 [1 kHz]			10.2	[20]	ex.
C_5H_{11}	SO_2CHF_2	Cr(119°C)I	20 [1 kHz]			13.1	[20]	ex.
C_5H_{11}	CF_3	Cr(123°C)N(124-125°C)I				12.9	[21]	ex. ZLI-1132

TABLE 1 continued

Group 8

R—⬡—⬡—X

Structure		Mesophases	Temp/°C (Tr.K)	$\tilde{\varepsilon}_\parallel$	$\tilde{\varepsilon}_\perp$	$\Delta\tilde{\varepsilon}$	Ref	Meas. conditions
R	X							
C_3H_7	OCF_3	Cr(90°C)SmB(129°C)N(151.4°C)I				8.9	[21]	ex. ZLI-1132
C_5H_{11}	OCF_3	Cr(43°C)SmB(128°C)N(147°C)I	20 [1 kHz]			8.7	[2]	ex. 10%
						7.9	[33]	ex. ZLI-4792
			20 [1 kHz]			8.9	[7]	ex. ZLI-1132
C_3H_7	OCH_2CF_3					6.3	[9]	ex.
C_5H_{11}	$COCF_3$	Cr(70°C)N(141.2°C)I	20 [1 kHz]			17.7	[20]	ex.
C_5H_{11}	SCF_3	Cr(60°C)SmB(78°C)N(105.2°C)I	20 [1 kHz]			9.4	[20]	ex.
C_5H_{11}	$SOCF_3$	Cr(123°C)I	20 [1 kHz]			12.5	[20]	ex.
C_5H_{11}	SO_2CF_3	Cr(125°C)I	20 [1 kHz]			22.9	[20]	ex.
C_5H_{11}	OCF_2Cl	Cr(96.0°C)Sm(112.5°C)N(123.0°C)I	20			9.2	[34]	ex. ZLI-1132

TABLE 1 continued

Group 9

Structure		Mesophases	Temp/°C (Tr.K)	$\tilde{\varepsilon}_\parallel$	$\tilde{\varepsilon}_\perp$	$\Delta\tilde{\varepsilon}$	Ref	Meas. conditions
R	X							
C_5H_{11}	CN	Cr(84°C)N(175.4°C)I	20 [1 kHz]			20.3	[7]	ex. ZLI-1132
C_3H_7	F	Cr(66-67.3°C) N(97.9-102.4°C)I	20 [1 kHz]			10	[7]	ex. ZLI-1132
			20			8.3	[29]	ex. 20% FB-01
						7.8	[2]	ex. 10%
			25			5.5	[30]	ex. 20% FB-01
C_5H_{11}	F	Cr(55°C)N(105-108.2°C)I	20 [1 kHz]			10	[32]	ex. 10% ZLI-1132
			20			7.3	[29]	ex. 20% FB-01
			25			5.3	[30]	ex. 20% FB-01
C_3H_7	$OCHF_2$	Cr(50.5°C)N(118.4°C)I	20			12.3	[24]	ex. ZLI-1132
			(98.8), Tr = 0.95	8.7	4.6	8.7	[9]	ex.
						4.1	[24]	ex.
C_3H_7	OCF_3					10.9	[35]	ex.
C_5H_{11}	OCF_3					9.4	[9]	ex.
C_3H_7	OCH_2CF_3					9.6	[9]	ex.
C_3H_7	Cl					7.4	[9]	ex.
C_3H_7	OCF_2Cl	Cr(43.0°C)N(75.0°C)I	20			11.5	[34]	ex. ZLI-1132

TABLE 1 continued

Group 10

Structure		Mesophases	Temp/°C (Tr.K)	$\tilde\varepsilon_\parallel$	$\tilde\varepsilon_\perp$	$\Delta\tilde\varepsilon$	Ref	Meas. conditions
R	X							
C₃H₇	F	Cr(40.7°C)(N33.2°C)I	20			12.8	[31]	ex. 20% FB-01
			20 [1 kHz]			12.6	[2]	ex. 10%
			20			6.8	[29]	ex. 20% FB-01
			25			4.4	[30]	ex. 20% FB-01
C₅H₁₁	F	Cr(30.4°C)N(58.0°C)I	20			11.3	[29,31]	ex. 20% FB-01
			25			6.1	[30]	ex. 20% FB-01
C₃H₇	OCHF₂					11.6	[9]	ex.
C₃H₇	OCF₃					14.8	[35]	ex.
C₃H₇	OCH₂CF₃					12.2	[9]	ex.

Group 11

Structure		Mesophases	Temp/°C (Tr.K)	$\tilde\varepsilon_\parallel$	$\tilde\varepsilon_\perp$	$\Delta\tilde\varepsilon$	Ref	Meas. conditions
R	X							
C₃H₇	F	Cr(66°C)N(94°C)I				15.2	[35]	ex.
						9.7	[28]	ex.
C₅H₁₁	F	Cr 63°C I				13.8	[33]	ex. ZLI-4792
C₃H₇	OCF₃					16.9	[35]	ex.

TABLE 1 continued

Group 12

Structure:

R	X	Mesophases	Temp/°C (Tr.K)	$\tilde{\varepsilon}_{\parallel}$	$\tilde{\varepsilon}_{\perp}$	$\Delta\tilde{\varepsilon}$	Ref	Meas. conditions
C$_3$H$_7$	OCF$_3$					12.5	[9,33]	ex. ZLI-4792
C$_5$H$_{11}$	OCF$_3$	Cr 86°C I				11.5-11.9	[9,33]	ex. ZLI-4792
CH$_3$OC$_2$H$_5$O	OCF$_3$					9.1	[9]	ex.

Group 13

Structure:

R	X	Mesophases	Temp/°C (Tr.K)	$\tilde{\varepsilon}_{\parallel}$	$\tilde{\varepsilon}_{\perp}$	$\Delta\tilde{\varepsilon}$	Ref	Meas. conditions
C$_3$H$_7$	F					15.6	[35]	ex.
C$_5$H$_{11}$	F					14	[2]	ex. 10%
C$_3$H$_7$	OCF$_3$	Cr 72°C I	20 [1 kHz]			12.4	[33]	ex. ZLI-4792
C$_5$H$_{11}$	OCF$_3$					16.7	[9]	ex.
		Cr 64°C I				15.1	[9]	ex.
						14.2	[33]	ex. ZLI-4792
C$_3$H$_7$	OCHFCF$_3$					19.3	[35]	ex.

TABLE 1 continued

Group 14

Structure		Mesophases	Temp/°C (Tr.K)	$\tilde{\varepsilon}_\parallel$	$\tilde{\varepsilon}_\perp$	$\Delta\tilde{\varepsilon}$	Ref	Meas. conditions
R	X							
C_3H_7	F					20.5	[9]	ex.
C_5H_{11}	F	Cr 114°C I				17.8	[33]	ex. ZLI-4792
C_3H_7	$OCHF_2$					16.1	[35]	ex.
C_5H_{11}	$OCHF_2$	Cr 77°C I				15.1	[33]	ex. ZLI-4792
C_3H_7	OCH_2CHF_2					14.3	[35]	ex.

Group 15

Structure		Mesophases	Temp/°C (Tr.K)	$\tilde{\varepsilon}_\parallel$	$\tilde{\varepsilon}_\perp$	$\Delta\tilde{\varepsilon}$	Ref	Meas. conditions
R	X							
C_3H_7	None	Cr(81°C)N(197°C)I				1.3	[28]	ex.
C_3H_7	CH_3	Cr(62°C)SmB(108°C)N(177°C)I	20 [1 kHz]			0.2	[7]	ex. ZLI-1132
C_3H_7	CN	Cr(79°C)N(241°C)I	20 [1 kHz]			14.8	[2]	ex. 10%
			20 [1 kHz]			13.2	[7]	ex. ZLI-1132
			(36.5), Tr = 0.8	9.68		6.48	[36]	ex. 15 ms% ZLI-4792
C_3H_7	F	Cr(54.1°C)Sm(96.6°C)N(155.2°C)I	20 [1 kHz]			7.3	[7]	ex. ZLI-1132
C_3H_7		Cr(90°C)N(158°C)I	20			3.8	[29]	ex. 20% FB-01
			20 [1 kHz]			3	[2]	ex. 10%

TABLE 1 continued

Group 15

R–⬡–⬡–◯–X

Structure		Mesophases	Temp/°C (Tr.K)	$\tilde\varepsilon_\parallel$	$\tilde\varepsilon_\perp$	$\Delta\tilde\varepsilon$	Ref	Meas. conditions
R	X							
C$_3$H$_7$	OCHF$_2$	Cr(50.8-52°C)Sm(69°C)N(172.2-173.6°C)I	25			4.6	[30]	ex. 20% FB-01
			(28.9), Tr = 0.8	7.77		4.78	[36]	ex. 15 ms% ZLI-4792
C$_3$H$_7$	CF$_3$	Cr(133°C)I	20			10.5	[21]	ex. ZLI-1132
						8.3	[24]	ex. ZLI-1132
						5.2	[9]	ex.
			(29.8), Tr = 0.8	8.16		5.03	[36]	ex. 15 ms% ZLI-4792
C$_3$H$_7$	OCF$_3$	Cr(38-39°C)SmB(68-69°C)N(149-153.7°C)I	(105.4), Tr = 0.85	5	2.8	2.2	[24]	ex. ZLI-1132
						13.2	[21]	ex. ZLI-1132
			(23.5), Tr = 0.8	8.82		5.79	[36]	ex. 15 ms% ZLI-4792
						7.6	[9]	ex.
C$_3$H$_7$	CO.CF$_2$CH$_3$	Cr(73°C)SmB(152°C)N(199°C)I	20 [1 kHz]			9.2	[7,21]	ex. ZLI-1132
			20 [1 kHz]			6.9	[2]	ex. 10%
			20			6.5	[37]	ex. 30% HM1
			(27.3), Tr = 0.8	8.21		5.19	[36]	ex. 15 ms% ZLI-4792
			20 [1 kHz]			7.4	[20]	ex.

TABLE 1 continued

Group 15

R	X	Mesophases	Temp/°C (Tr.K)	$\tilde\varepsilon_\parallel$	$\tilde\varepsilon_\perp$	$\Delta\tilde\varepsilon$	Ref	Meas. conditions
C₃H₇	Cl		20 [1 kHz]			3.9	[2]	ex. 10%
			(32.8), Tr = 0.8	7.85		4.87	[36]	ex. 15 ms% ZLI-4792
C₃H₇	OCF₂Cl	Cr(82.0°C)N(133.0°C)I	20			7.5	[34]	ex. ZLI-1132

Group 16

R	X	Mesophases	Temp/°C (Tr.K)	$\tilde\varepsilon_\parallel$	$\tilde\varepsilon_\perp$	$\Delta\tilde\varepsilon$	Ref	Meas. conditions
CH₃-CH=CH	CH₃	Cr(87°C)N(186°C)I	(176), T_{NI}-10			0.18	[27]	ex. ZLI-1132
C₃H₇	CN	Cr(53-54.3°C)N(203-204°C)I	20 [1 kHz]			19.4	[7,21]	
			(32.5), Tr = 0.8	10.47		7.28	[36]	ex. 15 ms% ZLI-4792
C₃H₇	NCS		(34.9), Tr = 0.8	9.09		5.9	[36]	ex. 15 ms% ZLI-4792
			20 [1 kHz]			9.3	[7]	ex. ZLI-1132
C₃H₇	F	Cr(44.2-46°C) N(118.0-124°C)I	20 [1 kHz]			6.4	[2]	ex. 10%
			20			5.8	[29]	ex. 20% FB-01
			(22.8), Tr = 0.8	8.51		5.31	[36]	ex. 15 ms% ZLI-4792
			25			4.4	[30]	ex. 20% FB-01

TABLE 1 continued

Group 16

Structure R	Structure X	Mesophases	Temp/°C (Tr.K)	$\tilde{\varepsilon}_\parallel$	$\tilde{\varepsilon}_\perp$	$\Delta\tilde{\varepsilon}$	Ref	Meas. conditions
CH$_3$-CH=CH	F	Cr(49°C)Sm(65°C)N(159°C)I	(149), T$_{NI}$-10			3.22	[27]	ex. ZLI-1132
C$_3$H$_7$	OCHF$_2$	Cr(33°C)N(144°C)I				11.7	[21]	ex.
						7.4	[9]	
C$_3$H$_7$	CF$_3$		(24.7), Tr = 0.8	8.61		5.36	[36]	ex. 15 ms% ZLI-4792
C$_3$H$_7$	OCF$_3$		(18.6), Tr = 0.8	9.5		6.25	[36]	ex. 15 ms% ZLI-4792
						9	[35]	ex.
C$_3$H$_7$	OCH$_2$CF$_3$		(22.8), Tr = 0.8	8.8		5.61	[36]	ex. 15 ms% ZLI-4792
			(24.8), Tr = 0.8	8.69		5.3	[36]	ex. 15 ms% ZLI-4792
C$_3$H$_7$	Cl		20 [1 kHz]			5.6	[2]	ex. 10%
			(22.5), Tr = 0.8	8.41		5.22	[36]	ex. 15 ms% ZLI-4792

TABLE 1 continued

Group 17

Structure R	Structure X	Mesophases	Temp/°C (Tr.K)	$\tilde{\varepsilon}_{\parallel}$	$\tilde{\varepsilon}_{\perp}$	$\Delta\tilde{\varepsilon}$	Ref	Meas. conditions
C_3H_7	CN	$T_{NI} = 102.1°C$	(27.1), Tr = 0.8	12.09		8.57	[36]	ex. 15 ms% ZLI-4792
C_3H_7	NCS	$T_{NI} = 106.3°C$	(30.4), Tr = 0.8	9.57		6.41	[36]	ex. 15 ms% ZLI-4792
C_3H_7	F	Cr(64.7°C)N(93.7°C)I	(19.6), Tr = 0.8	9.14		5.96	[36]	ex. 15 ms% ZLI-4792
			20 [1 kHz]			9.7	[2]	ex. 10%
			20			8.3	[29,31]	ex. 20% FB-01
			25			5.5	[30]	ex. 20% FB-01
C_3H_7	CHF_2					9.3	[9]	ex.
C_3H_7	$OCHF_2$	$T_{NI} = 92.8°C$	(19.6), Tr = 0.8	9.08		5.8	[36]	ex. 15 ms% ZLI-4792
C_3H_7	CF_3	$T_{NI} = 88.9°C$	20 [1 kHz]			8.8	[35]	ex.
			20 [1 kHz]			16.3	[2]	ex. 10%
			(16.5), Tr = 0.8	9.73		6.33	[36]	ex. 15 ms% ZLI-4792
C_3H_7	OCF_3	$T_{NI} = 91.8°C$	20 [1 kHz]			10.5	[2]	ex. 10%
			(18.8), Tr = 0.8	9.35		6.16	[36]	ex. 15 ms% ZLI-4792
						9.5	[9]	ex.
C_3H_7	OCH_2CF_3	$T_{NI} = 95.4°C$	(21.7), Tr = 0.8	9.08		5.75	[36]	ex. 15 ms% ZLI-4792

TABLE 1 continued

Group 17

Structure		Mesophases	Temp/°C (Tr.K)	$\tilde{\varepsilon}_{\parallel}$	$\tilde{\varepsilon}_{\perp}$	$\Delta\tilde{\varepsilon}$	Ref	Meas. conditions
R	X							
C$_3$H$_7$	OCHFCF$_3$	$T_{NI} = 95.7$°C	20 [1 kHz]			13.9	[35]	ex.
C$_3$H$_7$	Cl		(22.0), Tr = 0.8	8.93		9.4	[2]	ex. 10%
						5.81	[36]	ex. 15 ms% ZLI-4792

Group 18

Structure		Mesophases	Temp/°C (Tr.K)	$\tilde{\varepsilon}_{\parallel}$	$\tilde{\varepsilon}_{\perp}$	$\Delta\tilde{\varepsilon}$	Ref	Meas. conditions
R	X							
C$_3$H$_7$	OCH$_3$	Cr(207°C)SmB(211°C)I	20			-2.2	[37]	ex. 20% HM1
C$_3$H$_7$	OCF$_3$	Cr(44°C)SmX(112°C)	20			4.3	[37]	ex. 20% HM1
		SmB(147°C)N(189°C)I						

TABLE 1 continued

Group 19

Structure		Mesophases	Temp/°C	$\tilde\varepsilon_\parallel$	$\tilde\varepsilon_\perp$	$\Delta\tilde\varepsilon$	Ref	Meas. conditions
R	X							
C$_3$H$_7$	F	Cr(109.3°C)N(234.0°C)I	20			12.8	[31]	ex. 20% FB-01
	F	Cr(105.8°C)N(>250°C)I	20			12.6	[31]	ex. 20% FB-01
			20			5.3	[29]	ex. 20% FB-01
			25			6.5	[30]	ex. 20% FB-01
C$_5$H$_{11}$	F	Cr(87.8°C)N(>250°C)I	25			6.1	[30]	ex. 20% FB-01
			20			11.3	[29,31]	ex. 20% FB-01

Group 20

Structure		Mesophases	Temp/°C (Tr.K)	$\tilde\varepsilon_\parallel$	$\tilde\varepsilon_\perp$	$\Delta\tilde\varepsilon$	Ref	Meas. conditions
R	X							
C$_5$H$_{11}$	CN	Cr(96.0°C)(SmA93.5°C)N(109°C)I					[24]	
C$_5$H$_{11}$	SCHF$_2$	Cr(50.8)I	20			14.7	[24]	ex. ZLI-1132

TABLE 1 continued

Group 21

Structure		Mesophases	Temp/°C (Tr.K)	$\tilde{\varepsilon}_{\parallel}$	$\tilde{\varepsilon}_{\perp}$	$\Delta\tilde{\varepsilon}$	Ref	Meas. conditions
R	X							
C$_5$H$_{11}$	CN	Cr(70.5-71°C)(N52-53.3°C)I	20		7.6	43.5	[8]	ex. 10 wt% I-Mix.
			20		10	34	[24]	ex. ZLI-1132
			0.98 T$_{NI}$	31.3		21.3	[14]	
			Temp. dependence				[14,38]	
C$_3$H$_7$	SCHF$_2$	Cr(53.2°C)I	20			26.9	[24]	ex. ZLI-1132
C$_5$H$_{11}$	SCHF$_2$	Cr(43.1°C)I	20			22.7	[24]	ex. ZLI-1132
C$_3$H$_7$	OCF$_2$Cl	Cr(44.9°C)I	20			13.1	[34]	ex. ZLI-1132

Group 22

Structure		Mesophases	Temp/°C (Tr.K)	$\tilde{\varepsilon}_{\parallel}$	$\tilde{\varepsilon}_{\perp}$	$\Delta\tilde{\varepsilon}$	Ref	Meas. conditions
R	X							
C$_4$H$_9$	SCHF$_2$	Cr(69.1°C)Sm(111.1°C)N(131.4°C)I	20			14.4	[24]	ex. ZLI-1132

TABLE 1 continued

Group 23

Structure		Mesophases	Temp/°C (Tr.K)	$\tilde{\varepsilon}_{\parallel}$	$\tilde{\varepsilon}_{\perp}$	$\Delta\tilde{\varepsilon}$	Ref	Meas. conditions
R	X							
C$_5$H$_{11}$	CN	Cr(33.6°C)N(43.5°C)I	20			18.9	[24]	ex. ZLI-1132
C$_5$H$_{11}$	NCS	Cr(34.0°C)SmA(98.5°C)I	20			15.5	[24]	ex. ZLI-1132
C$_5$H$_{11}$	F	Cr(28.1°C)I	20			9.8	[24]	ex. ZLI-1132
C$_5$H$_{11}$	C$_6$F$_{13}$	Cr(81.1°C)(SmB74.6°C)I	20			0.7	[24]	ex. ZLI-1132
C$_5$H$_{11}$	OCHF$_2$	Cr(26.0°C)Sm(43.6°C)I	20			15.9	[24]	ex. ZLI-1132
C$_5$H$_{11}$	SCHF$_2$	Cr(2.5°C)I	20			22.9	[24]	ex. ZLI-1132
C$_5$H$_{11}$	OCF$_3$	Cr(18.6°C)SmB(38.5°C)SmA(52.4°C)I	20			11.8	[24]	ex. ZLI-1132

Group 24

Structure		Mesophases	Temp/°C (Tr.K)	$\tilde{\varepsilon}_{\parallel}$	$\tilde{\varepsilon}_{\perp}$	$\Delta\tilde{\varepsilon}$	Ref	Meas. conditions
R	X							
C$_4$H$_9$	OCHF$_2$	Cr(57.1°C)Sm(135.7°C)N(165.3°C)I	20			13.8	[24]	ex. ZLI-1132
C$_4$H$_9$	OCF$_3$	Cr(35.1°C)Sm(149.2°C)N(154.8°C)I	20			10.6	[24]	ex. ZLI-1132
C$_4$H$_9$	SCHF$_2$		Tr = 0.95	11.1	5.7	5.4	[24]	
C$_4$H$_9$	OCF$_2$Cl	Cr(79.5°C)Sm(121.7°C)N(130.3°C)I	20			12.2	[34]	ex. ZLI-1132

TABLE 1 continued

Group 25

Structure		Mesophases	Temp/°C (Tr.K)	$\tilde{\varepsilon}_{\parallel}$	$\tilde{\varepsilon}_{\perp}$	$\Delta\tilde{\varepsilon}$	Ref	Meas. conditions
R	X							
C$_5$H$_{11}$	OC$_9$H$_{19}$	Cr(48°C)(SmA41°C)N(56.5°C)I				0.02	[18]	
C$_5$H$_{11}$		Cr(56-57.0°C)(N48-52°C)I	20			32	[24]	ex. ZLI-1132
			20		8.2	29.6	[8]	ex. 10 wt% I-Mix.
C$_5$H$_{11}$	OCHF$_2$	Cr(23.0°C)(Sm4.0°C)(N8.0°C)I				23.8	[9]	ex.
			20			14.6	[24]	ex. ZLI-1132
C$_5$H$_{11}$	SCHF$_2$	Cr(42.5°C)I	20			16.1	[24]	ex. ZLI-1132
C$_5$H$_{11}$	OCF$_3$	Cr(23.6°C)SmB(34.9°C)I				11.8	[9]	ex.
			20			10	[24]	ex. ZLI-1132
C$_5$H$_{11}$	Cl					9.2	[9]	ex.

Group 26

Structure		Mesophases	Temp/°C (Tr.K)	$\tilde{\varepsilon}_{\parallel}$	$\tilde{\varepsilon}_{\perp}$	$\Delta\tilde{\varepsilon}$	Ref	Meas. conditions
R	X							
CH$_3$-CH=CH	F	Cr(111°C)N(160°C)I	(150), T$_{NI}$-10			6.8	[27]	

TABLE 1 continued

Group 27

Structure		Mesophases	Temp/°C (Tr.K)	$\tilde{\varepsilon}_\parallel$	$\tilde{\varepsilon}_\perp$	$\Delta\tilde{\varepsilon}$	Ref	Meas. conditions
R	X							
CH₃-CH=CH	F	Cr(96°C)N(125°C)I	(115), T_{NI}-10			12.4	[27]	

Group 28

Structure		Mesophases	Temp/°C (Tr.K)	$\tilde{\varepsilon}_\parallel$	$\tilde{\varepsilon}_\perp$	$\Delta\tilde{\varepsilon}$	Ref	Meas. conditions
R	X							
C₃H₇	C₅H₁₁	Cr(15°C)N(19°C)I	10.4	4.56	4.28	0.28	[39]	
C₅H₁₁	OC₄H₉	Cr(49°C)N(58°C)I	48.7	4.44	4.73	-0.28	[39]	
C₃H₇	CN	Cr(102°C)(N51-53°C)I	20			35.3	[9]	ex.
				38.7	7.4	35.1	[8]	ex. 10 wt% I-Mix.
					9.1	29.6	[40]	ex. 5-20 wt% Mix. A
C₅H₁₁	CN	Cr(64°C)(N57°C)I	Temp. dependence			25	[22]	ex.
						36		
C₃H₇	OCHF₂	Cr(48°C)I				17.5	[21]	ex. ZLI-1132
C₃H₇	CF₃	Cr(77°C)I				24.5	[21]	ex. ZLI-1132
C₃H₇	OCF₃	Cr(52°C)I				19.4	[21]	ex. ZLI-1132
C₅H₁₁	OCF₃	Cr(51°C)(SmA32°C)I				16.7	[21]	ex. ZLI-1132
						15	[9]	ex.
C₅H₁₁	CO.CF₃	Cr(27°C)I	20 [1 kHz]			26.7	[20]	ex.

TABLE 1 continued

Group 29

Structure		Mesophases	Temp/°C	$\tilde{\varepsilon}_{\parallel}$	$\tilde{\varepsilon}_{\perp}$	$\Delta\tilde{\varepsilon}$	Ref	Meas. conditions
R	X							
C$_7$H$_{15}$	CN	N(27.8°C)I	12.8, Tr = 0.950	62.52	13.6	48.9	[41]	
			18.8, Tr = 0.970	58.43	14.43	44	[41]	
			24.8, Tr = 0.990	51.7	16.14	35.6	[41]	
			26.3, Tr = 0.995	48.7	17.09	31.6	[41]	
C$_7$H$_{15}$	F					37.3	[9]	ex.

Group 30

Structure		Mesophases	Temp/°C (Tr.K)	$\tilde{\varepsilon}_{\parallel}$	$\tilde{\varepsilon}_{\perp}$	$\Delta\tilde{\varepsilon}$	Ref	Meas. conditions
R	X							
C$_5$H$_{11}$	C$_5$H$_{11}$	Cr(36.0°C)(Sm29.0°C) N(48.0°C)I	(41.6), Tr = 0.98			-0.4	[42]	
C$_5$H$_{11}$	CN	Cr(47.2°C)N(78.9-79.2°C)I	50			7	[43]	
			43.8, Tr = 0.900	13.86	5.67	8.19	[17]	
			50.0, Tr = 0.917	13.7	5.74	7.96	[17]	
			61.4, Tr = 0.950	13.23	5.92	7.31	[17]	
			70.0, Tr = 0.974	12.68	6.15	6.53	[17]	
			77.0, Tr = 0.994	11.83	6.57	5.26	[17]	
C$_5$H$_{11}$	OCHF$_2$	Cr(54°C)I				8	[21]	ex. ZLI-1132
C$_5$H$_{11}$	CF$_3$	Cr(63°C)(SmB59°C)I				8.6	[21]	ex. ZLI-1132

TABLE 1 continued

Group 30

Structure		Mesophases	Temp/°C (Tr.K)	$\tilde{\varepsilon}_{\parallel}$	$\tilde{\varepsilon}_{\perp}$	$\Delta\tilde{\varepsilon}$	Ref	Meas. conditions
R	X							
C$_3$H$_7$	OCF$_3$	Cr(32°C)(N26.1°C)I				7.5	[21]	ex. ZLI-1132
C$_5$H$_{11}$	OCF$_3$	Cr(37°C)(N34.8°C)I				6.4–6.55	[9,21]	ex. ZLI-1132
C$_5$H$_{11}$	OCH$_2$CF$_3$	Cr(84°C)I	20 [1 kHz]			6.8	[20]	ex.
C$_5$H$_{11}$	CO.CF$_3$	Cr(71°C)I	20 [1 kHz]			14	[20]	ex.

Group 31

Structure		Mesophases	Temp/°C (Tr.K)	$\tilde{\varepsilon}_{\parallel}$	$\tilde{\varepsilon}_{\perp}$	$\Delta\tilde{\varepsilon}$	Ref	Meas. conditions
R	X							
C$_5$H$_{11}$	None	Cr(87.5-88°C)N(114°C)I	(84), T$_{NI}$-30 [1592 Hz]	4.44	3.52	0.92	[44]	
C$_5$H$_{11}$	OCH$_3$	Cr(122-122.5°C)N(212°C)I	Temp. dependence (182), T$_{NI}$-30 [1592 Hz]	3.7	3.4	0.3	[44]	
C$_5$H$_{11}$	CH$_3$	Cr(106-107°C)N(176°C)I	(166), T$_{NI}$-10 [1592 Hz]	3.46	3.06	0.3	[44]	
			(146), T$_{NI}$-30 [1592 Hz]	3.65	3.13	0.52	[44]	
C$_5$H$_{11}$	C$_5$H$_{11}$	Cr(73.5-74°C)SmB(99.5°C) SmA(136°C)N(172.5°C)I	(162.5), T$_{NI}$-10 [1592 Hz]	3.18	2.98	0.2	[44]	

TABLE 1 continued

Structure:

R—[cyclohexane]—[benzene]—COO—[benzene]—X

Group 31

R	X	Mesophases	Temp/°C (Tr.K)	$\tilde{\varepsilon}_\parallel$	$\tilde{\varepsilon}_\perp$	$\Delta\tilde{\varepsilon}$	Ref	Meas. conditions
C_5H_{11}	CN	Cr(111–111.5°C)N(226°C)I	(142.5), T_{NI}-30	3.4	3.2	0.2	[44]	
			(196), T_{NI}-30 [1592 Hz]	24.4	7.5	16.9	[44]	
C_5H_{11}	CH_2CN	Cr(129–130°C)(SmE124°C)N(165.5°C)I	(156), T_{NI}-70 [1592 Hz]	24.4	7.5	16.9	[44]	
C_5H_{11}	CH_2CH_2CN	Cr(110–111°C)N(185°C)I	(135.5), T_{NI}-30 [1592 Hz]	15	9.6	5.4	[44]	
			Temp. dependence				[44]	
C_5H_{11}	NCS	Cr(118.5–119°C)SmA(129°C)N(235°C)I	(175), T_{NI}-10 [1592 Hz]	11	6.8	4.2	[44]	
			(155), T_{NI}-30 [1592 Hz]	12.3	7.1	5.2	[44]	
			(205), T_{NI}-30 [1592 Hz]	6.7	3.3	3.4	[44]	
C_5H_{11}	SCN	Cr(104–105°C)N(125°C)I	(165), T_{NI}-70 [1592 Hz]	9.3	3.8	5.5	[44]	
			(115), T_{NI}-10 [1592 Hz]	9.7	5.2	4.5	[44]	
C_5H_{11}	F	Cr(92–93°C)N(156°C)I	(126), T_{NI}-30 [1592 Hz]	8	4.1	3.9	[44]	
			Temp. dependence				[44]	

TABLE 1 continued

Structure: R—(cyclohexane)—(benzene)—COO—(benzene)—X

Group 31

R	X	Mesophases	Temp/°C (Tr.K)	$\tilde{\varepsilon}_\parallel$	$\tilde{\varepsilon}_\perp$	$\Delta\tilde{\varepsilon}$	Ref	Meas. conditions
C$_5$H$_{11}$	OCHF$_2$	Cr(87°C)(SmB69°C) SmA(96°C)N(177.9°C)I				14.5	[21]	ex. ZLI-1132
C$_5$H$_{11}$	OCF$_3$	Cr(106°C)(SmB84°C) SmA(131°C)N(167.9°C)I	(110), Tr = 0.85	12.1	4.8	7.3 15.3	[24] [21]	ex. ZLI-1132
C$_5$H$_{11}$	CO.CF$_3$	Cr(87°C)N(161.7°C)I	20 [1 kHz]			26	[20]	ex.
C$_5$H$_{11}$	Cl	Cr(104-104.5°C)N(191°C)I	(161), T$_{NI}$-30 [1592 Hz]	7.35	3.6	3.75	[44]	
C$_5$H$_{11}$	OCF$_2$Cl	Cr(113.0°C)N(150.0°C)I	Temp. dependence 20			13.8	[44]	
C$_5$H$_{11}$	Br	Cr(115.5-116°C)N(193°C)I	(163), T$_{NI}$-30 [1592 Hz]	6.2	3.4	2.8	[34] [44]	ex. ZLI-1132
C$_5$H$_{11}$	I	Cr(126.5-127.5°C) (SmA126°C)N(186°C)I	Temp. dependence (156), T$_{NI}$-30 [1592 Hz] Temp. dependence	6	3.3	2.7	[44] [44] [44]	

TABLE 1 continued

Group 32

Structure		Mesophases	Temp/°C (Tr.K)	$\tilde{\varepsilon}_{\parallel}$	$\tilde{\varepsilon}_{\perp}$	$\Delta\tilde{\varepsilon}$	Ref	Meas. conditions
R	X							
C₃H₇	F	Cr(98.3°C)(N90.1°C)I	20?			11.4	[29]	ex. 20% FB-01
			25			7.6	[30]	ex. 20% FB-01
C₅H₁₁	F	Cr(91.4°C)N(100.0°C)I	20?			21.9	[29]	ex. 20% FB-01
			25			8.4	[30]	ex. 20% FB-01

Group 33

Structure		Mesophases	Temp/°C (Tr.K)	$\tilde{\varepsilon}_{\parallel}$	$\tilde{\varepsilon}_{\perp}$	$\Delta\tilde{\varepsilon}$	Ref	Meas. conditions
R	X							
C₅H₁₁	OCHF₂	Cr(61°C)SmB(93°C)N(196.9°C)I				8.9	[21]	ex. ZLI-1132
C₅H₁₁	OCF₃	Cr(57°C)SmB(124°C)N(184.1°C)I	(126), Tr = 0.85	6.5	4.8	2.5	[24]	
						6.5	[9]	ex.
C₅H₁₁	OCH₂CF₃	Cr(84°C)SmB(178°C)N(197.4°C)I	20 [1 kHz]			7.4	[21]	ex. ZLI-1132
						7.9	[20]	ex.
C₅H₁₁	COCF₃	Cr(84°C)N(182.8°C)I	20 [1 kHz]			11.3	[20]	ex.

TABLE 1 continued

Group 34

Structure		Mesophases	Temp/°C (Tr.K)	$\tilde{\varepsilon}_\parallel$	$\tilde{\varepsilon}_\perp$	$\Delta\tilde{\varepsilon}$	Ref	Meas. conditions
R	X							
C₃H₇	F	Cr(56.8°C)N(117.5°C)I	20?			11.4	[29]	ex. 20% FB-01
			25			6.1	[30]	ex. 20% FB-01
C₅H₁₁	F	Cr(73.8°C)N(124.4°C)I	20?			10.4	[29]	ex. 20% FB-01
			25			5.9	[30]	ex. 20% FB-01

Group 35

Structure		Mesophases	Temp/°C (Tr.K)	$\tilde{\varepsilon}_\parallel$	$\tilde{\varepsilon}_\perp$	$\Delta\tilde{\varepsilon}$	Ref	Meas. conditions
R	X							
C₂H₅O	OCHF₂	Cr(37°C)I				14	[21]	ex. ZLI-1132
C₂H₅O	OCF₃	Cr(43°C)I				10.5	[21]	ex. ZLI-1132

552

TABLE 1 continued

Group 36

R—⬡—CH₂CH₂—⬢—X

Structure		Mesophases	Temp/°C (Tr.K)	$\tilde{\varepsilon}_\parallel$	$\tilde{\varepsilon}_\perp$	$\Delta\tilde{\varepsilon}$	Ref	Meas. conditions
R	X							
C$_5$H$_{11}$	OC$_2$H$_5$	Cr(18°C)N(46-46.6°C)I	(45), T$_{NI}$-10	2.67		-0.26	[45]	
C$_5$H$_{11}$	CN	Cr(31°C)N(52.5°C)I	(45.6), T$_{NI}$-10		2.98	-0.24	[23]	ex.
			22		4.4	10.8	[10]	
C$_5$H$_{11}$	OCHF$_2$	Cr(4°C)N(5.1°C)I	(42.5), T$_{NI}$-10		4.98	9.77	[10]	
C$_5$H$_{11}$	CF$_3$	Cr(33°C)I				7.8	[21]	ex. ZLI-1132
C$_5$H$_{11}$	OCF$_3$	Cr(20°C)I				9.1	[21]	ex. ZLI-1132
						7.2	[21]	ex. ZLI-1132

Group 37

R—⬡—CH₂CH₂—⬢(F)—X

Structure		Mesophases	Temp/°C (Tr.K)	$\tilde{\varepsilon}_\parallel$	$\tilde{\varepsilon}_\perp$	$\Delta\tilde{\varepsilon}$	Ref	Meas. conditions
R	X							
C$_5$H$_{11}$	F	Cr(1.5°C)I	20			3.4	[29]	ex. 20% FB-01
			25			4.5	[30]	ex. 20% FB-01

TABLE 1 continued

Group 38

Structure		Mesophases	Temp/°C (Tr.K)	$\tilde{\varepsilon}_{\parallel}$	$\tilde{\varepsilon}_{\perp}$	$\Delta\tilde{\varepsilon}$	Ref	Meas. conditions
R	X							
C$_5$H$_{11}$	F	Cr(-0.6°C)I	20			6.8	[29,31]	ex. 20% FB-01
			25			4.4	[30]	ex. 20% FB-01

Group 39

Structure		Mesophases	Temp/°C (Tr.K)	$\tilde{\varepsilon}_{\parallel}$	$\tilde{\varepsilon}_{\perp}$	$\Delta\tilde{\varepsilon}$	Ref	Meas. conditions
R	X							
C$_3$H$_7$	OCHF$_2$	Cr(131°C)I				12.5	[21]	ex. ZLI-1132
C$_3$H$_7$	OCF$_3$	Cr(97°C)SmB(137°C)I	20 [1 kHz]			9.7	[21]	ex. ZLI-1132
C$_3$H$_7$	Cl	Cr(111°C)N(123°C)I				7	[46]	ex. 10 wt% ZLI-4792

Group 40

Structure		Mesophases	Temp/°C (Tr.K)	$\tilde{\varepsilon}_{\parallel}$	$\tilde{\varepsilon}_{\perp}$	$\Delta\tilde{\varepsilon}$	Ref	Meas. conditions
R	X							
C$_3$H$_7$	None	Cr(67°C)N(82°C)I	25	3.5	1.38	2.12	[47]	ex. Mix. B
C$_3$H$_7$	OCH$_3$	Cr(96°C)N(175.6°C)I	20 [1 kHz]			1	[48]	ex. 10 wt% ZLI-4792
C$_3$H$_7$	SCH$_3$	Cr(92°C)SmA(124°C)N(165°C)I	20 [1 kHz]			2	[48]	ex. 10 wt% ZLI-4792

TABLE 1 continued

Group 40

R	X	Mesophases	Temp/°C (Tr.K)	$\tilde{\varepsilon}_{\parallel}$	$\tilde{\varepsilon}_{\perp}$	$\Delta\tilde{\varepsilon}$	Ref	Meas. conditions
C$_3$H$_7$	CN	Cr(74°C)N(188°C)I	20 [1 kHz]			13	[48]	ex. 10 wt% ZLI-4792
C$_3$H$_7$	F	Cr(76–76.7°C) N(125–126.7°C)I	20 [1 kHz]			3.6	[49]	ex. 10 wt% ZLI-4792
C$_3$H$_7$	Cl	Cr(76°C)N(125°C)I	25	6.32	1.43	4.89	[47]	ex. Mix. B
		Cr(100–101.2°C)N(158°C)I	20 [1 kHz]			1–3.5	[48,49]	ex. 10 wt% ZLI-4792
			25	8.83	1.33	7.5	[47]	ex. Mix. B
C$_3$H$_7$	Br	Cr(125°C)N(163°C)I	25	8.51	1.36	7.15	[47]	ex. Mix. B

Group 41

R	X	Mesophases	Temp/°C (Tr.K)	$\tilde{\varepsilon}_{\parallel}$	$\tilde{\varepsilon}_{\perp}$	$\Delta\tilde{\varepsilon}$	Ref	Meas. conditions
C$_3$H$_7$	CN	Cr(74.6°C)N(153.0°C)I	20 [1 kHz]			16.6	[49]	ex. 10 wt% ZLI-1132
C$_3$H$_7$	F	Cr(61.1°C)N(88.7–92.6°C)I	20			6.4	[29]	ex. 20% FB-01
			20 [1 kHz]			6.3	[49]	ex. 10 wt% ZLI-4792
C$_3$H$_7$	Cl	Cr(64.7°C)N(120.6°C)I	25			5.1	[30]	ex. 20% FB-01
			20 [1 kHz]			5.6	[49]	ex. 10 wt% ZLI-4792

555

TABLE 1 continued

Group 42

Structure		Mesophases	Temp/°C (Tr.K)	$\tilde{\varepsilon}_{\parallel}$	$\tilde{\varepsilon}_{\perp}$	$\Delta\tilde{\varepsilon}$	Ref	Meas. conditions
R	X							
C_3H_7	F	Cr(48.0°C)N(51.8°C)I	25			6.2	[30]	ex. 20% FB-01
			20			11.8	[29,31]	ex. 20% FB-01
C_4H_9	F	Cr(50.9°C)(N50.7°C)I	20			11.8	[31]	ex. 20% FB-01
C_5H_{11}	F	Cr(40.2°C)N(65.2°C)I	25			6	[30]	ex. 20% FB-01
			20			10.8	[29,31]	ex. 20% FB-01
C_3H_7	Cl	Cr(61.2°C)N(87.3°C)I	20 [1 kHz]			12.2	[49]	ex. 10 wt% ZLI-4792

Group 43

Structure		Mesophases	Temp/°C (Tr.K)	$\tilde{\varepsilon}_{\parallel}$	$\tilde{\varepsilon}_{\perp}$	$\Delta\tilde{\varepsilon}$	Ref	Meas. conditions
R	X							
C_3H_7	OCH_3	Cr(60°C)N(143°C)I	20 [1 kHz]			1	[48]	ex. 10 wt% ZLI-4792
C_3H_7	SCH_3	Cr(72°C)N(127°C)I	20 [1 kHz]			3	[48]	ex. 10 wt% ZLI-4792
C_3H_7	CN	Cr(63.1°C)N(159.7°C)I	20 [1 kHz]			20	[48]	ex. 10 wt% ZLI-4792
C_3H_7	F	Cr(47.0°C)N(88.7°C)I	20 [1 kHz]			5.1	[49]	ex. 10 wt% ZLI-4792

TABLE 1 continued

Group 43

Structure R	X	Mesophases	Temp/°C (Tr.K)	$\tilde{\varepsilon}_{\parallel}$	$\tilde{\varepsilon}_{\perp}$	$\Delta\tilde{\varepsilon}$	Ref	Meas. conditions
C$_3$H$_7$	CF$_3$	Cr(80.4°C)(N68.4°C)I	20 [1 kHz]			12.3	[49]	ex. 10 wt% ZLI-4792
C$_3$H$_7$	OCF$_3$	Cr(36.0°C)N(89.1°C)I	20 [1 kHz]			6.6	[49]	ex. 10 wt% ZLI-4792
C$_3$H$_7$	Cl	Cr(72°C)N(120°C)I	20 [1 kHz]			5	[48]	ex. 10 wt% ZLI-4792
			20 [1 kHz]			5.3	[49]	ex. 10 wt% ZLI-4792

Group 44

Structure R	X	Mesophases	Temp/°C (Tr.K)	$\tilde{\varepsilon}_{\parallel}$	$\tilde{\varepsilon}_{\perp}$	$\Delta\tilde{\varepsilon}$	Ref	Meas. conditions
C$_3$H$_7$	F	Cr(38.7°C)N(61.9°C)I	20 [1 kHz]			8.6	[49]	ex. 10 wt% ZLI-4792
C$_3$H$_7$	Cl	Cr(55.0°C)N(96.6°C)I	20 [1 kHz]			9	[49]	ex. 10 wt% ZLI-4792

TABLE 1 continued

Group 45

Structure		Mesophases	Temp/°C (Tr.K)	$\tilde{\varepsilon}_\parallel$	$\tilde{\varepsilon}_\perp$	$\Delta\tilde{\varepsilon}$	Ref	Meas. conditions
R	X							
C_3H_7	F	Cr(37.4°C)(N34.0°C)I	20 [1 kHz]			12.9	[49]	ex. 10 wt% ZLI-4792
C_3H_7	Cl	Cr(47.2°C)N(72.0°C)I	20 [1 kHz]			13.4	[49]	ex. 10 wt% ZLI-4792

Group 46

Structure		Mesophases	Temp/°C (Tr.K)	$\tilde{\varepsilon}_\parallel$	$\tilde{\varepsilon}_\perp$	$\Delta\tilde{\varepsilon}$	Ref	Meas. conditions
R	X							
C_5H_{11}	CN	Cr(123.0°C)N(131.0°C)I	20 [1 kHz]			22.4	[49]	ex. 10 wt% ZLI-1132
C_5H_{11}	F	Cr(91.0°C)(N67.6°C)I	20 [1 kHz]			6.1	[49]	ex. 10 wt% ZLI-4792
C_5H_{11}	CF_3	Cr(93.0°C)(N48.0°C)I	20 [1 kHz]			11.4	[49]	ex. 10 wt% ZLI-4792
C_5H_{11}	OCF_3	Cr(49.0°C)N(65.8°C)I	20 [1 kHz]			9.1	[49]	ex. 10 wt% ZLI-4792
C_5H_{11}	Cl	Cr(54.0°C)N(91.0°C)I	20 [1 kHz]			7.6	[49]	ex. 10 wt% ZLI-4792

TABLE 1 continued

Group 47

Structure		Mesophases	Temp/°C (Tr.K)	$\tilde{\varepsilon}_{\parallel}$	$\tilde{\varepsilon}_{\perp}$	$\Delta\tilde{\varepsilon}$	Ref	Meas. conditions
R	X							
C_5H_{11}	F	Cr(40.0°C)N(46.6°C)I	20 [1 kHz]			8.6	[49]	ex. 10 wt% ZLI-4792
C_5H_{11}	Cl	Cr(47.2°C)N(74.5°C)I	20 [1 kHz]			8.8	[49]	ex. 10 wt% ZLI-4792

Group 48

Structure		Mesophases	Temp/°C (Tr.K)	$\tilde{\varepsilon}_{\parallel}$	$\tilde{\varepsilon}_{\perp}$	$\Delta\tilde{\varepsilon}$	Ref	Meas. conditions
R	X							
C_5H_{11}	F	Cr(65.5°C)(N28.8°C)I	20 [1 kHz]			14.9	[49]	ex. 10 wt% ZLI-4792
C_5H_{11}	Cl	Cr(66.0°C)(N60.8°C)I	20 [1 kHz]			14.3	[49]	ex. 10 wt% ZLI-4792

559

TABLE 1 continued

Group 49

Structure		Mesophases	Temp/°C (Tr.K)	$\tilde{\varepsilon}_{\parallel}$	$\tilde{\varepsilon}_{\perp}$	$\Delta\tilde{\varepsilon}$	Ref	Meas. conditions
R	X							
C$_3$H$_7$	F	Cr(36.1°C)N(105.2°C)I	20			4	[29]	ex. 20% FB-01
			25			4.6	[30]	ex. 20% FB-01
C$_5$H$_{11}$	F	Cr(37.6°C)N(110.6°C)I	20			4.3	[29]	ex. 20% FB-01
			25			4.7	[30]	ex. 20% FB-01

Group 50

Structure		Mesophases	Temp/°C (Tr.K)	$\tilde{\varepsilon}_{\parallel}$	$\tilde{\varepsilon}_{\perp}$	$\Delta\tilde{\varepsilon}$	Ref	Meas. conditions
R	X							
C$_3$H$_7$	F	Cr(50.7°C)N(83.4°C)I	20			7.8	[29,31]	ex. 20% FB-01
			25			5.4	[30]	ex. 20% FB-01
C$_5$H$_{11}$	F	Cr(46.5°C)N(91.1°C)I	20			6.3	[29,31]	ex. 20% FB-01
			25			5.1	[30]	ex. 20% FB-01

Group 51

Structure		Mesophases	Temp/°C (Tr.K)	$\tilde{\varepsilon}_{\parallel}$	$\tilde{\varepsilon}_{\perp}$	$\Delta\tilde{\varepsilon}$	Ref	Meas. conditions
R	X							
C$_5$H$_{11}$	F	Cr(60°C)Sm(102°C)I				8	[32]	ex. 10 wt% ZLI-1132
C$_5$H$_{11}$	OCHF$_2$	Cr(133°C)I				11.5	[21]	ex. 10 wt% ZLI-1132

TABLE 1 continued

Group 51

Structure		Mesophases	Temp/°C (Tr.K)	$\tilde{\varepsilon}_\parallel$	$\tilde{\varepsilon}_\perp$	$\Delta\tilde{\varepsilon}$	Ref	Meas. conditions
R	X							
C_5H_{11}	CF_3	Cr(85°C)SmB(126°C)I				10.8	[21]	ex. 10 wt% ZLI-1132
C_5H_{11}	OCF_3	Cr(62°C)SmB(136°C)I				8.6	[21]	ex. 10 wt% ZLI-1132

Group 52

Structure		Mesophases	Temp/°C (Tr.K)	$\tilde{\varepsilon}_\parallel$	$\tilde{\varepsilon}_\perp$	$\Delta\tilde{\varepsilon}$	Ref	Meas. conditions
R	X							
C_5H_{11}	F	Cr(35°C)N(53°C)I				10	[32]	ex. 10 wt% ZLI-1132
C_5H_{11}	OCF_3	Cr(39°C)N(82°C)I				11	[32]	ex. 10 wt% ZLI-1132
C_5H_{11}	Cl	Cr(53°C)(SmB19°C)(SmA43°C)N(82°C)I	20 [1 kHz]			4	[46]	ex. 10 wt% ZLI-4792

Group 53

Structure		Mesophases	Temp/°C (Tr.K)	$\tilde{\varepsilon}_\parallel$	$\tilde{\varepsilon}_\perp$	$\Delta\tilde{\varepsilon}$	Ref	Meas. conditions
R	X							
C_3H_7	F					10.6	[9]	ex.

561

TABLE 1 continued

Group 54

Structure		Mesophases	Temp/°C (Tr.K)	$\tilde{\varepsilon}_{\parallel}$	$\tilde{\varepsilon}_{\perp}$	$\Delta\tilde{\varepsilon}$	Ref	Meas. conditions
R	X							
C$_3$H$_7$	OCHF$_2$	Cr(51°C)N(79.6°C)I				10.7	[21]	ex. 10 wt% ZLI-1132
C$_5$H$_{11}$	OCHF$_2$	Cr(32°C)SmB(49°C)N(90.2°C)I				10	[21]	ex. 10 wt% ZLI-1132
C$_3$H$_7$	OCF$_3$	Cr(68°C)I				8.4	[21]	ex. 10 wt% ZLI-1132
C$_5$H$_{11}$	OCF$_3$	Cr(47°C)SmB(68°C)N(73.7°C)I				7.2	[21]	ex. 10 wt% ZLI-1132
C$_5$H$_{11}$	Cl	Cr(128.8°C)(N106.5°C)I	20				[46]	ex. 10 wt% ZLI-4792

Group 55

Structure		Mesophases	Temp/°C (Tr.K)	$\tilde{\varepsilon}_{\parallel}$	$\tilde{\varepsilon}_{\perp}$	$\Delta\tilde{\varepsilon}$	Ref	Meas. conditions
R	X							
C$_3$H$_7$	OCHF$_2$	Cr(43°C)I				10.2	[33]	ex. 10 wt% ZLI-4792
C$_3$H$_7$	OCF$_3$	Cr(45°C)I				9.5	[33]	ex. 10 wt% ZLI-4792

562

TABLE 1 continued

Group 56

Structure		Mesophases	Temp/°C (Tr.K)	$\tilde\varepsilon_\parallel$	$\tilde\varepsilon_\perp$	$\Delta\tilde\varepsilon$	Ref	Meas. conditions
R	X							
C₃H₇	F					10.4	[9]	ex.

Group 57

Structure		Mesophases	Temp/°C (Tr.K)	$\tilde\varepsilon_\parallel$	$\tilde\varepsilon_\perp$	$\Delta\tilde\varepsilon$	Ref	Meas. conditions
R	X							
C₃H₇	CH₃	Cr(45°C)SmB(127°C)N(153°C)I				0	[22]	ex. ZLI-2857
C₃H₇	OCH₃	Cr(88°C)SmB(140°C)N(176°C)I				0	[22]	ex.
C₃H₇	CN	Cr(69°C)N(196°C)I				12	[22]	ex.
C₃H₇	F	Cr(45°C)SmB(83°C)N(134°C)I				6	[22]	ex.
C₃H₇	OCHF₂	Cr(19°C)SmB(105°C)N(153.6°C)I				7.1	[21]	ex. ZLI-1132
C₃H₇	CF₃	Cr(50°C)SmB(114°C)N(117°C)I	20 [1 kHz]			11.6	[7]	ex. ZLI-1132
C₅H₁₁	OCH₂CF₃	Cr(66°C)SmB(166°C)I	20 [1 kHz]			7.2	[20]	ex.
C₃H₇	OCF₃	Cr(24-60°C)SmG(76°C)N(134°C)I				8.1	[21,22]	ex. ZLI-1132
						7.4	[21]	ex. ZLI-1132
			20 [1 kHz]			8.1	[7]	ex. ZLI-1132

TABLE 1 continued

Group 58

Structure		Mesophases	Temp/°C (Tr.K)	$\tilde{\varepsilon}_{\parallel}$	$\tilde{\varepsilon}_{\perp}$	$\Delta\tilde{\varepsilon}$	Ref	Meas. conditions
R	X							
C$_3$H$_7$	OCH$_3$	Cr(93°C)SmB(109°C)N(160°C)I				-1	[22]	ex.
C$_3$H$_7$	CN	Cr(72°C)N(170°C)I				18	[22]	ex.
C$_3$H$_7$	F	Cr(20°C)SmB(50°C)N(117°C)I				12	[22]	ex.
C$_5$H$_{11}$	F	Cr(47°C)SmB(74°C)N(188°C)I	20 [1 kHz]			7.6	[7]	ex. ZLI-1132

Group 59

Structure		Mesophases	Temp/°C (Tr.K)	$\tilde{\varepsilon}_{\parallel}$	$\tilde{\varepsilon}_{\perp}$	$\Delta\tilde{\varepsilon}$	Ref	Meas. conditions
R	X							
C$_3$H$_7$	F	Cr(41.8°C)N(98.3°C)I	20			7.3	[29,31]	ex. 20% FB-01
			25			5.3	[30]	ex. 20% FB-01
C$_5$H$_{11}$	F	Cr(54.6°C)SmB(57.0°C)N(103.9°C)I	20			6.3	[29,31]	ex. 20% FB-01
			25			5.1	[30]	ex. 20% FB-01

TABLE 1 continued

Group 60

Structure		Mesophases	Temp/°C (Tr.K)	$\tilde{\varepsilon}_\parallel$	$\tilde{\varepsilon}_\perp$	$\Delta\tilde{\varepsilon}$	Ref	Meas. conditions
R	X							
C3H7	F	Cr(74°C)SmA(130°C)I				15.4	[9]	ex.
C7H15	OCHF2	Cr(87°C)SmB(111°C)SmA(146°C)I				17.9	[21]	ex. ZLI-1132
C7H15	CF3					21.6	[21]	ex. ZLI-1132
C7H15	OCF3	Cr(63°C)SmB(98°C)SmA(144°C)I				15.7	[21]	ex. ZLI-1132

Group 61

Structure		Mesophases	Temp/°C (Tr.K)	$\tilde{\varepsilon}_\parallel$	$\tilde{\varepsilon}_\perp$	$\Delta\tilde{\varepsilon}$	Ref	Meas. conditions
R	X							
C3H7	F	Cr(113°C)I				23.4	[33]	ex. ZLI-4792

Group 62

Structure		Mesophases	Temp/°C (Tr.K)	$\tilde{\varepsilon}_\parallel$	$\tilde{\varepsilon}_\perp$	$\Delta\tilde{\varepsilon}$	Ref	Meas. conditions
R	X							
C3H7	F	Cr(79.2°C)N(216.0°C)I	25			6.1	[30]	ex. 20% FB-01
			20			11.3	[29,31]	ex. 20% FB-01
C5H11	F	Cr(86.1°C)N(212.5°C)I	25			6	[30]	ex. 20% FB-01
			20			10.8	[29,31]	ex. 20% FB-01

565

TABLE 1 continued

Group 63

Structure		Mesophases	Temp/°C (Tr.K)	$\tilde{\varepsilon}_\parallel$	$\tilde{\varepsilon}_\perp$	$\Delta\tilde{\varepsilon}$	Ref	Meas. conditions
R	X							ex.
C_5H_{11}	OC_2H_5	Cr(61-62°C)N(89°C)I	20	3.2	3	0.2	[50,51]	
C_5H_{11}	$OCHF_2$	Cr(28.1°C)Sm(37.8°C)I	20			8.8	[24]	ex. ZLI-1132

Group 64

Structure		Mesophases	Temp/°C (Tr.K)	$\tilde{\varepsilon}_\parallel$	$\tilde{\varepsilon}_\perp$	$\Delta\tilde{\varepsilon}$	Ref	Meas. conditions
R	X							ex.
C_3H_7	C_4H_9	Cr(87°C)N(201°C)I	25			0	[50,52]	
C_3H_7	OC_2H_5	Cr(110°C)N(253°C)I	20				[51]	
C_3H_7	F	Cr(94°C)I				5.3	[33]	ex. ZLI-4792
C_3H_7	$OCHF_2$	Cr(87°C)SmB(108°C)SmA(132°C)N(212°C)I				11.6	[21]	ex. ZLI-1132
C_3H_7	OCF_3	Cr(88°C)SmB(126°C)SmA(163°C)N(198°C)I				9.7	[21]	ex. ZLI-1132

Group 65

Structure		Mesophases	Temp/°C (Tr.K)	$\tilde{\varepsilon}_\parallel$	$\tilde{\varepsilon}_\perp$	$\Delta\tilde{\varepsilon}$	Ref	Meas. conditions
R	X							
C_3H_7	F	Cr(76°C)I	20 [1 kHz]			12.1	[33]	ex. ZLI-4792
						8.6	[2]	ex. 10%

TABLE 1 continued

Group 66

R	R'	X	Y	Mesophases	Temp/°C (Tr.K)	$\tilde{\varepsilon}_\parallel$	$\tilde{\varepsilon}_\perp$	$\Delta\tilde{\varepsilon}$	Ref	Meas. conditions
C_3H_7	C_3H_7	None	4-CN	Cr(34°C)(N19°C)I				-8.7	[9]	ex.
C_5H_{11}	C_3H_7	None	4-CN	Cr(32°C)(SmB24°C)N(42°C)I				-8.2	[53]	ex. ZLI-2857
$CH_2{=}CH$	C_3H_7	None	4-F	Cr(29°C)SmB(58°C)N(69.1°C)I	20			-7.7	[51]	ex.
$CH_2{=}CH$	C_4H_9	None	4-F	Cr(16°C)SmB(77°C)I				-1.9	[54]	ex. ZLI-2857
$CH_2{=}CH$	$C_2H_4CH{:}CH_2$	None	4-F	Cr(13°C)SmB(37°C)N(79.8°C)I				-1.9	[54]	ex. ZLI-2857
C_3H_7	C_4H_9	None	4-F	Cr(30°C)SmB(112°C)I				-1.9	[54]	ex. ZLI-2857
C_3H_7	C_5H_{11}	None	4-F	Cr(46°C)SmB(115°C)I				-1.7	[54]	ex. ZLI-2857
C_3H_7	$C_2H_4CH{:}CH_2$	None	4-F	Cr(48°C)SmB(95°C)N(102.3°C)I				-2.1	[53]	ex. ZLI-2857
C_5H_{11}	C_3H_7	None	4-F	Cr(52°C)SmB(109°C)I				-1.7	[54]	ex. ZLI-2857
C_3H_7	C_5H_{11}	1-F	4-F	Cr(142°C)I				-1.7	[28]	ex. ZLI-2857
C_5H_{11}	C_5H_{11}	1-F	4-F	Cr(100°C)SmB(153°C)I				-1.5	[28]	ex. ZLI-2857

TABLE 1 continued

Group 67

R	R'	X1	X2	Mesophases	Temp/°C (Tr.K)	$\tilde{\varepsilon}_{\parallel}$	$\tilde{\varepsilon}_{\perp}$	$\Delta\tilde{\varepsilon}$	Ref	Meas. conditions
C$_3$H$_7$	OC$_2$H$_5$	F	F					-7	[55]	ex.
C$_5$H$_{11}$	OC$_2$H$_5$	F	F	Cr(49°C)(N12.9°C)I				-6.2	[53]	ex. ZLI-2857
C$_5$H$_{11}$	CH$_3$	F	F	Cr(14°C)I				-1.8	[53]	ex. ZLI-2857

Group 68

R	R'	X1, Y1, Z1	X2, Y2, Z2	Mesophases	Temp/°C (Tr.K)	$\tilde{\varepsilon}_{\parallel}$	$\tilde{\varepsilon}_{\perp}$	$\Delta\tilde{\varepsilon}$	Ref	Meas. conditions
C$_3$H$_7$	C$_3$H$_7$	H, H, F	H, H, F	Cr(132°C)N(148°C)I	20			-1.9	[51]	ex.
C$_5$H$_{11}$	OC$_2$H$_5$	H, H, F	H, H, F	Cr(105°C)SmC(135°C)N(185°C)I				-5.4	[53]	ex. ZLI-2857
C$_3$H$_7$	C$_3$H$_7$	H, F, H	H, F, H	Cr(96°C)N(132°C)I	20			-4.2	[51]	ex.
C$_5$H$_{11}$	C$_5$H$_{11}$	Cl, H, H	Cl, H, H	Cr(70.0°C)SmC(83.5°C)SmA(110.0°C)N(112.5°C)I	20	3.5		-1.7	[51]	ex.
								-1.65	[56]	

568

TABLE 1 continued

Group 69

R	R'	X1	X2	Mesophases	Temp/°C (Tr.K)	$\tilde{\varepsilon}_{\parallel}$	$\tilde{\varepsilon}_{\perp}$	$\Delta\tilde{\varepsilon}$	Ref	Meas. conditions
C_5H_{11}	OC_2H_5	F	CN	Cr(117°C)(SmA95°C) N(107°C)I				-9	[53]	ex. ZLI-2857
C_5H_{11}	C_2H_5	F	F					-2.2	[55]	ex.
C_5H_{11}	OC_2H_5	F	F	Cr(68-74°C) SmA(86-87°C) N(171-172°C)I	20	3.9	8	-4.1	[51,57]	ex.
C_3H_7	OC_2H_5	CF_3	F	Cr(80°C)I				-5.3	[55]	
								-7.3	[53]	ex. ZLI-2857

Group 70

R	R'	X1	X2	Mesophases	Temp/°C (Tr.K)	$\tilde{\varepsilon}_{\parallel}$	$\tilde{\varepsilon}_{\perp}$	$\Delta\tilde{\varepsilon}$	Ref	Meas. conditions
C_3H_7	OCH_3	F	F	Cr(67°C)N(145.3°C)I				-2.7	[53]	ex. ZLI-2857
C_3H_7	OC_2H_5	F	F	Cr(76°C)SmB(79°C) N(186°C)I	20			-4.4	[51]	ex.
C_5H_{11}	OC_2H_5	F	F	Cr(79°C)(SmB78°C) N(184.5°C)I				-5.9	[53]	ex. ZLI-2857
								-5.3	[55]	ex.

TABLE 1 continued

Group 71

R	R'	X	Mesophases	Temp/°C (Tr.K)	$\tilde\varepsilon_\parallel$	$\tilde\varepsilon_\perp$	$\Delta\tilde\varepsilon$	Ref	Meas. conditions
C₃H₇	OC₂H₅	F					-9.9	[55]	ex.
C₅H₁₁	OCH₃	F					-8.5	[55]	ex.

Group 72

R	R'	X	Mesophases	Temp/°C (Tr.K)	$\tilde\varepsilon_\parallel$	$\tilde\varepsilon_\perp$	$\Delta\tilde\varepsilon$	Ref	Meas. conditions
C₅H₁₁	C₃H₇	F	Cr(64°C)(SmA63°C)N(122°C)I		3.9	6.5	-2.6	[53,57]	ex. ZLI-2857
C₅H₁₁	OC₂H₅	F	Cr(72°C)N(147°C)I		3.3	3.8	-0.5	[53,57]	ex. ZLI-2857
C₅H₁₁	F	F	Cr(58°C)N(107°C)I		10.6	8.3	2.3	[57]	ex.

570

TABLE 1 continued

Group 73

Structure			Mesophases	Temp/°C (Tr.K)	$\tilde{\varepsilon}_{\parallel}$	$\tilde{\varepsilon}_{\perp}$	$\Delta\tilde{\varepsilon}$	Ref	Meas. conditions
R	R'	X							
C$_5$H$_{11}$	OC$_2$H$_5$	F	Cr(112°C)(SmA105°C)N(169.7-190°C)I		4.1	9.4	-5.3	[57]	ex.
							-7.4	[53]	ex. ZLI-2857
C$_5$H$_{11}$	C$_3$H$_7$	F	Cr(60°C)SmG(86°C)SmF(91°C)SmA(124°C)N(150.4°C)I		3.4	6.5	-3.1	[57]	ex.
							-4	[53]	ex. ZLI-2857

Group 74

Structure			Mesophases	Temp/°C (Tr.K)	$\tilde{\varepsilon}_{\parallel}$	$\tilde{\varepsilon}_{\perp}$	$\Delta\tilde{\varepsilon}$	Ref	Meas. conditions
R	R'	X							
C$_3$H$_7$	OC$_2$H$_5$	F	Cr(81°C)SmB(98°C)N(179°C)I				-6.1	[57]	ex.
							-8.4	[55]	ex.

TABLE 1 continued

Group 75

Structure				Temp/°C (Tr.K)	Mesophases	$\tilde{\varepsilon}_{\parallel}$	$\tilde{\varepsilon}_{\perp}$	$\tilde{\Delta\varepsilon}$	Ref	Meas. conditions
R	R'	X1	X2							
C_2H_5O	C_5H_{11}	F	F		Cr(61°C)(N51°C)I			-3.5	[57]	ex.
$C_8H_{17}O$	OC_8H_{17}	F	F	20	Cr(48°C)SmC(71°C)N(82°C)I			-2.7	[51]	ex.
$C_8H_{17}O$	C_8H_{17}	F	F	20	Cr(37°C)SmC(49°C)N(57°C)I			-2.7	[51]	ex.

Group 76

Structure				Temp/°C (Tr.K)	Mesophases	$\tilde{\varepsilon}_{\parallel}$	$\tilde{\varepsilon}_{\perp}$	$\tilde{\Delta\varepsilon}$	Ref	Meas. conditions
R	R'	X1	X2							
C_5H_{11}	OC_4H_9	CN	CN		Cr(93°C)(N64°C)I			-22	[42]	ex.
C_5H_{11}	C_5H_{11}	F	H	20	Cr(17-17.5°C)N(36.5-37°C)I			-0.9	[51]	ex.
C_5H_{11}	OC_2H_5	H	F	(21), Tr = 0.95				-0.8	[42]	ex.
C_5H_{11}	C_5H_{11}	F	F	20	Cr(49°C)N(59°C)I			-1.9	[51]	ex.
C_5H_{11}	OC_2H_5	F	F	20	Cr(13°C)N(29°C)I			-2.1	[51]	ex.
C_5H_{11}		F	F		Cr(51°C)N(63°C)I			-4.6	[57]	ex.

TABLE 1 continued

Group 77

R	X1	X2	Mesophases	Temp/°C (Tr.K)	$\tilde\varepsilon_\parallel$	$\tilde\varepsilon_\perp$	$\Delta\tilde\varepsilon$	Ref	Meas. conditions
	Structure								
R	R'	X1	X2						
C$_8$H$_{17}$O	C$_5$H$_{11}$	F	F	Cr(90°C)SmC(98°C)N(170°C)I	20			-2.1	[51]	ex.

Group 78

R	R'	X1	X2	Mesophases	Temp/°C (Tr.K)	$\tilde\varepsilon_\parallel$	$\tilde\varepsilon_\perp$	$\Delta\tilde\varepsilon$	Ref	Meas. conditions
C$_5$H$_{11}$	C$_7$H$_{15}$	CN	H	Cr(45°C)N(101°C)I				-3.7	[42]	ex.
C$_8$H$_{17}$O	C$_5$H$_{11}$	F	H	Cr(49°C)SmC(121°C)(Sm130°C)SmA(128°C)N(164.4°C)I	20 [1 kHz]			-0.047	[58]	ex. NH2
C$_8$H$_{17}$O	C$_7$H$_{15}$	F	H	Cr(34°C)SmC(115°C)N(138°C)I	20 [1 kHz]			-1.93	[58]	ex. NH2
C$_8$H$_{17}$O	OC$_7$H$_{15}$	F	F	Cr(60°C)SmC(150°C)SmA(155°C)N(157°C)I	20 [1 kHz]			-3.84	[58]	ex. NH2

TABLE 1 continued

Group 79

Structure				Mesophases	Temp/°C (Tr.K)	$\tilde{\varepsilon}_{\parallel}$	$\tilde{\varepsilon}_{\perp}$	$\Delta\tilde{\varepsilon}$	Ref	Meas. conditions
R	R'	X1	X2							
C₃H₇	OC₂H₅	CN	CN	Cr(160°C)N(175°C)I	20			-17	[53]	ex. ZLI-2857
C₃H₇	OC₂H₅	F	F	Cr(87°C)(SmB81°C)				-4.1	[51]	ex.
				SmA(98°C)N(222°C)I				-5.8	[53]	ex. ZLI-2857

Group 80

Structure		Mesophases	Temp/°C (Tr.K)	$\tilde{\varepsilon}_{\parallel}$	$\tilde{\varepsilon}_{\perp}$	$\Delta\tilde{\varepsilon}$	Ref	Meas. conditions
R	R'	X						
C₂H₅O	C₅H₁₁	F	Cr(88°C)I				-5	[57]	ex.

Group 81

Structure		Mesophases	Temp/°C (Tr.K)	$\tilde{\varepsilon}_{\parallel}$	$\tilde{\varepsilon}_{\perp}$	$\Delta\tilde{\varepsilon}$	Ref	Meas. conditions
R	R'	X						
C₅H₁₁	OC₂H₅	F	Cr(62°C)(N28°C)I				-1.1	[57]	ex.

TABLE 1 continued

Group 82

R	R'	X	Mesophases	Temp/°C (Tr.K)	ε̃∥	ε̃⊥	Δε̃	Ref	Meas. conditions
C₅H₁₁		OC₂H₅	Cr(58°C)(N23°C)I				-5.3	[57]	ex.
							-7.7	[55]	ex.

Group 83

R	R'	X1	X2	Mesophases	Temp/°C (Tr.K)	ε̃∥	ε̃⊥	Δε̃	Ref	Meas. conditions
C₃H₇	C₅H₁₁	F	F	Cr(62°C)N(154°C)I				-5	[55]	ex.
C₃H₇	OC₂H₅	F	F					-6	[57]	ex.

Group 84

R	R'	X	Y	Mesophases	Temp/°C (Tr.K)	ε̃∥	ε̃⊥	Δε̃	Ref	Meas. conditions
C₃H₇	C₃H₇	F	F	Cr(78°C)SmB(105°C)I				-4.6	[53]	ex. ZLI-2857
C₅H₁₁	C₃H₇	F	F	Cr(68°C)SmB(120°C)I				-4.2	[28]	ex. ZLI-2857
CH₂=CH	C₂H₄CH:CH₂	F	F	Cr(74°C)(SmB70°C)N(83°C)I				-3.8	[54]	ex. ZLI-2857

TABLE 1 continued

Group 85

Structure				Mesophases	Temp/°C (Tr.K)	$\tilde{\varepsilon}_{\parallel}$	$\tilde{\varepsilon}_{\perp}$	$\Delta\tilde{\varepsilon}$	Ref	Meas. conditions
R	R'	X1	X2							
C₃H₇	OC₂H₅	F	F					-5.7	[55]	ex.

Group 86

Structure				Mesophases	Temp/°C (Tr.K)	$\tilde{\varepsilon}_{\parallel}$	$\tilde{\varepsilon}_{\perp}$	$\Delta\tilde{\varepsilon}$	Ref	Meas. conditions
R	R'	X1	X2							
C₃H₇	OC₂H₅	F	F					-4.8	[55]	ex.

Group 87

Structure				Mesophases	Temp/°C (Tr.K)	$\tilde{\varepsilon}_{\parallel}$	$\tilde{\varepsilon}_{\perp}$	$\Delta\tilde{\varepsilon}$	Ref	Meas. conditions
R	R'	X1	X2							
C₃H₇	None	F	F	Cr(42°C)SmB(81°C) N(96°C)I	20			-1.3	[51]	ex.
C₃H₇	OC₂H₅	F	F	Cr(68°C)SmB(109°C) N(160°C)I	20			-3.7	[51]	ex.

TABLE 1 continued

Group 88

R	R'	X1	X2	Mesophases	Temp/°C (Tr.K)	$\tilde{\varepsilon}_\parallel$	$\tilde{\varepsilon}_\perp$	$\Delta\tilde{\varepsilon}$	Ref	Meas. conditions
C$_5$H$_{11}$	OC$_2$H$_5$	F	F	Cr(57°C)N(61°C)I	20			-4.4	[51]	ex.

Group 89

R	R'	X1	X2	Mesophases	Temp/°C (Tr.K)	$\tilde{\varepsilon}_\parallel$	$\tilde{\varepsilon}_\perp$	$\Delta\tilde{\varepsilon}$	Ref	Meas. conditions
C$_3$H$_7$	OC$_2$H$_5$	F	F	Cr(84°C)N(228.1-229°C)I	20			-4.1	[51]	ex.
								-5.2	[53]	ex. ZLI-2857

Group 90

R	R'	X1	X2	Mesophases	Temp/°C (Tr.K)	$\tilde{\varepsilon}_\parallel$	$\tilde{\varepsilon}_\perp$	$\Delta\tilde{\varepsilon}$	Ref	Meas. conditions
C$_3$H$_7$	C$_5$H$_{11}$	F	F	Cr(35°C)SmA(66°C)N(118°C)I	20			-3.4	[51]	ex.
C$_3$H$_7$	OC$_2$H$_5$	F	F	Cr(62°C)N(154°C)I	20			-6	[51]	ex.

TABLE 1 continued

ex: extrapolated.

Host mixtures:

ZLI-1132: cyano-based mixture from Merck, $\Delta\tilde{\varepsilon} = 11.23$.

I-Mix: ternary mixture of I-compounds (fluorobiphenylalkanes), $T_{NI} = 100$ °C, $\tilde{\varepsilon}_\parallel = 3.0$, $\tilde{\varepsilon}_\perp = 3.0$ [8].

FB-01: ternary mixture of fluoro-compounds, $T_{NI} = 112.8$°C, $\Delta\tilde{\varepsilon} = 4.8$ [31].

ZLI-4792: fluoro-based mixture from Merck, $T_{NI} = 92.8$°C, $\Delta\tilde{\varepsilon} = 5.1 - 5.3$.

HM1: binary mixture of alkenyl compounds, $T_{NI} = 116.7$°C, $\Delta\tilde{\varepsilon} = 4.8$ [18].

MIX.A: six-components mixture of ester compounds, $T_{NI} = 73.0$°C [30].

MIX.B: mixture of three cyano-PCHs and six ester components, $T_{NI} = 54.0$°C, $\Delta\tilde{\varepsilon} = 6.6$ [47].

ZLI-2857: negative $\Delta\tilde{\varepsilon}$ mixture from Merck, $T_{NI} = 82.3$°C, $\Delta\tilde{\varepsilon} = -1.4$.

NH2: non-polar nematic host [8].

As can be seen in TABLE 1, not all substances have a nematic phase but some of these compounds can also be important components of practical mixtures. That means, from an application point of view, it is necessary to characterise those non-nematic or non-liquid-crystalline substances by an effective or virtual dielectric anisotropy. Mainly for this reason, the dielectric properties are usually measured in mixtures, containing the corresponding substances at a fixed concentration, and the value characteristic of the component is obtained by extrapolation. This provides the main reasons for the scatter in the data, some of which can be attributed to the different host mixtures used. In this connection, too, dipole-dipole association is an important factor to make the microscopic investigation of the dielectric property of a nematic phase precise. The extrapolation conditions are noted in TABLE 1 and details can be found in the appropriate references.

Directly linked phenyl rings and/or cyclohexane rings are typical structures for the molecular core and also confer chemical stability. In order to increase the molecular dipole, some hydrogen atoms, which are attached to the core rings, are substituted by fluorine atoms. The properties of these substances with different terminal groups are listed as Groups 1-19 in TABLE 1.

Some heterocyclic rings such as pyrimidine, pyridine and dioxane are often introduced in the core to increase the molecular dipole, as shown in TABLE 1 (Groups 20-27). The ester linkage is also effective as an additional group moment, and is shown together with some other linkage structures (Groups 28-34). Ethanes (Groups 35-62) and tolanes (Groups 63-65) are examples of other typical links.

The structures of substances with negative dielectric anisotropy are presented as Groups 66-90 in TABLE 1. In cases when the polar group is located at the axial 4-position not of a benzene ring but of a cyclohexane ring, it can provide a dipole almost perpendicular to the molecular long axis, and this structure results in a negative dielectric anisotropy as given in Group 66.

Substitution of a hydrogen by a fluorine atom or a polar group at the 2-position of a benzene ring, preferably together with that at the 3-position, gives another typical molecular structure with a negative anisotropy (Groups 67-70). The dielectric anisotropy is enhanced when there is an ether linkage connecting a terminal alkyl chain on the adjacent 4-position carbon of the benzene ring (alkyloxy chains, for example).

Fluoropyridines are typical heterocyclic rings to be effectively used in the core of the molecules yielding a negative dielectric anisotropy (Groups 71-74). Some linkage groups similar to those for dielectrically positive substances are also introduced to the molecular core of this category (Groups 75-90).

REFERENCES

[1]   S. Naemura [to appear in *Mater. Res. Soc. Symp. Proc. (USA)* vol.559 (1999)]
[2]   M. Klasen, M. Bremer, A. Goetz, A. Manabe, S. Naemura, K. Tarumi [*Jpn. J. Appl. Phys. (Japan)* vol.37 (1998) p.L945-8]
[3]   D.A. Dunmur, K. Toriyama [*Mol. Cryst. Liq. Cryst. (UK)* vol.198 (1991) p.201-13]
[4]   S. Naemura [*Mater. Res. Soc. Symp. Proc. (USA)* vol.424 (1997) p.295-310]

[5] S. Naemura [presented at *5th IUMRS-ICAM'99* 13-18 June 1999, and to appear in *J. SID (USA)*]

[6] M. Bremer, S. Naemura, K. Tarumi [*Jpn. J. Appl. Phys. (Japan)* vol.37 (1998) p.L88-90]

[7] U. Finkenzeller, A. Kurmeier, E. Poetsch [*18 Freiburger Arbeitstagung Flussigkristalle* (1989) no.17]

[8] D.G. McDonnell, E.P. Raynes, R.A. Smith [*Liq. Cryst. (UK)* vol.6 (1989) p.515-23]

[9] M. Bremer, K. Tarumi [*Adv. Mater. (Germany)* vol.5 (1993) p.842-8]

[10] M. Schadt, M. Petrzilka, P.R. Gerber, A. Villiger [*Mol. Cryst. Liq. Cryst. (UK)* vol.122 (1985) p.241-60]

[11] D.A. Dunmur, A.E. Tomes [*Mol. Cryst. Liq. Cryst. (UK)* vol.97 (1983) p.241-53]

[12] P.G. Cummins, D.A. Dunmur, D.A. Laidler [*Mol. Cryst. Liq. Cryst. (UK)* vol.30 (1975) p.109-23]

[13] G.W. Gray, K.J. Harrison, J.A. Nash [*Electron. Lett. (UK)* vol.9 (1973) p.130-1]

[14] A. Boller, M. Cereghetti, M. Schadt, H. Scherrer [*Mol. Cryst. Liq. Cryst. (UK)* vol.42 (1977) p.215-31]

[15] D.A. Dunmur, M.R. Manterfield, W.H. Miller, J.K. Dunleavy [*Mol. Cryst. Liq. Cryst. (UK)* vol.45 (1978) p.127-44]

[16] P. Bordewijk, W.H. de Jeu [*J. Chem. Phys. (USA)* vol.58 (1978) p.116-8]

[17] Hp. Schad, M.A. Osman [*J. Chem. Phys. (USA)* vol.75 (1981) p.880-5]

[18] H.-M. Vorbrodt, S. Deresch, H. Kresse, A. Wiegeleben, D. Demus, H. Zaschke [*J. Prakt. Chem. (Germany)* vol.323 (1981) p.902-13]

[19] D. Lippens, J.P. Parneix, A. Chapoton [*J. Phys. (France)* vol.38 (1977) p.1465-71]

[20] E. Bartmann, D. Dorsch, U. Finkenzeller [*Mol. Cryst. Liq. Cryst. (UK)* vol.204 (1991) p.77-89]

[21] E. Bartmann, D. Dorsch, U. Finkenzeller, H.A. Kurmeier, R. Poetsch [*19 Freiburger Arbeitstagung Flussigkristalle* (1990)]

[22] B.S. Scheuble [*Proc. ITE Jpn. Annual Conf. 1989* (1989)]

[23] M. Schadt, R. Buchecker, F. Leenhouts, A. Boller, A. Villiger, M. Petrzilka [*Mol. Cryst. Liq. Cryst. (UK)* vol.139 (1986) p.1-25]

[24] A.I. Pavluchenko, N.I. Smirnova, V.F. Petrov, Y.A. Fialkov, S.V. Shelyazhenko, L.M. Yagupolsky [*Mol. Cryst. Liq. Cryst. (UK)* vol.209 (1991) p.225-35]

[25] L. Pohl, R. Eidenschink, G. Krause, D. Erdmann [*Phys. Lett. A (Netherlands)* vol.60 (1977) p.421-3]

[26] R. Buchecker, M. Schadt [*Mol. Cryst. Liq. Cryst. (UK)* vol.149 (1987) p.359-73]

[27] M. Schadt, R. Buchecker, A. Villiger [*Liq. Cryst. (UK)* vol.7 (1990) p.519-36]

[28] P. Kirsch, S. Naemura, K. Tarumi [*27 Freiburger Arbeitstagung Flussigkristalle* (1998)]

[29] E. Nakagawa et al [*Tech. Digest 1995 Asian Symp. Info. Display* (1995) p.235-40]

[30] H. Saito et al [*IEICE Trans. Electron. (Japan)* vol.E79-C (1996) p.1027-34]

[31] D. Demus, Y. Goto, S. Sawada, E. Nakagawa, H. Saito, R. Tarao [*Mol. Cryst. Liq. Cryst. (Switzerland)* vol.260 (1995) p.1-21]

[32] H.J. Plach, B. Rieger, E. Poetsch, V. Reiffenrath [*Eurodisplay* (1990) no.5.3]

[33] E. Bartmann, U. Finkenzeller, E. Poetsch, V. Reiffenrath, K. Tarumi [*22 Freiburger Arbeitstagung Flussigkristalle* (1993) Poster no.8]

[34] A.I. Pavluchenko et al [*Mol. Cryst. Liq. Cryst. (Switzerland)* vol.265 (1995) p.41-5]

[35] K. Tarumi, M. Bremer, T. Geelhaar [*Annu. Rev. Mater. Sci. (USA)* vol.27 (1997) p.423-41]

[36]   A. Beyer, B. Schuler, K. Tarumi [*22 Freiburger Arbeitstagung Flussigkristalle* (1993) no.13]

[37]   K. Kanie, Y. Tanaka, M. Shimizu, S. Takehara, T. Hiyama [*Chem. Lett. (Japan)* (1997) p.827-8]

[38]   M. Schadt [*Mol. Cryst. Liq. Cryst. (UK)* vol.165 (1988) p.405-38]

[39]   R.T. Klingbiel, D.J. Genova, T.R. Criswell, J.P. van Meter [*J. Am. Chem. Soc. (USA)* vol.96 (1974) p.7651-5]

[40]   M. Sasaki, K. Takeuchi, H. Sato, H. Takatsu [*Mol. Cryst. Liq. Cryst. (UK)* vol.109 (1984) p.169-78]

[41]   Hp. Schad, S.M. Kelly [*J. Phys. (France)* vol.46 (1985) p.1395-404]

[42]   M. Osman [*Mol. Cryst. Liq. Cryst. Lett. (UK)* vol.82 (1982) p.295-302]

[43]   D. Demus, H. Zaschke [*Mol. Cryst. Liq. Cryst. (UK)* vol.63 (1981) p.129-44]

[44]   R. Dabrowski, J. Dziaduszek, T. Szczucinski, Z. Raszewski [*Mol. Cryst. Liq. Cryst. (UK)* vol.107 (1984) p.411-43]

[45]   M. Schadt, M. Petrzilka, P.R. Gerber, A. Villiger, G. Trickes [*Mol. Cryst. Liq. Cryst. (UK)* vol.94 (1983) p.139-53]

[46]   M.J. Goulding, S. Greenfield, D. Coates, R. Clemitson [*Liq. Cryst. (UK)* vol.14 (1993) p.1397-408]

[47]   H. Takatsu, K. Takeuchi, H. Sato [*Mol. Cryst. Liq. Cryst. (UK)* vol.100 (1983) p.345-55]

[48]   M.J. Goulding, S. Greenfield, O. Parri, D. Coates [*Mol. Cryst. Liq. Cryst. (Switzerland)* vol.265 (1995) p.27-40]

[49]   S. Greenfield, D. Coates, M. Goulding, R. Clemitson [*Liq. Cryst. (UK)* vol.18 (1995) p.665-72]

[50]   H. Takatsu, K. Takeuchi, Y. Tanaka, M. Sasaki [*Mol. Cryst. Liq. Cryst. (UK)* vol.141 (1986) p.279-87]

[51]   V. Reiffenrath, J. Krause, H.J. Plach, G. Weber [*Liq. Cryst. (UK)* vol.5 (1989) p.159-70]

[52]   K. Takeuchi, Y. Tanaka, M. Sasaki, H. Takatsu [*11th Jpn. Domestic Liq. Cryst. Conf.* (1985) 1N01]

[53]   P. Kirsch, V. Reiffenrath, M. Bremer [*Synlett. (Germany)* (1999) p.389-96]

[54]   P. Kirsch, M. Heckmeier, K. Tarumi [*Liq. Cryst. (UK)* vol.26 (1999) p.449-52]

[55]   M. Bremer, K. Tarumi, H. Ichinose, H. Numata [*Digest Tech. Papers AMLCD 1995* (1995) p.105-8]

[56]   M. Hird, K. Toyne, P. Hindmarsh, J. Clifford, V. Minter [*Mol. Cryst. Liq. Cryst. (Switzerland)* vol.260 (1995) p.227-40]

[57]   V. Reiffenrath, M. Bremer [*Angew. Chem. (Germany)* vol.106 (1994) p.1435-8]

[58]   M. Chambers, R. Clemitson, D. Coates, S. Greenfield, J.A. Jenner, I.C. Sage [*Liq. Cryst. (UK)* vol.5 (1989) p.153-8]

11.3   Elastic properties of nematics for applications

H. Saito

November 1999

A   INTRODUCTION

The elastic constants of nematic liquid crystals are very important properties that determine electro-optical characteristics of liquid crystal displays (LCDs). They are physical quantities that fix the director deformed by an external force (e.g. magnetic field, electric field, influence of cell condition etc).

There are three bulk elastic constants, splay K_1, twist K_2, and bend K_3, in liquid crystals. In twisted nematic (TN) LCDs, the threshold voltage (V_c) is given by [1]

$$V_c = \pi [K/(\varepsilon_o \Delta \tilde{\varepsilon})]^{\frac{1}{2}} \tag{1}$$

where $\Delta \tilde{\varepsilon}$ is the dielectric anisotropy and $K = K_1 + (K_3 - 2K_2)/4$.

EQN (1) means that the threshold voltage can be decreased by reducing K, and that to make K_1 small is the most effective strategy. Further, the response time is given by [2]

$$\tau_r = \gamma_1 d^2 / [\varepsilon_o \Delta \tilde{\varepsilon} (V^2 - V_c^2)] \tag{2}$$

$$\tau_d = \gamma_1 d^2 / K\pi^2 \tag{3}$$

where τ_r is the rise time, τ_d is the decay time, and γ_1 is the rotational viscosity coefficient. It follows that the decay time can also be shortened by making K smaller but the rise time is not directly influenced by the elastic constant.

B   MEASUREMENT TECHNIQUES

The elastic constants are determined by fitting the data in accordance with theory, for example by measuring the change of light transmission when a cell is switched between homogeneous and homeotropic alignments by an electric or a magnetic field.

Liquid crystal cells are connected to a guard-ring electrode, and are treated to give homogeneous director alignment. The cell gap is set at about 20 μm, and is determined by a light interference method. Generally, the director pre-tilt angle is measured exactly by the crystal rotation method [3] determined by the symmetry point of the optical anisotropy. It can be also measured in practice by the magnetic null method [3] using a magnetic field. However, in order not to be affected by the pre-tilt, the cells used are aligned by oblique evaporation of silicon monoxide at 60° giving zero pre-tilt.

B1   Capacitance Method

(1)   The Freederiksz transition is used to determine the elastic constants by measuring the threshold voltage V_c which corresponds to a change in capacitance when the cells are subjected to an electric field. The elastic constant K_1 is calculated from the following equation [4]:

$$V_c = \pi \left(K_1 / \varepsilon_o \Delta\widetilde{\varepsilon} \right)^{\frac{1}{2}} \tag{4}$$

where $\Delta\widetilde{\varepsilon}$ is the dielectric anisotropy measured using other techniques.

(2)   When the liquid crystal in the same cell as described previously has a voltage applied greater than the Freederiksz transition threshold voltage, it induces a deformation of the director. Suppose the angle between the local optic axis defined at each point in the sample and the initial axis is defined as θ, and suppose θ_m is the maximum deformation angle in the centre of the cell; the following equation relates the applied voltage and θ [4]:

$$V / V_c = 2\pi^{-1} \left(1 + \gamma \sin^2 \theta_m \right)^{\frac{1}{2}} \times$$

$$\int_0^{\theta_m} \left[\left(1 + \kappa \sin^2 \theta \right) / \left\{ \left(1 + \gamma \sin^2 \theta \right) \left(\sin^2 \theta_m - \sin^2 \theta \right) \right\} \right]^{\frac{1}{2}} d\theta \tag{5}$$

where

$$\gamma = \Delta\widetilde{\varepsilon}/\varepsilon_\perp , \; \kappa = (K_3 - K_1)/K_1$$

V is the applied voltage and V_c is the threshold voltage.

The capacitance, C, of the liquid crystal cell is given by the following equation [4] under the same voltage condition:

$$C / C_o = 2V_c \left(\pi V \right)^{-1} \left(1 + \gamma \sin^2 \theta_m \right)^{\frac{1}{2}} \times$$

$$\int_0^{\theta_m} \left\{ \left(1 + \kappa \sin^2 \theta \right) \left(1 + \gamma \sin^2 \theta \right) / \left(\sin^2 \theta_m - \sin^2 \theta \right) \right\}^{\frac{1}{2}} d\theta \tag{6}$$

where C is the initial capacitance, $C_o = \varepsilon_\perp A/d$, and A is the electrode area while d is the cell gap.

This is the theoretical equation describing how the capacitance changes with the applied voltage. The capacitance change (C-V curve) can be fitted to the corresponding voltage changes by using values of γ and κ. The C-V curve is measured for the liquid crystal cell; this measured C-V curve is compared to the calculated C-V curve and is fitted to a suitable curve by using the variable κ. The elastic constant K_3 is calculated from the fitted κ.

Using computer simulation, many curves calculated in detail by changing κ are fitted to those which are measured in practice. Then, the exact value can be obtained.

(3)   Finally, a TN-LCD cell with a 90° twist is prepared. The threshold voltage (V_c) of this cell is given by

$$V_c = \pi \left(K / \varepsilon_o \Delta \widetilde{\varepsilon} \right)^{\frac{1}{2}}$$

where $K = K_1 + (K_3 - 2K_2)/4$. The elastic constant K_2 is obtained by using the values for K_1, K_3 and $\Delta \widetilde{\varepsilon}$. Many parameters are used in the calculation, and this method tends to give K_2 with decreased accuracy. If the director in the cells can be aligned uniformly, an elastic constant can be measured by determining only the threshold voltage on two kinds of cells.

B2   Birefringence Method

(1)   In this method, the homogeneous cell is used just as in the capacitance method. When the voltage is applied to this cell, the light transmission of polarised light changes with electric field. EQN (4) is used to determine K_1, as before.

(2)   When a magnetic field is applied to this cell perpendicular to the cell surface, the polarised light transmission changes with magnetic field. The threshold magnetic field B for this 'vertical' magnetic Freedericksz transition is given by [5]

$$B_{cv} = \pi \left(\mu_o K_1 / \Delta \widetilde{\chi} \right)^{\frac{1}{2}} / d \tag{7}$$

where $\Delta \widetilde{\chi}$ is the magnetic susceptibility anisotropy and can be calculated from EQN (7) by using the value of K_1.

Secondly, while the magnetic field is applied to this cell parallel to the cell surface, the light transmitted exhibits a change at the threshold magnetic field (B_{ch}: horizontal magnetic Freedericksz transition). B_{ch} is given by [5]

$$B_{ch} = \pi \left(\mu_o K_2 / \Delta \widetilde{\chi} \right)^{\frac{1}{2}} / d \tag{8}$$

where K_2 is calculated from this equation by using the value of $\Delta \widetilde{\chi}$.

(3)   When the same cell as above experiences an applied voltage greater than the Freedericksz transition voltage, a voltage/birefringence curve can be constructed because of the theoretical relationship between the change of the birefringence (Δ) and the electric field. The following equation relates Δ and applied voltage [4]:

$$\frac{\Delta}{n_e} = 1 - 2\left(\frac{V}{\pi V_c}\right) \int_0^{\theta_m} \left(1 + \kappa \sin^2 \theta\right)^{\frac{1}{2}} d\theta \bigg/ \left\{\left(1 + \Delta\tilde{n} \sin^2 \theta\right)\left(\sin^2 \theta_m - \sin^2 \theta\right)\right\}^{\frac{1}{2}} \qquad (9)$$

$\Delta\tilde{n} = \tilde{n}_e - \tilde{n}_o$, $\kappa = (K_3 - K_1)/K_1$

This is a theoretical equation which predicts that the change of birefringence is proportional to the applied voltage. The calculated birefringence change (Δ/V curve) can be fitted to voltage changes by adjusting the values of $\Delta\tilde{n}$ and κ. The Δ/V curve is practically measured for the liquid crystal cell; this measured Δ/V curve is compared to the calculated Δ/V one and can be fitted to a suitable curve by using the variable κ. The elastic constant K_3 is calculated from the fitted κ.

The temperature dependence of elastic constants can be measured by controlling the LC cell temperature. In practical liquid crystal displays, the elastic constant of the LC mixture has been controlled to match each display mode.

In the twisted nematic (TN) mode, the steepness of the T-V (transmission versus voltage) curve, proposed by Schadt and Gerber [6], is given by

$$\alpha = (V_{90} - V_{50})/V_{90} = 0.1330 + 0.0266\{(K_3/K_1) - 1\} + 0.0443 \{\ln(d\Delta n/2\lambda)\}^2$$

where V_{90} is the applied voltage changing 90% of light intensity, V_{50} is the applied voltage changing 50% of light intensity, d is the cell gap, $\Delta\tilde{n}$ is the optical anisotropy and λ is the wavelength.

In the TN mode, the steepness is decreased by the smaller elastic constant ratio K_3/K_1. In super twisted nematic (STN) mode, the elastic constant ratios K_3/K_1 and K_3/K_2 are decreased in order to improve the steepness.

C   RESULTS AND DATA

TABLE 1 shows the elastic constants for typical liquid crystals used for practical applications. The measurements were made at the given values of the reduced temperature, T/T_{NI}. The results for binary mixtures are given in TABLES 2 and 4. These were obtained in a host liquid crystal or host mixture.

TABLE 1 shows typical data for materials with positive dielectric anisotropy. Compared to the reference biphenyl compound 1d in TABLE 1, the compounds in which a cyclohexane ring is introduced tend to give smaller K_1 and K_2, and to increase the elastic constant ratio K_3/K_1 especially, for example, from 1d = 1.19 to 1e = 1.59, 1f = 1.66 in TABLE 1.

The elastic constant ratio, K_3/K_1, for the pyrimidine compound, 1c, in TABLE 1 is smaller than for other materials with positive dielectric anisotropy. Compounds having a double bond in the terminal chain show large values as for example 1i = 2.15, 1j = 2.02 in TABLE 1, and furthermore low polarity compounds without the cyano group generally show small values as for example 2i = 1.28, 2j = 0.85

in TABLE 2, and 3j = 0.75, 3k = 0.81 in TABLE 3. The compounds having an ethylenic double bond tend to have large values of K_3, for example 1h and 1j in TABLE 1. The fluorinated derivatives also have a large K_3 as we can see in TABLE 4, and have a large value of the elastic constant ratio K_3/K_1, for example 4d ~ 4f in TABLE 4.

TABLE 1 Typical materials with positive dielectric anisotropy.

Liquid crystal material	Elastic constant			Reduced temperature	
Molecular structure	$K_1/10^{-12}$ N	$K_2/10^{-12}$ N	$K_3/10^{-12}$ N	T/T_{NI}	Ref
1a	7.27	4.66	13.9	0.96	[9]
1b	5.53	3.34	9.06	0.96	[7]
1c	5.04	2.84	4.71	0.98	[7]
1d	8.55	4.43	10.2	0.95	[7]
1e	7.59	4.24	12.1	0.95	[7]
1f	5.93	3.10	9.85	0.95	[6]
1g	6.00	3.60	9.20	0.95	[11]
1h	8.85	4.60	16.1	0.97	[11]
1i	6.22	3.90	13.4	0.97	[11]
1j	9.18	5.28	18.5	0.97	[11]
1k	4.93	4.37	7.66	0.97	[14]
1l	6.71	2.93	7.38	0.97	[12]

TABLE 2 Typical binary mixtures.

	Liquid crystal material	Elastic constant			Reduced temperature	
	Molecular structure	$K_1/10^{-12}$ N	$K_2/10^{-12}$ N	$K_3/10^{-12}$ N	T/T_{NI}	Ref
2a		6.05	3.62	10.1	0.97	[10]
2b		6.25	3.16	8.70	0.97	[10]
2c		6.70	3.60	9.00	0.97	[10]
2d		5.15	3.10	6.40	0.97	[10]
2e		5.50	3.16	8.50	0.97	[10]
2f		6.75	3.30	10.5	0.97	[10]
2g		4.90	3.14	7.00	0.97	[10]
2h		7.00	3.86	8.10	0.97	[10]
2i		6.20	3.43	7.95	0.97	[10]
2j		7.95	3.89	6.75	0.97	[10]

R = C_5H_{11} and C_7H_{15}, R_1 = C_3H_7, R_2 = C_2H_5 and C_4H_9, R_3 = C_5H_{11}

Binary mixture: C_5/C_7: 40/60 mol%; C_2/C_4($:R_2$): 40/60 mol%

TABLE 3 Typical material with weak dielectric anisotropy.

Liquid crystal material		Elastic constant			Reduced temperature	
	Molecular structure	$K_1/10^{-12}$ N	$K_2/10^{-12}$ N	$K_3/10^{-12}$ N	T/T_{NI}	Ref
3a	CH₃O–⟨⟩–CH=N–⟨⟩–C₄H₉	5.06	3.19	6.85	0.96	[9]
3b	CH₃O–⟨⟩–COO–⟨⟩–C₅H₁₁	6.40	3.60	9.20	0.95	[8]
3c	C₅H₁₁–⟨⟩–COO–⟨⟩–OCH₃	6.80	3.70	8.20	0.95	[8]
3d	C₅H₁₁–⟨⟩–COO–⟨⟩–CH₃	5.10	2.70	5.40	0.98	[8]
3e	C₅H₁₁–⟨⟩–C₂H₄–⟨⟩–OC₂H₅	10.15	5.20	10.86	0.97	[13]
3f	⟨⟩–⟨⟩–OC₂H₅	6.32	3.53	8.22	0.97	[13]
3g	⟨⟩–⟨⟩–O	10.86	5.60	12.38	0.97	[13]
3h	⟨⟩–C₂H₄–⟨⟩–OC₂H₅	8.86	4.00	12.67	0.97	[13]
3i	C₅H₁₁–⟨⟩–COO–⟨⟩–OC₃H₇	6.23	3.08	5.79	0.97	[13]
3j	⟨⟩–COO–⟨⟩–OC₃H₇	5.43	2.51	4.07	0.97	[13]
3k	C₅H₁₁–⟨⟩–COO–⟨⟩–O	6.04	2.87	4.89	0.97	[13]
3l	C₅H₁₁–⟨⟩–COO–⟨⟩	6.22	2.60	6.17	0.97	[13]

TABLE 4 Binary mixtures with fluorinated materials.

Liquid crystal material		Elastic constant			Reduced temperature	
	Molecular structure	$K_1/10^{-12}$ N	$K_2/10^{-12}$ N	$K_3/10^{-12}$ N	T/T_{NI}	Ref
4a		8.31	4.10	10.10	0.97*[1]	[15]
4b		7.86	4.03	9.12	0.97*[1]	[15]
4c		7.86	4.15	9.20	0.97*[1]	[16]
4d		8.90	7.52	18.7	0.87*[2]	[16]
4e		8.11	7.10	17.0	0.88*[2]	[16]
4f		7.63	7.11	16.3	0.90*[2]	[16]
Base						
M1		8.82	4.45	10.2	0.97	[15]
M2		8.02	7.11	16.7	0.89	[16]

Condition of measurement
*1: binary mixture M1/material = 80/20 mol%
*2: binary mixture M2/material = 80/20 mol%
R = C_2H_5, C_3H_7, C_5H_{11}

The value of K_1 ranges from 5 to 9×10^{-12} N, K_2 from 2.8 to 5.3×10^{-12} N and smaller than the other elastic constants, and K_3 has a wide range of values from 4.7 to 18.5×10^{-12} N. The compounds having a pyrimidine ring are characterised by small values of all of the elastic constants.

Leenhouts and others [9] say that the elastic constant ratio K_3/K_1 is proportional to the length to breadth ratio of the molecular dimensions. This theory holds when a mesogenic molecule does not have a long chain length, because, in the same derivatives of molecular structure, there is a tendency that the longer an alkyl chain is, the smaller K_3/K_1 becomes.

There is generally an odd-even effect in the elastic constants. In addition, K_3/K_1 for weak polar nematic liquid crystals tends to be smaller than those of highly polar materials. Further, it has been reported that the K_3 value of alkenyl derivatives which possess a double bond is large due to the dependence on the position of the double bond.

With regard to liquid crystal mixtures, an elastic constant shows an average value of each component. The elastic constant in mixtures is then [8]

$$\left(K_i\right)^{\frac{1}{2}} = x_A \cdot \left(K_{i,A}\right)^{\frac{1}{2}} + x_B \cdot \left(K_{i,B}\right)^{\frac{1}{2}}$$

where x_A, x_B are the mole fractions of the A, B components, $K_{i,A}$, $K_{i,B}$ are the elastic constants of the two components and K_i is the elastic constant of the mixture. This equation is supported for both polar-polar mixtures and non-polar-non-polar mixtures. It also holds true except for K_3 for a polar-non-polar mixture. The equation is not applicable for K_3 of polar-non-polar mixtures because this elastic constant is considered to be affected by the dimer that is formed by polar molecules in a non-polar liquid crystal [17-19].

REFERENCES

[1]   M. Schadt, W. Helfrich [*Appl. Phys. Lett. (USA)* vol.18 (1971) p.127]

[2]   E. Jakeman, E.P. Raynes [*Phys. Lett. A (Netherlands)* vol.39 (1972) p.69]

[3]   T.J. Scheffer, J. Nehring [*J. Appl. Phys. (USA)* vol.48 no.5 (1977) p.1783]

[4]   H. Gruler, T.J. Scheffer, G. Meier [*Z. Nat.forsch. A (Germany)* vol.27 (1972) p.966]

[5]   A. Saupe [*Z. Nat.forsch. A (Germany)* vol.15 (1960) p.815]

[6]   M. Schadt, P.R. Gerber [*Proc. SID Inf. Display (USA)* vol.23 (1982) p.29]

[7]   Hp. Schad, M.A. Osman [*J. Chem. Phys. (USA)* vol.75 no.2 (1981) p.880]

[8]   Hp. Schad, M.A. Osman [*J. Chem Phys. (USA)* vol.79 no.11 (1983) p.5710]

[9]   F. Leenhouts, H.J. Roebers, A.J. Dekker, J.J. Jonker [*J. Phys. (France)* vol.40 no.C-3 (1979) pt.4 p.291]

[10]   M. Schadt, P.R. Gerber [*Z. Nat.forsch. A (Germany)* vol.37 (1982) p.165]

[11]   M. Schadt, M. Petrzilka, P.R. Gerber, A. Villiger [*Mol. Cryst. Liq. Cryst. (UK)* vol.122 (1985) p.241]

[12]   M. Schadt, R. Buchecker, F. Leenhouts, A. Boller, A. Villiger, M. Petrzilka [*Mol. Cryst. Liq. Cryst. (UK)* vol.139 (1986) p.1]

[13]   R. Buchecker, M. Schadt [*Mol. Cryst. Liq. Cryst. (UK)* vol.149 (1987) p.359]

[14]   M. Schadt, R. Buchecker, K. Muller [*Liq. Cryst. (UK)* vol.5 no.1 (1989) p.293]

[15]   M. Schadt, R. Buchecker, A. Villiger [*Liq. Cryst. (UK)* vol.7 no.4 (1990) p.519]

[16]   Chisso Corp. Data

[17]   M.A. Osman, Hp. Schad, H.R. Zeller [*J. Chem. Phys. (USA)* vol.78 (1981) p.906]

[18]   M.J. Bradshaw, E.P. Raynes [*Mol. Cryst. Liq. Cryst. (UK)* vol.91 (1983) p.145]

[19]   B.S. Scheuble, G. Baur, G. Meier [*Mol. Cryst. Liq. Cryst. (UK)* vol.68 (1981) p.1005-15]

11.4 Viscous properties of nematics for applications

K. Tarumi and M. Heckmeier

September 1999

A   INTRODUCTION

In this chapter the important physical features of nematic liquid crystals (LCs) such as the optical, dielectric and elastic properties have been discussed. These properties reflect the static behaviour of the nematic state, whereas the viscosity is related to the dynamics. The viscous property comes into play when a nematic is subjected to a certain deformation by an external field. The physical nature of this viscous property has been well elucidated in Chapter 8. Here we confine ourselves only to discussing the important feature of the viscous property that plays a dominant role in liquid crystal display (LCD) applications.

LCDs as optical devices should have the capability of changing or refreshing their pictures. To this end the corresponding change of the orientation of the nematic director has to be made by applying the electric field. This dynamic switching time is directly related to the viscosity. Several advanced LCDs [1-5] have recently been developed and the display performance been improved significantly not only in the optical performance such as contrast brightness and viewing angle but also in the switching times. However, these times still need to be improved in order to realise the so-called full moving pictures by LCDs. The investigation of the viscous property of nematics is therefore of major importance for the further improvement of LCDs.

The viscous feature of isotropic Newtonian fluids is defined as the constant of proportionality between the frictional force and the rate of strain tensor. However, in the case of nematics this coefficient is not a single constant but is dependent on the orientation of the nematic director. Taking symmetry into consideration the number of viscosity coefficients can be reduced to five. It is beyond our scope to discuss all of the viscosity coefficients here. Instead we focus on the viscosity coefficient that plays the most dominant role in LCD applications.

First, the relationship between the viscosity coefficients and the switching times of LCDs will be elucidated. Secondly, experimental data for nematics will be presented and the relationship to structure will be discussed. Further, the switching time characteristics of current LCD modes will be referred to briefly.

B   RELEVANT VISCOSITY COEFFICIENTS FOR LCD APPLICATIONS

Miesowicz introduced three viscosity coefficients, which are called Miesowicz viscosity coefficients (often denoted as η_1, η_2, η_3) where each of these is the proportional constant between the frictional force and the rate of stress tensor in three different director orientations. In addition there exists another coefficient η_{12} and the viscosity coefficient (often denoted as γ_1) in which the director is subjected to a rotational movement. The general and rigorous mathematical definition of these quantities is given in Chapter 8.

The most important and relevant viscosity coefficient in the application of LCDs is γ_1 and is called the rotational viscosity. The switching dynamics in twisted nematic cells (TN and STN) have been studied by perturbation theory and the switching times are found to be given by [6,7]:

$$\tau_{on} \propto \frac{\gamma_1 d^2}{K_{eff}\left(\left(\dfrac{V}{V_o}\right)^2 - 1\right)} \tag{1}$$

$$\tau_{off} \propto \frac{\gamma_1 d^2}{K_{eff}} \tag{2}$$

where τ_{on} and τ_{off} are the switching on and off times, d is the cell thickness, V is the applied voltage and V_o is the threshold voltage. K_{eff} is the effective elastic constant; for example, in the case of TN

$$K_{eff} = K_1 + \frac{(K_3 - 2K_2)}{4}$$

and here K_1, K_2, K_3 are the splay, twist, bend elastic constants, respectively. EQNS (1) and (2) are derived for the twisted LCD modes; however these equations turn out to be valid if K_{eff} is interpreted in a corresponding manner, for example $K_{eff} = K_2$ (IPS display), $K_{eff} = K_3$ (VA display). EQNS (1) and (2) show that the switching times are proportional to the rotational viscosity γ_1. It means the nematic director is subject to the rotational motion as a first approximation in the operation of typical LCDs. This first approximation must be modified when flow effects become important. Indeed, this is the case if the so-called back-flow effect needs to be taken into consideration [8]. The switching times of LCDs can no longer be described by the single viscosity coefficient γ_1, but the original equation of the motion for the director has the form of

$$\left(\gamma_1 - \gamma_{flow}\right)\frac{\partial \theta}{\partial t} = -\frac{\delta}{\delta \theta} F\left(\theta, \varphi; K_i\right) \tag{3}$$

with $\gamma_{flow} = \gamma_{flow}(\theta, \varphi; \eta_i)$.

Here ∂ and δ denote partial and functional derivatives, respectively, F is the Frank free energy, θ and φ are spherical polar coordinates for the director and γ_{flow} is the contribution of the flow effect which gives rise to a non-linearity of the rotational viscosity. If the flow effect plays a dominant role, all of the viscosity coefficients η_i should be taken into consideration. For instance, for TN γ_{flow} is not negligible compared with γ_1 when the operating voltage V is very high compared with the effective saturation voltage of the voltage-transmission curve. However, in the current TFT-TN devices V is designed to be close to this saturation voltage due to the limited power consumption of LCDs. It means that γ_{flow} plays only a minor role in the dynamics of LCDs. Altogether, we could assume that the switching times of representative LCDs are dominated by the rotational viscosity coefficient γ_1.

The measurement methods for the rotational viscosity coefficient γ_1 can be roughly categorised into two types, the relaxation method and the stationary method. The former measures a certain relaxation time of the rotational motion of directors by detecting, for example, optical or dielectric (capacitive) signals. This method is easy and not time consuming. But the elastic constants have to be measured in advance since the relaxation times are obtained as a combination of γ_1 and K_i. However, the measurement of elastic constants K_i is not straightforward and is time consuming. One of the representative methods is the transient current method [9,10]. The magnetic field rotational method is an example of the stationary method [11]. The rotational viscosity is calculated from the balance equation between the torque of magnetic force applied and the rotational frictional force. The advantage of this method is that the elastic constants are not needed for the evaluation of γ_1. The disadvantage of this method is that it is time consuming because it is necessary to wait until the system reaches a steady state.

The data for the rotational viscosity given in this paper were mainly measured by the magnetic field rotational method. We have recently introduced the transient current method and optimised this method so that the necessary dielectric and elastic information can be obtained in a parallel manner. Data obtained by both methods agree very well.

C   NEMATICS AND ROTATIONAL VISCOSITY

As we have shown in the previous section, the switching behaviour of nematics for display applications is basically governed by their rotational viscosity, γ_1. It is well-known, that a single nematogen does not match the many requirements for a real display application. For example, a reasonable broad nematic temperature range can only be obtained with mixtures of several different liquid crystals. This fact gives us a guideline for this Datareview. First, we will discuss briefly the relationship between the structure of a single mesogenic molecule and its rotational viscosity. Different examples will be presented in a tabular form. The relationship between rotational viscosity and other important parameters will be discussed qualitatively. Then, we turn to LC-mixtures and will describe the basic features with regard to their rotational viscosity. The impact of the rotational viscosity on other important physical LC-mixture parameters is discussed. This will lead to the last part, which deals with some typical LCD-modes and their demand for tailor-made LC-mixtures with specific rotational viscosities.

TABLE 1 gives an overview of different nematogens and their corresponding rotational viscosity. The principal focus of these data is the assessment of the materials with regard to their application in LC-mixtures. Hence, the reported values are not directly measured single values. Instead, a certain quantity of the compound is mixed with the commercially available host mixture ZLI-4792 and the single γ_1 value of the guest is obtained by extrapolating the measured γ_1 from the well-known value of ZLI-4792. To compare different compounds, the shift of the value of the nematic-isotropic transition temperature is taken into account.

TABLE 1 Various LCs and their rotational viscosity obtained
by extrapolation in a mixture with ZLI-4792.

Example	Molecular structure	γ_1/mPa s
A		42
B		233
C		703
D		117
E		151
F		191
G		180
H		252

TABLE 1 continued

Example	Molecular structure	γ_1/mPa s
I		171
J		173
K		112
L		413
M		99

Examples A, B and C demonstrate the most general rule, which exists between the molecular structure of an LC and its corresponding rotational viscosity; the longer a molecule the higher γ_1. Also by small changes in the overall molecular length this simple rule is nicely confirmed with the homologous series D-E-F. Increasing length of the terminal chain leads directly to a larger value of γ_1. However this is not a universal rule, which can be deduced from the values of γ_1 for G and H. Although the single fluorine substituent is shorter than the -OCF3 group, the value of γ_1 is not governed by this effect and is significantly lower for G than for H.

So far, we have focused on the relation between structural changes of the long axis of the nematogenic molecule and γ_1. We now extend this to present typical examples of how a change in the lateral structure, which means the structure-direction inclined to the molecular long axis, influences the rotational viscosity of a nematic. Compound I can be compared directly with compound E, only a cyclohexane ring is substituted by a phenyl ring which leads to a significant decrease of the rotational viscosity. This is a rare case, which can be easily understood with the microstructure of the molecule. The geometry of a phenyl ring is planar, while a cyclohexane ring exhibits a highly non-planar structure. Hence, a molecule consisting of phenyl rings rotates more easily. On the other hand, the curved structure of a cyclohexane ring leads to various steric interaction possibilities during a rotation, which results in an increased value of γ_1.

Another route to controlling the rotational viscosity is by tuning the microstructure of a molecule beyond its molecular long axis, as is evident with the molecules J and E. The introduction of a lateral fluorine atom leads to the expected increase in γ_1.

To complete the description of the influence of lateral substitutions on γ_1, we consider compounds K and L, two typical representatives of mesogenic molecules which can be used for mixtures with negative dielectric anisotropy, $\Delta\varepsilon$. To achieve a negative dielectric anisotropy, polar side groups have to be introduced. This necessarily increases γ_1, which can be seen by comparing compound K with A and L with B.

During the design of liquid crystal mixtures for display applications, various factors with regard to their rotational viscosity have to be taken into account. Due to the complex interaction of the different compounds in the mixture, the rotational viscosity of a mixture, consisting of N LCs with concentration C_i and rotational viscosity γ_i, is not the arithmetic mean of their compounds; hence the formula

$$\gamma_{Mix} = \sum_{i=1}^{N} C_i \gamma_i$$

can only serve as a very crude approximation with which to obtain a first idea of the rotational viscosity of the final mixture. A somewhat better correlation between γ_1 of an LC-mixture and its components is given by

$$\ln \gamma_{Mix} = \sum_{i=1}^{N} C_i \ln \gamma_i$$

which is based on an Arrhenius-like description of the rotational movement of a molecule in its environment, which leads to a geometric mean [12]. Although better than the simple linear description, the geometric mean can only be considered as an expedient due to the lack of a reliable microscopic description of the rotational viscosity of an LC mixture.

Another important point for the design of an LC mixture is the trade-off between γ_1 and other fundamental physical parameters of the mixture. The relationship between γ_1 and the nematic-isotropic transition temperature of the LC mixture and the connection between γ_1 and the dielectric anisotropy of the mixture will be described briefly.

As has been shown in TABLE 1, the larger a molecule, the higher on average is the corresponding γ_1. On the other hand, it is well-known that there is also an empirical relationship between the size of a molecule and its nematic-isotropic transition temperature. A four-ring mesogen with a nematic phase will almost always exhibit a higher transition temperature than a corresponding two-ring material. Since many molecular details are blurred out in an LC-mixture which consists of more than ten different liquid crystalline materials, these points lead to the following principle for creating LC-mixtures: the higher the T_{NI} of the mixture, the higher is the corresponding rotational viscosity. The impact of this quite general rule for the design and the application of LC-mixtures for displays will be discussed later in this section.

A second important trade-off relation is the link between the rotational viscosity and the dielectric anisotropy of an LC-mixture. Once again, this can be made plausible with an approximate argument stemming from the microscopic structure of the constituent molecules. High $\Delta\tilde{\varepsilon}$ - materials are obtained when the molecules have strongly polar end-groups, e.g. fluorine or cyano end-groups enhance the dielectric anisotropy significantly. As has already been argued, on average such strongly polar end-groups increase γ_1 of a molecule. Taking again the argument that many molecular details are washed out in an LC-mixture of about ten components, this leads straightforwardly to the following trade-off relation: the higher the polar nature of the components of the mixture, the higher the corresponding value of γ_1. We next briefly describe some relevant LCD-modes with regard to the LC-mixtures being used and their rotational viscosity.

The most fundamental mode for LCDs is still based on the twisted nematic cell (TN-cell). For such applications, the described trade-off relations between γ_1 and other important physical parameters lead to important consequences. This is illustrated with two examples.

(1)   For monitor use, the required nematic range has not to be as broad as for outdoor applications. The somewhat lower T_{NI} opens more possibilities in the LC-mixture development to decrease γ_1. These result in quite fast switching possibilities and TN-LCD-monitors with video rate switching capability have already been presented.

(2)   A low driving voltage might be not so important for LC-monitors as for mobile display applications. Since a higher available voltage level leaves room for mixtures with relatively low $\Delta\tilde{\varepsilon}$, according to the described trade-off, also mixtures with low γ_1 are available. Hence, these two trade-off relations indicate that, from basic physical principles, LC-mixtures support the fast switching requirement for moving pictures for LC-monitor applications. Compounds B, D, E, F, I and J are typical representatives which can be used for TN-TFT monitor applications.

Different types of LC-modes with homeotropically aligned nematics have been described [3]. This so-called VA (vertically aligned) mode intrinsically exhibits a very fast switching time and is a promising candidate for full moving picture LCDs. The VA-mode improves the switching time significantly as compared to the TN-case. LC-mixtures in this VA application field typically contain compounds like K or L.

A different approach has been demonstrated with the in-plane switching mode (IPS) [2]. As described previously, the switching time of this mode is governed by the elastic constant K_2, which leads to a somewhat larger switching time, as compared to the TN-mode. However, this point is compensated by the fact that IPS is the only TFT-LCD-mode that allows the use of LC-mixtures which contain cyano materials. For example, compound M [13] is a typical candidate, which can be used to improve the switching time of LC-mixtures for the IPS-mode, by use of a material class which is not available for other TFT-driven LCD-modes.

D   CONCLUSION

A brief and comprehensive review of the viscous properties of nematics with regard to their application in displays has been given. It has been shown that the rotational viscosity is the key viscous parameter for display applications of LCs. The relationship between this parameter and the switching time in an LCD-panel has been discussed.

Both from the single compound side and from the mixture side we have pointed out some important issues concerning the rotational viscosity of various liquid crystals. Basic rules and the link between the molecular structure of a nematogen and the resulting rotational viscosity have been presented. The further relationship between γ_1 and other important physical parameters of LC-mixtures has been outlined.

The results obtained have been discussed in the framework of real display modes. One of the future targets of LCDs will be to realise a full moving picture for monitor applications. Taking into account various improvement possibilities both from the LCD-manufacturer side (smaller cell gap, optimised voltage driving scheme, future pixel architecture) and from the LC-material research side (reduction of γ_1), we are optimistic that LCDs will be in a position to compete with the well-established cathode ray tubes.

REFERENCES

[1]   M. Hirata et al [*Proc. Int. Workshop on Active Matrix LCDs* (in association with IDW'96), Kobe, Japan, 27-29 Nov. 1996 (Japan Soc. Appl. Phys., Tokyo, Japan, 1996) p.193-6]

[2]   M. Oh-e, M. Ohta, S. Aratani, K. Kondo [*Proc. 15th Int. Display Research Conf.* (Asia Display 1995), Hamamatsu, Japan, 16-18 Oct. 1995 (USA, 1995) p.577]

[3]   A. Takeda et al [*Proc. SID 1998* Anaheim, CA, USA, 17-22 May 1998 (Soc. Inf. Display, Anaheim, CA, USA, 1998) vol.XXIX, Digest of Technical Papers, p.1070]

[4]   K.H. Kim, K. Lee, S.B. Park, J.K. Song, S. Kim, J.H. Souk [*Proc. Int. Display Research Conf.* (Asia Display 1998) (1998)]

[5]   S. Mizushima et al [*Proc. Int. Workshop on Active Matrix LCDs* (1999) p.177]

[6]   K. Tarumi, U. Finkenzeller, B. Schuler [*Jpn. J. Appl. Phys. (Japan)* vol.31 (1992) p.2829]

[7]   K. Tarumi et al [*Int. Display Research Conf.* (Japan Display 1992) (1992) p.587]

[8]   K. Tarumi, T. Jacob, H.-H. Graf [*Mol. Cryst. Liq. Cryst. (Switzerland)* vol.263 (1995) p.459]

[9]   M. Imai et al [*Jpn. J. Appl. Phys. (Japan)* vol.33 (1994) p.L119]

[10]   H. Ichinose, Y. Ikedo, D. Klement, A. Pausch, K. Tarumi [*Ekisho-Toronkai* (1997) p.212]

[11]   F.-J. Bock, H. Kneppe, F. Schneider [*Liq. Cryst. (UK)* vol.3 (1988) p.217]

[12]   I. Sage [in *Thermotropic Liquid Crystals* Ed. G.W. Gray (John Wiley & Sons, New York, 1987)]

[13]   D. Klement, K. Tarumi [*Proc. SID 1998* Anaheim, CA, USA, 17-22 May 1998 (Soc. Inf. Display, Anaheim, CA, USA, 1998) vol.XXIX, Digest of Technical Papers, p.393]

11.5   Material properties for nematic reflective displays

H. Molsen

October 1998

A   INTRODUCTION

This Datareview deals with the requirements of nematic liquid crystals specific to a variety of reflective liquid crystal displays (R-LCDs) for direct view or projection and also gives a brief description of their mode of operation. The focus will be on modes requiring a single or no linear polariser.

B   FUNDAMENTALS OF REFLECTIVE LIQUID CRYSTAL DISPLAYS

The purpose of R-LCDs is to reduce power consumption and weight compared to transmissive liquid crystal displays where ~75% of the power is used by the backlight. In other words, R-LCDs are targetted at portable equipment relying on battery operation.

R-LCDs are designed for reflecting ambient light which can vary in colour temperature between 2600 K for tungsten lighting and 6500 K for direct sunlight. At low ambient light level supplementary illumination can be provided by a frontlight or a backlight. The latter case is often referred to as a transflective liquid crystal display.

In general, R-LCDs can be divided into two categories according to their use of linear polarisers:

(1)   Birefringence mode R-LCDs relying on the interaction between linearly polarised light and the nematic, subdivided into two types:

      (a)   Dual polariser R-LCDs employing two linear polarisers, one between the viewer and the cell and a second between the mirror and the cell (see FIGURE 1).

      (b)   Single polariser R-LCDs with just one polariser between the viewer and the cell using an internal mirror (see FIGURE 2).

(2)   Zero polariser R-LCDs using

      (a)   voltage dependent absorption by dichroic dyes as a solute or in a nematic solvent;

      (b)   field dependent selective reflection in a highly twisted chiral nematic;

      (c)   scattering by mismatching refractive indices within the nematic layer.

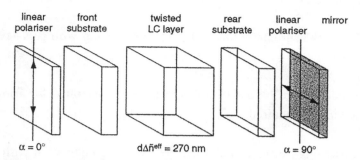

FIGURE 1 Schematic representation of a dual polariser reflective liquid crystal display.

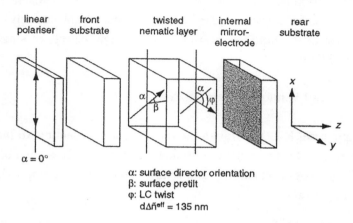

FIGURE 2 Schematic representation of a single polariser reflective liquid crystal display.

We discuss the materials requirements for all of these modes with the exception of two:

- Dual polariser R-LCDs are basically identical to transmissive liquid crystal displays in their mode of operation but restricted to low resolution due to parallax (optical cross-talk) and ~20% lower reflectivity compared to single polariser R-LCDs [1]. Apart from general demands on the nematic posed by mobile, in particular out-door, application on the range of operating temperature, the requirements are similar to non-reflective LCDs.

- Scattering modes in heterogeneous nematic media like polymer dispersed liquid crystals or polymer network liquid crystals, where a high ratio of back to forward scattering is a prerequisite for a good contrast ratio are not discussed here.

Apart from scattering, the electrooptics of reflective nematic modes can be calculated by solving the Jones matrix [2] or Berreman's 4×4 matrix [3] which are available as commercial software [4].

C   SINGLE POLARISER REFLECTIVE LIQUID CRYSTAL DISPLAYS

Depending on its polarisation efficiency, a single polariser transmits at best 45% of the incoming unpolarised light. Consequently, in any single polariser R-LCD this will be the maximum reflectance if other optical losses are ignored.

In order to produce an image on such an R-LCD two approaches are possible:

(1)   A nematic layer with achromatic black and white switching with grey-level capability can be used.

(2)   Birefringence colours can be used in an electrically controlled birefringence R-LCD if no grey-level capability is required.

The black and white switching nematic modes can be used in combination with a colour filter plate with broad transmission windows for each colour. The colour filter plate will reduce the reflectance by ~50%. Generally, the contrast requirements for R-LCDs are less stringent than for transmissive displays since in transmission the surface glare due to ambient light strongly deteriorates the dark state and thus the contrast of the liquid crystal display. Design aspects of single polariser R-LCDs have been reviewed by Tillin [1].

C1   Reflective Black and White Nematic Modes

The optics of a single polariser R-LCD with an internal mirror are identical to those of a transmissive LCD with parallel polarisers. To achieve the achromaticity required for a good black state resulting in high contrast ratio with grey-level capability and a large viewing angle, the R-LCD must operate in the first Gooch-Tarry minimum.

For a single light pass, the nematic switches between an effective retardation of 0.25 producing the dark state and $m(\Delta\tilde{n}^{eff}d_{LC}/2)$ producing the white state, where m is an integer, $\Delta\tilde{n}^{eff}$ is the effective birefringence for a particular wavelength and twist of the nematic layer and d_{LC} is the thickness of the nematic layer. The display can be operated in a normally white or in a normally black mode at zero field across the nematic layer. For a nematic with a positive dielectric anisotropy, $\Delta\tilde{\varepsilon} > 0$, m equals zero for the normally white mode or one for the normally black mode. The mode of operation is reversed for nematics with $\Delta\tilde{\varepsilon} < 0$. Since the entering light passes the nematic layer twice, for first minimum devices the birefringence $\Delta\tilde{n}$ of the nematic should be as small as possible to maintain a reasonable cell gap.

To eliminate residual retardation of the switched nematic state, to increase achromaticity, or improve the viewing angle, uniaxial or biaxial retarders can be added between the linear polariser and the cell.

Single polariser nematic modes can be separated into untwisted and twisted LC modes. The untwisted LC modes are Fréedericksz transition [5], optically compensated birefringence (OCB) [6], hybrid-aligned nematic (HAN, R-OCB) [7], and zenithal bistable device (ZBD) [8]. The twisted nematic modes include: twisted nematic (TN) [9,10], supertwisted nematic (STN) [11], bistable twisted nematic (BTN) [12], and the bistable surface controlled nematic (SCN) [13].

We will use FIGURE 2 to describe each reflective nematic mode in terms of surface pretilt β, surface director orientation α and the twist φ of the director. The surface director orientation is the angle between the projection of the director onto the surface (xy plane) and the polarisation direction of the light. The surface pretilt β is the angle between the surface director orientation and the director orientation at the surface along the z direction.

As TABLE 1 shows, R-LCDs, depending on their mode, require nematic materials of either positive or negative dielectric anisotropy with low birefringence to give cell gaps which are easy to manufacture. The best nematic mixtures with good stability to ultraviolet light and with a very broad temperature range achieve birefringence values of about 0.07. This is achieved by using fluorine substitution and non-conjugated ring systems like cyclohexyl. A further reduction of birefringence is desirable to increase the cell gap of R-LCDs in untwisted modes. A good black state requires reduced chromaticity, i.e. the dispersion of the nematic should be as low as possible and matched to that of the polymer used as a compensating retarder.

TABLE 1 Single polariser reflective nematic modes and their cell parameters.

Reflective nematic mode	Dielectric anisotropy $\Delta\widetilde{\varepsilon}$	Twist in LC layer $\varphi/°$	Surface director orientation $\alpha/°$	Surface pretilt at top/bottom substrate $\beta/°$
Passive matrix				
R-STN	>0	≤240		4/4
R-BTN	>0	$\varphi - 180°/\varphi + 180°$		2/2
R-ZBD	>0	0	45	90/90 on grating
R-SCN	>0	0/180	45	15/2 weak
Active matrix				
R-TN	>0; <0	≤80		2/2; 88/88
R-HAN (R-OCB)	>0; <0	0	45	2/90
OCB	>0	0	45	5/5
R-Fréedericksz	>0; <0	0	45	2/2; 88/88

C2   Electrically Controlled Birefringence Mode

Birefringence colours in the electrically controlled birefringence mode are produced by nematic layers with comparatively large retardations $d\Delta\tilde{n}^{\text{eff}}$. Untwisted, active matrix addressed and twisted supertwist type passive matrix addressed panels have been investigated.

Nowadays, the expensive active matrix addressing is considered unsuitable for direct view R-LCDs [14] because of the drawbacks of the electrically controlled birefringence mode (untwisted nematic with $\Delta\widetilde{\varepsilon} < 0$). As the reflected colour (or hue) is very sensitive to the applied voltage, this mode is susceptible to electrical cross-talk and does not provide grey-scale. The use of birefringence colours makes the reflected colour sensitive to the angle and source of illumination. Furthermore, creating a balanced red-green-blue colour triangle (especially red) in one panel is not possible. However, low-cost passive matrix displays with five colours are commercialised [15] using a dual polariser layout described previously. Their nematic requirements are similar to the transmissive supertwist mode but used with a larger cell gap.

D   ZERO POLARISER REFLECTIVE LIQUID CRYSTAL DISPLAYS

The requirements of reflective-nematic modes, which do not suffer from the absorption loss associated with the linear polariser, are discussed for the following cases:

(1)   Guest-host nematic.
(2)   Selective reflection from a chiral nematic.
(3)   Dynamic scattering mode.

D1   Guest-Host Reflective Liquid Crystal Displays

The guest-host mode relies on the voltage dependent absorption of dichroic dyes dissolved in the nematic [16]. According to the molecular structure the transition moment of the dichroic dyes lies either parallel (positive dichroic, pleochroic) or normal (negative dichroic) to the long molecular axis which in turn aligns parallel to the director. The dichroic ratio $D = \tilde{A}_{\parallel}/\tilde{A}_{\perp}$ defines the orientational order of the transition moment of the dye. In the common case of azo dyes, the absorption axis of the dye aligns parallel to the director, so absorption occurs when the director and hence the dye are aligned perpendicular to the optic axis, i.e. planar. The degree to which the absorption axes of the dye molecules are aligned parallel to the director is measured by the temperature-dependent order parameter of the transition moment $S = (A_{\parallel} - A_{\perp})/(A_{\parallel} + 2A_{\perp}) = (D - 1)/(D + 2)$. For positive dielectric anisotropy of the nematic, applying a voltage then results in a decreased absorption. The higher the voltage and the higher the order parameter of the dichroic dye, the lower this absorption will be. Complete solubility of the dye in the nematic at low temperature limits the maximum dye content to about 3% by weight. The limited absorption demands cell gaps larger than 5 μm even for reflective displays thus affecting switching times.

The guest-host modes can be distinguished in untwisted (uniform) [17] and twisted [18] (phase change) modes. The modes requiring a linear polariser will not be discussed for they are inferior to single polariser nematic modes in terms of brightness and contrast. Untwisted guest-host modes require a quarterwave retarder between the mirror and the nematic [19] to achieve the dark state. Alternatively, two layers of untwisted guest-host nematic can be oriented orthogonally and stacked on top of each other [20], separated by a thin polymer film [21].

The phase change guest-host mode as shown in FIGURE 3 can absorb the incoming light for all polarisations provided guiding of the plane of polarisation by the nematic is reduced to a minimum to operate outside the Mauguin limit.

The twist angle of the nematic host for active matrix driving is constrained by the optimum contrast ratio which requires twist angles larger than 220°, but twist angles must be small enough to avoid stripe domains and to achieve grey-scales without hysteresis (~300°) [22]. As for supertwist displays, the steepness of the curve is strongly dependent on the ratio K_3/K_1 with values closer to unity providing the required flat response.

For an active matrix addressing technology, in particular thin film diodes [23] or metal-insulator-metal switches which support data voltages up to eight volts, the contrast ratio can be as high as 10:1 while maintaining good reflectance values.

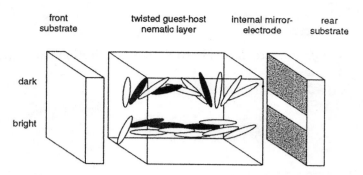

FIGURE 3 Schematic representation of a phase change guest-host reflective
liquid crystal display in its bright and dark state.

The nematic and the dichroic dye must have resistivities high enough to provide a sufficient voltage holding ratio, usually $>10^{11}$ Ω/cm. Normally white and normally black phase change guest-host modes are possible by using a nematic host with either negative or positive dielectric anisotropy in combination with a homeotropic or planar alignment, respectively.

Colour phase change guest-host displays can be achieved in two ways. The dye mixture absorbs across the visible spectrum for black and white switching and is combined with a colour filter plate [24,25]. Alternatively, the individual dyes are separated into two [26] or three layers [27] and switched independently.

For passive matrix addressing the nematic host should possess the essential properties required for the supertwist display discussed elsewhere (i.e. steep electrooptic curve, etc.) plus good solubility and high order parameter environment for the dichroic dye. However, passive matrix addressed phase change guest-host displays without a polariser suffer from very low contrast of less than 3:1.

D2   Selective Reflection from Chiral Nematic Liquid Crystals

A planar aligned highly twisted nematic, i.e. a chiral nematic, exhibits first order diffractive reflection of circularly polarised light of wavelength $\lambda = p(0.5n_o(\lambda) + 0.5n_e(\lambda))$ at normal incidence with p the pitch describing a rotation of the director by 180°. The spectral width of the reflection band is determined by the optical anisotropy of the nematic according to $\Delta\lambda = p(n_e(\lambda) - n_o(\lambda))$. With oblique incidence of light, the selective reflection waveband is shifted to shorter wavelengths and its analysis becomes complicated [28]. In practice, it is sought to disturb the planar orientation and create small domains with a distribution of orientation to overcome the problems associated with oblique incidence of light. The polydomain formation can be achieved by dispersion of a polymer [29], random alignment of domains [30,31] or the addition of surfactant and water [32].

Application of an electric field to a planar texture removes the selective reflection in favour of a slightly backscattering focal conic texture which forms the dark state if an absorber is applied to the rear substrate. The switching between the two stable states, the planar and the focal conic texture, of the monodomain chiral nematic [33] is exemplified in FIGURE 4. The two unstable states, the field-induced homeotropic orientation and the transient G* phase are required in order to switch from the focal conic to the planar state. The liquid crystal requirements for chiral nematic displays are apparent from the need to lower the threshold field

$$E_c \propto \sqrt{(K_2 / \Delta\widetilde{\varepsilon})}$$

Therefore, a low K_2 and high $\Delta\widetilde{\varepsilon}$ are desirable while broad reflection bands are brought about by a high $\Delta\tilde{n}$.

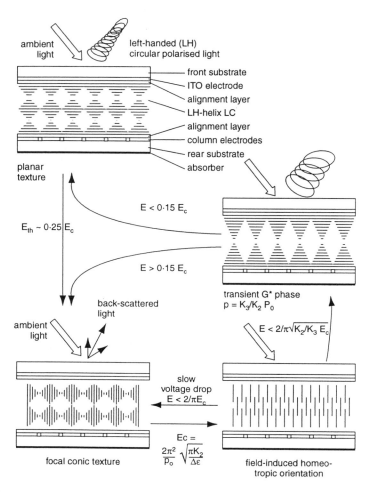

FIGURE 4 Schematic representation of the four states involved in switching a chiral nematic reflective liquid crystal display.

Colour can be achieved using three layers [34] of chiral nematic if addressed independently. For colour rendition by chiral nematics a complication occurs for the colour red since the side bands of the selective reflection towards the green tend to dominate the visual impression due to the photopic eye response [35]. In this case, the addition of a magenta dye to the chiral nematic is desirable.

D3   Bistable Dynamic Scattering Mode

The dynamic scattering mode relies on light scattering induced by electrohydrodynamic instabilities above a threshold in a nematic created by charge injection or ionic motion [36]. This mode combines a negative dielectric anisotropy $\Delta\widetilde{\varepsilon} < 0$ with a positive conductivity anisotropy $\Delta\widetilde{\sigma} > 0$ in the nematic ($\Delta\widetilde{\sigma}$ is always positive for calamitic nematics) [37]. The mathematic description of this current effect is complicated by non-linearities and defects.

A transparent state and a scattering state can be selected either by standard multiplexing or by frequency variation (dual frequency addressing). At low frequency, in the conduction regime, the ionic movement favouring a homeotropic orientation competes with the planar orientation favoured by the dielectric anisotropy. Above a cut-off frequency, the ionic response is practically zero and the negative dielectric anisotropy dominates, forcing the nematic into a planar orientation. If a twisted nematic is used the scattering state could be stable creating a bistable electrooptic effect in combination with dual frequency addressing.

The optical properties can be thought of as a first order diffraction with a random periodicity. This makes the dynamic scattering mode rather impractical because a visible backscattering would require a specular mirror with too restricted illumination conditions. Another drawback of this mode is its temperature dependence. The conductivity increases exponentially as does the cut-off frequency making correct ion doping for the complete temperature range impossible and increasing the current consumption. The dynamic scattering mode is now obsolete due to drawbacks in the optics and the temperature dependence of the effect itself.

E   PROJECTION

Reflective liquid crystal modes for projection are now in competition with the transmissive modes dominant so far. They combine standard silicon CMOS backplanes with either single polariser modes discussed previously, notably twisted nematic [38], R-HAN [39], R-Fréedericksz [40,41], or nematics in switchable polarisation independent diffraction gratings. Here, we will focus on the nematic phase gratings while a general overview on liquid crystal projection displays is discussed in [42].

A projection system based on diffraction gratings or, generally, scattering typically uses a Schlieren optical system. In a black pixel the light is scattered outside an aperture or the collection lens, and in a bright pixel the light remains undiffracted, which in a reflective system is equivalent to specular reflection. An inverted structure, i.e. the signal comes from the diffracted light, is also possible [43] thus eliminating problems with the diffracted light intensity leaking through the aperture.

The diffraction grating can be polarisation independent provided the phase change between two neighbouring nematic areas can be switched between zero and π. This can be achieved by interdigitated electrodes within one pixel [44], fringing fields [45], or patterned alignment [46,47]. The nematic here takes up an untwisted or a twisted director configuration and therefore the materials requirements for these diffractive modes are similar to those of the single polariser R-LCDs discussed previously.

An additional issue for high-resolution projection systems is the width of defect lines between two pixels due to, for example, reverse twist domains. Depending on the Schlieren optical system used, they can reduce the contrast ratio.

F   CONCLUSION

In direct view and projection R-LCDs the variety of nematic effects worth exploiting makes a unified requirement for the nematic material difficult, except for general properties like a low rotational viscosity, a broad nematic phase range, a high dielectric anisotropy, an excellent UV stability, and low ionic contamination. For single polariser nematic modes with low twist, the material challenge is to produce low birefringence material while maintaining a positive dielectric anisotropy. In addition to this requirement, for guest-host devices good solubility and high order parameter of the dye has to be added. For selective reflection from chiral nematics a high birefringence in combination with a high dielectric anisotropy while maintaining UV stability are required.

REFERENCES

[1]   M. Tillin [to be published in *Display and Imaging - International Edition (USA)* vol.1 (1999)]

[2]   R.C. Jones [*J. Opt. Soc. Am. (USA)* vol.32 (1942) p.486]

[3]   D.W. Berreman, T.J. Scheffer [*Mol. Cryst. Liq. Cryst. (UK)* vol.11 (1970) p.395]

[4]   [1DIMOS Software, Autronic-Melchers GmbH, Karlsruhe, Germany]

[5]   S.-T. Wu, C.-S. Wu [*Proc. Soc. Inf. Disp. (USA)* vol.29 (1998) p.770]

[6]   P.J. Bos, P.A. Johnson, K.R. Koehler/Beran [*Mol. Cryst. Liq. Cryst. (UK)* vol.263 (1984) p.329]

[7]   M. Shibazaki et al [*Proc. 18th Int. Display Research Conf.* (Society for Information Display, San Jose, 1998) p.51]

[8]   G.P. Bryan-Brown et al [*Proc. Soc. Inf. Disp. (USA)* vol.28 (1997) p.37]

[9]   E. Beynon, K. Saynor, M. Tillin, M. Towler [*Proc. 17th Int. Display Research Conf.* (Society for Information Display, Santa Ana, 1997) p.L-34]

[10]   S. Fujiwara, Y. Itoh, N. Kimura, S. Mizushima, F. Funada [*Proc. Int. Display Workshops 1997* Japan (1997) p.879]

[11]   M. Ohizumi, T. Hoshino, M. Kano, T. Miyashita, T. Uchida [*Proc. 18th Int. Display Research Conf.* (Society for Information Display, San Jose, 1998) p.1059]

[12]   Y.J. Kim, C.-J. Yu, S.-D. Lee [*Proc. 18th Int. Display Research Conf.* (Society for Information Display, San Jose, 1998) p.763]

[13]   I. Dozov, M. Nobili, G. Durand [*Appl. Phys. Lett. (USA)* vol.70 (1997) p.1179]

[14]   M.G. Pitt et al [*Proc. Soc. Inf. Disp. (USA)* vol.28 (1997) p.473]

[15]   M. Ozeki, H. Mori, E. Shidoji, Y. Hirai, H. Koh, M. Akatsuka [*Proc. Soc. Inf. Disp. (USA)* vol.27 (1996) p.107]

[16]   B. Bahadur [in *Liquid Crystals - Applications and Uses* vol.3, Ed. B. Bahadur (World Scientific, Singapore, 1992) p.65-208]

[17]   G.H. Heilmeier, L.A. Zanoni [*Appl. Phys. Lett. (USA)* vol.13 (1968) p.91]

[18]   D.L. White, G.N. Taylor [*J. Appl. Phys. (USA)* vol.45 (1974) p.4718]

[19]   H.S. Cole, R.A. Kashnow [*Appl. Phys. Lett. (USA)* vol.30 (1977) p.619]

[20]   T. Uchida, H. Seki, C. Shishido, M. Wada [*Proc. Soc. Inf. Disp. (USA)* vol.22 (1981) p.41]

[21]   A.C. Lowe, M. Hasegawa [*Proc. 17th Int. Display Research Conf.* (Society for Information Display, Santa Ana, 1997) p.250]

[22]   H. Suzuki et al [*Proc. 15th Int. Display Research Conf.* (Society for Information Display, Santa Ana, 1995) p.611]

[23]   R.A. Hartman [*Proc. Soc. Inf. Disp. (USA)* vol.26 (1995) p.7]

[24]   S. Mitsui et al [*Proc. Soc. Inf. Disp. (USA)* vol.23 (1992) p.437]

[25]   H. Ikeno, H. Kanoh, N. Ikeda, K. Yanai, H. Hayama, S. Kaneko [*Proc. Soc. Inf. Disp. (USA)* vol.28 (1997) p.1015]

[26]   H. Molsen, E. Beynon, M. Tillin [*Proc. Int. Display Workshops 1998* USA (1998) p.843]

[27]   Y. Nakai et al [*Proc. Soc. Inf. Disp. (USA)* vol.28 (1997) p.83]

[28]   V.A. Belyakov, V.E. Dmitrienko, V.P. Orlov [*Sov. Phys.-Usp. (USA)* vol.22 (1979) p.63]

[29]   D.-K. Yang, L.-C. Chien, J.W. Doane [*Proc. 11th Int. Display Research Conf.* (Society for Information Display, Santa Ana, 1991) p.44]

[30]   D.-K. Yang, J.L. West, L.-C. Chien, J.W. Doane [*J. Appl. Phys. (USA)* vol.76 (1994) p.1331]

[31]   Z.-J. Lu, W.D. St. John, X.-Y. Huang, D.-K. Yang, J.W. Doane [*Proc. Soc. Inf. Disp. (USA)* vol.26 (1995) p.172]

[32]   B.-G. Wu, H. Zhou, Y.-D. Ma [PCT patent application (1996) WO96/36905]

[33]   D.-K. Yang, Z.-J. Lu [*Proc. Soc. Inf. Disp. (USA)* vol.26 (1995) p.351]

[34]   M. Okada, T. Hatano, K. Hashimoto [*Proc. Soc. Inf. Disp. (USA)* vol.28 (1997) p.1019]

[35]   P. Kipfer, R. Klappert, J.-M. Künzi, J. Grupp, H.P. Herzig [*Proc. 27th Freiburg Liquid Crystal Workshop* Germany (1998) vol.27 poster P8]

[36]   R. Williams [*J. Chem. Phys. (USA)* vol.44 (1966) p.638]

[37]   B. Bahadur [in *Liquid Crystals - Applications and Uses* vol.1, Ed. B. Bahadur (World Scientific, Singapore, 1990) p.195-230]

[38]   R.L. Melcher, M. Ohhata, K. Enami [*Proc. Soc. Inf. Disp. (USA)* vol.29 (1998) p.25]

[39]   J. Glück, E. Lüder, T. Kallfass, H.-U. Lauer [*Proc. Soc. Inf. Disp. (USA)* vol.23 (1992) p.277]

[40]   H. Kurogane et al [*Proc. Soc. Inf. Disp. (USA)* vol.29 (1998) p.33]

[41]   F. Sato, Y. Yagi, K. Hanihara [*Proc. Soc. Inf. Disp. (USA)* vol.28 (1997) p.997]

[42]   S.E. Shields, W.P. Bleha [in *Liquid Crystals - Applications and Uses* vol.1, Ed. B. Bahadur (World Scientific, Singapore, 1990) p.437-91]

[43]   K.H. Yang, M. Lu [*IBM J. Res. Dev. (USA)* vol.42 (1998) p.401]

[44]   M. Fritsch, H. Wöhler, G. Haas, D. Mlynski [*IEEE Trans. Electron Devices (USA)* vol.36 (1989) p.1882]

[45]   H. Hatoh, Y. Hisatake, M. Sato, T. Ohyama, T. Watanabe [*Proc. SPIE (USA)* vol.2650 (1996) p.234]

[46]   P. Shannon, W. Gibbons, S. Sun, B. Swetlin [*Nature (UK)* vol.351 (1991) p.351]

[47]   C.M. Titus, P.J. Bos, C. Holton, W. Glenn [*Proc. Soc. Inf. Disp. (USA)* vol.28 (1997) p.769]

CHAPTER 12

COMPUTER MODELLING

12.1   Results of hard particle simulations

M.A. Bates

August 2000

A   INTRODUCTION

It is generally accepted that for atomic liquids far from the critical point, the microscopic organisation is dominated by the repulsive interactions between the atoms. The longer range attractive forces serve only to maintain the high density of the liquid phase and the temperature acts only as a mechanism with which to vary the density. With this in mind, the hard sphere fluid has become the standard starting point for liquid state theories in which the attractive forces can be introduced, for example, via a van der Waals approach. Hard sphere systems have been studied extensively using computer simulation, with the numerical data determined for the equation of state used as a substitute for an exactly solvable model for the liquid phase. This is in contrast to the gas and solid phases, for which the ideal gas and harmonic solid provide analytic models, respectively. Indeed, the lack of an analytic model for the liquid phase has meant that many of the current theories rely substantially on the insight obtained from the earliest simulations of the hard sphere fluid [1].

Given the dominant role of repulsive interactions in the organisation of atomic liquids, it seems entirely reasonable that repulsive interactions are also largely responsible for the molecular organisation within the nematic and other liquid crystalline phases. However, because of its isotropic shape, the hard sphere model does not provide a suitable reference point for common mesogens and so simulations of anisotropic hard body models are necessary. In contrast to the case of (spherical) atoms, the choice of reference model for a liquid crystal is neither unique nor straightforward. Indeed, there have been a number of hard body models introduced as reference models for liquid crystals. An in depth review of the phase behaviour and dynamical properties of rigid, hard body fluids based on theoretical and simulation results was given in 1993 [2] and little has changed in our understanding of the nematic phase since this time. However, attention has been paid to the accurate mapping of phase diagrams of, for example, hard spherocylinders as a function of their length-to-breadth ratio [3]. Attention has also been paid to how non-idealities, such as polydispersity in particle size [4,5] and shape [6], influence the stability of the liquid crystal phases exhibited by rod and disc shaped models. There have also been many simulations of hard particles with attractive potentials added to model, for example, electrostatic interactions; however, since these are not strictly hard body systems, these will not be discussed here.

B   HARD ELLIPSOIDS

The simplest anisometric generalisation of the hard sphere model is obtained by either stretching or compressing a sphere in a single direction to give a prolate or an oblate ellipsoid, respectively. Vieillard-Baron performed much of the groundwork for the simulation of hard ellipsoids in the 1970s, by, for example, determining an algorithm to test if two particles overlap, but was unable to observe the formation of a liquid crystalline phase in his simulations [7,8]. In the 1980s, Frenkel and Mulder

[9] performed a systematic study of hard ellipsoids as a function of their aspect or length-to-breadth ratio, from x = 1/3 (discotic particles) to x = 3 (calamitic particles). The single parameter x in this model, which characterises the shape of the ellipsoids, is defined as the ratio of the length of the particle along its symmetry axis divided by its breadth. The value x = 1 corresponds to the hard sphere model and the values x → ∞ and x → 0 correspond to the special cases of infinitely long rods and infinitely thin discs, respectively.

The phase behaviour of hard ellipsoids depends critically on the anisotropy parameter x, with up to three phases observed for any given anisotropy. The stability ranges of these phases is shown as a function of the aspect ratio in FIGURE 1. For strongly anisometric shapes (x > 2.5 for rods and x < 0.4 for discs) a nematic phase is observed. This phase undergoes a weak first order transition to an isotropic phase as the density is decreased and a stronger first order transition to the solid phase at higher densities. The density changes at these transitions are typically 2% and 10%, respectively. Although 2% for the nematic-isotropic transition is higher than the values observed for real mesogens, this is similar to values observed for the Gay-Berne model [10,11] and much less than the 25% density change predicted by the Onsager theory for infinitely long rods [12]. As the anisometry increases, so does the relative density range over which the nematic phase is stable. For calamitic ellipsoids, the number densities of the coexisting phases at the nematic-isotropic transition tend to zero, and thus to the limiting behaviour of the Onsager theory [12]. Similarly for discs, the nematic-isotropic coexistence densities also tend to zero, as the aspect ratio approaches the limiting case of infinitely thin platelets [13]. The nematic phase is not stable for 0.4 < x < 2.5, clearly indicating that a reasonably anisometric (molecular) shape is an important factor (although not necessarily the most important factor) in determining if the nematic phase is stable. For weakly anisometric ellipsoids, an

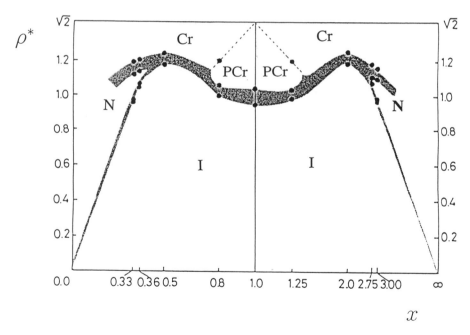

FIGURE 1 The phase behaviour of hard ellipsoids as a function of the aspect ratio x [2,9]. The reduced density ρ* is defined such that the density of regular close packing is equal to √2 for all x. The grey areas indicate the coexistence regions between the various phases, which are: isotropic (I), nematic (N), solid (Cr) and plastic solid (PCr). The solid points indicate the simulation results.

orientationally disordered, plastic solid phase is observed. Although the general form of the phase diagram is not in doubt, there is a dispute about the minimum anisometry needed to observe nematic behaviour. Simulations by Zarragoicoechea et al [14] indicate that the nematic phase observed in the simulations of Frenkel and Mulder [9] using a system of 100 ellipsoids with anisometry x = 3 may not be stable in larger systems of 256 particles. Thus it appears that the weak nematic-isotropic transition may be more sensitive to finite size effects than the nematic-solid transition. However, Allen and Mason [15] observed that the nematic for x = 3 is stable in systems of 108 - 576 particles. This dispute, which is based on rather small simulations, appears not to have been resolved; this is somewhat surprising given that simulations of much larger systems are now not only feasible but regularly performed for other, more computationally expensive models.

A surprising feature of the phase diagram, at least at first sight, is that prolate ellipsoids do not exhibit a smectic phase and oblate ones do not exhibit a columnar phase. This lack of translationally ordered mesophases was explained by Frenkel using the following simple scaling argument [16]. As there is high orientational order in these phases, the molecular symmetry axes can be taken to be parallel to the director. Since an affine transformation along this direction can convert the ellipsoid system into a hard sphere system, which exhibits only (isotropic) fluid and solid phases, parallel ellipsoids can exhibit only two phases, namely the nematic and solid. Thus it is unlikely that non-parallel but orientationally highly ordered ellipsoids will form either a smectic or a columnar phase. This does not rule out modelling of the translationally ordered phases, and there are a number of theories on why the smectic and columnar phases are thermodynamically stable. One theory is that attractive interactions between the molecular cores are important; models taking this into account are described in this volume by Zannoni [11]. An alternative mechanism is through the gain in entropy by melting the flexible alkyl tails attached to the mesogenic cores; models including this level of detail are described by Wilson [17]. A third important feature is the shape of the particle, since this can influence the local arrangement of the particles and, therefore, the packing entropy of the various phases. Two such models, namely spherocylinders and cut spheres, will be discussed in the following sections.

Before we leave ellipsoids, we note that it is also possible to stretch or compress along a second axis, to leave a biaxial ellipsoid. Allen has studied systems of biaxial ellipsoids, in which the molecular axes are all of differing lengths (a : b : c) [18]. At the rod limit (1 : 1 : 10) and disc limit (1 : 10 : 10), uniaxial nematic phases are observed, as expected. In between these limits (1 < b < 10), the molecule is biaxial (1 : b : 10); it is, therefore, possible that these model molecules could exhibit a biaxial nematic phase, in which the orientations of all three molecular axes become correlated at long range. The simulations provide evidence that such a system does indeed exhibit isotropic, uniaxial nematic and biaxial liquid crystalline phases. The biaxial phase is found to be most stable when

$$b = \sqrt{ac}$$

For molecules of this shape, the isotropic phase is found to undergo a transition directly into the biaxial phase on compression. A biaxial phase has also been observed in simulations of mixtures of rods and platelets, in which the different species are modelled ellipsoids with aspect ratios of x = 20 and x = 1/20, respectively [19]. However, here the range of the biaxial phase is severely limited by demixing into two coexisting uniaxial nematics, one rich in rods, the other rich in discs.

C   SPHEROCYLINDERS

Whilst hard ellipsoids represent a simple anisometric generalisation of the hard sphere model, another simple model exists for rod shaped particles. The hard spherocylinder is a model composed of a cylinder of length L and diameter D capped with two hemispheres of diameter D at both ends. The anisotropy of the particle is usually characterised by the ratio L/D, so that the limiting value L/D = 0 refers to the hard sphere system. Note that care should be taken when comparing the length-to-breadth ratio of spherocylinders with that of ellipsoids, since this is (L + D)/D and not L/D. The first simulations on freely rotating (as opposed to perfectly aligned) spherocylinders with length L/D = 5 performed by Frenkel et al [20] indicated that a smectic A phase, in addition to a nematic phase, was also possible in hard body systems. Since the spherocylinder provides a model with which to study the role of excluded volume effects in the formation of nematic and smectic phases, a number of studies have been performed to determine the phase diagram as a function of the aspect ratio. The long spherocylinder also provides a useful model for elongated colloidal particles such as boehmite (an aluminium oxide) [21], which are widely used in the ceramics industry, since the width of these particles is essentially constant along their length. Indeed, since systems of spherocylinders exhibit both nematic and smectic phases, the spherocylinder is often used as a reference system for both theoretical and experimental systems. Two major studies of spherocylinders were performed by Bolhuis and Frenkel [3] and McGrother et al [22]. In [21], the region 3 < L/D < 5 was studied in detail to determine the location of the onset of the liquid crystalline phases. The later study of Bolhuis and Frenkel [3] went on to determine the complete phase diagram from hard spheres at one extreme (L/D = 0) to infinitely long rods (L/D = ∞) at the other. The resulting phase diagram is shown in FIGURE 2. For very small elongations, up to L/D = 0.35, a face centred cubic rotator or plastic solid phase is observed, similar to hard dimers or dumbbells of the same aspect ratio [23]. For aspect ratios between L/D = 0.35 and L/D = 3.1, the only phases found are the isotropic liquid and crystalline phases. The smectic A phase is the first liquid crystal phase to enter the phase diagram, with the isotropic-smectic A-solid triple point occurring at L/D = 3.1 [3] or L/D = 3.2 [22]. The nematic phase only appears for longer rods, with the isotropic-nematic-smectic A triple point occurring at L/D = 3.7 [3] or L/D = 4.0 [22]. The differences in the location of these triple points arise presumably because in [22] they were not determined from free energy calculations, but estimated from the hysteresis region in the equation of state obtained from simulations in a cubic box; the imposition of a cubic box shape on a translationally ordered phase results in an increase of the free energy of that phase due to the generation of an anisotropic pressure. Note that the length-to-breadth ratio of a spherocylinder (L + D)/D has to be considerably longer than that of an ellipsoid (x) for the particle to exhibit a nematic phase. The nematic-smectic A transition for spherocylinders appears to be first order, and essentially independent of particle length, with a difference in coexistence densities of about 3%. Although the nematic-smectic A transition was initially thought to be continuous for infinitely long spherocylinders [3], further simulations using a special scaling technique, which allows simulation of infinitely long particles at finite volume fraction, indicate that this transition is first order for L/D = ∞ [24] and, presumably, for all elongations where both phases coexist.

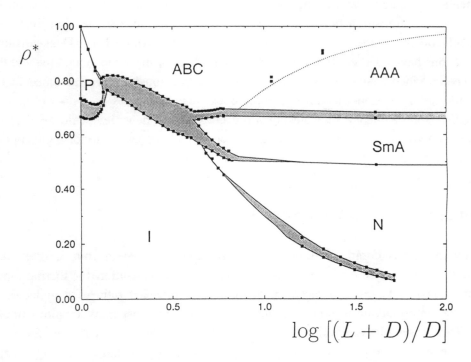

FIGURE 2 The phase behaviour of spherocylinders as a function of length-to-breadth ratio (L + D)/D [3]. The reduced density ρ* is defined as ρ/ρ_{CP}, where ρ_{CP} is the close packed density. The grey areas indicate the coexistence regions between the various phases, which are: isotropic (I), nematic (N), smectic A (SmA), solid (ABC and AAA stacking) and plastic solid (PCr). The solid points indicate the simulation results. Note that the aspect ratio is on a logarithmic scale.

D   CUT SPHERES

A change in shape from an ellipsoid to a spherocylinder can result in a remarkable difference in the stability of the (liquid crystalline) phases and, therefore, in the phase diagram of calamitic models. A model similar to the spherocylinder was devised for discotic systems by taking a sphere of diameter D and cutting off the top and bottom, to leave a tablet shaped particle of thickness L [25]; this particle is known as the cut sphere. This particle was used instead of a cylinder or a spheroplatelet (the disc equivalent of a spherocylinder, consisting of a short cylinder with a hemispherical rim) since the overlap test is computationally cheaper. The model again exhibits a variety of phases depending on the aspect ratio [26]. In addition to the isotropic and solid phases, a columnar phase is exhibited for particles with L/D = 0.1 and L/D = 0.2. For L/D = 0.1, a nematic phase is observed, whereas for L/D = 0.2 a cubatic phase is observed, in which the particles stack into short columns and these columns are oriented with cubic symmetry, but no positional order. Simulations of cut spheres in the limit L/D = 0 have also been performed [27], and indicate that the nematic-columnar transition is first order for all aspect ratios in the range, L/D = 0 to L/D = 0.1, with an anisotropy independent density change at the transition of approximately 10%.

The anisotropy dependence of the nematic-isotropic transition in systems of infinitely thin platelets has also been investigated, since this is important if we are to understand the behaviour of discs with a non-circular cross-section [6]. In addition to circular platelet systems, systems composed of hexagons, pentagons, squares and triangles of equal surface area were simulated and the location of the nematic-isotropic transition determined in each case. The coexistence densities are found to depend strongly on the shape, decreasing along the series as written [6]. This indicates that, for example, hexagons have a larger effective diameter than discs of the same facial area and this must be taken into account when analysing experimental systems; previously, it was assumed that hexagonal particles should have the same nematic-isotropic coexistence densities as discs of equal facial area [28]. Indeed, a simple theory indicates that the coexistence densities scale as the isotropically averaged excluded volume of a pair of particles, which is trivial to calculate for a given particle shape [6].

E   INFLUENCE OF POLYDISPERSITY

Polydispersity in size is common to many synthetic colloidal systems but is often neglected in simulation and theoretical studies. It is also an important consideration for thermotropic systems, since conformational changes in a mesogen can change the aspect ratio of a molecule [29]. The standard simulation technique to study the influence of the polydispersity employs the semi-grand ensemble. This ensemble is similar to the conventional canonical ensemble for monodisperse systems, but the size of the particles is imposed by a chemical potential or activity distribution; the particle sizes are allowed to change during the simulation and thus a continuous distribution is simulated. This also allows us to study the effect of size segregation more easily, although it does have the disadvantage that we do not know the size distribution until the simulation is finished.

The influence of polydispersity on the nematic-isotropic transition has been studied in systems of infinitely thin discs [5]. This model provides a useful system with which to study polydispersity since all simulations are carried out on particles of the same aspect ratio; no matter what the particle diameter, the aspect ratio L/D is always zero, and so the results from simulations with different activity distributions can be scaled for easy comparison. Similar simulations are not possible for rods since infinitely thin rods (of finite length) do not order at finite number density. The simulations indicate two main influences of polydispersity. The first is that, at the transition, the size distribution of the particles is different in the coexisting nematic and isotropic phases. In the nematic phase, the average diameter is larger than that in the isotropic phase, and this ratio depends on the square of the polydispersity or normalised standard deviation of the (Gaussian) size distribution. This means that, on compressing a polydisperse system to its coexistence region the larger particles will tend to be found in the nematic phase and the smaller ones in the isotropic; that is, size fractionation occurs at the nematic-isotropic transition. The other influence of polydispersity is that the biphasic gap or coexistence density region is increased with increasing polydispersity. It is also interesting to note that these simulations can provide data which would be very difficult to obtain from experiments. For example, the orientational distribution function in the nematic phase as a function of particle size is readily available, and this is found to be more peaked and narrower for larger particles than for smaller ones. This implies that the larger particles are essentially aligned with the director, whereas the smaller ones are free to tumble.

Polydispersity has been investigated in rod systems in the orientationally ordered nematic and smectic A phases [4]. In this case, the scaling technique used to study infinitely long rods [3,24] and infinitely thin discs [27] was used to study the relative stability of the phases as a function of polydispersity in the length of the rods, in the Onsager limit L = ∞. For weak to moderate polydispersity (standard deviation s < 0.08), the location of the nematic-smectic transition is unchanged. However, as the polydispersity is increased, the rods of differing lengths find it increasingly difficult to pack effectively into layers and so the coexistence densities are shifted to higher values. Eventually (s ≈ 0.18), the smectic phase becomes unstable with respect to the nematic, which undergoes a transition at higher densities to a columnar phase. We may wonder why the coexistence densities are not shifted for weakly polydisperse systems. A simple explanation given by Bates and Frenkel [4] is that, since the translational distribution in the smectic A phase is relatively broad and not a series of sharp peaks, the layering of the particles can accommodate a small range in particle lengths. However if the distribution becomes too wide, then the layering is destroyed.

F   INFLUENCE OF FLEXIBILITY

The influence of molecular flexibility on the stability of the nematic phase has been studied using the rattling sphere model, which is constructed from chains of hard spheres with constraints on the allowed distances between spheres along the chain [30]. As the molecular flexibility is increased, the location of the nematic-isotropic transition is shifted to higher densities. However, the nematic-smectic transition is also influenced by flexibility, and is also forced to higher densities, at the expense of the solid phase [30,31]. As the nematic phase is entered from the isotropic, there is a small change in anisotropy, with the particles in the nematic phase becoming slightly longer and thinner than their counterparts in the coexisting isotropic phase. This change in anisotropy occurs because non-linear conformations are less favoured in the nematic phase as the surrounding molecules tend to confine the chain to align along the director [30]. This anisotropy fractionation is similar to the size fractionation observed in polydisperse systems of discs at the nematic-isotropic transition [5].

G   TWO-DIMENSIONAL NEMATICS

Simulations have also been performed on two-dimensional (2D) systems, that is on mesogenic molecules confined to lie in a plane. One of the most interesting properties of 2D systems is the lack of true long range order. Although the correlation length in such a system may be macroscopically large, it is not infinite, as required for a true crystal. Straley has shown that, whilst true (orientational) long range order cannot exist for 2D nematics if the particles interact via a separable potential, this does not necessarily apply if the potential is not separable into positional and orientational parts [32]. This decoupling of the potential is not possible for almost all models in which particle shape is taken into account, and so it is interesting to know whether hard rod systems confined to a plane exhibit a 2D nematic phase.

The phase diagram for discorectangles (the 2D equivalent of spherocylinders) confined to a plane was determined using computer simulation by Bates and Frenkel [33] and is shown in FIGURE 3. Two types of phase behaviour are observed depending on the aspect ratio. Long rods (L/D > 7) behave in a very similar manner to infinitely thin needles in 2D [34], in that they exhibit a 2D nematic phase with algebraic decay in the orientational correlations between the particles. Note that it is possible for infinitely thin rods of finite length to order at finite density in 2D, in contrast to the case in 3D. Since the orientational correlations decay algebraically, this phase does not possess true long range order, but rather only quasi-long range order, and so is referred to as a 2D nematic. The 2D nematic phase undergoes a continuous Kosterlitz-Thouless disclination unbinding transition [35] to an isotropic phase as the density is lowered. Shorter rods (L/D < 7) do not exhibit the 2D nematic phase, but undergo a melting transition from a crystal to a phase dominated by chains of particles which align side-by-side, but are isotropically arranged; see FIGURE 4. Thus the most favourable way in which to maximise the entropy for short rods on melting is to retain both local positional and translational order of the crystal; compare this to the quasi-long range orientational order but lack of translational order observed for longer rods. This may at first seem to suggest that a 2D version of the cubatic phase, observed in 3D systems of cut spheres [26], could be observed for short rods, with short stacks of particles forming squares, which then assemble to form a phase with tetratic orientational order. However, the short rod systems are truly isotropic on melting and do not exhibit any long range (or quasi-long range) orientational order [33].

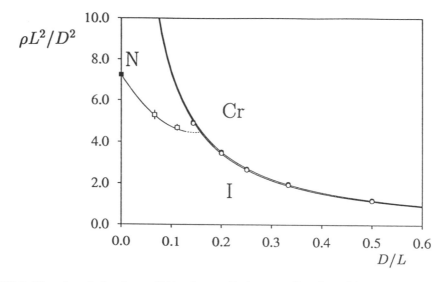

FIGURE 3 The phase behaviour of 2D spherocylinders as a function of inverse aspect ratio D/L and density ρ [33]. The phases exhibited are: isotropic (I), nematic (N) and crystal (Cr). The points indicate the simulation results.

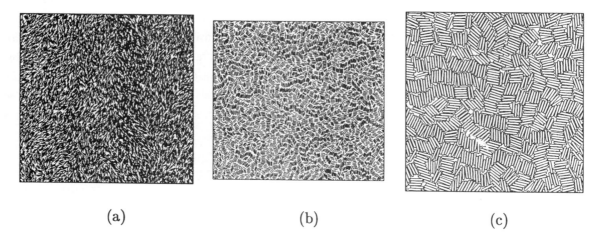

(a)   (b)   (c)

FIGURE 4 Snapshots from simulations of discorectangles [33]. (a) L/D = 15, N = 6400 in the nematic phase at a density just above the Kosterlitz-Thouless transition, (b) L/D = 5, N = 2400 in the isotropic phase at a density just below melting, and (c) L/D = 5, N = 840 at the same state point.

Simulations have also been performed using 2D ellipses of a limited number of aspect ratios and a stable 2D nematic is found for aspect ratios x = 4 and x = 6 [36]. For x = 6, the nematic-isotropic transition is found to be of the Kosterlitz-Thouless type, but for x = 4 it is apparently first order. Thus on decreasing the density, the nematic phase undergoes a first order transition before the point where it becomes unstable with respect to disclination unbinding.

H   FINAL REMARKS

Most of this volume is concerned with thermotropic nematic liquid crystals. The hard body models discussed in this brief review provide numerical data for the equation of state which can be used as an input to theories of liquid crystalline behaviour and so play an equivalent role to the hard sphere in the field of atomic fluids; however, we may wonder if the comparison of hard body systems to real (thermotropic) liquid crystals is relevant. Recall that, in a hard body simulation, we fix either the reduced pressure, $P^* = P/k_BT$, and measure the density, or fix the density and measure the reduced pressure. Of course, we should also recall that hard body models do not exhibit temperature dependence at constant volume like real thermotropic mesogens. However, a change in temperature T at fixed pressure P does lead to a change in density, even for hard body models. This is because the reduced pressure in a hard body system is temperature dependent (at fixed P) and we can, therefore, view an increase in P^* as an equivalent to a decrease in the temperature T at constant pressure P. Thus hard body models are not so unrelated to thermotropic mesogens as may first be thought; it is just that the phase boundaries in the temperature-density (T-ρ) phase diagram are vertical.

Another reason to study hard body models comes from the large recent expansion in experiments on colloidal liquid crystals. These are suspensions of large particles (length scales of the order of 10 - 1000 nm) which may be organic (cellulose, for example [37]), inorganic (such as boehmite [21], gold [38] and gibbsite [28]) or biological (including viruses, such as the tobacco mosiac virus [39]). Under the right solution conditions, the interactions between these particles are essentially hard body and so simulations of hard body systems provide excellent models for these. Whilst the behaviour of these systems on their own is interesting, the addition of smaller particles can lead to a very rich liquid crystalline phase diagram, since this causes a depletion-induced attraction between the larger particles [40]. The range and strength of these attractions can be modified by changing the size and concentration of the depletant. Experimental, simulation and theoretical work has shown the possibility for isotropic-isotropic coexistence in spherocylinder systems, similar to a liquid-gas phase separation [41]. For shorter range attractions (which cannot be modified systematically with thermotropic mesogens), nematic-nematic coexistence is observed. Recent studies of suspensions of hard body discs have shown the possibility of demixing within the isotropic, nematic and columnar phases, depending on the size of the depleting agent [42,43]. We conclude that simulation of hard body models provides a useful basis with which to study both thermotropic mesogens (at the level of short range repulsive interactions only) and colloidal suspensions of anisometric particles.

ACKNOWLEDGEMENT

I am grateful to Daan Frenkel and Bela Mulder for discussions on the simulations described in this review. The work of the FOM Institute for Atomic and Molecular Physics is part of the research programme of FOM (Stichting Fundamenteel Onderzoek der Materie) and is supported by the NWO (Nederlandse Organisatie voor Wetenschappelijk Onderzoek).

REFERENCES

[1]   J.-P. Hansen, I.R. McDonald [*Theory of Simple Liquids* (Academic Press, New York, 1986)]

[2]   M.P. Allen, G.T. Evans, D. Frenkel, B.M. Mulder [*Adv. Chem. Phys. (USA)* vol.86 (1993) p.1]

[3]   P. Bolhuis, D. Frenkel [*J. Chem. Phys. (USA)* vol.106 (1997) p.666]

[4]   M.A. Bates, D. Frenkel [*J. Chem. Phys. (USA)* vol.109 (1998) p.6193]

[5]   M.A. Bates, D. Frenkel [*J. Chem. Phys. (USA)* vol.110 (1999) p.6553]

[6]   M.A. Bates [*J. Chem. Phys. (USA)* vol.111 (1999) p.1732]

[7]   J. Vieillard-Baron [*J. Chem. Phys. (USA)* vol.56 (1972) p.4729]

[8]   J. Vieillard-Baron [*Mol. Phys. (UK)* vol.28 (1974) p.809]

[9]   D. Frenkel, B.M. Mulder [*Mol. Phys. (UK)* vol.55 (1985) p.1171]

[10]   M.A. Bates, G.R. Luckhurst [*Struct. Bond. (USA)* vol.94 (1998) p.65]

[11]   C. Zannoni [Datareview in this book: *12.2 Results of generic model simulations*]

[12]   L. Onsager [*Ann. New York Acad. Sci. (USA)* vol.51 (1949) p.627]

[13]   R. Eppenga, D. Frenkel [*Mol. Phys. (UK)* vol.52 (1984) p.1303]

[14]   G.J. Zarragoicoechea, D. Levesque, J.J. Weis [*Mol. Phys. (UK)* vol.74 (1984) p.629]

[15]   M.P. Allen, C.P. Mason [*Mol. Phys. (UK)* vol.86 (1995) p.467]

[16]   D. Frenkel [*Mol. Phys. (UK)* vol.60 (1987) p.1]

[17]   M.R. Wilson [Datareview in this book: *12.3 Calculations of nematic mesophase properties using realistic potentials*]

[18]   M.P. Allen [*Liq. Cryst. (UK)* vol.8 (1990) p.499]

[19]   P.J. Camp, M.P. Allen, P.G. Bolhuis, D. Frenkel [*J. Chem. Phys. (USA)* vol.106 (1997) p.9270]

[20]   D. Frenkel, H.N.W. Lekkerkerker, A. Stroobants [*Nature (UK)* vol.332 (1988) p.882]

[21]   P.A. Buining, C. Pathmamanoharan, J.B.H. Jansen, H.N.W. Lekkerkerker [*J. Am. Ceram. Soc. (USA)* vol.74 (1991) p.1303]

[22]   S.C. McGrother, D.C. Williamson, G. Jackson [*J. Chem. Phys. (USA)* vol.104 (1996) p.6755]

[23]   S.J. Singer, R. Mumaugh [*J. Chem. Phys. (USA)* vol.93 (1990) p.1278]

[24]   J.M. Polson, D. Frenkel [*Phys. Rev. E (USA)* vol.56 (1997) p.6260]

[25]   D. Frenkel [*Liq. Cryst. (UK)* vol.5 (1989) p.929]

[26]   J.A.C. Veerman, D. Frenkel [*Phys. Rev. A (USA)* vol.45 (1992) p.5633]

[27]   M.A. Bates, D. Frenkel [*Phys. Rev. E (USA)* vol.57 (1998) p.4824]

[28]   F.M. van der Kooij, H.N.W. Lekkerkerker [*J. Phys. Chem. (USA)* vol.102 (1998) p.7829]

[29]   T.J. Sluckin [*Liq. Cryst. (UK)* vol.6 (1989) p.111]

[30]   M.R. Wilson [*Mol. Phys. (UK)* vol.81 (1994) p.675]

[31]   P. Bladon, D. Frenkel [*J. Phys., Condens. Matter (UK)* vol.8 (1996) p.9445]

[32]   P. Straley [*Phys. Rev. A (USA)* vol.4 (1971) p.675]

[33]   M.A. Bates, D. Frenkel [*J. Chem. Phys. (USA)* vol.112 (2000) p.10034]

[34]   D. Frenkel, R. Eppenga [*Phys. Rev. A (USA)* vol.31 (1985) p.1776]

[35]   J.M. Kosterlitz, D. Thouless [*J. Phys. C (UK)* vol.6 (1973) p.1181]

[36]   J.A. Cuesta, D. Frenkel [*Phys. Rev. A (USA)* vol.42 (1990) p.2126]

[37]   X.M. Dong, T. Kimura, J.F. Revol, D.G. Gray [*Langmuir* vol.12 (1996) p.2076]

[38]   Y.Y. Yu, S.S. Chang, C.L. Lee, C.R.C. Wang [*J. Phys. Chem. B (USA)* vol.101 (1997) p.6661]

[39]   See, for example, S. Fraden [*Observation, Prediction and Simulation of Phase Transitions in Complex Fluids* Eds. M. Baus, L.F. Rull, J.-P. Ryckaert (Kluwer, Dordrecht, 1995)]

[40]   See, for example, D. Frenkel [*Observation, Prediction and Simulation of Phase Transitions in Complex Fluids* Eds. M. Baus, L.F. Rull, J.-P. Ryckaert (Kluwer, Dordrecht, 1995)]

[41]   See, for example, H.N.W. Lekkerkerker, P. Buining, J. Buitenhuis, G.J. Vroege, A. Stroobants [*Observation, Prediction and Simulation of Phase Transitions in Complex Fluids* Eds. M. Baus, L.F. Rull, J.-P. Ryckaert (Kluwer, Dordrecht, 1995)]

[42]   F.M. van der Kooij, H.N.W. Lekkerkerker [*Phys. Rev. Lett. (USA)* vol.84 (2000) p.781]

[43]   M.A. Bates, D. Frenkel [*Phys. Rev. E (USA)* in press]

12.2   Results of generic model simulations

C. Zannoni

March 2000

A   INTRODUCTION

The rationalisation of the physical properties and phase transitions of liquid crystals at a molecular level requires the setting up of appropriate models for mesogenic molecules and the computer simulation of the equilibrium state of a sufficiently large system N of these particles at certain thermodynamic conditions. The model system should be able to yield the phases of interest (isotropic, nematic, smectic etc.) and their transitions and to provide properties of particular interest, e.g. order parameters, for a certain choice of molecular features. This seemingly simple task carried out starting from standard atomistic models and force fields and employing Monte Carlo or molecular dynamics (MD) computer simulation techniques [1-4] turns out to be a particularly daunting one when a number of molecules sufficiently large to be able to simulate phase transitions is considered. Moreover, the modelling problem that has to be tackled is often that of designing molecules that have not yet been synthesised, and that are good candidates to yield mesophases with specific properties of interest for applications (such as ferroelectricity). Under these circumstances the emphasis can be on understanding trends rather than calculating in detail the properties of already known molecules.

In this Datareview we briefly discuss generic models that give up atomistic detail in favour of lower resolution, molecular level, models (see FIGURE 1). These are based on the idea that molecules that yield liquid crystals (mesogenic molecules) can be represented by a suitable attractive - repulsive centre as far as the prediction of the molecular organisation and phase transitions goes. In particular we concentrate on an overview of results for systems of particles interacting with Gay-Berne (GB) intermolecular potentials.

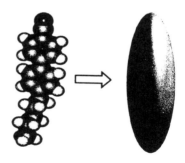

FIGURE 1 An atomistic representation for the mesogen 4-pentyl-4′-cyanobiphenyl (5CB) (left) and a generic ellipsoidal model (right).

B   THE GAY-BERNE POTENTIAL

The Gay-Berne potential in its uniaxial [5] and biaxial [6-9] versions can be regarded as a generalised anisotropic and shifted version of the Lennard-Jones interaction commonly used for simple fluids [1], with attractive and repulsive contributions that decrease as 6 and 12 inverse powers of distance. In the GB model [5] the Lennard-Jones strength, ε, and range, σ, parameters depend on the orientation vectors $\hat{\mathbf{u}}_i$, $\hat{\mathbf{u}}_j$ of the two particles and on their separation vector $\mathbf{r}$:

$$U(\hat{\mathbf{u}}_i,\hat{\mathbf{u}}_j,\mathbf{r})=4\varepsilon_0\varepsilon'^{\mu}\varepsilon^{\nu}\left[\left(\frac{\sigma_0}{r-\sigma(\hat{\mathbf{u}}_i,\hat{\mathbf{u}}_j,\hat{\mathbf{r}})+\sigma_0}\right)^{12}-\left(\frac{\sigma_0}{r-\sigma(\hat{\mathbf{u}}_i,\hat{\mathbf{u}}_j,\hat{\mathbf{r}})+\sigma_0}\right)^{6}\right] \tag{1}$$

where the cap indicates a unit vector and the anisotropic contact distance σ is given by

$$\sigma(\hat{\mathbf{u}}_i,\hat{\mathbf{u}}_j,\hat{\mathbf{r}})=\sigma_0\left\{1-\frac{\chi}{2}\left[\frac{(\hat{\mathbf{u}}_i\cdot\hat{\mathbf{r}}+\hat{\mathbf{u}}_j\cdot\hat{\mathbf{r}})^2}{1+(\hat{\mathbf{u}}_i\cdot\hat{\mathbf{u}}_j)\chi}+\frac{(\hat{\mathbf{u}}_i\cdot\hat{\mathbf{r}}-\hat{\mathbf{u}}_j\cdot\hat{\mathbf{r}})^2}{1-(\hat{\mathbf{u}}_i\cdot\hat{\mathbf{u}}_j)\chi}\right]\right\}^{-1/2} \tag{2}$$

χ is a shape anisotropy parameter related to the length σ_e and the breadth σ_s of the ellipsoid representing the molecule:

$$\chi=\frac{\sigma_e^2-\sigma_s^2}{\sigma_e^2+\sigma_s^2} \tag{3}$$

Similarly the interaction anisotropy is the product of two terms:

$$\varepsilon'^{\mu}\varepsilon^{\nu}=\left\{1-\frac{\chi'}{2}\left[\frac{(\hat{\mathbf{u}}_i\cdot\hat{\mathbf{r}}+\hat{\mathbf{u}}_j\cdot\hat{\mathbf{r}})^2}{1+(\hat{\mathbf{u}}_i\cdot\hat{\mathbf{u}}_j)\chi'}+\frac{(\hat{\mathbf{u}}_i\cdot\hat{\mathbf{r}}-\hat{\mathbf{u}}_j\cdot\hat{\mathbf{r}})^2}{1-(\hat{\mathbf{u}}_i\cdot\hat{\mathbf{u}}_j)\chi'}\right]\right\}^{\mu}\left[1-(\hat{\mathbf{u}}_i\cdot\hat{\mathbf{u}}_j)^2\chi^2\right]^{\nu/2} \tag{4}$$

where μ and ν are parameters used to tune the shape of the potential and

$$\chi'=\frac{(\varepsilon_s/\varepsilon_e)^{1/\mu}-1}{(\varepsilon_s/\varepsilon_e)^{1/\mu}+1} \tag{5}$$

reflects the anisotropy in the potential well depths for the side-by-side and end-to-end configurations. μ and ν were taken to be 2,1 in the original formulation [5], by fitting the interaction of two particles consisting of 4 Lennard-Jones sites in a row. The GB potential for a certain choice of parameters is conveniently indicated by the notation GB(χ,χ',μ,ν) suggested in [10]. It is convenient to work with scaled, dimensionless variables, that here we define for distance r, number density ρ, temperature T, energy U, pressure P, viscosity η, elastic constants K and time t and indicate with an asterisk:

$$r^* = r/\sigma_0, \quad \rho^* = N\sigma_0^3/V, \quad T^* = k_BT/\varepsilon_0, \quad U^* = U/\varepsilon_0, \quad P^* = \sigma_0^3P/\varepsilon_0,$$

$$\eta^* = \sigma_0^2\eta/(m\varepsilon_0)^{1/2}, \quad K^* = K\sigma_0/\varepsilon_0, \quad t^* = t(\varepsilon_0/m\sigma_0^2)^{1/2} \tag{6}$$

The simulation of the macroscopic properties and of the molecular organisation obtained for a system of N model molecules at a certain temperature and pressure (T, P) typically proceeds through one of the two current mainstream methods of computational statistical mechanics: molecular dynamics or Monte Carlo [1,2]. MD sets up and solves step by step the equations of motion for all the particles in the system and calculates properties as time averages from the trajectories obtained. MC calculates instead average properties from equilibrium configurations of the system obtained with an algorithm designed to generate sets of positions and orientations of the N molecules with a frequency proportional to their Boltzmann factor. Both methods, although quite different, proceed through repeated evaluations of the energy and thus of the intermolecular interactions in the sample (as well as of their derivatives to evaluate forces, at least in MD). The GB potential with its analytic formulation and its differentiability can be easily coded in both methods. Attention has however to be paid, especially in MC where molecules can perform fairly large jumps overcoming the increasing repulsion of two approaching particles, to avoid a spurious minimum of the potential near the origin, which could lead to unphysical trapping of particles.

Computer simulation of a set of GB particles with suitable parametrisation can lead to a rich variety of liquid crystal phases. We now briefly summarise results for calamitic and discotic mesophases.

C   CALAMITIC SYSTEMS

The Gay-Berne potential contains four parameters: σ_e/σ_s, $\varepsilon_s/\varepsilon_e$, μ and ν. Typical parameters used for simulating rod-like molecules (see FIGURE 1) are the length-to-breadth ratio $\sigma_e/\sigma_s = 3$ and well depth anisotropy $\varepsilon_s/\varepsilon_e = 5$. The corresponding potential is strongly anisotropic and favours a side-by-side alignment, as we can gather from the three sections of the potential surface obtained by moving one molecule around one taken at the origin shown in FIGURE 2.

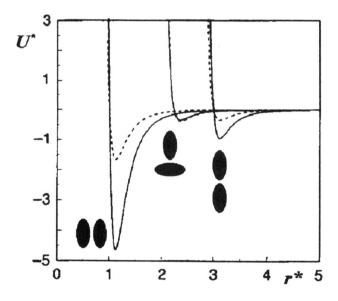

FIGURE 2 The GB potential as a function of the intermolecular separation for $\sigma_e/\sigma_s = 3$, $\varepsilon_s/\varepsilon_e = 5$ and energy parameters $\mu = 2$, $\nu = 1$ (dashed line) and $\mu = 1$, $\nu = 3$ (continuous line).

The original GB parametrisation with $\mu = 2$ and $\nu = 1$ is probably the most thoroughly studied one [5,11-16] and both Monte Carlo and molecular dynamics methods have been employed [11-20,64] to obtain the equilibrium phases generated under a variety of thermodynamic conditions and to construct, at least in part, its phase diagram [14-16]. Isotropic, nematic and smectic B phases have been found and their order and molecular organisation have been determined. Curiously, direct simulation of the 4 site Lennard-Jones potential which was originally fitted to yield this choice of GB parameters does not give a liquid crystal phase, although longer strings of these centres [21] do! Apart from the basic transition properties, a number of physical observables have been determined for GB(3,5,2,1) systems, including translational and rotational correlations [15-17], viscosity [18-20], elastic constants [22,23], thermal conductivity and diffusion coefficients [24].

Smondyrev et al [18] have determined the viscosities for the GB(3,5,2,1) and their temperature dependence for a fairly small (N = 256) system using MD. Simulations by Bennett and Hess [19] using the Gay-Berne potential and a director-based coordinate system show good agreement with [17] for η_2 and η_3, while a discrepancy exists for both η_1 and γ_1. Reference [19] also extends the calculation of viscosities to various densities and includes pre-smectic behaviour. The Miesowicz viscosity η_1 for the orientation parallel to the direction of flow and the Helfrich viscosity η_{12} are both found to increase rapidly as the phase transition is approached. In most of the cases investigated, a critical exponent, ν, of 1/3 was found. The viscosity coefficients η_2, η_3, γ_1 and γ_2 were found to vary regularly in the pre-smectic region.

The dimensionless Frank elastic constants for the GB(3,5,2,1) model have been estimated using MD by Allen et al [23] to be $K^*_1 = 0.7 \pm 0.07$, $K^*_2 = 0.72 \pm 0.07$, $K^*_3 = 2.43 \pm 0.11$ at $\rho^* = 0.33$ and $T^* = 1.00$ (where $\langle P_2 \rangle \sim 0.7$). These results are at variance with those of Stelzer et al [21]: $K^*_1 = 2.7 \pm 0.2$, $K^*_2 = 2.5 \pm 0.2$, $K^*_3 = 1 \pm 0.2$ at the same state point [23].

The nematic range of the GB(3,5,2,1) system is rather narrow and other parametrisations have been explored; some of them are summarised in TABLE 1. In particular, it is found that keeping the same elongation and potential well anisotropy, choosing energy parameters $\mu = 1$ and $\nu = 3$ makes the side-by-side interaction of two molecules stronger (see FIGURE 2) and generates nematics with a wider temperature range [25]. An MC simulation of $N = 10^3$ particles in canonical (constant NVT) conditions gave, at a density $\rho^* = 0.30$, a nematic-isotropic transition temperature T_{NI}^* of 3.55 ± 0.05 and a nematic-smectic transition at $T_{SmN}^* = 2.40 \pm 0.05$. An MD study on much larger samples of $N = 8000$ particles, which also investigated the pre-transitional behaviour of the model, placed T_{NI}^* between 3.45 and 3.50 [26]. Typical values for σ_0, ε_0 in real units could be $\sigma_0 = \sigma_s = 5$ Å, $\varepsilon_0/k_B = 100$ K.

In FIGURE 3 we show the second and fourth rank order parameters $\langle P_2 \rangle$, $\langle P_4 \rangle$ obtained for a system of $N = 10^3$ GB(3,5,1,3) particles using MC at a scaled density ρ^* of 0.30 [25]. The temperature dependence of $\langle P_2 \rangle$ is well-represented, after subtraction of the small residual order due to the finite sample size, by the Haller type expression often used to fit experimental data:

$$<P_2> \propto \left(1 - \frac{T^*}{T^*_{NI}}\right)^\beta \qquad (7)$$

627

TABLE 1 A summary of the phases observed for various Gay-Berne parametrisations.
Here I, N, Sm indicate isotropic, nematic, smectic.

GB parameters (χ,χ',μ,ν)	Phases observed	Method and comments	Ref
3,5,2,1	I,N, SmB	MD	[5]
3,5,1,2	I, N, Sm	MC, N = 256	[12]
3,5,1,3	I, N, SmA, SmB	MC (NVT), N = 1000	[24]
		MD, N = 8000	[25]
3.0,4,2,1; 3.2,4,2,1; 3.6,4,2,1; 3.8,4,2,1; 3,4,2,1	I, N, SmA , SmB	effect of changes in elongation	[27]
3,1,1,3; 3,1.25,1,3; 3,2.5,1,3; 3,5,1,3; 3,10,1,3; 3,25,1,3	I, N, SmA, SmB	effect of changing attractive well anisotropy MD, N = 256 - 864	[28]
4.4,20.0,1,1	I, N, SmA , SmB	N only observed at high P MC(NPT), MD(T)	[10]

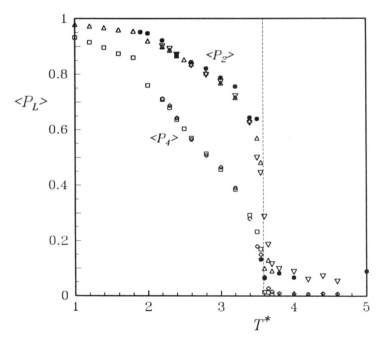

FIGURE 3 Order parameters of GB(χ,χ',1,3) systems at various temperatures in the crystal, smectic B, nematic and isotropic phase. Details are given in [25].

The exponent β found was about 0.17, which is in very good agreement with the values found for 4-methoxybenzylidene-4'-butylaniline (MBBA) ($\beta \cong 0.174$), other Schiff bases [29] ($0.17 \leq \beta \leq 0.22$), and for 5CB [30], $\beta \cong 0.172$. Order parameter data for GB(3,5,2,1) can still be fitted as in EQN (7) but with a larger exponent, e.g. for $\rho^* = 0.30$, $\beta \cong 0.37$ [13], corresponding to a much steeper variation of $\langle P_2 \rangle$ with temperature.

Properties determined for the (1,3) parametrisation include elastic constants [23] and at $\rho^* = 0.30$, $T^* = 3.40$ (where $\langle P_2 \rangle \sim 0.55$), these were found to be $K^*_1 = 2.17 \pm 0.01$, $K^*_2 = 1.71 \pm 0.01$ and $K^*_3 = 3.95 \pm 0.06$.

The molecular organisation at the nematic-isotropic coexistence for a GB(3,5,1,3) model was studied [31] with a specially developed MD method, where the two halves of the cell containing a sufficiently large (N = 12960) number of molecules were separately thermostated at temperatures slightly above and below the transition temperature. It was shown that in this case molecules align parallel to the interface. Experimentally this is what happens for some liquid crystals such as 4-methoxybenzylidene-4'-butylaniline (MBBA), although other types of alignment are found for other nematics. The same MD procedure gave a planar alignment also at the smectic-nematic interface [34].

The effect of changing the elongation ratio on the phase behaviour and on the dynamics has been studied in [27] and [17], while the effect of the attractive interactions and particularly of the potential well anisotropy χ' on the phase behaviour of the Gay-Berne liquid crystal model for a given value of the molecular elongation, χ, of 3 was investigated in [28]. It was found that smectic order is favoured at lower densities as χ' increases. When χ' is lowered, the smectic phase is pre-empted by the nematic phase which becomes increasingly stable at lower temperatures as χ' is decreased. The liquid-vapour coexistence region for different values of χ' was found using Gibbs ensemble and Gibbs-Duhem Monte Carlo techniques with evidence of a vapour-isotropic-nematic triple point for $\chi' = 1$ and 1.25 [28]. The nematic-vapour transition of the GB(2,1,3,1.25) model has been investigated [32,33] and molecules at the phase interface have been shown to be preferentially aligned parallel to it. Experimentally various types of behaviour are found at a free interface: e.g. planar for 4,4'-dimethoxyazoxybenzene (PAA) and perpendicular for the cyanobiphenyls (see references in [32,33]). A perpendicular alignment was also observed for the GB(2,5,1,2) system with shorter particles [34].

The interaction of GB liquid crystals with surfaces, particularly with the aim of investigating surface-induced ordering and the details of anchoring and structuring [35], has been explored for both generic [14,36-40] and specific substrates like graphite [41].

Very large GB systems (over 80,000 molecules) have also been recently studied to investigate some of the most distinctive features of liquid crystals: topological defects [27,28], until now simulated only with lattice models [29]. In particular, the twist grain boundary phase in smectics [27] and the formation of a variety of defects in nematics by rapid quenching [28] have been examined.

D   DISCOTIC SYSTEMS

Discotic mesogens [45,46] can be modelled as squashed GB ellipsoids with thickness σ_f and diameter σ_e, although comparatively fewer studies are available. Luckhurst and co-workers have shown that a GB model based on the dimensions of a triphenylene core, with $\sigma_f/\sigma_e = 0.345$, $\varepsilon_f/\varepsilon_e = 5$, $\mu = 1$, $\nu = 2$, gives an isotropic, nematic and columnar phase with rectangular structure [47]. The same parametrisation, but with a change of μ,ν to 1,3 (FIGURE 4) which has the effect of lowering the well depths of the face-to-face and side-by-side configurations, gives a hexagonal columnar structure [48], as often found in real discotic systems. A hexagonal columnar phase was also found [49] with a slightly modified version of the GB potential in [48], where the parameter σ_0 in EQN (1) was chosen as σ_f, instead of σ_e as in [47]. The formation of very ordered columnar structures makes the constant volume simulations troublesome, as indicated by the development of cavities inside the sample. A constant pressure algorithm [1] where the sample volume and aspect ratio can change has been employed by various authors to adjust the system to the equilibrium state density [49].

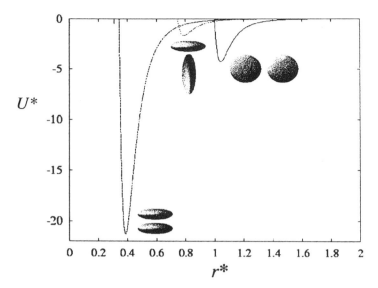

FIGURE 4 A GB potential for discotic mesogens with thickness-to-diameter ratio $\sigma_f/\sigma_e = 0.345$, well depth ratio $\varepsilon_f/\varepsilon_e = 5$ and parameters $\mu = 1$, $\nu = 3$ as a function of the intermolecular separation [48].

While order parameters, radial distributions and thermodynamic quantities have been calculated for GB discotics, very few other observables have been determined until now. An exception is the determination of elastic constants, via the direct pair correlation function route [50], where an ordering $K^*_3 < K^*_1 < K^*_2$ was found, in agreement with experiment.

E   **MODIFIED GB MODELS**

The GB model, although satisfactory in predicting many of the liquid crystalline phases, has various shortcomings. It lacks electrostatic contributions, flexibility, deviations from cylindrical symmetry and specific interactions. A number of investigations have been trying to improve the situation by adding suitable terms or generalising the model. While there is no possibility to cover fully these generalisations here, we shall recall at least a few of them.

Permanent dipole terms have been added to rod-shaped GB models [51-55] and the important modifications observed for the molecular organisation have been studied. The effect of a dipole on mesophase stability depends on its position, orientation and strength, so that a detailed analysis is required. In general a terminal axial dipole shifts the nematic-isotropic transition to higher temperatures [51] and a central axial dipole stabilises the smectic [52]. Particularly noticeable is the formation of dipolar domains for GB particles with an axial near terminal dipole [52], leading to a striped dipolar organisation, the so-called smectic Ã phase [58]. GB molecules with transverse dipoles [55] give rise to chains of dipoles in a plane perpendicular to the director, similarly to what has been found in hard spherocylinders with a transverse dipole [57].

Dipolar effects on GB discs have been studied for the case of axial dipoles [58], and coherent dipolar domains in the columns have been found, even if these mono-oriented domains do not comprise the whole column. This does not allow the formation of ferroelectric columnar systems [46], that indeed were not observed.

GB systems of elongated particles with parallel and perpendicular linear quadrupoles have been studied with MD [59] and smectic A, B, C phases have been observed.

Quadrupolar discotic GB particles have also been studied [60] both for pure systems and for mixtures of quadrupoles of opposite signs with a view to understanding possible mechanisms for the formation of chemically-induced columnar liquid crystals [46].

The range and strength parameters in the potential depend on the orientations of the molecules and that of the intermolecular vector. A generalised single site or Corner potential where ε and σ are expanded in rotational invariants has been proposed [61] as a way of obtaining non-ellipsoidal monosite interactions.

The cases discussed here have concerned simple one-site models. More complex molecular structures can be simulated by suitable combinations of various ellipsoidal Gay-Berne and spherical Lennard-Jones particles, for example to attempt modelling asymmetric molecules [62,63] or to include flexible chains [67,68].

F   CONCLUSION

The simple GB potential has proved capable of yielding the main liquid crystal phases and their properties and thus of constituting an attractive reference potential for investigating trends of variation in the order and organisation of the nematic and smectic phases under the effect of additional specific contributions. The simulation of model systems based on simple, molecular level, rather than atomistic, intermolecular potentials allows the identification of some of the physical features (e.g. molecular shape and attractive interaction anisotropy, biaxiality, electrostatic moments etc.) responsible for a certain collective behaviour, providing useful guidelines for the design of novel mesogenic molecules.

ACKNOWLEDGEMENTS

I wish to thank the University of Bologna, MURST (PRIN *Cristalli Liquidi*), EU TMR (CT) and NEDO (Japan) for their support and R. Berardi, A. Emerson, S. Orlandi and P. Pasini for useful discussions.

REFERENCES

[1]   M.P. Allen, D.J. Tildesley [*Computer Simulation of Liquids* (Clarendon Press,1987)]
[2]   P. Pasini, C. Zannoni (Eds.) [*Advances in the Computer Simulations of Liquid Crystals* (Kluwer, 2000)]

[3] J. Crain, A.V. Komolkin [*Adv. Chem. Phys. (USA)* vol.109 (1999) p.39]

[4] M.R. Wilson [Datareview in this book: *12.3 Calculations of nematic mesophase properties using realistic potentials*]

[5] J.G. Gay, B.J. Berne [*J. Chem. Phys. (USA)* vol.74 (1981) p.3316]

[6] R. Berardi, C. Fava, C. Zannoni [*Chem. Phys. Lett. (Netherlands)* vol.236 (1995) p.462 and vol.297 (1995) p.8]

[7] G. Ayton, G.N. Patey [*J. Chem. Phys. (USA)* vol.102 (1995) p.9040]

[8] D.J. Cleaver, C.M. Care, M.P. Allen, M.P. Neal [*Phys. Rev. E (USA)* vol.54 (1996) p.559]

[9] V.V. Ginzburg, M.A. Glaser, N.A. Clark [*Chem. Phys. Lett. (Netherlands)* vol.214 (1997) p.253]

[10] M.A. Bates, G.R. Luckhurst [*J. Chem. Phys. (USA)* vol.110 (1999) p.7087]

[11] D.J. Adams, G.R. Luckhurst, R.W. Phippen [*Mol. Phys. (UK)* vol.61 (1987) p.1575]

[12] G.R. Luckhurst, R.A. Stephens, R.W. Phippen [*Liq. Cryst. (UK)* vol.8 (1990) p.451]

[13] J.W. Emsley, G.R. Luckhurst, W.E. Palke, D.J. Tildesley [*Mol. Phys. (UK)* vol.11 (1992) p.519]

[14] M.K. Chalam, K.E. Gubbins, E. de Miguel, L.F. Rull [*Molec. Simul. (Switzerland)* vol.7 (1991) p.357]; E. de Miguel, L.F. Rull, M.K. Chalam, K.E. Gubbins [*Mol. Phys. (UK)* vol.74 (1991) p.405]; E. de Miguel, L.F. Rull, M.K. Chalam, K.E. Gubbins, F. van Swol [*Mol. Phys. (UK)* vol.72 (1991) p.593]

[15] E. de Miguel, L.F. Rull, K.E. Gubbins [*Phys. Rev. A (USA)* vol.45 (1992) p.3813]

[16] L.F. Rull [*Physica A (Netherlands)* vol.220 (1995) p.113]

[17] A. Perera, S. Ravichandran, M. Moreau, B. Bagchi [*J. Chem. Phys. (USA)* vol.106 (1997) p.1280]; S. Ravichandran, A. Perera, M. Moreau, B. Bagchi [*J. Chem. Phys. (USA)* vol.107 (1997) p.8469]

[18] A.M. Smondyrev, G.B. Loriot, R.A. Pelcovits [*Phys. Rev. Lett. (USA)* vol.75 (1995) p.2340]

[19] L. Bennett, S. Hess [*Phys. Rev. E (USA)* vol.60 (1999) p.1063]; S. Hess [in *Advances in The Computer Simulations of Liquid Crystals* Eds. P. Pasini, C. Zannoni (Kluwer, 2000)]

[20] S. Cozzini, L.F. Rull, G. Ciccotti, G.V. Paolini [*Physica A (Netherlands)* vol.240 (1997) p.173]

[21] G.V. Paolini, G. Ciccotti, M. Ferrario [*Mol. Phys. (UK)* vol.80 (1993) p.297]

[22] J. Stelzer, L. Longa, H.-R. Trebin [*J. Chem. Phys. (USA)* vol.103 (1995) p.3098 and vol.107 (1997) p.1295E]

[23] M.P. Allen, M.A. Warren, M.R. Wilson, A. Sauron, W. Smith [*J. Chem. Phys. (USA)* vol.105 (1996) p.2850]

[24] S. Sarman, D.J. Evans [*J. Chem. Phys. (USA)* vol.99 (1993) p.620]

[25] R. Berardi, A.P.J. Emerson, C. Zannoni [*J. Chem. Soc. Faraday Trans. (UK)* vol.89 (1993) p.4069]

[26] M.P. Allen, M.A. Warren [*Phys. Rev. Lett. (USA)* vol.78 (1997) p.1291]

[27] J.T. Brown, M.P. Allen, E. Martín del Río, E. de Miguel [*Phys. Rev. E (USA)* vol.57 (1998) p.6685]; J.T. Brown, M.P. Allen, M. Warren [*J. Phys., Condens. Matter (UK)* vol.8 (1996) p.9433]

[28] E. de Miguel, E.M. del Río, J.T. Brown, M.P Allen [*J. Chem. Phys. (USA)* vol.105 (1996) p.4234]

[29] F. Leenhouts, W.H. de Jeu, A.J. Dekker [*J. Phys. (France)* vol.40 (1979) p.989]

[30] S.T. Wu, R.J. Cox [*J. Appl. Phys. (USA)* vol.64 (1988) p.821]

[31]   M. Bates, C. Zannoni [*Chem. Phys. Lett. (Netherlands)* vol.280 (1997) p.40]

[32]   M. Bates [*Chem. Phys. Lett. (Netherlands)* vol.288 (1998) p.209]

[33]   A.P.J. Emerson, S. Faetti, C. Zannoni [*Chem. Phys. Lett. (Netherlands)* vol.271 (1997) p.241]; E. de Miguel, E.M. del Río [*Phys. Rev. E (USA)* vol.55 (1997) p.2916]

[34]   S.J. Mills, C.M. Care, M.P. Neal, D.J. Cleaver [*Phys. Rev. E (USA)* vol.58 (1998) p.3284]

[35]   E.M. del Río, M.M. Telo da Gama, E. de Miguel, L.F. Rull [*Phys. Rev. E (USA)* vol.52 (1995) p.5028]

[36]   Z.P. Zhang, A. Chakrabarti, O.G. Mouritsen, M.J. Zuckermann [*Phys. Rev. E. (USA)* vol.53 (1996) p.2461]

[37]   J. Stelzer, P. Galatola, G. Barbero, L. Longa [*Phys. Rev. E (USA)* vol.55 (1997) p.477]; J. Stelzer, L. Longa, H.-R. Trebin [*Phys. Rev. E (USA)* vol.55 (1997) p.7085]

[38]   G.D. Wall, D.J. Cleaver [*Phys. Rev. E (USA)* vol.56 (1997) p.4306]

[39]   T. Gruhn, M. Schoen [*Mol. Phys. (UK)* vol.93 (1998) p.681]; T. Gruhn, M. Schoen, D.J. Diestler [*J. Chem. Phys. (USA)* vol.109 (1998) p.301]

[40]   T. Miyazaki, H. Hayashi, M. Yamashita [*Mol. Cryst. Liq. Cryst. (Switzerland)* vol.330 (1999) p.1611]

[41]   V. Palermo, F. Biscarini, C. Zannoni [*Phys. Rev. E (USA)* vol.57 (1998) p.2519]

[42]   M.P. Allen, M.A. Warren, M.R. Wilson [*Phys. Rev. E (USA)* vol.57 (1998) p.5585]

[43]   J.L. Billeter, A.M. Smondyrev, G.B. Loriot, R.A. Pelcovits [*Phys. Rev. E (USA)* vol.60 (1999) p.6831]

[44]   C. Chiccoli, O.D. Lavrentovich, P. Pasini, C. Zannoni [*Phys. Rev. Lett. (USA)* vol.79 (1997) p.4401]

[45]   S. Chandrasekhar [*Liquid Crystals* 2nd Edition (Cambridge U.P., 1992)]

[46]   D. Guillon [*Struct. Bond. (USA)* vol.95 (1999) p.41]

[47]   A.P.J. Emerson, G.R. Luckhurst, S.G. Whatling [*Mol. Phys. (UK)* vol.82 (1994) p.113]

[48]   C. Bacchiocchi, C. Zannoni [*Phys. Rev. E (USA)* vol.58 (1998) p.3237]

[49]   M.A. Bates, G.R. Luckhurst [*J. Chem. Phys. (USA)* vol.104 (1996) p.6696]

[50]   J. Stelzer, M.A. Bates, L. Longa, G.R. Luckhurst [*J. Chem. Phys. (USA)* vol.107 (1997) p.7483]

[51]   K. Satoh, S. Mita, S. Kondo [*Chem. Phys. Lett. (Netherlands)* vol.255 (1996) p.99]

[52]   R. Berardi, S. Orlandi, C. Zannoni [*Chem. Phys. Lett. (Netherlands)* vol.261 (1996) p.357]

[53]   E. Gwozdz, A. Brodka, K. Pasterny [*Chem. Phys. Lett. (Netherlands)* vol.267 (1997) p.557]

[54]   M. Houssa, A. Oualid, L.F. Rull [*Mol. Phys. (UK)* vol.94 (1998) p.439]; M. Houssa, S.C. McGrother, L.F. Rull [*J. Chem. Phys. (USA)* vol.109 (1998) p.9529]; M. Houssa, S.C. McGrother, L.F. Rull [*Comput. Phys. Commun. (Netherlands)* vol.121-122 (1999) p.259]

[55]   R. Berardi, S. Orlandi, C. Zannoni [*Int. J. Mod. Phys. C (Singapore)* vol.10 (1999) p.477]

[56]   A.M. Levelut, R.J. Tarento, F. Hardouin, M.F. Achard, G. Sigaud [*Phys. Rev. A (USA)* vol.24 (1981) p.2180]

[57]   A. Gil-Vilegas, S. McGrother, G. Jackson [*Chem. Phys. Lett. (Netherlands)* vol.269 no.5-6 (1997) p.441]

[58]   R. Berardi, S. Orlandi, C. Zannoni [*J. Chem. Soc. Faraday Trans. (UK)* vol.93 (1997) p.1493]

[59]   M.P. Neal, A.J. Parker [*Mol. Cryst. Liq. Cryst. (Switzerland)* vol.330 (1999) p.1809]; M.P. Neal, A.J. Parker [*Chem. Phys. Lett. (Netherlands)* vol.294 (1998) p.277]

[60]   M.A. Bates, G.R. Luckhurst [*Liq. Cryst. (UK)* vol.24 (1998) p.229]

[61] H. Zewdie [*J. Chem. Phys. (USA)* vol.108 (1998) p.2117]; H. Zewdie [*Phys. Rev. E (USA)* vol.57 (1998) p.1793]

[62] M.P. Neal, A.J. Parker, C.M. Care [*Mol. Phys. (UK)* vol.91 (1997) p.603]

[63] J. Stelzer, R. Berardi, C. Zannoni [*Chem. Phys. Lett. (Netherlands)* vol.299 (1999) p.9]

[64] G. La Penna, D. Catalano, C.A. Veracini [*J. Chem. Phys. (USA)* vol.105 (1996) p.7097]

[65] M.R. Wilson [*J. Chem. Phys. (USA)* vol.107 (1997) p.8654]; C. McBride, M.R. Wilson [*Mol. Phys. (UK)* vol.97 (1999) p.511]

12.3 Calculations of nematic mesophase properties using realistic potentials

M.R. Wilson

August 1999

A   INTRODUCTION

This Datareview describes the simulation methods used to calculate structural and dynamical properties of nematic mesophases using realistic potentials, and summarises the progress made in determining these properties.

B   SIMULATION METHODS

Computer simulations of nematic phases rely on either molecular dynamics or molecular Monte Carlo simulation techniques. The molecular dynamics method [1,2] solves Newton's equations of motion for a system of interacting molecules. In contrast, molecular Monte Carlo methods [3,4] use random numbers to generate appropriate configurations from a Boltzmann distribution of states of the system. For calculations using realistic potentials both techniques require the use of classical force fields of the form

$$E_{tot} = \sum_{all\ bonds} E_{bond} + \sum_{all\ angles} E_{angle} + \sum_{all\ dihedrals} E_{torsion} + \sum_{i=1}^{N}\sum_{j<i}^{N}\left(E_{non\text{-}bond}^{ij} + E_{elec}^{ij}\right) \qquad (1)$$

Here, E_{bond}, E_{angle} and $E_{torsion}$ are the bond and bond angle distortion energies and the dihedral angle energy (torsional energy), respectively. $E_{non\text{-}bond}^{ij}$ and E_{elec}^{ij} represent the non-bonded and electrostatic interaction energies between two atomic sites i and j.

The force fields commonly used for modelling nematics [5-23] are shown in TABLE 1. They differ principally in the functional form used for $E_{non\text{-}bond}$ (an exponential form is more accurate for the repulsive part of the potential but is more expensive in terms of computer time), and in the inclusion/exclusion of explicit hydrogens attached to carbons. The most accurate all-atom force fields employ a number of cross-terms in addition to those shown in EQN (1). Most force fields have been designed to model a wide range of organic molecules, with particular emphasis on proteins and nucleic acids. Reference [23] (HFF), however, describes a new force field designed specifically for modelling liquid crystal molecules. Force fields such as OPLS-AA are parametrised to predict experimental heats of formation and liquid state densities for small molecules. Such force fields are extremely valuable for the simulation of nematic materials where it is essential to be able to predict the correct density of the fluid.

TABLE 1 Force fields for the simulation of liquid crystal systems.

Force field	Ref	UA/AA[a]	Cross terms[b]	LJ/exp[c]
GROMOS96	[5]	UA	No	LJ
AMBER	[6]	UA	No	LJ
AMBER94	[7]	AA	No	LJ
OPLS/AMBER	[8]	UA	No	LJ
OPLS-AA	[9,10]	AA	No	LJ
MM2	[11]	AA	Yes	exp
MM3	[12]	AA	Yes	exp
MM4	[13-17]	AA	Yes	exp
MMFF94	[18-22]	AA	Yes	exp
HFF	[23]	UA[d]	No	LJ

[a]UA and AA refer to united-atom and all-atom force fields. [b]Cross terms refer to the addition of coupled interactions (e.g. stretch/bend) to those shown in EQN (1). [c]Refers to the presence of a Lennard-Jones or an exponential form in the non-bonded interactions. [d]In the HFF force-field the united atom approximation is used for aliphatic but not aromatic hydrogens.

C   NEMATIC SIMULATIONS

A brief summary of the existing nematic simulation studies [24-41] is presented in TABLE 2. Most of these calculations use united atom force fields, and are limited to small systems (50 - 200 molecules) and fairly short simulation times. These limit the accuracy of the data obtained from these studies. Only in the most recent study [41] has a nematic phase been grown from a fully isotropic liquid. This process requires in the region of 6 - 12 ns of simulation time (for molecular dynamics), depending on the system and the proximity to the phase transition. Only when such simulation lengths can be applied to larger systems (several thousand atoms) will it be possible to move towards the prediction of accurate nematic-isotropic transition temperatures. The values of the non-bonded parameters in the molecular force field also have a crucial influence on transition temperatures. In addition, they also have a strong influence on the bulk density of a nematogen, and this provides a useful test of the accuracy of these parameters. TABLE 3 lists the calculated density in constant-NpT simulations of the nematic phase for a series of cyano-mesogens. The range of values, compared to experiment, is a reflection of the relative inaccuracy of many of the currently available molecular force fields.

TABLE 2 Atomistic simulations of nematic phases using Monte-Carlo (MC) or molecular dynamics (MD) methods. (N_m represents the number of molecules in the study and NVT/NpT represents the statistical mechanical ensemble used in the calculations.)

Acronym	Molecule	Ref	N_m	Force field	NVT/NpT	MD/MC
EBBA	4-ethoxybenzylidene-4'-butylaniline	[24]	60	AA	NpT	MC
5CB	4-pentyl-4'-cyanobiphenyl	[25]	64	UA	NpT	MD
		[26]	80	UA	NpT	
		[27]	75	UA	NVT	
		[27]	38	AA	NVT	
pHB	phenyl-4-(4-benzoyloxy-)benzoyloxybenzoate	[28]	8, 16	AA	NVT	MD
CCH5	4-(trans-4-pentylcyclohexyl)cyclohexylcarbonitrile	[29-31]	128	UA	NpT	MD
THE5	hexakis(pentyloxy)triphenylene	[32]	54	UA	NVT	MD
nOCB (n = 5 - 8)	4-alkoxy-4'-cyanobiphenyl	[33,34]	64, 125	UA	NpT	MD
5OCB		[35]	144	UA	NpT	
PCH5	4-(trans-4-pentylcyclohexyl)benzonitrile	[36-38]	50, 100	UA	NpT	MD
		[39]	200	UA	NpT	
HBA	tetramer of 4-hydroxybenzoic acid	[40]	125	AA	NpT	MD
5,5-BBCO	4,4'-di-pentyl-bibicyclo[2.2.2]octane	[41]	64, 125	UA	NVT	MD

TABLE 3 Calculated and experimental densities for some common cyano-mesogens.

Mesogen	T/K	ρ(calc.)/g cm^{-3}	ρ(exp.)/g cm^{-3}
5CB[ab]	300	1.01 [26]	1.01 [45] 0.98 [46]
CCH5	350	0.9161 [29]	1.11 [46]
	370	0.8819	
PCH5[a]	333	0.870 [39]	1.07 [46]
5OCB	330	1.020 [35]	-

[a]Approximate values obtained from quoted simulation box dimensions. [b]Non-bonded parameters of the aromatic carbons were adjusted to obtain good agreement with the experimental density of [45].

D   ORDER PARAMETERS

The second rank orientational order parameter <P$_2$> for a nematic liquid crystal is given by the largest eigenvalue obtained by diagonalising the ordering tensor

$$Q_{\alpha\beta} = \frac{1}{N_m} \sum_{j=1}^{N_m} \frac{3}{2} u_{j\alpha} u_{j\beta} - \frac{1}{2} \delta_{\alpha\beta}, \quad \alpha, \beta = x, y, z \tag{2}$$

For a flexible molecule the unit vector **u** can be assigned as the molecular long axis obtained by diagonalising the molecular inertia tensor [30] or the polarisability tensor [31]. Values of <P$_2$> calculated at a series of state points are given in TABLE 4. (Run lengths in the region of several nanoseconds are often required to check that the calculated values of the order parameter are constant.) Alternatively, individual bond order parameters can be calculated by associating **u** with a bond vector. TABLE 5 summarises calculated bond order parameters for three mesogens with pentyl chains. Each material exhibits a classic odd-even effect in the degree of ordering of alternate carbon-carbon bonds in the alkyl chain.

E   DIFFUSION CONSTANTS

In molecular dynamics simulations the translational diffusion constant D is available from the Einstein relation

$$D = \frac{1}{6t} \left\langle \left| \mathbf{r}_i(t + t_0) - \mathbf{r}_i(t_0) \right|^2 \right\rangle \tag{3}$$

calculated for the position vectors of the molecular centres of mass $\mathbf{r}_i(t)$ at time t. In a nematic liquid crystal D can be resolved into two components parallel (D$_\parallel$) and perpendicular (D$_\perp$) to the director with D = (D$_\parallel$ + 2D$_\perp$)/3. Results for D, D$_\parallel$ and D$_\perp$ are summarised in TABLE 4. Most studies have found diffusion coefficients in the 10^{-9} m^2 s^{-1} range with the ratio D$_\parallel$/D$_\perp$ between 1.59 and 3.73 for calamitic nematics. It should be stressed that absolute values of D$_\parallel$ and D$_\perp$ depend critically on the

TABLE 4 Calculated diffusion constants and uniaxial order parameters
from molecular dynamics simulations of calamitic nematics.

Mesogen	Ref	T/K	$D_\parallel/10^{-9}$ m^2 s^{-1}	$D_\perp/10^{-9}$ m^2 s^{-1}	$D_\parallel/D_\perp$	$<P_2>$
5CB[a]	[26]	300	1.12	0.30	3.73	0.60
5CB	[27]	300	0.250	0.10	2.5	0.72
CCH5	[29]	350	0.554	0.192	2.89	0.62
		370	1.076	0.517	2.08	0.38
PCH5	[36]	333	0.157	0.046	3.4	0.58
5OCB	[35]	330	0.064	0.021	3.05	0.61
5OCB	[34]c	331	0.36	0.188	1.91	0.53
6OCB	[34]	339	0.33	0.168	1.96	0.50
7OCB	[34]	337	0.316	0.176	1.80	0.50
8OCB	[34]	342	0.282	0.177	1.59	0.47

[a]Experimental values at 296.5 K: $D_\parallel = 0.053 \times 10^{-9}$ m^2 s^{-1}, $D_\perp = 0.041 \times 10^{-9}$ m^2 s^{-1}.

TABLE 5 Calculated bond order parameters for CCH5, 5CB and 5,5-BBCO.
(Bonds 1 - 5 are labelled according to their position
relative to the molecular core.)

Vector	CCH5 at 350 K [30]	5CB at 300 K [26]	5,5-BBCO at 300 K [41]
Molecule	0.66	0.60	0.64
C-N bond	0.57	-	-
Bond 1	0.57	0.57	0.58
Bond 2	0.38	-0.07	-0.01
Bond 3	0.48	0.50	0.41
Bond 4	0.21	-0.06	0.01
Bond 5	0.38	0.38	0.26

density of the fluid. In many cases intermolecular potentials are not yet good enough to predict accurately the density within a constant NpT simulation, and so calculated values of $D_\parallel$ and $D_\perp$ are often found to differ from experimental data.

F   MOLECULAR STRUCTURE

Short molecular dynamics simulations, of approximately 500 ps, seem sufficient in length to equilibrate the internal molecular structure of most mesogens provided that the barriers to internal rotation about bonds are not too high (<20 kJ/mol). A number of workers have studied dihedral angle distributions S(ϕ) and the relative proportions of different molecular conformers in the nematic phase [26,29,30,41]. Dihedral angle distributions S(ϕ) are linked to the effective torsional potential V(ϕ) by the following relation:

$$S(\phi) = C \exp\left[-\frac{V(\phi)}{k_B T}\right] \qquad (4)$$

where C is a normalisation factor. $V(\phi)$ can be split into contributions from the internal molecular structure, and external contributions from the surrounding molecules in the nematic. The nematic mean field is seen to favour linear conformers selectively by raising the internal energy of configurations with gauche conformers in that part of the molecule which lies at an angle to the director. This leads to an odd-even influence in the values of $S(\phi)$ and $V(\phi)$ as shown in FIGURE 1 for the molecule 4-(trans-4-pentylcyclohexyl)cyclohexylcarbonitrile (CCH5), and a change in the populations of different conformers in comparison with the isotropic liquid. The relative proportions of different chain conformers are shown in TABLE 6 for the mesogens CCH5 and 4-pentyl-4'-cyanobiphenyl (5CB). Typical changes in the gauche-trans energy difference

$$V_{gt} = V(\text{gauche}) - V(\text{trans}) \qquad (5)$$

on going from the isotropic to the nematic phase are of the order of 1 - 2 kJ/mol [26,30,42].

TABLE 6 Relative populations for different conformers of CCH5 and 5CB.
(Dihedral angles α, β, γ are the same as indicated in FIGURE 1.)

Dihedral α	Dihedral β	Dihedral γ	CCH5 [30] at 350 K /%	CCH5 [30] at 370 K /%	5CB [27] at 300 K /%	Multiplicity
t	t	t	49.2	44.4	44.7	1
t	$g^\pm$	t	12.75	12.65	15.6	2
t	t	$g^\pm$	5.75	7.7	8.1	2
$g^\pm$	t	t	2.35	2.95	1.6	2
t	$g^\pm$	$g^\pm$	1.55	2.0	0.8	2
$g^\pm$	t	$g^\mp$	0.9	0.75	0.6	2
$g^\pm$	$g^\pm$	t	0.8	1.1	1.0	2
t	$g^\pm$	$g^\mp$	0.3	0.1	0.0	2
$g^\pm$	t	$g^\pm$	0.35	0.45	0.0	2
$g^\pm$	$g^\mp$	t	0.3	0.05	0.0	2
$g^\pm$	$g^\pm$	g^+	0.1	0.0	0.0	2
$g^\pm$	$g^\pm$	$g^\pm$	0.1	0.0	0.0	2
$g^\pm$	g^+	g^+	0.05	0.0	0.0	2
$g^\pm$	$g^\mp$	$g^\pm$	0.05	0.0	0.0	2

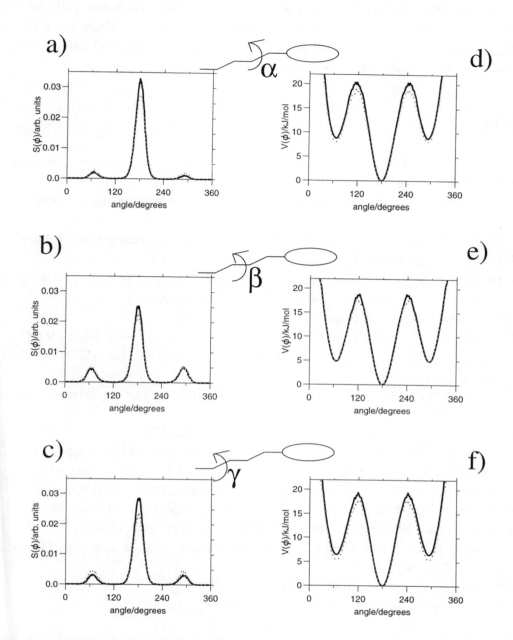

FIGURE 1 Dihedral angle distributions S(ϕ) and effective torsional potentials V(ϕ) for three dihedral angles in the alkyl chain of CCH5. Bold line - 350 K nematic phase (order parameter $\langle P_2 \rangle$ = 0.66), dotted grey line - isotropic phase at 390 K.

G   CONCLUSIONS

Realistic simulations of nematics cannot yet provide quantitative data for transition temperatures or for bulk properties of real systems. However, the work described provides some encouraging signs. The constant NpT work for CCH5 and other molecules demonstrates stable nematic phases at the right temperatures, and this same methodology can be extended to larger systems as computer power increases. With recent improvements in force field accuracy, this opens up the possibility of predicting nematic-isotropic transition temperatures. Accurate techniques for calculating material properties important for displays (elastic constants [43] and rotational viscosities [44]) have already been developed for simple systems (e.g. hard spherocylinders and the Gay-Berne potential). Given large enough systems (several thousand rather than several hundred molecules), these same techniques can also be applied to atomistic models.

REFERENCES

[1]   M.P. Allen, D.J. Tildesley [*Computer Simulation of Liquids* (Oxford University Press, Oxford, 1987)]

[2]   D.C. Rapaport [*The Art of Molecular Dynamics Simulation* (Cambridge University Press, 1995)]

[3]   D. Frenkel, B. Smit [*Understanding Molecular Simulation* (Academic Press, 1996)]

[4]   W.L. Jorgensen [*J. Chem. Phys. (USA)* vol.87 (1983) p.5304]

[5]   X. Daura, A.E. Mark, W.F. van Gunsteren [*J. Comput. Chem. (USA)* vol.19 (1998) p.535]

[6]   S.J. Weiner et al [*J. Am. Chem. Soc. (USA)* vol.106 (1984) p.765]

[7]   W.D. Cornell et al [*J. Am. Chem. Soc. (USA)* vol.117 (1995) p.5179]

[8]   W.L. Jorgensen, J. Tirado-Rives [*J. Am. Chem. Soc. (USA)* vol.110 (1988) p.1666]

[9]   W.L. Jorgensen, D.S. Maxwell, J. Tirado-Rives [*J. Am. Chem. Soc. (USA)* vol.118 (1996) p.11225]

[10]   W.L. Jorgensen, N.A. McDonald [*Theochem. - J. Molec. Structure (Netherlands)* vol.424 (1998) p.145]

[11]   N.L. Allinger [*J. Am. Chem. Soc. (USA)* vol.99 (1977) p.8127]

[12]   N.L. Allinger, Y.H. Yuh, J. Lii [*J. Am. Chem. Soc. (USA)* vol.111 (1989) p.8551]

[13]   N.L. Allinger, K.S. Chen, J.H. Lii [*J. Comput. Chem. (USA)* vol.17 (1996) p.642]

[14]   N. Nevins, K.S. Chen, N.L. Allinger [*J. Comput. Chem. (USA)* vol.17 (1996) p.669]

[15]   N. Nevins, J.H. Lii, N.L. Allinger [*J. Comput. Chem. (USA)* vol.17 (1996) p.695]

[16]   N. Nevins, N.L. Allinger [*J. Comput. Chem. (USA)* vol.17 (1996) p.730]

[17]   N.L. Allinger, K.S. Chen, G.M.A.J.A. Katzenellenbogen, S.R. Wilson [*J. Comput. Chem. (USA)* vol.17 (1996) p.747]

[18]   T.A. Halgren [*J. Comput. Chem. (USA)* vol.17 (1996) p.490]

[19]   T.A. Halgren [*J. Comput. Chem. (USA)* vol.17 (1996) p.520]

[20]   T.A. Halgren [*J. Comput. Chem. (USA)* vol.17 (1996) p.553]

[21]   T.A. Halgren, R.B. Nachbar [*J. Comput. Chem. (USA)* vol.17 (1996) p.587]

[22]   T.A. Halgren [*J. Comput. Chem. (USA)* vol.17 (1996) p.616]

[23]   E. Garcia, M.A. Glaser, N.A. Clark, D.M. Walba [*Theochem. - J. Molec. Structure (Netherlands)* vol.464 (1999) p.39]

[24]   A.V. Komolkin, Y.V. Molchanov, P.P. Yakutseni [*Liq. Cryst. (UK)* vol.6 (1989) p.39]

[25]   S.J. Picken, W.F. van Gunsteren, P.T. van Duijnen, W.H. de Jeu [*Liq. Cryst. (UK)* vol.6 (1989) p.357]

[26]   C.W. Cross, B. Fung [*J. Chem. Phys. (USA)* vol.101 (1994) p.6839]

[27]   A.V. Kolmolkin, A. Laaksonen, A. Maliniak [*J. Chem. Phys. (USA)* vol.101 (1994) p.4103]

[28]   B. Jung, B.L. Schürmann [*Mol. Cryst. Liq. Cryst. (UK)* vol.185 (1990) p.141]

[29]   M.R. Wilson, M.P. Allen [*Mol. Cryst. Liq. Cryst. (UK)* vol.198 (1991) p.465]

[30]   M.R. Wilson, M.P. Allen [*Liq. Cryst. (UK)* vol.12 (1992) p.157]

[31]   M.R. Wilson [*J. Mol. Liq. (Netherlands)* vol.68 (1996) p.23]

[32]   I. Ono, S. Kondo [*Mol. Cryst. Liq. Cryst. (UK)* vol.8 (1991) p.69]

[33]   I. Ono, S. Kondo [*Bull. Chem. Soc. Jpn. (Japan)* vol.65 (1992) p.1057]

[34]   I. Ono, S. Kondo [*Bull. Chem. Soc. Jpn. (Japan)* vol.66 (1993) p.633]

[35]   S. Hauptmann, T. Mosell, S. Reiling, J. Brickmann [*Chem. Phys. (Netherlands)* vol.208 (1996) p.57]

[36]   G. Krömer, D. Paschek, A. Geiger [*Ber. Bunsenges. Phys. Chem. (Germany)* vol.97 (1993) p.1188]

[37]   S.Y. Yakovenko, A.A. Muravski, G. Kromer, A. Geiger [*Mol. Phys. (UK)* vol.86 (1995) p.1099]

[38]   S.Y. Yakovenko, G. Kromer, A. Geiger [*Mol. Phys. (UK)* vol.275 (1996) p.91]

[39]   S.Y. Yakovenko, A.A. Muravski, F. Eikelschulte, A. Geiger [*Liq. Cryst. (UK)* vol.24 (1998) p.657]

[40]   J. Huth, T. Mosell, K. Nicklas, A. Sariban, J. Brickmann [*J. Phys. Chem. (USA)* vol.98 (1994) p.768]

[41]   C. McBride, M.R. Wilson, J.A.K. Howard [*Mol. Phys. (UK)* vol.93 (1998) p.955]

[42]   M.R. Wilson [*J. Chem. Phys. (USA)* vol.107 (1997) p.8654]

[43]   M.P. Allen, M.A. Warren, M.R. Wilson, A. Sauron, W. Smith [*J. Chem. Phys. (USA)* vol.105 (1996) p.2850]

[44]   S. Sarman, D.J. Evans [*J. Chem. Phys. (USA)* vol.99 (1993) p.9021]

[45]   T. Hanemann, W. Haase [*Ber. Bunsenges. Phys. Chem. (Germany)* vol.98 (1994) p.596]

[46]   A. Wüflinger, M. Sandman [Datareview in this book: *3.3 Equations of state for nematics*]

12.4 Material properties of nematics obtained from computer simulation

D.J. Cleaver

September 1999

A   INTRODUCTION

We present here a summary of the results obtained from and the techniques developed for computer simulation based calculations of certain bulk material properties of nematic liquid crystals. Calculations of such properties are of value since: (i) via comparison with experiment and theory, they provide a stringent test of simulation models and techniques; (ii) the results obtained from models at one length scale (e.g. molecular) can be used as inputs to more coarse grained models (e.g. finite element or lattice Boltzmann) and continuum theory calculations; and, most importantly, (iii) they offer insight into the molecular basis of significant material behaviour.

Here we restrict ourselves to two of the more widely measured material properties, the Frank elastic constants, and the viscosity coefficients. Whilst several other bulk properties have been simulated (e.g. the diffusion coefficients [1], helical twisting power [2] and flexoelectric coefficients [3]), these two have the merit of having been investigated for a significant range of models and are properties of considerable technological interest. We also note that very recently, significant developments have been made in the simulation of surface properties via computer simulation; surface tension [4], surface pretilt [5] and anchoring strengths [6] have all been calculated for model systems. That said, the field of surface property simulations is still rather in its infancy.

B   THE FRANK ELASTIC CONSTANTS

Due to thermal excitations, the director field, $\hat{\mathbf{n}}(\mathbf{r})$, in a bulk nematic liquid crystal is not spatially uniform. Following the well established theoretical treatments of this situation [7], the free energy associated with the distortions in $\hat{\mathbf{n}}(\mathbf{r})$ may be written as

$$\Delta F = \frac{1}{2}\int K_1\left(\nabla\cdot\hat{\mathbf{n}}\right)^2 + K_2\left(\hat{\mathbf{n}}\cdot\left(\nabla\times\hat{\mathbf{n}}\right)\right)^2 + K_1\left(\left(\hat{\mathbf{n}}\times\left(\nabla\times\hat{\mathbf{n}}\right)\right)\right)^2 d\mathbf{r} \tag{1}$$

where the coefficients K_1, K_2 and K_3 are, respectively, the splay, twist and bend Frank elastic constants [8]. These orientational elasticity coefficients are not actually constant, showing considerable variation with temperature and density, particularly near to the limits of nematic stability. A less thermodynamic, but more intuitive, expression describing this variation is the molecular field result $K_i \propto \langle P_2^2 \rangle$, where $\langle P_2 \rangle$ is the second rank nematic orientational order parameter.

Determinations of the Frank constants by computer simulation have, in the main, adopted routes closely related to pertinent experimental techniques, such as the field-induced Fréedericksz transition and light scattering. In the Fréedericksz transition, a nematic film sandwiched between strong anchoring plates is subjected to an aligning field applied perpendicular to the substrate induced director distribution. At small fields, the director profile is unaffected. However, at a critical K_i-dependent field, the director distribution becomes lobe-shaped, the main distortion occurring midway between the confining plates (see FIGURE 1(a)). The value of the index i is governed by the choice of set-up geometry.

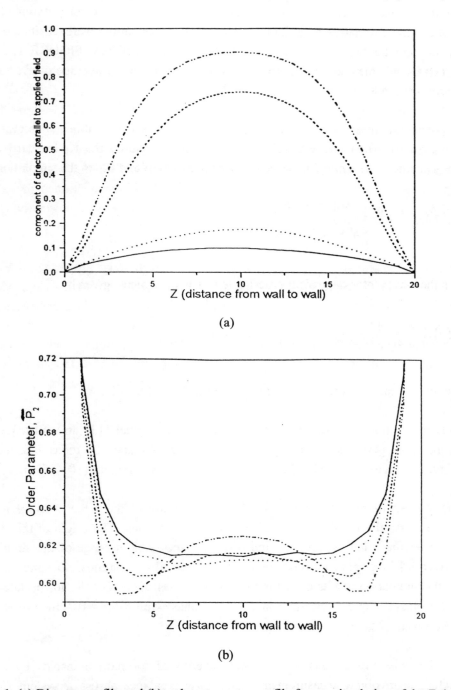

(a)

(b)

FIGURE 1 (a) Director profile and (b) order parameter profile from a simulation of the Fréedericksz transition performed using the Lebwohl-Lasher model [9]. Applied field varies from just over (solid line) to more than four times the critical field (dash-dot-dot line).

Since the Fréedericksz transition is second order, and so associated with divergent correlation times, it has attracted relatively little attention from simulators interested in determining the Frank constants. Indeed, such simulations have only been attempted using the lattice-based Lebwohl-Lasher model [10] (for which the three K_i are equal). Whilst the results obtained from these simulations [11,12] are consistent with those found using other methods, the Fréedericksz transition-based simulations were both less accurate (i.e. had larger error bars) and orders of magnitude more computationally demanding than those made with alternative techniques. The large uncertainties were attributed to the non-uniformity of the nematic order parameter across the simulated cell (see FIGURE 1(b)); the Frank constant obtained by this method was interpreted as being some sort of average over a range of $<P_2>$ (recall the approximation $K_i \propto <P_2^2>$).

An alternative technique used to calculate the Frank constants is the equilibrium fluctuation method, originally employed by Allen and Frenkel [13] when they measured the K_i for hard ellipsoid and spherocylinder systems. The method is based on measurements of and fits to the simulation average

$$\left\langle \hat{Q}_{\alpha 3}(\mathbf{q}) \hat{Q}_{\alpha 3}(-\mathbf{q}) \right\rangle = \frac{9 < P_2^2 > V k_B T}{4 \left(K_\alpha q_1^2 + K_3 q_3^2 \right)}, \quad \alpha = 1, 2 \tag{2}$$

where $\hat{Q}(\mathbf{q})$ is the wavevector-dependent ordering tensor with elements given by

$$\hat{Q}_{\alpha\beta}(\mathbf{q}) = \frac{V}{N} \sum_{j=1}^{N} \frac{3}{2} \left(e_{j\alpha} e_{j\beta} - \frac{1}{3} \delta_{\alpha\beta} \right) \exp\left(i \mathbf{q} \cdot \mathbf{r}_j \right)$$

where $\mathbf{e}_j$ is the unit vector defining the orientation of molecule j at position $\mathbf{r}_j$.

EQN (2) holds in the limit of small $q = |\mathbf{q}|$. However, the original [13] and subsequent [14] hard particle measurements, performed using systems of between 144 and 600 molecules, were unable to explore this limit systematically, and suffered from sizeable error estimates (10 - 15%).

These hard-body simulations gave values for the K_i which are of the same order of magnitude but consistently higher than those obtained from theoretical calculations. The origin of this discrepancy is thought to be either the small system size and restricted accessible q range of the simulations or the incorrect treatment of long-range correlations in the theoretical calculations. Despite this quantitative disagreement, the simulation results obey the usual inequalities: $K_3 > K_1 > K_2$ for prolate particles and $K_2 > K_1 > K_3$ for oblate. For 3:1 ellipsoids in a nematic phase with $<P_2> = 0.7$, the ratios $K_3/K_1 = 3.33$ and $K_3/K_2 = 3.66$ were obtained.

Subsequently, the twist Frank constant, K_2, was determined for hard ellipsoids by a more direct simulation technique involving measurement of the torque balance across a system simulated with twisted boundary conditions [15]. The results thus obtained are consistent with, but slightly lower than, those given by the equilibrium-fluctuation method. However, these small differences were attributed to technical factors (i.e. the system sizes and box shapes, and the effective restriction to a single $\mathbf{q}$ in the twisted system simulations) rather than being used to clarify the discrepancy between previous simulations and theory.

The extensive lattice simulations of [12] demonstrated that the basis of the equilibrium fluctuation method is sound and certainly not the source of the discrepancies mentioned previously. Here, system sizes of $32 \times 32 \times 32$ were accessible, leading to very small error estimates (typically of order 0.1%) and small system size effects (halving the system size resulted in a shift of only 0.7% in the measured K value). The results matched the molecular field prediction $K = 3 < P_2^2 >$ [16] exactly in the low temperature limit, though this was found to break down somewhat at higher temperatures (the proportionality constant rising by over 20% as T_{NI} was approached). In this case, the disagreement was demonstrated to originate from the theory since, for this model, all but the leading $< P_2^2 >$ term in the order parameter expansion of K can be shown to be zero [12]. This suggests that explaining experimental observations of the departure from $K_i \propto < P_2^2 >$ behaviour with molecular-field-based arguments may not be appropriate.

As well as performing equilibrium fluctuation based calculations, Cleaver and Allen [12] explored an alternative route to the Frank constants founded on the formal equivalence of the amplitudes of thermal fluctuations and responses to small perturbations. Thus, the Frank constant for the Lebwohl-Lasher model was determined via direct measurements of its director profile in the presence of a series of spatially modulating fields of wavevector $\mathbf{q}$. Whilst these linear response based simulations proved accurate and highly consistent with the equilibrium fluctuation results, they were found to represent a less efficient route to K since they required numerous moderate runs (several field values being considered for each $\mathbf{q}$) as opposed to a single extended run (from which data could be accumulated simultaneously for all wavevectors).

More recently, the equilibrium fluctuation method has been applied to simulations of the prolate Gay-Berne model ($\mu = 2$, $\nu = 1$, $\kappa' = 5$), using system sizes in the range 1000 - 8000 particles [17]. For most of the measurements made here, the efficiency of the method was enhanced through the use of a constrained-director implementation of molecular dynamics; this greatly simplified calculation of the components of $\hat{Q}(\mathbf{q})$, leading to significantly improved statistics. The numerical values thus obtained for the K_i and their ratios stand comparison with those found for hard ellipsoid systems. For 3:1 particles at $T^* = 1.0$ and $\rho = 0.33$, $<P_2> = 0.708$, $K_3/K_1 = 3.48$ and $K_3/K_2 = 3.38$ were obtained. Whilst both the director constraint and the periodic boundary conditions used can be expected to have had some damping effect on long wavelength fluctuations, for the reasonably large system sizes used in this study the results obtained from the constrained-director and free-director simulations were found to be consistent. This may be explained by the fact that in the equilibrium fluctuation method, the K_i values are determined via fits over a large spectrum of $\mathbf{q}$ points and so are relatively unaffected by such damping.

Reference [17] was not, however, the first publication on measurements of the elastic constants of this important model system. Previously, Stelzer et al [18] performed MD simulations of the Gay-Berne model to evaluate numerically spherical harmonic expansions of the system's direct correlation function. These expansions were then employed in formal expressions for the K_i based on integrals over the direct correlation function [19], Simpson's rule being used to perform the numerical integrations. The values thus obtained for the three K_i were surprisingly uniform (i.e. had ratios close to unity) and were very different from those found with the equilibrium fluctuation method. Subsequently, Stelzer et al discovered an error in their calculation - the corrected K_i values have larger ratios [18], but continue to differ markedly from the equilibrium fluctuation based results ($K_3/K_1 \approx 6.5$, $K_3/K_2 \approx 2.2$ and $<P_2> \approx 0.69$ for the $T^* = 1.0$ and $\rho = 0.33$ state point).

Whilst the system sizes used in these two sets of Gay-Berne model measurements differ somewhat, this does not appear sufficient to explain the discrepancy in the K_i values obtained. Similarly, the consistency between the results of the equilibrium fluctuation measurements of the Gay-Berne and hard ellipsoid systems suggests that the constrained-director MD used in the former is not the cause of this disagreement. A more significant difference may be the run-lengths used: Stelzer et al averaged their measurements over 40,000 timesteps whereas [17] used run-lengths of order 1,000,000 timesteps. Given that the outputs of Stelzer's short runs were used to calculate K_i values via numerical integration, problems due to poor statistics in the integrand data could be both significant and difficult to detect. Interestingly, a later paper describing measurements of the Gay-Berne model's direct correlation function states that '... runs of up to 450,000 steps were judged necessary ...' [20]. Recently, the direct correlation function integration scheme has been applied to a discotic variant of the Gay-Berne model [21]. Again, however, the simulation run lengths involved were rather short.

C   THE VISCOSITY COEFFICIENTS

Due to their broken orientational symmetry, nematic liquid crystals have a number of viscosity coefficients, which comprise the bulk viscosity, the three Miesowicz coefficients (η_1, η_2 and η_3), the Helfrich viscosity (η_{12}), the two Leslie viscosity coefficients (γ_1 and γ_2) and two further coupling parameters. Due to two Onsager relations relating various of these, this list can be expressed in terms of five independent coefficients [22].

Calculation of these coefficients via MD simulations can be achieved using one of two approaches: a given system can be simulated using standard equilibrium MD techniques, enabling the desired coefficient to be evaluated via an appropriate Green-Kubo integral [23], or a perturbative non-equilibrium MD (NEMD) method can be employed [24]. In the latter case, a planar Couette flow shear is imposed using Lees-Edwards (or 'sliding brick') boundary conditions with an appropriate NEMD scheme such as SLLOD [24].

The first simulation based determination of these coefficients was performed using NEMD on systems of perfectly aligned ellipsoidal particles with interaction potentials given by affine (i.e. coordinate stretching) transformations of Lennard-Jones and soft sphere potentials [25]. In this work, it was argued that the restriction to perfect orientational order should have little bearing on the measured Miesowicz viscosity ratios since these are usually temperature-independent away from phase transitions. For prolate ellipsoids, the inequalities $\eta_1 < \eta_3 < \eta_2$ and $\gamma_2 < 0$ were satisfied, as expected for a nematic. Good agreement was also achieved with analytical results for these idealised systems. The theory of this affine transformation approach was later extended to systems with imperfect orientational order [26]. The simulation technique was also later applied to the Gay-Berne model [27].

In 1993, Sarman and Evans derived the Green-Kubo relations needed to calculate the nematic viscosity coefficients and determined them using equilibrium MD [23]. As a comparison, they also performed equivalent calculations using the SLLOD NEMD algorithm. The model used for these simulations was a purely repulsive (and, therefore, relatively short ranged) variant of the Gay-Berne model. The results from the two sets of simulations generally compared very well, though run-lengths of several million time steps were found necessary. Subsequently, Sarman developed a constrained director form of MD [28]. He went on to apply this to NEMD based measurements of the viscosity coefficients, again using the repulsive Gay-Berne variant [29]. The results thus obtained were found to be consistent with those obtained using unconstrained NEMD, though the calculations involved were substantially simplified through the use of the director constraint.

Subsequently, the use of the director constraint was questioned by a group determining the viscosity coefficients of the full Gay-Berne model via equilibrium MD simulations [30]. The basis of this criticism arises from the observation that for the relatively small system sizes used in such simulations (typically just a few hundred particles due to the very long run-lengths needed) there is insufficient separation between the timescales of director motion and molecular motion. This effect was argued to be pertinent both to Sarman's thermodynamically constrained NEMD simulations and to an earlier equilibrium MD study which measured the viscosity coefficients of the full Gay-Berne model in the presence of a director-pinning field [31]. Given that this effect was not apparent in Sarman's previous comparisons of various constrained and unconstrained systems, however, its significance has yet to be established fully. The difference in the ranges of the models used may offer some explanation for this disagreement.

Equilibrium MD has also been used to calculate the viscosity coefficients of prolate hard ellipsoids [32]. Here, extended runs were found to be necessary to gain agreement with the predictions of kinetic theory. Such simulations are, however, restricted to the calculation of viscosity coefficients in the linear-response regime. In order to measure the coefficients corresponding to higher shear rates, an NEMD scheme with a thermalisation scheme able to take account of the angular velocity field induced by the shear flow (with molecules preferentially aligning along the flow direction) is necessary.

In simple fluid NEMD, such a scheme is called a 'positionally unbiased thermostat' (PUT) [24]. A relatively complicated molecular equivalent, developed by Travis et al [33], was recently applied to a hard ellipsoid system and found to produce significant improvements over the usual constant streaming angular velocity scheme [34]. In the same publication, a third thermostat, significantly simpler than the molecular PUT of Travis et al, was also tested and found to compare favourably.

D   CONCLUSIONS

Several computer simulation techniques have been developed to determine technologically relevant material properties of nematic liquid crystals. In general, the results obtained have been consistent with experiment and theory, supporting the validity of the relatively simple simulation models used. The significant time- and length-scales involved in determination of correlation based quantities has led to novel algorithmic developments such as constrained director MD. Further developments will be essential if properties such as the Frank elastic constants and viscosity coefficients are to be determined for more realistic simulation models.

REFERENCES

[1]   M.P. Allen [*Phys. Rev. Lett. (USA)* vol.65 (1990) p.2881]

[2]   M.P. Allen [*Phys. Rev. E (USA)* vol.47 (1993) p.4611]

[3]   J. Stelzer, R. Berardi, C. Zannoni [*Chem. Phys. Lett. (Netherlands)* vol.299 (1999) p.9]; M.P. Allen, P.J. Camp [unpublished]

[4]   D.J. Cleaver, M.P. Allen [*Mol. Phys. (UK)* vol.80 (1993) p.253]; E. Martín del Río, E. de Miguel [*Phys. Rev. E (USA)* vol.55 (1997) p.2916]; S.J. Mills, C.M. Care, M.P. Neal, D.J. Cleaver [*Phys. Rev. E (USA)* vol.58 (1998) p.3248]

[5]   Z. Zhang, A. Chakrabarti, O.G. Mouritsen, M.J. Zuckermann [*Phys. Rev. E (USA)* vol.53 (1998) p.3248]; G.D. Wall, D.J. Cleaver [*Phys. Rev. E (USA)* vol.56 (1997) p.4306]

[6]   M.P. Allen [*Mol. Phys. (UK)* vol.96 (1999) p.1391]

[7]   See, for example, P.G. de Gennes, J. Prost [*The Physics of Liquid Crystals* 2nd Edition (Oxford Science Publications, Oxford, 1993)]

[8]   F.C. Frank [*Discuss. Faraday Soc. (UK)* vol.25 (1958) p.19]

[9]   P.I.C. Teixeira, V. Yarmolenko, D.J. Cleaver [in preparation (1999)]

[10]   P.A. Lebwohl, G. Lasher [*Phys. Rev. A (USA)* vol.6 (1972) p.426]

[11]   P. Simpson [PhD Thesis, University of Southampton (1980)], though here a factor of 1.5 was omitted in relating the critical field to the elastic constant.

[12]   D.J. Cleaver, M.P. Allen [*Phys. Rev. A (USA)* vol.43 (1991) p.1918]

[13]   M.P. Allen, D. Frenkel [*Phys. Rev. A (USA)* vol.37 (1988) p.1813], though the value given for the K_i were subject to a missing factor of 2.25, as pointed out in the erratum [*Phys. Rev. A (USA)* vol.42 (1990) p.3641E]

[14]   B. Tjipto-Margo, G.T. Evans, M.P. Allen, D. Frenkel [*J. Phys. Chem. (USA)* vol.96 (1992) p.3942]

[15]   M.P. Allen, A.J. Masters [*Mol. Phys. (UK)* vol.79 (1993) p.277]

[16]   R.G. Priest [*Phys. Rev. A (USA)* vol.7 (1973) p.720]

[17]   M.P. Allen, M.A. Warren, M.R. Wilson, A. Sauron, W. Smith [*J. Chem. Phys. (USA)* vol.105 (1996) p.2850]

[18]   J. Stelzer, L. Longa, H.-R. Trebin [*J. Chem. Phys. (USA)* vol.103 (1995) p.3098], though several of the results given were affected by an error in the calculation of the total correlation function. Figures displaying corrected data are given in [*J. Phys. Chem. (USA)* vol.107 (1997) p.1295]

[19]   A. Poniewierski, J. Stecki [*Mol. Phys. (UK)* vol.38 (1979) p.1931]; M.D. Lipkin, S.A. Rice, U. Mohanty [*J. Chem. Phys. (USA)* vol.82 (1985) p.472]; M.P. Allen, C.P. Mason, E. de Miguel, J. Stelzer [*Phys. Rev. E (USA)* vol.52 (1995) p.R25]

[20]   M.P. Allen, M.A. Warren [*Phys. Rev. Lett. (USA)* vol.78 (1997) p.1291]

[21]   J. Stelzer, M.A. Bates, L. Longa, G.R. Luckhurst [*J. Chem. Phys. (USA)* vol.107 (1997) p.7483]

[22]   S. Hess [*J. Non-Equilib. Thermodyn. (Germany)* vol.11 (1986) p.175]

[23]   S. Sarman, D.J. Evans [*J. Chem. Phys. (USA)* vol.99 (1993) p.9021]

[24]   D.J. Evans, G.P. Morriss [*Statistical Mechanics of Non-Equilibrium Liquids* (Academic, London, 1990)]

[25]   D. Baalss, S. Hess [*Phys. Rev. Lett. (USA)* vol.57 (1986) p.86]

[26]   H. Ehrentraut, S. Hess [*Phys. Rev E (USA)* vol.51 (1995) p.2203]

[27]   S. Hess, C. Aust, L. Bennett, M. Kroger, C. Pereira Borgmeyer, T. Weider [*Physica A (Netherlands)* vol.240 (1997) p.126]

[28]   S. Sarman [*J. Chem. Phys. (USA)* vol.103 (1995) p.393]

[29]   S. Sarman [*J. Chem. Phys. (USA)* vol.103 (1995) p.10378]; S. Sarman [*Physica A (Netherlands)* vol.240 (1997) p.160]

[30]   S. Cozzini, L.F. Rull, G. Ciccotti, G.V. Paolini [*Physica A (Netherlands)* vol.240 (1997) p.173]

[31]   A.M. Smondyrev, G.B. Loriot, R.A. Pelcovits [*Phys. Rev. Lett. (USA)* vol.75 (1995) p.2340]

[32]   S. Tang, G.T. Evans, C.P. Mason, M.P. Allen [*J. Chem. Phys. (USA)* vol.102 (1995) p.3794]; M.P. Allen, P.J. Camp, C.P. Mason, G.T. Evans, A.J. Masters [*J. Chem. Phys. (USA)* vol.105 (1996) p.11175]

[33]   K.P. Travis, P.J. Davies, D.J. Evans [*J. Chem. Phys. (USA)* vol.103 (1995) p.1109 and p.10638]

[34]   X.-F. Yuan, M.P. Allen [*Physica A (Netherlands)* vol.240 (1997) p.145]

12.5 Simulation of macroscopic phenomena for nematics

S. Day and F.A. Fernandez

August 2000

A   INTRODUCTION

The successful operation of electro-optic effects in nematic liquid crystals requires the combination of orientation of the director in response to an electric field and reorientation when that field is changed or removed. Elastic forces, or restoring torques in the nematic prevent complete orientation along an applied field; the extent of the director orientation depends on the magnitude of the field and the elastic properties of the particular liquid crystal mixture. In addition, in displays, the elastic forces restore the nematic to the original state when the field has been removed. The accurate modelling of this behaviour is very important in the development of liquid crystal display technology.

In a liquid there are no elastic distortions in the classical sense because the liquid will deform or flow in response to an applied shear distortion. However, in a liquid crystal there are restoring torques which oppose any distortion or curvature in the director field, **n**. In nematics the distortions in the director field can be calculated successfully using the continuum theory. The origins of the theory can be traced to Oseen [1] and Zocher [2], followed by a more direct reformulation by Frank [3] and Ericksen [4]. The dynamic theory was completed by Ericksen and Leslie [5,6]. An alternative formulation has been given more recently by Leslie based on the balance laws for linear and angular momentum [7].

B   NUMERICAL MODELLING IN ONE DIMENSION

The free energy density of the liquid crystal is described in [8]:

$$F = \frac{1}{2}\left\{ K_1(\mathrm{div}\,\mathbf{n})^2 + K_2(\mathbf{n} \cdot \mathrm{curl}\,\mathbf{n})^2 + K_3\,|(\mathbf{n} \times \mathrm{curl}\,\mathbf{n})|^2 - \varepsilon_0 \Delta\tilde{\varepsilon}(\mathbf{n} \cdot \mathbf{E})^2 \right\} \tag{1}$$

This equation includes the electrostatic free energy density, which depends on the director, indicating that reorientation occurs under the influence of an applied electric field.

The director distribution can be found by minimising the integral of the free energy density (a variational expression), or by solving the corresponding Euler-Lagrange equation [10]. If the variation in the director occurs only in one direction then expressions in terms of the polar coordinates θ and ϕ can be integrated numerically to obtain the director distribution and the corresponding voltages applied across the liquid crystal layer [10]. The director **n** is a unit vector since it does not vary in length, only in orientation. This allows the use of the variables θ and ϕ, but the formulation can lead to difficulties when the director approaches $\theta = 0$, since the free energy becomes independent of ϕ.

This model is a static one and does not include any analysis of the dynamic switching of the liquid crystal. Nematic liquid crystals have five viscosity parameters and for most materials the values are not available, although they can be measured using a variety of techniques [11]. Often simulations are carried out using only one viscosity coefficient. The viscous torque is set equal to the elastic torque [12]. In addition to the viscosity coefficients, the liquid nature means that backflow can be significant in switching [13,14]. Another factor that is important in some circumstances is the conductivity of the liquid crystal, which may also be anisotropic [15].

The modelling in one dimension has proved very successful in many practical cases where the 1D approximation is valid and has been used extensively in the development of liquid crystal displays which use nematic liquid crystals [16].

C   NUMERICAL MODELLING IN TWO AND THREE DIMENSIONS

In complicated structures the one-dimensional approximation cannot be assumed and the modelling must be carried out in two or three dimensions. The nematic phase is characterised by a vector **n**, the director, but the head-to-tail symmetry of the phase is such that **n** is equal to -**n**. This symmetry is maintained in EQN (1), which is quadratic in **n**. However, the Euler equation, written in terms of the vector **n**, is no longer quadratic and using this vector approach to solve the free energy equation numerically does not maintain the correct symmetry. Instead, a tensor form of the Euler equation has been suggested [17] to preserve this symmetry. The orientational order is characterised by a second rank tensor,

$$Q_{\nu\mu} = S(n_\nu n_\mu - \delta_{\nu\mu}/3)$$

where $\nu,\mu = x,y,z$, $\delta_{\nu\mu}$ is the Kroeneker delta and S is the scalar order parameter. The tensor order parameter preserves the head-to-tail symmetry of the director and it can be used as a variable for the free energy. The resulting Euler equation, simplified by using a one elastic constant approximation and including dynamic effects, becomes

$$n_\nu \left(\gamma \frac{\partial}{\partial t}(n_\nu n_\mu) - K\nabla^2(n_\nu n_\mu) - \varepsilon_o \Delta\widetilde{\varepsilon} E_\nu E_\mu \right) = 0 \qquad (2)$$

where γ is the viscosity coefficient, K the elastic constant, ε_o the vacuum permittivity, $\Delta\widetilde{\varepsilon}$ the dielectric anisotropy, n_ν, n_μ and E_ν, E_μ the Cartesian components ($\nu,\mu = x,y,z$) of the director and electric fields, respectively, and the conventional tensor notation is used, i.e. summation over repeated suffixes. This expression for numerical solution preserves the head-to-tail symmetry of the nematic phase. This is also true for the corresponding expression including all three elastic constants, although the resulting equations for numerical solution are very complex.

For both the vector and the tensor approaches the calculation can be done as a time stepping process. At each time step the electric field and the director distribution can be calculated in turns, repeatedly, until the two are consistent [18,19]. FIGURE 1 shows this schematically.

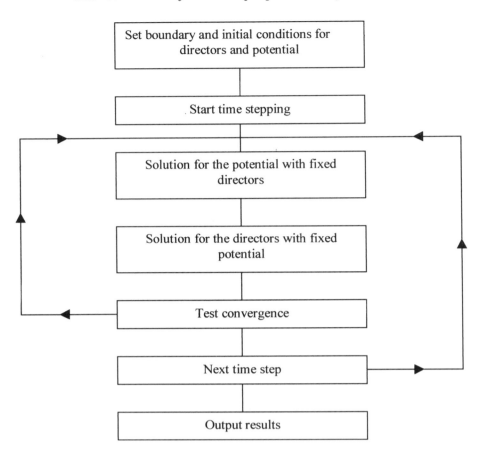

FIGURE 1 Time stepping process for the numerical calculation
of the electric field and director distributions.

If Cartesian coordinates are used in the calculation instead of the angles θ and ϕ, then the director must be normalised to ensure it remains of unit length. FIGURE 2 shows the results of such a 2D simulation for an IPS device [21]. The electrodes are shown as black lines. The potential is shown with shading, the director orientation as short lines whose length indicates twist out of the plane.

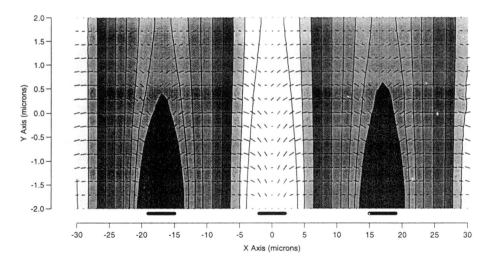

FIGURE 2 Potential and director distribution for a 2D tensor
method calculation using one elastic constant.

There are advantages and disadvantages of the two methods. In the case of the tensor approach the head-to-tail symmetry is preserved and so no assumptions need to be made as to the orientation of the vector. This is shown by Killian and Hess [17] for a square cell with homeotropic boundary conditions. The algorithm used for the vector formulation depends on the polarity of the vectors, so the boundary was chosen to have all the vectors pointing towards the centre of the cell. This gave only one solution with the director field escaping perpendicular to the plane in the centre. The tensor formulation gave a number of solutions which were very sensitive to the initial conditions. The solutions included different disclinations, as would be found in real nematics. The different solutions had different energies, but most had a lower energy than the solution found using the vector formulation. The tensor formulation has also been shown to give good results in inhomogeneous electric fields [20,21]. Neither the vector nor the tensor approach can model adequately defects and disclinations since at the scale of the disclination the continuum theory does not apply and the energy is not known. Around the disclination a microscopic theory would have to be used. It is needed over the region where the director no longer gives a good description of the liquid crystal structure.

It has been shown [22] that the numerical implementation of the tensor approach can give wrong results in a dynamic calculation if at one instant the mesh becomes too coarse to represent adequately a large variation of the director distribution. Of course, both the vector and tensor approaches will give inaccurate results in such a situation but if at one time the director distribution is such that a large variation (greater than 90°) occurs between mesh points, the tensor approach will not be able to recover the structure of the material once the director field relaxes and the mesh is again good enough. In the vector approach the sign of the vector will indicate that a highly twisted structure exists between mesh points and in subsequent time steps the correct structure can be recovered. This problem can be overcome by using adaptive meshing, thus preventing inaccurate representation of the director field. Some examples of this problem are given by Anderson et al [22]. Care must be taken in the interpretation of the result since in some circumstances, memory of the twist should be lost by the director field and the vector approach produces spurious results. For example, in the case of the Pi-cell modelling in [22], when the director field is switched into the V state the 180° twist initially present in the liquid crystal disappears. FIGURE 3 shows the V state and subsequent state with no

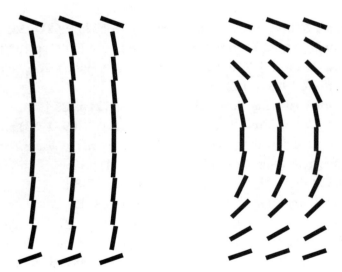

FIGURE 3 Fully switched V-state and subsequent unswitched state for the Pi-cell.

twist present, as is obtained using the tensor model. Experimentally, the twist is only restored after some considerable time and occurs by growth from defects in the cell. Modelling using the vector method retains the information about the twist in the V state and so when the electric field is removed, calculations show the director returning relatively quickly to the original twisted state. So while the vector method has advantages in terms of ease of implementation, particularly when all three elastic constants are included and in three dimensions, it is important that the initial directions of the vectors are chosen carefully to make sure that in cases where the vectors in adjacent cells are found to be in opposite directions, this is indeed a correct physical representation of the director field.

D   CONCLUSIONS

Modelling of nematic liquid crystal structures has always been important in the development of displays, starting with the one-dimensional model. Successful modelling of the liquid crystal structure in three dimensions is becoming even more technologically important as new structures and effects are being developed. Microstructures are being used, together with non-uniform electric fields between the substrates to improve the viewing angle and resolution of liquid crystal displays. Modelling allows improvements to be made more rapidly, without the need for complex fabrication of each structure. Modelling also allows optimisation of the liquid crystal parameters, without complex synthesis and mixture formulation of actual materials. It can also improve the understanding of the way in which switching of the liquid crystal occurs in more complex structures, since the director field is accessible in a way that is difficult experimentally.

REFERENCES

[1]   W.C. Oseen [*Trans. Faraday Soc. (UK)* vol.29 (1933) p.883]

[2]   H. Zocher [*Trans. Faraday Soc. (UK)* vol.29 (1933) p.945]

[3]   F.C. Frank [*Discuss. Faraday Soc. (UK)* vol.25 (1958) p.19]

[4]   J.L. Ericksen [*Arch. Ration. Mech. Anal. (Germany)* vol.9 (1962) p.371]

[5]   J.L. Ericksen [*Trans. Soc. Rheol. (USA)* vol.5 (1961) p.23]

[6]   F.M. Leslie [*Arch. Ration. Mech. Anal. (Germany)* vol.28 (1968) p.265]

[7]   F.M. Leslie [*Contin. Mech. Thermodyn. (Germany)* vol.4 (1992) p.167]

[8]   P.G. de Gennes, J. Prost [*The Physics of Liquid Crystals* 2nd edition (Oxford Science Publications, Oxford, 1993) p.102 and p.134]

[9]   H.J. Deuling [*Mol. Cryst. Liq. Cryst. (UK)* vol.19 (1972) p.123]

[10]   D.W. Berreman [*Philos. Trans. R. Soc. Lond. A (UK)* vol.309 (1983) p.203]

[11]   J.K. Moscicki [Datareview in this book: *8.2 Measurements of viscosities in nematics*]

[12]   D.W. Berreman [*Appl. Phys. Lett. (USA)* vol.25 (1974) p.12]

[13]   C.Z. van Doorn [*J. Appl. Phys. (USA)* vol.46 (1975) p.3738]

[14]   D.W. Berreman [*J. Appl. Phys. (USA)* vol.46 (1975) p.3746]

[15]   P. Vetter, B. Maximus, H. Pauwels [*J. Phys D, Appl. Phys. (UK)* vol.25 (1992) p.481]

[16]   C.M. Waters, E.P. Raynes, V. Brimmell [*Mol. Cryst. Liq. Cryst. (UK)* vol.123 (1985) p.303]

[17]   A. Killian, S. Hess [*Z. Nat.forsch. A (Germany)* vol.44 (1989) p.693]

[18]   J.B. Davies, S. Day, F. Di Pasquale, F.A. Fernandez [*Electron. Lett. (UK)* vol.32 (1996) p.582]

[19]   F. Di Pasquale, F.A. Fernandez, S. Day, J.B. Davies [*IEEE J. Sel. Top. Quantum Electron. (USA)* vol.2 (1996) p.128]

[20]   G. Haas, H. Wohler, M.W. Fritsch, D.A. Mlynski [*Mol. Cryst. Liq. Cryst. (UK)* vol.198 (1991) p.15]

[21]   F. Di Pasquale et al [*IEEE Trans. Electron Devices (USA)* vol.46 (1999) p.661]

[22]   J.E. Anderson, P. Watson, P.J. Bos [*SID Int. Symposium, Seminar and Exhibition* San Jose, USA, 21st May 1999, p.16.5]

Subject Index